Additive Manufacturing of Functional Polymers and Nanocomposites

AF696601

About the Series

The **Woodhead Publishing Series in Composites Science and Engineering** is a suite of professional reference books providing comprehensive coverage of recent developments in composite materials research for application now and in the future.

Editor-in-Chief
Professor Costas Soutis, Head of Aerospace Engineering, University of Manchester, UK

Series Editors
Professor Suresh Advani, Assoc. Director, Centre for Composite Materials, Univ. Delaware, United States
Professor Bodo Fiedler, Director Inst. Plastics and Composites, TU Hamburg, Germany
Professor Leif Asp, Division of Materials and Computational Mechanics, Chalmers, Sweden
Professor Chun H. Wang, Head of School of Mechanical Engineering, UNSW, Australia

Woodhead Publishing Series in Composites Science and Engineering

Additive Manufacturing of Functional Polymers and Nanocomposites

Recent Progress, Applications, Challenges and Future Opportunities

Edited by

Chaudhery Mustansar Hussain

Department of Chemistry and Environmental Science, Jordan Hu College of Science and Liberal Arts, New Jersey Institute of Technology, Newark, NJ, United States

Kalim Deshmukh

New Technologies - Research Centre, University of West Bohemia, Pilsen, Czech Republic

ELSEVIER

WP WOODHEAD PUBLISHING

An imprint of Elsevier

Woodhead Publishing is an imprint of Elsevier
50 Hampshire Street, 5th Floor, Cambridge, MA 02139, United States
125 London Wall, London EC2Y 5AS, United Kingdom

Copyright © 2025 Elsevier Ltd. All rights are reserved, including those for text and data mining, AI training, and similar technologies.

For accessibility purposes, images in electronic versions of this book are accompanied by alt text descriptions provided by Elsevier. For more information, see https://www.elsevier.com/about/accessibility.

Publisher's note: Elsevier takes a neutral position with respect to territorial disputes or jurisdictional claims in its published content, including in maps and institutional affiliations.

No part of this publication may be reproduced or transmitted in any form or by any means, electronic or mechanical, including photocopying, recording, or any information storage and retrieval system, without permission in writing from the publisher. Details on how to seek permission, further information about the Publisher's permissions policies and our arrangements with organizations such as the Copyright Clearance Center and the Copyright Licensing Agency, can be found at our website: www.elsevier.com/permissions.

This book and the individual contributions contained in it are protected under copyright by the Publisher (other than as may be noted herein).

Notices
Knowledge and best practice in this field are constantly changing. As new research and experience broaden our understanding, changes in research methods, professional practices, or medical treatment may become necessary.

Practitioners and researchers must always rely on their own experience and knowledge in evaluating and using any information, methods, compounds, or experiments described herein. In using such information or methods they should be mindful of their own safety and the safety of others, including parties for whom they have a professional responsibility.

To the fullest extent of the law, neither the Publisher nor the authors, contributors, or editors, assume any liability for any injury and/or damage to persons or property as a matter of products liability, negligence or otherwise, or from any use or operation of any methods, products, instructions, or ideas contained in the material herein.

ISBN: 978-0-443-26516-7 (print)
ISBN: 978-0-443-26517-4 (online)

For information on all Woodhead Publishing publications
visit our website at https://www.elsevier.com/books-and-journals

Publisher: Matthew Deans
Acquisitions Editor: Gwen Jones
Editorial Project Manager: Debarati Roy
Production Project Manager: Prem Kumar Kaliamoorthi
Cover Designer: Miles Hitchen

Typeset by MPS Limited, Chennai, India

Working together to grow libraries in developing countries
www.elsevier.com • www.bookaid.org

I would like to dedicate this book to
Our Extremely Beloved GOD
"Meray Pyarey Allah (Subhana Wa Ta Alla)"

Contents

Section 4 Theoretical modelling and machine learning approaches

Section 5 Challenges, environmental and health impact and commercialization aspects

List of contributors

Swati Agarwal Drumlins Water Technologies Pvt. Ltd., Jaipur, Rajasthan, India

Yanshi Agrawal Department of Bioscience and Biotechnology, Banasthali Vidyapith, Banasthali, Tonk, Rajasthan, India

Robert Aguirresarobe POLYMAT and Department of Polymers and Advanced Materials: Physics, Chemistry and Technology, Faculty of Chemistry, University of the Basque Country UPV/EHU, Donsotia-San Sebastian, Spain

Waleed Ahmed Engineering Requirements Unit, UAE University, Al-Ain, United Arab Emirates

Ali H. Al-Marzouqi Chemical and Petroleum Engineering Department, UAE University, Al-Ain, United Arab Emirates

Noura AlMazrouei Department of Chemical and Biological Engineering, Monash University, Clayton, Victoria, Australia

Abdullah N. Alodhayb Department of Physics and Astronomy, College of Science, King Saud University, Riyadh, Saudi Arabia

Thomas D. Anthopoulos King Abdullah University of Science and Technology (KAUST), KAUST Solar Center (KSC), Thuwal, Saudi Arabia; Department of Electrical and Electronic Engineering, Henry Royce Institute, The University of Manchester, United Kingdom; Department of Electrical and Electronic Engineering, Photon Science Institute, The University of Manchester, United Kingdom

Muhammad Tuoqeer Anwar Department of Mechanical Engineering, COMSATS University Islamabad, Sahiwal Campus, Sahiwal, Punjab, Pakistan

Saritha Appukuttan Department of Chemistry, Amrita Vishwa Vidyapeetham, Amritapuri, Kollam, Kerala, India

Muhammad Rehman Asghar Institute for Energy Research, Jiangsu University, Zhenjiang, Jiangsu, P.R. China

Punuri Jayasekhar Babu Biomaterials and Bioengineering Research Laboratory, Department of Biotechnology, Pachhunga University College, Mizoram University, Aizawl, Mizoram, India

Shalini Bhatt 2D Materials and LASER Actuation Laboratory, Centre of Excellence for Research, PP Savani University, Kosamba, Gujarat, India

Astha Bisht Department of Bioscience and Biotechnology, Banasthali Vidyapith, Banasthali, Rajasthan, India

Syed Masood Arif Bukhari Department of Mechanical Engineering, Faculty of Engineering and Technology, Bahauddin Zakariya University, Multan, Punjab, Pakistan

M. Itxaso Calafel POLYMAT and Department of Polymers and Advanced Materials: Physics, Chemistry and Technology, Faculty of Chemistry, University of the Basque Country UPV/EHU, Donsotia-San Sebastian, Spain

M. Virgínia Candal School of Engineering, Science and Technology, Valencian International University (VIU), Valencia, Spain

Sanjeevani Deshkar Department of Pharmaceutics, Dr. D.Y. Patil Institute of Pharmaceutical Sciences & Research, Pimpri, Pune, India

Hendrik Faber King Abdullah University of Science and Technology (KAUST), KAUST Solar Center (KSC), Thuwal, Saudi Arabia

Mustafa Gandhi Mechanical Engineering Department, Institute of Technology, Nirma University, Ahmedabad, Gujarat, India

Amol D. Gholap Department of Pharmaceutics, St. John Institute of Pharmacy and Research, Palghar, Maharashtra, India

Prabhanjan Giram Department of Pharmaceutical Sciences, The State University of New York, Buffalo, NY, United States

Mohammad Harun-Ur-Rashid Department of Chemistry, International University of Business Agriculture and Technology (IUBAT), Uttara Model Town, Uttara, Dhaka, Bangladesh

Yan Huang State Key Laboratory of Solid Lubrication, Lanzhou Institute of Chemical Physics, Chinese Academy of Sciences, Lanzhou, P.R. China; Center of Materials Science and Optoelectronics Engineering, University of Chinese Academy of Sciences, Beijing, P.R. China

Naveed Husnain Department of Mechanical Engineering, Faculty of Engineering and Technology, Bahauddin Zakariya University, Multan, Punjab, Pakistan

Muzamil Hussain Department of Mechanical Engineering, COMSATS University Islamabad, Sahiwal Campus, Sahiwal, Punjab, Pakistan

Shashikant J. Joshi Mechanical Engineering Department, Institute of Technology, Nirma University, Ahmedabad, Gujarat, India

Patrik Karlsson AiDEAS OU, Tallinn, Estonia

Suphiya Khan Shriram Institute for Industrial Research, Gurugram, Haryana, India

Ajit Khosla Department of Applied Chemistry, School of Advanced Materials and Nanotechnology, Xidian University, Xi'an, P.R. China

Dileep Kumar Department of Pharm Chemistry, Manipal College of Pharmaceutical Sciences, Manipal Academy of Higher Education, Manipal, Karnataka, India

Rakesh Kumar Department of Regulatory Affairs & Quality Assurance, Kaushik Orthopaedic Private Limited, Sonipat, Haryana, India

Santosh Kumar Department of Mechanical Engineering, Chandigarh Group of Colleges, Mohali, Punjab, India

Mahak Kushwaha Department of Bioscience and Biotechnology, Banasthali Vidyapith, Banasthali, Rajasthan, India; Drumlins Water Technologies Pvt. Ltd., Jaipur, Rajasthan, India

Zhangpeng Li State Key Laboratory of Solid Lubrication, Lanzhou Institute of Chemical Physics, Chinese Academy of Sciences, Lanzhou, P.R. China; Center of Materials Science and Optoelectronics Engineering, University of Chinese Academy of Sciences, Beijing, P.R. China; Yantai Zhongke Research Institute of Advanced Materials and Green Chemical Engineering, Yantai, P.R. China; Qingdao Center of Resource Chemistry & New Materials, Qingdao, P.R. China

T. Stanis Mangsianlun Biomaterials and Bioengineering Research Laboratory, Department of Biotechnology, Pachhunga University College, Mizoram University, Aizawl, Mizoram, India

Ameera F. Mohammad Engineering Requirements Unit, UAE University, Al-Ain, United Arab Emirates

Serafeim Moustakidis AiDEAS OU, Tallinn, Estonia

Alejandro J. Müller POLYMAT and Department of Polymers and Advanced Materials: Physics, Chemistry and Technology, Faculty of Chemistry, University of the Basque Country UPV/EHU, Donsotia-San Sebastian, Spain; IKERBASQUE, Basque Foundation for Science, Plaza Euskadi 5, Bilbao, Spain

Aisha Naveed Department of Mechanical Engineering, Faculty of Engineering and Technology, Bahauddin Zakariya University, Multan, Punjab, Pakistan

Muhammad Salman Nawaz Department of Mechanical Engineering, COMSATS University Islamabad, Sahiwal Campus, Sahiwal, Punjab, Pakistan

Diksha Nunse Department of Pharmaceutics, Sandip Institute of Pharmaceutical Science, Nashik, India

Kaushik M. Patel Mechanical Engineering Department, Institute of Technology, Nirma University, Ahmedabad, Gujarat, India

Kautilya S. Patel Mechanical Engineering Department, Institute of Technology, Nirma University, Ahmedabad, Gujarat, India

Rakshit Pathak 2D Materials and LASER Actuation Laboratory, Centre of Excellence for Research, PP Savani University, Kosamba, Gujarat, India

Ranjitsinh Pawar Department of Pharmaceutics, Poona College of Pharmacy, Bharti Vidyapeeth, Erandwane, Pune, India

Ricardo Pérez-Camargo POLYMAT and Department of Polymers and Advanced Materials: Physics, Chemistry and Technology, Faculty of Chemistry, University of the Basque Country UPV/EHU, Donsotia-San Sebastian, Spain

Mayank Punetha 2D Materials and LASER Actuation Laboratory, Centre of Excellence for Research, PP Savani University, Kosamba, Gujarat, India

Vinay Deep Punetha 2D Materials and LASER Actuation Laboratory, Centre of Excellence for Research, PP Savani University, Kosamba, Gujarat, India

Ravi Shankar Rai Automation and Robotics Engineering Department, Sandip Institute of Technology and Research Centre, Nashik, Maharashtra, India

Akhila Raman Department of Chemistry, Amrita Vishwa Vidyapeetham, Amritapuri, Kollam, Kerala, India

Tahir Rasheed Interdisciplinary Research Centre for Advanced Materials, King Fahad University of Petroleum and Minerals (KFUPM), Dhahran, Eastern Province, Saudi Arabia

Muhammad Rafi Raza Department of Mechanical Engineering, COMSATS University Islamabad, Sahiwal Campus, Sahiwal, Punjab, Pakistan

Ramisha Sajjad Department of Mechanical Engineering, COMSATS University Islamabad, Sahiwal Campus, Sahiwal, Punjab, Pakistan

Parag Sanghani 2D Materials and LASER Actuation Laboratory, Centre of Excellence for Research, PP Savani University, Kosamba, Gujarat, India

Dhaval B. Shah Mechanical Engineering Department, Institute of Technology, Nirma University, Ahmedabad, Gujarat, India

Syed Khawar Hussain Shah Department of Mechanical Engineering, COMSATS University Islamabad, Sahiwal Campus, Sahiwal, Punjab, Pakistan

Abhinav Sharma King Abdullah University of Science and Technology (KAUST), KAUST Solar Center (KSC), Thuwal, Saudi Arabia

Konstantinos Stergiou AiDEAS OU, Tallinn, Estonia

Laldinthari Suamte Biomaterials and Bioengineering Research Laboratory, Department of Biotechnology, Pachhunga University College, Mizoram University, Aizawl, Mizoram, India

Viswapriya Viswalingam Department of Pharmaceutical Chemistry, Poona College of Pharmacy, Bharati Vidyapeeth (Deemed to be University), Pune, Maharashtra, India

Jyotsna G. Vitore Department of Pharmaceutics, National Institute of Pharmaceutical Education and Research, Ahmedabad, Gujarat, India

Jinqing Wang State Key Laboratory of Solid Lubrication, Lanzhou Institute of Chemical Physics, Chinese Academy of Sciences, Lanzhou, P.R. China; Center of Materials Science and Optoelectronics Engineering, University of Chinese Academy of Sciences, Beijing, P.R. China

Essam Zaneldin Department of Civil and Environmental Engineering, College of Engineering, United Arab Emirates University, Al Ain, United Arab Emirates

Ke Zhang State Key Laboratory of Solid Lubrication, Lanzhou Institute of Chemical Physics, Chinese Academy of Sciences, Lanzhou, P.R. China; Center of Materials Science and Optoelectronics Engineering, University of Chinese Academy of Sciences, Beijing, P.R. China

About the editors

Chaudhery Mustansar Hussain, PhD, is an Adjunct Professor and Director of Laboratories in the Department of Chemistry & Environmental Sciences at the New Jersey Institute of Technology (NJIT), Newark, New Jersey, United States. His research is focused on sustainability, nanotechnology and advanced materials, environmental management, analytical chemistry, and other various industries. Dr. Hussain is the author of numerous papers in peer-reviewed journals as well as a prolific author and editor of around **200 books**, including scientific monographs and handbooks in his research areas. He has published with ELSEVIER, American Chemical Society, Royal Society of Chemistry, John Wiley & Sons, CRC Press, and Springer.

Kalim Deshmukh is a Senior Researcher at the New Technologies-Research Centre, University of West Bohemia, Pilsen, Czech Republic. He has over 15 years of research experience in the field of synthesis, characterization, and structure–property relationships in a wide variety of polymeric materials, polymer blends, and nanocomposites for various technological applications. His research interest is mainly focused on the synthesis, characterization, and property investigations of polymer nanocomposites reinforced with different nanofillers including nanoparticles and carbon allotropes such as carbon black, carbon nanotubes, graphene, and its derivatives for potential electronic applications. He has over 100 research publications in peer-reviewed journals and 36 book chapters with renowned international publishers and is currently coediting several books with international publishers. Moreover, he has actively participated and presented his research work at several international conferences and has been serving as a reviewer for several highly reputed international journals.

Preface

The advancement of additive manufacturing (AM) in functional polymers and nanocomposites has greatly enhanced the development of materials with unique properties tailored for specific applications. By merging the adaptability of polymers with the advanced characteristics of nanocomposites, these materials have catalyzed innovation across multiple fields, including health care, aerospace, and electronics. The ability to precisely control microlevel structure and composition allows for the production of materials with superior mechanical strength, electrical conductivity, and thermal stability. Nevertheless, challenges persist, particularly in achieving consistent quality and dimensional accuracy when incorporating nanomaterials, which can sometimes exhibit uneven dispersion and aggregation. Additionally, optimizing synthesis and processing methods is essential to improve scalability and cost-effectiveness. Environmental impacts, such as waste management and energy use during manufacturing, are also crucial considerations. Furthermore, the long-term durability and biocompatibility of these materials must be rigorously assessed, especially in medical and consumer contexts. This current book is dedicated to addressing these issues by focusing on sustainable material choices and improving AM efficiency. With continued progress, these materials hold great promise for transforming industries and fostering smarter, more efficient products. Overall, this book provides a comprehensive exploration of AM, focusing on advanced techniques, materials, and applications in functional polymers and nanocomposites.

To capture the entire topic of AM in functional polymers and nanocomposites and offer the reader a realm of understanding in a systematic manner the book is divided into five sections whereas every section comprises 24 chapters. Section 1 begins with foundational insights, covering AM of semicrystalline polymers for improved structural integrity, a roadmap for intellectual and industrial developments in the field, and an examination of mechanisms, materials, and applications of various techniques. This section also reviews the AM of smart polymeric materials and dives into both the opportunities and challenges associated with functional polymers and nanocomposites. Section 2 delves into innovative polymer nanocomposite synthesis for 3D printing and insights into sustainable additive manufacturing with composite plastics. It further explores the synthesis and functionalization methods of polymer nanocomposites and evaluates their thermal and mechanical properties, contributing to a thorough understanding of material performance. In Section 3, the book highlights diverse applications, from 3D-printed polymers and nanocomposites to advancements in environmental applications and soft robotics. It also discusses the use of functional materials in automotive, aerospace, and aviation industries, as well as in sports goods, electronics, optoelectronics, and electrochemical devices. The section concludes with the promising roles of these materials in health care and biomedical fields and environmental applications, demonstrating the far-reaching impact of additive manufacturing. Section 4 examines optimization techniques and the role of machine learning in improving the AM of functional polymers and nanocomposites, emphasizing data-driven approaches for enhanced efficiency and performance. Finally, Section 5 addresses critical topics such as future opportunities, life cycle assessments, and the environmental and health impacts of these manufacturing processes, alongside considerations for commercialization. This section underscores the challenges and opportunities that lie ahead in making AM more sustainable and commercially viable.

Overall, this book is designed to be a reference guide for researchers, scientists, and experts who are searching for new and modern developments in AM in functional polymers and nanocomposites. We extend our deepest gratitude to all the authors for their exceptional contributions to this book. We are also thankful to Drs. Gwen Jones, Debarati Roy, and the dedicated Elsevier team for their invaluable support and assistance throughout the publication process. Finally, we sincerely appreciate Elsevier for bringing this important edition to print.

Dr. Chaudhery Mustansar Hussain
Dr. Kalim Deshmukh

Acknowledgments

We would like to acknowledge Chaudhery Ghazanfar Hussain for his dedicated support during the compiling this book. We also would like to thank Dr. Vajiha Behranvand, Dr. Farbod Tabesh, Miss Fariba Sirous, and Miss Elham Azadi for their special support during the preparation of this book.

Section 1

Additive manufacturing techniques, material processing and design optimizations

1 Additive manufacturing of semicrystalline polymers: understanding structural integrity and dimensional accuracy

M. Itxaso Calafel[1], *Ricardo Pérez-Camargo*[1], *Robert Aguirresarobe*[1], *M. Virgínia Candal*[2], *and Alejandro J. Müller*[1,3]
[1]POLYMAT and Department of Polymers and Advanced Materials: Physics, Chemistry and Technology, Faculty of Chemistry, University of the Basque Country UPV/EHU, Donsotia-San Sebastian, Spain, [2]School of Engineering, Science and Technology, Valencian International University (VIU), Valencia, Spain, [3]IKERBASQUE, Basque Foundation for Science, Plaza Euskadi 5, Bilbao, Spain

1.1 Introduction

The exponential growth of additive manufacturing (AM) techniques has revolutionized the manufacturing of three-dimensional (3D) polymeric objects (Tamir et al., 2023). AM is based on manufacturing a 3D object by superimposing consecutive layers in a bottom-up concept (Gao et al., 2015; Ligon et al., 2017; MacDonald & Wicker, 2016). Although AM does not compete in productivity with traditional manufacturing techniques, its intrinsic versatility, the ability to produce complex geometries, and the scarce use of secondary tooling have allowed AM to compete against mass production technologies (e.g., injection molding) (Gao et al., 2015; Wendel et al., 2008). Thus, product customization processes have been highly favored, allowing the manufacturing of small batches at low costs.

The term AM does not respond to a single technology but to a manufacturing concept where the material is deposited layer by layer. According to the ASTM F2792-12a standard (ASTM F2792-12a Standard Terminology for Additive Manufacturing Technologies, 2015), these technologies can be classified into seven large groups, namely, Vat Photopolymerization (VP), Material Extrusion (MEX), Binder Jetting (BJ), Material Jetting (MJ), Powder Bed Fusion (PBF), Sheet Lamination (SL), and Direct Energy Deposition (DED). Each technology produces the final 3D object based on a different approach and utilizes various types of materials. Based on the materials used in each AM method, these groups can be classified as shown in Fig. 1.1. Although this classification gives a general overview, it is necessary to highlight that this field is very active in academia and industry, and new manufacturing technologies are continually appearing. The number of materials that each technology can print has been steadily increasing.

Additive Manufacturing of Functional Polymers and Nanocomposites. DOI: https://doi.org/10.1016/B978-0-443-26516-7.00001-8
© 2025 Elsevier Ltd. All rights are reserved, including those for text and data mining, AI training, and similar technologies.

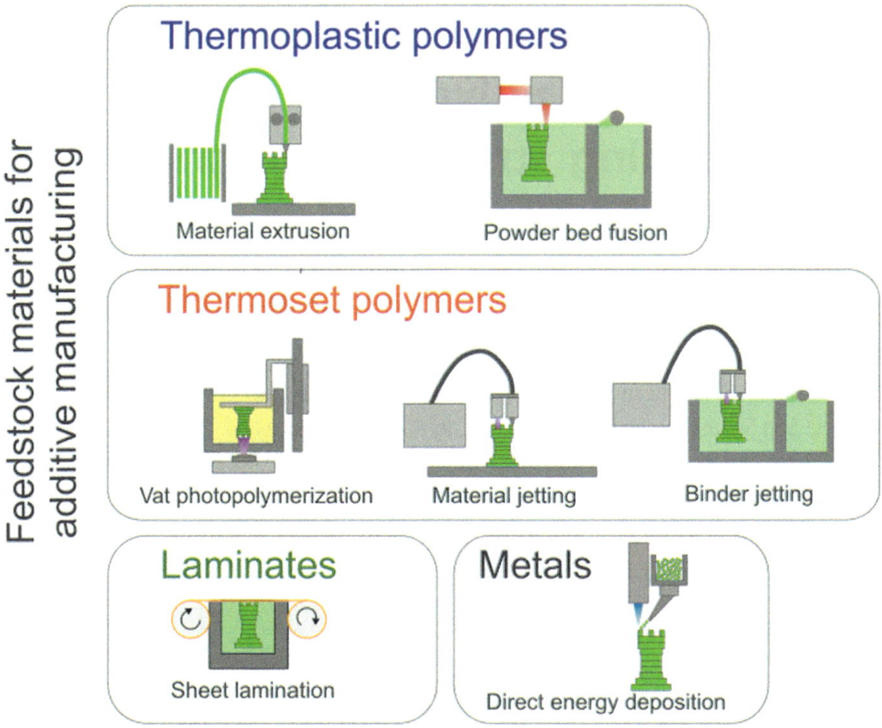

Figure 1.1 Classification of the seven leading AM technologies according to the type of material used for manufacturing.

As shown in Fig. 1.1, there is a big difference in the nature of the feedstock materials used in each technology. On the one hand, VP, MJ, and BJ technologies use resins capable of cross-linking when irradiated with visible and/or ultraviolet light to produce the final object. On the other hand, SL uses laminates of different natures that are cut in the correct shape and stick together using the appropriate adhesive. Finally, DED is a technology specifically developed for metal processing where a plasma beam melts material powder on top of a metal surface. Each of these technologies has extensively been reviewed (Elkaseer et al., 2022; Gibson et al., 2020; Zhang et al., 2021; Ziaee & Crane, 2019).

In this book chapter, we will focus on the printing of thermoplastic materials, more specifically, semicrystalline materials. Thermoplastic polymers are the most used polymeric materials, whose properties range from commodity materials to even engineering and high-performance polymers (Vaes & Van Puyvelde, 2021). From a topological point of view, these materials consist of linear or branched macromolecules capable of diffusing and flowing upon heating. In the specific case of semicrystalline polymers, these structures present a certain degree of regularity that permits aggregation into ordered crystalline domains, providing the printed part with unique properties. The formation of such domains depends not only on their chemical structure but also on the thermal treatments and shear occurring during processing, adding complexity to their printing compared with amorphous polymers. This book chapter reviews the most important AM technologies for semicrystalline thermoplastic materials. It includes the key aspects that one should consider for optimal printing, addressing the possible failures that can occur during the entire process. The chapter also introduces some basic polymer crystallization concepts, which are essential to understand before printing semicrystalline materials. Furthermore, the printing process is addressed from a rheological perspective, offering valuable insights into the different stages and viscoelastic parameters that govern them. Rheology will provide the framework for understanding how the material behaves during printing. Additionally, the chapter presents a brief overview of the mechanical properties of semicrystalline printed parts. Finally, the authors provide an outlook and perspectives on printing semicrystalline materials.

1.2 Additive manufacturing (AM) of thermoplastic materials

Thermoplastic materials, unlike thermosets, can flow when heated above specific thermal transitions. This ability to fluidize is the basis of AM techniques for thermoplastic materials. The main AM processes for thermoplastic materials are MEX and PBF.

1.2.1 Material extrusion-based technologies (MEX)

Extrusion-based technologies are based on passing the fluidized material through a heated printing nozzle that can move in three dimensions and deposit it on a printing base layer by layer. The consolidation of the successive layers of material occurs by cooling the material, giving rise to the 3D object (Cano-Vicent et al., 2021). These methods are commonly used for their versatility, cost-effectiveness, and suitability for a wide range of thermoplastics. The pioneering and most popular MEX technology is Fused Filament Fabrication (FFF), initially developed by Stratasys in 1989 (Crump, 1992) and commercialized as Fused Deposition Modeling (FDM). Fig. 1.2 shows the main MEX technologies. In all of them, the extrusion process is controlled by a computer-controlled system that precisely moves the nozzle along the *x*-, *y*-, and *z*-axes according to the printed object's digital design (CAD model). Although different MEX alternatives have been developed, relying on pressurizing the material using either pneumatic (Jain et al., 2020) or mechanic (Tseng et al., 2018) extrusion (Fig. 1.2A–C, respectively), FFF is the most used technology. In this case (Fig. 1.2D), the material is fed as a filament of 1.75 mm or 3 mm in diameter. The filament is usually housed on a spool and guided towards the liquefier through the feed mechanism (drive wheels or pinch roller). At the liquefier, also known as an extruder or hot end, the filament is heated to its melting or processing temperature to be extruded through the nozzle. Recently, the capabilities of EM were expanded by

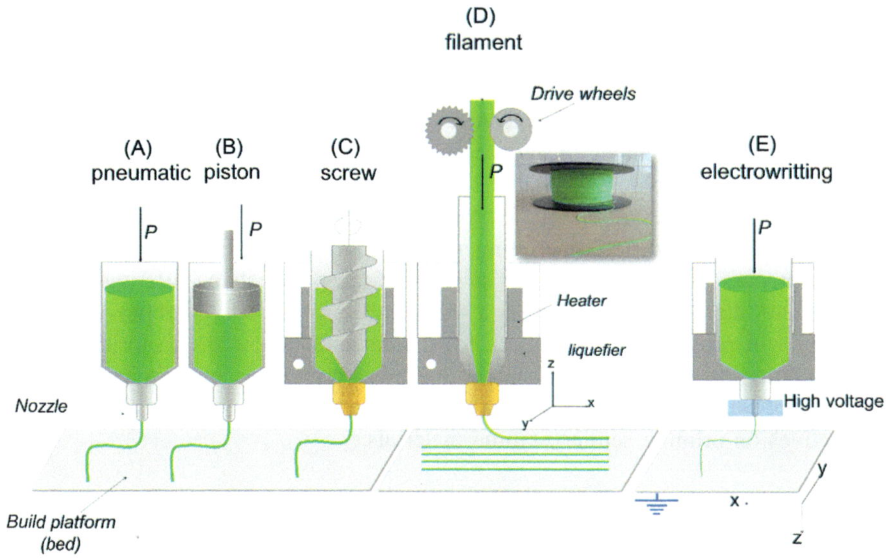

Figure 1.2 Schematic representation of the main MEX technologies: (A, B) pressure-driven extrusion, (C) screw-driven extrusion, (D) filament-driven extrusion, and (E) melt electrowriting.

incorporating electrospinning features, leading to a hybrid concept named melt electrowriting (Fig. 1.2E) (Dalton, 2017). This technology combines the on-demand 3D disposition of material with the production of thinner beads due to electrostatic elongation.

Extrusion AM systems are usually categorized into two main sub-groups: processes based on material melting, as shown in Fig. 1.2, and processes without material melting (Vaezi et al., 2018). Since the ones based on melting the material have already been briefly explained (Fig. 1.2D), this paragraph will outline those where material melting is unnecessary. These include Pressure-Assisted Microsyringe (PAM) (Vozzi et al., 2002), Low-temperature Deposition Manufacturing (LDM) (Y. Q. Zhao et al., 2020), 3D bioplotting (Landers & Mülhaupt, 2000), robocasting (Dave et al., 2023), direct-write assembly, and solvent-based extrusion free-forming (SEF) (Zhang et al., 2020). While, in material melting processes, filament-driven systems or mini-screw extruder systems were developed (Fig. 1.2C), in processes without material melting, four major nozzle designs have been exploited: pressure-actuated (Fig. 1.2A), volume-driven injection nozzles (usually using a stepper-motor) (Fig. 1.2B), solenoid, and piezoelectric-actuated (Fig. 1.2E). Nevertheless, this book chapter will not expand on these techniques without material melting, since, except for polycaprolactone (PCL) and its derivatives, most semicrystalline polymers are processed by melt processing techniques.

1.2.1.1 Key aspects of extrusion-based AM

Independent of the amorphous or semicrystalline nature of the feedstock, some key factors are a priori conditions for the success of the printing. The common failures during printing, particularly in FFF and FDM, can be related to the solid filament (before entering the heater), the extrusion process (flow), and the solid printed part, as summarized in Fig. 1.3.

Failures are associated with solid filament, its extrusion, and flow

When it comes to EM techniques, the filament and its features are of utmost importance. In fact, filament physical and mechanical characteristics play a vital role in preventing irregularities even before the polymer is molten. Among the requirements that the filament itself must possess to be suitable for FFF is to be strong and hard enough to avoid shearing due to drive wheel pinching and stiff enough to prevent buckling, that is, a sudden filament deflection between the drive wheels and the heating zone (Fig. 1.3C). Nevertheless, the filament must also be flexible enough to be easily stored and spooled in compact spaces, and it must be able to bend

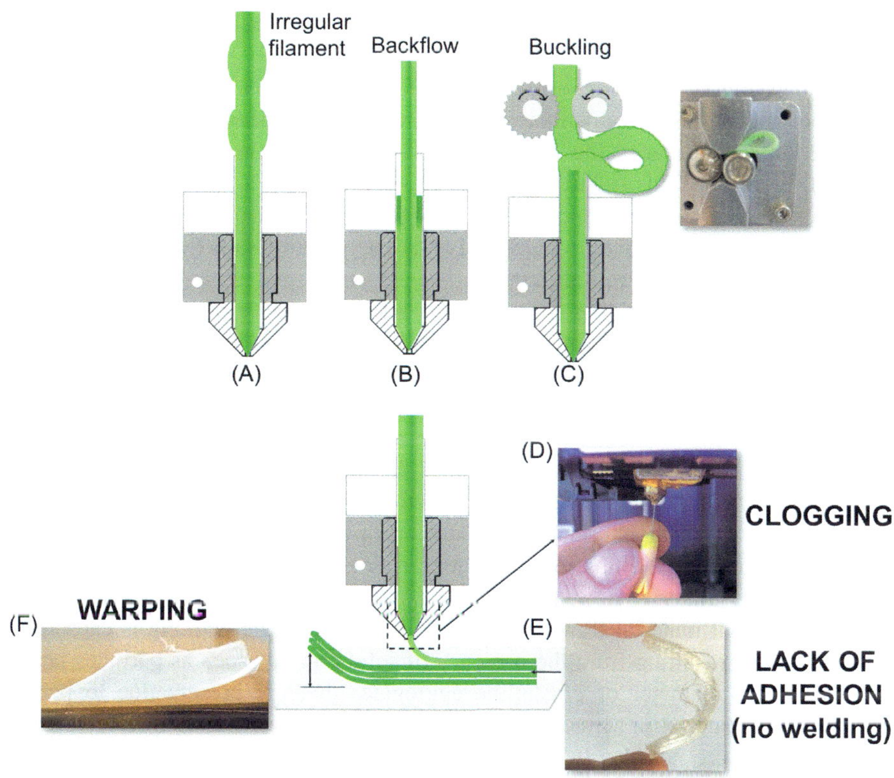

Figure 1.3 Schematic representation of the common failures during printing by FFF or FDM related to the solid filament, the flow (molten material), and the solid printed part.

between the spool and the printing head to match the printing patterns. The text below provides essential information on the basics of failures. Section 1.3 presents various methods of handling them, including quantification.

In MEX printing, the diameter of the filament is a crucial factor. A small diameter can cause buckling due to decreased critical buckling stress, while a large diameter restricts entry into the entrance (Fig. 1.3A). Commercial filaments have a tight and uniform diameter along the entire length of the filament. However, developing new polymeric materials involves the production of homemade filaments, usually by melt extrusion and drawing processes, whose dimensional homogeneity can be altered by many factors. Quality control is crucial in filament production to ensure consistency and reliability in printing.

Once the filament is molten, the polymer flow characteristic becomes critical to avoid failures. Higher temperatures decrease the melt viscosity of thermoplastic polymers, facilitating the flow and, therefore, the printability of the material. Nevertheless, temperatures that are too high can lead to polymer degradation and loss of dimensional stability. Filaments must be stiff and flexible with adequate flow properties at processing temperatures to achieve the best results. Annular backflow, buckling, and clogging are some of the most common failures associated with extrusion and flow (Fig. 1.3B–D, respectively).

Annular backflow is a common failure in FFF (Fig. 1.3B). It happens when the molten polymer flows upwards through the annular region between the filament and the liquefier wall, escapes the heated area, and solidifies in the low-temperature region of the heater. This failure is typical in polymers with a low activation energy for flow, which can experience a substantial drop in viscosity over a narrow temperature range. Sodium sulfonated polyethylene glycol (NaSPEG) is an example of such a polymer (Gilmer et al., 2018).

On the other hand, buckling (Fig. 1.3C) is a typical failure of flexible filaments. Ultimately, during FFF, the filament acts as a piston, exerting the necessary force to extrude the molten material through the nozzle. Inadequate stiffness and/or too high viscosity leads to filament buckling (Gibson et al., 2010). Different devices have been designed to prevent filament buckling; however, knowledge of the mechanical and flow properties of the material will be one of the most effective tools for predicting and controlling it, as will be discussed in Section 1.2.

Finally, it is worth mentioning that when carrying out the extrusion process, it is important to consider the possibility of nozzle clogging and choking. These issues can obstruct the material flow and lead to printing failure. Clogging occurs when there's insufficient material flow due to various reasons while choking is caused by nozzle blockage resulting from the presence of solid regions, such as particles larger than the nozzle diameter or crystalline aggregates that are present due to partial melting of the material in semicrystalline materials. However, in most studies, these two phenomena are not clearly differentiated, so the following discussion will only refer to "clogging."

Nozzle clogging is a common issue that occurs during AM. It can cause interruptions in the printing process, leading to the replacement of the nozzle and even a complete stoppage of the printing process (Fig. 1.3D). Unfortunately, nozzle clogging problems cannot be automatically resolved on FDM machines. They can only be detected by the operator through visual inspection. Many factors can cause nozzle clogging, such as the accumulation of dust inside the liquefier, the absence of space for extrusion due to insufficient layer height or warping of the printing part, and other issues (Tlegenov et al., 2018). This problem is particularly critical when printing with polymeric materials containing high filler content (between 45% and 60%) (Gonzalez-Gutierrez et al., 2018). Clogging can sometimes occur when the heat transfer from the liquefier to the feedstock material is inefficient. The feedstock material must be heated through direct thermal contact with the heated liquefier wall to reduce its viscosity and make it extrudable. However, it is widely reported in the literature that the actual extrusion temperature can be several tens of degrees lower than the set liquefier temperature (Phan et al., 2018; Seppala & Migler, 2016; Srinivas et al., 2018; Vaes et al., 2019; Vanaei et al., 2020). This difference is further exacerbated when the printing velocity is increased because the filament spends less time in the liquefier (Peng et al., 2018).

Failures associated with the 3D-printed part

After the filament is correctly extruded from the nozzle, it is deposited layer by layer to create the final 3D-printed object. However, since no molds or cooling devices are involved in this process, the success of the printing will be intimately related to the nature of the material and its specific behavior under different printing conditions (printing speed, layer height, printing temperature, and bed temperature). Interlayer adhesion and warping are the two main issues that can arise during this process.

Interlayer adhesion: general aspects

Another critical aspect of AM techniques is interlayer adhesion, often referred to as bonding, welding, or healing, between layers since it determines the actual performance of the printed specimens. Indeed, this aspect is not exclusive to extrusion-based techniques. This concept could be extended to all AM techniques that involve a layer-by-layer construction of the final part (taking into account the peculiarities of each one). Generally, 3D-printed parts obtained from commercial materials or engineering polymers exhibit inferior properties to their counterparts produced by conventional processes such as injection molding (Das et al., 2020; Gao et al., 2015) and compression molding (Goh et al., 2020). In traditional injection molding, high pressures are applied to the molten polymer, resulting in high compaction; however, during the FFF process, the molten polymer solidifies and bonds (or welds) with the deposited neighboring strands without the help of any external energy or force other than gravity or the extrusion force of the FFF machine (Coogan & Kazmer, 2017; Sun et al., 2008). Indeed, it has been reported that the strength along the build direction (z-direction) can reach only 40%–50% of that in other directions (Goh et al., 2020). Therefore, the quality of the bond interface between neighboring layers is an essential factor that will ultimately determine the structural integrity and mechanical properties of extrusion-based 3D-printed parts.

Interlayer adhesion between deposited strands is a thermally driven process that involves molecular interdiffusion and the reentanglement of the polymeric chains across the inter-filament interface (McIlroy & Olmsted, 2017). Interlayer bonding of isothermal amorphous interfaces has been conventionally described by three consecutive stages: (1) surface contact, (2) surface tension-driven neck growth, and (3) molecular diffusion and entanglement across the interface between filaments (Bellehumeur et al., 1998; Coogan & Kazmer, 2017). At this point, it should be noted that necking or neck formation is referred to the liquid bridge formed during the melting of neighboring filaments in the case of MEX (Bellehumeur et al., 2004), or particles in the case of PBF (Mazlan et al., 2023). During strand deposition, intimate contact is established between polymer interfaces as the material is deposited layer by layer. When neighboring filaments come into contact, coalescence begins instantaneously. Once surface contact has occurred, healing of the polymer interface proceeds to a mechanism of macromolecular chain diffusion across the interface (Wool and O'Connor, 1981). Coalescence and diffusion require molecular mobility, which means the temperature should remain sufficiently higher than the glass transition temperature (T_g). Thus, for amorphous polymers, interlayer adhesion can be improved by maintaining the layer-layer interface above the materials T_g for an extended period. Above T_g, macromolecular chains retain mobility, facilitating chain diffusion across the interface. Many studies have been conducted under this premise.

On the other hand, the effect of the different printing parameters on interlayer adhesion remains unclear. Several studies have confirmed the positive impact of printing at high liquefier temperatures (Abbott et al., 2018; Aliheidari et al., 2018). However, the effect of print velocity on interlayer adhesion is less clear. Some studies have reported no significant influence of print velocity on welding (Seppala et al., 2017), while others have observed both positive and negative effects of increasing print speed on interlayer bonding. Increasing the build plate temperature can also positively impact the interlayer bonding, although its effect is usually less pronounced than adjusting the printing temperature (Aliheidari et al., 2018). Nevertheless, it was found that the bond mechanism for the amorphous and semicrystalline polymers is slightly different. For amorphous polymers, interlayer bonding is dominated by interdiffusion and/or entanglements across the bond interface. Instead, for semicrystalline polymers, it is more complicated as the crystallization hinders the molecular diffusion since the polymer chain will then be embedded in the growing crystals, and entanglement and the partial melting of the strand will be the driving factor of the process. When printing semicrystalline polymers, neck growth occurs only at temperatures above the crystallization temperature (T_c) of the polymer (Shahriar et al., 2017). As the interfacial temperature decreases, the onset of crystallization induces a dramatic increase in melt viscosity and hinders neck growth, while molecular diffusion continues until the T_g of the polymer is reached. There are several studies concerning interlayer adhesion of semicrystalline feedstock polymers in FFF. These studies can be categorized into two main groups based on their approach to characterizing interlayer welding; however, they will be addressed in Section 1.4.

Warping: general aspects

Shrinkage, also referred to as "warping" or warpage, is a significant concern in polymer processing as it can result in a loss of dimensional stability of the final product. Shrinkage occurs due to decreased free volume between macromolecular chains and their vibrational volume when polymeric materials are cooled from the melt (liquid) state to temperatures below their T_g, where vitrification occurs. The extent of shrinkage is influenced by various factors such as polymer structure, crystallinity, and processing conditions. Semicrystalline polymers are more prone to shrinkage than amorphous polymers as they form highly dense crystalline structures during crystallization, leading to more significant volume contraction.

The shrinkage issue is particularly critical in AM, where the gradual cooling and solidification of the molten polymer can cause internal stresses within the printed part. These internal stresses can result in distortion, curling, or even printing failure. To ensure dimensional accuracy and minimize the impact of shrinkage, it is crucial to consider material properties, processing parameters, and part geometry.

1.2.2 Powder bed fusion (PBF)

Apart from most conventional MEX methods, another interesting AM alternative that uses semicrystalline polymers as feedstock is PBF, also called selective laser sintering (SLS). The technological concept implies the selective melting of powder material upon heating by a laser beam and the subsequent coalescence of particles layer by layer. Fig. 1.4 shows a schematic representation of the main steps of the process. The manufacturing process is based on different consecutive steps where the material found in the built area is heated (1) to the manufacturing temperature to subsequently apply the energy that will give rise to the 3D object fabrication; (2) finally, the manufacturing process finishes with a cooling down and consolidation

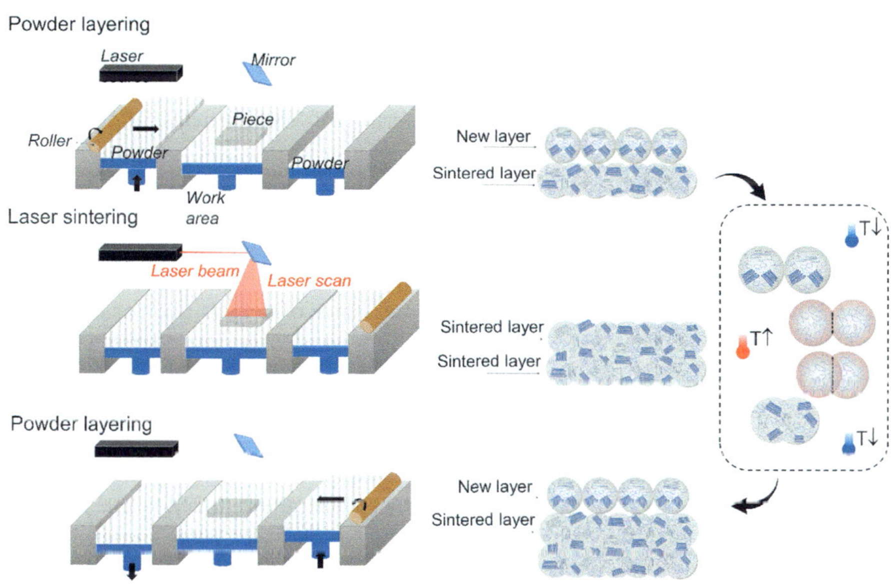

Figure 1.4 Schematic representation of the PBF technique and the process involved in forming the AM part.

step (Chatham, Long, et al., 2019). According to Chatham, Long, et al. (2019), the fabrication phase (2) can be divided into three subfunctions: powder recoating, energy input, and particle coalescence. Although this technological approach can use any thermoplastic material as feedstock, the industry has favored using semicrystalline materials as preferred polymers for PBF. In fact, polyamides (PA), mainly PA-12, are considered gold standard materials for this technology (Goodridge et al., 2012; Schmid et al., 2014).

One of the significant drawbacks of PBF compared with other AM technologies or formative manufacturing alternatives (i.e., injection molding) is the need to obtain the material in powder form. The particle morphology highly affects the recoating and fabrication phases (Chatham et al., 2019), and usually, rounded particles, typically around 100 μm, are desirable (Zou & Yu, 1996). Values related to powder size and flowability are the most frequently provided for feedstock materials (Kusoglu et al., 2021). The production of the powder is a complicated process, where its scale-up is highly difficult and, in most cases, protected under patent. Several alternatives have been presented for producing suitable semicrystalline powders for PBF, including material cryo-milling (Liang et al., 2002), spray drying (McAlea et al., 1998), or solvent precipitation (Lin et al., 1993). Other options, such as suspension polymerization (Dechet et al., 2020; Dechet et al., 2021), are being explored to expand the materials presented in this technology.

1.2.2.1 Key parameters for material development for PBF

In PBF manufacturing, the ability of the polymer chains to diffuse and subsequently stabilize the 3D object through crystallization governs the process (Fig. 1.4) (Chatham et al., 2019). In the right section of Fig. 1.4, delimited by a dashed line, the melting process of the crystalline structures due to the energy received from the laser, the subsequent interdiffusion and sintering of the particles and the new crystallization after cooling the system are schematized.

PBF is an intrinsically nonisothermal process that implies applying thermal and optical energy for long periods. Consequently, materials can suffer from several physicochemical processes, such as aging or degradation, which might alter their fabrication performance. Similarly, these processes influence the mechanical properties of 3D objects, their surface finish, or the appearance of defects, such as undesired curvature, shrinkage, or cracks. Three key points should be considered to avoid defects: material heating, particle coalescence, and material consolidation.

Material heating

The first point to consider concerns material heating. The material transformation occurs due to a laser beam's capacity to significantly heat the semicrystalline material above its processing/melting temperature. To reduce the laser beam's intensity and avoid material degradation, the feedstock material is heated to a temperature close to but below its T_m. Therefore, material heating occurs through a combination of thermal and optical energy that permits the selective melting of material in the building area. Further details are shown in Section 1.4.

Particle coalescence

Particle coalescence is the core of PBF. The driving force of particle coalescence is the reduction of the surface energy that occurs when increasing the particle size (and, therefore, reducing the surface area). Thus, the particle coalescence is controlled by the surface energy and the rheological behavior of the materials at the fabrication temperature. In contrast to conventional formative processes or MEX, particle coalescence occurs under very low shear rate conditions. Thus, the critical parameter that governs the molecular interdiffusion across the particle interphase is the zero-shear viscosity (η_0). It is worth noting that polymers present a high η_0 due to their high molecular weight. However, semicrystalline materials present a comparative advantage over their amorphous counterparts since they present a lower viscosity at processing temperatures.

Material consolidation

After light application and particle coalescence, the material is consolidated as the temperature cools down. This process is dominated by the feedstock material's crystallization characteristics and the final part's cooling conditions. It is to be noted that during the sintering processes, the material is kept at high temperatures in the building area with no control over the cooling of coalesced structures. In addition, the material subjected to radiation does not present the same thermal history across the particles. These two effects result in crystallization anisotropies that are worth analyzing.

The radiation is transformed into heat at the particle surface during particle coalescence, producing the material arrangement. However, as polymeric materials are poor thermal conductors, a thermal gradient is created between the core and the surface of particles; if insufficient thermal energy is applied to particles, an unsuccessful melting might occur, leading to core-shell structure in terms of material crystallinity (Zarringhalam et al., 2006). It is clear that an effective crystallization should lead to isotropic crystallization and a significant amount of interparticle crystals to provide enough mechanical support while avoiding anisotropic shrinkage (Chatham et al., 2019).

The effect of melt memory (a recent review is provided in Sangroniz et al., 2020) is one of the aspects that needs to be sufficiently investigated in the framework of PBF. As the heating and cooling during material irradiation is very fast, the melt may contain some remaining order (melt memory). Important crystallization and melt memory concepts are discussed in Section 1.3 to understand this effect.

To sum up, both MEX and PBF techniques require the material layers to interdiffuse in the molten state to produce robust interlayer adhesion and good mechanical properties. After that, the material should quickly consolidate to avoid spreading. Consequently, these manufacturing technologies are highly dependent on the rheological characteristics of the feedstock material (Das et al., 2021; Mackay, 2018) and its crystallization capabilities (McIlroy & Graham, 2018). Thus, this book chapter focuses on the most relevant aspects of AM of semicrystalline materials and their relationship with their rheological properties and crystallization capacity.

1.2.3 Other alternatives for Additive Manufacturing (AM) of semicrystalline polymers

MEX and PBF constitute the leading AM technologies for semicrystalline polymers. However, new synthetic approaches have recently been explored to incorporate semicrystalline materials into other technologies. Efforts have mainly focused on the VP due to its superior 3D resolution and high scalability. However, the necessity of polymerizing the material in situ through a photoactivated process and the cross-linking nature of the resulting material imposes a limitation for semicrystalline materials. Most synthetic approaches to confer crystallinity to objects produced by VP have incorporated crystallizable segments into flexible materials. For example, polycaprolactone dimethacrylate macromonomers were used as photoresins and conducting sensors (Zarek et al., 2016) or biological scaffolds (Samson et al., 2023). However, the crystallization degree in this approach was lower than in MEX and PBF.

The main drawback for materials to crystallize is the network character that limits the capacities of chains to rearrange in crystalline domains. To provide more chain mobility, Page et al. introduced crystallizable pendant groups into the resin formulation, leading to higher crystallinity (Rylski et al., 2023). In a different approach, Bowman et al. produced covalent adaptable networks capable of producing viscous dissipation and promoting VP object material crystallization (Kuenstler et al., 2023). Commisso et al. (2023) recently proposed a synthetic route to produce semicrystalline thermoplastic materials in situ by radical-mediated ring opening polymerization. Due to the interest in the field and the different synthetic approaches presented, the incorporation of semicrystalline materials into other AM techniques will grow in the following years. It will complement the most established MS and PBF technologies.

1.3 Semicrystalline polymers in AM

1.3.1 Types of semicrystalline polymers with printability capacity

Polymers have found extensive applications in ME, where both amorphous and semicrystalline thermoplastic polymers serve as primary feedstock:

1.3.1.1 Amorphous polymers

These materials are the most commonly recommended materials for MEX. They offer advantages such as a low thermal expansion coefficient and extrusion temperature, which help mitigate internal stresses during cooling, such as warpage. Acrylonitrile butadiene styrene (ABS) stands out as one of the most used and commercially available materials (Agarwal et al., 2022; Dusanapudi et al., 2023; He et al., 2021; Karad et al., 2023; Khabia & Jain, 2020; Mogan et al., 2024; Rodríguez et al., 2003; Shashikumar & Sreekanth, 2023). Its toughness and high impact, as well as its chemical and abrasion resistance, are well known. However, two disadvantages of the ABS are its shrinkage and warping after part production and the chemical fumes produced during its processing that could generate respiratory difficulties. Despite these drawbacks, ABS has found applications in different industries, including aerospace, automotive, and biomedicine. However, today, other researchers have studied other polymers, such as polycarbonate (PC)(Chinnasamy et al., 2021; Fang et al., 2020), polymethyl methacrylate (PMMA) (Lyu et al., 2024; Street et al., 2018), and polyetherimide (PEI) (Das et al., 2021; Wu et al., 2018).

1.3.1.2 Semicrystalline polymers

Semicrystalline polymers may exhibit a higher degree of anisotropy compared to amorphous polymers. Additionally, the study of interdiffusion across interfaces in semicrystalline polymers is more complex due to factors such as incipient crystallization, which can lead to part distortion. Semicrystalline polymers in ME can be used in many applications, such as electronics, biomedicine, automotive, and aerospace. The semicrystalline polymers used in ME are divided into three groups:

Commodity polymers

Polylactic acid (PLA) (Corapi et al., 2019; Costanzo et al., 2020; Hikmat et al., 2021; Kartikeyan et al., 2023; Lokesh et al., 2022; Mani et al., 2022; Marşavina et al., 2022; Maurya et al., 2022; Popa et al., 2023; Sahoo et al., 2023; Sandanamsamy et al., 2023; Singh et al., 2022; Yao et al., 2020; Zhao et al., 2019), PP (Charlon et al., 2021; Jin et al., 2020; Kumar et al., 2018; Petersmann et al., 2020; Spina & Cavalcante, 2020; Vidakis et al., 2022; Zander et al., 2019), PCL (Paetzold et al., 2022; Zhao et al., 2020), and high-density polyethylene (HDPE) (Poojitha et al., 2023; Schirmeister et al., 2019).

Engineering polymers

Thermoplastic polyurethane (TPU) (Candal et al., 2021; Domínguez-Robles et al., 2022; Hohimer et al., 2020; Kim et al., 2017; Lin et al., 2021; Yang et al., 2023), polyethylene terephthalate glycol (PETG) (Alarifi, 2023; Bhandari et al., 2019; García et al., 2022; Ginoux et al., 2023; Loskot et al., 2023; Vaught & Polycarpou, 2023), polyamides (PA) (Feroz Ali et al., 2022; Lupone et al., 2020; Peng et al., 2020; Qi et al., 2021; Qi et al., 2022; Sgrulletti et al., 2021; Su et al., 2022; Toro et al., 2019; Wang et al., 2024; Yang et al., 2024), polyethylene terephthalate (PET) (Chatham et al., 2019; O'Driscoll et al., 2024; Van de Voorde et al., 2022), poly (butylene succinate-co-adipate) (PBSA) (Candal et al., 2020), and poly(hydroxybutyrate-co-hydroxy hexanoate) (PHBH) (Caputo et al., 2023).

High-performance polymers

PEEK (Arif et al., 2020; Basgul et al., 2020; Ding et al., 2019; Greco et al., 2024; He et al., 2023; Lin et al., 2019; Pulipaka et al., 2023; Rodzeń et al., 2021; Saini et al., 2022; Sikder et al., 2022; Wang et al., 2020; Yao et al., 2024; Zhao et al., 2020), polyvinylidene fluoride (PVDF) (Kumar et al., 2021; Milenkovic et al., 2021; Mullaveettil et al., 2021; Simunec et al., 2023; Tao et al., 2022; Zhang et al., 2024), and polyphenylene sulfide (PPS) (Garmabi et al., 2022; Shuto et al., 2022).

In recent times, several researchers have tried to increase the number of materials suitable for ME. The manufacturing by FFF of semicrystalline polymers is not as simple as that of amorphous polymers due to the warpage and deformation that suffer during crystallization while processing (Brenken et al., 2019; Silva et al., 2017). However, it is still not known exactly what warpage process these materials undergo during their printing process. Only some authors have studied and tried to explain this process, but only with some particular materials. However, semicrystalline polymers are of particular interest since this class of materials generally outperforms amorphous polymers in applications where chemical resistance, resistance to environmental stress cracking, fatigue resistance, use temperature range, and barrier properties are essential (Dusunceli & Colak, 2008; Lagaron et al., 2004; Molnár et al., 2021; Şerban et al., 2013).

Regarding the PBF techniques, it has been previously mentioned that the industry has primarily favored using semicrystalline thermoplastic materials as the preferred feedstock. Among these materials, PA, particularly PA-12 (Nylon 12), is considered the gold standard for this technology (Goodridge et al., 2012; Schmid et al., 2014) because of its versatility, reliability, and suitability for producing high-quality 3D printed parts. However, other PAs such as PA-11, PA-6, and PA-66 are also utilized, as well as semicrystalline engineering polymers such as PEEK and poly aryl ether ketone (PEAK). Additionally, some attempts have been made with poly(butyl terephthalate) (PBT), PP, poly(methyl oxide) (POM), and PCL (Vasquez et al., 2014; Wegner, 2016).

Printing semicrystalline materials is a challenging task due to the various possible failures associated with crystallization. However, incorporating crystals in 3D-printed parts can provide unique properties that further boost AM processing growth. In this context, understanding polymer crystallization is crucial in AM processing, just as in other already mature manufacturing techniques. After reviewing the generalities of AM processing in the previous section, in this section, we introduce essential concepts on polymer crystallization before focusing on the particularities of printing semicrystalline materials presented in the following sections.

Semicrystalline polymers are integral to the plastic industry, with nearly 75% of all commercially available polymers being semicrystalline. These materials possess unique properties that are dependent on factors such as crystallinity degrees (X_c), T_m, and morphology and can be utilized across a wide range of temperatures between the T_g and T_m (Hadjichristidis et al., 2022). The anisotropic constitution of semicrystalline polymers, given by long-chain molecules regulating crystallization for kinetics reasons, is among their most significant features. This means these polymers will always have a portion of amorphous or unordered regions, while metals and ceramics can have 100% crystallinities. This feature ensures sufficient tensile strength given by the crystalline part and certain ductility thanks to the amorphous one (Nie et al., 2022). Thus, a deeper understanding of polymer crystallization, which has been studied over its 60-year history, is crucial for the manufacturing and processing of these materials. Crystallization is vital in the entire AM process, from obtaining pellets, powder, and filaments to printing and optimizing printing conditions. In traditional techniques such as injection molding and extrusion, the structuring process is well-understood, and this has been essential for achieving control and optimization. However, the understanding of the structuring process in AM still needs to be improved, and it is necessary to advance this understanding, particularly for printing semicrystalline materials. Doing so will significantly contribute to the development of AM. This section introduces some essential concepts that provide insights into polymer crystallization during printing.

1.3.2 Crystallization concepts

The crystallization process is a fascinating phenomenon that can transform matter from a molten disordered phase to a crystalline phase. This transformation process comprises different stages, the kinetics of which can be tailored, affecting the material's final properties. Thus, this transformation process can be controlled by using additives, like nucleating agents and plasticizers, that modify specific stages of the process or modify the material thermal history, for example, by changing the processing conditions. The crystallization process typically involves two stages: nucleation and growth, often referred to as "primary nucleation" and "secondary nucleation," respectively in literature. The "overall crystallization" process comprises both nucleation and growth and can be monitored by Differential Scanning Calorimetry (DSC) experiments. It is crucial to understand the different stages in the overall crystallization process. This section will briefly discuss these stages to understand the process better.

1.3.2.1 Primary nucleation or nucleation

Nucleation is a critical step in the crystallization of polymers, and it involves the formation of nuclei where none existed before. There are two types of nucleation: *homogeneous andheterogeneous nucleation.*

Homogeneous nucleation occurs when polymer chains spontaneously aggregate to form nuclei, while heterogeneous nucleation occurs at the surface of heterogeneities or impurities in the polymer (Pérez-Camargo et al., 2022). Homogeneous nucleation is a rare occurrence since the formation of the new phase depends solely on the fluctuation of the mother phase, that is, melt, without the contributions of foreign elements. This means that all the faces of the nucleus are in contact with the melt. However, homogeneous nucleation can play a significant role in the crystallization process at very large supercooling (ΔT of 50 to 100 K), often close to the polymer's T_g. Reaching such large ΔT without crystallizing during the cooling is only possible with ultrafast cooling rates such as the ones offered by Fast Chip Calorimetry or Flash DSC. Thus, on the one hand, Flash DSC experiments is one of the ways to study experimentally homogeneous nucleation, and on the other hand, to mimic real processing conditions, as in AM, that employ fast cooling rates.

Self-nucleation (*SN*) can be regarded as a unique form of homogeneous nucleation due to self-seeds or crystal fragments present in the material (due to partial melting) or chain segments with residual memory in a polymer. This nucleation is induced through a specific thermal treatment process. It involves subjecting the sample to a standard thermal history before heating it to a self-nucleation temperature, T_s, and thermally conditioning it for a set time, t_s, typically 5 minutes in a standard DSC. The T_s is usually set between $T_{m,\ onset}$, and the temperature required to erase the thermal

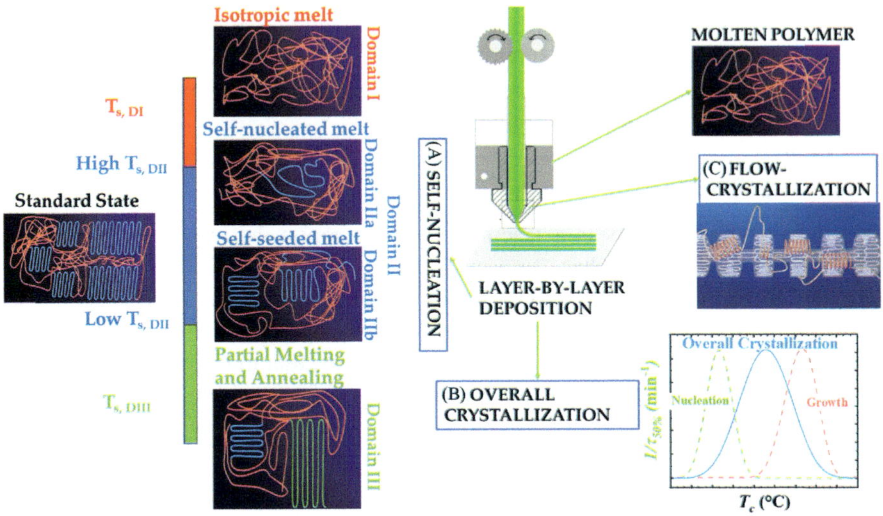

Figure 1.5 Schematic representation of polymer crystallization concepts applies to AM processing: (A) self-nucleation, (B) overall crystallization, and (C) flow crystallization.

history, that is, $T=T_m + 20$ (or 30°C). Depending on the T_s, the sample experiences different effects, defining the SN domains (Pérez-Camargo et al., 2022). The three main SN (see Fig. 1.5A) domains are as follows:

Domain I or complete melting domain (complete melting): With $T_s >> T_m$, the sample will be completely molten (isotropic state), and thus, its thermal behavior will be the same as the standard state, that is, $T_s=T$.

Domain II or exclusive SN domain (melting of most of the crystals): The T_s is high enough to melt most of the crystals but low enough to leave small crystal fragments (*Domain IIb or self-seeding Domain*) or ordered regions in the melt (*Domain IIa or melt memory Domain*) that act as self-nuclei. In more detail, *Domain IIb* is located at the lower temperature of *Domain II*, leaving small unmolten crystal fragments that represent self-seeds, however, incapable of annealing. *Domain IIa* starts at $T_s > T_{m, end}$ leaving certain ordered regions in the melt that act as self-nuclei upon the following cooling.

Domain III or self-nucleation and annealing domain (partial melting): The T_s is low enough to cause melting of a fraction of crystals and the annealing of unmolten crystals during the holding time at T_s.

In the AM process, using temperatures in *Domain II* might be beneficial due to the increase in nucleation density. Also, the memory effect can be significant in sintering, as mentioned in Section 1.2.2. However, these points are not often considered in AM processing. It is important to note that the layer-by-layer deposition process can cause cyclic heating-cooling profiles, resulting in a complex thermal history. This complex thermal history may cause inhomogeneity in the crystal morphology at the microstructure level (Baouch et al., 2024). Thus, the SN process may already be involved in the printing process rather than being solely an additional treatment to enhance nucleation density. Moreover, the core-shell-like structure that might be formed in PBF can also involve a temperature profile and an inhomogeneous crystallization. Altogether, this indicates that studying the SN process can be vital for AM.

Heterogeneous nucleation occurs at the surface of heterogeneities. These heterogeneities include extraneous solids, cavities, and container walls, among others, which enhance the formation of stable embryos. Heterogeneous nucleation is observed at relatively low ΔT and has been extensively studied in the literature. It occurs in most polymeric materials, either for inherent impurities of the synthesis or by nucleating agents purposely added to favor the nucleation.

1.3.2.2 Growth or secondary nucleation

After the formation of nuclei, the growth process begins as chain segments attach to the growth front. The most common morphology observed during polymer crystallization from the melt is the formation of spherulites—three-dimensional spherical aggregates of radially growing crystals. The growth or secondary nucleation is directly related to the radial growth of the spherulites. The process is considered complete when most of the spherulites impinge on one another at around half the crystallization time, $\tau_{50\%}$. However, even when the spherulites impinge, some chain attachments to the growth front events also occur. This is considered a different stage and is known as secondary crystallization. Overall, the growth process refers to the unrestrained growth of the material. Two opposing factors influence the rate (growth kinetics) at which a material grows during the crystallization process: diffusion and nucleation. As the temperature increases, the rate of molecular transport within the melt also increases due to diffusion, while the nucleation rate decreases.

During the process of *secondary crystallization*, which begins after the primary crystallization is completed (around $\tau_{50\%}$), the spherulites begin to impinge, triggering a series of inter- and intraspherulitic crystallization processes. As the process continues, the free spaces between the spherulites get filled with crystallites, and a densification process occurs within the spherulites, including a reorganization process that involves the thickening of the lamellae through an annealing effect and the formation of subsidiary lamellae.

The full solidification process is a complex process involving primary crystallization (nucleation and growth) and secondary crystallization. The overall crystallization rate, represented by the plot of the overall crystallization rate (usually expressed as the inverse of the half-crystallization time) versus isothermal crystallization temperature (see Fig. 1.5B), can be considered a combination or mathematical superposition of the bell-shaped behavior of nucleation and growth rates. This results in a bell-shaped curve where nucleation and diffusion factors compete. Thus, mobility increases with the increase in temperature, while the thermodynamic driving force decreases with temperature. These concepts are present in AM; recently, poly (ether ether ketone) (PEEK) crystallization was studied by simultaneously printing and taking synchrotron WAXS/SAXS patterns. It was observed that during the early deposition times (< ~10 seconds), the material near the print bed cools faster than at the polymer/air interface, so the temperature increases with height from the bed, resulting in a decrease in the driving force for crystallization and an increase in the time to crystallization, which causes a slope in the growth front (Ree et al., 2024). Similarly, increasing the bed temperature also increases the time to the onset of crystallization, and the crystallinity rises with an increase in bed temperature. This remarks the importance of crystallization in the AM process.

1.3.2.3 Flow-induced crystallization

The above crystallization concepts refer to the polymer under quiescent or static conditions. But, when the polymer is subjected to an external flow field, the crystallization is affected. Such an external flow field is present during processing. Processing techniques, such as fiber spinning, injection molding, and film blowing, all involve complex flow dynamics that can dramatically impact the formation of hierarchic structures and the final properties of semicrystalline polymers (Wang et al., 2016). Therefore, understanding the phenomenon of flow-induced crystallization for industrial applications is crucial. Flow-induced crystallization (see Fig. 1.5C) can be traced back to the stretch-induced crystallization of natural rubber conducted in 1805 (Wang et al., 2016). It is now widely recognized that flow can accelerate nucleation rate, modify nuclei morphology, and induce different crystalline structures (Graham, 2019; Nie et al., 2022). Flow dramatically enhances the rate at which polymer crystallizes and profoundly affects its morphology. By distorting the configuration of polymer chains, flow breaks down the kinetic barriers to crystallization and directs the resulting morphology, such as shish-kebab morphology. This flow-induced crystallization in polymers is a fascinating example of an externally driven, nonequilibrium phase transition controlled by kinetics (Graham, 2019).

The study of flow-induced crystallization is a highly complex and fascinating research field. When a phase transition occurs within a system, the rheological behavior of that system is significantly altered by the growing crystallites dispersed throughout the fluid matrix. In a polymer melt typically stable under quiescent conditions, crystals can be developed simply due to changes in chain conformation associated with an intense flow field. From a thermodynamic standpoint, an oriented amorphous polymer is in a lower entropy state than an unperturbed macromolecule in a random coil conformation. This reduction in melting entropy results in a deformation-dependent melting temperature, which increases with the extent of molecular orientation. As a result, the oriented polymer will experience higher ΔT than the quiescent system for a given T_c, leading to faster crystallization.

Understanding polymer flow-induced crystallization offers the potential to use processing conditions to control the crystallization kinetics and final morphology, leading to improved properties in a wide range of final products. As mentioned before, in AM, the influence of crystallization and flow-induced crystallization is somewhat limited compared to traditional techniques. Still, in polypropylene (PP), have been reported that the flow-induced crystalline morphology depends on the printing parameters: mild conditions (high nozzle temperature and low printing speed) lead to homogeneous morphology, while pronounced inhomogeneity in the crystalline morphology was produced at low extrusion temperatures and high printing speeds (Petersmann et al., 2020).

Recently, this phenomenon has been further highlighted by relating it to weld strength between layers. It has been highlighted that the polymer diffusion process at the filament-filament interface affects weld strength between layers. The interdiffusion of the polymer across this interface is crucial to ensure strength at the weld layers. In amorphous polymers, the diffusion process can be restricted by the T_g, which can lock the residual alignment and reduce mechanical properties. In semicrystalline polymers, once the temperature drops below the T_c, the polymer will nucleate, and spherulites will begin to grow. If the T_g is below RT, as in the case of PP, the crystallization will fill the space during cooling. The crystallization process slows the polymer dynamics, and interdiffusion will be arrested once complete space filling is achieved. Therefore, for semicrystalline polymers, the dominant mechanism restricting welding tends to be the crystallization rate, which is temperature-dependent rather than the time above T_g. Moreover, the spherulite's size and volume fraction developed during crystallization determines the mechanical properties chieved at solidification, making it inherently linked to the temperature profile.

Baouch et al. (2024) conducted a study to examine how PP feedstock's molecular weight and crystallization kinetics affect materials' welding properties in MEX. They discovered that samples with higher molecular weights had lower spherulite size and higher nucleation density than samples with lower molecular weights. This is due to the flow-enhanced nucleation that occurs during printing. Additionally, they found that applying shear rates similar to the shear at the nozzle increased the T_c (due to the increased nucleation density) in high molecular weight samples. However, no significant effect was observed in low molecular weight samples, even with shear rates higher than those experienced at the nozzle, which aligns with the morphological observations. Regarding mechanical properties, high molecular weight samples broke at lower stress levels and had a lower area under the stress-strain curve than low molecular weight samples. Furthermore, high molecular weight samples exhibited a brittle fracture pattern along the weld line. McIlroy et al. proposed a model to explain all the experimental findings. They found that the degree of interdiffusion decreases with molecular weight (and T_c) due to the reduced time available for welding. The model suggests that the flow-enhanced crystallization mechanism may further restrict welding. Therefore, flow-enhanced crystallization during ME can affect the spherulitic structure in the weld, impeding welding dynamics and leading to significantly reduced mechanical strength at the weld.

Printing semicrystalline polymers can be difficult, involving various concepts, as summarized in Fig. 1.5. On one hand, crystals can provide unique properties to the material. On the other hand, these crystals can affect certain properties, such as weld strength. Finding a balance between these opposing issues is key to increasing the use of semicrystalline materials in AM.

1.4 Understanding the AM of semicrystalline feedstock. An experimental and theoretical approach

1.4.1 Material extrusion-based technologies (MEX): Fused Filament Fabrication (FFF)

Fused Filament Fabrication (FFF) is one of the most popular MEX processes developed by Scott Crumb. Briefly recalling Section 1.2, in FFF, the material is supplied to the 3D printer as filaments, which are then fed into a control system that regulates the feed rate. The filament passes through a heating chamber or liquefier, where it is heated to processing temperature and, afterward, through a nozzle for deposition on the build platform. A horizontal layer is then constructed on this platform in relation to the extruder's horizontal *x*- and *y*-axes according to the digital object's predesign. Once the first layer is deposited, the extruder is moved vertically on the *z*-axis by a distance equivalent to the layer thickness, and a new layer is deposited (de Campos & dos Reis, 2024). This procedure is repeated until the final object is created. The mechanical and thermal controls within the extruder regulate the entire printing process and depend on the feedstock material properties (Al Rashid & Koç, 2021).

Many efforts have been made to identify the relevant physical parameters that govern the AM of semicrystalline polymers. As has been mentioned, AM of thermoplastic polymers implies at least three essential features: (a) adequate viscosity to ensure a liquid-like behavior in the nozzle to avoid filament buckling and clogging, (b) immediate welding of deposed polymer layers, (c) adequate cooling process to guarantee proper welding and minimize the shrinkage to obtain a good mechanical performance and dimensional accuracy.

Applying rheological knowledge to understand critical physics and its implications in all process aspects is essential in determining and identifying the correlation between material properties, processing variables, and the resulting mechanical properties. Most recent reviews, gathering numerous studies and detailed discussions on the subject, highlight the importance of temperature-dependent shear and extensional viscosity of the melt during the extrusion and the chain dynamic during solidification, as it will be intimately related to the degree of interlayer welding and, consequently, the mechanical performance of the final part. Therefore, evaluating the flow and viscoelastic characteristics of the materials in terms of polymer chain relaxation dynamics under the combined effect of shear rate and temperature profile is considered essential to get a deep understanding of the process.

In this section, critical aspects of the printing process of semicrystalline feedstocks will be analyzed from a rheological point of view. Fig. 1.6 will serve as a reference for the discussion. The process has been divided into three steps: extrusion (solid filament driving), flow (filament melting), and solidification (welding of the strands and subsequently crystallization). Therefore, Fig. 1.6 illustrates the key considerations for predicting material printability and comprehending the underlying physical principles of the process from a rheological perspective. It provides insight into the crucial determinations required for accurately characterizing the process. Additionally, some

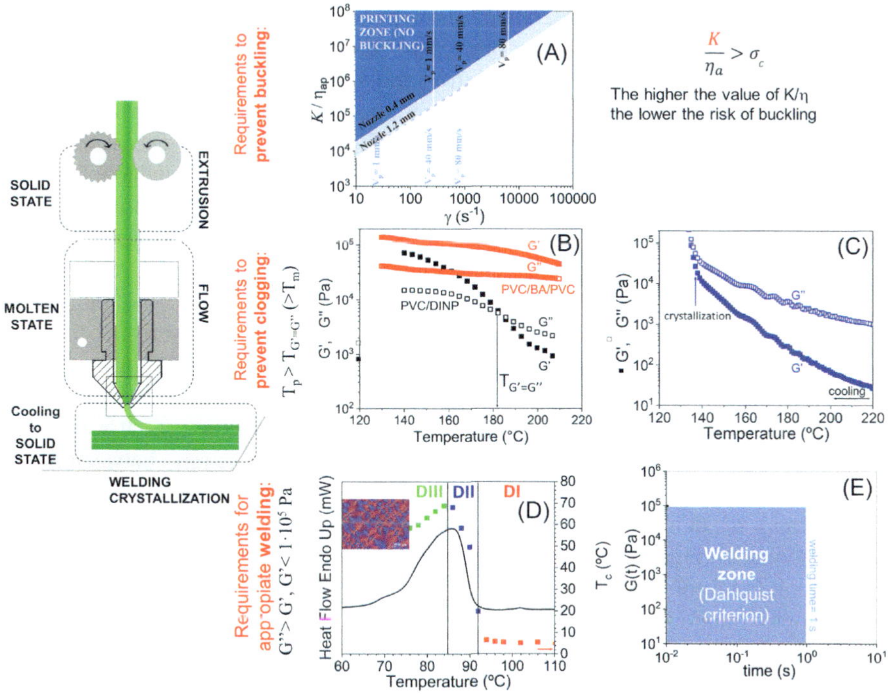

Figure 1.6 Fundamental rheological and thermal aspects that govern the FFF printing process: (A) Requirements to avoid buckling: *K*/*h* versus shear rate (s^{-1}). Colored area represents the adequate printing conditions to avoid buckling. (B, C) Requirements to ensure an appropriate flow avoiding clogging: Temperature scans under dynamic tests (SAOS), (B) refers to amorphous materials (plasticized PVC and PVC/PBA block copolymer) and (C) refers to semicrystalline materials (PLA); PBSA self-nucleation (SN) representation. Self-nucleation Domain (DII). (E) Stress relaxation test, showing colored welding zone according to Dahlquist criterion (see text).

guidelines will be given to predict the success or failure of the printing process depending on the material's flow and viscoelastic properties. Furthermore, some of the most successful simulations will be discussed and, finally, some tips related to process conditions (nozzle/bed temperature, printing velocity, patterns, layer height, etc.) will be given to improve printability, adhesion, and minimize warping, as well as to optimize the mechanical properties of the printed parts.

1.4.1.1 Extrusion process

Temperature-dependent shear and extensional viscosity at processing temperature play fundamental roles during the extrusion since they define the extrusion quality through the print nozzle and the dimensional stability of the strand in the region between the nozzle and the bed. Materials with lower viscosity flow more quickly through the nozzle, allowing for smoother extrusion and better print quality. Higher-viscosity materials may require higher extrusion temperatures to reach an appropriate viscosity for printing. However, high temperatures can also give rise to polymer degradation, even to a loss of dimensional stability in the case of excessively low viscosity. On the other hand, most polymers exhibit shear-thinning behavior, where the viscosity decreases as the shear rate increases due to the macromolecular and entangled nature of polymer chains. Shear-thinning behavior is advantageous in AM as it allows easier extrusion at high shear rates during printing.

Conversely, as it has been mentioned in Section 1.2, the suitability of a filament for FFF printing not only resides in the viscosity at processing temperatures. A filament should combine both stiffness and flexibility to avoid buckling. Another critical parameter determining the filament's efficiency in extruding without buckling is the ratio between compressive modulus, K, and the apparent viscosity, η_a. It was found that to ensure good filament flowability of ceramic filaments, the ratio between K and the melt viscosity, η, of the material at printing temperature should be higher than critical stress, σ_c, defined by dimensional parameters of the printing set-up, to avoid bucking (Venkataraman et al., 2000) (Fig. 1.6A). Calafel et al. extended this study to the printability of polyvinyl chloride (PVC)-based formulations. They observed that a better approximation was obtained when the viscosity value was corrected, considering the shear thinning behavior of the polymer melts during shear flow and the pressure effects related to short nozzle length (Calafel et al., 2020). Initially, unfilled semicrystalline polymers present hardly any problems related to buckling when they are FFF printed due to their medium/high crystallinity and relatively low compressibility modulus. Indeed, few studies have focused on the effect of printing temperature on semicrystalline filament viscosity and flowability. Most of them are oriented toward the impact of processing temperature on the mechanical properties of the printed parts (Benwood et al., 2018; Drummer et al., 2012; Gonzalez Ausejo et al., 2018; Kaygusuz & Özerinç, 2019; Senatov et al., 2016; Vanaei et al., 2020; Wang & Gardner, 2017). Gilmer et al. (2018) developed a rapid screening process for analyzing the extrudability of polymeric materials for filament extrusion-based AM by predicting extrusion failure. They predicted the buckling susceptibility of LDPE, suggesting that materials with high tensile moduli, that is, greater than 1000 MPa, have a lower probability of buckling under the conditions generated in an FFF printer. Such tensile modulus values would be expected in polymers with a T_g sufficiently above room temperature or in materials of lower T_g with sufficient crystallinity to impart substantial rubbery modules above T_g. Of the promising semicrystalline polymers for biomedical applications, PCL is the most susceptible to buckling due to its relatively high viscosity at melting temperatures and high ductility at room temperature. For this reason, different strategies have been opted for, such as the use of screw-based extrusion systems (Liu et al., 2018a; Liu et al., 2018b), the incorporation of fillers (Pires et al., 2018; Zhao et al., 2020), or tailoring the mechanical properties of filament through processing (Liu et al., 2018a; Liu et al., 2018b).

Concurrently, numerous analytical models have been developed to better understand the extrusion process in FFF 3D-printing (Osswald et al., 2018; Xia et al., 2019). Some interesting works focusing on the extrusion process have been collected in the review by Shaqour et al. (2021). It is generally well established that the extrusion process is highly nonisothermal, especially at high extrusion temperatures, in which the material's flow properties and viscoelasticity play a crucial role in the process (Mackay, 2018). Printing velocity, the pressure generated inside the nozzle, nozzle dimensions, and their effect on the flow behavior and die swelling (Xia et al., 2019) will be some of the main aspects addressed in these studies.

1.4.1.2 Flow behavior

The clogging of the nozzle was discussed in Section 1.2.1.1, and this section focuses on controlling it from a rheological perspective. When heated sufficiently above their T_g, the material presents a predominantly liquid behavior. At printing temperature, it is known that a predominantly liquid behavior is given by viscous moduli (G'') values equal to or higher than elastic moduli values (G'). Therefore, to prevent clogging, it is preferable to keep the printing temperature equal to or higher than the cross-over between G' and G'' (Fig. 1.6B). Nevertheless, this should not be a real problem in the case of semicrystalline polymers as their printing temperatures are always above T_m, so the material is in the terminal region of the viscoelastic spectrum ($G'' > G'$) (Fig. 1.6C). In those cases, as represented in Fig. 1.6C for the dynamic cooling test of a commercial PLA, the cross-over temperature is reached when the material starts to crystallize. Note, however, that this discussion is aimed at neat homopolymers, leaving aside blends of semicrystalline/amorphous materials whose behavior may differ significantly from that mentioned above. If the liquefier is not long enough so as not to heat the whole filament, the material may remain partially unmolten. These remaining unmolten crystallites can act as nucleation sites for crystallization, occurring when the strand is deposited and cooled down. Therefore, the SN phenomenon will be present and directly affect the deposited material's crystallization rate and crystalline morphology (Aliheidari et al., 2018; Candal et al., 2020, 2021; Caputo et al., 2023; Müller et al., 2015; Northcutt et al., 2018; Spoerk et al., 2017; Verbeeten et al., 2020; Wang et al., 2017; Wolszczak et al., 2018; Yang et al., 2017).

1.4.1.3 Solidification

Following extrusion, molten material strands are deposited layer by layer and subsequently solidify. The quality and mechanical performance of the part hinge directly on the degree of interlayer adhesion achieved during solidification. While the mechanism of interlayer adhesion in amorphous materials is well-documented, the process becomes more intricate in semicrystalline materials. Here, crystallization could interfere with or even hinder the reptation and movement of polymer chains minimizing welding (E). Moreover, the additional volume shrinkage resulting in crystallization and the consequent generation of internal stresses can lead to detaching of the part (F). Both phenomena will be discussed in this section.

1.4.1.4 Interlayer adhesion: rheological aspects

In Section 1.2.1.1, we discussed the fundamentals of interlayer adhesion. This section provides an overview of the research conducted on this issue, proposed solutions, and the rheological aspects of materials that should be considered to prevent such problems or failures.

Studies concerning interlayer adhesion of semicrystalline feedstock polymers in FFF could be categorized into two main groups based on their approach to characterizing interlayer welding. In the first set of studies, researchers employ a qualitative approach focused on observing the quality of the weld interface, assessing factors such as surface roughness, porosity, and neck dimensions. Most are focused on evaluating factors such as printing parameters, material properties, and processing conditions that influence interlayer adhesion (Abbott et al., 2018; Candal et al., 2020, 2021; Caputo et al., 2023; Spoerk et al., 2019; Verbeeten et al., 2020; Wang et al., 2017; Wolszczak et al., 2018). The second group of studies aims to incorporate the crystallization phenomenon into their analysis of the welding problem. Researchers in this group investigate the effect of crystallization on chain diffusion and focus on the crystalline morphology at the weld interface. They examine how the crystalline structure affects interlayer bonding and explore strategies to optimize printing conditions to mitigate the impact of crystallization on weld quality (Bhandari et al., 2019; Hertle et al., 2016; McIlroy et al., 2019; Nogales et al., 2019; Shahriar et al., 2017; Srinivas et al., 2018; Wittbrodt & Pearce, 2015).

Some authors have successfully studied and modeled interlayer welding of amorphous polymers in FFF at a more fundamental level based on molecular dynamics and rheological results (Bartolai et al., 2018; Coogan & Kazmer, 2017; McIlroy & Olmsted, 2017; Seppala et al., 2017). Results from simulations ratify the experimental results to a certain extent. These results suggest that increasing parameters such as build plate temperature, liquefier temperature, print speed, and layer width can improve weld strength and align with the fundamental principles of polymer chain diffusion across the interlayer interface: (a) increasing the build plate temperature enhances the initial layer adhesion by ensuring better contact between the deposited material and the build surface. This promotes stronger bonding at the bottom layers of the print, contributing to overall weld strength; (b) higher liquefier temperatures reduce the material's melt viscosity and improve fusion between adjacent layers. This facilitates polymer chain diffusion across the interlayer interface, leading to stronger interlayer bonding; (c) slower print speeds provide more time for the material to melt and flow, allowing for better intermixing of polymer chains at the interface. This enhances diffusion and stronger bonding between layers, improving weld strength. (d) Increasing the layer width provides a larger contact area between adjacent layers, which promotes more extensive interfacial bonding. These findings underscore the importance of optimizing printing parameters. To achieve optimal printing results, it is essential to consider a combination of parameters, including build plate temperature, printing temperature, print speed, and layer height.

From a rheological standpoint, conducting rheological tests and characterizations can provide valuable insights into a material's behavior under printing conditions, guiding the selection process and enhancing printing success. As discussed above, excellent interlayer adhesion is also favored by a predominantly viscous behavior of the material at the welding temperature, which is always lower than the printing temperature. For a successful bonding, the material's response at the welding temperature should correspond to the terminal or flow regime, that is, in the region where $G'' > G'$. However, polymers derived from PVC, with a predominant elastic behavior at low temperatures, have shown excellent welding (Calafel et al., 2020). Faced with this apparent contradiction, the Dahlquist criterion (Dahlquist, 1959; Voet & Geffken, 1951), initially developed for adhesives, can be considered for AM as a guideline for assessing interlayer adhesion. This criterion establishes a critical elastic modulus value of 1×10^5 Pa (corresponding to a compliance value, J_c, of 10^{-5} Pa^{-1}) below which a material exhibits good tack or instantaneous adhesion, irrespective of the dominance of G'' over G' state. While measuring the actual time and temperature at which welding occurs in the materials may be challenging, measuring the viscoelastic behavior at welding time less than 1 s, such as those provided by Seppala and Migler (2016) and Seppala et al. (2017), can be used as a reference point. Thus, using the Dahlquist criterion and conducting stress relaxation experiments at linear regions becomes a valuable tool for assessing welding behavior during printing (Fig. 1.6E).

1.4.1.5 Warping: thermophysical aspects

It has been pointed out that warping, see basics in Section 1.2.1.1 and the general changes in specific volume for amorphous and semicrystalline polymers that produce it in Fig. 1.7, is particularly critical in AM techniques as it leads to the detachment of the printing part from the built platform and, ultimately, to the printing failure. Extensive shrinkage has been reported in AM processes involving particle sintering (Gao et al., 2024; Yan et al., 2011). Several studies, encompassing experimental and numerical approaches, have been conducted to characterize the extent of shrinkage in FFF. These studies aim to understand the underlying mechanism causing shrinkage and develop mitigation strategies. Many experimental studies focus on systematically modifying printing conditions, such as printing velocity, cooling rate, and build environment (Candal et al., 2020; Caputo et al., 2023; Costanzo et al., 2020; Croccolo et al., 2023; Diederichs et al., 2019; Hu et al., 2019; Singh, 2018; Sreejith et al., 2023). All factors favoring a homogeneous cooling of the system and slow crystallization kinetics will lead to better warping control.

Additionally, the study conducted by Hämäläine (Hämäläinen, 2017) provides valuable insights into the effect of printed part geometry. It was found that cylindrical parts exhibit superior dimensional stability compared to cubic counterparts printed under the same conditions. This difference in dimensional stability was attributed to the distribution of internal stresses. In cubic parts, stress concentrations occur in the sharp corners, leading to distortion and detachment from the build plate during printing.

Numerous attempts have been made to minimize shrinkage. One of the most applied principles is the incorporation of fillers in the feedstock polymers, which has been extensively reported in the literature. Interesting results have been obtained with the addition of wood powder, calcium carbonate ($CaCO_3$) (Dong et al., 2019), talc (Bertolino et al., 2021; Prasong et al., 2020), glass fillers (Carneiro et al., 2015;

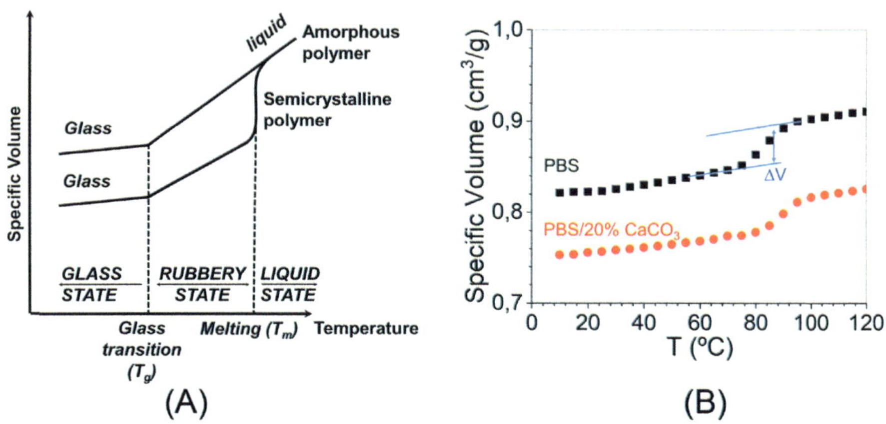

Figure 1.7 Specific volume temperature dependence.
(A) Schematic representation of the specific volume temperature dependence as a function of the amorphous or semicrystalline nature of the material. (B) PvT diagram for PBS and PBS with 20% $CaCO_3$ obtained during the cooling from the melt at 5 °C/min and 200 bar of pressure (unpublished data).

Sodeifian et al., 2019; Spoerk et al., 2019; Winter et al., 2022), and natural fibers (Stoof & Pickering, 2018) among others. In general terms, adding fillers reduces the coefficient of thermal expansion, effectively minimizing shrinkage. Conversely, filler can act as a nucleating agent, increasing the T_c. This early crystallization seems beneficial for controlling warpage in the printed part since it minimizes the abrupt change in specific volume and reduces the development of internal stresses. Nevertheless, excessive loading can lead to poorer performance due to particle aggregation (Dong et al., 2019), filament instability, or feeding problems during printing. Overall, the careful selection and optimization of fillers and loading levels are essential for achieving effective shrinkage reduction and improving the dimensional stability of printed parts in AM processes like FFF. Balancing the benefits of shrinkage reduction with potential challenges associated with excessive filler loading is crucial for achieving optimal print quality and performance.

Various techniques such as optical microscopy, 3D scanning, coordinate measuring machines (CMM), or digital image correlation (DIC) are used to quantify dimensional changes and deformations in printed parts. Pressure-volume-Temperature (PvT) measurements have been proven to be especially valuable in predicting the extent or degree of warping (Candal et al., 2020, 2021; Caputo et al., 2023). This PvT diagram finds extensive application in polymer injection molding for analyzing the volume reduction during crystallization. As an example, Fig. 1.7 shows the volume contraction related to the crystallization of PBS and a PBS/$CaCO_3$ blend during cooling. However, the precise alteration in specific volume throughout the AM process is typically overlooked. Some recent studies have studied the impact of pressure on flow and viscosity within the printer, employing the Tait model for analysis. Numerical simulations, such as finite element analysis (FEA) or computational fluid dynamics (CFD), are used to predict the warping phenomenon. These simulations involve modeling the thermal and mechanical processes occurring during printing, including material deposition rate, heating, cooling, and their effect on crystallization and internal stresses (Antony Samy et al., 2021; Gudadhe et al., 2019; Sreejith et al., 2023; Yang & Zhang, 2018). By varying simulation parameters, it is possible to analyze their impact and optimize printing conditions accordingly.

1.4.2 Mechanical properties: influence of printing parameters and testing

The mechanical properties of printed parts are important as these properties ultimately determine the functional performance of the parts. Various research works have focused on optimizing printing parameters to improve the mechanical properties of semicrystalline MEX 3D-printed parts. These efforts often involve experimental studies and numerical simulations to identify optimal process conditions. Some common strategies include adjusting nozzle temperature (Caputo et al., 2023; El Magri et al., 2021; Garmabi et al., 2022; He et al., 2023; Pulipaka et al., 2023; Rodríguez-Reyna et al., 2022; Vaezi & Yang, 2015; Yang et al., 2017; Yin et al., 2021), bed temperature (Candal et al., 2020; Caputo et al., 2023; He et al., 2023; Pulipaka et al., 2023), and printing velocity (Candal et al., 2020; Caputo et al., 2023; El Magri et al., 2021; Garmabi et al., 2022; He et al., 2021; Leary et al., 2013; Pulipaka et al., 2023; Yang et al., 2017) to optimize material flow and interlayer adhesion, modifying infill patterns (Greco et al., 2024), layer thickness/height (Bardiya et al., 2021a; Bardiya et al., 2021b; El Magri et al., 2021; Garmabi et al., 2022; Greco et al., 2024; He et al., 2023), layer orientation (Bardiya et al., 2021a; Bardiya et al., 2021b; Ding et al., 2019; Marşavina et al., 2022; Rahman et al., 2015; Sood et al., 2010; Sood et al., 2012), and density (Bardiya et al., 2021b; Singh et al., 2022) to enhance part strength and reduce warping, or implementing preheating or postprocessing techniques (El Magri et al., 2020; Greco et al., 2024; Yang et al., 2017) to minimize residual stress and improve part quality.

However, there are no standard test methods for determining the mechanical properties of 3D-printed plastic parts. ISO and ASTM test methods are used to evaluate them. These tests have been developed to evaluate reinforced and unreinforced polymers manufactured by injection, extrusion, blow molding, and other conventional manufacturing processes for polymer materials. Nowadays, those test methods are used to evaluate MEX printed polymers, even though they are parts with a lower density due to how they are manufactured, layer by layer. But recently, ASTM presented a guide (ASTM F3489-23) that established standards or adaptations of existing standards that could be used to evaluate particular static mechanical properties of polymeric parts produced through the MEX process.

Here is a brief overview of how mechanical properties are typically determined and their significance in MEX of semicrystalline polymers:

- *Tensile Testing*. Tensile testing by using the ASTM D638 standard is commonly used to evaluate the mechanical properties of MEX parts, including tensile strength, modulus of elasticity, and elongation at break. Test specimens are typically printed with different layer orientations and then subjected to controlled tension until failure occurs. A large volume of literature studied the mechanical properties of plastic parts fabricated with MEX process using tensile test. The importance of these works lies in knowing how strong this material is despite being less dense (presence of pores, anisotropy, poor interlayer adhesion) than manufactured by other conventional techniques such as injection and compression molding. Many of the authors report lower tensile properties for MEX specimens in comparison with those processed by traditional techniques (injection, compression, etc.) (Bardiya et al., 2021b; Candal et al., 2020; Caputo et al., 2023; Costanzo et al., 2020; Garmabi et al., 2022; Marşavina et al., 2022; Rahman et al., 2015; Rodzeń et al., 2021).

- *Flexural Testing*. Flexural testing, also known as 3-point bending testing, is done using the ASTM D790 standard and assesses the stiffness and strength of MEX parts under bending loads. Test specimens are subjected to bending forces until failure, and properties such as flexural strength and modulus of elasticity are determined. Different authors report that the flexural modulus from 3-point bending tests decreases with density, similar to Young's modulus from tensile tests (Bardiya et al., 2021a; Rahman et al., 2015).
- *Impact Testing*. Impact testing evaluates the ability of MEX parts to withstand sudden loading or shock. Charpy and Izod impact tests are commonly used to measure the energy absorbed by a specimen during fracture using the ASTM D256 standard. Some authors report that it is feasible to improve the impact resistance of MEX parts by changing the process parameters. Consequently, they have been exploring ways to broaden the utility of MEX parts (Bardiya et al., 2021a; Rahman et al., 2015).
- *Compression Testing*. Compression testing assesses MEX printed parts' compressive strength and deformation behavior under axial loading. Test specimens are subjected to compression until failure occurs, and properties such as compressive strength and modulus of elasticity are determined. The standard used is the ASTM D695 (Rahman et al., 2015).
- Other tests are not so common, but some authors have studied the fracture, torsion, and hardness properties of parts manufactured by MEX (Marşavina et al., 2022; Yin et al., 2021). Although reviewing the literature, it was found that the most commonly used test to verify the mechanical properties of parts manufactured by AM is the tension test, followed by compression and flexural tests. However, in more recent works, the adhesion between layers was studied using different ways to measure this property.

1.4.3 Powder bed fusion technologies: key parameters

In contrast to MEX technologies, PBF focuses on industrial-scale systems instead of lab-scale equipment, which restricts the access to researchers to this technology and, consequently, dismisses the understanding of the critical parameters influencing PBF from a material point of view. Few companies work on benchtop PBF, like Sinterint, Sharebot, or Sintratec. Still, their workflow is almost like a black box, and little or no parameters can be modified to better understand the physicochemical processes occurring during the fabrication. An interesting approach was made by Cardinaels et al. to visualize the particle neck formation in situ upon laser irradiation (Hejmady et al., 2019). The same authors upgraded this laser set-up incorporating thermography and X-ray diffraction accessories (Hejmady et al., 2022; Hejmady et al., 2019). This approach allowed for a better understanding of interparticle neck formation's rheological and crystallization implications.

This section will delve into the main material parameters that make semicrystalline polymers the most suitable materials for PBF based on their physicochemical and rheological characteristics. As explained above, the fabrication is based on a particle interdiffusion upon laser heating. As in interlayer adhesion, this process will be governed by the polymer reptation capabilities and surface rearrangement in a mechanism similar to the one proposed by Wool and O'Connor for self-healing materials (Wool and O'Connor, 1981). However, several other parameters must be considered to properly characterize the fabrication conditions of PBF. Here, we focus on material transformation during the fabrication phase. Previous warming, recoating, and cooling down phases will not be considered. The influence of the type of material during these stages was reviewed by Chatham et al. (2019). For clarity, different stages of the fabrication phase will be considered and discussed from a material perspective.

1.4.3.1 Material heating: processing window

The material transformation occurs due to the capacity of a laser beam to produce significant heating of the semicrystalline material above its T_m. Much of the efforts have centered on correlating the processing parameters and the material performance to establish a processing window for the material. In this sense, the most established methodology is related to the energy melt ratio (EMR), which incorporates the overall energy applied to the system and the energy required to produce the material melting (Vasquez et al., 2013). This EMR can be calculated as follows:

$$EMR = \frac{P \cdot S / y \cdot v_b \cdot z}{\left[c_p\left[T_m - T_b\right] + h_f\right] \cdot Q \cdot \Phi},$$

where P is the laser power, S is the scan count, y is the scan spacing, v_b is the beam velocity, z is the layer thickness, c_p is the specific heat, T_m is the melting temperature, T_b is the built temperature, h_f is the melting enthalpy, Q is the material density, and Φ is the packing density. Similarly, the same authors calculated the necessary energy to produce material degradation. Using these two parameters, they established the processing window as the one between EMR and EMR_{deg}. The main drawback of this methodology comes from assuming a complete transformation of optical energy to heat, which is not the case. In fact, researchers tried to enhance the heat transformation by incorporating photothermal sensitizers. Carbon black was used in laser sintering as a phothermic sensitizer. The results showed an unsatisfactory performance since the increase in absorbance required an increase in the applied energy, worsening the properties of the materials (Athreya et al., 2010). In an interesting approach, Powell et al. (2018) incorporated gold nanorods (GNRs) to produce plasmonic resonance, significantly increasing the heat produced in PA12 powder. Using this approach, they even printed bright-colored objects due to the effective heat transformation of these nanorods.

Despite the efforts, determining an accurate processing window based on processing and material parameters remains elusive. Thus, an accepted starting point for trial and error optimization establishes the chamber temperature to be between the onset of the melting of the material and the onset of the crystallization. Under these conditions, the material presents a rubbery behavior that permits interdiffusion upon minimal light exposure (permitting higher printing speeds) but avoids particle sintering in the building platform.

1.4.3.2 Understanding particle coalescence

As mentioned in Section 1.2, the driving force of particle coalescence is the reduction of the surface energy that occurs when increasing the particle size (and, therefore, reducing the surface area). Thus, the particle coalescence is controlled by the surface energy and the rheological behavior of the materials at the fabrication temperature.

In contrast to conventional formative processes or MEX methods, particle coalescence occurs under very low shear rate conditions. Thus, the key parameter that governs the molecular interdiffusion across the particle interphase is η_0. The Frenkel model is a widely accepted theory that describes the growing neck length, x, as a function of time, t (Hejmady et al., 2019).

$$\left(\frac{x}{r}\right)^2 = \frac{3\sigma t}{2r\eta_0},$$

where r, σ, and η_0 are the radii of coalescing particles, the surface tension, and zero-shear viscosity, respectively. This model presents the first approach to understanding the coalescence process and correlating the behavior with the material properties. This equation has been adjusted considering different radii for particles (Pokluda et al., 1997) or the viscoelastic nature of the polymer fluid (Bellehumeur et al., 1998). In fact, this effect is usually underestimated when considering the complete melt of material upon light application. According to this study, fluids' relaxation times highly affect the material's sintering rate. Further research should be conducted to determine the effect of the material viscoelasticity on the particle coalescence behavior.

1.4.3.3 Material consolidation: processing conditions and crystallization

The concepts of material consolidation were previously discussed in Section 1.2.2, where it was emphasized that the crystallization and cooling conditions of the material play an important role. Effective crystallization should result in uniform crystal growth and sufficient interparticle crystals to provide mechanical support while avoiding uneven shrinkage (Chatham et al., 2019). To better understand the molecular parameters affecting this process, Cardinaels et al. modified their optical setup to simulate the sintering process by incorporating X-ray and thermographic capacities (Hejmady et al., 2022; Hejmady et al., 2019; Hejmady et al., 2019). The authors examined how various processing conditions, such as laser power and chamber temperature, affect a material's rheological and crystallization properties.

They observed that particle coalescence and crystallization happen sequentially, leading to a uniform crystallization process. However, it's important to note that the influence of melt memory in this case cannot be dismissed.

The phenomenon of melt memory in the context of PBF has not been extensively studied. During material irradiation, the heating and cooling occur rapidly, which may result in some residual order in the melted material (known as melt memory). Fulchiron et al. investigated the impact of the remaining self-nuclei in polypropylene-ethylene random copolymers (Maulet et al., 2022). According to the research, when low laser energy is applied, the nuclei that remain in both the bottom layer and those that have not been erased contribute to faster crystallization kinetics during material processing. As a result, the microstructure exhibits a repeating pattern of oriented transcrystalline regions and isotropic spherulitic regions, which are related to the layer thickness.

1.5 Concluding remarks

The AM technique is a revolutionary way to create objects but has only been used with a limited variety of mostly amorphous polymeric materials, even when 75% of all industrially relevant polymers are semicrystalline. Expanding the range of 3D-printing materials to semicrystalline polymers can unlock AM's full potential. However, printing semicrystalline polymers is more complex due to the paradoxical nature of crystals, which provide unique mechanical properties but also make the AM process more difficult.

The AM processing of semicrystalline materials has been reviewed, and it has been noted that understanding the basics of crystallization and rheology is crucial to solving and improving the possible failures that can occur during printing. Understanding the variables involved in such failures is vital for progress in producing semicrystalline plastic parts by AM. By doing so, we can better understand the morphological self-assembly process of semicrystalline polymers during AM processing, as has already been done in other traditional manufacturing techniques. This will allow better control and optimization of the AM process. Being able to use a wide range of semicrystalline polymers with AM techniques will offer an unlimited range of potential applications, from aerospace parts to medical implants.

Acknowledgments

This research was supported by the projects PID2020–113045GB-C21 funded by MCIN/ AEI /10.13039/501100011033, KK-2023/00054 and KK-2024/00071 funded by Basque Goverment (ELKARTEK call) and by the Basque Government through grant IT1503-22. R.A.P-C is supported by the ADAGIO-H2020-MSCA-COFUND-2020 program (101034379). Itxaso Calafel and R. Aguirresarobe acknowledge the financial support received by ERCROS S.A. (Plastics and Bioplastics Division Monzón) through the UPV-ERCROS collaboration.

References

ASTM, ASTM F2792-12a Standard Terminology for Additive Manufacturing Technologies (2015).

Abbott, A. C., Tandon, G. P., Bradford, R. L., Koerner, H., & Baur, J. W. (2018). *Process-structure-property effects on ABS bond strength in fused filament fabrication. Additive Manufacturing, 19*, United States: Elsevier B.V, 29–38. https://www.journals.elsevier.com/additive-manufacturing. https://doi.org/10.1016/j.addma.2017.11.002.

Agarwal, K. M., Shubham, P., Bhatia, D., Sharma, P., Vaid, H., & Vajpeyi, R. (2022). Analyzing the Impact of print parameters on dimensional variation of ABS specimens printed using fused deposition modelling (FDM). *Sensors International, 3*, 100149. https://doi.org/10.1016/j.sintl.2021.100149.

Al Rashid, A., & Koç, M. (2021). Fused filament fabrication process: A review of numerical simulation techniques. *Polymers, 13*(20). https://doi.org/10.3390/polym13203534. https://www.mdpi.com/2073-4360/13/20/3534/pdf.

Alarifi, I. M. (2023). PETG/carbon fiber composites with different structures produced by 3D printing. *Polymer Testing, 120*, 107949. https://doi.org/10.1016/j.polymertesting.2023.107949.

Aliheidari, N., Christ, J., Tripuraneni, R., Nadimpalli, S., & Ameli, A. (2018). Interlayer adhesion and fracture resistance of polymers printed through melt extrusion additive manufacturing process. *Materials and Design, 156*, 351–361. https://doi.org/10.1016/j.matdes.2018.07.001.

Antony Samy, A., Golbang, A., Harkin-Jones, E., Archer, E., & McIlhagger, A. (2021). Prediction of part distortion in Fused Deposition Modelling (FDM) of semi-crystalline polymers via COMSOL: Effect of printing conditions. *IRP Journal of Manufacturing Science and Technology, 33*, 443–453. https://doi.org/10.1016/j.cirpj.2021.04.012. http://www.elsevier.com/wps/find/journaldescription.cws_home/714185/description#description.

Arif, M. F., Alhashmi, H., Varadarajan, K. M., Koo, J. H., Hart, A. J., & Kumar, S. (2020). Multifunctional performance of carbon nanotubes and graphene nanoplatelets reinforced PEEK composites enabled via FFF additive manufacturing. *Composites Part B: Engineering, 184*, 107625. https://doi.org/10.1016/j.compositesb.2019.107625.

Athreya, S. R., Kalaitzidou, K., & Das, S. (2010). Processing and characterization of a carbon black-filled electrically conductive Nylon-12 nanocomposite produced by selective laser sintering. *Materials Science and Engineering: A, 527*(10-11), 2637–2642. https://doi.org/10.1016/j.msea.2009.12.028. http://www.elsevier.com.

Baouch, Z., Vezzoli, R., Koster, J., Costanzo, A., Lanfranchi, A., Cavallo, D., & McIlroy, C. (2024). Polypropylene for material extrusion: Evidence that flow-enhanced crystallization restricts welding. *Additive Manufacturing, 83*, 104063. https://doi.org/10.1016/j.addma.2024.104063.

Bardiya, S., Jerald, J., & Satheeshkumar, V. (2021a). Effect of process parameters on the impact strength of fused filament fabricated (FFF) polylactic acid (PLA) parts. *Materials Today: Proceedings, 41*, 1103–1106. https://doi.org/10.1016/j.matpr.2020.08.066.

Bardiya, S., Jerald, J., & Satheeshkumar, V. (2021b). The impact of process parameters on the tensile strength, flexural strength and the manufacturing time of fused filament fabricated (FFF) parts. *Materials Today: Proceedings, 39*, 1362–1366. https://doi.org/10.1016/j.matpr.2020.04.691.

Bartolai, J., Simpson, T. W., & Xie, R. (2018). Predicting strength of additively manufactured thermoplastic polymer parts produced using material extrusion. *Rapid Prototyping Journal, 24*(2), 321–332. https://doi.org/10.1108/RPJ-02-2017-0026. http://www.emeraldinsight.com/info/journals/rpj/rpj.jsp.

Basgul, C., Yu, T., MacDonald, D. W., Siskey, R., Marcolongo, M., & Kurtz, S. M. (2020). Does annealing improve the interlayer adhesion and structural integrity of FFF 3D printed PEEK lumbar spinal cages? *Journal of the Mechanical Behavior of Biomedical Materials, 102.*

Bellehumeur, C., Li, L., Sun, Q., & Gu, P. (2004). Modeling of bond formation between polymer filaments in the fused deposition modeling process. *Journal of Manufacturing Processes, 6*(2), 170–178. https://doi.org/10.1016/S1526-6125(04)70071-7. http://www.elsevier.com/wps/find/journaldescription.cws_home/620379/description#description.

Bellehumeur, C. T., Kontopoulou, M., & Vlachopoulos, J. (1998). The role of viscoelasticity in polymer sintering. *Rheologica Acta, 37*(3), 270–278. https://doi.org/10.1007/s003970050114. http://link.springer.de/link/service/journals/00397/index.htm.

Benwood, C., Anstey, A., Andrzejewski, J., Misra, M., & Mohanty, A. K. (2018). Improving the impact strength and heat resistance of 3D printed models: Structure, property, and processing correlationships during fused deposition modeling (FDM) of poly(lactic acid). *ACS Omega, 3*(4), 4400–4411. https://doi.org/10.1021/acsomega.8b00129. http://pubs.acs.org/journal/acsodf.

Bertolino, M., Battegazzore, D., Arrigo, R., & Frache, A. (2021). Designing 3D printable polypropylene: Material and process optimisation through rheology. *Additive Manufacturing, 40*, 101944. https://doi.org/10.1016/j.addma.2021.101944.

Bhandari, S., Lopez-Anido, R. A., & Gardner, D. J. (2019). *Enhancing the interlayer tensile strength of 3D printed short carbon fiber reinforced PETG and PLA composites via annealing. Additive Manufacturing, 30.*

Brenken, B., Barocio, E., Favaloro, A., Kunc, V., & Pipes, R. B. (2019). Development and validation of extrusion deposition additive manufacturing process simulations. *Additive Manufacturing, 25*, 218–226. https://doi.org/10.1016/j.addma.2018.10.041.

Calafel, I., Aguirresarobe, R. H., Peñas, M. I., Santamaria, A., Tierno, M., Conde, J. I., & Pascual, B. (2020). Searching for rheological conditions for FFF 3D printing with PVC based flexible compounds. *Materials, 13*(1), 178. https://doi.org/10.3390/ma13010178. https://res.mdpi.com/d_attachment/materials/materials-13-00178/article_deploy/materials-13-00178.pdf.

de Campos, M. R., & dos Reis, A. C. (2024). Effect of post-processing on the mechanical properties of polymers printed by the fused filament fabrication method used as prosthodontic materials and dental biomaterials: a systematic review. *Polymer Bulletin, 81*(3), 2001–2021. https://doi.org/10.1007/s00289-023-04816-3. https://www.springer.com/journal/289.

Candal, M. V., Calafel, I., Aranburu, N., Fernández, M., Gerrica-Echevarria, G., Santamaría, A., & Müller, A. J. (2020). *Thermo-rheological effects on successful 3D printing of biodegradable polyesters.* Additive Manufacturing, *36*.

Candal, M. V., Calafel, I., Fernández, M., Aranburu, N., Aguirresarobe, R. H., Gerrica-Echevarria, G., Santamaría, A., & Müller, A. J. (2021). *Study of the interlayer adhesion and warping during material extrusion-based additive manufacturing of a carbon nanotube/biobased thermoplastic polyurethane nanocomposite. Polymer, 224.*

Cano-Vicent, A., Tambuwala, M. M., Hassan, S. S., Barh, D., Aljabali, A. A. A., Birkett, M., Arjunan, A., & Serrano-Aroca, Á. (2021). *Fused deposition modelling: Current status, methodology, applications and future prospects.* Additive Manufacturing, *47*.

Caputo, M. R., Fernández, M., Aguirresarobe, R., Kovalcik, A., Sardon, H., Candal, M. V., & Müller, A. J. (2023). Influence of FFF process conditions on the thermal, mechanical, and rheological properties of poly(hydroxybutyrate-co-hydroxy hexanoate). *Polymers, 15*(8). https://doi.org/10.3390/polym15081817. http://www.mdpi.com/journal/polymers.

Carneiro, O. S., Silva, A. F., & Gomes, R. (2015). Fused deposition modeling with polypropylene. *Materials and Design, 83*, 768–776. https://doi.org/10.1016/j.matdes.2015.06.053.

Charlon, S., Le Boterff, J., & Soulestin, J. (2021). Fused filament fabrication of polypropylene: Influence of the bead temperature on adhesion and porosity. *Additive Manufacturing, 38*, 101838. https://doi.org/10.1016/j.addma.2021.101838.

Chatham, C. A., Zawaski, C. E., Bobbitt, D. C., Moore, R. B., Long, T. E., & Williams, C. B. (2019). Semi-crystalline polymer blends for material extrusion additive manufacturing printability: A case study with poly(ethylene terephthalate) and polypropylene. *Macromolecular Materials and Engineering, 304*(5). https://doi.org/10.1002/mame.201800764. http://onlinelibrary.wiley.com/journal/10.1002/(ISSN)1439-2054.

Chatham, C. A., Long, T. E., & Williams, C. B. (2019). *A review of the process physics and material screening methods for polymer powder bed fusion additive manufacturing.* Progress in Polymer Science,, *93*.

Chinnasamy, A., Varalakshmi, L. M., Jims John Wessley, G., Malini, T., Sanjith, S., & Sheikh, M. F. A. (2021). IoT control and layer formation of FDM on polycarbonate based aluminium. *Materials Today: Proceedings, 45*, 2809–2811. https://doi.org/10.1016/j.matpr.2020.11.798.

Commisso, A. J., Sama, G. R., & Scott, T. F. (2023). Radical-mediated ring-opening photopolymerization for semicrystalline thermoplastic additive manufacturing. *Chemistry of Materials, 35*(10), 3825–3834. https://doi.org/10.1021/acs.chemmater.2c03352. http://pubs.acs.org/journal/cmatex.

Coogan, T. J., & Kazmer, D. O. (2017). Healing simulation for bond strength prediction of FDM. *Rapid Prototyping Journal, 23*(3), 551–561. https://doi.org/10.1108/RPJ-03-2016-0051. http://www.emeraldinsight.com/info/journals/rpj/rpj.jsp.

Corapi, D., Morettini, G., Pascoletti, G., & Zitelli, C. (2019). Characterization of a polylactic acid (PLA) produced by Fused Deposition Modeling (FDM) technology. *Procedia Structural Integrity*, 24, 289–295.

Costanzo, A., Spotorno, R., Candal, M. V., Fernández, M. M., Müller, A. J., Graham, R. S., Cavallo, D., & McIlroy, C. (2020). *Residual alignment and its effect on weld strength in material-extrusion 3D-printing of polylactic acid.* Additive Manufacturing, *36*.

Croccolo, D., De Agostinis, M., Fini, S., Mele, M., Olmi, G., & Campana, G. (2023). *Effects of infill temperature on the tensile properties and warping of 3D-printed polylactic acid. Progress in Additive Manufacturing*. Italy: Springer Science and Business Media Deutschland GmbH. https://www.springer.com/journal/40964. https://doi.org/10.1007/s40964-023-00492-x.

Crump, S. S. (1992). Apparatus and method for creating three-dimensional objects. US Patent US5121329A.

Dahlquist, C. A. (1959). An investigation into the nature of tack. *Adhesives Age, 2*, 25–29.

Dalton, P. D. (2017). Melt electrowriting with additive manufacturing principles. *Current Opinion in Biomedical Engineering, 2*, 49–57. https://doi.org/10.1016/j.cobme.2017.05.007.

Das, A., Chatham, C. A., Fallon, J. J., Zawaski, C. E., Gilmer, E. L., Williams, C. B., & Bortner, M. J. (2020). *Current understanding and challenges in high temperature additive manufacturing of engineering thermoplastic polymers. Additive Manufacturing, 34.*

Das, A., Gilmer, E. L., Biria, S., & Bortner, M. J. (2021). Importance of polymer rheology on material extrusion additive manufacturing: Correlating process physics to print properties. *ACS Applied Polymer Materials, 3*(3), 1218–1249. https://doi.org/10.1021/acsapm.0c01228. http://pubs.acs.org/journal/aapmcd.

Dave, H. K., Tank, Y. N., & Prajapati, A. R. (2023). *Synthesis and fabrication of bioactive glass for bone implant using 3D printing setup. Progress in additive manufacturing.* India: Springer Science and Business Media Deutschland GmbH. https://www.springer.com/journal/40964. https://doi.org/10.1007/s40964-023-00539-z.

Dechet, M. A., Demina, A., Römling, L., Gómez Bonilla, J. S., Lanyi, F. J., Schubert, D. W., Bück, A., Peukert, W., & Schmidt, J. (2020). *Development of poly(L-lactide) (PLLA) microspheres precipitated from triacetin for application in powder bed fusion of polymers. Additive Manufacturing, 32.*

Dechet, M. A., Gómez Bonilla, J. S., Grünewald, M., Popp, K., Rudloff, J., Lang, M., & Schmidt, J. (2021). *A novel, precipitated polybutylene terephthalate feedstock material for powder bed fusion of polymers (PBF): Material development and initial PBF processability. Materials and Design, 197.*

Diederichs, E. V., Picard, M. C., Chang, B. P., Misra, M., Mielewski, D. F., & Mohanty, A. K. (2019). Strategy to improve printability of renewable resource-based engineering plastic tailored for fdm applications. *ACS Omega, 4*(23), 20297–20307.

Ding, S., Zou, B., Wang, P., & Ding, H. (2019). Effects of nozzle temperature and building orientation on mechanical properties and microstructure of PEEK and PEI printed by 3D-FDM. *Polymer Testing, 78*, 105948. https://doi.org/10.1016/j.polymertesting.2019.105948.

Domínguez-Robles, J., Utomo, E., Cornelius, V. A., Anjani, Q. K., Korelidou, A., Gonzalez, Z., Donnelly, R. F., Margariti, A., Delgado-Aguilar, M., Tarrés, Q., & Larrañeta, E. (2022). *TPU-based antiplatelet cardiovascular prostheses prepared using fused deposition modelling. Materials and Design, 220.*

Dong, M., Zhang, S., Gao, D., & Chou, B. (2019). The study on polypropylene applied in fused deposition modeling. AIP Conference Proceedings. China: American Institute of Physics Inc. 2065. http://scitation.aip.org/content/aip/proceeding/aipcp.

Drummer, D., Cifuentes-Cuéllar, S., & Rietzel, D. (2012). Suitability of PLA/TCP for fused deposition modeling. *Rapid Prototyping Journal, 18*(6), 500–507. https://doi.org/10.1108/13552541211272045.

Dusanapudi, S. N., Krupakaran, R. L., Srinivas, A., Nikhil, K. S., & Vamshi, T. (2023). Optimization and experimental analysis of mechanical properties and porosity on FDM based 3D printed ABS sample. *Materials Today: Proceedings.* https://doi.org/10.1016/j.matpr.2023.10.096.

Dusunceli, N., & Colak, O. U. (2008). Modelling effects of degree of crystallinity on mechanical behavior of semicrystalline polymers. *International Journal of Plasticity, 24*(7), 1224–1242. https://doi.org/10.1016/j.ijplas.2007.09.003.

El Magri, A., El Mabrouk, K., Vaudreuil, S., Chibane, H., & Touhami, M. E. (2020). Optimization of printing parameters for improvement of mechanical and thermal performances of 3D printed poly(ether ether ketone) parts. *Journal of Applied Polymer Science, 137*(37). https://doi.org/10.1002/app.49087. http://onlinelibrary.wiley.com/journal/10.1002/(ISSN)1097-4628.

El Magri, A., El Mabrouk, K., Vaudreuil, S., & Ebn Touhami, M. (2021). Experimental investigation and optimization of printing parameters of 3D printed polyphenylene sulfide through response surface methodology. *Journal of Applied Polymer Science, 138*(1) https://doi.org/10.1002/app.49625.

Elkaseer, A., Chen, K. J., Janhsen, J. C., Refle, O., Hagenmeyer, V., & Scholz, S. G. (2022). Material jetting for advanced applications: A state-of-the-art review, gaps and future directions. *Additive Manufacturing, 60.*

Fang, L., Yan, Y., Agarwal, O., Seppala, J. E., Hemker, K. J., & Kang, S. H. (2020). Processing-structure-property relationships of bisphenol-A-polycarbonate samples prepared by fused filament fabrication. *Additive Manufacturing, 35.*

Feroz Ali, L., Raghul, R., Yogesh Muthu Ram, M., Harshavardhan Reddy, V., & Shakil Kanna, N. (2022). Evaluation of the polyamide's mechanical properties for varying infill percentage in FDM process. *Materials Today: Proceedings: 68*, (pp. 2509–2514). India: Elsevier Ltd. https://doi.org/10.1016/j.matpr.2022.09.227.

Gao, W., Zhang, Y., Ramanujan, D., Ramani, K., Chen, Y., Williams, C. B., Wang, C. C. L., Shin, Y. C., Zhang, S., & Zavattieri, P. D. (2015). The status, challenges, and future of additive manufacturing in engineering. *CAD Computer Aided Design, 69*, 65–89. https://doi.org/10.1016/j.cad.2015.04.001.

Gao, X., Okigami, F., Avedissian, N., & Zhang, W. (2024). An experimental and modeling study on warping in additively manufactured overhang structures. *Additive Manufacturing, 81*, 104017. https://doi.org/10.1016/j.addma.2024.104017.

García, E., Núñez, P. J., Caminero, M. A., Chacón, J. M., & Kamarthi, S. (2022). Effects of carbon fibre reinforcement on the geometric properties of PETG-based filament using FFF additive manufacturing. *Composites Part B: Engineering, 235*, 109766. https://doi.org/10.1016/j.compositesb.2022.109766.

Garmabi, M. M., Shahi, P., Tjong, J., & Sain, M. (2022). 3D printing of polyphenylene sulfide for functional lightweight automotive component manufacturing through enhancing interlayer bonding. *Additive Manufacturing, 56*, 102780. https://doi.org/10.1016/j.addma.2022.102780.

Gibson, I., Rosen, D. W., & Stucker, B. (2010). *Additive manufacturing technologies: Rapid prototyping to direct digital manufacturing.* Singapore: Springer US1–459. http://www.springerlink.com/openurl.asp?genre=book&isbn=978-1-4419-1119-3. https://doi.org/10.1007/978-1-4419-1120-9.

Gibson, I., Rosen, D., Stucker, B., & Khorasani, M. (2020). *Additive manufacturing technologies. Additive manufacturing technologies.* Netherlands: Springer International Publishing, 1–675. https://link.springer.com/book/10.1007/978-3-030-56127-7.

Gilmer, E. L., Miller, D., Chatham, C. A., Zawaski, C., Fallon, J. J., Pekkanen, A., Long, T. E., Williams, C. B., & Bortner, M. J. (2018). Model analysis of feedstock behavior in fused filament fabrication: Enabling rapid materials screening. *Polymer, 152*, 51–61.

Ginoux, G., Paux, J., & Allaoui, S. (2023). New preparation method of microstructurally and mechanically standardized PETG specimens by material extrusion additive manufacturing and machining. *Additive Manufacturing, 66*, 103471. https://doi.org/10.1016/j.addma.2023.103471.

Goh, G. D., Yap, Y. L., Tan, H. K. J., Sing, S. L., Goh, G. L., & Yeong, W. Y. (2020). Process–structure–properties in polymer additive manufacturing via material extrusion: A review. *Critical Reviews in Solid State and Materials Sciences, 45*(2), 113–133. https://doi.org/10.1080/10408436.2018.1549977. http://www.tandf.co.uk/journals/titles/10408436.asp.

Gonzalez Ausejo, J., Rydz, J., Musioł, M., Sikorska, W., Sobota, M., Włodarczyk, J., Adamus, G., Janeczek, H., Kwiecień, I., Hercog, A., Johnston, B., Khan, H. R., Kannappan, V., Jones, K. R., Morris, M. R., Jiang, G., Radecka, I., & Kowalczuk, M. (2018). A comparative study of three-dimensional printing directions: The degradation and toxicological profile of a PLA/PHA blend. *Polymer Degradation and Stability, 152*, 191–207. https://doi.org/10.1016/j.polymdegradstab.2018.04.024.

Gonzalez-Gutierrez, J., Cano, S., Schuschnigg, S., Kukla, C., Sapkota, J., & Holzer, C. (2018). Additive manufacturing of metallic and ceramic components by the material extrusion of highly-filled polymers: A review and future perspectives. *Materials, 11*(5), 840. https://doi.org/10.3390/ma11050840.

Goodridge, R. D., Tuck, C. J., & Hague, R. J. M. (2012). Laser sintering of polyamides and other polymers. *Progress in Materials Science, 57*(2), 229–267. https://doi.org/10.1016/j.pmatsci.2011.04.001.

Graham, R. S. (2019). Understanding flow-induced crystallization in polymers: A perspective on the role of molecular simulations. *Journal of Rheology, 63*(1), 203–214. https://doi.org/10.1122/1.5056170. http://scitation.aip.org/content/sor/journal/jor2.

Greco, A., De Luca, A., Sepe, R., & Gerbino, S. (2024). Investigation on tensile properties of FFF PEEK: Effects of printing parameters and post-processing treatment. *Structural Integrity Procedia, 53*, 178–184. https://www.sciencedirect.com/science/journal/24523216 .

Gudadhe, A., Bachhar, N., Kumar, A., Andrade, P., & Kumaraswamy, G. (2019). Three-dimensional printing with waste high-density polyethylene. *ACS Applied Polymer Materials, 1*(11), 3157–3164. https://doi.org/10.1021/acsapm.9b00813. http://pubs.acs.org/journal/aapmcd.

Hadjichristidis, N., Gnanou, Y., Matyjaszewski, K., & Muthukumar, M. (2022). *Polymer Crystallization macromolecular engineering: From precise synthesis to macroscopic materials and applications*. Wiley1–57 10.1002/9783527815562.mme0032.

He, F., Alshammari, Y. L. A., & Khan, M. (2021). The effect of printing parameters on crack growth rate of FDM ABS cantilever beam under thermo-mechanical loads. Procedia Structural Integrity, 34. http://www.journals.elsevier.com/procedia-structural-integrity.

He, Y., Shen, M., Wang, Q., Wang, T., & Pei, X. (2023). Effects of FDM parameters and annealing on the mechanical and tribological properties of PEEK. *Composite Structures, 313*, 116901. https://doi.org/10.1016/j.compstruct.2023.116901.

Hejmady, P., Cleven, L. C., Van Breemen, L. C. A., Anderson, P. D., & Cardinaels, R. (2019). A novel experimental setup for in situ optical and X-ray imaging of laser sintering of polymer particles. *Review of Scientific Instruments, 90*(8). https://doi.org/10.1063/1.5106406. http://scitation.aip.org/content/aip/journal/rsi.

Hejmady, P., Van Breemen, L. C. A., Anderson, P. D., & Cardinaels, R. (2019). Laser sintering of polymer particle pairs studied by in situ visualization. *Soft Matter, 15*(6), 1373–1387. https://doi.org/10.1039/c8sm02081g. http://pubs.rsc.org/en/journals/journal/sm.

Hejmady, P., van Breemen, L. C. A., Hermida-Merino, D., Anderson, P. D., & Cardinaels, R. (2022). *Laser sintering of PA12 particles studied by in-situ optical, thermal and X-ray characterization*. Additive Manufacturing, *52*.

Hertle, S., Drexler, M., & Drummer, D. (2016). Additive manufacturing of poly(propylene) by means of melt extrusion. *Macromolecular Materials and Engineering, 301*(12), 1482–1493.. https://doi.org/10.1002/mame.201600259. http://www3.interscience.wiley.com/journal/117930403/grouphome/home.html.

Hikmat, M., Rostam, S., & Ahmed, Y. M. (2021). Investigation of tensile property-based Taguchi method of PLA parts fabricated by FDM 3D printing technology. *Results in Engineering, 11*, 100264. https://doi.org/10.1016/j.rineng.2021.100264.

Hohimer, C. J., Petrossian, G., Ameli, A., Mo, C., & Pötschke, P. (2020). 3D printed conductive thermoplastic polyurethane/carbon nanotube composites for capacitive and piezoresistive sensing in soft pneumatic actuators. *Additive Manufacturing, 34*, 101281. https://doi.org/10.1016/j.addma.2020.101281.

Hu, B., Duan, X., Xing, Z., Xu, Z., Du, C., Zhou, H., Chen, R., & Shan, B. (2019). Improved design of fused deposition modeling equipment for 3D printing of high-performance PEEK parts. *Mechanics of Materials, 137*, 103139. https://doi.org/10.1016/j.mechmat.2019.103139.

Hämäläinen, J. P. (2017). *Semi-crystalline polyolefins in fused deposition modeling* (Master's thesis).

Jain, S., Fuoco, T., Yassin, M. A., Mustafa, K., & Finne-Wistrand, A. (2020). Printability and critical insight into polymer properties during direct-extrusion based 3D printing of medical grade polylactide and copolyesters. *Biomacromolecules, 21*(2), 388–396. https://doi.org/10.1021/acs.biomac.9b01112. http://pubs.acs.org/journal/bomaf6.

Jin, M., Neuber, C., & Schmidt, H. W. (2020). *Tailoring polypropylene for extrusion-based additive manufacturing.* Additive Manufacturing, *33*.

Karad, A. S., Sonawwanay, P. D., Naik, M., & Thakur, D. G. (2023). Experimental tensile strength analysis of ABS material through FDM technique. *Materials Today: Proceedings, 103, 506–512*. https://doi.org/10.1016/j.matpr.2023.09.216.

Kartikeyan, B., Ponshanmugakumar, A., Saravanan, G., BharathGanesh, S., & Hemamalini, V. (2023). Experimental and theoretical analysis of FDM AM PLA mechanical properties. *Materials Today: Proceedings*. https://doi.org/10.1016/j.matpr.2023.05.105.

Kaygusuz, B., & Özerinç, S. (2019). Improving the ductility of polylactic acid parts produced by fused deposition modeling through polyhydroxyalkanoate additions. *Journal of Applied Polymer Science, 136*(43). https://doi.org/10.1002/app.48154. http://onlinelibrary.wiley.com/journal/10.1002/(ISSN)1097-4628.

Khabia, S., & Jain, K. K. (2020). Comparison of mechanical properties of components 3D printed from different brand ABS filament on different FDM printers. *Materials Today: Proceedings, 26*, 2907–2914. https://doi.org/10.1016/j.matpr.2020.02.600.

Kim, K., Park, J., Suh, Jh, Kim, M., Jeong, Y., & Park, I. (2017). 3D printing of multiaxial force sensors using carbon nanotube (CNT)/thermoplastic polyurethane (TPU) filaments. *Sensors and Actuators, A: Physical, 263*, 493–500. https://doi.org/10.1016/j.sna.2017.07.020.

Kuenstler, A. S., Hernandez, J. J., Trujillo-Lemon, M., Osterbaan, A., & Bowman, C. N. (2023). Vat photopolymerization additive manufacturing of tough, fully recyclable thermosets. *ACS Applied Materials and Interfaces, 15*(8), 11111–11121. https://doi.org/10.1021/acsami.2c22081. http://pubs.acs.org/journal/aamick.

Kumar, N., Jain, P. K., Tandon, P., & Mohan Pandey, P. (2018). Experimental investigations on suitability of polypropylene (PP) and ethylene vinyl acetate (EVA) in additive manufacturing. *Materials Today: Proceedings, 5*, 4118–4127. https://www.sciencedirect.com/journal/materials-today-proceedings.

Kumar, R., Singh, R., Kumar, V., & Kumar, P. (2021). *On Mn doped ZnO nano particles reinforced in PVDF matrix for fused filament fabrication: Mechanical, thermal, morphological and 4D properties.* Journal of Manufacturing Processes, *62*, 817–832. http://www.elsevier.com/wps/find/journaldescription.cws_home/620379/description#description. https://doi.org/10.1016/j.jmapro.2021.01.003.

Kusoglu, I. M., Doñate-Buendía, C., Barcikowski, S., & Gökce, B. (2021). Laser powder bed fusion of polymers: quantitative research direction indices. *Materials, 14*(5), 1169. https://doi.org/10.3390/ma14051169.

Lagaron, J. M., Catalá, R., & Gavara, R. (2004). Structural characteristics defining high barrier properties in polymeric materials. *Materials Science and Technology, 20*(1), 1–7. https://doi.org/10.1179/026708304225010442.

Landers, R., & Mülhaupt, R. (2000). *Desktop manufacturing of complex objects, prototypes and biomedical scaffolds by means of computer-assisted design combined with computer-guided 3D plotting of polymers and reactive oligomers. Macromolecular Materials and Engineering, 282*, 17–21. http://onlinelibrary.wiley.com/journal/10.1002/(ISSN)1439-2054. https://doi.org/10.1002/1439-2054(20001001)282:1 %3C 17::AID-MAME17 %3E 3.0.CO;2-8.

Leary, M., Huang, S., Ataalla, T., Baxter, A., & Subic, A. (2013). Design of shape memory alloy actuators for direct power by an automotive battery. *Materials and Design, 43*, 460–466. https://doi.org/10.1016/j.matdes.2012.07.002.

Liang, S. B., Hu, D. P., Zhu, C., & Yu, A. B. (2002). Production of fine polymer powder under cryogenic conditions. *Chemical Engineering and Technology, 25*(4), 401–405. https://doi.org/10.1002/1521-4125(200204)25:4 %3C 401::AID-CEAT401 %3E 3.0.CO;2-S.

Ligon, S. C., Liska, R., Stampfl, J., Gurr, M., & Mülhaupt, R. (2017). Polymers for 3D printing and customized additive manufacturing. *Chemical Reviews, 117*(15), 10212–10290. https://doi.org/10.1021/acs.chemrev.7b00074. http://pubs.acs.org/journal/chreay.

Lin, L., Ecke, N., Huang, M., Pei, X. Q., & Schlarb, A. K. (2019). *Impact of nanosilica on the friction and wear of a PEEK/CF composite coating manufactured by fused deposition modeling (FDM). Composites Part B: Engineering, 177.*

Lin, T., Stickney, K. W., Rogers, M., Riffle, J. S., McGrath, J. E., Marand, H., Yu, T. H., & Davis, R. M. (1993). Preparation of submicrometre polyimide particles by precipitation from solution. *Polymer, 34*(4), 772–777. https://doi.org/10.1016/0032-3861(93)90362-E.

Lin, X., Gao, J., Wang, J., Wang, R., Gong, M., Zhang, L., Lu, Y., Wang, D., & Zhang, L. (2021). Desktop printing of 3D thermoplastic polyurethane parts with enhanced mechanical performance using filaments with varying stiffness. *Additive Manufacturing, 47*, 102267. https://doi.org/10.1016/j.addma.2021.102267.

Liu, F., Vyas, C., Poologasundarampillai, G., Pape, I., Hinduja, S., Mirihanage, W., & Bartolo, P. J. (2018a). Process-driven microstructure control in melt-extrusion-based 3D printing for tailorable mechanical properties in a polycaprolactone filament. *Macromolecular Materials and Engineering, 303*(8). https://doi.org/10.1002/mame.201800173. http://onlinelibrary.wiley.com/journal/10.1002/(ISSN)1439-2054.

Liu, F., Vyas, C., Poologasundarampillai, G., Pape, I., Hinduja, S., Mirihanage, W., & Bartolo, P. (2018b). Structural evolution of PCL during melt extrusion 3D printing. *Macromolecular Materials and Engineering, 303*(2). https://doi.org/10.1002/mame.201700494. http://www3.interscience.wiley.com/journal/117930403/grouphome/home.html.

Lokesh, N., Praveena, B. A., Sudheer Reddy, J., Vasu, V. K., & Vijaykumar S. (2022). Evaluation on effect of printing process parameter through Taguchi approach on mechanical properties of 3D printed PLA specimens using FDM at constant printing temperature. Materials Today: Proceedings, 52.

Loskot, J., Jezbera, D., Loskot, R., Bušovský, D., Barylski, A., Glowka, K., Duda, P., Aniołek, K., Voglová, K., & Zubko, M. (2023). Influence of print speed on the microstructure, morphology, and mechanical properties of 3D-printed PETG products. *Polymer Testing, 123*, 108055. https://doi.org/10.1016/j.polymertesting.2023.108055.

Lupone, F., Padovano, E., Veca, A., Franceschetti, L., & Badini, C. (2020). Innovative processing route combining fused deposition modelling and laser writing for the manufacturing of multifunctional polyamide/carbon fiber composites. *Materials & Design, 193*, 108869. https://doi.org/10.1016/j.matdes.2020.108869.

Lyu, Q., Dai, P., & Chen, A. (2024). Mechanical strengths and optical properties of translucent concrete manufactured by mortar-extrusion 3D printing with polymethyl methacrylate

(PMMA) fibers. *Composites Part B: Engineering, 268*, 111079. https://doi.org/10.1016/j.compositesb.2023.111079.

MacDonald, E., & Wicker, R. (2016). Multiprocess 3D printing for increasing component functionality. *Science (New York, N.Y.), 353*(6307). https://doi.org/10.1126/science.aaf2093. http://science.sciencemag.org/content/sci/353/6307/aaf2093.full.pdf.

Mackay, M. E. (2018). The importance of rheological behavior in the additive manufacturing technique material extrusion. *Journal of Rheology, 62*(6), 1549–1561. https://doi.org/10.1122/1.5037687. http://scitation.aip.org/content/sor/journal/jor2.

Mani, M., Karthikeyan, A. G., Kalaiselvan, K., Muthusamy, P., & Muruganandhan, P. (2022). Optimization of FDM 3-D printer process parameters for surface roughness and mechanical properties using PLA material. *Materials Today: Proceedings: 66*, (pp. 1926–1931). India: Elsevier Ltd. https://www.sciencedirect.com/journal/materials-today-proceedings. https://doi.org/10.1016/j.matpr.2022.05.422.

Marşavina, L., Vălean, C., Mărghitaş, M., Linul, E., Razavi, N., Berto, F., & Brighenti, R. (2022). Effect of the manufacturing parameters on the tensile and fracture properties of FDM 3D-printed PLA specimens. *Engineering Fracture Mechanics, 274*, 108766. https://doi.org/10.1016/j.engfracmech.2022.108766.

Maulet, T., Gimenez, J., Demé, F., Alcouffe, P., Vera, R., Barrès, C., & Fulchiron, R. (2022). Effect of self-nucleation on the crystallization kinetics and the microstructure of polypropylene parts manufactured by Laser Powder Bed Fusion. *Additive Manufacturing, 59*.

Maurya, S., Malik, B., Sharma, P., Singh, A., & Chalisgaonkar, R. (2022). Investigation of different parameters of cube printed using PLA by FDM 3D printer. *Materials Today: Proceedings, 64*, 1217–1222.

Mazlan, M. R., Jamadon, N. H., Rajabi, A., Sulong, A. B., Mohamed, I. F., Yusof, F., & Jamal, N. A. (2023). Necking mechanism under various sintering process parameters—A review. *Journal of Materials Research and Technology, 23*, 2189–2201.

McAlea, K. P., Forderhase, P. F., & Booth, R. B. (1998). Selective laser sintering of polymer powder of controlled particle size distribution, US5817206A.

McIlroy, C., & Graham, R. S. (2018). Modelling flow-enhanced crystallisation during fused filament fabrication of semi-crystalline polymer melts. *Additive Manufacturing, 24*, 323–340. https://doi.org/10.1016/j.addma.2018.10.018.

McIlroy, C., & Olmsted, P. D. (2017). Disentanglement effects on welding behaviour of polymer melts during the fused-filament-fabrication method for additive manufacturing. *Polymer, 123*, 376–391. http://www.journals.elsevier.com/polymer/. https://doi.org/10.1016/j.polymer.2017.06.051.

McIlroy, C., Seppala, J. E., & Kotula, A. P. (2019). Combining modeling and measurements to predict crystal morphology in material extrusion. *ACS Symposium Series, 1315*, 85–113. http://global.oup.com/academic/content/series/a/acs-symposium-series-acsss/?cc=in&lang=en. https://doi.org/10.1021/bk-2019-1315.ch006.

Milenkovic, S., Slavkovic, V., Fragassa, C., Grujovic, N., Palic, N., & Zivic, F. (2021). Effect of the raster orientation on strength of the continuous fiber reinforced PVDF/PLA composites, fabricated by hand-layup and fused deposition modeling. *Composite Structures, 270*, 114063. https://doi.org/10.1016/j.compstruct.2021.114063.

Mogan, J., Sandanamsamy, L., Harun, W. S. W., Ishak, I., Romlay, F. R. M., Kadirgama, K., & Ramasamy, D. (2024). Thermo-mechanical properties of ABS/stainless steel composite using FDM. *Materials Today: Proceedings, 109*, 120–125. https://doi.org/10.1016/j.matpr.2024.01.029.

Molnár, J., Zuba, Z., Sepsi, Ö., Ujhelyi, F., Erdei, G., Lenk, S., & Menyhárd, A. (2021). Structural investigation of semicrystalline polymers. *Polymer Testing, 95*, 107098. https://doi.org/10.1016/j.polymertesting.2021.107098.

Mullaveettil, F. N., Dauksevicius, R., & Wakjira, Y. (2021). Strength and elastic properties of 3D printed PVDF-based parts for lightweight biomedical applications. *Journal of the Mechanical Behavior of Biomedical Materials, 120*, 104603. https://doi.org/10.1016/j.jmbbm.2021.104603.

Müller, A. J., Michell, R. M., Pérez, R. A., & Lorenzo, A. T. (2015). Successive Self-nucleation and Annealing (SSA): Correct design of thermal protocol and applications. *European Polymer Journal, 65*, 132–154. https://doi.org/10.1016/j.eurpolymj.2015.01.015.

Nie, C., Peng, F., Cao, R., Cui, K., Sheng, J., Chen, W., & Li, L. (2022). Recent progress in flow-induced polymer crystallization. *Journal of Polymer Science, 60*(23), 3149–3175. https://doi.org/10.1002/pol.20220330. https://onlinelibrary.wiley.com/journal/26424169.

Nogales, A., Gutiérrez-Fernández, E., García-Gutiérrez, M. C., Ezquerra, T. A., Rebollar, E., Šics, I., Malfois, M., Gaidukovs, S., Gē Cis, E., Celms, K., & Bakradze, G. (2019). Structure development in polymers during fused filament fabrication (FFF): An in situ small- and wide-angle X-ray scattering study using synchrotron radiation. *Macromolecules, 52*(24), 9715–9723. https://doi.org/10.1021/acs.macromol.9b01620. http://pubs.acs.org/journal/mamobx.

Northcutt, L. A., Orski, S. V., Migler, K. B., & Kotula, A. P. (2018). *Effect of processing conditions on crystallization kinetics during materials extrusion additive manufacturing. Polymer (United Kingdom), 154*, 182–187. http://www.journals.elsevier.com/polymer/. https://doi.org/10.1016/j.polymer.2018.09.018.

O'Driscoll, C., Owodunni, O., & Asghar, U. (2024). Optimization of 3D printer settings for recycled PET filament using analysis of variance (ANOVA). *Heliyon, 10*(5), e26777. https://doi.org/10.1016/j.heliyon.2024.e26777.

Osswald, T. A., Puentes, J., & Kattinger, J. (2018). Fused filament fabrication melting model. *Additive Manufacturing, 22*. https://doi.org/10.1016/j.addma.2018.04.030.

Paetzold, R., Coulter, F. B., Singh, G., & Kelly, D. J. (2022). E.D. O'Cearbhaill, Fused filament fabrication of polycaprolactone bioscaffolds: Influence of fabrication parameters and thermal environment on geometric fidelity and mechanical properties. *Bioprinting, 27*, e00206. https://doi.org/10.1016/j.bprint.2022.e00206.

Peng, F., Vogt, B. D., & Cakmak, M. (2018). Complex flow and temperature history during melt extrusion in material extrusion additive manufacturing. *Additive Manufacturing, 22*, 197–206. https://doi.org/10.1016/j.addma.2018.05.015.

Peng, X., Zhang, M., Guo, Z., Sang, L., & Hou, W. (2020). Investigation of processing parameters on tensile performance for FDM-printed carbon fiber reinforced polyamide 6 composites. *Composites Communications, 22*, 100478. https://doi.org/10.1016/j.coco.2020.100478.

Petersmann, S., Spoerk-Erdely, P., Feuchter, M., Wieme, T., Arbeiter, F., & Spoerk, M. (2020). Process-induced morphological features in material extrusion-based additive manufacturing of polypropylene. *Additive Manufacturing, 35*, 101384. https://doi.org/10.1016/j.addma.2020.101384.

Phan, D. D., Swain, Z. R., & MacKay, M. E. (2018). Rheological and heat transfer effects in fused filament fabrication. *Journal of Rheology, 62*(5), 1097–1107. https://doi.org/10.1122/1.5022982. http://scitation.aip.org/content/sor/journal/jor2.

Pires, L. S. O., Fernandes, M. H. F. V., & de Oliveira, J. M. M. (2018). Crystallization kinetics of PCL and PCL–glass composites for additive manufacturing. *Journal of Thermal Analysis and Calorimetry, 134*(3), 2115–2125. https://doi.org/10.1007/s10973-018-7307-7. http://www.springer.com/sgw/cda/frontpage/0,11855,1-40109-70-35752391-0,00.html.

Pokluda, O., Bellehumeur, C. T., & Vlachopoulos, J. (1997). Modification of Frenkel's model for sintering. *AIChE Journal, 43*(12), 3253–3256. https://doi.org/10.1002/aic.690431213. http://www.interscience.wiley.com.

Poojitha, G., Talari, P., Banoth, S., & Kumar, A. (2023). Effects of combined infill angle with honeycomb pattern on the mechanical properties of HIPS and polypropylene in FDM process. *Materials Today: Proceedings*. https://doi.org/10.1016/j.matpr.2023.09.219, in press.

Popa, C. F., Krausz, T., Galatanu, S. V., Linul, E., & Marsavina, L. (2023). *Numerical and experimental study for FDM printed specimens from PLA under IZOD impact tests. Materials Today: Proceedings, 78*, 326–330.

Powell, A. W., Stavrinadis, A., De Miguel, I., Konstantatos, G., & Quidant, R. (2018). White and Brightly Colored 3D Printing Based on Resonant Photothermal Sensitizers. *Nano Letters, 18*(11), 6660–6664. https://doi.org/10.1021/acs.nanolett.8b01164. http://pubs.acs.org/journal/nalefd.

Prasong, W., Muanchan, P., Ishigami, A., Thumsorn, S., Kurose, T., & Ito, H. (2020). Properties of 3D printable poly(lactic acid)/poly(butylene adipate-co-terephthalate) blends and nano talc composites. *Journal of Nanomaterials, 2020*, 1–16. https://doi.org/10.1155/2020/8040517.

Pulipaka, A., Gide, K. M., Beheshti, A., & Bagheri, Z. S. (2023). *Effect of 3D printing process parameters on surface and mechanical properties of FFF-printed PEEK. Journal of Manufacturing Processes, 85*.

Pérez-Camargo, R. A., Cavallo, D., & Müller, A. J. (2022). Recent applications of the successive self-nucleation and annealing thermal fractionation technique. *Frontiers in Soft Matter, 2*. https://doi.org/10.3389/frsfm.2022.1003500.

Pérez-Camargo, R. A., Liu, G. M., Wang, D. J., & Müller, A. J. (2022). Experimental and data fitting guidelines for the determination of polymer crystallization kinetics. *Chinese Journal of Polymer Science (English Edition), 40*(6), 658–691. https://doi.org/10.1007/s10118-022-2724-2. http://www.springerlink.com/content/0256-7679.

Qi, S., Gao, X., Su, Y., Dong, X., Cavallo, D., & Wang, D. (2021). Correlation between welding behavior and mechanical anisotropy of long chain polyamide 12 manufactured with fused filament fabrication. *Polymer, 213*, 123318. https://doi.org/10.1016/j.polymer.2020.123318.

Qi, S., Gao, X., Su, Y., Zhou, Y., Dong, X., & Wang, D. (2022). Effect of carbon nanotubes on mechanical properties of polyamide 12 parts by fused filament fabrication. *Polymer, 247*, 124784. https://doi.org/10.1016/j.polymer.2022.124784.

Rahman, K. M., Letcher, T., & Reese, R. (2015). Mechanical properties of additively manufactured peek components using fused filament fabrication. ASME International Mechanical Engineering Congress and Exposition, Proceedings (IMECE). United States: American Society of Mechanical Engineers (ASME). http://www.asmedl.org/journals/doc/ASMEDL-home/proc/ 2.

Ree, B. J., Zheng, B., Abbott, A., Smieska, L. M., Page, K. A., Woll, A. R., Renwick, Z., & Koerner, H. (2024). Mapping crystallization kinetics during 3D printing of poly(ether ether ketone). *Macromolecules, 57*(6), 2810–2817. https://doi.org/10.1021/acs.macromol.3c02327. http://pubs.acs.org/journal/mamobx.

Rodríguez, J. F., Thomas, J. P., & Renaud, J. E. (2003). Mechanical behavior of acrylonitrile butadiene styrene fused deposition materials modeling. *Rapid Prototyping Journal, 9*(4), 219–230. https://doi.org/10.1108/13552540310489604.

Rodríguez-Reyna, S. L., Mata, C., Díaz-Aguilera, J. H., Acevedo-Parra, H. R., & Tapia, F. (2022). Mechanical properties optimization for PLA, ABS and Nylon + CF manufactured by 3D FDM printing. *Materials Today Communications, 33*, 104774. https://doi.org/10.1016/j.mtcomm.2022.104774.

Rodzeń, K., Harkin-Jones, E., Wegrzyn, M., Sharma, P. K., & Zhigunov, A. (2021). Improvement of the layer-layer adhesion in FFF 3D printed PEEK/carbon fibre composites. *Composites Part A: Applied Science and Manufacturing, 149*, 106532. https://doi.org/10.1016/j.compositesa.2021.106532.

Rylski, A. K., Maraliga, T., Wu, Y., Recker, E. A., Arrowood, A. J., Sanoja, G. E., & Page, Z. A. (2023). Digital light processing 3D printing of soft semicrystalline acrylates with localized shape memory and stiffness control. *ACS Applied Materials and Interfaces, 15*(28), 34097–34107. https://doi.org/10.1021/acsami.3c07172. http://pubs.acs.org/journal/aamick.

Sahoo, S., Sutar, H., Senapati, P., Shankar Mohanto, B., Ranjan Dhal, P., & Kumar Baral, S. (2023). *Experimental investigation and optimization of the FDM process using PLA.* Materials Today: Proceedings, *74.*

Saini, A., Elhattab, K., Gummadi, S. K., Nadkarni, G. R., & Sikder, P. (2022). *Fused filament fabrication-3D printing of poly-ether-ether-ketone (PEEK) spinal fusion cages.* Materials Letters, *328.*

Samson, K. D. G., Hidalgo-Alvarez, V., Dargaville, T. R., & Melchels, F. P. W. (2023). Tough, resorbable polycaprolactone-based bimodal networks for vat polymerization 3D printing. *Advanced Functional Materials, 33*(25). https://doi.org/10.1002/adfm.202213797. http://onlinelibrary.wiley.com/journal/10.1002/(ISSN)1616-3028.

Sandanamsamy, L., Mogan, J., Rajan, K., Harun, W. S. W., Ishak, I., Romlay, F. R. M., Samykano, M., & Kadirgama, K. (2023). *Effect of process parameter on tensile properties of FDM printed PLA. Materials Today: Proceedings, 109.*

Sangroniz, L., Cavallo, D., & Müller, A. J. (2020). Self-nucleation effects on polymer crystallization. *Macromolecules, 53*(12), 4581–4604. https://doi.org/10.1021/acs.macromol.0c00223. http://pubs.acs.org/journal/mamobx.

Schirmeister, C. G., Hees, T., Licht, E. H., & Mülhaupt, R. (2019). 3D printing of high density polyethylene by fused filament fabrication. *Additive Manufacturing, 28*, 152–159. https://doi.org/10.1016/j.addma.2019.05.003.

Schmid, M., Amado, A., & Wegener, K. (2014). Materials perspective of polymers for additive manufacturing with selective laser sintering. *Journal of Materials Research, 29*(17), 1824–1832. https://doi.org/10.1557/jmr.2014.138. https://www.springer.com/journal/43578.

Senatov, F. S., Niaza, K. V., Zadorozhnyy, M. Y., Maksimkin, A. V., Kaloshkin, S. D., & Estrin, Y. Z. (2016). *Mechanical properties and shape memory effect of 3D-printed PLA-based porous scaffolds. Federation Journal of the Mechanical Behavior of Biomedical Materials, 57*, 139–148. http://www.elsevier.com/wps/find/journaldescription.cws_home/711005/description#description. https://doi.org/10.1016/j.jmbbm.2015.11.036.

Seppala, J. E., & Migler, K. D. (2016). Infrared thermography of welding zones produced by polymer extrusion additive manufacturing. *Additive Manufacturing, 12*, 71–76. https://doi.org/10.1016/j.addma.2016.06.007.

Seppala, J. E., Hoon Han, S., Hillgartner, K. E., Davis, C. S., & Migler, K. B. (2017). Weld formation during material extrusion additive manufacturing. *Soft Matter, 13*(38),

6761–6769. https://doi.org/10.1039/c7sm00950j. http://www.rsc.org/Publishing/Journals/sm/Article.asp?Type=CurrentIssue.

Şerban, D. A., Weber, G., Marşavina, L., Silberschmidt, V. V., & Hufenbach, W. (2013). Tensile properties of semi-crystalline thermoplastic polymers: Effects of temperature and strain rates. *Polymer Testing, 32*(2), 413–425. https://doi.org/10.1016/j.polymertesting.2012.12.002.

Sgrulletti, M., Bragaglia, M., Giarnetti, S., Paleari, L., & Nanni, F. (2021). Understanding the impact of Fused Filament Fabrication conditions on the microstructure and tensile properties of polyamide 6 by thermal and optical live monitoring. *Materials Today Communications, 28*, 102679. https://doi.org/10.1016/j.mtcomm.2021.102679.

Shahriar, B. B., France, C., Valerie, N., Arthur, C., & Christian G. (2017). Toward improvement of the properties of parts manufactured by FFF (fused filament fabrication) through understanding the influence of temperature and rheological behaviour on the coalescence phenomenon. AIP Conference Proceedings. France: American Institute of Physics Inc. http://scitation.aip.org/content/aip/proceeding/aipcp.

Shaqour, B., Abuabiah, M., Abdel-Fattah, S., Juaidi, A., Abdallah, R., Abuzaina, W., Qarout, M., Verleije, B., & Cos, P. (2021). Gaining a better understanding of the extrusion process in fused filament fabrication 3D printing: A review. *International Journal of Advanced Manufacturing Technology, 114*(5–6), 1279–1291. https://doi.org/10.1007/s00170-021-06918-6. http://www.springerlink.com/content/0268-3768.

Shashikumar, S., & Sreekanth, M. S. (2023). The effect of printing parameters on tensile properties of thermoplastics prepared by fused deposition modeling (FDM) based additive manufacturing technique. Materials Today: Proceedings, 90.

Shuto, R., Norimatsu, S., Arola, D. D., & Matsuzaki, R. (2022). Effect of the nozzle temperature on the microstructure and interlaminar strength in 3D printing of carbon fiber/polyphenylene sulfide composites. *Composites Part C: Open Access, 9*, 100328. https://doi.org/10.1016/j.jcomc.2022.100328.

Sikder, P., Challa, B. T., & Gummadi, S. K. (2022). *A comprehensive analysis on the processing-structure-property relationships of FDM-based 3-D printed polyetheretherketone (PEEK) structures. Materialia, 22.*

Silva, A. F., Carneiro, O. S., & Gomes, R. (2017). AIP Conference Proceedings. American Institute of Physics Inc. Portugal 3D printing of polypropylene using the fused filament fabrication technique, 1896. doi: 10.1063/1.5008040 15517616; http://scitation.aip.org/content/aip/proceeding/aipcp.

Simunec, D. P., Breedon, M., Muhammad, F. U. R., Kyratzis, L., & Sola, A. (2023). *Electrical capability of 3D printed unpoled polyvinylidene fluoride (PVDF)/thermoplastic polyurethane (TPU) sensors combined with carbon black and barium titanate. Additive Manufacturing, 73.*

Singh, J., Kumar Goyal, K., & Kumar, R. (2022). *Effect of filling percentage and raster style on tensile behavior of FDM produced PLA parts at different build orientation. Materials Today: Proceedings, 63*, 433–439.

Singh, K. (2018). Experimental study to prevent the warping of 3D models in fused deposition modeling. *International Journal of Plastics Technology, 22*(1), 177–184. https://doi.org/10.1007/s12588-018-9206-y. http://www.springer.com/chemistry/polymer+science/journal/12588.

Sodeifian, G., Ghaseminejad, S., & Yousefi, A. A. (2019). *Preparation of polypropylene/short glass fiber composite as Fused Deposition Modeling (FDM) filament. Results in Physics, 12*, 205–222.

Sood, A. K., Ohdar, R. K., & Mahapatra, S. S. (2010). Parametric appraisal of mechanical property of fused deposition modelling processed parts. *Materials and Design, 31*(1), 287–295. https://doi.org/10.1016/j.matdes.2009.06.016.

Sood, A. K., Ohdar, R. K., & Mahapatra, S. S. (2012). Experimental investigation and empirical modelling of FDM process for compressive strength improvement. *Journal of Advanced Research, 3*(1), 81–90. https://doi.org/10.1016/j.jare.2011.05.001.

Spina, R., & Cavalcante, B. (2020). Preliminary analysis of extruded PP filaments for FFF. Procedia Manufacturing, 47.

Spoerk, M., Arbeiter, F., Cajner, H., Sapkota, J., & Holzer, C. (2017). Parametric optimization of intra- and inter-layer strengths in parts produced by extrusion-based additive manufacturing of poly(lactic acid). *Journal of Applied Polymer Science, 134*(41). https://doi.org/10.1002/app.45401. http://onlinelibrary.wiley.com/journal/10.1002/(ISSN)1097-4628.

Spoerk, M., Savandaiah, C., Arbeiter, F., Sapkota, J., & Holzer, C. (2019). Optimization of mechanical properties of glass-spheres-filled polypropylene composites for extrusion-based additive manufacturing. *Polymer Composites, 40*(2), 638–651. https://doi.org/10.1002/pc.24701. http://onlinelibrary.wiley.com/journal/10.1002/(ISSN)1548-0569.

Sreejith, P., Kannan, K., & Rajagopal, K. R. (2023). A thermodynamic framework for the additive manufacturing of crystallizing polymers. Part I: A theory that accounts for phase change, shrinkage, warpage and residual stress. *International Journal of Engineering Science, 183*, 103789. https://doi.org/10.1016/j.ijengsci.2022.103789.

Srinivas, V., van Hooy-Corstjens, C. S. J., & Harings, J. A. W. (2018). *Correlating molecular and crystallization dynamics to macroscopic fusion and thermodynamic stability in fused deposition modeling; a model study on polylactides. Polymer, 142*, 348–355.

Stoof, D., & Pickering, K. (2018). Sustainable composite fused deposition modelling filament using recycled pre-consumer polypropylene. *Composites Part B: Engineering, 135*, 110–118. https://doi.org/10.1016/j.compositesb.2017.10.005.

Street, D. P., Mah, A. H., Patterson, S., Pickel, D. L., Bergman, J. A., Stein, G. E., Messman, J. M., & Kilbey, S. M. (2018). *Interfacial interactions in PMMA/silica nanocomposites enhance the performance of parts created by Fused Filament Fabrication. Polymer, 157*, 87–94.

Su, N., Pierce, R. S., Rudd, C., & Liu, X. (2022). Comprehensive investigation of reclaimed carbon fibre reinforced polyamide (rCF/PA) filaments and FDM printed composites. *Composites Part B: Engineering, 233*, 109646. https://doi.org/10.1016/j.compositesb.2022.109646.

Sun, Q., Rizvi, G. M., Bellehumeur, C. T., & Gu, P. (2008). Effect of processing conditions on the bonding quality of FDM polymer filaments. *Rapid Prototyping Journal, 14*(2), 72–80. https://doi.org/10.1108/13552540810862028.

Tamir, T. S., Xiong, G., Shen, Z., Leng, J., Fang, Q., Yang, Y., Jiang, J., Lodhi, E., & Wang, F. Y. (2023). 3D printing in materials manufacturing industry: A realm of Industry 4.0. *Heliyon, 9*(9). https://doi.org/10.1016/j.heliyon.2023.e19689. http://www.journals.elsevier.com/heliyon/.

Tao, R., Shi, J., Rafiee, M., Akbarzadeh, A., & Therriault, D. (2022). Fused filament fabrication of PVDF films for piezoelectric sensing and energy harvesting applications. *Materials Advances, 3*(12), 4851–4860. https://doi.org/10.1039/d2ma00072e. http://www.rsc.org/journals-books-databases/about-journals/materials-advances/.

Tlegenov, Y., Hong, G. S., & Lu, W. F. (2018). Nozzle condition monitoring in 3D printing. *Robotics and Computer-Integrated Manufacturing, 54*, 45–55. https://doi.org/10.1016/j.rcim.2018.05.010.

Toro, E. V.de, Sobrino, J. C., Martínez, A. M., & Eguía, V. M. (2019). Analysis of the influence of the variables of the Fused Deposition Modeling (FDM) process on the mechanical properties of a carbon fiber-reinforced polyamide. *Procedia Manufacturing, 41*, 731–738. https://doi.org/10.1016/j.promfg.2019.09.064.

Tseng, J. W., Liu, C. Y., Yen, Y. K., Belkner, J., Bremicker, T., Liu, B. H., Sun, T. J., & Wang, A. B. (2018). Screw extrusion-based additive manufacturing of PEEK. *Materials and Design, 140*, 209–221. https://doi.org/10.1016/j.matdes.2017.11.032.

Vaes, D., & Van Puyvelde, P. (2021). Semi-crystalline feedstock for filament-based 3D printing of polymers. *Progress in Polymer Science, 118*, 101411. https://doi.org/10.1016/j.progpolymsci.2021.101411.

Vaes, D., Coppens, M., Goderis, B., Zoetelief, W., & Puyvelde, P. V. (2019). Assessment of crystallinity development during fused filament fabrication through fast scanning chip calorimetry. *Applied Sciences, 9*(13), 2676. https://doi.org/10.3390/app9132676.

Vaezi, M., & Yang, S. (2015). Extrusion-based additive manufacturing of PEEK for biomedical applications. *Virtual and Physical Prototyping, 10*(3), 123–135. https://doi.org/10.1080/17452759.2015.1097053. http://www.tandf.co.uk/journals/titles/17452759.asp.

Vaezi, M., Zhong, G., Kalami, H., & Yang, S. (2018). *Extrusion-based 3D printing technologies for 3D scaffold engineering. Functional 3D tissue engineering scaffolds: Materials, technologies, and applications*. United Kingdom: Elsevier235–254. http://www.sciencedirect.com/science/book/9780081009796. https://doi.org/10.1016/B978-0-08-100979-6.00010-0.

Van de Voorde, B., Katalagarianakis, A., Huysman, S., Toncheva, A., Raquez, J. M., Duretek, I., Holzer, C., Cardon, L., Bernaerts, K. V., Van Hemelrijck, D., Pyl, L., & Van Vlierberghe, S. (2022). *Effect of extrusion and fused filament fabrication processing parameters of recycled poly(ethylene terephthalate) on the crystallinity and mechanical properties. Additive Manufacturing, 50.*

Vanaei, H., Shirinbayan, M., Deligant, M., Raissi, K., Fitoussi, J., Khelladi, S., & Tcharkhtchi, A. (2020). Influence of process parameters on thermal and mechanical properties of polylactic acid fabricated by fused filament fabrication. *Polymer Engineering and Science, 60*(8), 1822–1831. https://doi.org/10.1002/pen.25419. http://onlinelibrary.wiley.com/journal/10.1002/(ISSN)1548-2634.

Vasquez, G. M., Majewski, C. E., Haworth, B., & Hopkinson, N. (2014). A targeted material selection process for polymers in laser sintering. *Additive Manufacturing, 1*, 127–138. https://doi.org/10.1016/j.addma.2014.09.003.

Vasquez, M., Haworth, B., & Hopkinson, N. (2013). Methods for quantifying the stable sintering region in laser sintered polyamide-12. *Polymer Engineering and Science, 53*(6), 1230–1240. https://doi.org/10.1002/pen.23386.

Vaught, L. O., & Polycarpou, A. A. (2023). Investigating the effect of fused deposition modelling on the tribology of PETG thermoplastic. *Wear, 524–525*, 204736. https://doi.org/10.1016/j.wear.2023.204736.

Venkataraman, N., Rangarajan, S., Matthewson, M. J., Harper, B., Safari, A., Danforth, S. C., Wu, G., Langrana, N., Guceri, S., & Yardimci, A. (2000). Feedstock material property–process relationships in fused deposition of ceramics (FDC). *Rapid Prototyping Journal, 6*(4), 244–253. https://doi.org/10.1108/13552540010373344.

Verbeeten, W. M. H., Lorenzo-Bañuelos, M., & Arribas-Subiñas, P. J. (2020). *Anisotropic rate-dependent mechanical behavior of Poly(Lactic Acid) processed by Material Extrusion Additive Manufacturing. Additive Manufacturing, 31.*

Vidakis, N., Petousis, M., Velidakis, E., Mountakis, N., Fischer-Griffiths, P. E., Grammatikos, S. A., & Tzounis, L. (2022). *Fused filament fabrication 3D printed polypropylene/alumina nanocomposites: Effect of filler loading on the mechanical reinforcement. Polymer Testing, 109.*

Voet, A., & Geffken, C. F. (1951). The nature of tack. *Industrial & Engineering Chemistry, 43*(7), 1614–1624. https://doi.org/10.1021/ie50499a041.

Vozzi, G., Flaim, C. J., Bianchi, F., Ahluwalia, A., & Bhatia, S. (2002). Microfabricated PLGA scaffolds: A comparative study for application to tissue engineering. *Materials Science and Engineering C, 20*(1-2), 43–47. https://doi.org/10.1016/S0928-4931(02)00011-5.

Wang, C., He, Y., Lin, Z., Zhao, X., Sun, C., Guo, R., Wang, X., & Zhou, F. (2024). Mechanical and tribological properties of FDM-printed polyamide. *Tribology International, 191*, 109198. https://doi.org/10.1016/j.triboint.2023.109198.

Wang, L., & Gardner, D. J. (2017). *Effect of fused layer modeling (FLM) processing parameters on impact strength of cellular polypropylene. Polymer, 113*, 74–80.

Wang, L., Gramlich, W. M., & Gardner, D. J. (2017). *Improving the impact strength of poly (lactic acid) (PLA) in fused layer modeling (FLM). Polymer, 114.*

Wang, P., Zou, B., Ding, S., Huang, C., Shi, Z., Ma, Y., & Yao, P. (2020). Preparation of short CF/GF reinforced PEEK composite filaments and their comprehensive properties evaluation for FDM-3D printing. *Composites Part B: Engineering, 198*, 108175. https://doi.org/10.1016/j.compositesb.2020.108175.

Wang, Z., Ma, Z., & Li, L. (2016). Flow-induced crystallization of polymers: Molecular and thermodynamic considerations. *Macromolecules, 49*(5), 1505–1517. https://doi.org/10.1021/acs.macromol.5b02688. http://pubs.acs.org/journal/mamobx.

Wegner, A. (2016). New polymer materials for the laser sintering process: Polypropylene and others. Physics Procedia, 83.

Wendel, B., Rietzel, D., Kühnlein, F., Feulner, R., Hülder, G., & Schmachtenberg, E. (2008). Additive processing of polymers. *Macromolecular Materials and Engineering, 293*(10), 799–809. https://doi.org/10.1002/mame.200800121Germany. http://www3.interscience.wiley.com/cgi-bin/fulltext/121402200/PDFSTART.

Winter, K., Wilfert, J., Häupler, B., Erlmann, J., & Altstädt, V. (2022). Large scale 3D printing: Influence of fillers on warp deformation and on mechanical properties of printed polypropylene components. *Macromolecular Materials and Engineering, 307*(1)https://doi.org/10.1002/mame.202100528http://onlinelibrary.wiley.com/journal/10.1002/(ISSN)1439-2054.

Wittbrodt, B., & Pearce, J. M. (2015). The effects of PLA color on material properties of 3-D printed components. *Additive Manufacturing, 8*, 110–116. https://doi.org/10.1016/j.addma.2015.09.006.

Wolszczak, P., Lygas, K., Paszko, M., & Wach, R. A. (2018). Heat distribution in material during fused deposition modelling. *Rapid Prototyping Journal, 24*(3), 615–622. https://doi.org/10.1108/RPJ-04-2017-0062. http://www.emeraldinsight.com/info/journals/rpj/rpj.jsp.

Wool, R. P., & O'Connor, K. M. (1981). A theory of crack healing in polymers. *Journal of Applied Physics, 52*(10), 5953–5963. https://doi.org/10.1063/1.328526

Wu, H., Sulkis, M., Driver, J., Saade-Castillo, A., Thompson, A., & Koo, J. H. (2018). Multi-functional ULTEM™1010 composite filaments for additive manufacturing using Fused Filament Fabrication (FFF). *Additive Manufacturing, 24*, 298–306. https://doi.org/10.1016/j.addma.2018.10.014.

Xia, H., Lu, J., & Tryggvason, G. (2019). *A numerical study of the effect of viscoelastic stresses in fused filament fabrication. Computer Methods in Applied Mechanics and Engineering, 346*, 242–259.

Yan, C., Shi, Y., & Hao, L. (2011). Investigation into the differences in the selective laser sintering between amorphous and semi-crystalline polymers. *International Polymer Processing, 26*(4), 416–423. https://doi.org/10.3139/217.2452UnitedKingdom. http://www.polymer-process.com/web/archiv/get_doc_abo.asp?url=get_doc_abo.asp&bin_id=21181191633-27156&o_id=2531143156-51.

Yang, C., Tian, X., Li, D., Cao, Y., Zhao, F., & Shi, C. (2017). Influence of thermal processing conditions in 3D printing on the crystallinity and mechanical properties of PEEK material. *Journal of Materials Processing Technology, 248*, 1–7. https://doi.org/10.1016/j.jmatprotec.2017.04.027.

Yang, H., & Zhang, S. (2018). Numerical simulation of temperature field and stress field in fused deposition modeling. *Journal of Mechanical Science and Technology, 32*(7), 3337–3344. https://doi.org/10.1007/s12206-018-0636-4. http://www.springerlink.com/content/1738-494X.

Yang, L., Wang, Y., Xiao, Y., Zhang, G., Pan, B., Li, S., Li, J., Gao, D., & Wang, Y. (2024). Fabrication of a sensor based on the FDM-printed CNT/PA6 dielectric layer with Hilbert fractal microstructure. *Sensors and Actuators A: Physical, 369*, 115190. https://doi.org/10.1016/j.sna.2024.115190.

Yang, T., Hu, J., Wang, P., Edeleva, M., Cardon, L., & Zhang, J. (2023). Two-step approach based on fused filament fabrication for high performance graphene/thermoplastic polyurethane composite with segregated structure. *Composites Part A: Applied Science and Manufacturing, 174*, 107719. https://doi.org/10.1016/j.compositesa.2023.107719.

Yao, J., Duongthipthewa, A., Xu, X., Liu, M., Xiong, Y., & Zhou, L. (2024). Interlayer bonding improvement of PEEK and CF-PEEK composites with laser-assisted fused deposition modeling. *Composites Communications, 45*, 101819. https://doi.org/10.1016/j.coco.2024.101819.

Yao, T., Ye, J., Deng, Z., Zhang, K., Ma, Y., & Ouyang, H. (2020). Tensile failure strength and separation angle of FDM 3D printing PLA material: Experimental and theoretical analyses. *Composites Part B: Engineering, 188*, 107894. https://doi.org/10.1016/j.compositesb.2020.107894.

Yin, B., He, Q., & Ye, L. (2021). Effects of deposition speed and extrusion temperature on fusion between filaments in single-layer polymer films printed with FFF. *Advanced Industrial and Engineering Polymer Research, 4*(4), 270–276. https://doi.org/10.1016/j.aiepr.2021.07.002. http://www.keaipublishing.com/en/journals/advanced-industrial-and-engineering-polymer-research/.

Zander, N. E., Gillan, M., Burckhard, Z., & Gardea, F. (2019). Recycled polypropylene blends as novel 3D printing materials. *Additive Manufacturing, 25*, 122–130. https://doi.org/10.1016/j.addma.2018.11.009.

Zarek, M., Layani, M., Cooperstein, I., Sachyani, E., Cohn, D., & Magdassi, S. (2016). 3D printing of shape memory polymers for flexible electronic devices. *Advanced Materials, 28*(22), 4449–4454. https://doi.org/10.1002/adma.201503132. http://www3.interscience.wiley.com/journal/119030556/issue.

Zarringhalam, H., Hopkinson, N., Kamperman, N. F., & de Vlieger, J. J. (2006). *Effects of processing on microstructure and properties of SLS Nylon 12. Materials Science and Engineering: A, 435–436*, 172–180. http://www.elsevier.com. https://doi.org/10.1016/j.msea.2006.07.084.

Zhang, B., Cristescu, R., Chrisey, D. B., & Narayan, R. J. (2020). Solvent-based extrusion 3D printing for the fabrication of tissue engineering scaffolds. *International Journal of Bioprinting, 6*(1), 28–42. https://doi.org/10.18063/ijb.v6i1.211. http://ijb.whioce.com/index.php/int-j-bioprinting/article/download/211/pdf.

Zhang, F., Zhu, L., Li, Z., Wang, S., Shi, J., Tang, W., Li, N., & Yang, J. (2021). *The recent development of vat photopolymerization: A review.* Additive Manufacturing, *48*.

Zhang, H., Xiao, J., Chen, Y., & Wang, H. (2024). Fused deposition modeling to predict interlayer weld strength of polyvinylidene fluoride and its MWCNTs-reinforced composites. *Composite Structures, 328*, 117715. https://doi.org/10.1016/j.compstruct.2023.117715.

Zhao, Y., Chen, Y., & Zhou, Y. (2019). Novel mechanical models of tensile strength and elastic property of FDM AM PLA materials: Experimental and theoretical analyses. *Materials & Design, 181*, 108089. https://doi.org/10.1016/j.matdes.2019.108089.

Zhao, Y., Zhao, K., Li, Y., & Chen, F. (2020). Mechanical characterization of biocompatible PEEK by FDM. *Journal of Manufacturing Processes, 56*, 28–42. https://doi.org/10.1016/j.jmapro.2020.04.063. http://www.elsevier.com/wps/find/journaldescription.cws_home/620379/description#description.

Zhao, Y. Q., Yang, J. H., Ding, X., Ding, X., Duan, S., & Xu, F. J. (2020). Polycaprolactone/polysaccharide functional composites for low-temperature fused deposition modelling. *KeAi Communications Co., China Bioactive Materials, 5*(2), 185–191. https://doi.org/10.1016/j.bioactmat.2020.02.006. http://www.keaipublishing.com/en/journals/bioactive-materials/.

Ziaee, M., & Crane, N. B. (2019). *Binder jetting: A review of process, materials, and methods. Additive Manufacturing*, *28*, 781–801.

Zou, R. P., & Yu, A. B. (1996). Evaluation of the packing characteristics of mono-sized non-spherical particles. *Powder Technology, 88*(1), 71–79. https://doi.org/10.1016/0032-5910(96)03106-3. http://www.elsevier.com/locate/powtec.

Additive manufacturing: a roadmap toward intellectual and industrial developments

2

Viswapriya Viswalingam[1], *and Dileep Kumar*[2]
[1]Department of Pharmaceutical Chemistry, Poona College of Pharmacy, Bharati Vidyapeeth (Deemed to be University), Pune, Maharashtra, India, [2]Department of Pharm Chemistry, Manipal College of Pharmaceutical Sciences, Manipal Academy of Higher Education, Manipal, Karnataka, India

2.1 Introduction

As pressing societal issues and ambitious objectives like the Sustainable Development Goals gain traction, the Science, Technology, and Innovation (STI) policy community is increasingly giving attention to roadmaps and road mapping techniques (Miedzinski et al., 2022). STI policy roadmaps are instrumental in crafting, documenting, strategizing, and executing public policies that align with long-term sustainability aims (Miedzinski et al., 2019). Road mapping, known for its sophisticated approach to strategic planning, is viewed as suitable for driving sustainability transitions (Miedzinski et al., 2022). Over the past three decades, the practice and study of road mapping have evolved in sync with shifting landscapes driven by advancements in technology and the rapid evolution of society (Vinayavekhin et al., 2023).

Within the industrial sphere, employing roadmaps entails plotting the trajectory of a product, establishing links between the fundamental business strategy and the development of features, costs, and essential technologies. Implementing this approach across a pivotal sector yields multiple benefits (Albright & Kappel, 2003). Within each product line, roadmaps serve as a crucial link that connects market strategy to product and technology strategies. The roadmaps at the product level form the foundation for technology planning across the industry, fostering a shared understanding of requirements, gaps, strengths, and weaknesses throughout the company. These roadmaps also emphasize long-term planning, enhance communication and accountability, and channel the team's focus toward the most critical priorities at every stage of the planning process (Park et al., 2020). The Road mapping methodology has been widely embraced and disseminated across organizations of diverse scales, ranging from small businesses to expansive government initiatives. Its objective is to seamlessly align an organization's objectives with the technology associated with its products and ensure critical success factors for numerous enterprises (de Alcantara & Martens, 2019).

Additive Manufacturing of Functional Polymers and Nanocomposites. DOI: https://doi.org/10.1016/B978-0-443-26516-7.00017-1
© 2025 Elsevier Ltd. All rights are reserved, including those for text and data mining, AI training, and similar technologies.

With the rapid advancement of AM technology within the past three decades, AM has moved beyond its initial use for prototyping and has now evolved into the advanced manufacturing of functional components across various industries (Mahmood et al., 2022; Tofail et al., 2018). Consequently, the potential hindrances to the widespread industrial implementation of AM technology in the future lie in the intellectualization and industrialization of processes and equipment. Intelligent AM technology shows promise in achieving a fully automated, highly precise, and stable manufacturing process through close-loop control (Fidan et al., 2023; Ghaffar et al., 2018; Zhironkina & Zhironkin, 2023). Developing a process model for AM allows for automatic adjustments of relevant parameters using a process database and advanced big data analysis technology, which is projected to be a critical pathway for AM's development. Despite being a rapidly emerging and evolving technology, AM still falls short of meeting the mass production requirements of traditional industries, particularly in terms of standardization and industrialization (Bayat et al., 2023; Ligon et al., 2017; Yang et al., 2021). However, it holds potential for achieving single-part production, small batch production, and mass customization in future industrialization, especially in extreme manufacturing scenarios related to the environment and scale (Bohr & Memarzadeh, 2020; Dash et al., 2019). This chapter provides a concise overview of AM technology, its history, industrial applications, challenges, and future prospects, with a focus on case studies utilizing AM.

2.2 Road mapping for AM

AM (AM) technology emerged through advancements in diverse technological domains. Much like in numerous manufacturing technologies, progress in computing capabilities and the reduction in mass storage expenses facilitated the processing of vast volumes of data inherent in contemporary 3D Computer-Aided Design (CAD) models within manageable time frames (Johnson et al., 2021; Kumar et al., 2023; Vafadar et al., 2021). In our present era, we've grown accustomed to the presence of robust computers and intricate automated machinery. It's sometimes challenging to envision the challenges faced by the early innovators in creating the first AM machines.

AM (AM) technology offers the ability to create personalized products while reducing development costs, shortening lead times, minimizing energy usage in production, and reducing material waste (Mehrpouya et al., 2019). Given the maturity of the technology, the extensive possibilities of 3D printing, and the institutional support, AM is on track to become a prominent technology in various sectors soon. A crucial aspect of Industry 4.0 is primarily constricted toward 3D printing, enabling the fabrication of intricate parts and empowering companies to reduce inventory, produce on-demand items, establish localized manufacturing setups, and streamline supply chains (Javaid et al., 2022; Javaid, Haleem, Singh, et al., 2021; Prashar et al., 2022). Consequently, the integration of intelligent technology into production systems, or in other words, AM's promotion of Industry 4.0, assumes great significance in addressing key requirements of the fourth industrial revolution (Zhong et al., 2017). AM is a prospective paradigm for futuristic production systems, and Industry 4.0 stands to leverage its potential to achieve crucial objectives. Today, AM finds applications across a diverse range of industries, including aerospace, healthcare, and consumer goods (Phuyal et al., 2020).

AM (AM) technology offers the ability to create personalized products while reducing development costs, shortening lead times, minimizing energy usage in production, and reducing material waste. Given the maturity of the technology, the extensive possibilities of 3D printing, and the institutional support, AM is on track to become a prominent technology in various sectors soon. A crucial aspect of Industry 4.0 centers on 3D printing, enabling the fabrication of intricate parts and empowering companies to reduce inventory, produce on-demand items, establish localized manufacturing setups, and streamline supply chains (Raja Santhi & Muthuswamy, 2023). Consequently, the integration of intelligent technology into production systems, or in other words, AM's promotion of Industry 4.0, assumes great significance in addressing key requirements of the fourth industrial revolution. AM is a prospective paradigm for futuristic production systems, and Industry 4.0 stands to leverage its potential to achieve crucial objectives (Rosário & Dias, 2022). Today, AM finds applications across a diverse range of industries, including aerospace, healthcare, and consumer goods.

AM stands apart from conventional manufacturing methods in its core approach. Yet, to harness the inherent advantages of AM effectively, it's vital to measure its fundamental characteristics in contrast to traditional techniques (Rouf et al., 2022; Wang et al., 2017). With this comprehension, it becomes evident that tailored design tools addressing the technical and economic facets of AM are essential. Recognizing the importance of specialized Design for AM (DFAM) capabilities (Durakovic, 2018), this chapter delineates current DFAM capabilities concerning their appropriateness for design objectives while pinpointing shortcomings in existing DFAM tools. These insights form a foundation for productive DFAM research strategies.

2.3 A room for 3D printing in 21st century

Digital fabrication technology, also known as AM or 3D printing, fabricates physical objects based on a geometric representation by incrementally adding materials. The rapid growth of 3D printing technology is evident, and it has become widespread across various industries. Three-Dimensional Printing (3DP) is a fabrication technology developed at the Massachusetts Institute of Technology that constructs 3D structures by depositing a liquid binder solution onto a powder bed, utilizing a variety of materials in either solid or liquid states (Chia & Wu, 2015; Cima et al., 1994). The process involves layering fine powder material evenly, selectively depositing the binder droplets onto the powder layer to create a desired 2D pattern and repeating this cycle until the entire part is completed. The technology offers precise control over material composition and microstructure, enabling the incorporation of biological agents and living cells, making it promising for tissue engineering. However, it has limitations such as resolution constraints and challenges with organic solvent compatibility in printheads, necessitating further advancements in printing techniques to overcome these drawbacks (Hartmann et al., 2022; Sachs et al., 2003).

3DP presents notable advantages, including its flexibility in material selection due to room temperature processing and use of materials in powder form, as well as the ability to print complex structures, overhangs, and internal architecture with controlled microstructure (Chakraborty et al., 2023; Jamróz et al., 2018; Ni et al., 2019). On the other hand, it faces challenges such as restrictions on using organic solvents as binders, potentially dissolving commercial printheads, and difficulties in removing unbound powder from intricate channels. Balancing these advantages and challenges is crucial for optimizing and expanding the application of 3DP in various fields (Cui et al., 2021). The applications now extend to mass customization and the production of diverse open-source designs in agriculture, healthcare, automotive, locomotive, and aviation sectors. Utilizing CAD models, 3D printing technology constructs an object layer by layer through material deposition (Jadhav & Jadhav, 2022; Shahrubudin et al., 2019).

AM technology emerged through advancements in diverse technological domains. Much like in numerous manufacturing technologies, progress in computing capabilities and the reduction in mass storage expenses facilitated the processing of vast volumes of data inherent in contemporary 3D CAD models within manageable time frames (Javaid, Haleem, Pratap Singh & Khan, & Suman, 2021; Xu et al., 2021). In our present era, we've grown accustomed to the presence of robust computers and intricate automated machinery. It's sometimes challenging to envision the challenges faced by the early innovators in creating the first AM machines (Fig. 2.1).

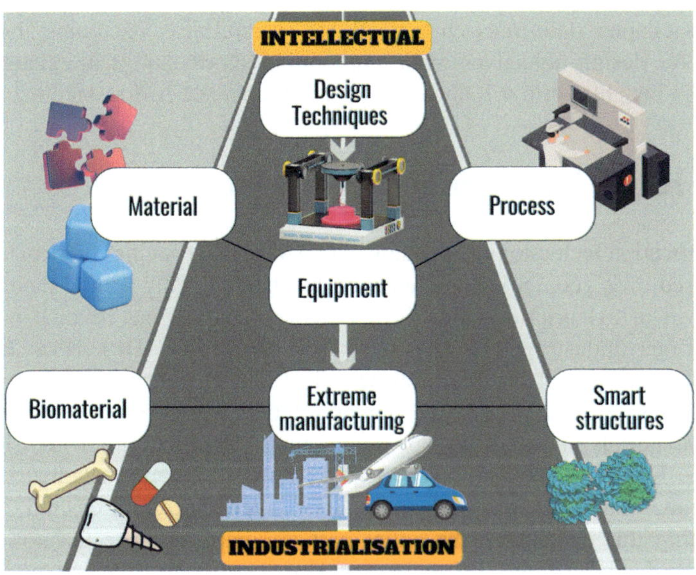

Figure 2.1 AM can be a revolutionary technology in connecting the principles of design with material processing and thereby their purpose in the spheres of biomaterials, extreme manufacturing, and smart structures.

2.3.1 3D printing in biomedicine

Since the early 2000s, 3D printing has found its way into the medical field. Initially, it was employed to manufacture dental implants and personalized prosthetics (Cui et al., 2012; Gross et al., 2014). Over time, advancements have significantly expanded the scope of medical applications for 3D printing (Schubert et al., 2014). Recent research outline a plethora of applications, including generating bones, ears, exoskeletons, windpipes, jawbones, eyeglasses, cell cultures, stem cells, blood vessels, vascular networks, tissues, organs, as well as innovative dosage forms and devices for drug delivery (Banks, 2013; Klein et al., 2013). The current utilization of 3D printing in medicine can be categorized into key areas: fabricating tissues and organs, crafting prosthetics, implants, and anatomical models, and conducting pharmaceutical research related to drug exploration, delivery methods, and dosage structures.

2.3.1.1 Tailored drug delivery

The utilization of 3D printing in pharmaceutical research and creation is already underway and holds the promise of a revolutionary impact (Ursan et al., 2013). Key advantages of 3D printing include precise control over droplet size and dosage, consistent reproducibility, and the capability to manufacture dosage forms with intricate drug-release characteristics (Banks, 2013; Khaled et al., 2014). Moreover, integration of 3D printing into the production of complex drugs could simplify and standardize the manufacturing processes. Additionally, this technology is highly significant in the advancement of personalized medicine (Khaled et al., 2014).

The integration of 3D printing in pharmaceutical processes is essential for transforming drug development, especially in achieving personalized medication goals (Khaled et al., 2014). Wherein, the primary focus is to enhance drug efficacy and reduce adverse reactions, particularly through the production of tailored drugs using 3D printing technology. Currently, oral tablets are the popular drug dosage form due to their ease of manufacture, precise dosing, and high patient compliance. But due to their lack of personalization in the dosage, traditional manufacturing processes, quicker drug degradation and lowered batch success often arise as a hindrance in their therapeutic effectiveness.

3D printing has the potential to address these challenges by simplifying complex drug manufacturing processes and offering a way to tailor medications to individual patients based on their pharmacogenetic profile, age, race, or gender. This could lead to the production of personalized drugs in novel formulations, such as pills containing multiple active ingredients or multireservoir tablets. Moreover, 3D printing enables the creation of diverse and intricate dosage forms, challenging traditional drug manufacturing approaches. Key 3D printing technologies in pharmaceutical production involve inkjet-based or inkjet powder-based methods, each offering distinct advantages in fabricating dosage forms. The potential to develop medications with complex drug-release profiles is a major area of research in 3D printing, enabling the creation of porous and multidrug-loaded dosage forms, as well as implantable drug delivery devices with precise drug-release mechanisms. These advancements have the potential to significantly impact drug manufacturing and personalized medicine, offering new possibilities for tailored drug treatments and improving patient compliance and outcomes.

2.3.1.2 Patient specific surgical planning

The intricacies and unique variations in human anatomy make 3D-printed models highly advantageous for surgical preparation (Klein et al., 2013). Providing surgeons with a tangible replica of a patient's anatomy for study or surgical simulation is more effective than relying solely on 2D MRI or CT scans displayed on a flat screen, which may not offer as much clarity (Gross et al., 2014). Moreover, utilizing 3D-printed models for surgical training is more favorable compared to training on cadavers due to challenges related to cadaver availability and cost. Cadavers often lack the specific pathology required for surgical training, focusing more on imparting anatomy knowledge rather than simulating a surgical patient.

While primarily in the experimental stage, 3D-printed representations have been effectively employed in numerous instances to comprehend a patient's unique anatomy prior to medical interventions. Pioneering surgeons at Kobe University Hospital in Japan have leveraged 3D-printed models for strategic planning in liver transplants. These models, constructed from affordable materials like partially transparent acrylic resin or polyvinyl alcohol, closely mimic living tissues, aiding surgeons in meticulously carving donor livers to minimize tissue loss and accommodate the recipient's abdominal cavity. In a different context, surgeons have utilized a 3D-printed replica of a calcified aorta for planning plaque removal procedures. Additionally, a premature infant's airway was replicated using 3D printing to study aerosol drug delivery to the lungs. An orthopedic surgery trainee utilized CT image scans and 3D modeling software to generate print files simulating a patient's bones, which were then sent to Shapeways for custom model printing, significantly reducing costs and speeding up the process compared to traditional methods. Beyond surgical planning, 3D-printed models have diverse applications, such as the creation of a polypeptide chain model capable of folding into secondary structures, aiding in the understanding of biological and biochemical structures. Studies indicate that students exhibit enhanced comprehension of molecular structures when utilizing these 3D models in pre- and post-comprehension studies.

2.3.1.3 Tissue and organ printing

3D bioprinting employs three main approaches: biomimicry, autonomous self-assembly, and mini tissue building blocks. Biomimicry involves replicating cellular and extracellular components of tissues or organs by mimicking biological patterns and structures. Autonomous self-assembly draws inspiration from embryonic organ development, allowing cellular components to organize themselves to mirror the desired biological micro-architecture and function. Mini tissues, the smallest functional building blocks of organs and tissues, are critical components in both biomimicry and autonomous self-assembly, and they can be assembled into larger constructs through careful design or self-assembly processes.

Combining these approaches is crucial to successfully 3D print intricate biological structures with diverse components and properties. The primary steps in the bioprinting process encompass imaging and design, material and cell selection, followed by the actual printing of the tissue construct. The resulting construct can then be transplanted or utilized for in vitro analysis, either directly or after a period of in vitro maturation (Murphy & Atala, 2014).

3D bioprinting is being employed in regenerative medicine to fulfill the demand for transplantable tissues and organs. Unlike nonbiological printing, 3D bioprinting introduces additional intricacies, such as material selection, cell varieties, growth factors, differentiation considerations, and technical challenges pertaining to the delicacy of living cells and tissue construction. Addressing these complexities necessitates the integration of technologies from engineering, biomaterials science, cell biology, physics, and medicine. Already, 3D bioprinting has proven effective in creating and implanting various tissues, including complex structures like multi-layered skin, bone, vascular grafts, tracheal splints, heart tissue, and cartilage. Applied functions extend to the development of high-throughput 3D-bioprinted tissue models for laboratory research, drug discovery, and toxicology (Xia et al., 2018).

2.3.2 Building blocks of architecture

The field of architecture has witnessed significant advantages through AM (AM) in two primary domains: modeling and construction (Xu et al., 2017). AM has been utilized for creating models of historic sites to even planning settlements on celestial bodies like the moon (Labeaga-Martínez et al., 2017). Architects find AM particularly beneficial for models, enabling them to enhance design details on a small scale and optimize their architectural plans. This technology has the potential to automate construction processes, enhance design and construction efficiencies, minimize waste, and promote sustainable development (Prasad et al., 2023). Moreover, AM contributes to the construction sector by transforming crucial aspects, including reducing production time and costs while enhancing adaptability. For the manufacturing of building components using materials such as polymers, metals, or aggregates, sophisticated systems like gantries and large robotic arms have been devised. A notable example is the Digital Construction Platform developed by the Massachusetts Institute of Technology, utilizing a mobile base with a robot arm to prevent collisions with existing structures. In concrete printing, Eindhoven University has pioneered a 3D printing machine for concrete, employing a nozzle to deposit fresh cementitious material along a predetermined trajectory, ultimately constructing a substrate layer by layer (Paolini et al., 2019).

2.3.3 Bringing together 3D printing and aeronautical

Complex components for engines, replacements, or restoration purposes are frequently utilized in aircraft. AM (AM) enables the creation of such components with significantly reduced weight and lower life-cycle costs, crucial for both space shuttles and airplanes. Conventional production methods make weight optimized designs excessively costly to manufacture. This is particularly evident with Sentinel

Satellites produced by Oerlikon AM, achieving a 40% weight reduction through topology optimization. AM's versatility facilitated the creation of aircraft structures that exceeded expectations by ensuring well-distributed stress, ultimately enhancing performance. Additionally, AM contributes to enhanced fuel efficiency, especially for machining high-temperature materials like Ni and similar alloys, which is challenging using traditional methods. In conventional approaches, intricate cooling channels for high combustion temperatures necessitated challenging machining. AM eliminated many of these manufacturing obstacles.

Furthermore, for astronauts to manufacture parts for space station repair or maintenance, it's imperative to have on-demand and on-site fabrication capabilities. Producing small batches of parts with intricate geometries necessary for airflow and heat dissipation functions is achievable with AM techniques, making it suitable for aircraft component production (Kumar & Krishnadas Nair, 2017). AM finds applications ranging from brackets and instrument housings in helicopters to battery compartments and fuselage structures in UAVs.

2.3.4 3D printing precision for food technology

To achieve the desired quality and functional attributes in 3D-printed food, it is crucial to comprehend and regulate various factors. A critical factor is the printability of the food ink, which relies on its composition, structure, and physicochemical properties. Existing research does not adequately elaborate on the precise specifications and operation of 3D printers, making it challenging to compare outcomes. Moreover, fundamental physicochemical attributes commonly used in traditional food, such as moisture, water activity, and microbial contamination, are inconsistently reported for 3D inks. This inconsistency hampers the practical applicability of the findings. Hence, we emphasize the key elements influencing 3D ink formulation and performance and propose guidelines to standardize and enhance the production of 3D-printed foods. Research in the realm of fats and proteins remains relatively constrained. There is potential for in-depth investigation into conventional protein sources like chicken and beef, alongside exploring plant-based protein alternatives like black beans, fava beans, and chickpeas (Severini et al., 2018). Incorporating these proteins through printing enriches food inks with vital nutrients and enhances the adaptability of customization, thanks to distinctive properties like gelling and denaturation. However, employing fruits and vegetables in 3D applications poses a challenge due to their high moisture levels and limited structural stability. 3D food printing represents a novel and inventive domain with the potential to tailor the design, nutritional content, and composition of food items. Numerous intricate and diverse shapes like butterflies, Mickey Mouse figures, apples, and pyramids have already been successfully printed (Keerthana et al., 2020; Yang et al., 2017; Zheng et al., 2019). Nonetheless, a gap exists between conventionally manufactured food and 3D-printed food concerning safety, efficacy, development, large-scale production, and reproducibility. Various properties, including moisture content, water activity, color, and different textural attributes, require further comprehensive data collection.

2.3.4.1 Erecting the pillars of 4D and 5D printing technologies

3D printing, pioneered by Japanese researcher Hideo Kodama in 1981, is an AM process that operates based on the stereolithographic method, encompassing three-dimensional axes: *x*, *y*, and *z* (Dommati et al., 2019). However, a notable drawback of 3D printing is its time-consuming nature, impacting the efficiency of the technology. To address these limitations, 4D printing emerged with Skylar Tibbits introducing the concept in February 2013 (Demoly et al., 2021). 4D printing involves the use of smart materials, integrating time as the fourth dimension, enabling properties like self-repair and self-assembly while significantly reducing printing time by 70%–90%. This advancement finds applications in diverse fields such as fashion, medicine, automobile manufacturing, and industrial polymer components (Javaid, Haleem, Pratap Singh, & Suman, 2021).

Nevertheless, 4D printing faces a challenge in printing complex shapes with curved surfaces. This limitation led to the development of 5D printing, which aims to overcome the constraints of 4D printing. Presently, the manufacturing industry is increasingly focusing on 4D and 5D printing due to their enhanced advantages. The fundamental idea behind 5D AM involves the rotation of the extruder head and the print bed, enabling printing along five different axes (Anas et al., 2021). Compared to 3D and 4D printing, 5D printing is remarkably stronger, approximately three to four times. This technology excels at producing complex, curved surfaces and finds widespread applications in biomedical, automobile, and aerospace components.

Incorporating 4D technology is also utilized to improve the tissue retention properties of drug delivery systems. An instance is the research by Melocchi et al., where they investigated the potential of 4D printing in creating a retentive device for intravesical drug delivery. This device is engineered to gradually release drugs into the bladder and relies on a water-induced shape memory response using polyvinyl alcohol (PVA). The device was fabricated using an FDM-based printer, whose shape memory features were assessed. The findings revealed that the 4D-printed device displayed commendable shape fixity and retrieval, ensuring its retention in the bladder. These outcomes indicate the feasibility of PVA-based retentive devices for intravesical drug delivery, holding promise for treating bladder conditions (Melocchi et al., 2019). Similarly, Han et al. have detailed bioinspired microneedle array development using PμSLA printing technology. These microneedles are known for possession rear-facing barbs for enhanced tissue adhesion, sustained drug delivery and minimally invasive procedures. The researchers employed a novel approach, integrating 3D printing and electrospinning to produce the microneedles from a biocompatible polymer. The study showcased the potential of employing a 4D printing strategy to create bioinspired microneedle arrays, offering substantial prospects in Table 2.1 realms of drug delivery and biomedical engineering (Han et al., 2020).

2.4 How can AM influence Industry 4.0?

The fourth industrial revolution, referred to as Industry 4.0, signifies the contemporary trend of employing advanced automation technology. Within this emerging

Table 2.1 A succinct comparison between 3D, 4D and 5D printing techniques.

Criteria	3D printing	4D printing	5D printing	References
Process of development	Constructs a design by building it layer by layer, starting from the bottom and progressing upward	Based on the principles of 3D printing	Creates products utilizing five approaches, involving movement of both the print object and the printer head during the printing process	(Chia & Wu, 2015; Vasiliadis et al., 2022)
Product durability	Often distorted due to layered construction	Alters its form in response to external triggers as time progresses	The product contains curvature that displays strength level three to five times higher than that of a 3D-printed object	(Sajjad et al., 2023; Vasiliadis et al., 2022)
Materials used for construction	Thermoplastic acrylonitrile styrene, polycarbonate, resins, photopolymers,	Shape memory polymers (SMPs), hydrogels	Polycarbonate, acrylonitrile styrene, photopolymers	(Vasiliadis et al., 2022)

phase, the integration of modern manufacturing capabilities in conjunction with innovative information technologies significantly influences economic competitiveness (Javaid, Haleem, Pratap Singh, & Suman, 2021; Yang & Gu, 2021). This era of industrial advancement is promoting the convergence of smart manufacturing systems and cutting-edge information technologies. Within this evolving landscape, AM (AM) stands as a pivotal component (Fanoro et al., 2021; Morgan et al., 2021). The fusion of cyber-physical systems enables the creation of highly efficient smart factories capable of producing superior-quality customized goods. On one hand, the progress in information technology has expedited the shift toward the upcoming industrial era (Javaid et al., 2023; Lee et al., 2018).

Material science is crucial for comprehending advancements in AM technologies, with researchers showing a keen inclination toward identifying new materials suitable for 3D printing that can eliminate the defects of previous generations of industrial revolution (Monfared et al., 2023; Shahrubudin et al., 2019).

The paradigm shift brought about by AM (AM) is reshaping industries across the board. At the forefront of this revolution is Metal AM (MAM), a field leveraging the exceptional mechanical properties of metals to engineer alternative components (Wei et al., 2023). Metals have always been a cornerstone in engineering due to their strength and versatility. The advent of 3D printing technologies has spurred the growth of MAM, enabling the utilization of materials like aluminum, titanium, and stainless steel (Awasthi et al., 2021). However, this growth is not without its challenges. Achieving cost-effectiveness, increasing production speed, enhancing mechanical properties, ensuring surface quality, and maintaining consistent microstructures are ongoing goals. Progress in MAM hinges on the fusion of advancements in material science and evolving MAM processes, a critical step toward unlocking MAM's full potential within the context of Industry 4.0 (Al-Mangour & Yang, 2016; Armstrong et al., 2022; Száva et al., 2023).

Concurrently, the integration of intelligent materials is amplifying the capabilities of AM. Smart structures, often referred to as 4D printing materials, possess the remarkable ability to dynamically adjust their shape or material properties in response to external stimuli or changing conditions (Zhang et al., 2019; Zhou et al., 2023). This introduces exciting prospects, such as self-assembly and stimuli-responsive mechanisms, adding a new layer of adaptability and functionality. Shape memory alloys (SMAs) and shape memory polymers (SMP) play pivotal roles in this domain, known for their unique properties (Sun et al., 2012). SMAs exhibit superelasticity and thermal shape recovery, making them suitable for a wide range of applications, from biomedical implants to micro-electromechanical devices. SMPs, on the other hand, respond to various external stimuli like light, humidity, and temperature gradients, opening doors to applications in medical engineering and beyond (Stachiv et al., 2021; Wen et al., 2018). The intelligent materials space within AM is a realm of ongoing exploration and discovery, with potential implications across multiple sectors.

Moreover, the incorporation of printable hydraulics and electronics helps to testify the versatility and potential of AM. Printable hydraulics involve the simultaneous printing of solid and liquid materials, effectively creating hydraulically actuated working mechanisms (Aguirre-Cortés et al., 2023; Hayes et al., 2022; Li et al., 2020). Soft robotic grippers, for instance, can be manufactured in a single stage without the need for additional assembly steps. On the electronics front, the ability to embed conductive substances within AM products allows for the seamless integration of electronic circuitry within printed objects, a concept known as embedded electronics (Goh et al., 2023; Tawk et al., 2022; Terrile et al., 2021). This breakthrough holds immense promise, enabling innovative designs and the full integration of electronics into objects, paving the way for smarter, more functional end-products.

In addition to these advancements, special materials have been gaining traction within the realm of AM (Betz et al., 2023). Concrete, for instance, is being explored for additive construction, with an eye toward creating future buildings and infrastructures. The clothing and jewelry industries are increasingly adopting AM practices, leveraging advancements in textile printing for rapid design processes, reduced fabrication times, and cost savings in packaging and transportation (Abdulhameed et al., 2022; Khajavi et al., 2021; Liu et al., 2023). Edible constituents within the food industry present yet another intriguing dimension of AM. Recent studies have begun exploring various applications, leveraging extrusion-based AM techniques to craft foods with desired surface textures and diverse nutritional content (Pinna et al., 2016; Xie et al., 2023). Even space exploration is not untouched by the potential of AM, with research delving into the 3D printing of Moondust for potential construction on the Moon, a glimpse into a potential future where AM may facilitate the building of habitats and infrastructures beyond Earth (Cesaretti et al., 2014).

As we step into the era of Industry 4.0, ongoing research and development will be key to fully harnessing the diverse potential of AM and its associated materials. The journey involves addressing challenges, refining processes, discovering new material combinations, and enhancing design paradigms (Khan et al., 2021; Popov et al., 2022). Through persistent exploration and innovation, AM is set to play an increasingly transformative role across industries, propelling us into a future where the possibilities of creation are limited only by our imagination.

Likely, advancements in AM technology will lead to the development of more sophisticated processes. However, the majority of these processes are primarily designed for printing typical materials such as polymers, commonly utilized in nonindustrial applications. With the rise of Industry 4.0 and the increasing demand for heavy engineering applications, specific AM processes have emerged (Jandyal et al., 2022; Rayna & Striukova, 2016). Given that metals are the prevalent choice in the industry, the issue of Metal AM has gained significant attention in this modern era. Moreover, the trajectory of manufacturing in the future is expected to steer toward the integration of these processes. Referred to as hybrid manufacturing, this burgeoning field involves combining subtractive manufacturing methods with additive ones to produce superior products with enhanced surface quality and tensile strength (Mohd Yusuf et al., 2019). Presently, the burgeoning interest in hybrid manufacturing has led to the exploration of various combinations of manufacturing processes that go beyond traditional AM methods (Fig. 2.2).

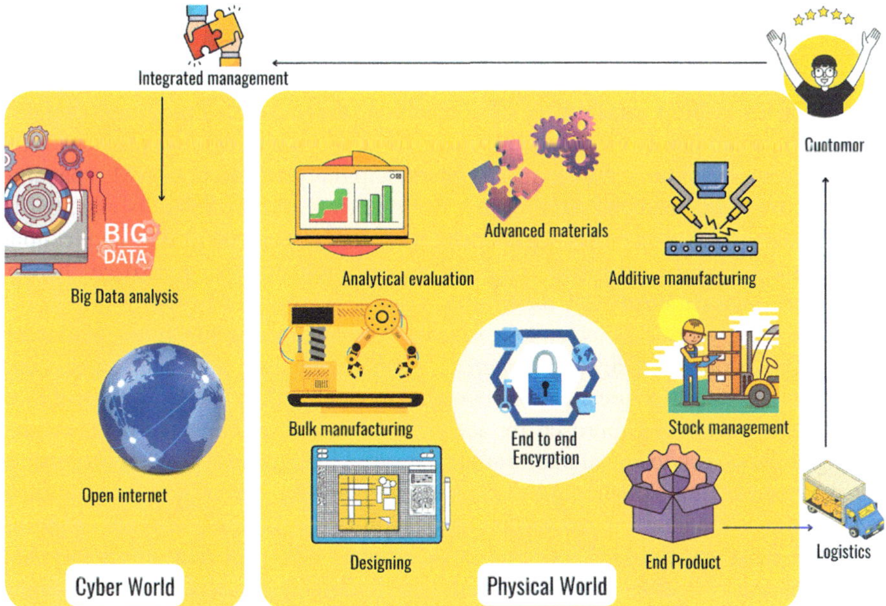

Figure 2.2 A descriptive flow of AM in producing the final output. In the process of overall industrialization, an adequate role of AM and data analytics play a vital role.

2.5 Future prospects and challenges

In the context of Industry 4.0, the success of AM hinges on how well printed components perform in real service conditions and their transformation into useful end products. Early on, AM was primarily used for prototyping and visualization, but advancements in materials and printers have expanded its applications to factory tooling, end-use products, and spare parts (Khorasani et al., 2022). While 63% of industries use AM for prototyping and 21% for manufacturing challenging products, there are still challenges hindering broader integration into production processes. These challenges encompass material-related limitations, the need for standardization, IT integration complexities, design paradigm shifts, skill gaps, and financial considerations (Chowdhury, 2023). Looking ahead, AM (AM) is poised for further advancements in intellectualization and industrialization. AM represents a highly intricate system, encompassing a multitude of factors across different levels and scales, interconnecting materials, structures, and various physical and chemical domains. To understand and optimize this intricate system, it is imperative to explore it in conjunction with big data and artificial intelligence, striving for a breakthrough in the fundamental principles and methods of multifunctional integrated design optimization for AM (Aldoseri et al., 2023). The aim is to develop intelligent AM technology with the capability to actively control shapes, laying a robust scientific and technical foundation for significant enhancements in material properties, processes, structural designs, product quality, and service efficiency within the realm of future AM technology. The crux lies in creating intelligent AM equipment endowed with in-built capabilities of acquisition, diagnosis, learning and decision making, to establish a crucial foundation for the widespread adoption of AM technology in the future. Furthermore, interdisciplinary research in AM, spanning materials, software, artificial intelligence, life sciences, and medicine, is essential to drive substantial original technological advancements. The scope of AM applications should broaden and concentrate on sectors like new energy, aerospace, health, architecture, cultural creativity, navigation, nuclear power, and beyond.

In the coming years, AM technology is set to evolve, focusing on the manufacturing of four-dimensional intelligent structures, living organisms, and components that integrate materials, structures, and functions (Liu et al., 2020, 2021). This evolution presents a new technical approach for achieving controllable shape and performance, offering a robust technological platform for industrial innovation and entrepreneurship. The development of AM should adhere to the guiding ideology of "application development as the guide, technological innovation as the driving force, and industrial development as the goal" (Mormina, 2019; Taalbi, 2017). A well-structured AM industry standard system needs to be established, promoting comprehensive innovation and application of AM processes by integrating emerging technologies and intelligent manufacturing systems including cloud manufacturing, big data, and the Internet of Things. This integration holds immense significance in propelling a leap forward in manufacturing technology (Li et al., 2017; Soori et al., 2023).

AM faces several challenges that must be addressed to achieve widespread integration into Industry 4.0. Historically used for prototyping, AM needs to surpass traditional design constraints and address technical challenges related to materials and processing (Stavropoulos et al., 2023). Assurance of material properties meeting industry standards is critical for economic success. AM's potential for sustainability and recyclability needs further research, particularly for polymers (Joseph et al., 2023; Moshood et al., 2022). Despite its ability to create complex structures, AM still lacks process stability, reliability, and cost-effectiveness, necessitating improvements in quality, metrological tools, and methodologies. Integrating AM into Industry 4.0 requires addressing IT integration challenges, design paradigm shifts, capability challenges, and financial considerations (Kassab et al., 2023). Bridging the skill gap and managing the financial aspects of AM will be essential for its successful transition into mainstream manufacturing in Industry 4.0. A four-step approach—evaluating AM fit, calculating the business case, developing a roadmap, and fostering organizational acceptance—can guide industries in navigating the challenges and maximizing the potential of AM in Industry 4.0 (Ghobakhloo, 2018; Ras et al., 2017).

Certainly, AM has revolutionized various industries like medical, aerospace, and automotive by streamlining processes and offering cost-effective, time-saving benefits, including rapid prototyping and decentralized production (Calignano & Mercurio, 2023). Despite its advantages, AM faces obstacles such as production time, size constraints, material challenges, and high machine and production costs (Bradshaw, 2015; Ekins & Zenghelis, 2021; Javaid, Haleem, Singh, Suman, et al., 2021).

This manufacturing method stands as an ecofriendly and efficient approach, significantly reducing material waste and fuel consumption (El-Haggar, 2007a, 2007b). Looking ahead, the future of AM appears promising, aiming for increased affordability, speed, and application diversification beyond industrial sectors. To reach its potential, further developments are necessary, focusing on cost reduction, faster and more precise printing methods, and seamless integration into smart factories, aligning with the broader evolution of manufacturing technologies.

2.6 Conclusion

AM, within the context of Industry 4.0, presents promising prospects in design and manufacturing. Undoubtedly, the capabilities of AM play a vital role in the fourth industrial revolution by facilitating the production of tailored and personalized items directly at the location. Intelligent manufacturing, in actuality, offers a remarkably adaptable manufacturing procedure capable of swiftly shifting from personalized mass production to the creation of top-tier products. This shift will significantly transform the roles and duties of customers, manufacturers, and designers in the future landscape of production.

AM is expected to revolutionize production, akin to the impact of assembly-line factories in the 20th century. Experts predict a shift toward smaller value chains and accessible online designs, enabling competition with traditional manufacturing. The United States, already a leader in AM, is poised to benefit, potentially enhancing manufacturing efficiency and competitiveness. Standardization and strategic policies are seen as key to accelerating the widespread adoption of this transformative technology.

References

Abdulhameed, O., Mian, S. H., Moiduddin, K., Al-Ahmari, A., Ahmed, N., & Aboudaif, M. K. (2022). A multi-part orientation planning schema for fabrication of non-related components using additive manufacturing. *Micromachines, 13*(10). https://doi.org/10.3390/mi13101777.

Aguirre-Cortés, J. M., Moral-Rodríguez, A. I., Bailón-García, E., Davó-Quiñonero, A., Pérez-Cadenas, A. F., & Carrasco-Marín, F. (2023). 3D printing in photocatalysis: Methods and capabilities for the improved performance. *Applied Materials Today, 32*, 101831. https://doi.org/10.1016/j.apmt.2023.101831.

Albright, R. E., & Kappel, T. A. (2003). Roadmapping in the corporation. *Research-Technology Management, 46*(2), 31–40. https://doi.org/10.1080/08956308.2003.11671552.

de Alcantara, D. P., & Martens, M. L. (2019). Technology Roadmapping (TRM): A systematic review of the literature focusing on models. *Technological Forecasting and Social Change, 138*, 127–138. https://doi.org/10.1016/j.techfore.2018.08.014.

Aldoseri, A., Al-Khalifa, K. N., & Hamouda, A. M. (2023). Re-thinking data strategy and integration for artificial intelligence: Concepts, opportunities, and challenges. *Applied Sciences, 13*(12). https://doi.org/10.3390/app13127082.

Al-Mangour, B., & Yang, J.-M. (2016). Improving the surface quality and mechanical properties by shot-peening of 17-4 stainless steel fabricated by additive manufacturing. *Materials & Design, 110*. https://doi.org/10.1016/j.matdes.2016.08.037.

Anas, S., Khan, M., Rafey, M., & Faheem, K. (2021). Concept of 5D printing technology and its applicability in the healthcare industry. *Materials Today: Proceedings, 56*. https://doi.org/10.1016/j.matpr.2021.10.391.

Armstrong, M., Mehrabi, H., & Naveed, N. (2022). An overview of modern metal additive manufacturing technology. *Journal of Manufacturing Processes, 84*, 1001–1029. https://doi.org/10.1016/j.jmapro.2022.10.060.

Awasthi, A., Saxena, K., & Arun, V. (2021). Sustainable and smart metal forming manufacturing process. *Materials Today: Proceedings, 44*. https://doi.org/10.1016/j.matpr.2020.12.177.

Banks, J. (2013). Adding value in additive manufacturing: Researchers in the United Kingdom and Europe look to 3D printing for customization. *IEEE Pulse, 4*(6), 22–26. https://doi.org/10.1109/MPUL.2013.2279617.

Bayat, M., Zinovieva, O., Ferrari, F., Ayas, C., Langelaar, M., Spangenberg, J., Salajeghe, R., Poulios, K., Mohanty, S., Sigmund, O., & Hattel, J. (2023). Holistic computational design within additive manufacturing through topology optimization combined with multiphysics multi-scale materials and process modelling. *Progress in Materials Science, 138*, 101129. https://doi.org/10.1016/j.pmatsci.2023.101129.

Betz, U. A. K., Arora, L., Assal, R. A., Azevedo, H., Baldwin, J., Becker, M. S., Bostock, S., Cheng, V., Egle, T., Ferrari, N., Schneider-Futschik, E. K., Gerhardy, S., Hammes, A., Harzheim, A., Herget, T., Jauset, C., Kretschmer, S., Lammie, C., Kloss, N., ... Zhao, G. (2023). Game changers in science and technology—Now and beyond. *Technological Forecasting and Social Change, 193*, 122588. https://doi.org/10.1016/j.techfore.2023.122588.

Bohr, A., & Memarzadeh, K. (2020). The rise of artificial intelligence in healthcare applications. *Artificial Intelligence in Healthcare,* 25–60. https://doi.org/10.1016/B978-0-12-818438-7.00002-2.

Bradshaw, E. A. (2015). Blockadia rising: Rowdy greens, direct action and the keystone XL pipeline. *Critical Criminology, 23*(4), 433–448. https://doi.org/10.1007/s10612-015-9289-0.

Calignano, F., & Mercurio, V. (2023). An overview of the impact of additive manufacturing on supply chain, reshoring, and sustainability. *Cleaner Logistics and Supply Chain, 7*, 100103. https://doi.org/10.1016/j.clscn.2023.100103.

Cesaretti, G., Enrico, D., De Kestelier, X., Colla, V., & Pambaguian, L. (2014). Building Components for an outpost on the lunar soil by means of a novel 3D printing technology. *Acta Astronautica, 93*, 430–450. https://doi.org/10.1016/j.actaastro.2013.07.034.

Chakraborty, P., Eqbal, M. D., & Ahmed, J. (2023). Three-dimensional printing and its application to legume proteins: A review. *Legume Science, 5*(2), e172. https://doi.org/10.1002/leg3.172.

Chia, H. N., & Wu, B. M. (2015). Recent advances in 3D printing of biomaterials. *Journal of Biological Engineering, 9*, 4. https://doi.org/10.1186/s13036-015-0001-4.

Chowdhury, H. (2023). *Circular economy integration in additive manufacturing*. https://doi.org/10.20944/preprints202310.0087.v1.

Cima, M.J., Sachs, E., Cima, L.G., Yoo, J., Khanuja, S., Borland, S.W., Wu, B., & Giordano, R.A. (1994). *Computer-derived microstructures by 3D Printing: Sio- and Structural Materials*. https://doi.org/10.15781/T2959CS8Q.

Cui, M., Pan, H., Su, Y., Fang, D., Qiao, S., Ding, P., & Pan, W. (2021). Opportunities and challenges of three-dimensional printing technology in pharmaceutical formulation development. *Acta Pharmaceutica Sinica. B, 11*(8), 2488–2504. https://doi.org/10.1016/j.apsb.2021.03.015.

Cui, X., Boland, T., D'Lima, D. D., & Lotz, M. K. (2012). Thermal inkjet printing in tissue engineering and regenerative medicine. *Recent Patents on Drug Delivery & Formulation, 6*(2), 149–155.

Dash, S., Shakyawar, S. K., Sharma, M., & Kaushik, S. (2019). Big data in healthcare: Management, analysis and future prospects. *Journal of Big Data, 6*(1), 54. https://doi.org/10.1186/s40537-019-0217-0.

Demoly, F., Dunn, M. L., Wood, K. L., Qi, H. J., & André, J.-C. (2021). The status, barriers, challenges, and future in design for 4D printing. *Materials & Design, 212*, 110193. https://doi.org/10.1016/j.matdes.2021.110193.

Dommati, H., Ray, S. S., Wang, J.-C., & Chen, S.-S. (2019). A comprehensive review of recent developments in 3D printing technique for ceramic membrane fabrication for water purification. *RSC Advances, 9*(29), 16869. https://doi.org/10.1039/c9ra00872a.

Durakovic, B. (2018). Design for additive manufacturing: Benefits, trends and challenges. *Periodicals of Engineering and Natural Sciences, 6*, 179–191. https://doi.org/10.21533/pen.v6i2.224.

Ekins, P., & Zenghelis, D. (2021). The costs and benefits of environmental sustainability. *Sustainability Science, 16*(3), 949–965. https://doi.org/10.1007/s11625-021-00910-5.

El-Haggar, S. M. (2007a). Chapter 2—Cleaner production. In In. S. M. El-Haggar (Ed.). *Sustainable industrial design and waste management* (pp. 21–84). Academic Press. https://doi.org/10.1016/B978-012373623-9/50004-6.

El-Haggar, S. M. (2007b). Chapter 10—Sustainability of industrial waste management. In S. M. El-Haggar (Ed.). *Sustainable industrial design and waste management* (pp. 307–369). Academic Press. https://doi.org/10.1016/B978-012373623-9/50012-5.

Fanoro, M., Božanić, M., & Sinha, S. (2021). A review of 4IR/5IR enabling technologies and their linkage to manufacturing supply chain. *Technologies, 9*(4). https://doi.org/10.3390/technologies9040077, Article 4.

Fidan, I., Huseynov, O., Ali, M. A., Alkunte, S., Rajeshirke, M., Gupta, A., Hasanov, S., Tantawi, K., Yasa, E., Yilmaz, O., Loy, J., Popov, V., & Sharma, A. (2023). Recent inventions in additive manufacturing: Holistic review. *Inventions, 8*(4). https://doi.org/10.3390/inventions8040103, Article 4.

Ghaffar, S., Corker, J., & Fan, M. (2018). Additive manufacturing technology and its implementation in construction as an eco-innovative solution. *Automation in Construction, 93*. https://doi.org/10.1016/j.autcon.2018.05.005.

Ghobakhloo, M. (2018). The future of manufacturing industry: A strategic roadmap toward Industry 4.0. *Journal of Manufacturing Technology Management, 29*. https://doi.org/10.1108/JMTM-02-2018-0057.

Goh, G. L., Goh, G. D., Nguyen, V. P., Toh, W., Lee, S., Li, X., Sunil, B. D., Lim, J. Y., Li, Z., Sinha, A. K., Yeong, W. Y., Campolo, D., Chow, W. T., Ng, T. Y., & Han, B. S. (2023). A 3D printing-enabled artificially innervated smart soft gripper with variable joint stiffness. Advanced Materials Technologies, 8(24), 2301426. https://doi.org/10.1002/admt.202301426.

Gross, B. C., Erkal, J. L., Lockwood, S. Y., Chen, C., & Spence, D. M. (2014). Evaluation of 3D printing and its potential impact on biotechnology and the chemical sciences. *Analytical Chemistry, 86*(7), 3240–3253. https://doi.org/10.1021/ac403397r.

Han, D., Morde, R. S., Mariani, S., La Mattina, A. A., Vignali, E., Yang, C., Barillaro, G., & Lee, H. (2020). 4D printing of a bioinspired microneedle array with backward-facing barbs for enhanced tissue adhesion. *Advanced Functional Materials, 30*(11), 1909197. https://doi.org/10.1002/adfm.201909197.

Hartmann, C., van den Bosch, L., Spiegel, J., Rumschöttel, D., & Günther, D. (2022). Removal of stair-step effects in binder jetting additive manufacturing using grayscale and dithering-based droplet distribution. *Materials, 15*(11). https://doi.org/10.3390/ma15113798, Article 11.

Hayes, B., Hainsworth, T., & MacCurdy, R. (2022). Liquid-solid co-printing of multi-material 3D fluidic devices via material jetting. *Additive Manufacturing, 55*, 102785. https://doi.org/10.1016/j.addma.2022.102785.

Jadhav, A., & Jadhav, V. S. (2022). A review on 3D printing: An additive manufacturing technology. *Materials Today: Proceedings, 62*, 2094–2099. https://doi.org/10.1016/j.matpr.2022.02.558.

Jamróz, W., Szafraniec, J., Kurek, M., & Jachowicz, R. (2018). 3D printing in pharmaceutical and medical applications—Recent achievements and challenges. *Pharmaceutical Research, 35*(9), 176. https://doi.org/10.1007/s11095-018-2454-x.

Jandyal, A., Chaturvedi, I., Wazir, I., Raina, A., & Ul Haq, M. I. (2022). 3D printing—A review of processes, materials and applications in industry 4.0. *Sustainable Operations and Computers, 3*, 33–42. https://doi.org/10.1016/j.susoc.2021.09.004.

Javaid, M., Haleem, A., Pratap Singh, R., Khan, S., & Suman, R. (2021). Blockchain technology applications for Industry 4.0: A literature-based review. *Blockchain: Research and Applications, 2*(4), 100027. https://doi.org/10.1016/j.bcra.2021.100027.

Javaid, M., Haleem, A., Pratap Singh, R., & Suman, R. (2021). Significance of Quality 4.0 towards comprehensive enhancement in manufacturing sector. *Sensors International, 2*, 100109. https://doi.org/10.1016/j.sintl.2021.100109.

Javaid, M., Haleem, A., Singh, R. P., & Suman, R. (2021). Substantial capabilities of robotics in enhancing industry 4.0 implementation. *Cognitive Robotics, 1*, 58–75. https://doi.org/10.1016/j.cogr.2021.06.001.

Javaid, M., Haleem, A., Singh, R. P., & Suman, R. (2023). An integrated outlook of Cyber–Physical Systems for Industry 4.0: Topical practices, architecture, and applications. *Green Technologies and Sustainability, 1*(1), 100001. https://doi.org/10.1016/j.grets.2022.100001.

Javaid, M., Haleem, A., Singh, R. P., Suman, R., & Gonzalez, E. S. (2022). Understanding the adoption of Industry 4.0 technologies in improving environmental sustainability. *Sustainable Operations and Computers, 3*, 203–217. https://doi.org/10.1016/j.susoc.2022.01.008.

Javaid, M., Haleem, A., Singh, R. P., Suman, R., & Rab, S. (2021). Role of additive manufacturing applications towards environmental sustainability. *Advanced Industrial and Engineering Polymer Research, 4*(4), 312–322. https://doi.org/10.1016/j.aiepr.2021.07.005.

Johnson, K. B., Wei, W., Weeraratne, D., Frisse, M. E., Misulis, K., Rhee, K., Zhao, J., & Snowdon, J. L. (2021). Precision medicine, AI, and the future of personalized health care. *Clinical and Translational Science, 14*(1), 86–93. https://doi.org/10.1111/cts.12884.

Joseph, T. M., Unni, A. B., Joshy, K. S., Kar Mahapatra, D., Haponiuk, J., & Thomas, S. (2023). Emerging Bio-based polymers from lab to market: Current strategies. *Market Dynamics and Research Trends. C, 9*(1). https://doi.org/10.3390/c9010030, Article 1.

Kassab, A., Al Nabhani, D., Mohanty, P., Pannier, C., & Ayoub, G. Y. (2023). Advancing plastic recycling: Challenges and opportunities in the integration of 3D printing and distributed recycling for a circular economy. *Polymers, 15*(19). https://doi.org/10.3390/polym15193881, Article 19.

Keerthana, K., Anukiruthika, T., Moses, J. A., & Anandharamakrishnan, C. (2020). Development of fiber-enriched 3D printed snacks from alternative foods: A study on button mushroom. *Journal of Food Engineering, 287*, 110116. https://doi.org/10.1016/j.jfoodeng.2020.110116.

Khajavi, S., Tetik, M., Mohite, A., Peltokorpi, A., Li, M., Weng, Y., & Holmström, J. (2021). Additive manufacturing in the construction industry: The comparative competitiveness of 3D concrete printing. *Applied Sciences, 11*, 3865. https://doi.org/10.3390/app11093865.

Khaled, S. A., Burley, J. C., Alexander, M. R., & Roberts, C. J. (2014). Desktop 3D printing of controlled release pharmaceutical bilayer tablets. *International Journal of Pharmaceutics, 461*(1–2), 105–111. https://doi.org/10.1016/j.ijpharm.2013.11.021.

Khan, I. S., Ahmad, M. O., & Majava, J. (2021). Industry 4.0 and sustainable development: A systematic mapping of triple bottom line, Circular Economy and Sustainable Business

Models perspectives. *Journal of Cleaner Production, 297*, 126655. https://doi.org/10.1016/j.jclepro.2021.126655.

Khorasani, M., Loy, J., Ghasemi, A. H., Sharabian, E., Leary, M., Mirafzal, H., Cochrane, P., Rolfe, B., & Gibson, I. (2022). A review of Industry 4.0 and additive manufacturing synergy. *Rapid Prototyping Journal, 28*(8), 1462–1475. https://doi.org/10.1108/RPJ-08-2021-0194.

Klein, G. T., Lu, Y., & Wang, M. Y. (2013). 3D printing and neurosurgery—Ready for prime time? *World Neurosurgery, 80*(3–4), 233–235. https://doi.org/10.1016/j.wneu.2013.07.009.

Kumar, L. J., & Krishnadas Nair, C. G. (2017). Current trends of additive manufacturing in the aerospace industry. In In. D. I. Wimpenny, P. M. Pandey, & L. J. Kumar (Eds.). *Advances in 3D printing & additive manufacturing technologies* (pp. 39–54). Springer. https://doi.org/10.1007/978-981-10-0812-2_4.

Kumar, S., Gopi, T., Harikeerthana, N., Gupta, M. K., Gaur, V., Krolczyk, G. M., & Wu, C. (2023). Machine learning techniques in additive manufacturing: A state of the art review on design, processes and production control. *Journal of Intelligent Manufacturing, 34*(1), 21–55. https://doi.org/10.1007/s10845-022-02029-5.

Labeaga-Martínez, N., Sanjurjo-Rivo, M., Díaz-Álvarez, J., & Martínez-Frías, J. (2017). Additive manufacturing for a Moon village. *Procedia Manufacturing, 13*, 794–801. https://doi.org/10.1016/j.promfg.2017.09.186.

Lee, M., Yun, J. J., Pyka, A., Won, D., Kodama, F., Schiuma, G., Park, H., Jeon, J., Park, K., Jung, K., Yan, M.-R., Lee, S., & Zhao, X. (2018). How to respond to the fourth industrial revolution, or the second information technology revolution? Dynamic new combinations between technology, market, and society through open innovation. *Journal of Open Innovation: Technology, Market, and Complexity, 4*(3), 21. https://doi.org/10.3390/joitmc4030021.

Li, B., Hou, B., Yu, W., Lu, X., & Yang, C. (2017). Applications of artificial intelligence in intelligent manufacturing: A review. *Frontiers of Information Technology & Electronic Engineering, 18*(1), 86–96. https://doi.org/10.1631/FITEE.1601885.

Li, J., Wu, C., Chu, P. K., & Gelinsky, M. (2020). 3D printing of hydrogels: Rational design strategies and emerging biomedical applications. *Materials Science and Engineering: R: Reports, 140*, 100543. https://doi.org/10.1016/j.mser.2020.100543.

Ligon, S. C., Liska, R., Stampfl, J., Gurr, M., & Mülhaupt, R. (2017). Polymers for 3D Printing and customized additive manufacturing. *Chemical Reviews, 117*(15), 10212–10290. https://doi.org/10.1021/acs.chemrev.7b00074.

Liu, G., He, Y., Liu, P., Chen, Z., Chen, X., Wan, L., Li, Y., & Lu, J. (2020). Development of bioimplants with 2D, 3D, and 4D additive manufacturing materials. *Engineering, 6*(11), 1232–1243. https://doi.org/10.1016/j.eng.2020.04.015.

Liu, G., Zhang, X., Chen, X., He, Y., Cheng, L., Huo, M., Yin, J., Hao, F., Chen, S., Wang, P., Yi, S., Wan, L., Mao, Z., Chen, Z., Wang, X., Cao, Z., & Lu, J. (2021). Additive manufacturing of structural materials. *Materials Science and Engineering: R: Reports, 145*, 100596. https://doi.org/10.1016/j.mser.2020.100596.

Liu, W., Liu, X., Liu, Y., Wang, J., Evans, S., & Yang, M. (2023). Unpacking additive manufacturing challenges and opportunities in moving towards sustainability: An exploratory study. *Sustainability, 15*(4). https://doi.org/10.3390/su15043827, Article 4.

Mahmood, A., Akram, T., Chen, H., & Chen, S. (2022). On the evolution of additive manufacturing (3D/4D printing) technologies: Materials, applications, and challenges. *Polymers, 14*(21). https://doi.org/10.3390/polym14214698, Article 21.

Mehrpouya, M., Dehghanghadikolaei, A., Fotovvati, B., Vosooghnia, A., Emamian, S. S., & Gisario, A. (2019). The potential of additive manufacturing in the smart factory Industrial 4.0: A review. *Applied Sciences, 9*(18). https://doi.org/10.3390/app9183865, Article 18.

Melocchi, A., Uboldi, M., Inverardi, N., Briatico-Vangosa, F., Baldi, F., Pandini, S., Scalet, G., Auricchio, F., Cerea, M., Foppoli, A., Maroni, A., Zema, L., & Gazzaniga, A. (2019). Expandable drug delivery system for gastric retention based on shape memory polymers: Development via 4D printing and extrusion. *International Journal of Pharmaceutics, 571*, 118700. https://doi.org/10.1016/j.ijpharm.2019.118700.

Miedzinski, M., Mcdowall, W., Fahnestock, J., Muller, G., & Diaz Lopez, F.J. (2019). Science, technology and innovation policy roadmaps for the SDGs: A guide for design and implementation. Innovation for Sustainable Development Network.

Miedzinski, M., McDowall, W., Fahnestock, J., Rataj, O., & Papachristos, G. (2022). Paving the pathways towards sustainable future? A critical assessment of STI policy roadmaps as policy instruments for sustainability transitions. *Futures, 142*, 103015. https://doi.org/10.1016/j.futures.2022.103015.

Mohd Yusuf, S., Cutler, S., & Gao, N. (2019). Review: The impact of metal additive manufacturing on the aerospace industry. *Metals, 9*(12). https://doi.org/10.3390/met9121286, Article 12.

Monfared, V., Ramakrishna, S., Nasajpour-Esfahani, N., Toghraie, D., Hekmatifar, M., & Rahmati, S. (2023). Science and technology of additive manufacturing progress: Processes, materials, and applications. *Metals and Materials International*. . https://doi.org/10.1007/s12540-023-01467-x.

Morgan, J., Halton, M., Qiao, Y., & Breslin, J. G. (2021). Industry 4.0 smart reconfigurable manufacturing machines. *Journal of Manufacturing Systems, 59*, 481–506. https://doi.org/10.1016/j.jmsy.2021.03.001.

Mormina, M. (2019). Science, technology and innovation as social goods for development: Rethinking research capacity building from Sen's capabilities approach. *Science and Engineering Ethics, 25*(3), 671–692. https://doi.org/10.1007/s11948-018-0037-1.

Moshood, T. D., Nawanir, G., Mahmud, F., Mohamad, F., Ahmad, M. H., & AbdulGhani, A. (2022). Biodegradable plastic applications towards sustainability: A recent innovations in the green product. *Cleaner Engineering and Technology, 6*, 100404. https://doi.org/10.1016/j.clet.2022.100404.

Murphy, S. V., & Atala, A. (2014). 3D bioprinting of tissues and organs. *Nature Biotechnology, 32*(8), 773–785. https://doi.org/10.1038/nbt.2958.

Ni, J., Ling, H., Zhang, S., Wang, Z., Peng, Z., Benyshek, C., Zan, R., Miri, A. K., Li, Z., Zhang, X., Lee, J., Lee, K.-J., Kim, H.-J., Tebon, P., Hoffman, T., Dokmeci, M. R., Ashammakhi, N., Li, X., & Khademhosseini, A. (2019). Three-dimensional printing of metals for biomedical applications. *Materials Today Bio, 3*, 100024. https://doi.org/10.1016/j.mtbio.2019.100024.

Paolini, A., Kollmannsberger, S., & Rank, E. (2019). Additive manufacturing in construction: A review on processes, applications, and digital planning methods. *Additive Manufacturing, 30*, 100894. https://doi.org/10.1016/j.addma.2019.100894.

Park, H., Phaal, R., Ho, J.-Y., & O'Sullivan, E. (2020). Twenty years of technology and strategic roadmapping research: A school of thought perspective. *Technological Forecasting and Social Change, 154*, 119965. https://doi.org/10.1016/j.techfore.2020.119965.

Phuyal, S., Bista, D., & Bista, R. (2020). Challenges, opportunities and future directions of smart manufacturing: A state of art review. *Sustainable Futures, 2*, 100023. https://doi.org/10.1016/j.sftr.2020.100023.

Pinna, C., Ramundo, L., Sisca, F., Angioletti, C., Taisch, M., & Terzi, S. (2016). Additive Manufacturing applications within Food industry: An actual overview and future opportunities.

Popov, V. V., Kudryavtseva, E. V., Kumar Katiyar, N., Shishkin, A., Stepanov, S. I., & Goel, S. (2022). Industry 4.0 and digitalisation in healthcare. *Materials, 15*(6), 2140. https://doi.org/10.3390/ma15062140.

Prasad, K. V., Vasugi, V., & Senthil Kumaran, G. (2023). Application of 3D printing concepts in the Architecture Engineering and Construction (AEC) industry—A scientometric review. *Materials Today: Proceedings*. https://doi.org/10.1016/j.matpr.2023.02.158.

Prashar, G., Vasudev, H., & Bhuddhi, D. (2022). Additive manufacturing: Expanding 3D printing horizon in industry 4.0. *International Journal on Interactive Design and Manufacturing (IJIDeM)*. https://doi.org/10.1007/s12008-022-00956-4.

Raja Santhi, A., & Muthuswamy, P. (2023). Industry 5.0 or industry 4.0S? Introduction to industry 4.0 and a peek into the prospective industry 5.0 technologies. *International Journal on Interactive Design and Manufacturing (IJIDeM), 17*(2), 947–979. https://doi.org/10.1007/s12008-023-01217-8.

Ras, E., Wild, F., Stahl, C., & Baudet, A. (2017). Bridging the Skills Gap of Workers in Industry 4.0 by Human Performance Augmentation Tools: Challenges and Roadmap. 428–432. https://doi.org/10.1145/3056540.3076192.

Rayna, T., & Striukova, L. (2016). From rapid prototyping to home fabrication: How 3D printing is changing business model innovation. *Technological Forecasting and Social Change, 102*, 214–224. https://doi.org/10.1016/j.techfore.2015.07.023.

Rosário, A. T., & Dias, J. C. (2022). Industry 4.0 and marketing: Towards an integrated future research agenda. *Journal of Sensor and Actuator Networks, 11*(3). https://doi.org/10.3390/jsan11030030, Article 3.

Rouf, S., Malik, A., Singh, N., Raina, A., Naveed, N., Siddiqui, M. I. H., & Haq, M. I. U. (2022). Additive manufacturing technologies: Industrial and medical applications. *Sustainable Operations and Computers, 3*, 258–274. https://doi.org/10.1016/j.susoc.2022.05.001.

Sachs, E. M., Cima, M. J., Caradonna, M. A., Grau, J., Serdy, J. G., Saxton, P. C., Uhland, S. A., & Moon, J. (2003). Jetting layers of powder and the formation of fine powder beds thereby. United States Patent US6596224B1. https://patents.google.com/patent/US6596224B1/en.

Sajjad, R., Chauhdary, S. T., Anwar, M. T., Zahid, A., Khosa, A. A., Imran, M., & Sajjad, M. H. (2023). A review of 4D printing—Technologies, shape shifting, smart polymer based materials, and biomedical applications. *Advanced Industrial and Engineering Polymer Research*. https://doi.org/10.1016/j.aiepr.2023.08.002.

Schubert, C., van Langeveld, M. C., & Donoso, L. A. (2014). Innovations in 3D printing: A 3D overview from optics to organs. *The British Journal of Ophthalmology, 98*(2), 159–161. https://doi.org/10.1136/bjophthalmol-2013-304446.

Severini, C., Derossi, A., Ricci, I., Caporizzi, R., & Fiore, A. (2018). Printing a blend of fruit and vegetables. New advances on critical variables and shelf life of 3D edible objects. *Journal of Food Engineering, 220*, 89–100. https://doi.org/10.1016/j.jfoodeng.2017.08.025.

Shahrubudin, N., Te Chuan, L., & Ramlan, R. (2019). An overview on 3D printing technology: Technological, materials, and applications, 35, 1286–1296. https://doi.org/10.1016/j.promfg.2019.06.089.

Soori, M., Arezoo, B., & Dastres, R. (2023). Internet of things for smart factories in industry 4.0, a review. *Internet of Things and Cyber-Physical Systems, 3*, 192–204. https://doi.org/10.1016/j.iotcps.2023.04.006.

Stachiv, I., Alarcon, E., & Lamac, M. (2021). Shape memory alloys and polymers for mems/nems applications: Review on recent findings and challenges in design, preparation, and characterization. *Metals and Materials (Institute of Metals), 11*, 415. https://doi.org/10.3390/met11030415.

Stavropoulos, P., Foteinopoulos, P., Stavridis, J., & Bikas, H. (2023). Increasing the industrial uptake of additive manufacturing processes: A training framework. *Advances in Industrial and Manufacturing Engineering, 6*, 100110. https://doi.org/10.1016/j.aime.2022.100110.

Sun, L., Huang, W. M., Ding, Z., Zhao, Y., Wang, C., Purnawali, H., & Tang, C. (2012). Stimulus-responsive shape memory materials: A review. *Materials & Design, 33*, 577–640. https://doi.org/10.1016/j.matdes.2011.04.065.

Száva, I., Vlase, S., Scutaru, M. L., Asztalos, Z., Gálfi, B.-P., oica, A., & oica, S. (2023). Dimensional methods used in the additive manufacturing process. *Polymers, 15*(18). https://doi.org/10.3390/polym15183694, Article 18.

Taalbi, J. (2017). What drives innovation? Evidence from economic history. *Research Policy, 46*(8), 1437–1453. https://doi.org/10.1016/j.respol.2017.06.007.

Tawk, C., Mutlu, R., & Alici, G. (2022). A 3D printed modular soft gripper integrated with metamaterials for conformal grasping. *Frontiers in Robotics and AI, 8*, 799230. https://doi.org/10.3389/frobt.2021.799230.

Terrile, S., Argüelles, M., & Barrientos, A. (2021). Comparison of different technologies for soft robotics grippers. *Sensors, 21*(9). https://doi.org/10.3390/s21093253, Article 9.

Tofail, S. A. M., Koumoulos, E. P., Bandyopadhyay, A., Bose, S., O'Donoghue, L., & Charitidis, C. (2018). Additive manufacturing: Scientific and technological challenges, market uptake and opportunities. *Materials Today, 21*(1), 22–37. https://doi.org/10.1016/j.mattod.2017.07.001.

Ursan, I. D., Chiu, L., & Pierce, A. (2013). Three-dimensional drug printing: A structured review. *Journal of the American Pharmacists Association: JAPhA, 53*(2), 136–144. https://doi.org/10.1331/JAPhA.2013.12217.

Vafadar, A., Guzzomi, F., Rassau, A., & Hayward, K. (2021). Advances in metal additive manufacturing: A review of common processes, industrial applications, and current challenges. *Applied Sciences, 11*(3). https://doi.org/10.3390/app11031213, Article 3.

Vasiliadis, A. V., Koukoulias, N., & Katakalos, K. (2022). From three-dimensional (3D)- to 6D-printing technology in orthopedics: Science fiction or scientific reality? *Journal of Functional Biomaterials, 13*(3), 101. https://doi.org/10.3390/jfb13030101.

Vinayavekhin, S., Phaal, R., Thanamaitreejit, T., & Asatani, K. (2023). Emerging trends in roadmapping research: A bibliometric literature review. *Technology Analysis & Strategic Management, 35*(5), 558–572. https://doi.org/10.1080/09537325.2021.1979210.

Wang, Y., Blache, R., & Xu, X. (2017). Selection of additive manufacturing processes. *Rapid Prototyping Journal, 23*, 434–447. https://doi.org/10.1108/RPJ-09-2015-0123.

Wei, Q., Han, C., & Kazantseva, N. (2023). Additive manufacturing of metallic materials: Structures, properties and methodologies. *Metals, 13*(7). https://doi.org/10.3390/met13071258, Article 7.

Wen, C., Yu, X., Zeng, W., Zhao, S., Wang, L., Wan, G., Huang, S., Grover, H., & Chen, Z. (2018). Mechanical behaviors and biomedical applications of shape memory materials: A review. *AIMS Materials Science, 5*(4), 559–590. https://doi.org/10.3934/matersci.2018.4.559.

Xia, Z., Jin, S., & Ye, K. (2018). Tissue and organ 3D bioprinting. *SLAS Technology, 23*(4), 301–314. https://doi.org/10.1177/2472630318760515.

Xie, Y., Liu, Q., Zhang, W., Yang, F., Zhao, K., Dong, X., Prakash, S., & Yuan, Y. (2023). Advances in the potential application of 3D food printing to enhance elderly nutritional dietary intake. *Foods, 12*(9). https://doi.org/10.3390/foods12091842, Article 9.

Xu, J., Ding, L., & Love, P. E. D. (2017). Digital reproduction of historical building ornamental components: From 3D scanning to 3D printing. *Automation in Construction, 76*, 85–96. https://doi.org/10.1016/j.autcon.2017.01.010.

Xu, Y., Liu, X., Cao, X., Huang, C., Liu, E., Qian, S., Liu, X., Wu, Y., Dong, F., Qiu, C.-W., Qiu, J., Hua, K., Su, W., Wu, J., Xu, H., Han, Y., Fu, C., Yin, Z., Liu, M., ... Zhang, J. (2021). Artificial intelligence: A powerful paradigm for scientific research. *The Innovation, 2*(4), 100179. https://doi.org/10.1016/j.xinn.2021.100179.

Yang, F., & Gu, S. (2021). Industry 4.0, a revolution that requires technology and national strategies. *Complex & Intelligent Systems, 7*(3), 1311–1325. https://doi.org/10.1007/s40747-020-00267-9.

Yang, F., Zhang, M., & Bhandari, B. (2017). Recent development in 3D food printing. *Critical Reviews in Food Science and Nutrition, 57*(14), 3145–3153. https://doi.org/10.1080/10408398.2015.1094732.

Yang, H., Rao, P., Simpson, T., Lu, Y., Witherell, P., Nassar, A. R., Reutzel, E., & Kumara, S. (2021). Six-Sigma quality management of additive manufacturing. In: *Proceedings of the IEEE. Institute of Electrical and Electronics Engineers, 109*(4). https://doi.org/10.1109/JPROC.2020.3034519.

Zhang, Z., Demir, K. G., & Gu, G. X. (2019). Developments in 4D-printing: A review on current smart materials, technologies, and applications. *International Journal of Smart and Nano Materials, 10*(3), 205–224. https://doi.org/10.1080/19475411.2019.1591541.

Zheng, L., Teng, F., Wang, N., Zhang, X.-N., Regenstein, J. M., Liu, J.-S., Li, Y., & Wang, Z.-J. (2019). Addition of salt ions before spraying improves heat- and cold-induced gel properties of soy protein isolate (SPI). *Applied Sciences, 9*(6). https://doi.org/10.3390/app9061076, Article 6.

Zhironkina, O., & Zhironkin, S. (2023). Technological and intellectual transition to Mining 4.0: A review. *Energies, 16*(3). https://doi.org/10.3390/en16031427, Article 3.

Zhong, R. Y., Xu, X., Klotz, E., & Newman, S. T. (2017). Intelligent manufacturing in the context of Industry 4.0: A review. *Engineering, 3*(5), 616–630. https://doi.org/10.1016/J.ENG.2017.05.015.

Zhou, X., Ren, L., Song, Z., Li, G., Zhang, J., Li, B., Wu, Q., Li, W., Ren, L., & Liu, Q. (2023). Advances in 3D/4D printing of mechanical metamaterials: From manufacturing to applications. *Composites Part B: Engineering, 254*, 110585. https://doi.org/10.1016/j.compositesb.2023.110585.

Additive manufacturing techniques: mechanism, materials, and applications

3

Rakesh Kumar[1], and Santosh Kumar[2]

[1]Department of Regulatory Affairs & Quality Assurance, Kaushik Orthopaedic Private Limited, Sonipat, Haryana, India, [2]Department of Mechanical Engineering, Chandigarh Group of Colleges, Mohali, Punjab, India

3.1 Introduction

The additive manufacturing (AM) or rapid Prototyping (RP) method is also known as "3D printing." It is the process of developing three-dimensional parts from a digital file by adding material in layer by layer (Jiménez et al., 2019). The common step of every AM printing process involves creating three-dimensional objects by using a CAD model to capture the required product shape. Thereafter, convert CAD model into an STL file, transfer the STL file to the machine by slicing into horizontal layers (thin type), set up the machine, and deposition of material in a controlled manner by selecting one of the 3D printer processes (fused deposition modeling, stereo lithography, selective laser sintering, inkjet printing, electron beam melting, etc.) and build an object in a layer according to sliced model, use support structure for temporary support in case of complex shape, remove the part from the 3D printer and do the post-processing (teat treatment, surface finishing) if required. In this way, the final 3D part is ready for use. The process flow diagram of AM is shown in Fig. 3.1 (Dev & Srivastava, 2024; Kumar, Kumar, Chohan, 2021).

Hence AM technology has been being developed for more than 30 years, but without removing or replacing materials. In addition, the timeline for AM technology is shown in Fig. 3.2.

3.1.1 Comparison of AM versus traditional manufacturing methods

The AM is a highly attractive process due to its several advantages; hence, it is classified in different ways as shown in Fig. 3.3.

The rapid proliferation of additive manufacturing (AM) in the last 50 years has seen the developing manufacturing sector integrated into design and modeling as a rapid prototyping technique (Gao et al., 2015). The comparison of AM versus conventional manufacturing process methods (Da Silveira et al., 2001; Dutta & Froes, 2017; Eastwood, 1996; Joffe, 2003; Korpela et al., 2020; Kumar, 2023; Levy et al., 2003; Mehrpouya et al., 2019; Pereira et al., 2019; Pine, 1993; Tseng et al., 1996; Unger et al., 2005; Upadhyay et al., 2017; Ventola, 2014) is shown in Table 3.1.

Additive Manufacturing of Functional Polymers and Nanocomposites. DOI: https://doi.org/10.1016/B978-0-443-26516-7.00003-1
© 2025 Elsevier Ltd. All rights are reserved, including those for text and data mining, AI training, and similar technologies.

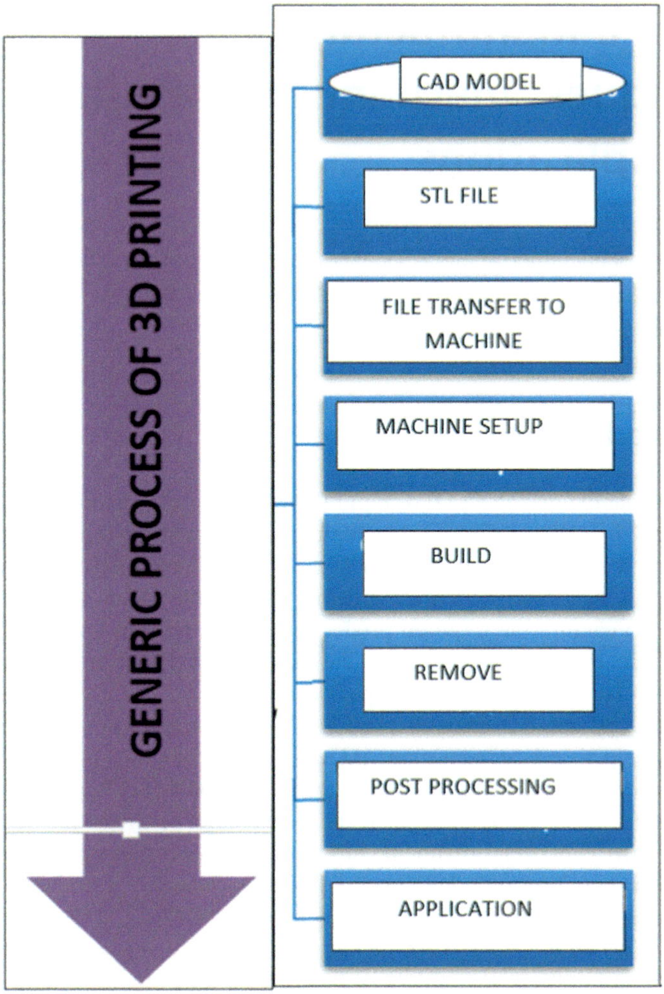

Figure 3.1 Generic process of AM technology.

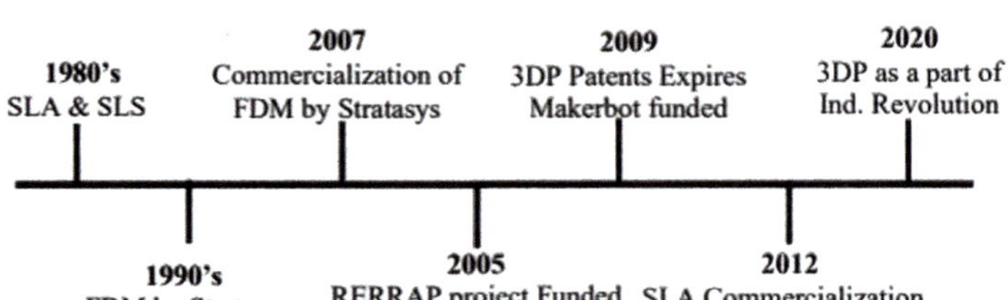

Figure 3.2 Timeline of 3D printing technology.

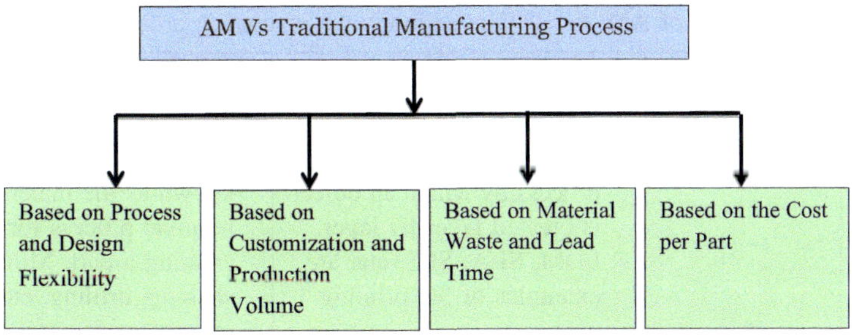

Figure 3.3 Different ways of classifying the AM and conventional manufacturing process.

Furthermore, AM's layer-by-layer manufacturing technique enables the material to be placed just where it is needed, saving material and producing lighter components (Ngo et al., 2018). Because the joints may be incorporated into the component, AM lessens the requirement to link pieces. This implies that intricate plumbing may be constructed all at once rather than having to create individual pipes and then manually fasten them together with bolts or welding afterward. However, because of the restricted construction volumes available, AM does not always eliminate the requirement to link pieces. AM offers a greater degree of freedom than traditional manufacturing techniques thus, allowing for the creation of more distinctive forms and goods. Additionally, AM enables the adjustment of a part's flow and heat properties. Large surface areas can be added to heat exchangers, for example, to produce enhanced heat-flow characteristics. Additionally, AM enables the manufacturing of spare parts for which the original molds, production equipment, or tooling are no longer in use. AM refers to the capacity to regulate the heat treatments used in the process to generate the required microstructure as needed (Attaran, 2017; Du Plessis et al., 2021; Laleh et al., 2022; Yang et al., 2017). In addition, AM is claimed to be 100–1000 times faster than the traditional metal AM. The framework of hybrid RP is categorized by energy source and mechanism utilized. Properties: mechanical consists of tensile strength, hardness, residual stress, yield strength, elongation etc. Microstructural properties: grain size, porosity, grain texture, dendrite spacing. Functional properties: surface roughness, geometric flexibility. Mechanisms include surface evaluation, stress state, thermal gradients, melt pool dynamics, and microstructure growth. Energy source: mechanical, electrical, thermal, plasma, magnetic. Hybrid RP processes: machining that includes milling, shaping, heating/cooling: air cooling nozzle, induction heating, mechanical compressing (rolling), laser (laser assisted machining), etc.

Some researchers (Kumar, 2020) reported that the porosity of the as-sprayed object degrades its mechanical properties during metal additive manufacturing and can prevent the technology from being used further. The study revealed that the possibility of enhancing keyhole stability to lessen or even prevent the creation of keyhole pores is demonstrated by the simulation conducted at low ambient pressure.

Table 3.1 Comparison of AM versus conventional manufacturing process methods.

S. No.	Feature	AM process	Conventional process
1.	Process	AM is a material addition process by which an object is created layer by layer. FDM, SLA, SLS, etc. are examples of 3D printing processes.	Conventional processes are known as the material removal process for creating a part. Milling, shaping, drilling, etc. are examples of conventional manufacturing processes.
2.	Flexibility	The design flexibility in the case of AM is very high because it allows for complex geometrical features.	The design flexibility in the case of the conventional process method is low, owing to tool capabilities.
3.	Production	AM is suitable for manufacturing less quantity of objects, specialized or prototyping objects.	It is suitable for the production of large quantities where no. of identical objects are manufactured.
4.	Wastage	AM minimize waste by the application of only the material needed for part manufacturing.	As the material is removed during processing, it generates more material wastage.
5.	Customization	AM process can easily create customized parts because of flexibility. Thus it is ideal for developing unique objects.	In the case of conventional methods customization is difficult, and more costly for individual pieces.
6.	Time	AM manufacturing methods in the case of complex manufacturing take less time, mainly for prototyping and for the production of less quantity of parts.	Conventional manufacturing methods in the case of complex manufacturing take more time, and also require no. of processes and tools during part development.
7.	Skill	A less skilled operator is required to operate the AM printer.	A highly skilled operator is required to operate and maintain the conventional machines.
8.	Cost	The cost per part in the case of the AM method is less by increasing volume.	The cost per part in the case of the conventional method is less in the case of mass production (i.e., no. of the same items).

3.1.2 Advantages and disadvantages of AM process methods

The merits and demerits of distinct AM methods (Kumar, Kumar, Chohan, et al., 2022; Kumar, Kumar, Kumar, et al., 2022) are shown in Table 3.2.

3.2 Additive manufacturing materials

The materials used in 3D printing are just as diverse as the products that come from the process. Because of this, 3D printing is flexible enough to let producers alter the shape, feel, and strength of a product. The best part is that a lot fewer steps are required to achieve these qualities than are typically required in traditional production

Table 3.2 Merits and demerits of AM processes.

Process method	Merits	Demerits
FDM	Material flexibility, easy to use, owing to the lesser cost of printers and wide availability, the lead times are minimal and cheaper as compared to other AM processes.	Rough surface finish, ABS produces warping during printing, allowing for single color printing. Support structure required.
SLA	High strength & flexibility, faster printing speed than other printing methods, no support structure required thus saving material and time.	Higher cost, limited material selection, required postprocessing (cleaning and curing in UV light).
SLS	Excellent strength, stiffness, dimensional stability, no need for support structure, etc.	High cost of the printer as compared to SLA and FDM printer, high cost of material than other AM printers, limited color options, potential health issue owing to fine powder particles. Postprocessing needed.
SLM	High strength and durability, direct metal printing, lightweight potential.	Extensively high cost of printer, slightly rough surface, safety issue owing to high temperature.
EBM	Outstanding mechanical properties like SLM, good fatigue resistance, direct metal printing, producing complex shapes, reducing cracking, and warping.	High cost, limited material selection, slightly rough surface, high accuracy.
LOM	Less cost of printer than SLA, SLS, SLM; used a distinct range of materials.	Large volume produces rough surface, warping and delamination, etc.

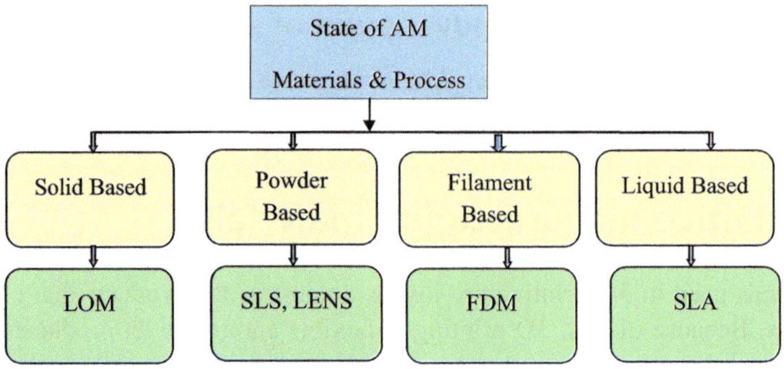

Figure 3.4 States of AM processes and printing methods.

processes. Since the dawn of technology, the materials accessible for 3D printing have advanced significantly (Kumar, 2022). Currently, a wide variety of materials in various stages are available (powder, filament, pellets, granules, resin etc.). More and more, customized materials with material properties that better suit the application are manufactured for specialized platforms carrying out specialized applications (such as in the aerospace, medical, and dental sectors). A wide range of materials are used in the additive manufacturing (AM) process, including metals, alloys, and metal composites. Materials that are frequently used in metal AM include stainless steel, titanium and titanium alloys, aluminum and aluminum alloys, nickel and nickel alloys, and cobalt-chromium alloys. Other materials that can be used in metal AM include copper, gold, and platinum, as well as metal composites, which combine two or more materials for improved properties. The distinct states of AM processes and printing methods (Kumar & Sathiya, 2020; Webster et al., 2021) are shown in Fig. 3.4.

These days, additive manufacturing (AM) uses a variety of materials, including ceramic, polymers, metals, plastics, and their mixtures. Advanced materials are still being improved. The overview of some polymer materials is described below.

3.2.1 Polymers materials

Plastic is also used in multiple products and distinct sectors. Plastic has three types natural plastics, semisynthetic plastics, synthesis plastics (thermoplastic) and (thermosetting plastic). Except, thermosetting type of plastics all the plastic used in FDM. Polymers known as thermoplastics may be molded and softened by heat, and when cooled, they will revert to their original form. They will soften if heat is given to them once more. Styrene and acrylic are two of the most commonly used polymers in-school workshops; they are examples of thermoplastics. Polymer components, from prototypes to functional buildings with intricate geometry, are produced via 3D printing. As per Sapkota et al. (2024), using fused deposition modeling (FDM) may produce a three-dimensional printed item by depositing successive layers of extruded thermoplastic filament, such as polylactic acid (PLA), acrylic butadiene styrene (ABS), polypropylene (PP), or polyethene (PE). Lately, 3D printing materials made of thermoplastic filaments with high melting temperatures, such as PEEK and PMMA, have been utilized (Caminero et al., 2018). Polymers were widely employed as inert materials in biomaterials and medical device products, which helped the devices, work properly and supported numerous orthopedic implants mechanically. The following substances are commonly used to make plastics. The examples of some plastic materials and their application are as follows:

Polylactic acid (PLA): Because it is biodegradable and made from natural resources like sugar cane and corn starch, it is among the greenest 3D printing materials available. In the upcoming years, plastic acid-based plastics—which come in soft and hard forms—are anticipated to rule the 3D printing market. However, hard PLA is a more resilient material that may be used for a greater range of purposes. Materials and products for medical devices both support the mechanical operation of numerous orthopedic implants and help the devices work well. Acrylonitrile butadiene styrene (ABS): Because of its durability and protection, acrylonitrile butadiene styrene, or ABS, is a material that is frequently used in home-based 3D printers. Pasta-like threads give ABS, popularly referred to as "LEGO plastic," its toughness and adaptability. Because it is multicolored, it is perfect for products like toys and stickers. Polyvinyl alcohol plastic (PVA): often used in inexpensive home printers, PVA is a dissolvable support material. PVA isn't the best option for things that require great strength, but it can be a cheap substitute for items that are only meant to be used once. Polyamide (Nylon): because of its strength and adaptability, polyamide (Nylon) provides a high level of detail on a 3D-printed object. The material works especially well for connecting and locking sections in a 3D-printed object.

Polyamide is used to print anything, from handles and fasteners to toy vehicles and figurines. The good balance of chemical and mechanical qualities found in polyamides, which are composed of semicrystalline structures, results in exceptional stability, stiffness, durability, and shock resistance. Polypropylene (PP) is a thermoplastic that finds widespread application in the automotive sector, specialty textiles, and the production of several daily items. PP is well known for its relative stiffness and flexibility, as well as its resistance to abrasion and shock absorption. Unfortunately, there are drawbacks to the material, such as its low-temperature tolerance and vulnerability to UV radiation, which can cause it to grow. Consequently, several producers have created simili-propilenos, a variation of polypropylene that possesses greater mechanical and physical strength. Polycarbonate (PC): Although less prevalent than other plastic types, PC is utilized in 3D printers with high-temperature capabilities and nozzle designs. Among other things, molding trays and inexpensive plastic fasteners are made of polycarbonate. HIPS stands for high performance polymers (PEEK, PEKK, ULTEM): Thanks to rigorous material testing brought about by the development of 3D printing technology, a wide range of high-performance filaments with mechanical qualities similar to those of metals can now be produced. Some examples of high-performance polymers for 3D printing are PEEK, PEKK, and ULTEM. They fall into one of two categories: polyetherimides (PEI) or polyaryletherketones (PAEK). These filaments are heavier than metals, but they also have great mechanical and thermal resistance. Because of their characteristics, not all FDM machines on the market can print on high-performance polymers. Biomaterials, medical device goods, and orthopedic implants are the primary applications for polymer materials. Yet, polymers are also employed in tissue engineering, cartilage restoration, and other applications (Bedi et al., 2019; Berretta et al., 2017; Joseph et al., 2023; Park & Fu, 2021; Sabir et al., 2009; Tümer & Erbil, 2021; Williams et al., 2005; Xin et al., 2017).

3.2.2 Metals and alloys

The metal used in additive manufacturing (AM) offers flexibility in creating complex shapes and has exceptional physical properties, including high compressive strength and great fatigue resistance. It is also utilized in the printing of human organs for aeronautical components. Cobalt-based alloys, aluminum-based alloys, nickel-based alloys, stainless steels, titanium alloys, etc. are a few significant examples of materials. The cobalt-based alloy has a high recovery capacity, more precise stiffness, elongation, and resilience while being highly recommended (Uhlmann et al., 2015). It is therefore most frequently utilized in dental applications.

3.2.3 Smart materials

The shape, geometry, and configuration of things may be altered by smart materials in response to environmental factors such as heat and water. Self-evolving structures and soft robotics systems are two prominent examples of additively built parts made using clever materials.4DP materials are a subset of smart materials (Lee et al., 2017). Among them are alloys (Ni-Ti) and shape memory polymers, which have been used in microelectromechanical devices and biomedical implants (Kumar, Kumar, Chohan & Kumar, 2021).

3.2.4 Special materials

The AM application in food industries for special materials (chocolate, pizza) and fashion industries (jewelry products) that needed less time to manufacture and very low packaging cost etc. are the major example of special materials (Jideani et al., 2020; Kumar, Kumar, Chohan, 2021, 2021; Sun et al., 2018). In addition, the commercially available materials are polymers (thermosetting, thermoplastic), metals, and materials that are under R&D stage are ceramics (alumina, suspension of zircona), composites (polymer metals, polymer ceramics, short fiber-reinforced composites), etc. (Banjanin et al., 2018).

3.2.5 Composites

When two or more constituent materials with noticeably different physical or chemical properties remain separate and distinct at the macroscopic or microscopic scale within the finished structure, they form composites, which are artificially created or naturally occurring materials that exhibit properties that cannot be achieved by any of the materials acting alone. A composite can be made of materials mixed uniformly to produce a homogeneous compound (a uniform composite) or nonuniformly to produce an inhomogeneous compound (functionally graded materials, for example) where the composition varies gradually over volume to produce corresponding changes in the composite material's properties. Composite are classified into three groups: particulate reinforced (includes large particle and dispersion particles), fiber-reinforced (includes continuous and discontinuous particles), and structural (laminate and sandwich panels). Particle reinforced composite (PRC): Materials with a minimum of two constituents are known as PRCs. To construct the composite, the reinforcing components are laid down in a matrix. Fiber-reinforced composites, or FRCs, are composed of three components: (a) the fibers, which act as an irregular or dispersed stage; (b) the matrix, which acts as a continuous stage; and (c) the fine interphase zone, sometimes referred to as the interface.

Composite structure (SC): Reinforced composite materials provide completely new possibilities in a variety of contemporary applications as compared to older materials. When speed, low weight, great performance, and exceptional mechanical and chemical qualities are essential, it makes sense to consider using composite material segments or components. The most common material used by the distinct AM printing methods by various researchers is shown in Table 3.3.

3.3 Additive manufacturing techniques

The different types of AM methods such as stereolithography (SLA), fusion deposition modeling (FDM), selective laser sintering (SLS), Selective laser melting (SLM), electron beam melting (EBM), and inkjet printing (IJP) have own advantages and limitations also (Jiménez et al., 2019; Kumar & Kumar, 2020; Lantada & Morgado, 2012; Sidhu & Kumar, 2020; Srivastava et al., 2022; Wendel et al., 2008; Yakout et al., 2018). The most widely used AM process comparisons are described in Table 3.4.

In addition, a list of equipment manufacturers and their products (Frazier, 2014) is shown in Table 3.5.

3.4 Additive manufacturing applications

Nowadays, the AM method is being used in many different industries, including the automotive, medical, aerospace, fashion, fabric, electronics, building, architectural,

Table 3.3 Different additive manufacturing processes and materials used.

Material	Printing process	References
PLA	FDM	Guo & Leu (2013)
Ti6Al4V	FDM	Nune et al. (2014)
Ti6Al4V	SLM	Al-Rubaie et al. (2020)
AlSi7Mg alloys	SLM	Wang et al. (2018)
Ti6Al4V (implant)	DMLS	Delikanli & Kayacan (2019)
17-4Ph SS	FFF	Godec et al. (2020)

Table 3.4 AM processes comparison.

Printing methods	FDM	SLA	Material jetting	LOM	SLS
Technology used	Material extrusion	Vat photo polymerization	Material-Jetting process	Sheet lamination	Powder bed fusion
Working process	The required heated polymer material is used to create an object	Photosensitive thermoset polymers are cured by exposing them to ultraviolet laser light	Uses several linear nozzles connected to a heated print head to deposit plastic resins, which are then cured by a UV lamp	Uses adhesives to join layers of material in sheet shapes, such as metal, plastic, or paper, to construct a component. Well-liked for prototypes	The fuses material layer by layer without ever reaching the melting temperature by using a laser beam as a heat source
Part finishing quality	Standard finish	Excellent surface finish is achieved	Standard finish	Wood-like characteristics and can be treated similarly	Standard finish
Materials	Thermoplastic materials such as PLA, ABS, etc.	ABS (semiflexible materials)	ABS, PP, rubber, and thermosetting photopolymers same as those utilized in SLA	Polymers and ceramics	Nylon and ceramics
Field of applica-tions	Toys, prototypes, home applications, light weight Implants, casings	Biomedical and prototyping, dental applications	Nonfunctional prototypes	Paper U foundry, nonfunctional prototypes	Aerospace, biomedical military and complex shape object, heat resistant
Accuracy (μm)	Good accuracy	25–150	More than 35	Less accuracy	300
Process	Filament	Liquid	-	Solid sheet	Powder
Inventor	Scott Crump	Charles W. Hull	Israeli	Helisys	CarlDeckard
Year of Invention	1990	1986	1998	1991	1989

Table 3.5 The equipment sources and specifications.

Printing method	System	Source of energy	Build Vol. in mm
PTAS FFF	Wire feed: MER plasma transferred arc selected FFF, Manufacturer: USA	Using two 350 A DC power supplies, plasma was transported	610 × 610 × 5182
LD	Powder feed: Trumpf, Manufacturer: Germany	CO_2 laser cladding	Length: 600 × 1000
LENS	Powder feed: LENS 850-R, Manufacturer: USA	IPG fiber laser (1–2 KW)	900 × 1500 × 900
DMD	Powder feed: POM DMD, Manufacturer: Canada	Fiber diode (1–5 KW)	1500×1000×8000
SLM	Powder bed: Realizer (SLM 250), Manufacturer: Germany	Yb fiber laser (400 W)	250 × 250 × 220
EBM	Powder bed: ARCAM (A2), Manufacturer: Sweden	Electron beam (7 kW)	200 × 200 × 350
SLM	Powder bed: Matsuura	Yb fiber laser (400 W); Hybrid AM or conventional system	Diameter: 250 × 250

construction, medicinal, and food industries. The different sectors of AM process applications are described below.

3.4.1 Biomedical application

The use of AM techniques in the biomedical field has expanded to include products like biologic chips medical devices (temporary and permanent types called implants and instruments), artificial organs, tissue scaffolds, biomaterials, biologic sciences, and biomedicine. In the case of orthopedic and dental applications, AM creates intricate shapes and geometries, provides a rough, surface (that promotes bone integration), and customization, and helps to meet specific requirements of each patient (Chang et al., 2010; Liu et al., 2005). Vandenbroucke and Kruth (2007) utilized the SLM printing method for the development of implant parts by using biocompatible metal alloys such as Ti6Al4Vand CoCrMo for biomedical applications. Among the uses of additive manufacturing in medicine are medical models, implants, biomanufacturing, medical aids, supporting guides, splints, prostheses, and tools, equipment, and parts for medical devices.

3.4.2 Aerospace application

Aerospace components are typically constructed of sophisticated materials, such as titanium alloys, nickel superalloys, special steels, or ultrahigh-temperature ceramics, which are difficult, expensive, and time-consuming to manufacture. These components frequently have complicated geometries. Aerospace production runs are also typically minimal, with a cap of a few thousand pieces. Consequently, AM technology is excellent for aircraft applications. The AM technology such as LENS for the aerospace part fabrication (satellites), EBM for functional components, and SLM for turbine blades, fixtures, brackets, accessories, molds, dies, etc. were used (Boyer et al., 2015; Katz-Demyanetz et al., 2019; Samal et al.,2022; Sinha et al., 2022; Yap et al.,2015; Zhou et al., 2024)

3.4.3 Improved 3D-printed biomaterials for tissue engineering

Additive manufacturing, or 3D printing, has allowed for the creation of multiscale biomimetic structures, medical devices, and scaffolds tailored to individual patients (Arif et al., 2021; Arif et al., 2022). As more patients require transplants, there is an increasing need for 3D-printed replacement organs. Due to its potential to completely transform the area of tissue engineering, 3D printing is now the center of attention for many researchers in the field. Both polylactide and polycaprolactone have garnered significant attention as prospective biomaterials for use in tissue engineering.

3.4.4 Architectural and jewelry industry

Models for architecture and jewelry need a lot of labor-intensive handwork and time due to the intricate forms needed to translate artistic concepts into tangible reality. When it comes to handling these intricate models, AM offers far higher resolution than other methods, for example, artistic models fabricated by the RP method (Petrovic et al., 2010; Wong & Hernandez, 2012).

3.4.5 Food industry

Patients, kids, athletes, pregnant women, and other groups require nutritious meals in particular because they require a certain amount of nutrients, which may be obtained by reducing the number of components that are deemed superfluous and boosting the number of nutritious substances (Dankar et al., 2018). This method allows certain materials to be treated and mixed into the required size and form (Liu et al., 2017). Pizza, chocolate, sugar, pasta, and other foods are a few examples of food products that may be used to create new food items with the required shape (Sun et al., 2018).

3.4.6 Furniture industry

The entire sector is seeing a shift toward 3D printing these days. It also has a fantastic furniture industry. In reality, by creating new types of furniture, connections, replacement components, etc., 3D printing is revolutionizing the furniture sector.

As is common knowledge, furniture—such as mattresses, coffee tables, chairs, and other accessories—plays a significant part in the décor of homes, hotels, offices, and other spaces. 3D printing is a terrific way to build fashionable or embellished furniture. Furthermore, it saves time, money, and other resources by doing away with or cutting down on assembly work, which was labor-intensive when done the old fashioned way (Ceschin & Gaziulusoy, 2016; Murmura & Bravi, 2017; Petrova & Jivkov, 2024).

3.4.7 Automobile industry

Automotive fast prototyping has been widely used recently to create bumpers, windbreakers, cooling vents, valves, pumps, and other components. Furthermore, it is anticipated that shortly, the automotive AM will be used to produce superalloys that are resistant to corrosion (Yang et al., 2024). In addition, AM is used for 3D-printed structures such as Bloom Pavilion, footprint, 3D-printed homes in China, Winsun, the first 3D-printed office in Dubai, woven concrete benches, etc. (Ibrahim et al., 2022). The distinct printing methods and fields of application are given in Table 3.6.

3.5 Challenges

There are several obstacles, including an understanding of AM process systems, materials, defects, cost, and productivity. The primary difficulties are as follows:

- Comparing components printed by various AM processes is proving to be challenging for academics and producers as a result of the absence of test standards for AM-processed products (Wu et al., 2020)
- Injection-based AM methods require higher resolution, which adds to processing time and energy costs. Thus, you must concentrate.
- The main issues to be addressed in the fabrication of hollow and thin-walled structures are sagging and regulating deformation from thermal stresses. The key to achieving acceptable microstructural properties in printed components is managing process variables such as cooling rates, laser beam energy, scanning velocity, and scanning route strategy. The ability to fabricate thin-walled objects using powder-based AM methods is restricted by low surface quality (Travitzky et al., 2014)
- The components created by 3D printing are required to have higher quality to fulfill the criteria of sophisticated applications. Researchers' growing levels of expertise can be used to conduct more thorough studies on postprocessing methods (heat treatment and surface finish). Future studies may concentrate on the cautious removal of supporting evidence.
- To combine the design, process parameters, process restrictions, material behavior, and development of analysis tools to enable accurate forecasts for extended design flexibilities are the primary issues in the 3DP of buildings and construction applications (Rashid et al., 2020).
- It is advised that future studies on the industrial side of additive manufacturing as well as its applications in the fields of medicine and dentistry use standardized language. This would make it possible to compare the use of various AM techniques more methodically (Chhaya et al., 2015)

Table 3.6 Summary of distinct AM methods used and field of applications.

S. No.	Printing method	Field of application	References
1.	LOM, FDM, SLA etc.	Industries: aviation industry, automobile, health care, electronics industry, etc.	Lifton et al. (2014); Nampoothiri et al. (2010); Rouf et al. (2022); Wojtyła et al. (2017)
2.	Binder Jetting, LOM, etc.	Industries: automotive defense, aerospace aviation, etc.	Duda and Raghavan (2016); Ngo et al. (2018)
3.	SLS, SLA, FDM, Binder Jetting, Polyjet printing etc.	Industries: aerospace, health care sector,	Dawood et al. (2015); Garcia et al. (2017); Rouf et al. (2022)
4.	FFF SLA, LDM, FFF, LOM, SLS, etc.	Industries: electronics, automobiles, health care, etc.	Carrico et al. (2015); Jiang et al. (2020); Singh et al. (2019); Valino et al. (2019)
5.	FDM, concrete printing, Binder Jetting etc.	Industries: food industry, biomedical, construction, etc.	Gibbons et al. (2009); Shakor et al. (2019); Tay et al. (2017)
	FDM, IJP, SLA, EBM, DLP, DPP, SLM, SLS etl.	Industries: health care, robotics, aviation, automotive defense, etc.	Javaid and Haleem (2020); Voon et al. (2019); Zhang et al. (2019)
6.	FDM, IJP, SLS, SLA, EBM, SLM	Surgical tools, dental; scaffolds for tissue engineering, maxillofacial reconstruction; implants: crowns, dental bridges, implants: crowns, dental bridges etc.; crowns, dental bridges, etc.	Cai et al. (2023); Hwang et al. (2014); Mekonnen et al. (2016); Mikler et al. (2017); Mohamed et al. (2017); Olivier et al. (2017); Schmidt et al. (2017); Travitzky et al. (2014); Wang and Wang (2017)

- The issues that require more research and investigation include the part size limit, anisotropic mechanical properties, the creation of overhang surfaces, high costs, low manufacturing efficiency, poor accuracy, warping, pillowing, stringing, gaps in the upper layers, underextrusion, layer misalignment, overextrusion, elephant foot, mass production, and material limitations. The following are AM's drawbacks and difficulties : void formation, anisotropic in mechanical properties, stair stepping, small build volume, fabrication of weapons or drugs for crime purposes, and compliance with food and drug administration safety standards. Hence, researchers need to focus on all the listed challenges for improvement in technology (Kumar & Kumar, 2022a, 2022b; Kumar, Kumar, Kumar, et al., 2020; Kumar, Kumar, Chohan, et al., 2022; Kumar, Kumar, Kumar, et al., 2022; Kumar, Kushawaha, et al., 2022).

3.6 Conclusions

The commercialization of AM in the future will be contingent upon several elements, including robust process stability, online quality control protocols, continuous certification, a database containing AM material attributes, and the accessibility of design standards. To ensure that manufacturers don't under or overinvest in QA technology, it's imperative to evaluate the level of quality assurance required for each item produced utilizing AM technologies. The growing usage of additive manufacturing (AM) techniques in the aerospace, automotive, biomedical, and other engineering sectors may be attributed to its benefits, which include the capacity to produce freeform parts, material savings, mass customization, and cost-effective prototyping capabilities. This may indicate that sales of AM equipment are rising in the markets and are expected to increase more in the future. Researchers and workers with access to technology need to be more knowledgeable about the latest technical advancements. A greater comprehension of AM processes, innovations, the process ability of materials, and problems/defects relating to process parameters, metallurgy, and thermal and mechanical properties are the synergistic aspects of the knowledge base on AM. An analysis of the AM's operations, features, and procedures was provided in this research along with insights into the application materials and the difficulties in meeting the criteria.

References

Arif, M. M., Khan, S. M., Gull, N., Tabish, T. A., Zia, S., Khan, R. U., Awais, S. M., & Butt, M. A. (2021). Polymer-based biomaterials for chronic wound management: Promises and challenges. *International Journal of Pharmaceutics, 598*, 120270. https://doi.org/10.1016/j.ijpharm.2021.120270.

Al-Rubaie, K. S., Melotti, S., Rabelo, A., Paiva, J. M., Elbestawi, M. A., & Veldhuis, S. C. (2020). Machinability of SLM-produced Ti6Al4V titanium alloy parts. *Journal of Manufacturing Processes, 57*, 768–786. https://doi.org/10.1016/j.jmapro.2020.07.035.

Arif, Z. U., Khalid, M. Y., Zolfagharian, A., & Bodaghi, M. (2022). 4D bioprinting of smart polymers for biomedical applications: Recent progress, challenges, and future perspectives. *Reactive and Functional Polymers, 179*, 105374. https://doi.org/10.1016/j.reactfunctpolym.2022.105374.

Attaran, M. (2017). The rise of 3 D printing: The advantages of additive manufacturing over traditional manufacturing. *Business Horizons, 60*(5), 677–688. https://doi.org/10.1016/j.bushor.2017.05.011.

Banjanin, B., Vladic, G., Pál, M., Balos, S., Dramicanin, M., Rackov, M., & Knezevic, I. (2018). Consistency analysis of mechanical properties of elements produced by FDM additive manufacturing technology. *Matéria (Rio De Janeiro), 23*(4)https://doi.org/10.1590/s1517-707620180004.0584.

Bedi, T. S., Kumar, S., & Kumar, R. (2019). Corrosion performance of hydroxyapaite and hydroxyapaite/titania bond coating for biomedical applications. *Materials Research Express, 7*(1), 015402. https://doi.org/10.1088/2053-1591/ab5cc5.

Berretta, S., Davies, R., Shyng, Y., Wang, Y., & Ghita, O. (2017). Fused deposition modelling of high temperature polymers: Exploring CNT PEEK composites. *Polymer Testing, 63*, 251–262. https://doi.org/10.1016/j.polymertesting.2017.08.024.

Boyer, R., Cotton, J., Mohaghegh, M., & Schafrik, R. (2015). Materials considerations for aerospace applications. *MRS Bulletin, 40*(12), 1055–1066. https://doi.org/10.1557/mrs.2015.278.

Cai, H., Xu, X., Lu, X., Zhao, M., Jia, Q., Jiang, H., & Kwon, J. (2023). Dental materials applied to 3D and 4D printing technologies: A review. *Polymers, 15*(10), 2405. https://doi.org/10.3390/polym15102405.

Caminero, M., Chacón, J., García-Moreno, I., & Rodríguez, G. (2018). Impact damage resistance of 3D printed continuous fibre reinforced thermoplastic composites using fused deposition modelling. *Composites Part B Engineering, 148*, 93–103. https://doi.org/10.1016/j.compositesb.2018.04.054.

Carrico, J. D., Traeden, N. W., Aureli, M., & Leang, K. K. (2015). Fused filament 3D printing of ionic polymer-metal composites (IPMCs). *Smart Materials and Structures, 24*(12), 125021. https://doi.org/10.1088/0964-1726/24/12/125021.

Ceschin, F., & Gaziulusoy, I. (2016). Evolution of design for sustainability: From product design to design for system innovations and transitions. *Design Studies, 47*, 118–163. https://doi.org/10.1016/j.destud.2016.09.002.

Chang, R., Emami, K., Wu, H., & Sun, W. (2010). Biofabrication of a three-dimensional liver micro-organ as an in vitro drug metabolism model. *Biofabrication, 2*(4), 045004. https://doi.org/10.1088/1758-5082/2/4/045004.

Chhaya, M. P., Poh, P. S., Balmayor, E. R., Van Griensven, M., Schantz, J., & Hutmacher, D. W. (2015). Additive manufacturing in biomedical sciences and the need for definitions and norms. *Expert Review of Medical Devices, 12*(5), 537–543. https://doi.org/10.1586/17434440.2015.1059274.

Dankar, I., Haddarah, A., Omar, F. E., Sepulcre, F., & Pujolà, M. (2018). 3D printing technology: The new era for food customization and elaboration. *Trends in Food Science & Technology, 75*, 231–242. https://doi.org/10.1016/j.tifs.2018.03.018.

Dawood, A., Marti, B. M., Sauret-Jackson, V., & Darwood, A. (2015). 3D printing in dentistry. *BDJ, 219*(11), 521–529. https://doi.org/10.1038/sj.bdj.2015.914.

Delikanli, Y. E., & Kayacan, M. C. (2019). Design, manufacture, and fatigue analysis of lightweight hip implants. *Journal of Applied Biomaterials & Functional Materials, 17*(2) https://doi.org/10.1177/2280800019836830.

Dev, S., & Srivastava, R. (2024). Experimental investigation and optimization of the additive manufacturing process through AI-based hybrid statistical approaches. *Progress in Additive Manufacturing*. https://doi.org/10.1007/s40964-024-00606-z.

Du Plessis, A., Razavi, N., Benedetti, M., Murchio, S., Leary, M., Watson, M., Bhate, D., & Berto, F. (2021). Properties and applications of additively manufactured metallic cellular materials: A review. *Progress in Materials Science, 125*, 100918. https://doi.org/10.1016/j.pmatsci.2021.100918.

Duda, T., & Raghavan, L. V. (2016). 3D metal printing technology. *IFAC-PapersOnLine, 49*(29), 103–110. https://doi.org/10.1016/j.ifacol.2016.11.111.

Dutta, B., & Froes, F. (2017). The additive manufacturing (AM) of titanium alloys. *Metal Powder Report, 72*(2), 96–106. https://doi.org/10.1016/j.mprp.2016.12.062.

Eastwood, M. A. (1996). Implementing mass customization. *Computers in Industry, 30*(3), 171–174. https://doi.org/10.1016/0166-3615(96)00010-3.

Frazier, W. E. (2014). Metal additive manufacturing: A review. *Journal of Materials Engineering and Performance, 23*(6), 1917–1928. https://doi.org/10.1007/s11665-014-0958-z.

Gao, W., Zhang, Y., Ramanujan, D., Ramani, K., Chen, Y., Williams, C. B., Wang, C. C., Shin, Y. C., Zhang, S., & Zavattieri, P. D. (2015). The status, challenges, and future of additive manufacturing in engineering. *Computer-Aided Design, 69*, 65–89. https://doi.org/10.1016/j.cad.2015.04.001.

Garcia, J., Yang, Z., Mongrain, R., Leask, R. L., & Lachapelle, K. (2017). 3D printing materials and their use in medical education: A review of current technology and trends for the future. *BMJ Simulation & Technology Enhanced Learning, 4*(1), 27–40. https://doi.org/10.1136/bmjstel-2017-000234.

Gibbons, G. J., Williams, R., Purnell, P., & Farahi, E. (2009). 3D printing of cement composites. *Advances in Applied Ceramics Structural Functional and Bioceramics, 109*(5), 287–290. https://doi.org/10.1179/174367509x12472364600878.

Godec, D., Cano, S., Holzer, C., & Gonzalez-Gutierrez, J. (2020). Optimization of the 3D printing parameters for tensile properties of specimens produced by fused filament fabrication of 17-4PH stainless steel. *Materials, 13*(3), 774. https://doi.org/10.3390/ma13030774.

Guo, N., & Leu, M. C. (2013). Additive manufacturing: technology, applications and research needs. *Front. Mech. Eng. 8*, 215–243. https://doi.org/10.1007/s11465-013-0248-8.

Hwang, S., Reyes, E. I., Moon, K., Rumpf, R. C., & Kim, N. S. (2014). Thermo-mechanical characterization of metal/polymer composite filaments and printing parameter study for fused deposition modeling in the 3D printing process. *Journal of Electronic Materials, 44*(3), 771–777. https://doi.org/10.1007/s11664-014-3425-6.

Ibrahim, I., Eltarabishi, F., Abdalla, H., & Abdallah, M. (2022). 3D printing in sustainable buildings: Systematic review and applications in the United Arab Emirates. *Buildings, 12*(10), 1703. https://doi.org/10.3390/buildings12101703.

Javaid, M., & Haleem, A. (2020). Exploring smart material applications for COVID-19 pandemic using 4D printing technology. *Journal of Industrial Integration and Management, 05*(04), 481–494. https://doi.org/10.1142/s2424862220500219.

Jiang, L., Peng, X., & Walczyk, D. (2020). 3D printing of biofiber-reinforced composites and their mechanical properties: A review. *Rapid Prototyping Journal, 26*(6), 1113–1129. https://doi.org/10.1108/rpj-08-2019-0214.

Jideani, A. I. O., Mutshinyani, A. P., Maluleke, N. P., Mafukata, Z. P., Sithole, M. V., Lidovho, M. U., Ramatsetse, E. K., & Matshisevhe, M. M. (2020). Impact of industrial revolutions on food machinery—An ōverview. *Journal of Food Research, 9*(5), 42. https://doi.org/10.5539/jfr.v9n5p42.

Jiménez, M., Romero, L., Domínguez, I. A., Del Mar Espinosa, M., & Domínguez, M. (2019). Additive manufacturing technologies: An overview about 3D printing methods and future prospects. *Complexity, 2019*, 1–30. https://doi.org/10.1155/2019/9656938.

Jiménez, M., Romero, L., Domínguez, I. A., Del Mar Espinosa, M., & Domínguez, M. (2019). Additive Manufacturing Technologies: An Overview about 3D Printing Methods and Future Prospects. *Complexity, 2019*(1)https://doi.org/10.1155/2019/9656938.

Joffe, L. (2003). Invisalign®: Early experiences. *Journal of Orthodontics, 30*(4), 348–352. https://doi.org/10.1093/ortho/30.4.348.

Joseph, T. M., Kallingal, A., Suresh, A. M., Mahapatra, D. K., Hasanin, M. S., Haponiuk, J., & Thomas, S. (2023). 3D printing of polylactic acid: Recent advances and opportunities. *The International Journal of Advanced Manufacturing Technology, 125*(3–4), 1015–1035. https://doi.org/10.1007/s00170-022-10795-y.

Katz-Demyanetz, A., Popov, V. V., Kovalevsky, A., Safranchik, D., & Koptyug, A. (2019). Powder-bed additive manufacturing for aerospace application: Techniques, metallic and metal/ceramic composite materials and trends. *Manufacturing Review, 6*(5)https://doi.org/10.1051/mfreview/2019003.

Korpela, M., Riikonen, N., Piili, H., Salminen, A., & Nyrhilä, O. (2020). Additive Manufacturing—Past, Present, and the Future. In M. Collan, & KE. Michelsen (Eds.). *Technical, Economic and Societal Effects of Manufacturing 4.0* Cham: Palgrave Macmillan. https://doi.org/10.1007/978-3-030-46103-4_2.

Kumar, S. (2020). *Additive manufacturing processes*. Springer. https://doi.org/10.1007/978-3-030-45089-2.

Kumar, S. (2022). Influence of processing conditions on the mechanical, tribological and fatigue performance of cold spray coating: a review. *Surface Engineering, 38*(4), 324–365. https://doi.org/10.1080/02670844.2022.2073424.

Kumar, S. (2023). Comprehensive review on high entropy alloy-based coating. *Surface and Coatings Technology, 477*, 130327. https://doi.org/10.1016/j.surfcoat.2023.130327.

Kumar, M. B., & Sathiya, P. (2020). Methods and materials for additive manufacturing: A critical review on advancements and challenges. *Thin-Walled Structures, 159*, 107228. https://doi.org/10.1016/j.tws.2020.107228.

Kumar, R., Kumar, M., & Chohan, J. S. (2021). Material-specific properties and applications of additive manufacturing techniques: A comprehensive review. *Bulletin of Materials Science, 44*(3). https://doi.org/10.1007/s12034-021-02364-y.

Kumar, R., Kumar, M., Chohan, J. S., & Kumar, S. (2021). Overview on metamaterial: History, types and applications. *Materials Today Proceedings, 56*, 3016–3024. https://doi.org/10.1016/j.matpr.2021.11.423.

Kumar, R., Kumar, R., Kumar, S., & Goyal, N. (2020). Trending applications and mechanical properties of 3D printing: A review. *I Manager Journal on Mechanical Engineering, 11*(1), 22–39. https://doi.org/10.26634/jme.11.1.17627.

Kumar, R., & Kumar, S. (2020). Trending applications of 3D printing: A study. *Asian Journal of Engineering and Applied Technology, 9*(1), 1–12. https://doi.org/10.51983/ajeat-2020.9.1.1085.

Kumar, R., & Kumar, S. (2022a). Overview of 3D and 4D printing techniques and their emerging applications in medical sectors. *Current Materials Science, 16*(2), 143–170. https://doi.org/10.2174/2666145416666221019105748.

Kumar, R., & Kumar, S. (2022b). *Overview of 3D-printing technology. Historical, types, applications, and materials. CRC Press eBooks*265–289. https://doi.org/10.1201/9781003301066-15.

Kumar, R., Kumar, S., Kumar, M., & Luthra, G. (2022). *Conventional and 3D printing technology for the manufacturing of metal-matrix composite. CRC Press eBooks*197–223. https://doi.org/10.1201/9781003194897-9.

Kumar, R., Kumar, M., Chohan, J. S., & Kumar, S. (2022). Effect of process parameters on surface roughness of 316L stainless steel coated 3D printed PLA parts. *Materials Today Proceedings, 68*, 734–741. https://doi.org/10.1016/j.matpr.2022.06.004.

Kumar, S., Kushawaha, M. K., & Kumar, R. (2022). Comprehensive study on additive manufacturing process methods, cost comparison, challenges and industrial applications.

I-manager's. *Journal on Future Engineering and Technology, 17*(4), 31. https://doi.org/10.26634/jfet.17.4.18909.

Laleh, M., Sadeghi, E., Revilla, R. I., Chao, Q., Haghdadi, N., Hughes, A. E., Xu, W., De Graeve, I., Qian, M., Gibson, I., & Tan, M. Y. (2022). Heat treatment for metal additive manufacturing. *Progress in Materials Science, 133*, 101051. https://doi.org/10.1016/j.pmatsci.2022.101051.

Lantada, A. D., & Morgado, P. L. (2012). Rapid prototyping for biomedical engineering: current capabilities and challenges. *Annual Review of Biomedical Engineering, 14*(1), 73–96. https://doi.org/10.1146/annurev-bioeng-071811-150112.

Lee, J., An, J., & Chua, C. K. (2017). Fundamentals and applications of 3D printing for novel materials. *Applied Materials Today, 7*, 120–133. https://doi.org/10.1016/j.apmt.2017.02.004.

Levy, G. N., Schindel, R., & Kruth, J. (2003). Rapid manufacturing and rapid tooling with layer manufacturing (lm) technologies, state of the art and future perspectives. *CIRP Annals, 52*(2), 589–609. https://doi.org/10.1016/s0007-8506(07)60206-6.

Lifton, V. A., Lifton, G., & Simon, S. (2014). Options for additive rapid prototyping methods (3D printing) in MEMS technology. *Rapid Prototyping Journal, 20*(5), 403–412. https://doi.org/10.1108/rpj-04-2013-0038.

Liu, Q., Leu, M. C., & Schmitt, S. M. (2005). Rapid prototyping in dentistry: Technology and application. *The International Journal of Advanced Manufacturing Technology, 29*(3–4), 317–335. https://doi.org/10.1007/s00170-005-2523-2.

Liu Z., Zhang M., Bhandari B. and Wang Y. 2017 Trends Food Sci. Technol. 2-12.

Mehrpouya, M., Dehghanghadikolaei, A., Fotovvati, B., Vosooghnia, A., Emamian, S. S., & Gisario, A. (2019). The potential of additive manufacturing in the smart factory Industrial 4.0: A review. *Applied Sciences, 9*(18), 3865. https://doi.org/10.3390/app9183865.

Mekonnen, B. G., Bright, G., & Walker, A. (2016). A study on state of the art technology of laminated object manufacturing (LOM). *Lecture notes in mechanical engineering*, 207–216. https://doi.org/10.1007/978-81-322-2740-3_21.

Mikler, C. V., Chaudhary, V., Borkar, T., Soni, V., Jaeger, D., Chen, X., Contieri, R., Ramanujan, R. V., & Banerjee, R. (2017). Laser additive manufacturing of magnetic materials. *JOM, 69*(3), 532–543. https://doi.org/10.1007/s11837-017-2257-2.

Mohamed, O. A., Masood, S. H., & Bhowmik, J. L. (2017). Experimental investigation of time-dependent mechanical properties of PC-ABS prototypes processed by FDM additive manufacturing process. *Materials Letters, 193*, 58–62. https://doi.org/10.1016/j.matlet.2017.01.104.

Murmura, F., & Bravi, L. (2017). Additive manufacturing in the wood-furniture sector. *Journal of Manufacturing Technology Management, 29*(2), 350–371. https://doi.org/10.1108/jmtm-08-2017-0175.

Nampoothiri, K. M., Nair, N. R., & John, R. P. (2010). An overview of the recent developments in polylactide (PLA) research. *Bioresource Technology, 101*(22), 8493–8501. https://doi.org/10.1016/j.biortech.2010.05.092.

Ngo, T. D., Kashani, A., Imbalzano, G., Nguyen, K. T., & Hui, D. (2018). Additive manufacturing (3D printing): A review of materials, methods, applications and challenges. *Composites Part B Engineering, 143*, 172–196. https://doi.org/10.1016/j.compositesb.2018.02.012.

Nune, K. C., Misra, R. D. K., Gaytan, S. M., & Murr, L. E. (2014). Interplay between cellular activity and three-dimensional scaffold-cell constructs with different foam structure processed by electron beam melting. *Journal of Biomedical Materials Research. Part A, 103*(5), 1677–1692. https://doi.org/10.1002/jbm.a.35307.

Olivier, D., Travieso-Rodriguez, J. A., Borros, S., Reyes, G., & Jerez-Mesa, R. (2017). Influence of building orientation on the flexural strength of laminated object manufacturing specimens. *Journal of Mechanical Science and Technology, 31*(1), 133–139. https://doi.org/10.1007/s12206-016-1212-4.

Park, S., & Fu, K. (2021). Polymer-based filament feedstock for additive manufacturing. *Composites Science and Technology, 213*, 108876. https://doi.org/10.1016/j.compscitech.2021.108876.

Pereira, T., Kennedy, J. V., & Potgieter, J. (2019). A comparison of traditional manufacturing vs additive manufacturing, the best method for the job. *Procedia Manufacturing, 30*, 11–18. https://doi.org/10.1016/j.promfg.2019.02.003.

Petrova, B., & Jivkov, V. (2024). Application of 3D printing technology in furniture construction. *Materials, 17*(19), 4848. https://doi.org/10.3390/ma17194848.

Petrovic, V., Gonzalez, J. V. H., Ferrando, O. J., Gordillo, J. D., Puchades, J. R. B., & Griñan, L. P. (2010). Additive layered manufacturing: Sectors of industrial application shown through case studies. *International Journal of Production Research, 49*(4), 1061–1079. https://doi.org/10.1080/00207540903479786.

Pine, B. J., II (1993). Making mass customization work. . *Harvard Business Review, 71*(5), 108–111. . https://hbr.org/1993/09/making-mass-customization-work.

Rashid, A. A., Khan, S. A., Al-Ghamdi, S. G., & Koç, M. (2020). Additive manufacturing: Technology, applications, markets, and opportunities for the built environment. *Automation in Construction, 118*, 103268. https://doi.org/10.1016/j.autcon.2020.103268.

Rouf, S., Malik, A., Singh, N., Raina, A., Naveed, N., Siddiqui, M. I. H., & Haq, M. I. U. (2022). Additive manufacturing technologies: Industrial and medical applications. *Sustainable Operations and Computers, 3*, 258–274. https://doi.org/10.1016/j.susoc.2022.05.001.

Sabir, M. I., Xu, X., & Li, L. (2009). A review on biodegradable polymeric materials for bone tissue engineering applications. *Journal of Materials Science, 44*(21), 5713–5724. https://doi.org/10.1007/s10853-009-3770-7.

Samal, S. K., Vishwanatha, H. M., Saxena, K. K., Behera, A., Nguyen, T. A., Behera, A., Prakash, C., Dixit, S., & Mohammed, K. A. (2022). 3D-printed satellite brackets: Materials, manufacturing and applications. *Crystals, 12*(8), 1148. https://doi.org/10.3390/cryst12081148.

Sapkota, A., Ghimire, S. K., & Adanur, S. (2024). A review on fused deposition modeling (FDM)-based additive manufacturing (AM) methods, materials and applications for flexible fabric structures. *Journal of Industrial Textiles, 54*. https://doi.org/10.1177/15280837241282110.

Schmidt, M., Merklein, M., Bourell, D., Dimitrov, D., Hausotte, T., Wegener, K., Overmeyer, L., Vollertsen, F., & Levy, G. N. (2017). Laser based additive manufacturing in industry and academia. *CIRP Annals, 66*(2), 561–583. https://doi.org/10.1016/j.cirp.2017.05.011.

Shakor, P., Nejadi, S., Paul, G., & Malek, S. (2019). Review of emerging additive manufacturing technologies in 3D printing of cementitious materials in the construction industry. *Frontiers in Built Environment, 4*. https://doi.org/10.3389/fbuil.2018.00085.

Sidhu, H. S., & Kumar, S. (2020). Experimental investigation on design and analysis of prosthetic leg. *Journal of Xidian University, 14*(5). https://doi.org/10.37896/jxu14.5/491.

Da Silveira, G., Borenstein, D., & Fogliatto, F. S. (2001). Mass customization: Literature review and research directions. *International Journal of Production Economics, 72*(1), 1–13. https://doi.org/10.1016/s0925-5273(00)00079-7.

Singh, S., Ramakrishna, S., & Berto, F. (2019). 3D Printing of polymer composites: A short review. *Material Design & Processing Communications, 2*(2)https://doi.org/10.1002/mdp2.97.

Sinha, A., Swain, B., Behera, A., Mallick, P., Samal, S. K., Vishwanatha, H. M., & Behera, A. (2022). A review on the processing of aero turbine blade using 3D print techniques. *Journal of Manufacturing and Materials Processing, 6*(1), 16. https://doi.org/10.3390/jmmp6010016.

Srivastava, M., Rathee, S., Patel, V., Kumar, A., & Koppad, P. G. (2022). A review of various materials for additive manufacturing: Recent trends and processing issues. *Journal of Materials Research and Technology, 21*, 2612–2641. https://doi.org/10.1016/j.jmrt.2022.10.015.

Sun, J., Zhou, W., Huang, D., & Yan, L. (2018). 3D Food Printing: Perspectives. In T. Gutiérrez (Ed.). *Polymers for Food Applications* Cham: Springer. https://doi.org/10.1007/978-3-319-94625-2_26.

Tay, Y. W. D., Panda, B., Paul, S. C., Mohamed, N.a N., Tan, M. J., & Leong, K. F. (2017). 3D printing trends in building and construction industry: A review. *Virtual and Physical Prototyping, 12*(3), 261–276. https://doi.org/10.1080/17452759.2017.1326724.

Travitzky, N., Bonet, A., Dermeik, B., Fey, T., Filbert-Demut, I., Schlier, L., Schlordt, T., & Greil, P. (2014). Additive manufacturing of ceramic-based materials. *Advanced Engineering Materials, 16*(6), 729–754. https://doi.org/10.1002/adem.201400097.

Tseng, M. M., Jiao, J., & Merchant, M. E. (1996). Design for mass customization. *CIRP Annals, 45*(1), 153–156. https://doi.org/10.1016/s0007-8506(07)63036-4.

Tümer, E. H., & Erbil, H. Y. (2021). Extrusion-based 3D printing applications of PLA composites: A review. *Coatings, 11*(4), 390. https://doi.org/10.3390/coatings11040390.

Uhlmann, E., Kersting, R., Klein, T. B., Cruz, M. F., & Borille, A. V. (2015). Additive manufacturing of titanium alloy for aircraft components. *Procedia CIRP, 35*, 55–60. https://doi.org/10.1016/j.procir.2015.08.061.

Unger, A. S., Lewis, R. J., & Gruen, T. (2005). Evaluation of a porous tantalum uncemented acetabular cup in revision total hip arthroplasty. *The Journal of Arthroplasty, 20*(8), 1002–1009. https://doi.org/10.1016/j.arth.2005.01.023.

Upadhyay, M., Sivarupan, T., & Mansori, M. E. (2017). 3D printing for rapid sand casting—A review. *Journal of Manufacturing Processes, 29*, 211–220. https://doi.org/10.1016/j.jmapro.2017.07.017.

Valino, A. D., Dizon, J. R. C., Espera, A. H., Chen, Q., Messman, J., & Advincula, R. C. (2019). Advances in 3D printing of thermoplastic polymer composites and nanocomposites. *Progress in Polymer Science, 98*, 101162. https://doi.org/10.1016/j.progpolymsci.2019.101162.

Vandenbroucke, B., & Kruth, J. (2007). Selective laser melting of biocompatible metals for rapid manufacturing of medical parts. *Rapid Prototyping Journal, 13*(4), 196–203. https://doi.org/10.1108/13552540710776142.

Ventola, C.L. (2014). Medical Applications for 3D Printing: Current and Projected Uses. PubMed. https://pubmed.ncbi.nlm.nih.gov/25336867.

Voon, S. L., An, J., Wong, G., Zhang, Y., & Chua, C. K. (2019). 3D food printing: A categorised review of inks and their development. *Virtual and Physical Prototyping, 14*(3), 203–218. https://doi.org/10.1080/17452759.2019.1603508.

Wang, F., & Wang, F. (2017). Liquid resins-based additive manufacturing. *Journal of Molecular and Engineering Materials, 05*(02), 1740004. https://doi.org/10.1142/s2251237317400044.

Wang, M., Song, B., Wei, Q., Zhang, Y., & Shi, Y. (2018). Effects of annealing on the microstructure and mechanical properties of selective laser melted AlSi7Mg alloy. *Materials Science and Engineering A, 739*, 463–472. https://doi.org/10.1016/j.msea.2018.10.047.

Webster, S., Lin, H., Carter, F. M., III, Ehmann, K., & Cao, J. (2021). Physical mechanisms in hybrid additive manufacturing: A process design framework. *Journal of Materials Processing Technology, 291*, 117048. https://doi.org/10.1016/j.jmatprotec.2021.117048.

Wendel, B., Rietzel, D., Kühnlein, F., Feulner, R., Hülder, G., & Schmachtenberg, E. (2008). Additive processing of polymers. *Macromolecular Materials and Engineering, 293*(10), 799–809. https://doi.org/10.1002/mame.200800121.

Williams, J. M., Adewunmi, A., Schek, R. M., Flanagan, C. L., Krebsbach, P. H., Feinberg, S. E., Hollister, S. J., & Das, S. (2005). Bone tissue engineering using polycaprolactone scaffolds fabricated via selective laser sintering. *Biomaterials, 26*(23), 4817–4827. https://doi.org/10.1016/j.biomaterials.2004.11.057.

Wojtyła, S., Klama, P., & Baran, T. (2017). Is 3D printing safe? Analysis of the thermal treatment of thermoplastics: ABS, PLA, PET, and nylon. *Journal of Occupational and Environmental Hygiene, 14*(6), D80–D85. https://doi.org/10.1080/15459624.2017.1285489.

Wong, K. V., & Hernandez, A. (2012). A review of additive manufacturing. *ISRN Mechanical Engineering, 2012*, 1–10. https://doi.org/10.5402/2012/208760.

Wu, H., Fahy, W., Kim, S., Kim, H., Zhao, N., Pilato, L., Kafi, A., Bateman, S., & Koo, J. (2020). Recent developments in polymers/polymer nanocomposites for additive manufacturing. *Progress in Materials Science, 111*, 100638. https://doi.org/10.1016/j.pmatsci.2020.100638.

Xin, Y., Wu, J., Yao, Z., Guan, Q., Aleman, A., & Luo, Y. (2017). The relationship between personality and the response to acute psychological stress. *Scientific Reports, 7*(1)https://doi.org/10.1038/s41598-017-17053-2.

Yakout, M., Elbestawi, M., & Veldhuis, S. C. (2018). A review of metal additive manufacturing technologies. *Diffusion and Defect Data, Solid State Data. Part B, Solid State Phenomena/Solid State Phenomena, 278*, 1–14. https://doi.org/10.4028/www.scientific.net/ssp.278.1.

Yang, J., Li, B., Liu, J., Tu, Z., & Wu, X. (2024). Application of additive manufacturing in the automobile industry: A mini review. *Processes, 12*(6), 1101. https://doi.org/10.3390/pr12061101.

Yang, L., Hsu, K., Baughman, B., Godfrey, D., Medina, F., Menon, M., & Wiener, S. (2017). Additive manufacturing of metals: The technology, materials, design and production. *Springer series in advanced manufacturing*. https://doi.org/10.1007/978-3-319-55128-9.

Yap, C. Y., Chua, C. K., Dong, Z. L., Liu, Z. H., Zhang, D. Q., Loh, L. E., & Sing, S. L. (2015). Review of selective laser melting: Materials and applications. *Applied Physics Reviews, 2*(4)https://doi.org/10.1063/1.4935926.

Zhang, Z., Demir, K. G., & Gu, G. X. (2019). Developments in 4D-printing: A review on current smart materials, technologies, and applications. *International Journal of Smart and Nano Materials, 10*(3), 205–224. https://doi.org/10.1080/19475411.2019.1591541.

Zhou, L., Miller, J., Vezza, J., Mayster, M., Raffay, M., Justice, Q., Tamimi, Z. A., Hansotte, G., Sunkara, L. D., & Bernat, J. (2024). Additive manufacturing: A comprehensive review. *Sensors, 24*(9), 2668. https://doi.org/10.3390/s24092668.

Additive manufacturing of smart polymeric materials: recent developments and future perspectives

4

Yan Huang[1,2], *Ke Zhang*[1,2], *Zhangpeng Li*[1,2,3,4], *and Jinqing Wang*[1,2]
[1]State Key Laboratory of Solid Lubrication, Lanzhou Institute of Chemical Physics, Chinese Academy of Sciences, Lanzhou, P.R. China, [2]Center of Materials Science and Optoelectronics Engineering, University of Chinese Academy of Sciences, Beijing, P.R. China, [3]Yantai Zhongke Research Institute of Advanced Materials and Green Chemical Engineering, Yantai, P.R. China, [4]Qingdao Center of Resource Chemistry & New Materials, Qingdao, P.R. China

4.1 Introduction

Intelligent manufacturing is currently a burgeoning trend in the development of new materials and devices, promoting the research, development, and application of smart materials. One such material is the smart polymer, which has the function of self-adaptation and self-regulation similar to natural biological systems. These properties enable dynamic adjustments in physical and chemical responses to the stimulation of the external environment (Wang et al., 2023). The stimuli that drive these responses include temperature (Luo et al., 2024), pH (Ma, Zhang, Javed, et al., 2023), mechanical force (Li et al., 2019), light (Shao et al., 2023), magnetic field (Lv et al., 2024), electric field (Alkahtani et al., 2023), enzymes (Xing et al., 2023), etc. Under the corresponding signals, the smart polymer undergoes changes in phase (Wang, Zhao, et al., 2023), shape (Xie et al., 2022), volume (Seo et al., 2020), color (Xu & Feringa, 2023), etc. Consequently, it can be applied in fields such as biomedicine (Kumar et al., 2023), sensors (Savastru et al., 2023), intelligent bionics (Gu, Bi, et al., 2023), aerospace (Wang et al., 2022), and other fields (Hu et al., 2012; Lai, 2023; Sayyed et al., 2018). However, traditional manufacturing technology has limitations such as low preparation efficiency, complex preparation process, low material utilization rate, and simplified product performance, which no longer meets the requirements of the rapid development of intelligent manufacturing on the structure and performance of smart polymers (BG et al., 2023; Detamornrat et al., 2022; Mi et al., 2018).

AM technology, also known as 3D printing, is defined as "process of joining materials to make parts from 3D model data, usually layer upon layer, in contrast to subtractive manufacturing and formative manufacturing methodologies" (ISO/ASTM 52900:2015(en), 2015). The emergence of AM technology has brought great innovation to the traditional manufacturing field (Khan et al., 2021). It offers the

Additive Manufacturing of Functional Polymers and Nanocomposites. DOI: https://doi.org/10.1016/B978-0-443-26516-7.00004-3
© 2025 Elsevier Ltd. All rights are reserved, including those for text and data mining, AI training, and similar technologies.

advantages of rapid manufacturing of complex structures in single pieces and small batches (Crapnell et al., 2023; Montanari et al., 2023), and solves the limitations of difficult or impossible processing of traditional material subtractive manufacturing technologies. Furthermore, its manufacturing cost does not increase greatly with the increase in the complexity of parts (Dong et al., 2023). Despite its potential advantages, AM technology, being an emerging and transformative technology, also presents challenges related to the flexibility and insufficient environmental adaptability of the printed materials (Subash & Kandasubramanian, 2020).

4D printing technology is based on the development of smart materials and AM technology (Gardan, 2019; Kumar et al., 2022). 4D printing augments the time dimension compared to 3D printing. This allows the shape of the physical structure can change with the external environmental stimulation (Momeni & Ni, 2020; Pingale et al., 2023). The emergence of 4D printing makes it possible to prepare 3D smart material structures with arbitrary complex shapes, and further expands the application range of smart materials. At present, research momentum on 4D printing is booming, and it has shown significant potential in fields such as biomedical equipment (Arif et al., 2022; Chiesa et al., 2023; Faruque et al., 2023), soft robotics

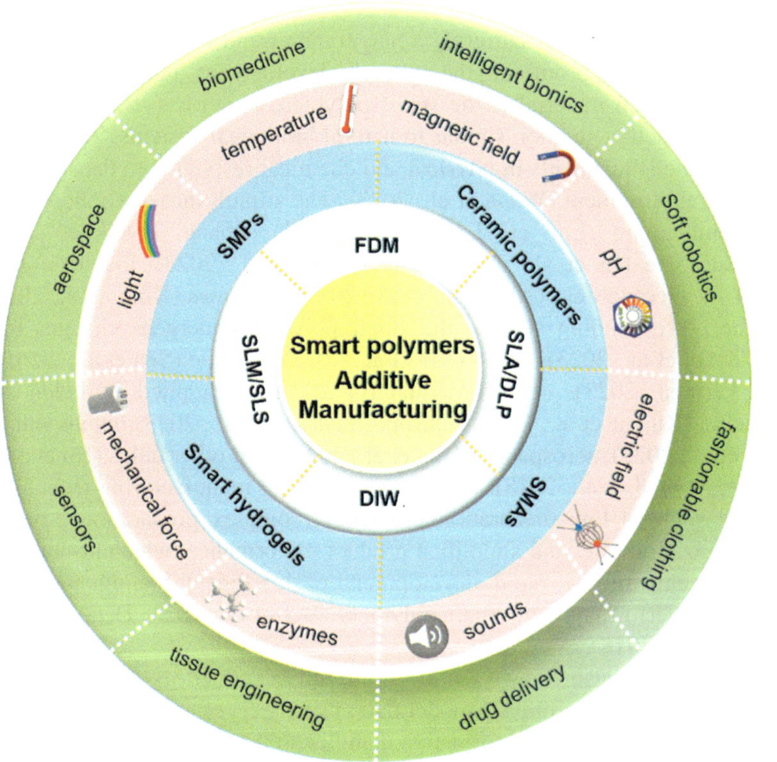

Figure 4.1 Preparation techniques, material classification, response conditions, and applications of smart polymers for AM.

(Soleimanzadeh et al., 2023; Zhang et al., 2023), tissue engineering (Chen, Su, et al., 2023; Chen, Wang, et al., 2023; Zeenat et al., 2023), aerospace (Liu et al., 2023; Ntouanoglou et al., 2018), and other fields (Ghazal et al., 2023; Raina et al., 2021; Zouboulis et al., 2022).

As the demand for materials with tailored properties continues to grow, the synergistic integration of AM and smart polymers emerges as a promising avenue for innovation across various domains. As shown in Fig. 4.1, the technology and latest research findings in intelligent polymer AM, the classification and preparation of intelligent materials are summarized. The application direction and challenges of intelligent polymer AM are outlined, as well as the potential trajectories for future advancements in this dynamic field are envisaged.

4.2 Classification of 4D printing techniques

At present, most 4D printing technologies are built on the basis of 3D printing. In this section, the latest applications and significance of different printing technologies in 4D printing are discussed.

4.2.1 *Fused deposition modeling (FDM)*

FDM is a kind of forming technology in which thermoplastic polymer materials are fused and extruded and piled up layer by layer to produce complex structure parts, as shown in Fig. 4.2A. This technology has specific requirements for the mechanical properties, thermal properties, melt viscosity, shrinkage, and so on of the printed polymer material. The main operation process involves heating the molten material using the printing nozzle, which then moves along the section profile of the part and the filling track. As the molten material is extruded into a filamentous shape, it rapidly cures and bonds with the surrounding material (Jiang et al., 2020). Each layer is stacked on top of the previous layer, acting as a positioning and support for the current layer. With the increase of height, the area and shape of the lamellar profile will change. When there is a suspended part in the model, it is necessary to design some auxiliary structures to provide positioning and support for the suspended layer to ensure the smooth realization of the forming process.

Because of its convenient manufacturing process, simple material preparation, and cost effectiveness (Cano-Vicent et al., 2021), FDM can manufacture the required complex parts in a variety of environments. This has led to its emergence as a research hotspot in the fields of industrial manufacturing (Parada et al., 2019), biomedicine (Kökcü et al., 2023), building manufacturing (Esposito Corcione et al., 2018), culture and art (Higueras et al., 2023). In the realm of 4D printing technology, FDM is frequently used to print high performance and shape memory thermosetting plastics. Wang et al. used multivelocity FDM to embed gradient built-in strain into planar precursor patterns to achieve fast and flexible deformation under thermal stimulation, as shown in Fig. 4.2B and C (Wang, Luo, et al., 2023). By systematically optimizing the processing parameters of FDM technology, complex three-dimensional structures can be constructed directly from planar precursor patterns with both

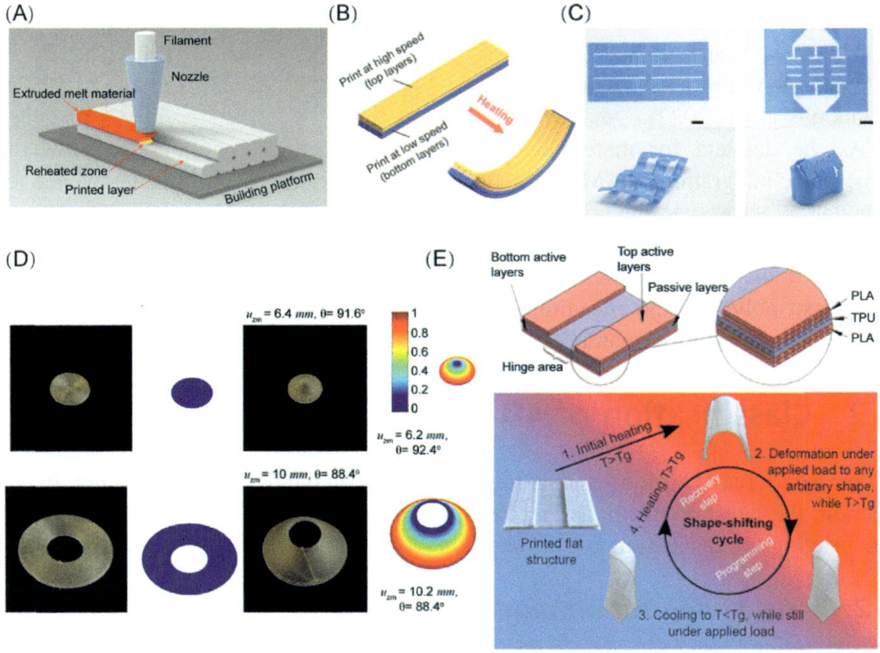

Figure 4.2 (A) Schematic diagram of FDM extrusion system (Hu & Bodaghi, 2023). (B) Printing thermal bending diagrams of PLA samples and (C) 4D printing examples of multibending component structures at different speeds, Scale bar: 10 mm (Wang, Luo, et al., 2023). Copyright from Wiley 2022. (D) Reliability of SMP programming during FDM printing: the configuration of the self-bending structure before and after thermal activation (Hu et al., 2017). Copyright from IOP 2022. (E) Design of multimaterial deformation structure of "hinge" structure (Kačergis et al., 2019).

flexibility and high fidelity. This significantly reduces the construction time and process complexity, making the 4D printing method more efficient and economical.

Hu et al. (2017) used FDM as a filament-based printing method for programming shape memory polymers (SMPs) during the deposition of materials. This 4D printing method is capable of fabricating complex polymer structures with self-bending characteristics and is expected to be a tool for fabricating and simultaneously programming SMP structures in future studies. Because FDM is a layer-by-layer assembly technique, SMP materials are selected for programming during the material deposition process. A prestrained planar two-dimensional structure is manufactured in the manufacturing process; one-dimensional to two-dimensional deformation is achieved through a self-bending mechanism, and then transformed into a three-dimensional complex structure by simple heating through free strain recovery (Fig. 4.2D). The prestrain value depends directly on the mechanical forcing stage.

The influence of FDM process parameters on the transformation of deformed structures in 4D printing was also explored (Kačergis et al., 2019). FDM allows structures to be preprogrammed by changing the print parameters during printing, and they can be manipulated to obtain various curvatures. In Kačergis's study, 4D-printed structures made of polylactic acid (PLA) and thermoplastic polyurethane (TPU) were investigated. These structures, initially printed flat with an FDM-type 3D printer, bend into a U-shape when the environmental temperature higher than T_g (higher than 60°C) (Kačergis et al., 2019). As shown in Fig. 4.2E, structures in U shape can be manually deformed or twisted into arbitrary shape and can be cooled down to maintain that new shape. Repeated heating above T_g will cause the structure to deform back to its original U-shape. The influence of printing parameters on the U-shape (bending angle) of the structure was experimentally investigated. By changing these parameters, a multimaterial deformation structure can be designed. The combination of these parameters provides possibilities for the creation of larger structures using different programming hinges.

However, there are also some limitations of FDM, such as insufficient strength of the printed parts (Wu et al., 2018), poor surface finish (Chohan & Singh, 2017), the printed object has low interlayer adhesion (Chen, Xu, et al., 2022), and other problems, which are still important challenges restricting the further development of this technology.

4.2.2 Direct ink writing (DIW)

Similar to FDM, DIW is an extrusion-based printing technology that uses a special nozzle to continuously extrude functional slurry, stacking it layer by layer to obtain specific geometries (Shahzad & Lazoglu, 2021). The main difference between the two is that FDM requires heating to melt the printing wire, while DIW can print liquid materials with viscosity at room temperature. The outstanding advantages of DIW technology include the lower material requirements (Yang et al., 2021), with high sensitivity (Guo et al., 2019) and dimensional controllability of the periodic structure (Zhang et al., 2020). The ink based on DIW 4D printing technology has shear thinning properties and fast pseudoplastic expansion recovery properties. So far, DIW has successfully printed various smart polymer materials (Mazeeva et al., 2023), such as SMPs (Yang et al., 2023), liquid crystal elastomers (Zhang et al., 2019), stimulation-response hydrogels (Narupai et al., 2021), and more (Hao et al., 2021; Revelo & Colorado, 2018).

At present, although DIW technology has been greatly developed, there are still some limitations. Compared with other printing technologies, manufacturing efficiency will be affected due to the low printing speed and poor bonding between interfaces resulting from the layer-by-layer deposition process. Therefore, in the future, the overall design of the equipment, the functionality of the hardware and software need to be further studied to improve the printing accuracy and efficiency.

4.2.3 Stereolithography (SLA)/digital light processing (DLP)

Both SLA and DLP are liquid resin-based printing technologies. SLA is the world's first printing technology, which prints objects on the basis of light curing reaction, selectively irradiates the liquid surface through ultraviolet light, and moves the lifting platform accordingly, so that the resin can be cured layer by layer (Razzaq et al.,

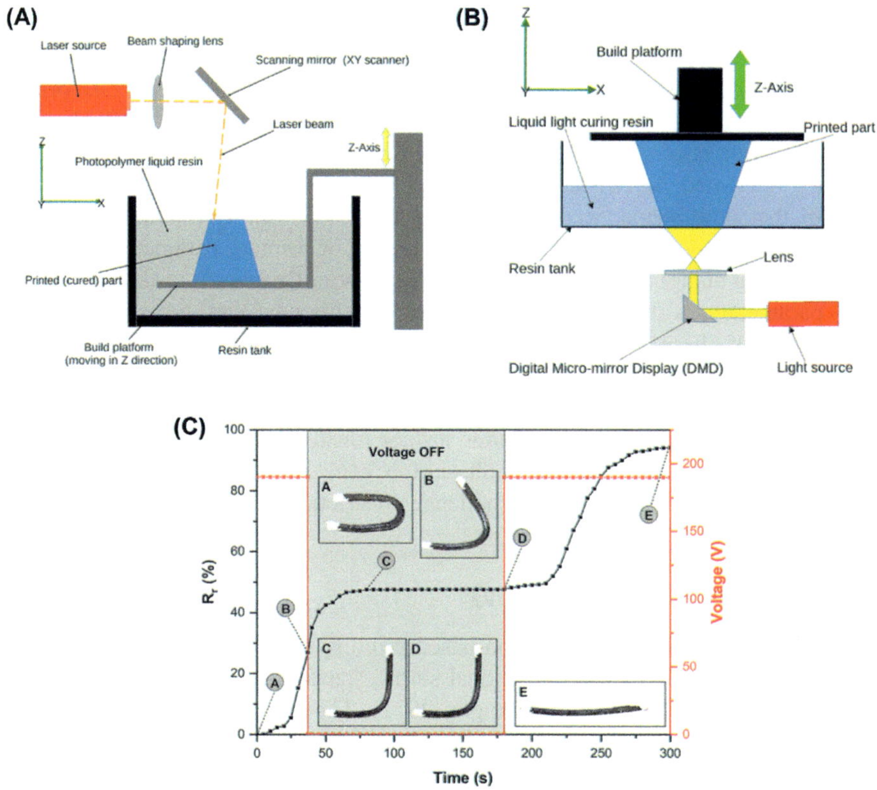

Figure 4.3 Schematic representation of the (A) SLA and (B) DLP (Ramezani & Ripin, 2023). (C) Shape recovery of PS by turning on and off voltage drives (Cortés et al., 2021).

2022), as shown in Fig. 4.3A. Currently, the SLA technology is relatively mature. It is widely used because of its high degree of automation in the molding process, high molding accuracy, excellent surface finish, and ability to manipulate complex structures (Fu et al., 2019; Yao et al., 2020). By selecting suitable materials, the process can produce polymers with shape memory properties. Zhao et al. (2018) prepared a shape memory polyurethane using a photopolymer by SLA technique. Initially, a polyurethane acrylate was synthesized, and then photoinitiated with epoxy acrylate and isobornyl acrylate to obtain a photopolymer with high ultraviolet (UV) curing activity and high printing accuracy. The printed object exhibited excellent shape memory and good mechanical properties. Lu et al. (2017) developed a field-assisted projection stereolithography technique in which magnetic particles are deposited in a liquid polymer and guided by an applied magnetic field to form a specific pattern, which finally solidifies into the desired distribution pattern. In this way, a variety of smart polymer composites can be printed.

Unlike SLA, as illustrated in Fig. 4.3B, DLP technology uses a high-resolution digital processor projector instead of the SLA laser for layer-by-layer light-curing liquid polymerization. This allows the entire graphic of a layer to be processed in a single exposure before forming a three-dimensional object, thereby increasing the efficiency of printing (Sun et al., 2021). Using DLP technology can also print photocurable SMP, manufacturing 4D printing structures with complex geometry and high resolution (Alam et al., 2023; Shin & Chang, 2021; Zhang, Li et al., 2021). Zhang, Li et al. (2021) developed a large-deformation-fatigue-resistant shape memory polymer photosensitive resin for DLP printing with ultrahigh mechanical properties. It can meet the comprehensive requirements of high fracture strain, high fatigue resistance, high load-bearing capacity, and adaptability to advanced manufacturing processes of advanced smart materials for various 4D printing engineering applications. But converting liquid resins into solids with extremely high resolution using photochemical processes remains a challenge. Cortés et al. (2021) reported a novel electrically activated shape memory polymer composites (SMPCs) consisting of a photocured and DLP-printable SMP matrix embedded with multiwalled carbon nanotubes (Fig. 4.3C). This method can print SMPCs directly through DLP technology, and has significant electrical trigger response, which provides a good solution for remote and selectively controllable intelligent device 4D printing.

SLA and DLP 3D printing technologies require raw materials to be photosensitive resins, which are less diverse and still difficult to replace existing engineering plastics. Although DLP technology has high accuracy, it is still difficult to print some large-size objects, and it is mainly used in fields such as dentistry (Han, Liu, et al., 2023) and jewelry casting (Saorin et al., 2021).

4.2.4 Selective laser sintering/melting (SLS/SLM)

SLS and SLM are the predominant application technologies in metal printing, and their molding raw materials are mainly in powder form (Papadakis et al., 2018). As shown in Fig. 4.4A, SLS is based on the basic principle of high-temperature sintering of powder materials under laser irradiation, with precise positioning achieved by a computer-controlled light source positioning device, followed by layer-by-layer sintering (Launhardt et al., 2016). SLM uses a high-energy, high-power laser beam to melt the powder at the surface and then makes the melted powder solidify and superimpose. The main distinction between SLS and SLM lies in the fact that while

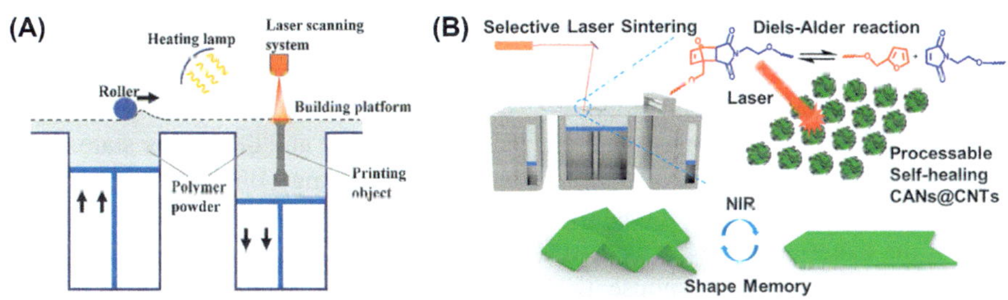

Figure 4.4 (A) Schematic illustration of SLS (Sun et al., 2020). Copyright from Elsevier 2020. (B) SLS 4D printing dynamic polymer example (Ouyang et al., 2022). Copyright from American Chemical Society 2019.

sintering in SLS, the strike beam completely melts the powder in SLM. Thus, the density and mechanical properties of SLM forming parts are better. The SLS is typically used to print polymers and composites, while SLM is more suitable for printing complex metal products.

The 4D printing smart materials using SLS is still a huge challenge due to the limited development of suitable materials (Ouyang et al., 2022). Ouyang et al. (2022) printed a polyurethane with a covalent adaptive network via SLS technology; the introduction of carbon nanotubes (CNTs) provides good photothermal and electrical responsiveness to composites, enabling it to exhibit near-infrared (NIR)-triggered self-healing and shape memory functions (Fig. 4.4B). SLM technology is usually used to synthesize some smart alloys. de Wild et al. (2020) used SLM to fabricate a biologically compatible NiTi shape memory alloy, resulting in an intelligently programmed structure with custom properties. This allows for spatiotemporal programming of 3D-printed responsive surgical instruments or implants.

4.3 Materials for 4D printing

Smart materials are an essential part of 4D printing, which can self-evolve under different environmental conditions. In other words, smart materials are able to respond to external stimuli with the ability to self-perceive, self-repair, self-drive, self-diagnose, and deform (Gobin et al., 1996; Montoya et al., 2023; Nicolay et al., 2023; Wang, Lee, et al., 2023). The current 4D printing smart polymeric materials can be categorized into polymers and hydrogels. Each of these categories will be discussed.

4.3.1 Polymer material

SMPs are currently among the most widely used and diverse smart materials for 4D printing (Yue et al., 2023). Unlike hydrogels, changes in SMPs can be programmed after printing. Compared to shape memory alloys (SMAs), SMPs materials are more processable, possess superior shape memory properties, and require lower manufacturing costs (Alshebly et al., 2021; Carlson & Li, 2020; Monzón et al., 2017).

The shape memory effect of SMPs depends on two coexisting phase transitions. The stationary phase can be formed by cross-linking, partial crystallization and winding of ultrahigh molecular chains, which have great influence on shape memory and recovery of SMPs (Zheng et al., 2019). The reversible phase is usually the phase structure of the reversible transition between the glass state and the rubber state or the partial crystalline phase that produces the reversible change of crystallization and crystallization melting, which has a greater influence on the deformation characteristics of the material (Zhang et al., 2016). Depending on the driving mode, SMPs can be divided into thermal response (Zhang et al., 2023a), magnetic response (Liu, Wang, et al., 2023), electrical response (Zhang et al., 2023b), optical response (Guo et al., 2023), microwave response (Koh & Sutradhar, 2022), and ultrasound responsive (Wang et al., 2022) SMPs materials.

4.3.1.1 Thermal response SMPs

Thermal response SMPs materials are driven by direct heat generated by temperature changes, leading to shape changes. When the temperature surpasses the transition temperature (T_t), the material transitions into a rubber state and exhibits proneness to deformation. When the temperature falls below T_t, the material can maintain its heat-induced shape and reverts to its original state when the temperature rises again (Gopinath et al., 2023). Currently, it is the most widely used material in 4D printing. Gu, Bi et al. (2023) developed a cellulose nanofiber (NF) reinforced polycaprolactone (PCL)/polybutylene adipate-butylene terephthalate (PBAT) thermoresponsive shape memory polymer composite. By introducing CNF, they addressed the issues of PCL's nonfixed initial deformation shape incomplete recovery. Using shape memory effect, climbing plant bionic model, and 3D printing technology, the 4D printing thermal response self-deformation structure is designed. This structure is capable of rapid response and stable winding through thermal induction, presenting great potential for development in other fields such as biomedical devices, soft robotics, and intelligent actuators. In addition to focusing on the deformation behavior and structural design of 4D-printed structures, Chen et al. have developed an UV-curable and thermochromic (UVT) SMPs system to enable color-variable 4D printing (Fig. 4.5A) (Chen et al., 2021). The thermochromic properties of SMPs are endowed by the addition of thermochromic microcapsules. When this material is used as a bionic gripper, it can produce changes in both shape and color, and thus can be used to predetermine the deformation of polymers. Moreover, high-resolution multilayered security patterns printed using this material can switch between visible and invisible states within a certain temperature range. Therefore, this SMPs system holds promising prospects for practical applications in the fields of intelligent anticounterfeiting and bionic devices.

4.3.1.2 Magnetic response SMPs

In addition to the above direct drive that generates heat through temperature change, the others are driven by indirect heat generation. SMPs that can be driven by remote and contactless stimuli have a wider range of applications. Magnetically actuated SMPs are mainly realized by adding nanoparticles with magnetic properties to a thermally responsive SMPs matrix (Wang & Zhao, Luo, et al., 2023). By adding magnetic nanoparticles such as Fe_3O_4 (Gu et al., 2016), Fe_2O_3 (Pourmadadi et al., 2022), NdFeB (Ha et al., 2021), and nickel zinc ferrite (Kumari et al., 2016), the magnetized SMPs generate induced heat under the action of alternating magnetic fields, and deformation occurs when the temperature surpasses the transition temperature of the SMPs matrix. Fe_3O_4 nanoparticles are a kind of material with good biocompatibility and can participate in human metabolism through normal pathways (Du et al., 2023), so they have received high attention in biomedical fields such as magnetothermal therapy (Galarreta-Rodriguez et al., 2023) and drug controlled release (Li, Kong, et al., 2023). Because the magnetic response SMPs can achieve remote controlled local precise heating, and the process is safe, effective, and highly penetrating, it is suitable for noncontact drives in closed environments or

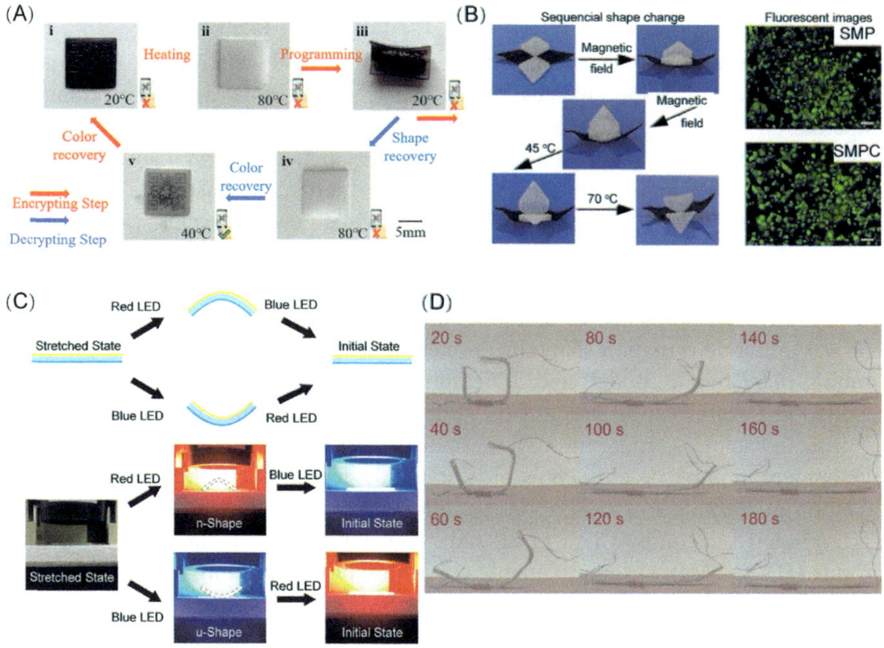

Figure 4.5 (A) The QR code structure is encrypted, decrypted, and re-encrypted through shape–color recovery brought about by temperature changes (Chen et al., 2021). Copyright from American Chemical Society 2021. (B) The 4D-printed multi-SMP composite has both thermal and magnetic response shape change capability and good biocompatibility (Wan et al., 2022). Copyright from Elsevier 2022. (C) Bending behavior of the multicolor sample (Jeong et al., 2020). (D) Sequential shape recovery process of E-SMPs samples under current stimulation (Zhang et al., 2023b). Copyright from Elsevier 2023.

confined spaces (such as the human body) (Liu, et al., 2023; Sánchez et al., 2022). Magnetic braking is usually combined with other braking methods to flexibly achieve multistimulus control and multistage deformation in SMPs systems. For example, Wan et al. optimized the composition of SMPs using DIW technology to obtain multiple shape memory effects, and added Fe_3O_4 to prepare poly (D,L-lactide-co-trimethylene carbonate) (PLMC)/poly (trimethylene carbonate) (PTMC)/Fe_3O_4 composite materials (Wan et al., 2022). This approach simultaneously endows the material's thermal response, remote magnetic response, and greater freedom of shape change (Fig. 4.5B), offering an example for the development of selective multi-stimulus materials for 4D printing. However, the addition of magnetic nanoparticles such as Fe_3O_4 often affects the inherent strength and toughness of SMPs materials, and the sacrifice of the mechanical properties of the materials will limit its further practical application. To address this issue, Gu et al. designed a sacrificial network of "hydrogen bond-metal-phenolic." By grafting epoxy groups onto PCL and subsequently coating it with polydopamine, Fe_3O_4 was uniformly dispersed in the SMP matrix (Gu, Ji, et al., 2023). The resulting composite exhibited a tensile strength of 18.98 MPa, an elongation at break of 897%, and a toughness of 121.38 MJ/m^3. It can realize shape programming and recovery by remote operation of NIR light and magnetic stimulation, which has potential application in intelligent soft robot handling of hazardous chemicals.

4.3.1.3 Light response SMPs

Similar to magnetic drive, light-driven SMPs also have the advantages of controllable operating space and precise local control. The introduction of photothermal fillers into the SMPs matrix facilitates the absorption of light energy, which subsequently convert into heat energy to induce the shape memory process (Zhang et al., 2019). Alternatively, by adding light-sensitive particles, the corresponding cross-linking reaction is generated within the material after absorbing light energy, so as to achieve the macroscopic shape change of the material (Fang et al., 2018). Photothermal fillers such as gold nanoparticles (Wang & Ding, 2018), black phosphorus (BP) (Chen, Sun, et al., 2022), carbon black (CB) (Liu et al., 2022), CNTs (Li, Li, et al., 2023), and graphene oxide (GO) (Bayar et al., 2023) absorb NIR and convert it into heat to induce the shape memory process. It is worth noting that long-wave NIR is a benign form of energy in the human body, able to penetrate tissue without causing other harm (Jia et al., 2018). In the 4D printing process, light-driven SMPs materials are often preferentially printed using the FDM method. This is because SLA and DLP methods require high-intensity ultraviolet light sources in the printing process. If these photoresponsive groups and photothermal conversion materials are exposed to ultraviolet light for a long time, they can undergo conformational transformations or structural changes, or release a lot of heat, thereby affecting the properties of the material itself and the final printing effect. In Jeong et al.'s (2020) study, remote drives with light were achieved by using color-dependent selective light absorption and heating in multicolor SMPs composites. As seen in Fig. 4.5C, when illuminated with a red light-emitting diode, the material bends, while when illuminated with blue light, the material returns to its flat structure. Different shapes can be obtained by controlling the color of the light and the duration of the illumination. Relevant simulations and analytical calculations are also provided to theoretically model the structural changes of multicolor composites. Cui et al. (2019) team built on this by designing a NIR responsive nanocomposite to create a proof-of-concept 4D-printed brain model, introducing graphene into a thermally responsive SMP matrix to evaluate the ability to control 4D transformation and the feasibility of photothermal stimulation to modulate neural stem cell behavior. The position and time of shape change of the composite model can be controlled by NIR light to produce progressive and targeted shape change. This is the first study to propose remote and dynamically controllable 4D printing technology.

4.3.1.4 Electrical response SMPs

Electroinduced SMPs involve the introduction of metal nanoparticles (Koo et al., 2019), CNTs (Zhang, Chen, et al., 2021), CB (Kang et al., 2023), and other conductive particles into the SMPs material matrix by doping, embedding, and coating. This process forms an internal conductive network, endowing the SMPs with the corresponding conductivity (Rosales et al., 2018). Joule heat is generated when an external current is applied, which induces deformation in the material. Most 4D printing studies on electroresponse-driven SMPs incorporate carbon-based materials into SMPs, typically requiring high drive voltages and long heating times. On the other hand, because the carbon-based material is easy to absorb the energy of ultraviolet light, it affects the formation of SMP matrix to a certain extent. Zhang et al. (2023b) used a double-layer silver nanowire (AgNW) electrothermal layer (ETL) embedded in a photocurable printed SMPs to generate heat through electrical response to achieve shape change and recovery (Fig. 4.5D). The double-layer ETL has more faster and reliable shape recovery, while also avoiding thermal runaway by introducing cooling channels to regulate heat production. These electroinduced SMPs provide a new idea for the development of photocurable 4D printing SMPs in actuator applications. In addition to single shape deformation and control, triple shape polymers have been studied because they can remember two independent shapes during controlled recovery. Razzaq et al. (2023) prepared electroactive triple shape composite materials containing three conductive fillers by melting processing, including polyester urethane (PEU), PLA, and multiwall carbon nanotubes (MWCNTs). In practice, the triple shape effect is achieved by gradually increasing the voltage. This paves the way for the preparation of more complex 4D printing shapes and applications in soft robotics, actuators, and smart textiles.

4.3.2 Hydrogels

Hydrogels used in 4D printing are one of the hot spots in today's research (Champeau et al., 2020). Most of them are polymers with unique structure and excellent properties, thus occupying a rather important position in 4D printing. Intelligent materials are the prerequisite for realizing the intelligent evolution of 4D-printed products. Among these, intelligent hydrogels stand out due to their simple preparation process, high stability, adjustable physical and chemical properties, good biocompatibility, and easy control of response behavior (Bakarich et al., 2015; Liu et al., 2022; Pudkon et al., 2021). These smart hydrogels, also known as stimulus-responsive hydrogels, are capable of changing their physical and chemical properties in response to various exogenous or endogenous stimuli (Fang et al., 2023), such as heat, electricity, light, mechanical force, pH, magnetic field, concentration of biomolecules (Norioka et al., 2017), and so on. The potential applications of hydrogel 4D printing have already been demonstrated in tissue engineering (Gugulothu & Chatterjee, 2023), soft actuator (Li, Yang, et al., 2023), and drug release applications (Dong et al., 2020). Li et al. developed a general-purpose 4D printing technique that, unlike the soluble/dispersible MXene sol pattern/structure in the solvent produced by traditional 3D printing. This technique transforms the MXene sol into a cross-linked MXene hydrogel with good mechanical strength after a self-assembly process induced by simple thermal stimulation (Li et al., 2022). Moreover, the prepared MXene hydrogels showed high capacity of pseudocapacitance storage.

4.3.2.1 Thermally stimulated responsive hydrogels

Among the hydrogels printed in 4D, thermally stimulated responsive hydrogels are most frequently responsible for shape changes. Poly(N-isopropylacrylamide) (PNIPAm) is the most commonly used polymer for the fabrication of thermally printed structures due to its unique ability to respond to temperature (Solis & Czekanski, 2022). This material exhibits at a low critical solution temperature (LCST) similar to that of the human body, with the polymer chain segments being hydrophilic below LCST and hydrophobic crumpling above LCST (Singh et al., 2017). The hydrophilic–hydrophobic transition induces an internal volume change, which is reversible. Owing to its good biocompatibility, it is widely utilized in drug delivery (Wang, Jiang, et al., 2023), bioengineering (Ciarleglio et al., 2023) and other fields. Zu et al. developed a core-shell structure hydrogel capsule using 4D printing method, as shown in Fig. 4.6A (Zu et al., 2022). In this model, the drug serves as the nucleus while the PNIPAM hydrogel acts as the shell. The temperature functions as a

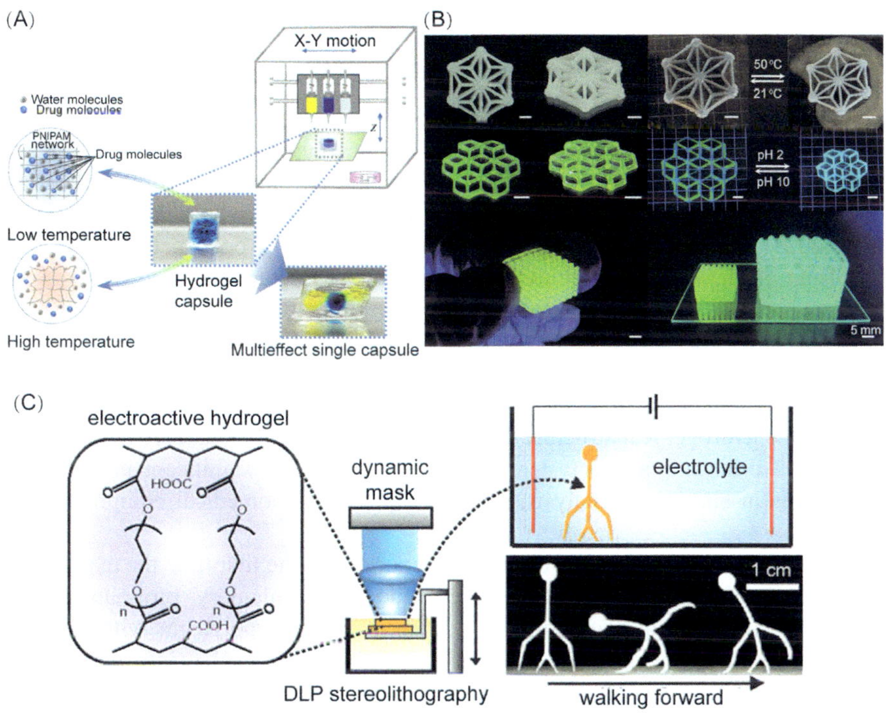

Figure 4.6 (A) Temperature-sensitive controlled drug release process for 4D-printed hydrogel capsules (Zu et al., 2022). Copyright from Zhejiang University Press 2022. (B) Shape or volume changes of hydrogels induced by temperature, pH, and enzymatic degradation of 3D-printed structures (Narupai et al., 2021). Copyright from Wiley 2021. (C) Diagram of the electroactive hydrogel bending or changing shape with electrolyte concentration and the strength of an electric field (Han et al., 2018). Copyright from American Chemical Society 2018.

"switch" controlling the drug release rate; as the temperature rises, the hydrogel network shrinks, and the pore size of the hydrogel decreases, thereby slowing down the drug release rate. This multimaterial extrusion 4D printing technology prepared by the capsule can be loaded with long-term therapeutic drugs, achieving controlled release of drugs.

Heat-driven smart gels are also often used in conjunction with other drives to meet the higher requirements of stimulation hydrogels for 4D printing. Narupai et al. (2021) developed protein-based hydrogels prepared using DIW technology and applied them to programmable structural changes under the action of temperature, pH, or enzymes. As depicted in Fig. 4.6B, temperature and pH responsiveness are introduced into the printed object by N-isopropylacrylamide (NIPAAm) (Temp-Ink) and 2-dimethylaminoethyl methacrylate (DMAEMA) (pH-Ink) to change the material shape reversibly, and irreversible shape is obtained by enzymatic degradation of protein-based hydrogels. In other words, different shapes can be obtained through programming under multiple stimuli of temperature, pH, and enzymes, providing conditions for the preparation of smart biomaterials.

4.3.3 pH stimulated responsive hydrogels

pH-driven hydrogels will undergo structural or chemical alterations due to changes in the pH of the chemical environment, resulting in volume changes (Falahati et al., 2020). For the 4D printing of these pH-responsive hydrogels, raw materials containing acidic groups are generally used, such as polyacrylic acid (Abdullah & Okay, 2023), acrylic acid ester (Selianitis & Pispas, 2023), poly(methacrylic acid), etc. (Teng et al., 2024). Hu et al. (2020) used femtosecond lasers to produce 3D hydrogel samples with nanoscale spatial resolution, ultralow thermal effects, and excellent geometric design. The expansion, contraction, and torsion of hydrogel are realized by the pH-triggered response mechanism, thus forming shape transformation with multiple degrees of freedom. When the pH value of the liquid environment is changed, the hydrogel leaf structure will expand and contract relatively. This material has a small print size and a fast response rate, and holds significant potential for applications in single-cell analysis, micromanipulation, biomedical equipment, and other fields.

In a separate study, Pan and Goto (2023) reported for the first time the exploration of nanoassembly topology-related drivers and shape memory programming in biomimetic 4D printing. Topologically different block copolymers with multiple stimulus responses are designed, and the polymers self-assemble into vesicles for DLP 4D printing. When these polymers are expanded in an acidic solution, the double-layer brake can bend downward to grip and lift objects. Conversely, when expanded in an alkaline solution, their ability to clamp upward motion provides a prototype for 4D printing technology to create unique smart materials.

4.3.3.1 Electroresponsive hydrogels

Under the action of applied electric field, the concentration gradient difference exists between the ions flowing inside the hydrogel and the solution, which leads to the change of osmotic pressure and the macroscopic deformation of the hydrogel (Yang et al., 2017). At present, electroresponsive hydrogels are mainly divided into two types, one is the hydrogels with ionic groups (Miao et al., 2023), and another is the hydrogels with electroactivity by adding conductive particles or intrinsically conductive polymer networks (Enshaei et al., 2022; Li, Yun, et al., 2023). The majority of current research on electrically driven hydrogels for 4D printing is focused on the former type. Zolfagharian et al. printed a polyelectrolyte hydrogel soft actuator using a bioprinter, realizing maximum free bending deformation of the flexible actuator through topology optimization design. They also observed a higher bending deformation rate and driving rate under electrical stimulation (Zolfagharian et al., 2020). Han et al. (2018) used DLP micro-3D printing technology to prepare an electroactive hydrogel. During the 3D printing process, light is shone on a photosensitive solution to turn it into a gel. As depicted in Fig. 4.6C, the hydrogel is immersed in a saline (or electrolyte) solution, and motion is triggered with two thin wires: walking forward, reversing course, grabbing and moving objects. The hydrogel will bend or change shape with the concentration of the brine (or electrolyte) solution and the strength of the electric field. Therefore, the actuator model will move and change shape when activated by electricity.

Similar to shape memory polymers, the 4D printing of photoresponsive and magnetically responsive hydrogels is mainly achieved by adding nanoparticles with specific functions to the polymer network (Hu et al., 2022; Zheng et al., 2020). This aspect will not be further elaborated in this chapter.

4.4 Current and potential applications of 4D printing

4D printing, an emerging manufacturing technology, opens up unprecedented opportunities across various fields by printing smart materials that can self-assemble (Liu et al., 2021), self-repair (Ma, Zhang, Li, et al., 2023), self-deform (Joshi et al., 2023), or respond to external stimuli (Wang & Guo, 2023). Through unremitting efforts, researchers have developed a variety of 4D printing materials, which have a wide range of applications in aerospace (Zhang, Li, et al., 2021), biomedics (Ding et al., 2022), soft robotics (Wang et al., 2023), engineering (Rivera-Tarazona et al., 2022), information storage (Chen et al., 2023), fashion and clothing (Zarek et al., 2016).

In the aerospace field, materials are required to be lightweight properties, high temperature resistance, corrosion resistance, high strength and durability, self-healing capabilities and adaptability (Getachew et al., 2023; Li, Yue, et al., 2023). 4D printing products can realize the rapid prototyping of complex parts and products have multifunctional evolution capabilities. Consequently, it is expected to produce more aerospace parts that meet these requirements through 4D printing. For example, Li et al. were inspired by the dispersal mechanisms of *Viola verecunda* fruit seeds. They developed a catapultusing 4D printing that can achieve high-speed ejection under the logical stimuli such as external force, temperature, light, humidity, or electricity. This breaks the limit of movement speed and is expected to play a key role in the rapid unlocking and deployment of aerospace equipment (Fig. 4.7A) (Li et al., 2023). Smart

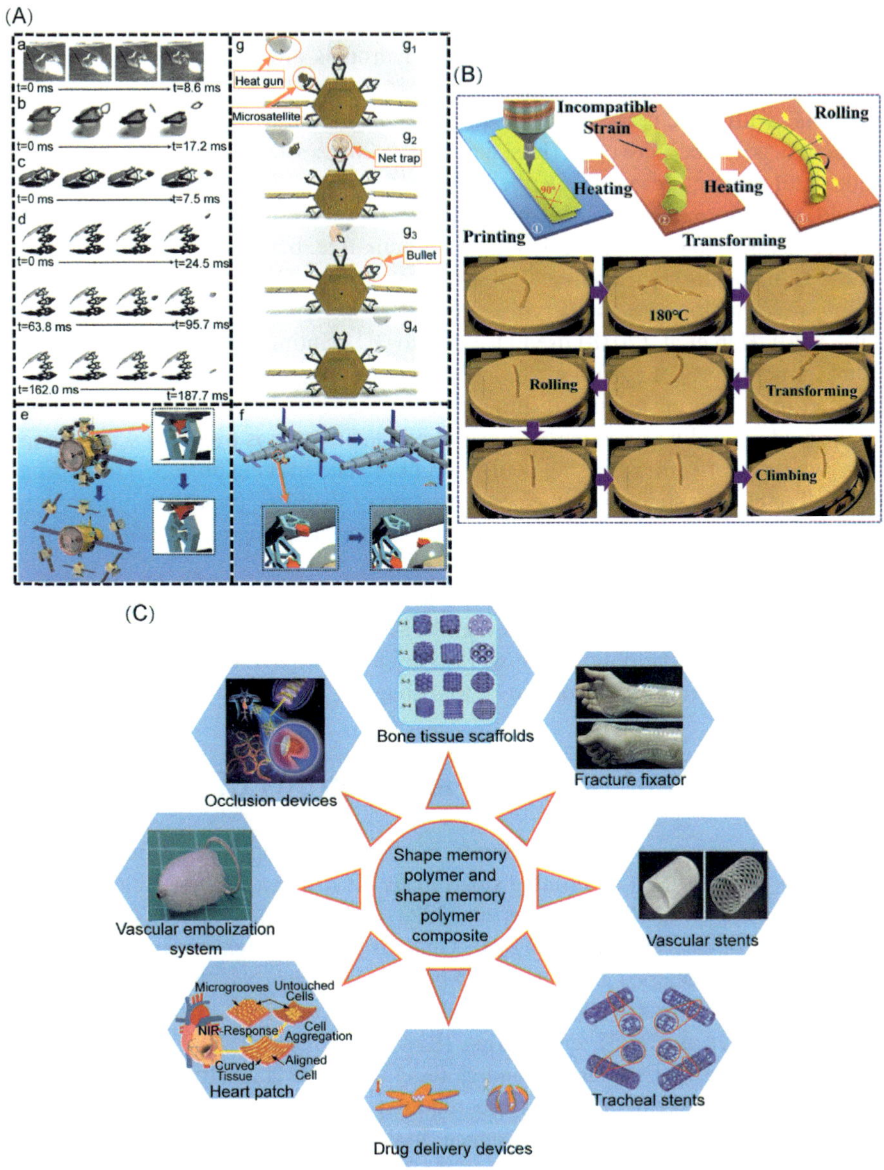

Figure 4.7 (A) Experimental water bath heating trigger, infrared heating trigger, and electrical trigger of a 4D-printed bionic catapult and prospects for aerospace applications (Li et al., 2023). Copyright from Wiley 2023. (B) Tactile perception of 4D printing soft robot (Zhai et al., 2021). Copyright from Elsevier 2021. (C) Biomedical applications of SMP (Zhao et al., 2023). Copyright from Wiley 2022.

materials have the potential to replace actuators and hinges, thereby reducing heavier deformation compared to traditional materials. At present, 4D-printed parts are being more and more widely used in aerospace, but also face challenges such as slow mass production speed, the size limitions imposed by additive manufacturing machines. Researchers are constantly exploring solutions to these problems. In the future, metamaterials designed using 4D printing capable of deforming in response to changing aerodynamic conditions, improving the performance and efficiency of aircraft and spacecraft. The wing shape can be dynamically adjusted during flight to optimize fuel consumption. The 4D-printed materials will also play an increasingly important role in the aerospace field.

The combination of 4D printing and soft robotics breaks through the traditional concept of robot as motor gear coupled with control. The soft robot formed by 4D printing of intelligent materials does not need the complex drive system and transmission structure of the traditional rigid body robot. They can realize the autonomous movement of the robot only by directly giving the external signal stimulation and change, thereby achieving the simplification and high efficiency of the drive (Khalid, Arif, & Ahmed, 2022; Soleimanzadeh et al., 2023). Zhai et al. (2021) used 4D printing to develop a single-material soft robot with unrestricted movement without any other machining procedures (Fig. 4.7B). This liquid crystal elastomer-based robot deforms into a small tube when heated above 160°C, enabling cordless self-walking rolling, and carrying loads up to 40 times its own weight. The robot can be applied to cargo transportation and intelligent exploration of unconventional environments in outer space. Cecchini et al. (2023) created a novel 4D-printed biodegradable soft robot that designed to mimic the shape of a seed and can change its shape in response to changes in humidity, capable of navigating through soil. The device has great potential as a new environment monitoring tool. The application of 4D printing to the direction of soft robot research is in a booming trend, but most of them are in the experimental stage, and it will take time for widespread practical applications. In the future, 4D printing software robots need to improve the speed of movement, the diversity of complex actions, responsiveness and intelligence, and service time, etc. Furthermore, the breakthroughs in 4D printing within the domain of software robots are anticipated to propel the evolution of artificial intelligence.

4D printing materials, especially shape memory materials, have shown application potential in the field of biological devices such as minimally invasive surgery, tissue stents, and orthopedic devices due to their programmable structure, multistimulus drive, deployment adaptability, biodegradability, and biocompatibility (Khalid, Arif & Noroozi, Zolfagharian, et al., 2022; Yan et al., 2023). 4D-printed smart implants can be adapted to the unique anatomy of a patient, ensuring precise placement and minimizing the risk of complications, such as dental implants (Singh et al., 2023), intestinal stents (Lin et al., 2023), vascular stents (Han, Saiding, et al., 2023), etc. (Fig. 4.7C) In addition, 4D-printed smart materials can precisely control the release of drugs, thereby reducing side effects and improving treatment efficiency. Liu, Lou, et al. (2023) pioneered the usage of 4D printing technology to embed the arranged cell sheets on a deformable hydrogel containing biophysical signals and spatially adjustable physical properties. The hydrogel expands with the absorption of water, and the bionic periosteum morphs into a three-dimensional macro structure preset in the 4D printing process. This active macrogeometric change maintains the healing microenvironment and further promotes bone regeneration at the affected area, which can be used in various areas of bone defects. Since the most widely used in the biomedical field is the responsive 4D printing material, future research will likely focus on the improvement of the durability of the stimulus response and the precision of the noncontact spatiotemporal control mode. 4D printing and smart materials will also play an important role in the development of smart products for future biomedical devices.

4.5 Development status

According to the principles of the 4D printing, a variety of materials have been developed and they have applied in various fields. For example, some people will cause irreversible damage of bones due to natural or accidents, which far exceeds the self-healing ability of bone cells. Therefore, it is necessary to customize biocompatible bone materials according to different positions. 4D printing offers a new method to solve some specific situation (Chen et al., 2022). Similarly, 4D printing can also be applied to heart valve tower bridge, blood pressure monitoring, and other areas where surgery is often challenging due to complex structures. Shape memory polymer 4D printing allows us to prepare structures with predetermined shapes after a specific stimulation response in the body, thereby achieving good therapeutic effect. It greatly reduces the disadvantage that the mechanical properties of the postoperative structure are not obvious (Zhao et al., 2023). Furthermore, the reactions of each part of the body during movement are different; some require frequent bending while others may not deform at all. Traditional 3D printing struggles with these tasks, but 4D printing adeptly addresses them. 4D materials have already played a huge role in shape memory polymers as well as soft polymers; it is important to note that hard material (such as ceramics) is difficult to change the shape. Wang, Liu, et al. (2023) prepared two layers ZrO_2 ceramics with different ink solid content through the 4D printing direct writing molding technology, and after sintering, the ceramics can self-shape complex shapes.

4D printing has injected new impetus into the development of biological tissue engineering. Annually, a number of individuals succumb to congenital or acquired diseases (Chen, Wang, et al., 2023). Similar to patients suffering from heart disease, these individuals can improve their condition through surgical interventions such as heart bypass surgery. The application of 4D materials can greatly improve the mechanical properties of heart bypass surgery materials, thereby prolonging their service life and minimizing surgical risks through biomimetic means.

The application of soft robots in the healthcare field has been facilitated by the use of 3D materials, which have enabled the miniaturization of soft robots possible. However, 4D-printed soft materials has paved the way for minimally invasive technology (Ntouanoglou et al., 2018) and is projected to play a crucial role in the field of soft robots in the future. Soft robots can serve as carriers for drugs, greatly improving their curative efficacy. For cancer treatment, soft robots can accurately locate cancer cells and achieve precise strikes. This addresses the current issue of the immune system being compromised after chemotherapy, thereby offering the potential for eradicating cancer.

The emergence of 4D materials not only promotes the innovation of molding processes but also plays a role as a catalyst for the emergence of new materials. Future research and development efforts will focus on creating more intelligent materials with better mechanical properties, self-healing capabilities, and longer service life. The application of 4D printing technology can greatly improve the efficiency of molding and provide more intelligent and adaptive structures in the construction field. With the further deepening of research, 4D printing will play an increasingly role in construction, medical, military, and other fields. At the same time, the vigorous development of 4D printing will also provide a solid foundation for the emergence of 5D and 6D printing technologies.

Despite its rapid development, 4D printing technology still has several shortcomings and bottlenecks. The primary issue is the cost, which limits the widespread application of 4D printing. 4D printing starts with the material, but as a substrate it is often more expensive. Optimizing experimental conditions to reduce costs is the key factor that must be considered in the future research. Additionally, the 4D printing process itself is complex, and the adjustment of its process parameters often requires rich experience and proficient skills, and there are relatively few talents with these two aspects. 4D printing process will inevitably cause some material waste, which may cause serious pollution to the environment. Some polymer particles may accumulate in the food chain, causing irreversible harm. Therefore, the development of new environmentally friendly products is the top priority for future research.

4.6 Conclusion

4D printing has become a research hotspot due to its incomparable advantages, but the current research on 4D printing is still in its nascent stage. The increasing demand for smart materials in high-tech industries such as aerospace, soft robotics, biomedicine, and engineering will further promote the development and improvement of 4D printing technology. From the vigorous development of 3D printing, it is reasonable to predict that 4D printing will also have unlimited possibilities in the future. However, it also faces challenges including slow and inaccurate drive, lack of control of deformation intermediate state, and limited material availability. By designing multifunctional stimulus response materials with precision regulation, long-term stability, and immediacy of response, it is possible to meet the higher demands for 4D printing materials. However, the availability of 4D printing materials is largely limited by the tools and software used for printing. Therefore, optimizing printing tools and manufacturing processes is also a top priority. In summary, the additive manufacturing technology of smart materials will play an increasingly important role in human production and life. Continuous research and innovation in this field will bring substantial benefits.

Acknowledgments

This work was supported by the Strategic Priority Research Program of the Chinese Academy of Sciences (Grant No. XDB 0470301), CAS "Light of West China" Program, and Natural Science Foundation of Gansu Province (22JR5RA102).

References

Abdullah, T., & Okay, O. (2023). 4D printing of body temperature-responsive hydrogels based on poly(acrylic acid) with shape-memory and self-healing abilities. *ACS Applied Bio Materials, 6*(2), 703–711. https://doi.org/10.1021/acsabm.2c00939. http://pubs.acs.org/journal/aabmcb.

Alam, F., Ubaid, J., Butt, H., & El-Atab, N. (2023). Swift 4D printing of thermoresponsive shape-memory polymers using vat photopolymerization. *NPG Asia Materials, 15*(1). https://doi.org/10.1038/s41427-023-00511-x.

Alkahtani, M. E., Elbadawi, M., Chapmen, C. A., Green, R. A., Gaisford, S., Orlu, M., & Basit, A. W. (2023). Electro-active polymers for on-demand drug release. *Advanced Healthcare Materials,* 2301759. https://doi.org/10.1002/adhm.202301759.

Alshebly, Y. S., Nafea, M., Ali, M. S. M., & Almurib, H. A. F. (2021). Review on recent advances in 4D printing of shape memory polymers. *European Polymer Journal, 159,* 110708. https://doi.org/10.1016/j.eurpolymj.2021.110708.

Arif, Z. U., Khalid, M. Y., Zolfagharian, A., & Bodaghi, M. (2022). 4D bioprinting of smart polymers for biomedical applications: recent progress, challenges, and future perspectives. *Reactive and Functional Polymers, 179*. https://doi.org/10.1016/j.reactfunctpolym.2022.105374. https://www.journals.elsevier.com/reactive-and-functional-polymers.

Bakarich, S. E., Gorkin, R., Panhuis, M. I. H., & Spinks, G. M. (2015). 4D printing with mechanically robust, thermally actuating hydrogels. *Macromolecular Rapid Communications, 36*(12), 1211–1217. https://doi.org/10.1002/marc.201500079. http://www3.interscience.wiley.com/journal/117932056/grouphome.

Bayar, E. Y., Getiren, B., Soysal, F., Çiplak, Z., Yildiz, N., & Bayraktar, E. (2023). Graphene oxide/polyaniline/silver nanocomposite synthesis and photothermal performance. *Materials research bulletin, 166,* 112352. https://doi.org/10.1016/j.materresbull.2023.112352.

BG, P. K., Mehrotra, S., Marques, S. M., Kumar, L., & Verma, R. (2023). 3D printing in personalized medicines: A focus on applications of the technology. *Materials Today Communications, 35*. https://doi.org/10.1016/j.mtcomm.2023.105875. http://www.journals.elsevier.com/materials-today-communications/.

Cano-Vicent, A., Tambuwala, M. M., Hassan, S. S., Barh, D., Aljabali, A. A. A., Birkett, M., Arjunan, A., & Serrano-Aroca, Á. (2021). Fused deposition modelling: Current status, methodology, applications and future prospects. *Additive Manufacturing, 47*. https://doi.org/10.1016/j.addma.2021.102378. https://www.journals.elsevier.com/additive-manufacturing.

Carlson, M., & Li, Y. (2020). Development and kinetic evaluation of a low-cost temperature-sensitive shape memory polymer for 4-dimensional printing. *International Journal of Advanced Manufacturing Technology, 106*(9–10), 4263–4279. https://doi.org/10.1007/s00170-020-04927-5. http://www.springerlink.com/content/0268-3768.

Cecchini, L., Mariani, S., Ronzan, M., Mondini, A., Pugno, N. M., & Mazzolai, B. (2023). 4D printing of humidity-driven seed inspired soft robots. *Advanced Science, 10*(9). https://doi.org/10.1002/advs.202205146. http://onlinelibrary.wiley.com/journal/10.1002/(ISSN)2198-3844.

Champeau, M., Heinze, D. A., Viana, T. N., de Souza, E. R., Chinellato, A. C., & Titotto, S. (2020). 4D printing of Hydrogels: A review. *Advanced Functional Materials, 30*(31). https://doi.org/10.1002/adfm.201910606. http://onlinelibrary.wiley.com/journal/10.1002/(ISSN)1616-3028.

Chen, A., Wang, W., Mao, Z., He, Y., Chen, S., Liu, G., Su, J., Feng, P., Shi, Y., & Yan, C. (2023). Multi-material 3D and 4D bioprinting of heterogeneous constructs for tissue engineering. *Advanced Materials for Additive Manufacturing, 36*, 2307686. https://doi.org/10.1002/adma.202307686.

Chen, A., Su, J., Li, Y., Zhang, H., Shi, Y., Yan, C., & Lu, J. (2023). 3D/4D printed bio-piezoelectric smart scaffolds for next-generation bone tissue engineering. *International Journal of Extreme Manufacturing, 5*(3), 032007. https://doi.org/10.1088/2631-7990/acd88f.

Chen, F., Xu, Q., Huang, F., Xie, Z., & Fang, H. (2022). Effect of nozzle vibration at different frequencies on surface structures and tensile properties of PLA parts printed by FDM. *Materials Letters, 325*, 132612. https://doi.org/10.1016/j.matlet.2022.132612.

Chen, H., Sun, R., Zeng, T., Zheng, J., Yoshitomi, T., Kawazoe, N., Yang, Y., & Chen, G. (2022). Stepwise photothermal therapy and chemotherapy by composite scaffolds of gold nanoparticles, BP nanosheets and gelatin immobilized with doxorubicin-loaded thermo-sensitive liposomes. *Biomaterials Science, 10*(24), 7042–7054. https://doi.org/10.1039/d2bm01155g. http://pubs.rsc.org/en/journals/journal/bm.

Chen, L., Zhang, Y., Ye, H., Duan, G., Duan, H., Ge, Q., & Wang, Z. (2021). Color-Changeable four-dimensional printing enabled with ultraviolet-curable and thermochromic shape memory polymers. *ACS Applied Materials and Interfaces, 13*(15), 18120–18127. https://doi.org/10.1021/acsami.1c02656. http://pubs.acs.org/journal/aamick.

Chen, M., Hou, Y., An, R., Qi, H. J., & Zhou, K. (2023). 4D printing of reprogrammable liquid crystal elastomers with synergistic photochromism and photoactuation. *Advanced Materials for Additive Manufacturing, 36,* 2303969. https://doi.org/10.1002/adma.202303969.

Chen, X., Han, S., Wu, W., Wu, Z., Yuan, Y., Wu, J., & Liu, C. (2022). Harnessing 4D printing bioscaffolds for advanced orthopedics. *Small (Weinheim an der Bergstrasse, Germany), 18*(36). https://doi.org/10.1002/smll.202106824. http://onlinelibrary.wiley.com/journal/10.1002/(ISSN)1613-6829.

Chiesa, I., Ceccarini, M. R., Bittolo Bon, S., Codini, M., Beccari, T., Valentini, L., & De Maria, C. (2023). 4D printing shape-morphing hybrid biomaterials for advanced bioengineering applications. *Materials, 16*(20). https://doi.org/10.3390/ma16206661. http://www.mdpi.com/journal/materials.

Chohan, J. S., & Singh, R. (2017). Pre and post processing techniques to improve surface characteristics of FDM parts: A state of art review and future applications. *Rapid Prototyping Journal, 23*(3), 495–513. https://doi.org/10.1108/RPJ-05-2015-0059. http://www.emeraldinsight.com/info/journals/rpj/rpj.jsp.

Ciarleglio, G., Toto, E., & Santonicola, M. G. (2023). Conductive and thermo-responsive composite hydrogels with poly(N-isopropylacrylamide) and carbon nanotubes fabricated by two-step photopolymerization. *Polymers, 15*(4). https://doi.org/10.3390/polym15041022. http://www.mdpi.com/journal/polymers.

Cortés, A., Cosola, A., Sangermano, M., Campo, M., González Prolongo, S., Pirri, C. F., Jiménez-Suárez, A., & Chiappone, A. (2021). DLP 4D-printing of remotely, modularly, and selectively controllable shape memory polymer nanocomposites embedding carbon nanotubes. *Advanced Functional Materials, 31*(50). https://doi.org/10.1002/adfm.202106774. http://onlinelibrary.wiley.com/journal/10.1002/(ISSN)1616-3028.

Crapnell, R. D., Kalinke, C., Silva, L. R. G., Stefano, J. S., Williams, R. J., Abarza Munoz, R. A., Bonacin, J. A., Janegitz, B. C., & Banks, C. E. (2023). Additive manufacturing electrochemistry: An overview of producing bespoke conductive additive manufacturing filaments. *Materials Today, 71*. https://doi.org/10.1016/j.mattod.2023.11.002. https://www.sciencedirect.com/science/journal/13697021.

Cui, H., Miao, S., Esworthy, T., Lee, Sj, Zhou, X., Hann, S. Y., Webster, T. J., Harris, B. T., & Zhang, L. G. (2019). A novel near-infrared light responsive 4D printed nanoarchitecture with dynamically and remotely controllable transformation. *Nano Research*. https://doi.org/10.1007/s12274-019-2340-9. http://www.springer.com/materials/nanotechnology/journal/12274.

Detamornrat, U., McAlister, E., Hutton, A. R. J., Larrañeta, E., & Donnelly, R. F. (2022). The role of 3D printing technology in microengineering of microneedles. *Small (Weinheim an der Bergstrasse, Germany), 18*(18). https://doi.org/10.1002/smll.202106392. http://onlinelibrary.wiley.com/journal/10.1002/(ISSN)1613-6829.

Ding, A., Lee, S. J., Tang, R., Gasvoda, K. L., He, F., & Alsberg, E. (2022). 4D cell-condensate bioprinting. *Small (Weinheim an der Bergstrasse, Germany), 18*(36). https://doi.org/10.1002/smll.202202196. http://onlinelibrary.wiley.com/journal/10.1002/(ISSN)1613-6829.

Do, X. H., Nguyen, T. D., Le, T. T. H., To, T. T., Bui, T. V. K., Pham, N. H., Lam, K., Hoang, T. M. N., & Ha, P. T. (2023). High biocompatibility, mri enhancement, and dual chemo- and thermal-therapy of curcumin-encapsulated alginate/Fe3O4 nanoparticles. *Pharmaceutics, 15*(5). https://doi.org/10.3390/pharmaceutics15051523. http://www.mdpi.com/journal/pharmaceutics.

Dong, L., Bai, J., Wan, X., Wu, Z., Zhao, H., Li, Y., Guo, T., Zhang, K., Wang, J., Chu, Y., Zhang, J., & Peng, G. D. (2023). Effect of sintering condition on the attenuation of silica glasses fabricated by additive manufacturing. *Ceramics International, 49*(23), 39400–39408. https://doi.org/10.1016/j.ceramint.2023.09.285. https://www.journals.elsevier.com/ceramics-international.

Dong, Y., Wang, S., Ke, Y., Ding, L., Zeng, X., Magdassi, S., & Long, Y. (2020). 4D printed hydrogels: Fabrication, materials, and applications. *Advanced Materials Technologies, 5*(6). https://doi.org/10.1002/admt.202000034. http://onlinelibrary.wiley.com/journal/10.1002/(ISSN)2365-709X.

Enshaei, H., Molina, B. G., Puiggalí-Jou, A., Saperas, N., & Alemán, C. (2022). Polypeptide hydrogel loaded with conducting polymer nanoparticles as electroresponsive delivery system of small hydrophobic drugs. *European Polymer Journal, 173*, 111199. https://doi.org/10.1016/j.eurpolymj.2022.111199.

Esposito Corcione, C., Palumbo, E., Masciullo, A., Montagna, F., & Torricelli, M. C. (2018). Fused deposition modeling (FDM): An innovative technique aimed at reusing Lecce stone waste for industrial design and building applications. *Construction and Building Materials, 158*, 276–284. https://doi.org/10.1016/j.conbuildmat.2017.10.011.

Falahati, M., Ahmadvand, P., Safaee, S., Chang, Y. C., Lyu, Z., Chen, R., Li, L., & Lin, Y. (2020). Smart polymers and nanocomposites for 3D and 4D printing. *Materials Today, 40*, 215–245. https://doi.org/10.1016/j.mattod.2020.06.001. http://www.journals.elsevier.com/materials-today/.

Fang, T., Cao, L., Chen, S., Fang, J., Zhou, J., Fang, L., Lu, C., & Xu, Z. (2018). Preparation and assembly of five photoresponsive polymers to achieve complex light-induced shape deformations. *Materials and Design, 144*, 129–139. https://doi.org/10.1016/j.matdes.2018.02.029.

Fang, Z., Chen, P., Ji, Q., Yan, C., & Gong, A. (2023). Stimuli-responsive hydrogel for disease therapy. *Polymer Bulletin*. https://doi.org/10.1007/s00289-023-04862-x. https://www.springer.com/journal/289.

Faruque, M. O., Lee, Y., Wyckoff, G. J., & Lee, C. H. (2023). Application of 4D printing and AI to cardiovascular devices. *Journal of Drug Delivery Science and Technology, 80*. https://doi.org/10.1016/j.jddst.2023.104162. http://www.editionsdesante.fr/category.php?id_category=48.

Fu, X., Zou, B., Xing, H., Li, L., Li, Y., & Wang, X. (2019). Effect of printing strategies on forming accuracy and mechanical properties of ZrO2 parts fabricated by SLA technology. *Ceramics International, 45*(14), 17630–17637. https://doi.org/10.1016/j.ceramint.2019.05.328. https://www.journals.elsevier.com/ceramics-international.

Galarreta-Rodriguez, I., Etxebeste-Mitxeltorena, M., Moreno, E., Plano, D., Sanmartín, C., Megahed, S., Feliu, N., Parak, W. J., Garaio, E., Gil de Muro, I., Lezama, L., Ruiz de Larramendi, I., & Insausti, M. (2023). Preparation of selenium-based drug-modified polymeric ligand-functionalised Fe3O4 nanoparticles as multimodal drug carrier and magnetic hyperthermia inductor. *Pharmaceuticals, 16*(7). https://doi.org/10.3390/ph16070949. https://www.mdpi.com/journal/pharmaceuticals.

Gardan, J. (2019). Smart materials in additive manufacturing: State of the art and trends. *Virtual and Physical Prototyping, 14*(1), 1–18. https://doi.org/10.1080/17452759.2018.1518016. http://www.tandf.co.uk/journals/titles/17452759.asp.

Getachew, M. T., Shiferaw, M. Z., & Ayele, B. S. (2023). The current state of the art and advancements, challenges, and future of additive manufacturing in aerospace applications. *Advances in Materials Science and Engineering, 2023*. https://doi.org/10.1155/2023/8817006. http://www.hindawi.com/journals/amse/.

Ghazal, A. F., Zhang, M., Mujumdar, A. S., & Ghamry, M. (2023). Progress in 4D/5D/6D printing of foods: Applications and R&D opportunities. *Critical Reviews in Food Science and Nutrition, 63*(25), 7399–7422. https://doi.org/10.1080/10408398.2022.2045896. https://www.tandfonline.com/loi/bfsn20.

Gobin, P.-F., Gutnin, G., Morin, M., Salvia, M., & Tatibouet, J. (1996). Smart materials: A future for composites. *Journal of intelligent material systems and structures, 7*(3), 353–357. https://doi.org/10.1177/1045389x9600700318.

Gopinath, S., Adarsh, N. N., Radhakrishnan Nair, P., & Mathew, S. (2023). Recent trends in thermo-responsive elastomeric shape memory polymer nanocomposites. *Polymer Composites, 44*(8), 4433–4458. https://doi.org/10.1002/pc.27464. http://onlinelibrary.wiley.com/journal/10.1002/(ISSN)1548-0569.

Gu, S. Y., Jin, S. P., Gao, X. F., & Mu, J. (2016). Polylactide-based polyurethane shape memory nanocomposites (Fe3O4/PLAUs) with fast magnetic responsiveness. *Smart Materials and Structures, 25*(5). https://doi.org/10.1088/0964-1726/25/5/055036. http://iopscience.iop.org/article/10.1088/0964-1726/25/5/055036/pdf.

Gu, T., Bi, H., Sun, H., Tang, J. F., Ren, Z., Zhou, X., & Xu, M. (2023). Design and development of 4D-printed cellulose nanofibers reinforced shape memory polymer composites: Application for self-deforming plant bionic soft grippers. *Additive Manufacturing, 70*, 103544. https://doi.org/10.1016/j.addma.2023.103544.

Gu, T., Ji, T., Bi, H., Ding, K. J., Sun, H., Zhai, W., Ren, Z., Wei, Y., & Xu, M. (2023). 4D printed and multi-stimulus responsive shape memory polymer nanocomposites developed on hydrogen bonding–metal-phenolic sacrificial network: Application for hazardous chemical operations soft robots. *Applied Materials Today, 35*, 102009. https://doi.org/10.1016/j.apmt.2023.102009.

Gugulothu, S. B., & Chatterjee, K. (2023). Visible light-based 4D-bioprinted tissue scaffold. *ACS Macro Letters, 12*(4), 494–502. https://doi.org/10.1021/acsmacrolett.3c00036. http://pubs.acs.org/journal/amlccd.

Guo, W., Zhou, B., Zou, Y., & Lu, X. (2023). 4D printed poly(l-lactide)/(FeCl3–TA/MgO) composite scaffolds with near-infrared light-induced shape-memory effect and antibacterial properties. *Advanced Engineering Materials*. https://doi.org/10.1002/adem.202301381. http://onlinelibrary.wiley.com/journal/10.1002/(ISSN)1527-2648.

Guo, Z., Xu, J., Chen, Y., Guo, Z., Yu, P., Liu, Y., & Zhao, J. (2019). High-sensitive and stretchable resistive strain gauges: Parametric design and DIW fabrication. *Composite Structures, 223*, 110955. https://doi.org/10.1016/j.compstruct.2019.110955.

Ha, M., Cañón Bermúdez, G. S., Liu, J. A. C., Oliveros Mata, E. S., Evans, B. A., Tracy, J. B., & Makarov, D. (2021). Reconfigurable magnetic origami actuators with on-board sensing for guided assembly. *Advanced Materials, 33*(25). https://doi.org/10.1002/adma.202008751. http://onlinelibrary.wiley.com/journal/10.1002/(ISSN)1521-4095.

Han, D., Farino, C., Yang, C., Scott, T., Browe, D., Choi, W., Freeman, J. W., & Lee, H. (2018). Soft robotic manipulation and locomotion with a 3D printed electroactive hydrogel. *ACS Applied Materials and Interfaces, 10*(21), 17512–17518. https://doi.org/10.1021/acsami.8b04250. http://pubs.acs.org/journal/aamick.

Han, X., Saiding, Q., Cai, X., Xiao, Y., Wang, P., Cai, Z., Gong, X., Gong, W., Zhang, X., & Cui, W. (2023). Intelligent vascularized 3D/4D/5D/6D-printed tissue scaffolds. *Nano-Micro Letters, 15*, 239. https://doi.org/10.1007/s40820-023-01187-2.

Han, Z., Liu, S., Qiu, K., Liu, J., Zou, R., Wang, Y., Zhao, J., Liu, F., Wang, Y., & Li, L. (2023). The enhanced ZrO2 produced by DLP via a reliable plasticizer and its dental application. *Journal of the Mechanical Behavior of Biomedical Materials, 141*, 105751. https://doi.org/10.1016/j.jmbbm.2023.105751.

Hao, L., Tang, D., Sun, T., Xiong, W., Feng, Z., Evans, K. E., & Li, Y. (2021). Direct Ink writing of mineral materials: A review. *International Journal of Precision Engineering and Manufacturing—Green Technology, 8*(2), 665–685. https://doi.org/10.1007/s40684-020-00222-6. http://www.springer.com/engineering/production+engineering/journal/40684.

Higueras, M., Collado-Montero, F. J., & Medina, V. J. (2023). Colorimetric evaluation of 3D printing polymers exposed to accelerated aging for Cultural Heritage applications. *Color Research and Application, 48*(3), 283–295. https://doi.org/10.1002/col.22846. http://onlinelibrary.wiley.com/journal/10.1002/(ISSN)1520-6378.

Hu, G. F., Damanpack, A. R., Bodaghi, M., & Liao, W. H. (2017). Increasing dimension of structures by 4D printing shape memory polymers via fused deposition modeling. *Smart Materials and Structures, 26*(12), 125023. https://doi.org/10.1088/1361-665x/aa95ec.

Hu, G., & Bodaghi, M. (2023). Direct fused deposition modeling 4D printing and programming of thermoresponsive shape memory polymers with autonomous 2D-to-3D shape transformations. *Advanced Engineering Materials, 25*(19), 2300334. https://doi.org/10.1002/adem.202300334.

Hu, J., Meng, H., Li, G., & Ibekwe, S. I. (2012). A review of stimuli-responsive polymers for smart textile applications. *Smart Materials and Structures, 21*(5), 053001. https://doi.org/10.1088/0964-1726/21/5/053001.

Hu, X., Ge, Z., Wang, X., Jiao, N., Tung, S., & Liu, L. (2022). Multifunctional thermo-magnetically actuated hybrid soft millirobot based on 4D printing. *Composites Part B: Engineering, 228*, 109451. https://doi.org/10.1016/j.compositesb.2021.109451.

Hu, Y., Wang, Z., Jin, D., Zhang, C., Sun, R., Li, Z., Hu, K., Ni, J., Cai, Z., Pan, D., Wang, X., Zhu, W., Li, J., Wu, D., Zhang, L., & Chu, J. (2020). Botanical-inspired 4D printing of hydrogel at the microscale. *Advanced Functional Materials, 30*(4). https://doi.org/10.1002/adfm.201907377. http://onlinelibrary.wiley.com/journal/10.1002/(ISSN)1616-3028.

Jeong, H. Y., Woo, B. H., Kim, N., & Jun, Y. C. (2020). Multicolor 4D printing of shape-memory polymers for light-induced selective heating and remote actuation. *Nature Research, South Korea Scientific Reports, 10*(1). https://doi.org/10.1038/s41598-020-63020-9. http://www.nature.com/srep/index.html.

Jia, S., Fong, W. K., Graham, B., & Boyd, B. J. (2018). Photoswitchable molecules in long-wavelength light-responsive drug delivery: From molecular design to applications. *Chemistry of Materials, 30*(9), 2873–2887. https://doi.org/10.1021/acs.chemmater.8b00357. http://pubs.acs.org/journal/cmatex.

Jiang, Z., Diggle, B., Tan, M. L., Viktorova, J., Bennett, C. W., & Connal, L. A. (2020). Extrusion 3D printing of polymeric materials with advanced properties. *Advanced Science, 7*(17). https://doi.org/10.1002/advs.202001379. http://onlinelibrary.wiley.com/journal/10.1002/(ISSN)2198-3844.

Joshi, A., Choudhury, S., Baghel, V. S., Ghosh, S., Gupta, S., Lahiri, D., Ananthasuresh, G. K., & Chatterjee, K. (2023). 4D printed programmable shape-morphing hydrogels as intraoperative self-folding nerve conduits for sutureless neurorrhaphy. *Advanced Healthcare Materials, 12*(24). https://doi.org/10.1002/adhm.202300701. http://onlinelibrary.wiley.com/journal/10.1002/(ISSN)2192-2659.

Kačergis, L., Mitkus, R., & Sinapius, M. (2019). Influence of fused deposition modeling process parameters on the transformation of 4D printed morphing structures. *Smart Materials and Structures, 28*(10), 105042. https://doi.org/10.1088/1361-665x/ab3d18.

Kang, D. J., An, K. H., Seo, S. B., Kim, Y. A., Park, Y. K., Lee, G. G., & Jung, S. C. (2023). Evaluation of carbon black produced with hydrogen by the liquid phase plasma method as a conductive material for the anode of a supercapacitor. *Journal of Power Sources, 580*. https://doi.org/10.1016/j.jpowsour.2023.233346. https://www.journals.elsevier.com/journal-of-power-sources.

Khalid, M. Y., Arif, Z. U., & Ahmed, W. (2022). 4D Printing: Technological and manufacturing renaissance. *Macromolecular Materials and Engineering, 307*, 2200003. https://doi.org/10.1002/adem.202300334.

Khalid, M. Y., Arif, Z. U., Noroozi, R., Zolfagharian, A., & Bodaghi, M. (2022). 4D printing of shape memory polymer composites: A review on fabrication techniques, applications, and future perspectives. *Journal of Manufacturing Processes, 81*, 759–797. https://doi.org/10.1016/j.jmapro.2022.07.035. http://www.elsevier.com/wps/find/journaldescription.cws_home/620379/description#description.

Khan, S. A., Kumar, H., & Arora, P. K. (2021). Retrospective investigation on emergence and development of additive manufacturing. *Indian Journal of Engineering and Materials Sciences, 28*(2), 115–124. http://nopr.niscair.res.in/bitstream/123456789/57663/1/IJEMS-115-124.pdf.

Koh, T. Y., & Sutradhar, A. (2022). Untethered selectively actuated microwave 4D printing through ferromagnetic PLA. *Additive Manufacturing, 56*, 102866. https://doi.org/10.1016/j.addma.2022.102866.

Kökcü, İ., Eryildiz, M., Altan, M., Ertuğrul, M.İ., & Odabaş, S. (2023). Scaffold fabrication from drug loaded HNT reinforced polylactic acid by FDM for biomedical applications. *Polymer Composites, 44*(4), 2138–2152. https://doi.org/10.1002/pc.27231. http://onlinelibrary.wiley.com/journal/10.1002/(ISSN)1548-0569.

Koo, W. T., Kim, S. J., Jang, J. S., Kim, D. H., & Kim, I. D. (2019). Catalytic metal nanoparticles embedded in conductive metal–organic frameworks for chemiresistors: Highly active and conductive porous materials. *Advanced Science, 6*(21). https://doi.org/10.1002/advs.201900250. http://onlinelibrary.wiley.com/journal/10.1002/(ISSN)2198-3844.

Kumar, D., Sahu, B., & Banerjee, S. (2023). Amino acid-derived smart and functional polymers for biomedical applications: current status and future perspectives. Macromolecular Chemistry and Physics, 224, 2300207.

Kumar, S., Singh, R., Batish, A., & Singh, T. P. (2022). Additive manufacturing of smart materials exhibiting 4-D properties: A state of art review. *Journal of Thermoplastic Composite Materials, 35*(9), 1358–1381. https://doi.org/10.1177/0892705719895052.

Kumari, V., Dey, K., Giri, S., & Bhaumik, A. (2016). Magnetic memory effect in self-assembled nickel ferrite nanoparticles having mesoscopic void spaces. *RSC Advances, 6*(51), 45701–45707. https://doi.org/10.1039/C6RA05483H.

Lai, W. F. (2023). Design and application of self-healable polymeric films and coatings for smart food packaging. *npj Science of Food, 7*(1). https://doi.org/10.1038/s41538-023-00185-3. https://www.nature.com/npjscifood/about.

Launhardt, M., Wörz, A., Loderer, A., Laumer, T., Drummer, D., Hausotte, T., & Schmidt, M. (2016). Detecting surface roughness on SLS parts with various measuring techniques. *Polymer Testing, 53*, 217–226. https://doi.org/10.1016/j.polymertesting.2016.05.022.

Li, G., Yang, S., Wu, W., Chen, F., Li, X., Tian, Q., Wang, K., Zhao, Y., Wang, J., Liu, Q., Ren, L., Han, Z., & Ren, L. (2023). Biomimetic 4D printing catapult: From biological prototype to practical implementation. *Advanced Functional Materials, 33*(32). https://doi.org/10.1002/adfm.202301286. http://onlinelibrary.wiley.com/journal/10.1002/(ISSN)1616-3028.

Li, K., Zhao, J., Zhussupbekova, A., Shuck, C. E., Hughes, L., Dong, Y., Barwich, S., Vaesen, S., Shvets, I. V., Möbius, M., Schmitt, W., Gogotsi, Y., & Nicolosi, V. (2022). 4D printing of MXene hydrogels for high-efficiency pseudocapacitive energy storage. *Nature Research, Ireland Nature Communications, 13*(1). https://doi.org/10.1038/s41467-022-34583-0. https://www.nature.com/ncomms/.

Li, M., Zhang, H., Gao, F., Tang, Z., Zeng, D., Pan, Y., Su, P., Ruan, Y., Xu, Y., & Weng, W. (2019). A cyclic cinnamate dimer mechanophore for multimodal stress responsive and mechanically adaptable polymeric materials. *Polymer Chemistry, 10*(7), 905–910. https://doi.org/10.1039/c8py01654b.

Li, S., Yang, H., Chen, G., Zheng, J., Wang, W., Ren, J., Zhu, C., Yang, Y., Cong, Y., & Fu, J. (2023). 4D printing of biomimetic anisotropic self-sensing hydrogel actuators. *Chemical Engineering Journal, 473*, 145444. https://doi.org/10.1016/j.cej.2023.145444.

Li, S., Li, Y., Tan, Y., Li, J., Wang, D., Yuan, D., & Zhang, J. (2023). A sustainable superhydrophobic and photothermal coatings for anti-icing application on concrete with a simple method for CNTs/SiO_2 modification. *Sustainability, 15*(22), 15865. https://doi.org/10.3390/su152215865.

Li, S. S., Yue, X., Li, Q. Y., Peng, H. L., Dong, B. X., Liu, T. S., Yang, H. Y., Fan, J., Shu, S. L., Qiu, F., & Jiang, Q. C. (2023). Development and applications of aluminum alloys for aerospace industry. *Journal of Materials Research and Technology, 27*, 944–983. https://doi.org/10.1016/j.jmrt.2023.09.274. http://www.elsevier.com/journals/journal-of-materials-research-and-technology/2238-7854.

Li, Y., Kong, J., Zhao, H., & Liu, Y. (2023). Synthesis of multi-stimuli responsive Fe3O4 coated with diamonds nanocomposite for magnetic assisted chemo-photothermal therapy. *Molecules, 28*(4), 1784. https://doi.org/10.3390/molecules28041784.

Li, Z., Yun, H., Yan, Y., Yuan, M., Zhao, Y., & Zhao, F. (2023). Electro-responsive breathing transition of conductive hydrogel for broadband kinetic energy harvesting. *Advanced Materials*. https://doi.org/10.1002/adma.202305837. http://onlinelibrary.wiley.com/journal/10.1002/(ISSN)1521-4095.

Lin, C., Huang, Z., Wang, Q., Zou, Z., Wang, W., Liu, L., Liu, Y., & Leng, J. (2023). Mass-producible near-body temperature-triggered 4D printed shape memory biocomposites and their application in biomimetic intestinal stents. *Composites Part B: Engineering, 256*, 110623. https://doi.org/10.1016/j.compositesb.2023.110623.

Liu, C., Lou, Y., Sun, Z., Ma, H., Sun, M., Li, S., You, D., Wu, J., Ying, B., Ding, W., Yu, M., & Wang, H. (2023). 4D printing of personalized-tunable biomimetic periosteum with anisotropic microstructure for accelerated vascularization and bone healing. *Advanced*

Healthcare Materials, 12(22). https://doi.org/10.1002/adhm.202202868. http://onlinelibrary.wiley.com/journal/10.1002/(ISSN)2192-2659.

Liu, H., Wang, F., Wu, W., Dong, X., & Sang, L. (2023). 4D printing of mechanically robust PLA/TPU/Fe3O4 magneto-responsive shape memory polymers for smart structures. *Composites Part B: Engineering, 248*, 110382. https://doi.org/10.1016/j.compositesb.2022.110382.

Liu, J., Xu, L., Li, Y., Zhao, J., Jia, X., Chao, J., Lv, B., & Zhao, Y. (2022). Construction of novel biomass-based solar evaporator with asymmetric dual-layer structure for water desalination. *Advanced Sustainable Systems, 6*(1). https://doi.org/10.1002/adsu.202100274. www.advsustainsys.com.

Liu, S., Dong, X., Wang, Y., Xiong, J., Guo, R., Xiao, J., Sun, C., Zhai, F., & Wang, X. (2023). 4D printing of shape memory epoxy for adaptive dynamic components. *Advanced Materials Technologies, 8*(12). https://doi.org/10.1002/admt.202202004. http://onlinelibrary.wiley.com/journal/10.1002/(ISSN)2365-709X.

Liu, X., Gao, M., Chen, J., Guo, S., Zhu, W., Bai, L., Zhai, W., Du, H., Wu, H., Yan, C., Shi, Y., Gu, J., Qi, H. J., & Zhou, K. (2022). Recent advances in stimuli-responsive shape-morphing hydrogels. *Advanced Functional Materials*. https://doi.org/10.1002/adfm.202203323. http://onlinelibrary.wiley.com/journal/10.1002/(ISSN)1616-3028.

Liu, X., Wei, M., Wang, Q., Tian, Y., Han, J., Gu, H., Ding, H., Chen, Q., Zhou, K., & Gu, Z. (2021). Capillary-force-driven self-assembly of 4D-printed microstructures. *Advanced Materials, 33*(22). https://doi.org/10.1002/adma.202100332. http://onlinelibrary.wiley.com/journal/10.1002/(ISSN)1521-4095.

Lu, P., Guo, & Pan, Y. (2017). Magnetic-field-assisted projection stereolithography for three-dimensional printing of smart structures. *Journal of Manufacturing Science and Engineering, Transactions of the ASME, 139*(7). https://doi.org/10.1115/1.4035964. http://asmedl.aip.org/Manufacturing.

Luo, X., Liu, X., Wang, Y., Chen, M., Zhao, M., Wang, Y., & Zhang, Q. (2024). Thermoresponsive membrane based on UCST-type organoboron polymer for smart gating and self-cleaning. *Journal of Membrane Science, 693*, 122343. https://doi.org/10.1016/j.memsci.2023.122343.

Lv, M., Wang, L., Ding, W., Tan, J., Zhang, N., Chen, L., Zhai, Y., Liu, X., Xie, H., & Wang, J. (2024). Synthesis of magneto-thermal biresponsive temperature-sensitive copolymer/inorganic hybrid green nanoemulsifier with tunable LCST behavior and simulation and calculation of LCST transition behavior of its copolymer. *Journal of Molecular Liquids, 393*, 123686. https://doi.org/10.1016/j.molliq.2023.123686.

Ma, B., Zhang, Y., Li, J., Chen, D., Liang, R., Fu, S., & Li, D. (2023). 4D printing of multi-stimuli responsive rigid smart composite materials with self-healing ability. *Chemical Engineering Journal, 466*, 143420. https://doi.org/10.1016/j.cej.2023.143420.

Ma, X., Zhang, J., Javed, M., Wu, J., Hu, Y., Yin, S., Zhu, Y., Wu, W., & Liu, F. (2023). Chitosan based smart polymer composites: Fabrication and pH-responsive behavior for bio-medical applications. *Environmental Research, 221*, 115286. https://doi.org/10.1016/j.envres.2023.115286.

Mazeeva, A., Masaylo, D., Razumov, N., Konov, G., & Popovich, A. (2023). 3D Printing technologies for fabrication of magnetic materials based on metal–polymer composites: A review. *Materials, 16*(21), 6928. https://doi.org/10.3390/ma16216928.

Mi, S., Du, Z., Xu, Y., & Sun, W. (2018). The crossing and integration between microfluidic technology and 3D printing for organ-on-chips. *Royal Society of Chemistry, China Journal of Materials Chemistry B, 6*(39), 6191–6206. https://doi.org/10.1039/c8tb01661e. http://pubs.rsc.org/en/journals/journal/tb.

Miao, Z., Tan, H., Gustavsson, L., Zhou, Y., Xu, Q., Ikkala, O., & Peng, B. (2023). Gustation-inspired dual-responsive hydrogels for taste sensing enabled by machine learning. *Small (Weinheim an der Bergstrasse, Germany)*. https://doi.org/10.1002/smll.202305195. http://onlinelibrary.wiley.com/journal/10.1002/(ISSN)1613-6829.

Momeni, F., & Ni, J. (2020). Laws of 4D printing. *Engineering, 6*(9), 1035–1055. https://doi.org/10.1016/j.eng.2020.01.015. http://www.journals.elsevier.com/engineering/.

Montanari, R., Palombi, A., Richetta, M., & Varone, A. (2023). Additive manufacturing of aluminum alloys for aeronautic applications: Advantages and problems. *Metals, 13*(4), 716. https://doi.org/10.3390/met13040716.

Montoya, C., Roldan, L., Yu, M., Valliani, S., Ta, C., Yang, M., & Orrego, S. (2023). Smart dental materials for antimicrobial applications. *Bioactive Materials, 24*, 1–19. https://doi.org/10.1016/j.bioactmat.2022.12.002. http://www.keaipublishing.com/en/journals/bioactive-materials/.

Monzón, M. D., Paz, R., Pei, E., Ortega, F., Suárez, L. A., Ortega, Z., Alemán, M. E., Plucinski, T., & Clow, N. (2017). 4D printing: Processability and measurement of recovery force in shape memory polymers. *The International Journal of Advanced Manufacturing Technology, 89*(5–8), 1827–1836. https://doi.org/10.1007/s00170-016-9233-9.

Narupai, B., Smith, P. T., & Nelson, A. (2021). 4D printing of multi-stimuli responsive protein-based hydrogels for autonomous shape transformations. *Advanced Functional Materials, 31*(23). https://doi.org/10.1002/adfm.202011012. http://onlinelibrary.wiley.com/journal/10.1002/(ISSN)1616-3028.

Nicolay, P., Schlögl, S., Thaler, S. M., Humbert, C., & Filipitsch, B. (2023). Smart materials for green (Er) cities: A short review. *Applied Sciences, 13*(16), 9289. https://doi.org/10.3390/app13169289.

Norioka, C., Okita, K., Mukada, M., Kawamura, A., & Miyata, T. (2017). Biomolecularly stimuli-responsive tetra-poly(ethylene glycol) that undergoes sol–gel transition in response to a target biomolecule. *Polymer Chemistry, 8*(41), 6378–6385. https://doi.org/10.1039/C7PY01370A.

Ntouanoglou, K., Stavropoulos, P., & Mourtzis, D. (2018) 4D printing prospects for the aerospace industry: A critical review. Procedia Manufacturing, 18, 120–129.

Ouyang, H., Li, X., Lu, X., & Xia, H. (2022). Selective laser sintering 4D printing of dynamic cross-linked polyurethane containing Diels-Alder bonds. *ACS Applied Polymer Materials, 4*(5), 4035–4046. https://doi.org/10.1021/acsapm.2c00565. pubs.acs.org/journal/aapmcd.

Pan, H. M., & Goto, A. (2023). Topology-dependent pH-responsive actuation and shape memory programming for biomimetic 4D printing. *Macromolecular Rapid Communications, 44*(9). https://doi.org/10.1002/marc.202300074. http://onlinelibrary.wiley.com/journal/10.1002/(ISSN)1521-3927.

Papadakis, L., Chantzis, D., & Salonitis, K. (2018). On the energy efficiency of pre-heating methods in SLM/SLS processes. *The International Journal of Advanced Manufacturing Technology, 95*(1-4), 1325–1338. https://doi.org/10.1007/s00170-017-1287-9.

Parada, L. R., Mayuet, P. F., & Gámez, A. J. (2019). Industrial product design: Study of FDM technology for the manufacture of thermoformed prototypes. *Procedia Manufacturing, 41*, 587–593. https://doi.org/10.1016/j.promfg.2019.09.046.

Pingale, P., Dawre, S., Dhapte-Pawar, V., Dhas, N., & Rajput, A. (2023). Advances in 4D printing: From stimulation to simulation. *Drug Delivery and Translational Research, 13*(1), 164–188. https://doi.org/10.1007/s13346-022-01200-y. https://www.springer.com/journal/13346.

Pourmadadi, M., Ahmadi, M., Dinani, H. S., Ajalli, N., & Dorkoosh, F. (2022). Theranostic applications of stimulus-responsive systems based on Fe2O3. *Pharmaceutical*

Nanotechnology, 10(2), 90–112. https://doi.org/10.2174/2211738510666220210105113. http://www.benthamscience.com/pnt/index.htm.

Pudkon, W., Laomeephol, C., Damrongsakkul, S., Kanokpanont, S., & Ratanavaraporn, J. (2021). Comparative study of silk fibroin-based hydrogels and their potential as material for 3-dimensional (3D) printing. *Molecules, 26*(13), 3887. https://doi.org/10.3390/molecules26133887.

Raina, A., Haq, M. I. U., Javaid, M., Rab, S., & Haleem, A. (2021). 4D printing for automotive industry applications. *Journal of The Institution of Engineers (India): Series D, 102*(2), 521–529. https://doi.org/10.1007/s40033-021-00284-z. http://www.springer.com/journal/40033/about.

Ramezani, M., & Ripin, M. (2023). 4D printing in biomedical engineering: Advancements, challenges, and future directions. *Journal of Functional Biomaterials, 14*(7), 347. https://doi.org/10.3390/jfb14070347.

Razzaq, M. Y., Gonzalez-Gutierrez, J., Farhan, M., Das, R., Ruch, D., Westermann, S., & Schmidt, D. F. (2023). 4D printing of electroactive triple-shape composites. *Polymers, 15*(4). https://doi.org/10.3390/polym15040832. http://www.mdpi.com/journal/polymers.

Razzaq, M. Y., Gonzalez-Gutierrez, J., Mertz, G., Ruch, D., Schmidt, D. F., & Westermann, S. (2022). 4D printing of multicomponent shape-memory polymer formulations. *Applied Sciences (Switzerland), 12*(15). https://doi.org/10.3390/app12157880. http://www.mdpi.com/journal/applsci/.

Revelo, C. F., & Colorado, H. A. (2018). 3D printing of kaolinite clay ceramics using the Direct Ink Writing (DIW) technique. *Ceramics International, 44*(5), 5673–5682. https://doi.org/10.1016/j.ceramint.2017.12.219.

Rivera-Tarazona, L. K., Shukla, T., Singh, K. A., Gaharwar, A. K., Campbell, Z. T., & Ware, T. H. (2022). 4D printing of engineered living materials. *Advanced Functional Materials, 32*(4). https://doi.org/10.1002/adfm.202106843. http://onlinelibrary.wiley.com/journal/10.1002/(ISSN)1616-3028.

Rosales, C. A. G., Duarte, M. F. G., Kim, H., Chavez, L., Hodges, D., Mandal, P., Lin, Y., & Tseng, T.-L. B. (2018). 3D printing of shape memory polymer (SMP)/carbon black (CB) nanocomposites with electro-responsive toughness enhancement. *Materials Research Express, 5*(6), 065704. http://doi.org/10.1088/2053-1591/aacd53.

Sánchez, C. P., Jérôme, C., Noels, L., & Vanderbemden, P. (2022). Review of thermo-responsive electroactive and magnetoactive shape memory polymer nanocomposites. *ACS Omega, 7*(45), 40701–40723. https://doi.org/10.1021/acsomega.2c05930. http://pubs.acs.org/journal/acsodf.

Saorin, J. L., Diaz-Alemán, M. D., De La Torre-Cantero, J., Meier, C., & Pérez Conesa, I. (2021). Design and validation of an open source 3D printer based on digital ultraviolet light processing (DLP), for the improvement of traditional artistic casting techniques for microsculptures. *Applied Sciences (Switzerland), 11*(7). https://doi.org/10.3390/app11073197. https://www.mdpi.com/2076-3417/11/7/3197/pdf.

Savastru, R., Miclos, S., Baschir, L., Savastru, D., & Lancranjan, I. I. (2023). Long period grating fibre sensor for detection of impurities infesting smart polymer composites. *Composite Structures, 312*, 116877. https://doi.org/10.1016/j.compstruct.2023.116877.

Sayyed, M. I., AlZaatreh, M. Y., Matori, K. A., Sidek, H. A. A., & Zaid, M. H. M. (2018). Comprehensive study on estimation of gamma-ray exposure buildup factors for smart polymers as a potent application in nuclear industries. *Results in Physics, 9*, 585–592. https://doi.org/10.1016/j.rinp.2018.01.057.

Selianitis, D., & Pispas, S. (2023). Thermo- and pH-responsive poly[(diethylene glycol methyl ether methacrylate)- co -(2-diisopropylamino ethyl methacrylate)] hyperbranched copolymers: self-assembly and drug-loading. *Polymer Chemistry, 14*(5), 587–599. https://doi.org/10.1039/d2py01447e.

Seo, S. W., Rostami Azmand, H., & Song, Y. (2020). A fiber optic sensor platform for smart hydrogel event detection. *Optical Fiber Technology, 58*. https://doi.org/10.1016/j.yofte.2020.102246. http://www.journals.elsevier.com/optical-fiber-technology/.

Shahzad, A., & Lazoglu, I. (2021). Direct ink writing (DIW) of structural and functional ceramics: Recent achievements and future challenges. *Composites Part B: Engineering, 225*, 109249. https://doi.org/10.1016/j.compositesb.2021.109249.

Shao, Z., Hakobyan, K., Xu, J., Chen, R., Kumar, N., Willcox, M., & Wong, E. H. H. (2023). Photoinduced unveiling of cationic amine: Toward smart photoresponsive antimicrobial polymers. *ACS Applied Polymer Materials, 5*(10), 8735–8743. https://doi.org/10.1021/acsapm.3c01904. pubs.acs.org/journal/aapmcd.

Shin, C. S., & Chang, Y. C. (2021). Fabrication and compressive behavior of a micro-lattice composite by high resolution dlp stereolithography. *Polymers, 13*(5), 1–17. https://doi.org/10.3390/polym13050785. https://www.mdpi.com/2073-4360/13/5/785/pdf.

Singh, R., Das, A., & Anand, A. (2023). On 3D printing of customized multi-root dental implants for the strategic tooth of canine by direct metal laser sintering. *Rapid Prototyping Journal, 29*(3), 447–459. https://doi.org/10.1108/RPJ-04-2022-0112. http://www.emeraldinsight.com/info/journals/rpj/rpj.jsp.

Singh, R., Deshmukh, S. A., Kamath, G., Sankaranarayanan, S. K. R. S., & Balasubramanian, G. (2017). Controlling the aqueous solubility of PNIPAM with hydrophobic molecular units. *Computational Materials Science, 126*, 191–203. https://doi.org/10.1016/j.commatsci.2016.09.030.

Soleimanzadeh, H., Rolfe, B., Bodaghi, M., Jamalabadi, M., Zhang, X., & Zolfagharian, A. (2023). Sustainable robots 4D printing. *Advanced Sustainable Systems, 7*(12), 2300289. https://doi.org/10.1002/adsu.202300289.

Solis, D. M., & Czekanski, A. (2022). The effect of the printing temperature on 4D DLP printed pNIPAM hydrogels. *Soft Matter, 18*(17), 3422–3429. https://doi.org/10.1039/d2sm00201a. http://pubs.rsc.org/en/journals/journal/sm.

ISO/ASTM 52900:2015(en). (2015). Standard terminology for additive manufacturing–general principles–part 1: terminology. ASTM International.

Subash, A., & Kandasubramanian, B. (2020). 4D printing of shape memory polymers. *European Polymer Journal, 134*, 109771. https://doi.org/10.1016/j.eurpolymj.2020.109771.

Sun, Y. C., Boero, G., & Brugger, J. (2021). Stretchable conductors fabricated by stencil lithography and centrifugal force-assisted patterning of liquid metal. *ACS Applied Electronic Materials, 3*(12), 5423–5432. https://doi.org/10.1021/acsaelm.1c00884. pubs.acs.org/journal/aaembp.

Sun, S., Gan, X., Wang, Z., Fu, D., Pu, W., & Xia, H. (2020). Dynamic healable polyurethane for selective laser sintering. *Additive Manufacturing, 33*, 101176.

Teng, X., Zhang, M., Adhikari, B., & Ma, Y. (2024). 4D printed pH-responsive labels of methacrylic anhydride grafted konjac glucomannan for detecting quality changes in respiring climacteric fruits. *Food Hydrocolloids, 149*, 109543. https://doi.org/10.1016/j.foodhyd.2023.109543.

Wan, X., He, Y., Liu, Y., & Leng, J. (2022). 4D printing of multiple shape memory polymer and nanocomposites with biocompatible, programmable and selectively actuated properties. *Additive Manufacturing, 53*, 102689. https://doi.org/10.1016/j.addma.2022.102689.

Wang, B., Lee, T. L., & Qin, Y. (2023). Advances in smart materials and structures. *Materials, 16*, 7206. https://doi.org/10.3390/ma16227206.

Wang, F., Luo, F., Huang, Y., Cao, X., & Yuan, C. (2023). 4D printing via multispeed fused deposition modeling. *Advanced Materials Technologies, 8*(2). https://doi.org/10.1002/admt.202201383. http://onlinelibrary.wiley.com/journal/10.1002/(ISSN)2365-709X.

Wang, F., Liu, C., Yang, H., Wang, H., Zhang, H., Zeng, X., Wang, C., Zhang, W., Lv, W., Zhu, P., & Li, B. (2023). 4D printing of ceramic structures. *Additive Manufacturing, 63*, 103411. https://doi.org/10.1016/j.addma.2023.103411.

Wang, H., & Guo, J. (2023). Recent advances in 4D printing hydrogel for biological interfaces. *China International Journal of Material Forming. 16*(5). https://doi.org/10.1007/s12289-023-01778-9. https://www.springer.com/journal/12289.

Wang, H., Zhao, J., Luo, Z., & Li, Z. (2023). Recent research developments of 4D printing technology for magnetically controlled smart materials: A review. *Magnetochemistry, 9*(8), 204. https://doi.org/10.3390/magnetochemistry9080204.

Wang, L., Zhang, F., Liu, Y., Du, S., & Leng, J. (2022). Thermal, mechanical and shape fixity behaviors of shape memory cyanate under γ-ray radiation. *Smart Materials and Structures, 31*(4), 045010. https://doi.org/10.1088/1361-665x/ac5538.

Wang, Q., Tian, X., Zhang, D., Zhou, Y., Yan, W., & Li, D. (2023). Programmable spatial deformation by controllable off-center freestanding 4D printing of continuous fiber reinforced liquid crystal elastomer composites. *Nature Communications, 14*(1), 3869. https://doi.org/10.1038/s41467-023-39566-3. https://www.nature.com/ncomms/.

Wang, S., & Ding, T. (2018). Photothermal-assisted optical stretching of gold nanoparticles. *ACS nano, 13*(1), 32–37. https://doi.org/10.1021/acsnano.8b06087.

Wang, X., He, Y., Liu, Y., & Leng, J. (2022). Advances in shape memory polymers: Remote actuation, multi-stimuli control, 4D printing and prospective applications. *Materials Science and Engineering: R: Reports, 151*, 100702. https://doi.org/10.1016/j.mser.2022.100702.

Wang, Y., Zhao, Y., Yan, Z., Li, T., Wang, S., & Dong, W. (2023). Photothermal-driven polymer for a smart window with adaptive solar modulation. *ACS Applied Materials and Interfaces, 15*(36), 43038–43047. https://doi.org/10.1021/acsami.3c10848. http://pubs.acs.org/journal/aamick.

Wang, Z., Jiang, Y., Shang, H., Qiao, N., Sun, X., Li, Q., Wang, X., Wu, Y., & Ma, H. (2023). Preparation and evaluation of controllable drug delivery system: A pH and temperature responsive nanosphere based on chitosan copolymer/mesoporous silica. *Journal of Polymer Research, 30*(6). https://doi.org/10.1007/s10965-023-03599-5. https://www.springer.com/journal/10965.

de Wild, M., Dany, S., John, C., & Schuler, F. (2020). Smart 4D-printed implants and instruments. *Current Directions in Biomedical Engineering, 6*(3), 209–212. https://doi.org/10.1515/cdbme-2020-3053.

Wu, W., Jiang, J., Jiang, H., Liu, W., Li, G., Wang, B., Tang, M., & Zhao, J. (2018). Improving bending and dynamic mechanics performance of 3D printing through ultrasonic strengthening. *Materials Letters, 220*, 317–320. https://doi.org/10.1016/j.matlet.2018.03.048. http://www.journals.elsevier.com/materials-letters/.

Xie, X., Xu, X., Zhu, Q., Lu, S., Li, Y., & Bai, Y. (2022). Thermo- and near infrared light-induced reversible multi-shape memory materials for actuators and sensors. *Materials Chemistry Frontiers, 6*(14), 1973–1981. https://doi.org/10.1039/d2qm00346e. http://rsc.li/frontiers-materials.

Xing, Y., Qiu, L., Liu, D., Dai, S., & Sheu, C. L. (2023). The role of smart polymeric biomaterials in bone regeneration: A review. *Frontiers in Bioengineering and Biotechnology, 11*. https://doi.org/10.3389/fbioe.2023.1240861. http://journal.frontiersin.org/journal/bioengineering-and-biotechnology#archive.

Xu, F., & Feringa, B. L. (2023). Photoresponsive supramolecular polymers: from light-controlled small molecules to smart materials. *Advanced Materials, 35*(10). https://doi.org/10.1002/adma.202204413. http://onlinelibrary.wiley.com/journal/10.1002/(ISSN)1521-4095.

Yan, S., Zhang, F., Luo, L., Wang, L., Liu, Y., & Leng, J. (2023). Shape memory polymer composites: 4D printing, smart structures, and applications. *Research: A Journal of Science and Its Applications, 6*, 0234. https://doi.org/10.34133/research.0234. https://spj.science.org/doi/epdf/10.34133/research.0234.

Yang, C., Liu, Z., Chen, C., Shi, K., Zhang, L., Ju, X. J., Wang, W., Xie, R., & Chu, L. Y. (2017). Reduced graphene oxide-containing smart hydrogels with excellent electro-response and mechanical properties for soft actuators. *ACS Applied Materials and Interfaces, 9*(18), 15758–15767. https://doi.org/10.1021/acsami.7b01710. http://pubs.acs.org/journal/aamick.

Yang, G., Sun, Y., Qin, L., Li, M., Ou, K., Fang, J., & Fu, Q. (2021). Direct-ink-writing (DIW) 3D printing functional composite materials based on supra-molecular interaction. *Composites Science and Technology, 215*, 109013. https://doi.org/10.1016/j.compscitech.2021.109013.

Yang, W., Gao, Y., Hu, R., Li, M., Zhao, F., Jiang, H., & Zhang, X. (2023). Construction of smart propellant with multi-morphologies. *Defence Technology, 33*, 180–185. https://doi.org/10.1016/j.dt.2023.06.015. https://www.journals.elsevier.com/defence-technology.

Yao, L., Hu, P., Wu, Z., Liu, W., Lv, Q., Nie, Z., & Zhengdi, H. (2020). Comparison of accuracy and precision of various types of photo-curing printing technology. *Journal of Physics: Conference Series, 1549*(3), 032151. https://doi.org/10.1088/1742-6596/1549/3/032151.

Yue, L., Sun, X., Yu, L., Li, M., Montgomery, S. M., Song, Y., Nomura, T., Tanaka, M., & Qi, H. J. (2023). Cold-programmed shape-morphing structures based on grayscale digital light processing 4D printing. *Nature Communications, 14*(1), 5519. https://doi.org/10.1038/s41467-023-41170-4. https://www.nature.com/ncomms/.

Zarek, M., Layani, M., Eliazar, S., Mansour, N., Cooperstein, I., Shukrun, E., Szlar, A., Cohn, D., & Magdassi, S. (2016). 4D printing shape memory polymers for dynamic jewellery and fashionwear. *Virtual and Physical Prototyping, 11*(4), 263–270. https://doi.org/10.1080/17452759.2016.1244085. http://www.tandf.co.uk/journals/titles/17452759.asp.

Zeenat, L., Zolfagharian, A., Sriya, Y., Sasikumar, S., Bodaghi, M., & Pati, F. (2023). 4D Printing for vascular tissue engineering: Progress and challenges. *Advanced Materials Technologies, 8*(23), 2300200. https://doi.org/10.1002/admt.202300200.

Zhai, F., Feng, Y., Li, Z., Xie, Y., Ge, J., Wang, H., Qiu, W., & Feng, W. (2021). 4D-printed untethered self-propelling soft robot with tactile perception: Rolling, racing, and exploring. *Matter, 4*(10), 3313–3326. https://doi.org/10.1016/j.matt.2021.08.014. http://www.cell.com/matter.

Zhang, B., Li, H., Cheng, J., Ye, H., Sakhaei, A. H., Yuan, C., Rao, P., Zhang, Y. F., Chen, Z., Wang, R., He, X., Liu, J., Xiao, R., Qu, S., & Ge, Q. (2021). Mechanically robust and UV-curable shape-memory polymers for digital light processing based 4D printing. *Advanced Materials, 33*(27). https://doi.org/10.1002/adma.202101298. http://onlinelibrary.wiley.com/journal/10.1002/(ISSN)1521-4095.

Zhang, C., Chen, G., Zhang, R., Wu, Z., Xu, C., Man, H., & Che, R. (2021). Charge modulation of CNTs-based conductive network for oxygen reduction reaction and microwave absorption. *Carbon, 178*, 310–319. https://doi.org/10.1016/j.carbon.2021.03.030. http://www.journals.elsevier.com/carbon/.

Zhang, C., Lu, X., Fei, G., Wang, Z., Xia, H., & Zhao, Y. (2019). 4D printing of a liquid crystal elastomer with a controllable orientation gradient. *ACS Applied Materials and Interfaces, 11*(47), 44774–44782. https://doi.org/10.1021/acsami.9b18037. http://pubs.acs.org/journal/aamick.

Zhang, H., Huang, S., Sheng, J., Wang, C., Zhang, J., Zhu, M., Agyenim-Boateng, E., Liang, C., Xue, B., Yang, H., Zhou, J., & Lu, J. (2023a). 4D printing of shape memory polymer nanocomposites for enhanced performances and tunable response behavior. *European Polymer Journal, 201*, 112568. https://doi.org/10.1016/j.eurpolymj.2023.112568.

Zhang, H., Huang, S., Sheng, J., Wang, C., Zhang, J., Zhu, M., Agyenim-Boateng, E., Liang, C., Xue, B., Yang, H., Zhou, J., & Lu, J. (2023b). Regulation of electrically responsive shape recovery behavior of 4D printed polymers for application in actuators. *Additive Manufacturing, 78*, 103836. https://doi.org/10.1016/j.addma.2023.103836.

Zhang, L., Huang, X., Cole, T., Lu, H., Hang, J., Li, W., Tang, S. Y., Boyer, C., Davis, T. P., & Qiao, R. (2023). 3D-printed liquid metal polymer composites as NIR-responsive 4D printing soft robot. *Nature Communications, 14*(1). https://doi.org/10.1038/s41467-023-43667-4. https://www.nature.com/ncomms/.

Zhang, X., Wu, Y., & Gao, Y. (2020). 3D printed cellulose-based high-efficiency oil-water separation mesh. *IOP Conference Series: Earth and Environmental Science, 514*(2), 022088. https://doi.org/10.1088/1755-1315/514/2/022088.

Zhang, Y., Jiang, X., Wu, R., & Wang, W. (2016). Multi-stimuli responsive shape memory polymers synthesized by using reaction-induced phase separation. *Journal of Applied Polymer Science, 133*(24). https://doi.org/10.1002/app.43534. http://onlinelibrary.wiley.com/journal/10.1002/(ISSN)1097-4628.

Zhang, Y., Zhou, S., Chong, K. C., Wang, S., & Liu, B. (2019). Near-infrared light-induced shape memory, self-healable and anti-bacterial elastomers prepared by incorporation of a diketopyrrolopyrrole-based conjugated polymer. *Materials Chemistry Frontiers, 3*(5), 836–841. https://doi.org/10.1039/c9qm00104b. http://rsc.li/frontiers-materials.

Zhao, T., Yu, R., Li, X., Cheng, B., Zhang, Y., Yang, X., Zhao, X., Zhao, Y., & Huang, W. (2018). 4D printing of shape memory polyurethane via stereolithography. *European Polymer Journal, 101*, 120–126. https://doi.org/10.1016/j.eurpolymj.2018.02.021.

Zhao, W., Yue, C., Liu, L., Liu, Y., & Leng, J. (2023). Research progress of shape memory polymer and 4D printing in biomedical application. *Advanced Healthcare Materials, 12*(16). https://doi.org/10.1002/adhm.202201975. http://onlinelibrary.wiley.com/journal/10.1002/(ISSN)2192-2659.

Zheng, X., Zhou, B., & Xue, S. (2019). A viscoelastic-plastic constitutive model of shape memory polymer. *Journal of Mechanics, 35*(5), 601–611. https://doi.org/10.1017/jmech.2018.56. http://www.jmech.org.tw/.

Zheng, Y., Chen, Z., Jiang, Q., Feng, J., Wu, S., & Del Campo, A. (2020). Near-infrared-light regulated angiogenesis in a 4D hydrogel. *Nanoscale, 12*(25), 13654–13661. https://doi.org/10.1039/d0nr02552f. http://pubs.rsc.org/en/journals/journal/nr.

Zolfagharian, A., Denk, M., Bodaghi, M., Kouzani, A. Z., & Kaynak, A. (2020). Topology-optimized 4D printing of a soft actuator. *Acta Mechanica Solida Sinica, 33*(3), 418–430. https://doi.org/10.1007/s10338-019-00137-z. http://www.springer.com/engineering/mechanics/journal/10338.

Zouboulis, P., Koumoulos, E. P., & Karatza, A. (2022). Employment of 3D-printed bilayer structures with embedded continuous fibers for thermal management applications: An axial cooling 4D-printed fan application case study. *Polymers, 14*(19). https://doi.org/10.3390/polym14193952. http://www.mdpi.com/journal/polymers.

Zu, S., Zhang, Z., Liu, Q., Wang, Z., Song, Z., Guo, Y., Xin, Y., & Zhang, S. (2022). 4D printing of core–shell hydrogel capsules for smart controlled drug release. *Bio-Design and Manufacturing, 5*(2), 294–304. https://doi.org/10.1007/s42242-021-00175-y.

Additive manufacturing of functional polymers and nanocomposites: opportunities and challenges

5

Yanshi Agrawal[1], *Mahak Kushwaha*[2], *Swati Agarwal*[2], *and Suphiya Khan*[3]
[1]Department of Bioscience and Biotechnology, Banasthali Vidyapith, Banasthali, Tonk, Rajasthan, India, [2]Drumlins Water Technologies Pvt. Ltd., Jaipur, Rajasthan, India, [3]Shriram Institute for Industrial Research, Gurugram, Haryana, India

5.1 Introduction

In today's world, additive manufacturing (AM), also known as three-dimensional printing (3DP), is shaking up industries by changing how we design, engineer, and produce goods. It's a game-changer, driving innovation, cutting costs, and reducing waste across different sectors. One exciting area where AM is gaining traction is in the realm of polymer composites (Wimpenny et al., 2017). These composites offer a myriad of benefits, providing flexibility in product properties and design options. They're not only commercially available but also cost-effective, making them a top choice even for critical applications such as aerospace and medical devices in Fig. 5.1 (Kumar et al., 2019).

AM operates differently from traditional manufacturing methods, building objects layer by layer, which offers unmatched versatility. Functional polymer devices are created using a variety of 3D printing techniques, including selective laser sintering (SLS) and fused deposition modeling (FDM). Nevertheless, limitations persist in terms of material diversity and scalability, posing challenges for large-scale production. Thus, there is a pressing need to innovate and develop new materials to unlock the full potential of large-scale 3D printing production.

Polymer-based nanocomposites, with their remarkable strength-to-weight ratio and customizable properties, are also making waves (Gibson et al., 2021). The fusion of 3D printing and functional polymers holds immense promise for revolutionizing manufacturing (Han, 2017). The increasing number of studies focusing on 3D-printed functional polymeric materials underscores their growing importance (Beniak et al., 2020). However, significant challenges remain, including limited material options and durability issues stemming from defects between layers (Komissarenko et al., 2021). Addressing these challenges requires in-depth research on materials and the development of new ones (Liu et al., 2014.)

Additive Manufacturing of Functional Polymers and Nanocomposites. DOI: https://doi.org/10.1016/B978-0-443-26516-7.00005-5
© 2025 Elsevier Ltd. All rights are reserved, including those for text and data mining, AI training, and similar technologies.

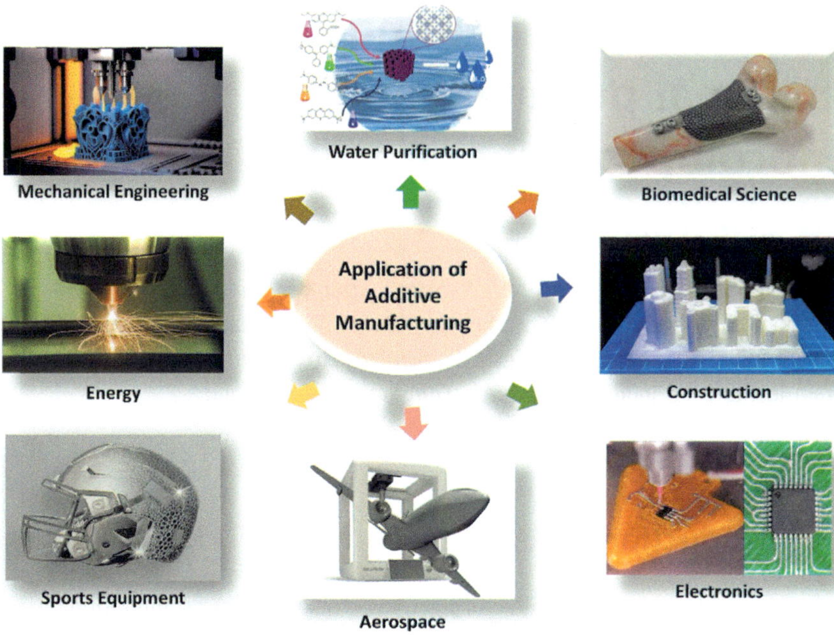

Figure 5.1 Application of additive manufacturing.

The recent advancements in additive manufacturing in mentioned Fig. 5.2 have ushered in a new era of production, offering unprecedented design freedom, faster manufacturing, waste reduction, and enhanced precision at lower costs. Using AM to maximize the flexibility of polymer composites makes multifunctional, quick, and affordable manufacturing more possible (Goh et al., 2017).

In this chapter, we will explore recent innovations in functional polymers and 3D-printed materials in greater detail, including their manufacturing processes, properties, and potential applications mentioned in Table 5.1 (Savastano et al., 2016). Specifically, we will examine their conductivity, heat resistance, interference shielding capabilities, and energy storage capacities. Moreover, we will delve into the challenges facing 3D printing and discuss strategies for overcoming them to pave the way for the development of superior functional materials and devices in the future.

5.2 Types of additively manufactured functional polymers and nanocomposites

5.2.1 Polylactic acid (PLA)–based polymer nanocomposites

Polylactic acid (PLA) is a remarkable biodegradable polymer sourced from renewable materials like corn starch and sugarcane (Gironi & Piemonte, 2011). This eco-friendly polymer is produced through the fermentation of carbohydrates, making it a sustainable alternative to traditional petroleum-based plastics. PLA

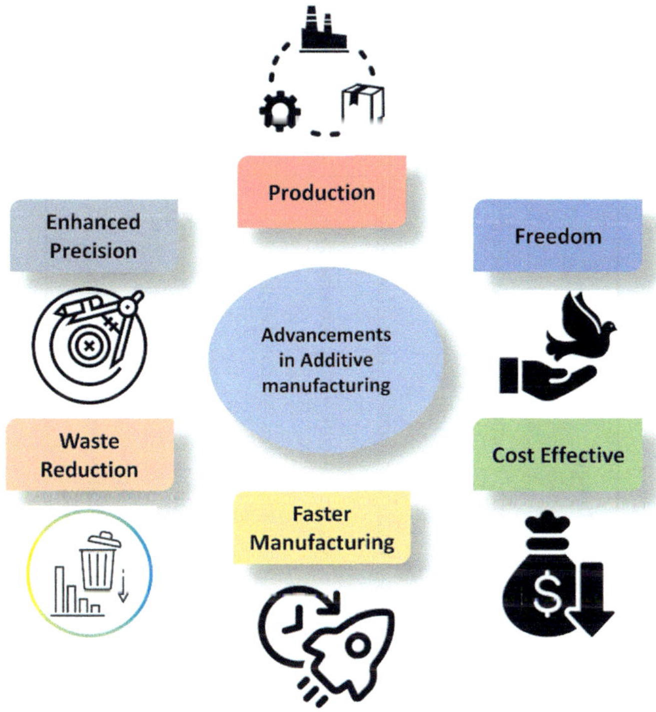

Figure 5.2 Advancements in additive manufacturing.

emerges as a promising biopolymer due to its renewable, recyclable, biodegradable, and compostable nature (Sha et al., 2016). Derived from crops like maize and potatoes, mentioned in Table 5.2 PLA polymers exhibit unique properties resulting from bacterial fermentation of carbohydrates (Koh et al., 2018). Researchers have enhanced PLA's properties by incorporating nanofillers like multiwalled carbon nanotubes (MWCNTs) into PLA matrices. Utilizing the thermo mechanical melt mixing method, they produce 3D printing PLA/MWCNTs Nanocomposite filaments, which are versatile, cost-effective, and environmentally friendly compared to solvent-based methods (Musial et al., 2020). The multifunctional things of PLA/MWCNT Nanocomposites make them suitable for various industrial applications (Farah et al., 2016).

5.2.2 Polycaprolactone–based polymer nanocomposites

Polycaprolactone (PCL) is a versatile biodegradable polyester belonging to the aliphatic polyester's family, synthesized from caprolactone monomers through ring-opening polymerization. Renowned for its exceptional biocompatibility, slow degradation rate, and adaptability, PCL finds significant use in various fields, especially biomedical engineering. Its low melting point makes it a preferred choice for applications such as 3D printing and scaffold materials in tissue

Table 5.1 Additive manufacturing potential application and its challenges.

S. no.	Additive manufacturing properties	Potential application	Challenges	References
1.	Rapid prototyping, high precision	Medical devices: prosthetics, implants, surgical tools	Material biocompatibility, sterilization, regulatory compliance	Francis et al. (2018)
2.	Design flexibility, lightweight structures	Aerospace components: turbine blades, structural elements	Material strength, certification	Blakey-Milner et al. (2021)
3.	Customization, on-demand production	Automotive parts: interior components, engine parts	Material durability, cost, scalability	Dalpadulo et al. (2022)
4.	Material diversity, electrical and thermal properties	Electronics: circuit boards, sensors, housings	Material compatibility, conductivity	Rao et al. (2022)
5.	Complex geometries, intricate designs	Consumer goods: fashion accessories, home appliances	Printing speed, surface finish	Jiang, Ma (2020)
6.	Layer-by-layer control, internal structures	Biomedical implants: dental prosthetics, tissue scaffolds	Material biocompatibility, mechanical strength	da Silva et al. (2021)
7.	Functional gradients, enhanced performance	Energy storage devices: batteries, fuel cells	Electrochemical stability, energy density	De et al. (2024)
8.	Multimaterial integration, enhanced functionality	Chemical processing: catalysts, membranes, filtration systems	Material compatibility, chemical resistance	Nunes et al. (2020)

Table 5.2 Classification of additive manufacturing.

Based on the process used to create the product	Based on the kind of material utilized	Based on the environment used for processing
Stereolithography (SLA), digital light processing (DLP), continuous liquid interface production (CLIP), continuous digital light manufacturing (CDLM)	Thermoplastics: ABS, PLA, PETG, etc. Photopolymers: acrylic, epoxy, polyurethane, etc.	Liquid resin bath: resin is stored in a vat, and a platform slowly lifts as each layer is cured by UV light. UV-curing chamber: entire layers or sections are cured simultaneously using a UV light source.
Fused deposition modeling (FDM), fused filament fabrication (FFF)	Thermoplastics: ABS, PLA, PETG, nylon, etc. Composites: carbon fiber, glass fiber, metal-filled filaments, etc.	Heated build chamber: the build environment is heated to prevent warping and improve layer adhesion. Ambient air: standard environmental conditions are used, with no additional heating or cooling.
Selective Laser sintering (SLS), selective laser melting (SLM), laser powder bed fusion (LPBF), direct metal laser sintering (DMLS), direct metal deposition (DMD)	Metals: stainless steel, aluminum, titanium, etc. Ceramics: alumina, zirconia, silicon carbide, etc. Polymers: nylon, TPU, PEKK, etc.	Inert gas atmosphere: uses inert gases like argon or nitrogen to prevent oxidation during processing. High-temperature environment: operating temperatures are elevated to ensure proper sintering or melting of materials.
Inkjet-based 3D printing, binder jetting, material jetting, drop on demand (DOD) printing	Ceramics: porcelain, clay-based materials, zirconia, etc. Metals: stainless steel, bronze, cobalt chrome, etc. Polymers: nylon, polyamide, etc.	Powder bed: powdered material is spread in a thin layer, and the binder is selectively applied where needed. Liquid binder: a liquid binder is jetted onto the powder to bind it together and form solid layers.
Electron beam melting (EBM), wire arc additive manufacturing (WAAM)	Metals: titanium alloys, aluminum alloys, stainless steel, etc. Alloys: nickel-based alloys, cobalt chrome alloys, etc.	High vacuum: maintains a vacuum environment to prevent contamination and ensure proper material fusion Controlled atmosphere: uses controlled gases like argon to create a specific atmosphere for optimal metal processing.

(Continued)

Table 5.2 (Continued)

Based on the process used to create the product	Based on the kind of material utilized	Based on the environment used for processing
Powder bed fusion with heat transfer (PBF-HT), powder bed fusion with infrared radiation (PBF-IR)	Photopolymers: flexible resins, tough resins, biocompatible resins, etc. Resins: epoxy resins, polyurethane resins, silicone resins, etc. Composites: carbon fiber composites, glass fiber composites, etc.	Continuous liquid resin bath: a continuous flow of resin allows for continuous curing as the object is pulled out. Controlled UV light exposure: UV light exposure is managed continuously across the entire build area for continuous curing.
Vat polymerization, continuous vat polymerization (CVP), Bottom-up projection, top-down projection	Photopolymers: clear resins, colored resins, rubber-like resins, etc. Ceramics: porcelain, silica-based ceramics, alumina ceramics, etc. Metals: silver, gold, platinum, etc.	Multiple material reservoirs: different materials are loaded into separate reservoirs and jetted as needed. Precise droplet placement: droplets are placed with high precision, ensuring accurate deposition of materials.
Laminated object manufacturing (LOM), sheet lamination, ultrasonic additive manufacturing (UAM)	Metals: stainless steel, titanium, aluminum, nickel alloys, etc. Alloys: cobalt chrome alloys, inconel alloys, etc. Composites: metal matrix composites, ceramic matrix composites, etc.	Powder bed: Metal powder is spread in a thin layer, and the laser selectively melts areas to create solid parts. High-energy laser beam: a powerful laser source is used to melt metal powders or deposit material through a nozzle.

engineering, supporting cell growth and tissue regeneration (Vidal et al., 2022). In tissue engineering, PCL stands out, particularly in stent manufacturing, due to its superb biocompatibility and ability to retain shape. Its low melting point makes it ideal for 3D printing, while its scaffold role has been extensively researched, aiding in cell attachment, proliferation, and differentiation (Heimowska et al., 2017). Notably, PCL scaffolds have diverse applications in tissue engineering, including bone regeneration, boasting advantages over alternative materials. PCL exhibits commendable properties; it may not induce bone formation alone at defect sites. However, its compatibility with various biomaterials opens avenues for enhancement. Combining PCL with substances like calcium phosphate compounds and ceramics such as hydroxyapatite can promote ossification, enhancing its efficacy in tissue regeneration (Khan et al., 2017).

5.2.3 High-density polyethylene–based polymer nanocomposites

High-density polyethylene (HDPE) is a versatile thermoplastic polymer known for its remarkable strength-to-density ratio, toughness, chemical resistance, and heat tolerance. This makes it a preferred choice in industries ranging from packaging, piping, and bottles to automotive and construction due to its durability and adaptability (Gao et al., 2019). In consumer industries, HDPE stands out as the most used polymer, offering superior strength-to-density performance compared to LDPE. Further enhancements in HDPE composites were explored using carbon nanotubes (CNTs), leading to improved electrical conductivity. Different reinforcement levels were applied, demonstrating suitability for electroconductive applications (Gao et al., 2012). These advancements highlight HDPE's continual evolution and its vital role in meeting diverse industrial needs.

5.2.4 Acrylonitrile butadiene styrene (ABS)–based polymer nanocomposites

The thermoplastic polymer known as acrylonitrile butadiene styrene (ABS) is made up of three monomers: styrene, butadiene, and acrylonitrile (Wang, 2011). Its exceptional mechanical qualities, like as hardness, impact resistance, and heat resistance, make it widely employed in many sectors. Since ABS is so adaptable, it may be used for a wide range of products, including toys, building materials, electronics, home appliances, automobile parts, and 3D printing. Research endeavors have extensively investigated the mechanical behavior of ABS in 3D printing, encompassing various tests such as strain rate analysis, tensile tests, compression, and creep studies (Schmitz et al., 2018). These investigations, coupled with the optimization of 3D-printing parameters using advanced modeling tools, have provided insights into enhancing part performance (Jeong et al., 2020). Furthermore, the ABS matrix has been fortified with carbon-based additives to exploit their remarkable electrical transport and storage properties, leading to the development of conductive polymer nanocomposites via additive manufacturing processes (Hsu & Lin, 2000).

5.2.5 Polyvinylidene fluoride–based polymer nanocomposites

Polyvinylidene fluoride (PVDF) stands out as a thermoplastic fluoropolymer sourced from vinylidene fluoride monomers, prized for its remarkable chemical resistance, purity, and robust mechanical attributes like toughness and flexibility. Its versatility finds applications across diverse sectors such as chemical processing, electronics, aerospace, and construction, given its ability to endure harsh conditions, extreme temperatures, and chemical exposure while retaining its integrity. PVDF is a highly favored material in situations that require corrosion resistance and durability because to its low permeability to gases and liquids. An alternative research direction, headed by Chiu, Chen, et al. (2014), examined the PVDF's piezoelectric response for FFF using MWCNT and barium titanate (BT) nanoparticles. Piezoelectric efficiency was increased by MWCNTs by enabling a larger concentration of b-phase polymers.

Furthermore, employing a dual-extrusion head allowed simultaneous 3D printing of structural and sensor components, streamlining the sensor fabrication process (Chiu et al., 2018).

5.2.6 Thermoplastic polyurethane–based polymer nanocomposites

Thermoplastic polyurethane (TPU) belongs to the class of thermoplastic elastomers, formed by the reaction of diisocyanates with polyols (Patton, Chen, et al., 2016). Its unique property is the ability to soften when heated and harden when cooled, making it versatile for various applications. With elasticity, durability, and chemical resistance, TPUs find use in footwear, automotive parts, medical devices, clothing, and flexible packaging, where a blend of flexibility and strength is needed. Biocompatible thermoplastic polyurethane (TPU) is frequently utilized in biomedical implants; for certain uses, it can be further reinforced using nanoparticle reinforcement. MWCNTs reinforced TPU nanocomposites exhibit improved elastic properties with higher concentrations, aiding in applications like sensors. Studies show that interlayer adhesion between printed layers enhances electrical conductivity, offering tunable piezo sensitivity based on reinforcement content. Additionally, investigations into mechanical properties reveal that printing orientation affects interfacial adhesion, with axially printed specimens showing better mechanical strength compared to transversely printed ones (Bhushan, 2003).

5.2.7 Polybutylene terephthalate–based polymer nanocomposites

Polybutylene terephthalate (PBT) is a thermoplastic polymer belonging to the polyester family, created through condensation polymerization (Zhang et al., 2017). It boasts excellent mechanical properties like stiffness, strength, and dimensional stability, alongside resistance to heat, chemicals, and abrasion. This versatility makes PBT ideal for a wide range of applications in the automotive, electronics, appliances, and consumer goods industries (Zheng et al., 2017). As per the study conducted by Gnanasekaran et al. (2017). PBT nanocomposites were 3D printed with graphene and carbon nanotubes integrated by a solution-mixing technique. When compared to CNT/PBT filaments, graphene/PBT showed a rougher surface and more brittle behavior. The electrical conductivity percolation threshold was determined to be 0.49 wt% for CNT and 5.2 wt% for graphene (Gong et al., 2018). These findings align with experimental results, although the presence of abrasive carbon nanoparticles led to significant nozzle wear.

5.2.8 Nylon-based polymer nanocomposites

Nylon-based polymer nanocomposites are advanced materials that combine the inherent strengths of nylon polymers with the enhanced properties provided by nanofillers. Nylon, recognized for its durability, strength, and resistance to abrasion, becomes even more versatile and robust when integrated with nanoparticles or nanotubes (Chen et al., 2006). These nanocomposites are perfect for demanding applications in sectors such as automotive, aerospace, textiles, and electronics because they exhibit superior mechanical, thermal, and barrier qualities when compared to pure nylon (Su et al., 2012). Nylon 6 and Nylon 12 (PA12) are widely available variants, with PA12 being extensively used in powder bed fusion (PBF) processes due to its higher melting point and impact strength (González et al., 2008). Researchers have explored the incorporation of graphene into PA12 nanocomposites, revealing improved thermal resistance and mechanical performance. The optimal graphene content for tensile strength, thermal stability, and compatibility with the FFF process was found to be around 6 wt% (Li et al., 2017).

5.2.9 Photopolymers and epoxies–based polymer nanocomposites

Photopolymers and epoxies represent two distinct classes of polymers with unique curing mechanisms and properties. Photopolymers undergo photopolymerization upon exposure to specific wavelengths of light, typically ultraviolet (UV) light (Lan & Pinnavaia, 1994). This process is initiated by photoinitiators present in the polymer formulation, leading to cross-linking and rapid solidification of the liquid or semi-liquid photopolymer. Due to their quick curing and ability to achieve high-resolution details, photopolymers are extensively used in applications such as 3D printing (stereolithography), dental materials, coatings, and adhesives (Wang & Pinnavaia, 1994). In contrast, epoxies are thermosetting polymers that cure through exposure to heat or a curing agent, often referred to as a hardener or catalyst. The chemical reaction between the epoxy resin and the hardener during curing results in cross-linking and the formation of a rigid and durable polymer network (Wang et al., 1996). Epoxies are widely used in adhesives, coatings, composite materials, and electronics encapsulation due to their remarkable adhesive qualities, chemical resistance, and mechanical strength.

5.2.10 Hybrid polymers–based polymer nanocomposites

Hybrid polymers are composite materials that blend two or more different types of polymers or polymer systems, whether from the same chemical family or different ones, to produce a novel substance with improved qualities. This blending process aims to capitalize on the strengths of each polymer component while offsetting their weaknesses. One common application of hybrid polymers is in the development of hybrid polymer-based nanocomposites (Sathishkumar et al., 2014). These involve combining polymers with nanofillers to achieve a blend of desirable properties such as strength, flexibility, and thermal stability. Industries like automotive, aerospace, electronics, and healthcare often rely on these nanocomposites to meet specific performance requirements. Researchers have explored various approaches to enhance polymer properties through hybridization.

5.2.11 Shape memory polymer composites

Shape memory polymer composites (SMPCs) are advanced materials that exhibit a unique ability to recall and revert to their original shape after significant deformation. These composites are typically formed by combining shape memory polymers (SMPs) with various materials such as fillers, fibers, or nanoparticles to enhance their mechanical, thermal, or functional properties (Zhou & Sheiko, 2016). The characteristic behavior of SMPCs stems from the reversible phase transition of SMPs between two distinct states: a transient structure when distorted at low temperatures (below their transition temperature), and an initial shape when heated over the transition temperature (Hager et al., 2015). This phase transition is often initiated by external factors like temperature changes, light exposure, or electrical stimulation. SMPCs are especially valuable in applications where shape adaptability and recovery are critical, such as in biomedical devices, smart textiles, robotics, and aerospace structures. Furthermore, the use of SMPCs in biological applications such as tissue engineering and patient-specific implants has been investigated. With advancements in additive manufacturing processes like fused filament fabrication (FFF), biocompatible scaffolds can be precisely fabricated with interconnected networks, offering potential alternatives for bone implant materials (Momeni et al., 2017).

5.3 Opportunities of additively manufactured functional polymers and nanocomposites

5.3.1 Water purification

Opportunities for additively manufactured functional polymers and nanocomposites are extensive, finding notable utility in water purification systems. Specifically, in the context of membrane separation systems, 3D-printed structures like spacers and filtration membranes play a pivotal role in removing pollutants and contaminants from various environments (Fijoł et al., 2022). These structures enhance mass transfer by increasing local shear rate and fluid velocity, thereby minimizing concentrated polarization effects.

In wastewater treatment, biofilm reactors with moving beds rely on biocarriers to enrich microorganisms and enhance biomass in reactors. Due to their wide surface-to-volume ratio and rough surface, laser-sintered biocarriers with fullerene-like structures show excellent microbial activity and adhesion ability (Ghosal et al., 2022). However, challenges persist in preparing membranes for submicroscale filtration, including resolution constraints during printing and cost factors compared to conventional methods. Future research is needed to explore candidate polymers and composites for optimal membrane performance in water purification applications (Yusoff et al., 2022).

5.3.2 Biomedical sciences

In biomedical sciences, additive manufacturing (AM) plays a vital role, especially in creating customized tools, therapies, medicines, and organ substitutions. Biomedical applications require a lot of polymer-based materials, which are further divided into soft and hard polymers and biodegradable and nonbiodegradable materials (Shahrubudin et al., 2020). Biodegradable polymers like polylactic acid (PLA) and polycaprolactone (PCL) are used in tissue growth applications, degrading in the body after use. On the other hand, nonbiodegradable polymers such as polyaryletherketones (PAEKs) serve as structural implants (Ngo et al., 2018).

Hard synthetic biodegradable polymers are used in prosthesis, dental implants, surgical models, and implants. Because of their special thermal characteristics, soft polymers like polyethylene glycol (PEG) and acrylate-based hydrogels are used in tissue engineering, bioprinting, drug delivery systems, and soft surgical models. Developed scaffolds using polycaprolactone and carbon nanotubes (CNTs) for cardiac tissue engineering, showcasing the versatility of AM in creating complex biomedical structures (Tan et al., 2020).

5.3.3 Electronics

In the realm of electronics, additive manufacturing presents promising applications, particularly in embedded electronics. There are two basic strategies used in this field. In the first method, discrete components are prefabricated and then laser-transferred directly into a substrate; in the second method, functional components such as interconnects, actives, and passives are laser printed into the substrate (Lamichhane et al., 2020). Researchers like have investigated the use of carbon nanotube-based materials for 3D printing in order to produce functioning electrical components. Developed flexible supercapacitors with polypyrrole MnO_2 carbon fiber hybrid structures using additive manufacturing, showcasing the immense potential of composite active materials in energy management. Wang et al. contributed by creating thin, flexible pressure sensors with a carbon black and silicone rubber nanocomposite.

Furthermore, introduced stretchable and flexible high-strain sensors using graphite flakes or carbon nanotubes (CNTs) dispersed randomly on a natural rubber substrate. investigated if conductive ABS might be used to create flexible 3D antennas using additive printing (Cardoso et al., 2020). These advancements underscore the versatility and innovative applications of additive manufacturing in the electronics sector.

5.3.4 Construction

The advancement of additive manufacturing (AM) processes within the construction industry has sparked the development of novel materials, designs, and procedures. Researchers are exploring polymers and their composites as viable alternatives to conventional materials and reinforcements (Ngo et al., 2018). Several types of reinforcement designs were printed and tested in both uniaxial and 4-point loading scenarios. To confirm the experimental findings, finite element simulations were also carried out (Mehrpouya et al., 2019). The results showed that reinforcement designs led to the strain hardening behavior that was intended for cementitious composites, which enhanced the tensile and flexural strengths of composites reinforced with structured polymers. These results were further corroborated by finite element simulation outcomes.

Additionally, there are still issues that need to be resolved despite these developments, such as overcoming reinforcement issues and automating the complete 3D concrete printing process. Salazar and colleagues contributed to this field by developing innovative lattice structures for concrete reinforcement. The material was poured into the lattice structures after PLA and ABS polymer materials were used to 3D print these design components. According to Zhang et al. (2020), the resultant structures showed strain-hardening behavior and better ductility when compared to pure concrete. However, the application of AM processes in construction faces hindrances, particularly in integrating conventional reinforcements into 3D-printed structures. Overcoming these challenges will require further research and development efforts to fully realize the potential of additive manufacturing in revolutionizing the construction industry.

5.3.5 Aerospace

In addition to its capacity to create complicated geometries and lower development costs, additive manufacturing (AM) techniques like laser-based 3D printing are widely employed in the aerospace and military sectors. A Service Level Agreement (SLA), Flight Data Monitoring (FDM), and jet-type processes like polyjet, etc. are among the most widely utilized polymer-based AM techniques in aerospace applications (Najmon et al., 2019). SLA-based products are ideally suited for casting applications in the aerospace sector since, according to studies, SLA-made components are more durable than FDM-made ones (Van Grootel et al., 2020).

The aircraft industry needs parts that are constructed from materials that are strong and light at the same time. AM makes this feasible and allows for on-demand manufacturing. About 70 pieces of the Desert RATS rover were created by NASA using additive manufacturing. These components include camera mounts, big pod doors, front bumpers, flame-retardant vents and housings, intricate electronics, etc. The FDM technique was utilized in the fabrication of ABS, PC, and PCABS parts (Nickels, 2015). We have looked at AM's potential uses for unmanned vehicles. The potential of AM in unmanned vehicles includes multifunctional constructions, on-site, on-demand printing, printing with numerous materials, including smart materials. Graphene, carbon nanotubes, graphene oxide, and clay are the most often employed nanocomposites in aeronautical applications. These nanocomposites do not, however, match the strength standards needed for aircraft constructions. Consequently, multiscale composites—a hybrid composite that combines nanocomposites and conventional composites—are created.

5.3.6 Energy

Energy and sustainable energy have become paramount global concerns due to the escalating depletion of fossil fuels (Kabeyi & Olanrewaju, 2022). Numerous studies on material- and nanomaterial-based polymers and their composites for a range of energy-related applications have been prompted by this increased awareness. Battery, solar cell, supercapacitor, home tool, car, fuel cell, biomedical, and surgical items are just a few of the many applications that these technologies may be used for. Conducting polymers, a subset of organic polymers capable of conducting electricity, has garnered significant attention in the energy sector (Ramasubramanian et al.,

2022). These polymers were created about 20 years ago and are also referred to as typically conducting polymers or electroactive conjugated polymers. Delocalized molecular orbitals are what give them their electrical conductivity.

The economic significance, environmental stability, electrical conductivity, and multifunctional properties of conducting polymers have made them highly sought-after. Their unique nanostructures with a greater specific capacitance make them a viable substitute for the creation of next-generation energy storage devices (Liu et al., 2021). Ionic conducting polymers, intrinsically conducting polymers (ICPs), also referred to as synthetic metals, and conducting polymer composites are the three main categories into which conducting polymers are usually divided. Each group possesses unique properties and potential applications in the realm of energy storage and production (Hatchett & Josowicz, 2008).

5.3.7 Sports equipment

Sports spectacles, golf putters, rackets, bicycles, and helmets are among the many sporting goods that are in great demand for lightweight and strong frames. In addition to conducting in-depth research on the mechanical characteristics of printed components, such as strength, toughness, and ductility, a major emphasis is on customization of these products (Yuan et al., 2019). They produced laser-sintered composites with remarkable specific mechanical strength and toughness, for instance, by developing CNT-coated PA12 powders for SLS. According to Biehler (2018), carbon fiber-reinforced polymers have proven to have more mechanical strength than other commercial composites. One example of this is microcarbon fiber/PEEK via SLS.

Carbon composites, due to their properties, hold immense potential for lightweight and customizable sports gear. Furthermore, the production of textiles and shoes has seen a rise in the usage of pliable, soft materials like TPE. Products made of polyurethane (PU) based on esters are especially noteworthy for their excellent flexibility, resistance to oils and solvents, and wear resistance. Researchers have assessed the nanosilica/TPU powders' processability and optimized settings to print complicated auxetic lattices using SLS (Bai et al., 2017). Given that these printed soft-material products are intended to be used in protective sports equipment and footwear, their damping and energy-dissipating properties are important factors to consider.

5.3.8 Mechanical engineering

Polymers are rarely, directly used in structural or thermal applications in mechanical engineering because of their poor mechanical characteristics. Different composites have been created and evaluated for their mechanical and thermal properties to overcome this constraint. For low-strength engineering applications, such as architectural modeling, interior design, and indoor paneling, for example, HDPE/cardboard dust composites provide an affordable and sustainable option (Li et al., 2023). The use of 3D printing (3DP) processes in thermal sciences has garnered attention due to its ability to manufacture complex geometries. This fabrication method allows for a high volume-to-area ratio, enabling the creation of compact and efficient heat exchangers.

Researchers are actively employing numerical and experimental tools to evaluate the thermal performance of 3D-printed polymer composites, enhancing their applicability in thermal engineering (Liu et al., 2021). These materials, even with tiny amounts of reinforcement, show promise for robotic applications. Furthermore, because of their strong form memory force and recovery capabilities, the 4D-printed CNT-reinforced PLA shape memory composites show promise for use in actuators (Zaharia et al., 2023).

5.4 Challenges of additively manufactured functional polymers and nanocomposites

5.4.1 Cost

Cost is a significant challenge associated with additively manufactured (AM) functional polymers and nanocomposites, impacting their widespread adoption across industries (Al Rashid et al., 2021). The initial investment in AM technology, including the procurement of high-quality printers, materials, and postprocessing equipment, can be substantial, posing financial barriers for smaller businesses and startups. Moreover, the cost of raw materials used in AM, especially specialized polymers, and nanocomposites, tends to be higher than traditional manufacturing materials due to factors such as material purity, customization requirements, and limited supplier availability (Ligon et al., 2017).

Beyond initial setup costs, operational expenses in the AM, including energy consumption, maintenance of equipment, and labor costs, contribute to the overall production expenditure. The complexity of AM processes, such as parameter optimization, support structure generation, and postprocessing steps like surface finishing and quality inspection, adds to the cost per part produced (Peng et al., 2021). Additionally, the need for skilled personnel proficient in AM techniques further adds to the operational expenses for training and workforce management. Efforts to address cost challenges in AM of functional polymers and nanocomposites include research into cost-effective material formulations, process optimization for material efficiency, and the development of cost estimation models tailored for AM workflows (Walia et al., 2021). Collaborative initiatives among industry players, academia, and government bodies are essential to drive down costs through economies of scale, standardization of AM practices, and incentivizing innovation in cost-effective AM technologies.

5.4.2 Reusability

Reusability poses a significant challenge in the domain of additively manufactured (AM) functional polymers and nanocomposites. Unlike conventional manufacturing methods where unused materials can often be recycled or repurposed, AM processes generate substantial material waste due to support structures, failed prints, and unutilized build volume. This excess waste not only escalates production costs but also raises environmental concerns regarding material disposal and the depletion of resources.

Moreover, the reusability of AM materials is impeded by factors such as material degradation during printing, postprocessing requirements that render materials unsuitable for reuse, and limitations in recycling technologies for complex polymer-based structures (Singh & Singh, 2017). Efforts to tackle these reusability challenges are underway, focusing on the development of recyclable polymers and nanocomposites tailored specifically for AM processes.

Additionally, there is a push for the implementation of efficient material recovery and recycling methods, alongside the adoption of closed-loop manufacturing systems to minimize waste and maximize material utilization. Collaboration among stakeholders across the AM ecosystem, including material suppliers, manufacturers, recycling facilities, and regulatory bodies, is crucial to advancing sustainable practices and enhancing the reusability of AM materials (Govindan et al., 2016).

5.4.3 Reliability

Reliability is a paramount challenge faced in the domain of additively manufactured (AM) functional polymers and nanocomposites, impacting their usability in critical applications (Shanthar et al., 2023). Variability in mechanical properties across batches, potential degradation over time, and uncertainties in meeting regulatory standards pose significant concerns regarding the consistency and durability of AM-produced parts. Factors such as build orientation, layer adhesion, and material homogeneity contribute to variations in mechanical strength and fatigue resistance within printed components (Wu et al., 2018).

Moreover, environmental factors like UV radiation and moisture can adversely affect the long-term performance of polymer-based materials, further complicating reliability issues in safety-critical applications (Al-Maharma et al., 2020). To tackle these challenges, ongoing research focuses on enhancing material consistency through improved processing techniques and material formulations. Optimization of AM processes to reduce defects and enhance structural integrity is another area of active investigation.

Additionally, the development of robust quality control measures tailored for AM materials and processes is imperative to ensure reliability. Efforts among industry stakeholders, researchers, and regulatory bodies are essential to establish standardized certification protocols that validate the reliability and performance of AM-produced functional polymers and nanocomposites. By advancing materials development, refining quality assurance techniques, and fostering industry-wide cooperation, the reliability of AM-produced functional polymers and nanocomposites can be significantly improved, paving the way for their widespread adoption across various sectors (Ferracane, 2011).

5.4.4 Material waste

Material waste presents a significant challenge in the realm of additively manufactured (AM) functional polymers and nanocomposites. Unlike traditional subtractive manufacturing processes, where excess material is primarily in the form of chips or shavings, AM often generates considerable waste in the form of unused powder, support structures, and failed prints (Sola & Trinchi, 2023). This surplus waste not only increases production costs but also has environmental repercussions, including resource depletion and landfill accumulation.

Furthermore, the disposal of AM waste materials requires careful consideration due to their composite nature and potential hazards, necessitating specialized recycling and waste management strategies (Faludi et al., 2017). Addressing the issue of material waste in AM involves innovative approaches across the entire manufacturing lifecycle. This includes optimizing print parameters to reduce support material usage, implementing design strategies that minimize material consumption, and exploring alternative materials with improved recyclability. Additionally, advancements in AM technologies such as powder recovery systems and postprocessing techniques like reusing excess material can contribute to mitigating material waste (El-Saadony et al., 2023).

5.4.5 Scale up

Scale-up in additive manufacturing (AM) of functional polymers and nanocomposites refers to the challenge of transitioning from small-scale production to larger volumes while maintaining quality and cost-effectiveness. This process involves managing increased material consumption, optimizing production time, and ensuring consistency and uniformity across larger print sizes (Wang et al., 2021). Challenges also arise in scaling up postprocessing steps such as curing, finishing, and inspection to meet the demands of mass production. To address the challenges of scale-up in AM, researchers have focused on developing scalable printing technologies and optimizing production workflows. Advanced printing techniques such as large-scale extrusion-based AM systems and high-speed powder bed fusion systems have been developed to facilitate the production of larger parts (Colledani et al., 2014).

Additionally, process monitoring and control systems have been implemented to ensure quality and reliability during large-scale production runs. Collaborative efforts between industry experts, research institutions, and technology providers are essential in advancing scalable AM solutions for functional polymers and nanocomposites. By integrating innovative technologies and best practices, scale-up challenges can be effectively addressed to enable the widespread adoption of AM in industrial manufacturing (Carrola et al., 2021).

5.4.6 Process optimization

Process optimization in additive manufacturing (AM) of functional polymers and nanocomposites involves enhancing efficiency, quality, and cost-effectiveness throughout the production process. One of the key challenges in process optimization is the need to fine-tune printing parameters such as layer thickness, print speed, and material deposition rates to achieve optimal part quality and mechanical properties (Baturynska et al., 2018). This optimization requires extensive experimentation and analysis to determine the ideal process settings for different materials and applications. Advanced process optimization techniques, including design of experiments (DOE), computational modeling, and machine learning algorithms, have been employed to streamline the AM process (Champa-Bujaico et al., 2022). These methods help in identifying the optimal combination of printing parameters while reducing trial and error iterations, thus accelerating the development and production of functional polymers and nanocomposites (Erps et al., 2021).

5.5 Conclusion

Expanding the material performance range through the development of polymer nanocomposites for additive manufacturing allows for the fabrication of multifunctional parts with more design flexibility. Rather than taking the place of traditional production methods entirely, additive manufacturing (AM) should enhance and strengthen them. Multifunctional polymer composite production is made possible by 3D printing, and the resulting device has a wide range of applications, from aeronautics to bioelectronics. Designers must develop new design processes to fully benefit from AM. Numerous industries, including but not limited to the oil and gas, aerospace, automotive, and medical sectors, may benefit from this. It is anticipated that material optimization, which includes process development and material formulation, will greatly aid in improving the mechanical, electrical, electrochemical, thermal, and optical properties of composites. Fundamental knowledge will be advanced and eventually, breakthroughs in energy, the environment, manufacturing, healthcare, mechanics, and acoustics will be made possible by structural topology. It is anticipated that the latest advancements in polymer materials for AM will stimulate more interest in and advancements in this area of study. It is just a matter of time until additive manufacturing (AM) emerges as a viable technology that supports conventional manufacturing methods, given the speed at which technology is developing.

The challenges for AM-printed materials and structures form the basis for progress in creating information that has the potential to revolutionize technology. It is strongly anticipated that new capacities in function and performance will be made possible by the design of novel materials and our growing understanding of their structures and properties. In the future, multiprocess printing could be able to address issues with functional limits, material selection, precision control, and other issues by effectively integrating AM techniques with other manufacturing and treatment processes. To improve the quality of products made through recycling, future research should concentrate on creating safer and more robust polymeric materials for additive manufacturing (AM).

Acknowledgments

The study was supported by the DST major project "Low Cost-Renewable Energy Driven (LC-RED) Water Treatment Solution Centre" (Ref No. DST/TM/WTI/WIC/2K17/124), the DBT

BIG-BIRAC Ref No. BIRAC/CCAMP0723/BIG-13/18, SERB-POWER major project entitled "Upgradation in MVP with Improved Adsorption, Regeneration, and Fluoride Sensing Capacities" File No. 'SPF/2022/000086' and B.B. and I.D. acknowledge the support of the ISF-NSFC joint research program (grant No. 3106/19).

References

Al Rashid, A., Khan, S. A., Al-Ghamdi, S. G., & Koç, M. (2021). Additive manufacturing of polymer nanocomposites: Needs and challenges in materials, processes, and applications. *Journal of Materials Research and Technology, 14*, 910–941.

Al-Maharma, A. Y., Patil, S. P., & Markert, B. (2020). Effects of porosity on the mechanical properties of additively manufactured components: a critical review. *Materials Research Express, 7*(12), 122001.

Bai, J., Yuan, S., Shen, F., Zhang, B., Chua, C. K., Zhou, K., & Wei, J. (2017). Toughening of polyamide 11 with carbon nanotubes for additive manufacturing. *Virtual and Physical Prototyping, 12*(3), 235–240.

Baturynska, I., Semeniuta, O., & Martinsen, K. (2018). Optimization of process parameters for powder bed fusion additive manufacturing by combination of machine learning and finite element method: A conceptual framework. *Procedia Cirp, 67*, 227–232.

Beniak, J. U. R. A. J., Krizan, P. E. T. E. R., & Matus, M. I. L. O. S. (2020). Conductive material properties for FDM additive manufacturing. *MM Science Journal, 2020*(1), 3846–3851.

Bhushan, B. (2003). Adhesion and stiction: Mechanisms, measurement techniques, and methods for reduction. *Journal of Vacuum Science & Technology B: Microelectronics and Nanometer Structures Processing, Measurement, and Phenomena, 21*(6), 2262–2296.

Biehler J. The future of 3D printed sports equipment, 2017 [accessed June 2018].

Blakey-Milner, B., Gradl, P., Snedden, G., Brooks, M., Pitot, J., Lopez, E., & Du Plessis, A. (2021). Metal additive manufacturing in aerospace: A review. *Materials & Design, 209*, 110008.

Cardoso, R. M., Kalinke, C., Rocha, R. G., Dos Santos, P. L., Rocha, D. P., Oliveira, P. R., & Munoz, R. A. (2020). Additive-manufactured (3D-printed) electrochemical sensors: A critical review. *Analytica Chimica Acta, 1118*, 73–91.

Carrola, M., Asadi, A., Zhang, H., Papageorgiou, D. G., Bilotti, E., & Koerner, H. (2021). Best of both worlds: Synergistically derived material properties via additive manufacturing of nanocomposites. *Advanced Functional Materials, 31*(46), 2103334.

Champa-Bujaico, E., García-Díaz, P., & Díez-Pascual, A. M. (2022). Machine learning for property prediction and optimization of polymeric nanocomposites: a state-of-the-art. *International Journal of Molecular Sciences, 23*(18), 10712.

Chen, L., Phang*, I. Y., Wong, S. C., Lv, P. F., & Liu, T. (2006). Embrittlement mechanisms of nylon 66/organoclay nanocomposites prepared by melt-compounding process. *Materials and Manufacturing Processes, 21*(2), 153–158.

Chiu, F. C., Chen, C. C., & Chen, Y. J. (2014). Binary and ternary nanocomposites based on PVDF, PMMA, and PVDF/PMMA blends: polymorphism, thermal, and rheological properties. *Journal of Polymer Research, 21*, 1–12.

Chiu, F. C., Chuang, Y. C., Liao, S. J., & Chang, Y. H. (2018). Comparison of PVDF/PVAc/GNP and PVDF/PVAc/CNT ternary nanocomposites: Enhanced thermal/electrical properties and rigidity. *Polymer Testing, 65*, 197–205.

Colledani, M., Tolio, T., Fischer, A., Iung, B., Lanza, G., Schmitt, R., & Váncza, J. (2014). Design and management of manufacturing systems for production quality. *Cirp Annals, 63*(2), 773–796.

Dalpadulo, E., Petruccioli, A., Gherardini, F., & Leali, F. (2022). A review of automotive spare-part reconstruction based on additive manufacturing. *Journal of Manufacturing and Materials Processing, 6*(6), 133.

da Silva, L. R. R., Sales, W. F., Campos, F. D. A. R., de Sousa, J. A. G., Davis, R., Singh, A., & Borgohain, B. (2021). A comprehensive review on additive manufacturing of medical devices. *Progress in Additive Manufacturing, 6*(3), 517–553.

De, A., Ramasubramian, B., Ramakrishna, S., & Chellappan, V. (2024). Advances in additive manufacturing techniques for electrochemical energy storage. *Advanced Materials Technologies, 9*(4), 2301439.

El-Saadony, M. T., Saad, A. M., El-Wafai, N. A., Abou-Aly, H. E., Salem, H. M., Soliman, S. M., & AbuQamar, S. F. (2023). Hazardous wastes and management strategies of landfill leachates: A comprehensive review. *Environmental Technology & Innovation*103150.

Erps, T., Foshey, M., Luković, M. K., Shou, W., Goetzke, H. H., Dietsch, H., & Matusik, W. (2021). Accelerated discovery of 3D printing materials using data-driven multiobjective optimization. *Science Advances, 7*(42), eabf7435.

Faludi, J., Cline-Thomas, N., & Agrawala, S. (2017). 3D printing and its environmental implications. The Next Production Revolution Implications for *Governments and Business, Organization for Economic Cooperation and* Development (OECD). Paris: OECD Publishing.

Farah, S., Anderson, D. G., & Langer, R. (2016). Physical and mechanical properties of PLA, and their functions in widespread applications—A comprehensive review. *Advanced Drug Delivery Reviews, 107*, 367–392.

Ferracane, J. L. (2011). Resin composite—state of the art. *Dental Materials, 27*(1), 29–38.

Fijoł, N., Aguilar-Sánchez, A., & Mathew, A. P. (2022). 3D-printable biopolymer-based materials for water treatment: A review. *Chemical Engineering Journal, 430*, 132964.

Francis, M. P., Kemper, N., Maghdouri-White, Y., & Thayer, N. (2018). *Additive manufacturing for biofabricated medical device applications.* Additive manufacturing. Butterworth-Heinemann311–344.

Gao, D., Zhang, Y., Lyu, B., Wang, P., & Ma, J. (2019). Nanocomposite based on poly (acrylic acid)/attapulgite towards flame retardant of cotton fabrics. *Carbohydrate Polymers, 206*, 245–253.

Gao, J., Zhang, Q., Wang, K., Fu, Q., Chen, Y., Chen, H., & Rego, J. M. (2012). Effect of shearing on the orientation, crystallization and mechanical properties of HDPE/attapulgite nanocomposites. *Composites Part A: Applied Science and Manufacturing, 43*(4), 562–569.

Ghosal, P., Gupta, B., Ambekar, R. S., Rahman, M. M., Ajayan, P. M., Aich, N., & Tiwary, C. S. (2022). 3D printed materials in water treatment applications. *Advanced Sustainable Systems, 6*(3), 2100282.

Gibson, I., Rosen, D., Stucker, B., Khorasani, M., Gibson, I., Rosen, D., & Khorasani, M. (2021). Design for additive manufacturing. *Additive Manufacturing Technologies*, 555–607.

Gironi, F., & Piemonte, V. (2011). Life cycle assessment of polylactic acid and polyethylene terephthalate bottles for drinking water. *Environmental Progress & Sustainable Energy, 30*(3), 459–468.

Gnanasekaran, K., Heijmans, T., Van Bennekom, S., Woldhuis, H., Wijnia, S., De With, G., & Friedrich, H. (2017). 3D printing of CNT-and graphene-based conductive polymer nanocomposites by fused deposition modeling. *Applied Materials Today, 9*, 21–28.

Goh, G. D., Agarwala, S., Goh, G. L., Dikshit, V., Sing, S. L., & Yeong, W. Y. (2017). Additive manufacturing in unmanned aerial vehicles (UAVs): Challenges and potential. *Aerospace Science and Technology, 63*, 140–151.

Gong, X., Chen, X., & Zhou, Y. (2018). *Advanced weaving technologies for high-performance fabrics. High-performance apparel.* Woodhead Publishing75–112.

González, T. V., Salazar, C. G., De la Rosa, J. R., & González, V. G. (2008). Nylon 6/ organoclay nanocomposites by extrusion. *Journal of Applied Polymer Science, 108*(5), 2923–2933.

Govindan, K., Jha, P. C., & Garg, K. (2016). Product recovery optimization in closed-loop supply chain to improve sustainability in manufacturing. *International Journal of Production Research, 54*(5), 1463–1486.

Hager, M. D., Bode, S., Weber, C., & Schubert, U. S. (2015). Shape memory polymers: Past, present and future developments. *Progress in Polymer Science, 49*, 3–33.

Han, P. (2017). Additive design and manufacturing of jet engine parts. *Engineering, 3*(5), 648–652.

Hatchett, D. W., & Josowicz, M. (2008). Composites of intrinsically conducting polymers as sensing nanomaterials. *Chemical Reviews, 108*(2), 746–769.

Heimowska, A., Morawska, M., & Bocho-Janiszewska, A. (2017). Biodegradation of poly (ε-caprolactone) in natural water environments. *Polish Journal of Chemical Technology, 19*(1), 120–126.

Hsu, Y. G., & Lin, F. J. (2000). Organic–inorganic composite materials from acrylonitrile–butadiene–styrene copolymers (ABS) and silica through an in situ sol-gel process. *Journal of Applied Polymer Science, 75*(2), 275–283.

Jeong, S., Song, Y. S., & Lim, E. (2020). Fabrication and characterization of aluminum nanoparticle-reinforced composites. *Polymers, 12*(12), 2772.

Jiang, J., & Ma, Y. (2020). Path planning strategies to optimize accuracy, quality, build time and material use in additive manufacturing: a review. *Micromachines, 11*(7), 633.

Kabeyi, M. J. B., & Olanrewaju, O. A. (2022). Sustainable energy transition for renewable and low carbon grid electricity generation and supply. *Frontiers in Energy Research, 9*.

Khan, I., Dutta, J. R., & Ganesan, R. (2017). Lactobacillus sps. lipase mediated poly (ε-caprolactone) degradation. *International Journal of Biological Macromolecules, 95*, 126–131.

Koh, J. J., Zhang, X., & He, C. (2018). Fully biodegradable Poly (lactic acid)/Starch blends: A review of toughening strategies. *International Journal of Biological Macromolecules, 109*, 99–113.

Komissarenko, D. A., Sokolov, P. S., Evstigneeva, A. D., Slyusar, I. V., Nartov, A. S., Volkov, P. A., & Dosovitsky, A. E. (2021). DLP 3D printing of scandia-stabilized zirconia ceramics. *Journal of the European Ceramic Society, 41*(1), 684–690.

Kumar, L. J., Pandey, P. M., & Wimpenny, D. I. (Vol. Eds.), (2019). *3D Printing and additive manufacturing technologies: Vol. 311*. Singapore: Springer.

Lamichhane, T. N., Sethuraman, L., Dalagan, A., Wang, H., Keller, J., & Paranthaman, M. P. (2020). Additive manufacturing of soft magnets for electrical machines—A review. *Materials Today Physics, 15*, 100255.

Lan, T., & Pinnavaia, T. J. (1994). Clay-reinforced epoxy nanocomposites. *Chemistry of Materials, 6*(12), 2216–2219.

Li, C., Xiang, M., Zhao, X., & Ye, L. (2017). In situ synthesis of monomer casting nylon-6/ graphene-polysiloxane nanocomposites: intercalation structure, synergistic reinforcing, and friction-reducing effect. *ACS Applied Materials & Interfaces, 9*(38), 33176–33190.

Li, Y., Ren, X., Zhu, L., & Li, C. (2023). Biomass 3D printing: Principles, materials, post-processing and applications. *Polymers, 15*(12), 2692.

Ligon, S. C., Liska, R., Stampfl, J., Gurr, M., & Mülhaupt, R. (2017). Polymers for 3D printing and customized additive manufacturing. *Chemical Reviews, 117*(15), 10212–10290.

Liu, C., Luan, P., Li, Q., Cheng, Z., Xiang, P., Liu, D., & Zhu, H. (2021). Biopolymers derived from trees as sustainable multifunctional materials: A review. *Advanced Materials, 33*(28), 2001654.

Liu, P., Huang, S. H., Mokasdar, A., Zhou, H., & Hou, L. (2014). The impact of additive manufacturing in the aircraft spare parts supply chain: Supply chain operation reference (SCOR) model based analysis. *Production Planning & Control, 25*(13-14), 1169–1181.

Mehrpouya, M., Dehghanghadikolaei, A., Fotovvati, B., Vosooghnia, A., Emamian, S. S., & Gisario, A. (2019). The potential of additive manufacturing in the smart factory industrial 4.0: A review. *Applied Sciences, 9*(18), 3865.

Momeni, F., Liu, X., & Ni, J. (2017). A review of 4D printing. *Materials & Design, 122*, 42–79.

Musial, J., Krakowiak, R., Mlynarczyk, D. T., Goslinski, T., & Stanisz, B. J. (2020). Titanium dioxide nanoparticles in food and personal care products—What do we know about their safety? *Nanomaterials, 10*(6), 1110.

Najmon, J. C., Raeisi, S., & Tovar, A. (2019). Review of additive manufacturing technologies and applications in the aerospace industry. *Additive Manufacturing for the Aerospace Industry*, 7–31.

Ngo, T. D., Kashani, A., Imbalzano, G., Nguyen, K. T., & Hui, D. (2018). Additive manufacturing (3D printing): A review of materials, methods, applications and challenges. *Composites Part B: Engineering, 143*, 172–196.

Nickels, L. (2015). AM and aerospace: an ideal combination. *Metal Powder Report, 70*(6), 300–303.

Nunes, S. P., Culfaz-Emecen, P. Z., Ramon, G. Z., Visser, T., Koops, G. H., Jin, W., & Ulbricht, M. (2020). Thinking the future of membranes: Perspectives for advanced and new membrane materials and manufacturing processes. *Journal of Membrane Science, 598*, 117761.

Patton, S. T., Chen, C., Hu, J., Grazulis, L., Schrand, A. M., & Roy, A. K. (2016). Characterization of thermoplastic polyurethane (TPU) and Ag-carbon black TPU nanocomposite for potential application in additive manufacturing. *Polymers, 9*(1), 6.

Peng, X., Kong, L., Fuh, J. Y. H., & Wang, H. (2021). A review of post-processing technologies in additive manufacturing. *Journal of Manufacturing and Materials Processing, 5*(2), 38.

Ramasubramanian, B., Sundarrajan, S., Rao, R. P., Reddy, M. V., Chellappan, V., & Ramakrishna, S. (2022). Novel low-carbon energy solutions for powering emerging wearables, smart textiles, and medical devices. *Energy & Environmental Science, 15*(12), 4928–4981.

Rao, C. H., Avinash, K., Varaprasad, B. K. S. V. L., & Goel, S. (2022). A review on printed electronics with digital 3D printing: fabrication techniques, materials, challenges and future opportunities. *Journal of Electronic Materials, 51*(6), 2747–2765.

Sathishkumar, T. P., Naveen, J. A., & Satheeshkumar, S. (2014). Hybrid fiber reinforced polymer composites—Ā review. *Journal of Reinforced Plastics and Composites, 33*(5), 454–471.

Savastano, M., Amendola, C., D′ Ascenzo, F., & Massaroni, E. (2016). *3-D printing in the spare parts supply chain: An explorative study in the automotive industry. Digitally supported innovation: A multi-disciplinary view on enterprise, public sector and user innovation.* Springer International Publishing153–170.

Schmitz, D. P., Ecco, L. G., Dul, S., Pereira, E. C. L., Soares, B. G., Barra, G. M. O., & Pegoretti, A. (2018). Electromagnetic interference shielding effectiveness of ABS carbon-based composites manufactured via fused deposition modelling. *Materials Today. Communications, 15*, 70–80.

Sha, L., Chen, Z., Chen, Z., Zhang, A., & Yang, Z. (2016). Polylactic acid based nanocomposites: Promising safe and biodegradable materials in biomedical field. *International Journal of Polymer Science,* 2016.

Shahrubudin, N., Koshy, P., Alipal, J., Kadir, M. H. A., & Lee, T. C. (2020). Challenges of 3D printing technology for manufacturing biomedical products: A case study of Malaysian manufacturing firms. *Heliyon, 6*(4).

Shanthar, R., Chen, K., & Abeykoon, C. (2023). Powder-based additive manufacturing: A critical review of materials, methods, opportunities, and challenges. *Advanced Engineering Materials, 25*(19), 2300375.

Singh, R. L., & Singh, P. K. (2017). Global environmental problems. Principles and applications of environmental biotechnology for a sustainable. *Future,* 13–41.

Sola, A., & Trinchi, A. (2023). Recycling as a key enabler for sustainable additive manufacturing of polymer composites: A critical perspective on fused filament fabrication. *Polymers, 15*(21), 4219.

Su, L., Tao, Q., He, H., Zhu, J., & Yuan, P. (2012). Locking effect: A novel insight in the silylation of montmorillonite surfaces. *Materials Chemistry and Physics, 136*(2-3), 292–295.

Tan, L. J., Zhu, W., & Zhou, K. (2020). Recent progress on polymer materials for additive manufacturing. *Advanced Functional Materials, 30*(43), 2003062.

Van Grootel, A., Chang, J., Wardle, B. L., & Olivetti, E. (2020). Manufacturing variability drives significant environmental and economic impact: The case of carbon fiber reinforced polymer composites in the aerospace industry. *Journal of Cleaner Production, 261*, 121087.

Vidal, J. L., Yavitt, B. M., Wheeler, M. D., Kolwich, J. L., Donovan, L. N., Sit, C. S., & Kerton, F. M. (2022). Biochar as a sustainable and renewable additive for the production of Poly (ε-caprolactone) composites. *Sustainable Chemistry and Pharmacy, 25*, 100586.

Walia, K., Khan, A., & Breedon, P. (2021). Polymer-based additive manufacturing: process optimisation for low-cost industrial robotics manufacture. *Polymers, 13*(16), 2809.

Wang, J., Zhang, Y., Aghda, N. H., Pillai, A. R., Thakkar, R., Nokhodchi, A., & Maniruzzaman, M. (2021). Emerging 3D printing technologies for drug delivery devices: Current status and future perspective. *Advanced Drug Delivery Reviews, 174*, 294–316.

Wang, M. S., & Pinnavaia, T. J. (1994). Clay-polymer nanocomposites formed from acidic derivatives of montmorillonite and an epoxy resin. *Chemistry of Materials, 6*(4), 468–474.

Wang, X. (2011). Investigation of electromagnetic shielding effectiveness of nanostructural carbon black/ABS composites. *Journal of Electromagnetic Analysis and Applications,* 2011.

Wang, Z., Lan, T., & Pinnavaia, T. J. (1996). Hybrid organic–inorganic nanocomposites formed from an epoxy polymer and a layered silicic acid (Magadiite). *Chemistry of Materials, 8*(9), 2200–2204.

Wimpenny, D. I., Pandey, P. M., & Kumar, L. J. (Vol. Eds.), (2017). *Advances in 3D printing & additive manufacturing technologies: Vol. 1*, (pp. 1–184). Singapore: Springer.

Wu, B., Pan, Z., Ding, D., Cuiuri, D., Li, H., Xu, J., & Norrish, J. (2018). A review of the wire arc additive manufacturing of metals: Properties, defects and quality improvement. *Journal of Manufacturing Processes, 35*, 127–139.

Yuan, S., Shen, F., Chua, C. K., & Zhou, K. (2019). Polymeric composites for powder-based additive manufacturing: Materials and applications. *Progress in Polymer Science, 91*, 141–168.

Yusoff, N. H. M., Teo, L. R. I., Phang, S. J., Wong, V. L., Cheah, K. H., & Lim, S. S. (2022). Recent advances in polymer-based 3D printing for wastewater treatment application: An overview. *Chemical Engineering Journal, 429*, 132311.

Zaharia, S. M., Pop, M. A., Cosnita, M., Croitoru, C., Matei, S., & Spîrchez, C. (2023). Sound absorption performance and mechanical properties of the 3D-printed bio-degradable panels. *Polymers, 15*(18), 3695.

Zhang, P., Wang, Z., Li, J., Li, X., & Cheng, L. (2020). From materials to devices using fused deposition modeling: A state-of-art review. *Nanotechnology Reviews, 9*(1), 1594–1609.

Zhang, W., Li, J., Shang, Y., Li, H., Jiang, S., & An, L. (2017). Temperature dependence of tensile behavior in poly (butylene terephthalate) with different crystallinity. *Materials & Design, 129*, 143–150.

Zheng, C., Zhao, G., Shen, Y., Zhu, B., & He, Y. (2017). Effect of a polyetheramine additive on the melt-flowability of poly (butylene terephthalate). *Polymer Testing, 61*, 191–196.

Zhou, J., & Sheiko, S. S. (2016). Reversible shape-shifting in polymeric materials. *Journal of Polymer Science Part B: Polymer Physics, 54*(14), 1365–1380.

Further reading

Adekanye, A., Mahamood, R.M., Akinlabi, E.T., & Owolabi, M.G. (2017). Additive manufacturing: the future of manufacturing. *Materials and Technologies*, 51(5), 709–715.

Aris, M. H., Annuar, M. S. M., & Ling, T. C. (2016). Lipase-mediated degradation of poly-ε-caprolactone in toluene: Behavior and its action mechanism. *Polymer Degradation and Stability, 133*, 182–191.

Chiu, F. C. (2014). Comparisons of phase morphology and physical properties of PVDF nanocomposites filled with organoclay and/or multi-walled carbon nanotubes. *Materials Chemistry and Physics, 143*(2), 681–692.

Di Cataldo, S., Vinco, S., Urgese, G., Calignano, F., Ficarra, E., Macii, A., & Macii, E. (2021). Optimizing quality inspection and control in powder bed metal additive manufacturing: Challenges and research directions. *Proceedings of the IEEE, 109*(4), 326–346.

Fu, Y. Q., Huang, W. M., Luo, J. K., & Lu, H. (2015). *Polyurethane shape-memory polymers for biomedical applications. Shape memory polymers for biomedical applications*. Woodhead Publishing167–195.

Gao, F., Son, S., Park, K., Biggs, D., Andrews, C., Mockensturm, E. M., & Goulbourne, N. C. (2011). *On a novel self-regulating shape memory polymer composite. Behavior and mechanics of multifunctional materials and composites 2011*, *Vol. 7978*, SPIE314–325.

Goh, G. D., Yap, Y. L., Tan, H. K. J., Sing, S. L., Goh, G. L., & Yeong, W. Y. (2020). Process–structure–properties in polymer additive manufacturing via material extrusion: A review. *Critical Reviews in Solid State and Materials Sciences, 45*(2), 113–133.

González-López, M. E., del Campo, A. S. M., Robledo-Ortíz, J. R., Arellano, M., & Pérez-Fonseca, A. A. (2020). Accelerated weathering of poly (lactic acid) and its biocomposites: A review. *Polymer Degradation and Stability, 179*, 109290.

Krueger, H. (2017). Standardization for additive manufacturing in aerospace. *Engineering, 3*(5), 585.

Kumar, A., Gori, Y., Dutt, N., Singla, Y. K., & Maurya, A. (Eds.). (2021). *Advanced computational methods in mechanical and materials engineering*. CRC Press.

Ma, Q., Shi, K., Su, T., & Wang, Z. (2020). Biodegradation of polycaprolactone (PCL) with different molecular weights by *Candida antarctica* lipase. *Journal of Polymers and the Environment, 28*, 2947–2955.

Omidvarkarjan, D., Rosenbauer, R., Klahn, C., & Meboldt, M. (2023). *Implementation of additive manufacturing in industry. Springer handbook of additive manufacturing*. Cham: Springer International Publishing55–71.

Patton, S. T., Frasca, A. J., Hu, J., Varshney, V., Phillips, B. S., Roy, A. K., & Voevodin, A. A. (2016). Multiphysics characterization of multi-walled carbon nanotube thermoplastic polyurethane polymer nanocomposites during compression. *Carbon, 98*, 638–648.

Rao, & Pochan, J. M. (2007). Mechanics of polymer–clay nanocomposites. *Macromolecules, 40*(2), 290–296.

Shi, H., Lan, T., & Pinnavaia, T. J. (1996). Interfacial effects on the reinforcement properties of polymer–organoclay nanocomposites. *Chemistry of Materials, 8*(8), 1584–1587.

Shin, D. G., Kim, T. H., & Kim, D. E. (2017). Review of 4D printing materials and their properties. *International Journal of Precision Engineering and Manufacturing-Green Technology, 4*, 349–357.

Tsou, C. H., Guo, J., Lei, J. A., De Guzman, M. R., & Suen, M. C. (2020). Characterizing attapulgite-reinforced nanocomposites of poly (lactic acid). *Polymer Science, Series A, 62*, 732–743.

Vicente, C. M., Sardinha, M., Reis, L., Ribeiro, A., & Leite, M. (2023). Large-format additive manufacturing of polymer extrusion-based deposition systems: Review and applications. *Progress in Additive Manufacturing, 8*(6), 1257–1280.

Wu, H., Fahy, W. P., Kim, S., Kim, H., Zhao, N., Pilato, L., & Koo, J. H. (2020). Recent developments in polymers/polymer nanocomposites for additive manufacturing. *Progress in Materials Science, 111*, 100638.

Yakacki, C. M., Shandas, R., Lanning, C., Rech, B., Eckstein, A., & Gall, K. (2007). Unconstrained recovery characterization of shape-memory polymer networks for cardiovascular applications. *Biomaterials, 28*(14), 2255–2263.

Yang, J., Wang, J., Zhang, Q., Chen, F., Deng, H., Wang, K., & Fu, Q. (2011). Cooperative effect of shear and nanoclay on the formation of polar phase in poly (vinylidene fluoride) and the resultant properties. *Polymer, 52*(21), 4970–4978.

Zarna, C., Rodríguez-Fabià, S., Echtermeyer, A. T., & Chinga-Carrasco, G. (2022). Preparation and characterisation of biocomposites containing thermomechanical pulp fibres, poly (lactic acid) and poly (butylene-adipate-terephthalate) or poly (hydroxyalkanoates) for 3D and 4D printing. *Additive Manufacturing, 59*, 103166.

Jiang, L., Lam, Y. C., Tam, K. C., Chua, T. H., Sim, G. W., & Ang, L. S. (2005). Strengthening acrylonitrile-butadiene-styrene (ABS) with nano-sized and micron-sized calcium carbonate. *Polymer, 46*(1), 243–252.

Section 2

Synthesis, properties and characterizations of additively manufactured functional polymers and nanocomposites

Innovative approaches to polymer nanocomposite synthesis for 3D printing

6

Aisha Naveed[1], Syed Masood Arif Bukhari[1], Naveed Husnain[1], Muhammad Tuoqeer Anwar[2], Muhammad Rehman Asghar[3], and Tahir Rasheed[4]
[1]Department of Mechanical Engineering, Faculty of Engineering and Technology, Bahauddin Zakariya University, Multan, Punjab, Pakistan, [2]Department of Mechanical Engineering, COMSATS University Islamabad, Sahiwal Campus, Sahiwal, Punjab, Pakistan, [3]Institute for Energy Research, Jiangsu University, Zhenjiang, Jiangsu, P.R. China, [4]Interdisciplinary Research Centre for Advanced Materials, King Fahad University of Petroleum and Minerals (KFUPM), Dhahran, Eastern Province, Saudi Arabia

6.1 Introduction

In the recent times of technical advancements, the use of Additive Manufacturing (AM) techniques has increased significantly. Initially introduced by Charles Hull in 1986, 3D printing is one of the AM techniques widely used for synthesis of complex 3D geometries and complicated parts using the printing of layers of the required part one by one with the aid of 3D-modeled design. There are numerous advantages associated with the use of 3D printing. Some of these advantages include fabrication of complex parts, low material wastage, ease of process, liberty of design, etc. Despite several advantages, the adaption of this technique at industrial level is not so fast due to several reasons. Major reasons include high cost, material limitations, barrier in construction of mega structures, mass production, etc. (Ngo et al., 2018). Moreover, the manufactured parts have poor surface integrity, porosity and cracking defects (Bukhari, 2023). Generally, metals, polymers (plastics), etc. are used for fabrication process in 3D printing. Some major factors that must be taken care of while fabricating some parts through 3D printing include strength and mechanical properties of printed parts which are directly linked with the material being used. It has been observed that the parts produced using a single material tend to lack strength.

A most favorable way is to add certain nanoparticles in the already used base polymer with the intention to improve the strength of the printed parts (Clarissa et al., 2022). Nanoparticles have the tendency to reduce the wear loss and surface roughness (Lee et al., 2009). When the nanoparticles are used as additives with the polymer, the properties of the nanoparticles are also included in the composite. The resulting nanocomposite polymer may have varied strength, porosity, surface roughness, and surface charge based on nanoparticles being used (Lee, 2019). Polymeric, metallic, ceramic, semiconductor, magnetic, lipid, and carbonic nanoparticles are general

Additive Manufacturing of Functional Polymers and Nanocomposites. DOI: https://doi.org/10.1016/B978-0-443-26516-7.00006-7
© 2025 Elsevier Ltd. All rights are reserved, including those for text and data mining, AI training, and similar technologies.

categories of nanoparticles out of which, metallic, carbonic and semiconductor nanoparticles are mostly preferred as 3D printing additives to polymers (Khan et al., 2019). There are some limitations to the use of nanoparticles in the 3D printing polymers. In theoretical aspects, the higher the nanoparticles, higher will be the performance and strength of the fabricated part. But some studies revealed that if a certain limit of nanoparticles is crossed, the nanoparticles can cause blockage in the nozzle of 3D printer (Shrestha et al., 2018). So, concentration of nanoparticles must be carefully selected to avoid blockage as well as to enhance the mechanical properties of polymer nanocomposite.

Traditionally, polymer nanocomposites are prepared by a simple approach known as "solution mixing." In this technique, first of all, a solvent is selected which is compatible with both the nanoparticles and base polymer in terms of solubility. Then nanoparticles are added in the solvent to make the solution. After that, polymer is also mixed in the same solution. Due to intercalation, the chains of polymer take the place of solvent and solvent is evaporated (Rane et al., 2018). This result in the formation of polymer nanocomposite. The spread of nanoparticles in the solution containing nanoparticles and polymer is often accompanied by ultrasonication. In this process, ultrasonic energy in the form of high frequency ultrasound waves is used to disperse the nanoparticles within the solution. Due to high energy waves, nanoparticles are separated from each other completely after which solvent evaporates (Akpan et al., 2019). This method is usually not favorable due to its environmental affects as large amount of solvent is needed which ultimately evaporates into the environment and this removal becomes issue. Moreover, this method is usually used at the lab scale and it is very difficult to scale this method to industrial level (Deshmukh et al., 2017).

Different innovative approaches to synthesize polymer nanocomposites for 3D printing are systematically analyzed and summarized in this chapter.

6.2 Innovative approaches

6.2.1 Shear mixing

The limitations of solution mixing can be reduced through shear mixing technique. This technique can be scaled to industrial level. Magnetic stirring is the general approach for shear mixing. Magnetic stirrer rotates in the solution containing nanoparticles and polymers. The rotation of stirrer creates shear forces which can disperse the nanofillers. But the shear force created from magnetic stirrer is not so strong and that's why reagglomeration occurs. To overcome this challenge, high-speed shear mixers are used. The main component of these mixers is rotor blade, which has the capacity to rotate at high velocities. This high velocity blades tend to pull the nanoparticles solution in the center and then redistributing due to centrifugal forces which can help to break the agglomerates. Here too, the solvent must be rightly selected as correct compatibility would require less shear force (Akpan et al., 2019). Moreover, this approach provides uniform dispersion of nanoparticles in the solution.

Recently, a shift toward research on high shear mixers has been noted. As per literature, the erosion is the major factor, which helps in the breakage of nanoparticles agglomerates while used in high shear mixers for synthesis of polymer nanocomposites (Özcan-Taşkin et al., 2009). Moreover, the breaking trend of nanoparticles can be modified by altering the process parameters and is also dependent on the properties of nanoparticles (Vashisth et al., 2021). Polylactide (PLA) can be mixed with cellulose nanocrystals using this approach to form polymer nanocomposite with certain desirable characteristics. It has been observed that high shear mixing of PLA with cellulose nanoparticles yield a nanocomposite, which exhibits increased strength, heat resistance, and toughness. Moreover, increasing the percentage of cellulose nanoparticles in PLA can result in increase of modulus of elasticity, which shows the enhancement of tensile behavior as well (Oguz et al., 2021). Sinnott and Cleary attempted to investigate the effect of nanoparticles' shape on the mixing behavior of high shear mixer. The shape of the particles showed negligible effect on the mixing behavior (Sinnott & Cleary, 2016).

6.2.2 Melt blending

Melt blending is another innovative approach to synthesize polymer nanocomposites for 3D printing. Unlike solution mixing, this approach is environment friendly as there is no solvent involved. This process is generally used for thermosetting polymers. As its name implies, melting of base polymer of 3D printing is required. First of all, the polymer is melted at elevated temperatures. After that, nanoparticles are added in the melted polymer matrix under shear. At those high temperatures, the polymer and nanoparticles get mixed. The chains of polymers diffuse into the nanoparticles to form polymer nanocomposite to be used in 3D printing. Injection molding and extrusion-based printing use this approach at high temperatures. This complete process can be controlled by several variables such as time, mixing environment, temperature of the melt pool, and shear. Melt blending at optimized variables and correctly selected polymers and nanoparticles can result in polymer nanocomposites of various desirable properties. Another merit of this process is that no toxins are involved in this (Deshmukh et al., 2017; Verma & Goh, 2019).

Polymer nanocomposite of PLA can be made using the nanofillers of Hydroxyapatite through melt blending approach. Ten percent hydroxyapatite by weight in the melted mixture of PLA can reduce the required printing pressure and can enhance the thermomechanical properties of the printed parts by 12%. Use of this polymer composite provides better printability and accuracy and printed parts can be widely used in medical applications (Backes et al., 2020). Melt blending approach can also mix polyethylene and PLA with carbon nanotubes to form polymer nanocomposite, which can be used in 3D printing to fabricate parts having electroconductive applications (Mora et al., 2020). Another nanocomposite that can be formed using melt blending is a blend of PLA with reduced grapheme oxide. The printed polymer nanocomposite can exhibit excellent mechanical and electrical properties, which can add its application in electronics (Zhang et al., 2016). Various researches reported the formulation of polymer nanocomposite of acrylonitrile butadiene styrene (ABS) and carbon nanotubes using melt blending approach. The proportions of nanotubes can be varied to get distinguished properties of 3D-printed parts. These polymer nanocomposites tend to have increased mechanical, thermal, and electrical properties (Dorigato et al., 2017; Dul et al., 2018; Meng et al., 2017).

6.2.3 Mechanical mixing

The mixing of powders of different components (whose mixture is desired) with the aid of ball mills and other mixers can be termed as mechanical mixing approach. It is a common method to produce polymer nanocomposite using a variety of powders of different materials for 3D printing. Filler powder and polymer powder are mixed using 3D mixer, kneader, or ball mill, and resulting polymer composite can be utilized in 3D printing process. High viscosity and too smaller particle size can result in nonuniform dispersion, which can affect the performance of printed part (Yan et al., 2021). For the milling process, the powders are not mixed homogeneously in the beginning. At the end of milling, powders become ultrafine and a homogeneous composite powder is obtained (El-Eskandarany, 2020). Other techniques are also followed to ensure the homogeneity of the polymer nanocomposite powder, one of which is the use of mechanical roller. The resulting composite powder can be placed on mechanical roller for a day or night which can help in proper dispersion of filler powder into the polymer powder (Pierantozzi et al., 2020).

Gregor-Svetec et al. utilized mechanical mixing technique to make a polymer nanocomposite of HDPE polymer and cardboard dust particles for 3D printing. The resulting composite was low cost but mechanical properties of composite also decreased. This study provided an idea about printability of cardboard based composite filament but the optimization of percentage of cardboard dust should be done (Gregor-Svetec et al., 2020). In another study, magnetic microparticles were mechanically mixed with photopolymer to form a composite for 3D printing. $SrFe_{12}O_{19}$ filler particles were added to UV prepolymer and mechanically mixed using an impeller. Through this polymer composite, the yield point was improved and viscosity was also enhanced. This type of composite is suitable for electronic applications (Nagarajan et al., 2019). Titanium dioxide ceramic nanopowder can be mixed with photoreactive polymer through mechanical mixing. The resulting nanocomposite can exhibit higher dielectric permittivity and shear rate as compared to unfilled photopolymer and the printing parts can be effectively used in high frequency electromagnetic applications (Malas et al., 2019).

Choong et al. printed novel shape memory polymer composite for 4D printing. They utilized nanosilica filler in shape memory polymer resin using mechanical mixing process and developed a novel polymer composite. They controlled the polymer and particle interaction to the optimized value. The resultant composite exhibited improved mechanical properties. Moreover, the printing time was reduced from 4 seconds per layer to 0.7 seconds per layer and polymer nanocomposite showed great shape recovery capability. This discovery created a high speed polymer composite for 3D printing technology (Choong et al., 2020). Montmorillonite nanoparticles can be mechanically mixed with UV 6105 photopolymer to create a composite which can be used to fabricate functional prototypes due to increase in glass transition temperature (Esposito Corcione et al., 2015). A bend of PLA and calcium carbonate powder can also be made using mechanical mixing process. Gayer et al. prepared a polymer composite of calcium carbonate powder in PLA polymer using mechanical milling process. 3D-printed part of the resultant composite showed decreased microporosity and increased strength as compared to unfilled polymer, which can be effectively used in bone tissue applications (Gayer et al., 2019).

6.2.4 Sol-gel method

Researchers started to prepare hybrid polymers using sol-gel technique back in 1980. Recently, this approach is being utilized to form polymer nanocomposites for 3D printing (Al Rashid et al., 2021; Oliveira & Machado, 2013). As its name implies, there are basically two steps involved in this method, that is, sol and gel. In the first step, a suspension is made that contains solid particles in a liquid. Whereas in second step, an interconnected network is created. In this method, the precursor is dissolved in the solvent initially. Then it is subjected to stirring and hydrolysis to convert the solution into the gel form. Finally, solvent is removed from the wet gel to obtain the resulting product. The drying method of the wet is dependent on the application of the final product as there are various techniques to remove the solvent from the gel (Bokov et al., 2021). The process of sol-gel technique is shown in Fig. 6.1

The first polymer nanocomposite using sol-gel technique was made by Wu and Liao. They used polyethylene-octene elastomer as a base polymer and added silica nanoparticles using sol-gel technique. Solution of silica nanoparticles, water, and catalyst was mixed into the molten polymer. Si-O-C covalent bonds replaced the already existing hydrogen bonds between polymer and silica matrices due to which the properties of resulting composite were better than base polymer (Wu & Liao, 2003). Ma et al. synthesized a polymer nanocomposite by adding silica nanoparticles into the polyacrylate polymer using sol-gel method. Ultraviolet absorbency was significantly improved as compared to polyacrylate (Ma et al., 2007). Sol-gel can make polymer nanocomposite of polypropylene and aluminosilicate nanoparticles. The synthesized nanocomposite exhibit superior mechanical properties (Moncada et al., 2007). In a study, sol-gel technique was utilized to prepare a nanocomposite of poly(ethyl methacrylate-*co*-hydroxyethyl acrylate) and silica. The concentration of silica was varied. The results in this case also followed the already-mentioned pattern, that is, enhancement of mechanical properties. Moreover, the presence of silica also

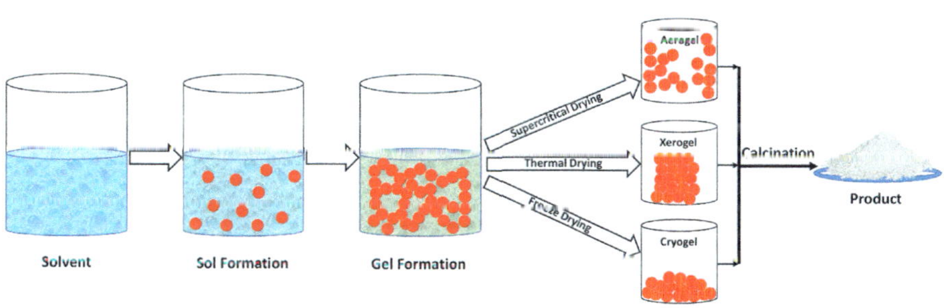

Figure 6.1 Process of sol-gel approach.

improved the surface wettability of the polymer nanocomposite (Vallés-Lluch et al., 2010). PLA polymer can be mixed with hydroxyapatite nanoparticles with different concentrations using sol-gel technique. About 19% increase in compressive strength and around 30% increase in impact strength can be observed for the synthesized polymer nanocomposite and it can be used in biomedical applications (Petrovskaya et al., 2018).

6.2.5 Stereolithography

Stereolithography is one of the mostly used technique of 3D printing. It is a laser-based process in which liquid resin is converted into solid. First of all, liquid resin is poured into a container, which contains a platform for the final part to be build. UV light or laser is passed over the liquid resin through programming. This UV light or laser converts the liquid resin into a solid shape of the required part through photopolymerization (Pham & Ji, 2000). Stereolithography uses thermoplastics such as PLA, polyamide (PA), and ABS. The only limitation of this method is poor mechanical properties of printed parts. Due to layer-by-layer fabrication, the tensile behavior is also poor due to which the failure of these parts occur, which is brittle in nature. The potential solution to these issues is the addition of nanofillers in the liquid resin solution, which has the capability to enhance the properties of resulting part (Shirvan et al., 2021). The process of stereolithography is shown in Fig. 6.2

Effect of different nanoparticles on the mechanical properties of 3D-printed parts through stereolithography is being explored. Carbon nanotubes can offer unique and enhanced properties. Sandoval et al. attempted to mix multiwalled carbon nanotubes

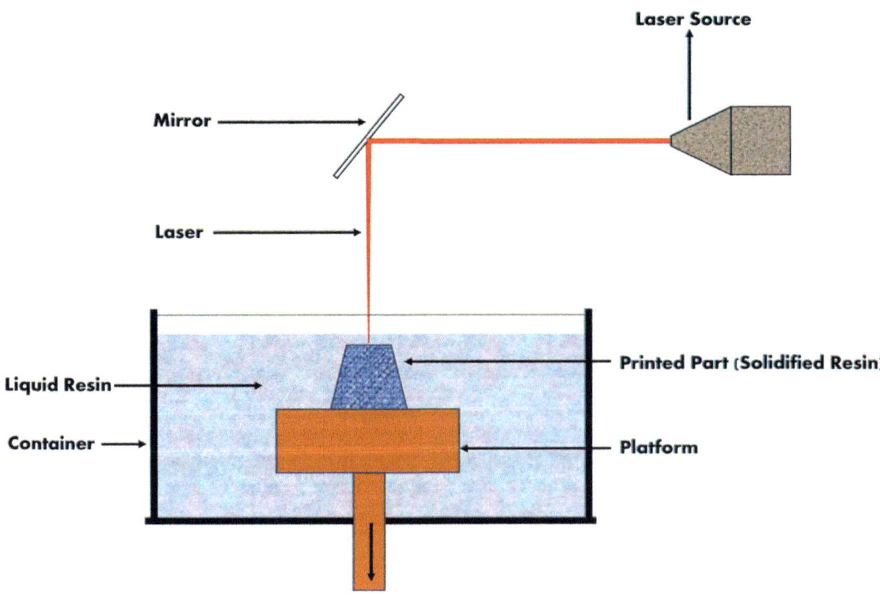

Figure 6.2 Stereolithography process.

in the epoxy resin solution for stereolithography. The results demonstrated that carbon nanotubes can potentially enhance the mechanical properties of the 3D printed parts. Moreover, increasing the percentage of carbon nanotubes can further enhance the properties as compared to based epoxy resin (Hector Sandoval & Wicker, 2006). Silica nanoparticles can also be added to the stereolithography resin through ultrasonic dispersion. These nanoparticles have the potential advantages of increasing penetration depth of laser, enhancing thermal stability and enhancing the mechanical properties as compared to untreated stereolithography resin (Liu & Mo, 2010). The viscosity of the resin plays a key role in determining the agglomeration capacity of particles. It has been seen that increasing the viscosity can support the easy dispersion of nanoparticles and reduce the particle agglomeration. This can be done by adding agents that have a property to increase the viscosity. Moreover, the viscosity must not be too high as it will create issues when a new layer is to be made (Manapat et al., 2017).

Clay is another filler whose nanoparticles can be added to polymer resin during stereolithography. Although the elongation capacity of polymer nanocomposite made of using clay nanoparticles is reduced with tensile strength remaining almost same, correct alignment and orientation of clay nanofillers can improve the elongation as well. In a study, montmorillonite clay nanoparticles were added into acrylate-based resin. The percentage of nanofillers on the tensile behavior was also observed. Bottom-up method was used to print the part with optimized orientation. The mechanical properties were enhanced whereas elongation was improved by 100% (Eng et al., 2017).

Due to supreme mechanical properties of titanium alloy, titanium dioxide nanoparticles are another potential nanofillers to be used in synthesis of polymer nanocomposites using stereolithography. Addition of titanium dioxide nanoparticles into stereolithographic epoxy resin can enhance the Young's modulus by 20%. Along with this enhancement, the brittleness of polymer nanocomposite also increases with the increase in the content of titanium dioxide nanoparticles. This may be due to the microporosities and agglomerations that can be observed with Scanning Electron Microscopy (SEM) (Aktitiz et al., 2021). Another study presents the effectiveness of titanium dioxide nanoparticles in the stereolithography technique in which these nanoparticles were dispersed in gray resin solution using sonication process. Toughness, yield strength, and compressive stress were significantly improved of the resulting nanocomposite. Moreover, the fabricated nanocomposite was also light in weight (Ávila-López et al., 2022). Stereolithography can also be used to print silver-based polymer nanocomposites with enhanced mechanical properties. One percent by weight silver nanoparticles can be added to the acrylic resin and part can be printed using stereolithography. Physical and mechanical properties of the polymer nanocomposite of silver can be increased significantly as compared to base acrylic polymer (Sciancalepore et al., 2017).

Graphene oxide-based polymer nanocomposite showed promise over last few years but its use is limited due to challenge of synthesis. Stereolithography technique can smoothly synthesize graphene oxide nanocomposite. Graphene oxide nanoparticles can be added to stereolithography elastomer resin to make polymer nanocomposite and ultimately 3D-printed parts. The resulting polymer nanocomposite reported to have reduced mechanical and thermal properties due to improper dispersion and agglomeration (Tsang et al., 2019). Chitin nanowhiskers can be blended with methacrylate polymer to synthesize polymer nanocomposite using stereolithography technique. A study investigated the properties of chitin nanowhiskers-based polymer nanocomposite and compared those properties with base methacrylate resin. Chitin nanowhiskers were obtained from crab shells and were added to methacrylate matrix using stereolithography. The resulting nanocomposite had approximately 80% enhanced tensile strength and stiffness (Maalihan et al., 2022). Stereolithography can also be employed to synthesize the polymer nanocomposite consisting gold nanoparticles. The macroscopic properties of gold nanoparticles can be embedded on the base resin of stereolithography. The resulting composite can be used in plasmonic applications (Cianni, 2023).

6.2.6 Extrusion-based printing

Extrusion-based 3D printing is a process of fabricating parts layer by layer through the deposition of polymer extruded from nozzle. The polymer from spool is fed into the extrusion head where it gets heated and is forced to flow through a nozzle on a platform and nozzle makes the required product (as programmed through CAD data) layer by layer. The motion in 3 axis is accompanied through nozzle head and platform (Mahamood, 2021). The mostly used technology of extrusion-based 3D printing is Fused Filament Fabrication (FFF), which utilizes thermosetting polymer filament. FFF was initially invented by Scott Crump. FFF is also known as Fused Deposition Modeling (FDM). FFF is being used widely in the commercial production of biomedical implants these days. Like other AM techniques, this too produces parts with reduced mechanical properties due to layer by layer deposition. Another demerit is the creation of voids between fibers. The quick solidification of layers doesn't allow adjacent layers to fuse properly into each other due to which voids are created (Bouzaglou et al., 2023). In this scenario also, the need of specific nanoparticles is required to enhance the mechanical properties of printed parts. The process of extrusion-based 3D printing is shown in Fig. 6.3

There are several studies that utilize the extrusion-based 3D printing technique to synthesize and fabricate polymer nanocomposites in order to obtain the superior mechanical properties and performance of printed parts. For instance, Kumar et al. used FFF approach to incorporate ZnO nanoparticles doped by graphene and manganese in polyvinylidene fluoride polymer. The motive behind this study was to develop a polymer nanocomposite having the capability of self-healing. It was concluded that the composite may be used in repairing the damages of building structures (Kumar et al., 2022). PLA can be blended with cellulose nanoparticles with the aid of extrusion for the synthesis of polymer nanocomposite. The resultant composite has the capacity of higher storage modulus as compared to that of other techniques (Frone et al., 2020). Carbon black nanoparticles can be incorporated in polyamide 12 polymer using FFF. The printed polymer nanocomposite exhibits superior mechanical properties. Moreover, electrical conductivity and antibacterial

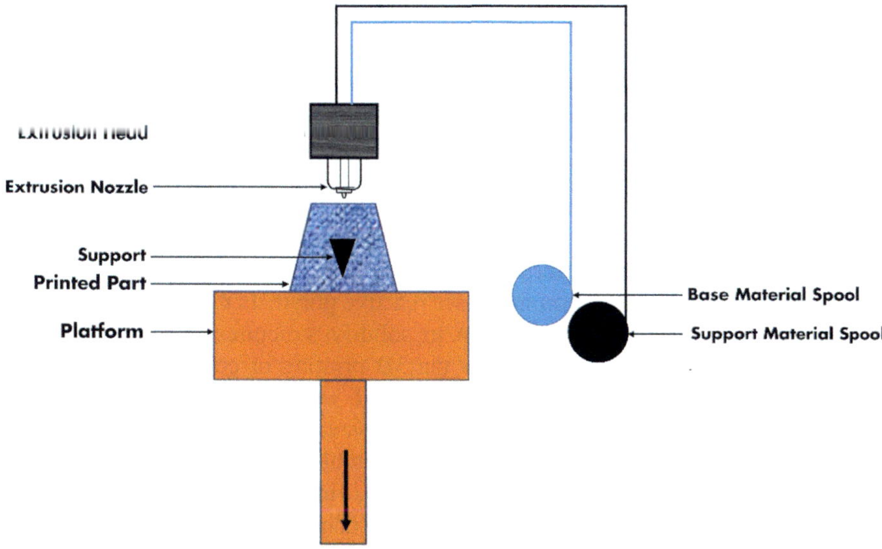

Figure 6.3 Extrusion-based 3D printing.

performance can also be improved subject to optimized percentage of nanoparticles inside the polymer matrix. To ease this fabrication, melt blending approach can be utilized (Vidakis et al., 2023).

In a study, Yamamoto et al. used FFF in an attempt to enhance the strength of thermosetting plastic. They selected ABS as base polymer and used graphene oxide nanofillers in it. The mechanical behavior of the resulting polymer nanocomposite was tested. The results depicted that the composite had superior strength, toughness, and strain-to-failure as compared to base ABS polymer whereas the stiffness of the composite was maintained (Yamamoto et al., 2019). Polypropylene polymer can be mixed with aluminum oxide nanoparticles through extrusion 3D printing based on melt blending. The composite exhibits enhanced tensile and thermomechanical properties (Vidakis, Petousis, Velidakis, Mountakis, et al., 2022). In another study, Vidakis et al. optimized the concentration of nanoparticles using extrusion printing to produce nanocomposite parts. Polypropylene polymer was taken and titanium oxide nanoparticles were added into it using melt blending. The polymer nanocomposite yielded enhanced properties. Moreover, it was observed that the properties become more superior as the amount of nanoparticles is increased (Vidakis, Petousis, Velidakis, Tzounis, et al., 2021).

Cellulose nanofibers can be mixed with polycarbonate polymer with the aid of melt blending in extrusion-based 3D printing. But for this blend, optimum amount is necessary as going below or above the optimized amount would reduce the performance of parts instead of enhancing it. For this specific polymer nanocomposite, the optimized percentage of nanoparticles is 0.5% by weight. At this ratio, the printed parts exhibit superior properties and performance (Vidakis, Petousis, Velidakis, Spiridaki, et al., 2021). Small amounts of clay nanoparticles have a significant effect on strength and performance of parts manufactured by AM. Nanoclay can be added to the epoxy resin for extrusion 3D printing. The resulting polymer nanocomposite exhibit significantly high flexural strength (Hmeidat et al., 2018). VIdakis et al. synthesized a polymer nanocomposite of silver doped antibacterial nanopowder and polyamide 12 using FFF technique. Different ratios of filler nanoparticles were added into the base polymer. The antibacterial performance of the composite was pretty high. Moreover optimized ratio of nanoparticles was 2% by weight on which increased tensile behavior by 27% (Vidakis, Petousis, Velidakis, Korlos, et al., 2022).

Extrusion-based reactive 3D printing uses reactive polymers that undergo curing or crosslinking immediately after extrusion to solidify structures in a layer-by-layer additive manner. This technique enables the 3D printing of complicated structures with living cells and biomolecules encapsulated within the deposited precursor solutions or suspensions (Malda et al., 2013). However, extensive work was needed to develop polymers and composites that are suitable for extrusion-based reactive printing as they must exhibit appropriate rheology, viscosity, and gelation properties to facilitate high-resolution deposition and curing (Zhu et al., 2016).

One of the earliest polymers used was pure alginate, a naturally derived polysaccharide (Billiet et al., 2014). However, alginate formulations showed limited resolution and structural integrity. Photocrosslinkable gelatin methacryloyl (GelMA) hydrogels were among the first synthetic polymers developed for extrusion-based reactive printing with improved properties (Khademhosseini & Langer, 2007). Tailoring the properties of individual polymers led to further improvements.

The addition of nano- and microscale composite reinforcing elements has significantly improved the mechanical properties of 3D-printed constructs. Commonly used nanocomposite reinforcements include clay (Monfared et al., 2021), hydroxyapatite, and carbon nanotubes (George et al., 2022). Ceramic microparticles like tricalcium phosphate (Birkholz et al., 2016; Bose & Tarafder, 2012) and calcium phosphate (Denry & Kuhn, 2016) have also been incorporated. Encapsulating viable cells within hydrogel microparticles prior to mixing them into bioinks enabled on-demand delivery of multiple cell types. Growth factors and signaling molecules could also be loaded into polymeric microcapsules serving as reinforcement elements with controlled release profiles. More recently, platelets and platelet-rich plasma have been 3D printed as cellular and acellular reinforcing components to modulate inflammation and enhance tissue regeneration (Andia et al., 2015; Daikuara et al., 2021; Grzelak et al., 2024).

6.2.7 Digital light processing

Digital light processing (DLP) projects a dynamic image pattern onto a vat of photocurable resin and solidifies the entire layer simultaneously rather than scanning a laser (Chaudhary et al., 2023). It provides faster printing speeds than SLA and mostly utilized in printing objects where fine details are needed (Zhang et al., 2019). DLP also commonly uses free radical or cationic photopolymerization of acrylate or epoxy monomers (Ding et al., 2022). Multimaterial and sequential curing are possible by altering the exposure pattern and resin composition over time.

Traditional resins include acrylated oligomers and monomers which rapidly polymerize when exposed to UV light. Advancements in resin chemistry led to the development of the resins with reduced shrinkage, enhanced mechanical properties and print speeds, and lower material costs (Chyr & DeSimone, 2023; Jacobs, 1992; Wen, 2007). Early resins used in DLP printing were based on acrylated epoxies and urethanes, which provided good mechanical properties but high shrinkage on curing (Calvert, 2001). To reduce shrinkage, new resins have utilized lower shrinkage chemistries like acrylated polyesters and polyethers (Zhang et al., 2022). Other approaches include blending rigid and flexible segments into one resin formulation to balance mechanical properties and shrinkage (González-Henríquez et al., 2019). New resins have also improved mechanical properties for load-bearing applications. Double-cure resins first partially cure layers with UV light, then fully cure later with heat, achieving higher modulus and strength than UV cure alone (Al-Fakih et al., 2023). Similarly, resins with increased functionality (e.g., more reactive acrylate groups) and hybrid chemistries copolymerizing acrylates with high Tg monomers provide parts with enhanced heat resistance and toughness (Jacobs, 1992). Improved print speeds have been achieved through developing resins with enhanced photoinitiation efficiency. Initiators like bisacylphosphine oxide have improved curing depth and print speeds over traditional arylketone initiators (Ramachandran, 1996). Extending the resin working curve where features can be rapidly cured without overexposure also enables higher build rates (Raki et al., 2010). Research has led to resins with improved flow properties and layer adhesion for better geometrical accuracy of fine features. For example, resins incorporating reactive surfactants providing a more uniform surface tension enhanced fine layer consolidation (Smay et al., 2002). Photoinitiators have also been paired with coinitiators and synergists to optimize initiation efficiency over the exposure region and accelerate curing (Rosiak et al., 2021).

Incorporating nanoparticles into photo curable resins can further enhance their desirable properties for DLP printing. Recently, some notable work has been conducted on the versatility of the bio-based resin formulation by the addition of nanocomposites. For example, Tao (2022) used a mixture of aqueous dextran microdroplets and gelatin methacryloyl solution (which was further stabilized by β-lactoglobulin nanoparticles) to prepare a nanoparticle-stabilized emulsion bioink. A microporous structure was created by the printed tissues due to the removal of dextran, which was responsible for providing favorable biophysical properties (Tao, 2022). Another study used lignin nanoparticles, which were surface-embedded with Ag nanoparticles (LNP@Ags) as a high-performance antimicrobial reagent to produce a bio-based antimicrobial resin by engaging it with methacrylated O-acetyl-galactoglucomannan (GGMMA) as a photocrosslinkable polymeric matrix (Wang et al., 2022). It was found that GGMMA/LNP@Ag-based resin facilitated a hydrogel construction having honeycomb and crosshatch structure stability with excellent printing fidelity.

Carbon nanotubes could be a suitable candidate to incorporate with photocurable resin to produce electrically conductive objects of polymer nanocomposites. Mu et al. (2017) have made a mixture of multiwalled carbon nanotubes (MWCNTs) with photocurable resin to make the ink for DLP printing. The concentrations of MWCNTs and the printing parameters were examined to achieve optimal conductivity and print quality. They have also shown that adding the MWCNT could increase the modulus and ultimate tensile stress; however, it slightly decreases the ultimate stretch. Similarly, Cortés et al. (2020) has also explored that post curing treatment has a significant effect on the mechanical properties such as Young's modulus; however, their effect on electrical properties was not so significant. Similarly, some other studies have discussed the benefits of combining CNTs with photocurable resin to obtain better properties; however, their utilization in DLP printing is yet to be explored (Farahani et al., 2012; Gardner et al., 2016; Postiglione et al., 2015). Besides CNTs, graphene oxide is another suitable option to be used with photocurable resin for DPL printing. Edaugal et al. have taken commercial photopolymer acrylate resin and combined it with poly(N-vinyl carbazole)-graphene oxide dispersion (PVK-GO) as a nanofiller to make a nanocomposite for DLP printing. With this nanocomposite-based coating, they have registered superior thermomechanical properties and elevated antibacterial activity (Edaugal et al., 2023).

6.2.8 Two-photon polymerizations

Two-photon polymerization (TPP) 3D printing is an additive manufacturing technique that utilizes a femtosecond pulsed laser to initiate a photopolymerization reaction within a photocurable resin. By focusing the laser beam into the resin and scanning it in three dimensions, arbitrary 3D structures can be directly fabricated layer by layer with resolution down to 100 nm (Tovar-Lopez, 2023). It uses near-infrared femtosecond laser pulses to initiate polymerization via two-photon absorption within the focal volume (Nguyen & Narayan, 2017). The setup of a typical TPP system consists of a laser source, focusing optics, and a photocurable resin.

It achieves unprecedented spatial resolution (< 100 nm) for applications such as photonics, microfluidics, and tissue engineering scaffolds (Faraji Rad et al., 2021). Common resins are acrylates, while the polymerization mechanism depends on the initiator system and can involve free radical or cationic processes. However, TPP is limited by slow printing speeds (Bednarczyk et al., 2021; Garra et al., 2019). TPP relies on simultaneous absorption of two low-energy photons to trigger polymerization (Deubel et al., 2004). Short pulsed near-infrared lasers such as femtosecond Ti:Sapphire lasers are commonly used light sources due to their high peak power and ability to induce nonlinear effects (Johnson et al., 2022; Maruo et al., 1997). The probability of two-photon absorption is directly proportional to the square of the laser intensity (Ramasubramanian et al., 2022). Hence, photopolymerization only occurs at the ultrasmall focal volume where laser intensity is highest, allowing structures to be created layer by layer with micrometer or submicrometer resolution in the *x*, *y*, and *z* axes (Fischer & Wegener, 2011; Käpylä, 2014) as shown in Fig. 6.4

Due to its highly localized photochemistry, TPP enables the fabrication of complex 3D micro- and nanostructures that are not possible with conventional lithographic or 3D printing techniques (Shi et al., 2020). Other studies have functionalized natural polymers such as alginate (Drury & Mooney, 2003), polycaprolactone (Thompson et al., 2019), polyurethanes (Mastrodonato, 2017), and polysaccharides such as hyaluronic acid (Chang et al., 2021) to yield biocompatible and biodegradable materials for tissue engineering applications. For example, TPP 3D-printed polycaprolactone scaffolds supported the growth and proliferation of human dermal fibroblasts and endothelial cells better than commercial resins (Thompson et al., 2019). In summary, these polymer advancements have expanded the material options for TPP 3D printing while improving properties such as mechanical strength, resolution, biocompatibility, and biodegradability.

The incorporation of inorganic nanoparticles into photopolymers has become a major area of research, resulting in advanced composite materials for TPP 3D printing. Silver nanoparticles added to acrylate resins were shown to introduce antibacterial properties to printed structures. Graphene oxide and gold nanoparticles improved mechanical properties and introduced electrical conductivity (Kelly, 2020; Sagdic et al., 2023). Ceramic nanoparticles such as TiO_2, SiO_2, and ZrO_2 increased hardness and reduced shrinkage of printed parts (Min & Jang, 2011). Several studies

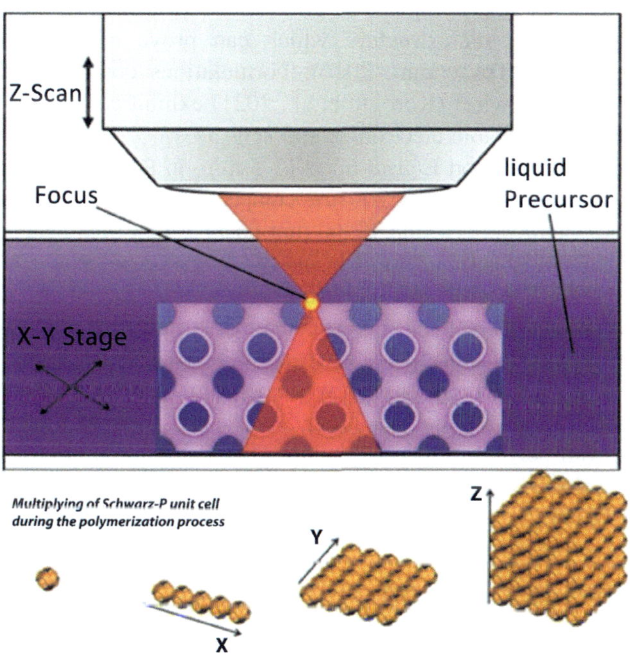

Figure 6.4 Layer-by-layer fabrication process (Zabidi et al., 2019).

demonstrated biomedical composites by adding hydroxyapatite or bioactive glass to enhance bone tissue regeneration (Wang et al., 2018; Xu et al., 2018). The incorporation of various inorganic nanoparticles into photopolymers has enabled new functional composite materials for TPP 3D printing. For example, addition of silver nanoparticles at 1–5 wt% into acrylate resins were shown to introduce efficient antibacterial properties, killing over 99.9% of E. coli within 2 hours compared to no killing for pristine resins (Lima et al., 2000). Graphene oxide (0.1–0.5 wt%) and gold nanoparticles (0.05–0.2 wt%) improved compressive strength by 20–30% and reduced shrinkage during curing, while also imparting electrical conductivity (> 100 S/m) (Bauld et al., 2018).

6.2.9 Material jetting

Material jetting works on the principle of inkjet printing where liquid photopolymer materials are selectively deposited in the form of droplets and UV light is used to cure the layers (Gülcan et al., 2021). Commonly used photopolymers include acrylates, epoxies and polyamides (García-Collado et al., 2022). To enhance the functionality of printed parts, polymer nanocomposites are being printed through material jetting approach. Jabari et al. printed graphene-based polymer nanocomposite using material jetting. The resulting nanocomposite exhibited low resistivity due to which it was made possible to make smart electronic gadgets using graphene-based polymer nanocomposite through this technique (Jabari et al., 2020). Another research presented the utilization of graphene and silver nanoparticles in material jetting ink to synthesize polymer nanocomposites. The printed part has low resistivity, high flexibility and uniform microstructure which can prove be vital in electronics applications (Jabari & Toyserkani, 2016). Formulations containing 1–5 vol% of graphene or silver nano-wires (Koroglu et al., 2021) exhibit conductivities sufficient for applications like wearable electronics and sensors.

Glass fiber composites: Short E-glass fibers (2–5 mm) in PA6 and PC enhance tensile strength by 15–30% and stiffness by 20–40% compared to unfilled polymers (Liu et al., 2022). Long fiber thermoplastics with 10–60 mm fibers exhibit even better properties (Li et al., 2022). Continuous carbon fiber/PEEK and PPSU composites show 2–3× increase in stiffness and 40–60% higher strength than short fiber formulations (Hamza et al., 2022). Composites reinforced with alumina, zirconia, and Si3N4 micro- or nano-particles exhibit improved wear resistance, hardness, and corrosion behavior (Khatkar et al., 2023). Metal nanocomposites can be material jetted at low temperatures, which can ease the process of fabricating plastic electronics (Ko et al., 2010). Sowade et al. prepared metal nanoparticles ink and used it with polymer in inkjet printing. They prepared an electronic device and concluded that major advantage of this technique is the utilization of different materials (Sowade et al., 2019).

6.3 Powder bed fusion

Powder bed fusion (PBF) 3D printing, also known as selective laser sintering (SLS), is an additive manufacturing technique used to fabricate complex geometries from thermoplastics and polymer composites (Fina et al., 2018). In PBF, a high-power laser selectively fuses powdered material layered onto a build plate (Lupone, 2021). Over the past decade, major advancements have been made in polymers and composites used for PBF 3D printing, enabling an even broader range of applications.

One of the earliest and most commonly used thermoplastics for PBF is polyamide 12 (PA12), commercialized under the trade name DuraForm (Bain, 2019). While PA12 offers acceptable mechanical properties and good layer-to-layer bonding, its high melting temperature of 178°C is close to decomposition (Morano & Pagnotta, 2023). In recent years, advanced amorphous polymers with lower melting points have been commercialized that offer improved printability compared to PA12. For example, polycarbonate (PC) melts at 135°C and maintains its strength after repeated exposures to the laser beam (Pérez et al., 2010). Arnite TPU 95A, an elastomeric polyurethane-based material, melts at only 110°C while offering rubber-like flexibility (Hart et al., 2020).

To improve the efficiency and thermal conductivity, Pannitz et al. prepared polymer nanocomposite of stainless steel 1.4404 and different nanoparticles including graphene and silicon carbide for PBF process. The resulting polymer nanocomposite exhibited enhanced thermal conductivity but some defects in the printed part with silicon carbide were also present (Pannitz et al., 2021). Lüddecke et al. also coated stainless steel, tool steel, and aluminum alloy powder with graphene, silicon carbide, and iron oxide black nanoparticles. The resulting polymer nanocomposite demonstrated increased density, energy absorption, and flow characteristics (Lüddecke et al., 2021). Addition of carbon nanoparticles can increase the functionality of PBF powders. Goodridge et al. coated PA12 powder with carbon nanofibers. The printed part had enhanced magnetic properties (Goodridge et al., 2023). Carbon fiber–reinforced PA12 (CFR-PA12) demonstrated a near doubling in flexural modulus and strength (Kudžma, 2021). Since then, a wide range of long-glass and long-carbon fiber composites have been commercialized and studied for PBF including polyamide 6 (PA6), PC, and high-density polyethylene (HDPE) (Ketola, 2014).

6.4 Conclusion

3D printing has offer various advantages over traditional manufacturing techniques but some limitations restrict its usage at mega level. A promising way to address those limitations is the addition of nanoparticles into the polymer used for 3D printing resulting in polymer nanocomposite. Traditionally, polymer nanocomposite was prepared through solution mixing, which had several limitations including use of large amount of solvent, which evaporates and makes issues for the environment. Secondly, it is a lab-scale approach and cannot be expanded at the industrial level. This chapter critically analyzed and summarized the other approaches of polymer nanocomposite synthesis, which can address the limitations of solution mixing as well as the limitations of 3D printing. Unlike solution mixing, these innovative approaches can be scaled to industrial level and are user-friendly as well. Several polymer nanocomposites are analyzed in this chapter along with their mechanical properties and applications. To unleash the full potential of this field, other nanoparticles should also be incorporated in the 3D printing polymers using the innovative approaches to check their effectiveness. The 3D printing technology can be further improved based on further research in polymer nanocomposite field.

References

Akpan, E. I., Shen, X., Wetzel, B., & Friedrich, K. (2019). *Design and synthesis of polymer nanocomposites*. Elsevier BV47–83. https://doi.org/10.1016/b978-0-12-814064-2.00002-0.

Aktitiz, İ., Aydın, K., & Topcu, A. (2021). Characterization of TiO2 nanoparticle–Reinforced polymer nanocomposite materials printed by stereolithography method. *Journal of Materials Engineering and Performance, 30*(7), 4975–4980. https://doi.org/10.1007/s11665-021-05574-x. http://www.springerlink.com/content/1059-9495/.

Al Rashid, A., Alim Khan, S., Al-Ghamdi, S. G., & Koç, M. (2021). Additive manufacturing of polymer nanocomposites: Needs and challenges in materials, processes, and applications. *Journal of Materials Research and Technology, 14*, 910–941. https://doi.org/10.1016/j.jmrt.2021.07.016.

Al-Fakih, A., Odeh, A., Abdul, M., Mahamood, A., Al-Shugaa, M. A., Al-Osta, M. A., & Ahmad, S. (2023). Review of the properties of sustainable cementitious systems incorporating ceramic waste. *Buildings, 13*(8), 2105. https://doi.org/10.3390/buildings13082105.

Andia, I., Rubio-Azpeitia, E., Martin, J. I., & Abate, M. (2015). *Current concepts and translational uses of platelet rich plasma biotechnology*. InTech. https://doi.org/10.5772/59954.

Ávila-López, M. A., Bonilla-Cruz, J., Méndez-Nonell, J., & Lara-Ceniceros, T. E. (2022). Strong and lightweight stereolithographically 3D-printed polymer nanocomposites with low friction and high toughness. *Polymers, 14*(17). https://doi.org/10.3390/polym14173628http://www.mdpi.com/journal/polymers.

Backes, E. H., Pires, L. D. N., Beatrice, C. A. G., Costa, L. C., Passador, F. R., & Pessan, L. A. (2020). Fabrication of Biocompatible composites of poly(lactic acid)/hydroxyapatite envisioning medical applications. *Polymer Engineering and Science, 60*(3), 636–644. https://doi.org/10.1002/pen.25322. http://onlinelibrary.wiley.com/journal/10.1002/(ISSN)1548-2634.

Bain, E. D. (2019). Polymer powder bed fusion additive manufacturing: recent developments in materials, processes, and applications. *ACS Symposium Series, 1315*, 7–36. https://doi.org/10.1021/bk-2019-1315.ch002. http://global.oup.com/academic/content/series/a/acs-symposium-series-acsss/?cc=in&lang=en.

Bauld, R., Choi, D. Y. W., Bazylewski, P., Divigalpitiya, R., & Fanchini, G. (2018). Thermo-optical characterization and thermal properties of graphene-polymer composites: A review. *Journal of Materials Chemistry C, 6*(12), 2901–2914. https://doi.org/10.1039/c7tc01719g. http://pubs.rsc.org/en/journals/journal/tc.

Bednarczyk, P., Irska, I., Gziut, K., & Ossowicz-Rupniewska, P. (2021). Novel rization. *Polymers, 13*(11), 1718. https://doi.org/10.3390/polym13111718.

Billiet, T., Gevaert, E., De Schryver, T., Cornelissen, M., & Dubruel, P. (2014). The 3D printing of gelatin methacrylamide cell-laden tissue-engineered constructs with high cell viability. *Biomaterials, 35*(1), 49–62. https://doi.org/10.1016/j.biomaterials.2013.09.078.

Birkholz, M.-N., Agrawal, G., Bergmann, C., Schröder, R., Lechner, S. J., Pich, A., & Fischer, H. (2016). Calcium phosphate/microgel composites for 3D powderbed printing of ceramic materials. *Biomedical Engineering / Biomedizinische Technik, 61*(3), 267–279. https://doi.org/10.1515/bmt-2014-0141.

Bokov, D., Jalil, A. T., Chupradit, S., Suksatan, W., Ansari, M. J., Shewael, I. H., Valiev, G. H., Kianfar, E., & Wang, Z. (2021). Nanomaterial by sol-gel method: Synthesis and application. *Advances in Materials Science and Engineering, 2021*, 1–21. https://doi.org/10.1155/2021/5102014.

Bose, S., & Tarafder, S. (2012). Calcium phosphate ceramic systems in growth factor and drug delivery for bone tissue engineering: A review. *Acta Biomaterialia, 8*(4), 1401–1421. https://doi.org/10.1016/j.actbio.2011.11.017. http://www.journals.elsevier.com/acta-biomaterialia.

Bouzaglou, O., Golan, O., & Lachman, N. (2023). Process design and parameters interaction in material extrusion 3D printing: A review. *Polymers, 15*(10). https://doi.org/10.3390/polym15102280, http://www.mdpi.com/journal/polymers.

Bukhari, M. A. (2023). Effect of laser surface remelting on Microstructure, mechanical properties and tribological properties of metals and alloys: A review. *Optics & Laser Technology, 165*, 2023.

Calvert, P. (2001). Inkjet printing for materials and devices. *Chemistry of Materials, 13*(10), 3299–3305. https://doi.org/10.1021/cm0101632.

Chang, D., Ma, Y., Xu, X., Xie, J., & Ju, S. (2021). Stimuli-responsive polymeric nanoplatforms for cancer therapy. *Frontiers in Bioengineering and Biotechnology, 9*. https://doi.org/10.3389/fbioe.2021.707319, http://journal.frontiersin.org/journal/bioengineering-and-biotechnology#archive.

Chaudhary, R., Fabbri, P., Leoni, E., Mazzanti, F., Akbari, R., & Antonini, C. (2023). Additive manufacturing by digital light processing: A review. *Progress in Additive Manufacturing, 8*(2), 331–351. https://doi.org/10.1007/s40964-022-00336-0.

Choong, Y. Y. C., Maleksaeedi, S., Eng, H., Yu, S., Wei, J., & Su, P. C. (2020). High speed 4D printing of shape memory polymers with nanosilica. *Applied Materials Today, 18*, 100515. https://doi.org/10.1016/j.apmt.2019.100515.

Chyr, G., & DeSimone, J. M. (2023). Review of high-performance sustainable polymers in additive manufacturing. *Green Chemistry, 25*(2), 453–466. https://doi.org/10.1039/d2gc03474c.

Cianni. (2023). Polymer nanocomposites for plasmonics: In situ synthesis of gold nanoparticles after additive manufacturing. *Polymer Testing, 117*, 2023.

Clarissa, W. H. Y., Chia, C. H., Zakaria, S., & Evyan, Y. C. Y. (2022). Recent advancement in 3-D printing: Nanocomposites with added functionality. *Progress in Additive Manufacturing, 7*(2), 325–350. https://doi.org/10.1007/s40964-021-00232-z. https://link.springer.com/journal/40964.

Cortés, A., Sánchez Romate, X. F., Jiménez-Suárez, A., Campo, M., Ureña, A., & Prolongo, S. G. (2020). Mechanical and strain-sensing capabilities of carbon nanotube reinforced composites by digital light processing 3D printing technology. *Polymers, 12*(4), 975. https://doi.org/10.3390/polym12040975.

Daikuara, L. Y., Yue, Z., Skropeta, D., & Wallace, G. G. (2021). In vitro characterisation of 3D printed platelet lysate-based bioink for potential application in skin tissue engineering. *Acta Biomaterialia, 123*, 286–297. https://doi.org/10.1016/j.actbio.2021.01.021. http://www.journals.elsevier.com/acta-biomaterialia.

Denry, I., & Kuhn, L. T. (2016). Design and characterization of calcium phosphate ceramic scaffolds for bone tissue engineering. *Dental Materials, 32*(1), 43–53. https://doi.org/10.1016/j.dental.2015.09.008.

Deshmukh, K., Basheer Ahamed, M., Deshmukh, R. R., Khadheer Pasha, S. K., Bhagat, P. R., & Chidambaram, K. (2017). *Biopolymer composites with high dielectric performance: interface engineering biopolymer composites in electronics (pp. 27–128)*. India: Elsevier Inc. http://www.sciencedirect.com/science/book/9780128092613. 10.1016/B978-0-12-809261-3.00003-6.

Deubel, M., von Freymann, G., Wegener, M., Pereira, S., Busch, K., & Soukoulis, C. M. (2004). Direct laser writing of three-dimensional photonic-crystal templates for telecommunications. *Nature Materials, 3*(7), 444–447. https://doi.org/10.1038/nmat1155.

Ding, H., Dong, M., Zheng, Q., & Wu, Z. L. (2022). Digital light processing 3D printing of hydrogels: A minireview. *Molecular Systems Design & Engineering, 7*(9), 1017–1029. https://doi.org/10.1039/d2me00066k.

Dorigato, A., Moretti, V., Dul, S., Unterberger, S. H., & Pegoretti, A. (2017). Electrically conductive nanocomposites for fused deposition modelling. *Synthetic Metals, 226*, 7–14. https://doi.org/10.1016/j.synthmet.2017.01.009.

Drury, J. L., & Mooney, D. J. (2003). Hydrogels for tissue engineering: Scaffold design variables and applications. *Biomaterials, 24*(24), 4337–4351. https://doi.org/10.1016/S0142-9612(03)00340-5. http://www.journals.elsevier.com/biomaterials/.

Dul, S., Fambri, L., & Pegoretti, A. (2018). Filaments production and fused deposition modelling of ABS/carbon nanotubes composites. *Nanomaterials, 8*(1)https://doi.org/10.3390/nano8010049, http://www.mdpi.com/2079-4991/8/1/49/pdf.

Edaugal, J. P., Ribeiro, E. L., Mitchell, M. K., Cheng, X., Buckner, E. M., Chen, J., Ivanov, I. N., Ellermann, M., & Advincula, R. C. (2023). Digital light processing (DLP): 3D printing of polymer-based graphene oxide nanocomposites—Efficient antimicrobial material for biomedical devices. *MRS Communications, 13*(4), 594–602. https://doi.org/10.1557/s43579-023-00390-x. https://www.springer.com/journal/43579.

El-Eskandarany, M. S. (2020). *Fabrication of and consolidation of hard nanocomposite materials*. Elsevier BV193–228. https://doi.org/10.1016/b978-0-12-818180-5.00008-x.

Eng, H., Maleksaeedi, S., Yu, S., Choong, Y. Y. C., Wiria, F. E., Tan, C. L. C., Su, P. C., & Wei, J. (2017). 3D stereolithography of polymer composites reinforced with orientated nanoclay. *Procedia Engineering, 216*, 1–7. https://doi.org/10.1016/j.proeng.2018.02.080.

Esposito Corcione, C., Striani, R., Montagna, F., & Cannoletta, D. (2015). Organically modified montmorillonite polymer nanocomposites for stereolithography building process. *Polymers for Advanced Technologies, 26*(1), 92–98. https://doi.org/10.1002/pat.3425.

Farahani, R. D., Dalir, H., Le Borgne, V., Gautier, L. A., El Khakani, M. A., Lévesque, M., & Therriault, D. (2012). Reinforcing epoxy nanocomposites with functionalized carbon nanotubes via biotin-streptavidin interactions. *Composites Science and Technology, 72*(12), 1387–1395. https://doi.org/10.1016/j.compscitech.2012.05.010.

Faraji Rad, Z., Prewett, P. D., & Davies, G. J. (2021). High-resolution two-photon polymerization: The most versatile technique for the fabrication of microneedle arrays. *Microsystems and Nanoengineering, 7*(1)https://doi.org/10.1038/s41378-021-00298-3http://www.nature.com/micronano/.

Fina, F., Gaisford, S., & Basit, A. W. (2018). Powder bed fusion: The working process, current applications and opportunities. *AAPS Advances in the Pharmaceutical Sciences Series, 31*, 81–105. https://doi.org/10.1007/978-3-319-90755-0_5. http://www.springerlink.com/content/2210-7371/.

Fischer, J., & Wegener, M. (2011). Three-dimensional direct laser writing inspired by stimulated-emission-depletion microscopy [Invited]. *Optical Materials Express, 1*(4), 614–624. https://doi.org/10.1364/OME.1.000614. http://www.opticsinfobase.org/

DirectPDFAccess/4B45F77E-A579-9396-2E5808CEF8709AE2_220288.pdf?da=1&id=220288&seq=0&mobile=no.

Frone, A. N., Batalu, D., Chiulan, I., Oprea, M., Gabor, A. R., Nicolae, C. A., Raditoiu, V., Trusca, R., & Panaitescu, D. M. (2020). Morpho-structural, thermal and mechanical properties of PLA/PHB/cellulose biodegradable nanocomposites obtained by compression molding, extrusion, and 3d printing. *Nanomaterials, 10*(1)https://doi.org/10.3390/nano10010051, https://www.mdpi.com/2079-4991/10/1/51/pdf.

García Collado, A., Blanco, J. M., Gupta, M. K., & Dorado-Vicente, R. (2022). Advances in polymers based multi-material additive-manufacturing techniques: State-of-art review on properties and applications. *Additive Manufacturing, 50*, 102577. https://doi.org/10.1016/j.addma.2021.102577.

Gardner, J. M., Sauti, G., Kim, J. W., Cano, R. J., Wincheski, R. A., Stelter, C. J., Grimsley, B. W., Working, D. C., & Siochi, E. J. (2016). 3-D printing of multifunctional carbon nanotube yarn reinforced components. *Additive Manufacturing, 12*, 38–44. https://doi.org/10.1016/j.addma.2016.06.008.

Garra, P., Dietlin, C., Morlet-Savary, F., Dumur, F., Gigmes, D., Fouassier, J.-P., & Lalevée, J. (2019). Redox two-component initiated free radical and cationic polymerizations: Concepts, reactions and applications. *Progress in Polymer Science, 94*, 33–56. https://doi.org/10.1016/j.progpolymsci.2019.04.003.

Gayer, C., Ritter, J., Bullemer, M., Grom, S., Jauer, L., Meiners, W., Pfister, A., Reinauer, F., Vučak, M., Wissenbach, K., Fischer, H., Poprawe, R., & Schleifenbaum, J. H. (2019). Development of a solvent-free polylactide/calcium carbonate composite for selective laser sintering of bone tissue engineering scaffolds. *Materials Science and Engineering: C, 101*, 660–673. https://doi.org/10.1016/j.msec.2019.03.101.

George, S. M., Nayak, C., Singh, I., & Balani, K. (2022). Multifunctional hydroxyapatite composites for orthopedic applications: A review. *ACS Biomaterials Science and Engineering, 8*(8), 3162–3186. https://doi.org/10.1021/acsbiomaterials.2c00140. http://pubs.acs.org/journal/abseba.

González-Henríquez, C. M., Sarabia-Vallejos, M. A., & Rodriguez-Hernandez, J. (2019). Polymers for additive manufacturing and 4D-printing: Materials, methodologies, and biomedical applications. *Progress in Polymer Science, 94*, 57–116. https://doi.org/10.1016/j.progpolymsci.2019.03.001. http://www.sciencedirect.com/science/journal/00796700.

Goodridge, R. D., Herreros-Lucas, C., & del Carmen Gimenez-Lopez, M. (2023). Adding functionality to powder bed fusion materials: Creating magnetic polymers using hybridized hollow carbon nanofibres. *Additive Manufacturing, 69*. https://doi.org/10.1016/j.addma.2023.103518, https://www.journals.elsevier.com/additive-manufacturing.

Gregor-Svetec, D., Leskovšek, M., Vrabič Brodnjak, U., Stankovič Elesini, U., Muck, D., & Urbas, R. (2020). Characteristics of HDPE/cardboard dust 3D printable composite filaments. *Journal of Materials Processing Technology, 276*, 116379. https://doi.org/10.1016/j.jmatprotec.2019.116379.

Grzelak, A., Hnydka, A., Higuchi, J., Michalak, A., Tarczynska, M., Gaweda, K., & Klimek, K. (2024). Recent achievements in the development of biomaterials improved with platelet concentrates for soft and hard tissue engineering applications. *International Journal of Molecular Sciences, 25*(3), 1525. https://doi.org/10.3390/ijms25031525.

Gülcan, O., Günaydın, K., & Tamer, A. (2021). The state of the art of material jetting—A critical review. *Polymers, 13*(16), 2829. https://doi.org/10.3390/polym13162829.

Hamza, H. M., Deen, K. M., Khaliq, A., Asselin, E., & Haider, W. (2022). Microstructural, corrosion and mechanical properties of additively manufactured alloys: A review. *Critical*

Reviews in Solid State and Materials Sciences, 47(1), 46–98. https://doi.org/10.1080/10408436.2021.1886044. www.tandf.co.uk/journals/titles/10408436.asp.

Hart, K. R., Dunn, R. M., & Wetzel, E. D. (2020). Tough, additively manufactured structures fabricated with dual-thermoplastic filaments. *Advanced Engineering Materials, 22*(4).

Hmeidat, N. S., Kemp, J. W., & Compton, B. G. (2018). High-strength epoxy nanocomposites for 3D printing. *Composites Science and Technology, 160*, 9–20. https://doi.org/10.1016/j.compscitech.2018.03.008. http://www.journals.elsevier.com/composites-science-and-technology/.

Jabari, E., & Toyserkani, E. (2016). Aerosol-Jet printing of highly flexible and conductive graphene/silver patterns. *Materials Letters, 174*, 40–43. https://doi.org/10.1016/j.matlet.2016.03.082. http://www.journals.elsevier.com/materials-letters/.

Jabari, E., Liravi, F., Davoodi, E., Lin, L., & Toyserkani, E. (2020). High speed 3D material-jetting additive manufacturing of viscous graphene-based ink with high electrical conductivity. *Additive Manufacturing, 35*, 101330. https://doi.org/10.1016/j.addma.2020.101330.

Jacobs, P. F. Rapid prototyping & manufacturing: fundamentals of stereolithography. 1992: Society of Manufacturing Engineers.

Johnson, J. E., Chen, Y., & Xu, X. (2022). Model for polymerization and self-deactivation in two-photon nanolithography. *Optics Express, 30*(15), 26824–26840. https://doi.org/10.1364/OE.461969. https://opg.optica.org/abstract.cfm?uri=oe-30-15-26824.

Kelly, A. (2020). Fabrication of biomimetic electrodes by laser processing and two photon lithography. NUI Galway.

Ketola, J. J. Cold sprayed coatings in biomedicine. (2014), 2014.

Khademhosseini, A., & Langer, R. (2007). Microengineered hydrogels for tissue engineering. *Biomaterials, 28*(34), 5087–5092. https://doi.org/10.1016/j.biomaterials.2007.07.021.

Khan, I., Saeed, K., & Khan, I. (2019). Nanoparticles: Properties, applications and toxicities. *Arabian Journal of Chemistry, 12*(7), 908–931. (ScienceDirect https://doi.org/10.1016/j.arabjc.2017.05.011http://colleges.ksu.edu.sa/Arabic%20Colleges/CollegeOfScience/ChemicalDept/AJC/default.aspxhttp://www.sciencedirect.com/science/journal/18785352.

Khatkar, S.K., Kumar, P., & Gupta, A. (2023). Reinforcement philosophy and effect of nano sized materials on mechanical, corrosive and wear behaviour of magnesium nanocomposites—A critical review. Materials Today: Proceedings.

Ko, S. H., Chung, J., Hotz, N., Nam, K. H., & Grigoropoulos, C. P. (2010). Metal nanoparticle direct inkjet printing for low-temperature 3D micro metal structure fabrication. *Journal of Micromechanics and Microengineering, 20*(12), 125010. https://doi.org/10.1088/0960-1317/20/12/125010.

Koroglu, L., Ayas, E., & Ay, N. (2021). 3D printing of polyvinylidene fluoride based piezoelectric nanocomposites: An overview. *Macromolecular Materials and Engineering, 306*(10)https://doi.org/10.1002/mame.202100277, http://onlinelibrary.wiley.com/journal/10.1002/(ISSN)1439-2054.

Kudžma, A. (2021). Effect of low oxidation graphene oxide on physical, dynamic and mechanical properties of MCC type refractory castable. In ICCS24—24th International Conference on Composite Structures (pp. 14–16). Faculty of Engineering.

Kumar, V., Singh, R., & Ahuja, I. S. (2022). On 3D printing of electro-active PVDF-Graphene and Mn-doped ZnO nanoparticle-based composite as a self-healing repair solution for heritage structures. *Proceedings of the Institution of Mechanical Engineers, Part B: Journal of Engineering Manufacture, 236*(8), 1141–1154. https://doi.org/10.1177/09544054211060912. https://journals.sagepub.com/home/PIB.

Käpylä, E. (2014). Direct laser writing of polymer-ceramic and hydrogel microstructures by two-photon polymerization. Tampere University of Technology.

Lee, J. (2019). *Carbon nanotube-based membranes for water purification nanoscale materials in water purification*. Australia: Elsevier309–331. https://www.sciencedirect.com/book/9780128139264. https://doi.org/10.1016/B978-0-12-813926-4.00017-3.

Lee, K., Hwang, Y., Cheong, S., Choi, Y., Kwon, L., Lee, J., & Kim, S. H. (2009). Understanding the role of nanoparticles in nano-oil lubrication. *Tribology Letters, 35*(2), 127–131. https://doi.org/10.1007/s11249-009-9441-7.

Li, L., Liu, W., & Sun, L. (2022). Mechanical characterization of 3D printed continuous carbon fiber reinforced thermoplastic composites. *Composites Science and Technology, 227*, 109618. https://doi.org/10.1016/j.compscitech.2022.109618.

Lima, S. M., Sampaio, J. A., Catunda, T., Bento, A. C., Miranda, L. C. M., & Baesso, M. L. (2000). Mode-mismatched thermal lens spectrometry for thermo-optical properties measurement in optical glasses: A review. *Journal of Non-Crystalline Solids, 273*(1-3), 215–227. https://doi.org/10.1016/s0022-3093(00)00169-1.

Liu, H., & Mo, J. (2010). Study on nanosilica reinforced stereolithography resin. *Journal of Reinforced Plastics and Composites, 29*(6), 909–920. https://doi.org/10.1177/0731684409102838.

Liu, X., Shan, Z., Liu, J., Xia, H., Ao, X., Zou, A., & Wu, S. (2022). Mechanical and electrical properties of additive manufactured high-performance continuous glass fiber reinforced PEEK composites. *Composites Part B: Engineering, 247*, 110292. https://doi.org/10.1016/j.compositesb.2022.110292.

Lupone, F. (2021). Process phenomena and material properties in selective laser sintering of polymers: A review. *Materials, 15*(1).

Lüddecke, A., Pannitz, O., Zetzener, H., Sehrt, J. T., & Kwade, A. (2021). Powder properties and flowability measurements of tailored nanocomposites for powder bed fusion applications. *Materials & Design, 202*, 109536. https://doi.org/10.1016/j.matdes.2021.109536.

Ma, J. Z., Hu, J., & Zhang, Z. J. (2007). Polyacrylate/silica nanocomposite materials prepared by sol-gel process. *European Polymer Journal, 43*(10), 4169–4177. https://doi.org/10.1016/j.eurpolymj.2007.06.051.

Maalihan, R. D., Chen, Q., Tamura, H., Joseph Rey, J. R. H., Pajarito, B. B., Caldona, E. B., & Advincula, R. C. (2022). Mechanically and thermally enhanced 3D-printed photocurable polymer nanocomposites containing functionalized chitin nanowhiskers by stereolithography. *ACS Applied Polymer Materials, 4*(4), 2513–2526. https://doi.org/10.1021/acsapm.1c01816. pubs.acs.org/journal/aapmcd.

Mahamood, R. M. (2021). *Role of additive manufacturing in the era of Industry 4.0. In Chapter 6)*. Woodhead Publishing, 107–126.

Malas, Isakov, Couling, & Gibbons (2019). Fabrication of high permittivity resin composite for vat photopolymerization 3D printing: Morphology, thermal, dynamic mechanical and dielectric properties. *Materials, 12*(23), 3818. https://doi.org/10.3390/ma12233818.

Malda, J., Visser, J., Melchels, F. P., Jüngst, T., Hennink, W. E., Dhert, W. J. A., Groll, J., & Hutmacher, D. W. (2013). 25th anniversary article: Engineering hydrogels for biofabrication. *Advanced Materials, 25*(36), 5011–5028. https://doi.org/10.1002/adma.201302042.

Manapat, J. Z., Chen, Q., Ye, P., & Advincula, R. C. (2017). 3D printing of polymer nanocomposites via stereolithography. *Macromolecular Materials and Engineering, 302*(9). https://doi.org/10.1002/mame.201600553http://www3.interscience.wiley.com/journal/117930403/grouphome/home.html.

Maruo, S., Nakamura, O., & Kawata, S. (1997). Three-dimensional microfabrication with two-photon-absorbed photopolymerization. *Optics Letters, 22*(2), 132–134. https://doi.org/10.1364/OL.22.000132. http://ol.osa.org/Issue.cfm.

Mastrodonato, C.M. (2017). Elaboration of fluorescent molecular probes and molecular-based nanoparticles for bioimaging purposes. Organic chemistry. Université de Bordeaux.

Meng, S., He, H., Jia, Y., Yu, P., Huang, B., & Chen, J. (2017). Effect of nanoparticles on the mechanical properties of acrylonitrile–butadiene–styrene specimens fabricated by fused deposition modeling. *Journal of Applied Polymer Science, 134*(7). https://doi.org/10.1002/app.44470.

Min, B.-H., & Jang, J.W. (2011). Method for manufacturing a porous three-dimensional scaffold using powder from animal tissue, and porous three-dimensional scaffold manufactured by same. Google Patents.

Moncada, E., Quijada, R., & Retuert, J. (2007). Nanoparticles prepared by the sol-gel method and their use in the formation of nanocomposites with polypropylene. *Nanotechnology, 18*(33). https://doi.org/10.1088/0957-4484/18/33/335606.

Monfared, V., Bakhsheshi-Rad, H. R., Ramakrishna, S., Razzaghi, M., & Berto, F. (2021). A brief review on additive manufacturing of polymeric composites and nanocomposites. *Micromachines, 12*(6), 704. https://doi.org/10.3390/mi12060704.

Mora, A., Verma, P., & Kumar, S. (2020). Electrical conductivity of CNT/polymer composites: 3D printing, measurements and modeling. *Composites Part B: Engineering, 183*, 107600. https://doi.org/10.1016/j.compositesb.2019.107600.

Morano, C., & Pagnotta, L. (2023). Additive manufactured parts produced by selective laser sintering technology: Porosity formation mechanisms. *Journal of Polymer Engineering, 43*, 537–555.

Mu, Q., Wang, L., Dunn, C. K., Kuang, X., Duan, F., Zhang, Z., Qi, H. J., & Wang, T. (2017). Digital light processing 3D printing of conductive complex structures. *Additive Manufacturing, 18*, 74–83. https://doi.org/10.1016/j.addma.2017.08.011.

Nagarajan, B., Arshad, M., Ullah, A., Mertiny, P., & Qureshi, A. J. (2019). Additive manufacturing ferromagnetic polymers using stereolithography—Materials and process development. *Manufacturing Letters, 21*, 12–16. https://doi.org/10.1016/j.mfglet.2019.06.003.

Ngo, T. D., Kashani, A., Imbalzano, G., Nguyen, K. T. Q., & Hui, D. (2018). Additive manufacturing (3D printing): A review of materials, methods, applications and challenges. *Composites Part B: Engineering, 143*, 172–196. https://doi.org/10.1016/j.compositesb.2018.02.012. https://www.journals.elsevier.com/composites-part-b-engineering.

Nguyen, A. K., & Narayan, R. J. (2017). Two-photon polymerization for biological applications. *Materials Today, 20*(6), 314–322. https://doi.org/10.1016/j.mattod.2017.06.004. http://www.journals.elsevier.com/materials-today/.

Oguz, O., Candau, N., Demongeot, A., Citak, M. K., Cetin, F. N., Stoclet, G., Michaud, V., & Menceloglu, Y. Z. (2021). Poly(lactide)/cellulose nanocrystal nanocomposites by high-shear mixing. *Polymer Engineering & Science, 61*(4), 1028–1040. https://doi.org/10.1002/pen.25621.

Oliveira, M., & Machado, A. (2013). Preparation of polymer-based nanocomposites by different routes. .

Özcan-Taşkin, N. G., Padron, G., & Voelkel, A. (2009). Effect of particle type on the mechanisms of break up of nanoscale particle clusters. *Chemical Engineering Research and Design, 87*(4), 468–473. https://doi.org/10.1016/j.cherd.2008.12.012.

Pannitz, O., Lüddecke, A., Kwade, A., & Sehrt, J. T. (2021). Investigation of the in situ thermal conductivity and absorption behavior of nanocomposite powder materials in laser powder bed fusion processes. *Materials & Design, 201*, 109530. https://doi.org/10.1016/j.matdes.2021.109530.

Petrovskaya, T. S., Toropkov, N. E., Mironov, E. G., & Azhmi, F. (2018). 3D printed biocompatible polylactide-hydroxyapatite based material for bone implants. *Materials and Manufacturing Processes, 33*(16), 1899–1904. https://doi.org/10.1080/10426914.2018.1476764.

Pham, D. T., & Ji, C. (2000). Design for stereolithography. *Proceedings of the Institution of Mechanical Engineers, Part C: Journal of Mechanical Engineering Science, 214*(5), 635–640. https://doi.org/10.1243/0954406001523650.

Pierantozzi, D., Scalzone, A., Jindal, S., Stīpniece, L., Šalma-Ancāne, K., & Dalgarno, K. (2020). Piergiorgio Gentile, Elena Mancuso, 3D printed Sr-containing composite scaffolds: Effect of structural design and material formulation towards new strategies for bone tissue engineering. *Composites Science and Technology, 191*, 108069. https://doi.org/10.1016/j.compscitech.2020.108069.

Postiglione, G., Natale, G., Griffini, G., Levi, M., & Turri, S. (2015). Conductive 3D microstructures by direct 3D printing of polymer/carbon nanotube nanocomposites via liquid deposition modeling. *Composites Part A: Applied Science and Manufacturing, 76*, 110–114. https://doi.org/10.1016/j.compositesa.2015.05.014.

Pérez, J. M., Vilas, J. L., Laza, J. M., Arnáiz, S., Mijangos, F., Bilbao, E., Rodríguez, M., & León, L. M. (2010). Effect of reprocessing and accelerated ageing on thermal and mechanical polycarbonate properties. *Journal of Materials Processing Technology, 210*(5), 727–733. https://doi.org/10.1016/j.jmatprotec.2009.12.009.

Raki, L., Beaudoin, J., Alizadeh, R., Makar, J., & Sato, T. (2010). Cement and Concrete Nanoscience and Nanotechnology. *Materials, 3*(2), 918–942. https://doi.org/10.3390/ma3020918.

Ramachandran, V. S. (1996). *Concrete admixtures handbook: Properties, science and technology*. William Andrew1996.

Ramasubramanian, B., Sundarrajan, S., Rao, R. P., Reddy, M. V., Chellappan, V., & Ramakrishna, S. (2022). Novel low-carbon energy solutions for powering emerging wearables, smart textiles, and medical devices. *Energy & Environmental Science, 15*(12), 4928–4981. https://doi.org/10.1039/d2ee02695c.

Rane, A. V., Kanny, K., Abitha, V. K., & Thomas, S. (2018). *Methods for synthesis of nanoparticles and fabrication of nanocomposites. Synthesis of inorganic nanomaterials: Advances and key technologies (pp. 121–139).* South Africa: Elsevier. https://www.sciencedirect.com/book/9780081019757. https://doi.org/10.1016/B978-0-08-101975-7.00005-1.

Rosiak, P., Latanska, I., Paul, P., Sujka, W., & Kolesinska, B. (2021). Modification of alginates to modulate their physic-chemical properties and obtain biomaterials with different functional properties. *Molecules (Basel, Switzerland), 26*(23), 7264. https://doi.org/10.3390/molecules26237264.

Sagdic, K., Fernández-Lavado, E., Mariello, M., Akouissi, O., & Lacour, S. P. (2023). Hydrogels and conductive hydrogels for implantable bioelectronics. *MRS Bulletin, 48*(5), 495–505. https://doi.org/10.1557/s43577-023-00536-1.

Sandoval, J. H., & Wicker, R. B. (2006). Functionalizing stereolithography resins: effects of dispersed multi-walled carbon nanotubes on physical properties. *Rapid Prototyping Journal, 12*(5), 292–303. https://doi.org/10.1108/13552540610707059.

Sciancalepore, C., Moroni, F., Messori, M., & Bondioli, F. (2017). Acrylate-based silver nanocomposite by simultaneous polymerization–reduction approach via 3D stereolithography. *Composites Communications, 6*, 11–16. https://doi.org/10.1016/j.coco.2017.07.006.

Shi, S., Peng, Z., Jing, J., Yang, L., & Chen, Y. (2020). 3D printing of delicately controllable cellular nanocomposites based on polylactic acid incorporating graphene/carbon nanotube hybrids for efficient electromagnetic interference shielding. *ACS Sustainable Chemistry & Engineering, 8*(21), 7962–7972. https://doi.org/10.1021/acssuschemeng.0c01877.

Shirvan, A. R., Nouri, A., & Wen, . C. (2021). *Structural polymer biomaterials. InStructural biomaterials: Properties, characteristics and selection(pp. 395–439).* Woodhead Publishing.

Shrestha, C., Walker, J.S., Arnold, J.M., Wong, M., Blanco, A., Carrigee, L.A., Granier, M., & Sutherland, E. (2018). A study of 3D printed silver-polymer composite structures. Honors Thesis.

Sinnott, M. D., & Cleary, P. W. (2016). The effect of particle shape on mixing in a high shear mixer. *Computational Particle Mechanics, 3*(4), 477–504. https://doi.org/10.1007/s40571-015-0065-4. http://www.springer.com/engineering/mechanics/journal/40571.

Smay, J. E., Cesarano, J., & Lewis, J. A. (2002). Colloidal inks for directed assembly of 3-D periodic structures. *Langmuir: The ACS Journal of Surfaces and Colloids, 18*(14), 5429–5437. https://doi.org/10.1021/la0257135.

Sowade, E., Polomoshnov, M., Willert, A., & Baumann, R. R. (2019). Toward 3D-printed electronics: Inkjet-printed vertical metal wire interconnects and screen-printed batteries. *Advanced Engineering Materials, 21*(10). https://doi.org/10.1002/adem.201900568, http://onlinelibrary.wiley.com/journal/10.1002/(ISSN)1527-2648.

Tao, J. (2022). Nanoparticle-stabilized emulsion bioink for digital light processing based 3D bioprinting of porous tissue constructs. *Advanced Healthcare Materials , 11*(12), e2102810.

Thompson, J. R., Worthington, K. S., Green, B. J., Mullin, N. K., Jiao, C., Kaalberg, E. E., Wiley, L. A., Han, I. C., Russell, S. R., Sohn, E. H., Guymon, C. A., Mullins, R. F., Stone, E. M., & Tucker, B. A. (2019). Two-photon polymerized poly(caprolactone) retinal cell delivery scaffolds and their systemic and retinal biocompatibility. *Acta Biomaterialia, 94*, 204–218. https://doi.org/10.1016/j.actbio.2019.04.057. http://www.journals.elsevier.com/acta-biomaterialia.

Tovar-Lopez, F. J. (2023). Recent Progress in micro-and nanotechnology-enabled sensors for biomedical and environmental challenges. *Sensors, 23*(12), 2023.

Tsang, C. H. A., Zhakeyev, A., Leung, D. Y. C., & Xuan, J. (2019). GO-modified flexible polymer nanocomposites fabricated via 3D stereolithography. *Frontiers of Chemical Science and Engineering, 13*(4), 736–743. https://doi.org/10.1007/s11705-019-1836-x. https://link.springer.com/journal/11705.

Vallés-Lluch, A., Costa, E., Gallego Ferrer, G., Monleón Pradas, M., & Salmerón-Sánchez, M. (2010). Structure and biological response of polymer/silica nanocomposites prepared by sol–gel technique. *Composites Science and Technology, 70*(13), 1789–1795. https://doi.org/10.1016/j.compscitech.2010.07.008.

Vashisth, V., Nigam, K. D. P., & Kumar, V. (2021). Design and development of high shear mixers: Fundamentals, applications and recent progress. *Chemical Engineering Science, 232*, 116296. https://doi.org/10.1016/j.ces.2020.116296.

Verma, D., & Goh, K. L. (2019). *Functionalized graphene-based nanocomposites for energy applications functionalized graphene nanocomposites and their derivatives: synthesis, processing and applications (pp. 219–243).* India: Elsevier. . https://www.sciencedirect.com/book/978012814548710.1016/B978-0-12-814548-7.00011-8.

Vidakis, N., Petousis, M., Velidakis, E., Mountakis, N., Fischer-Griffiths, P. E., Grammatikos, S. A., & Tzounis, L. (2022). Fused filament fabrication 3D printed polypropylene/alumina nanocomposites: Effect of filler loading on the mechanical reinforcement. *Polymer Testing, 109*, 107545. https://doi.org/10.1016/j.polymertesting.2022.107545.

Vidakis, N., Petousis, M., Velidakis, E., Korlos, A., Kechagias, J. D., Tsikritzis, D., & Mountakis, N. (2022). Medical-grade polyamide 12 nanocomposite materials for enhanced mechanical and antibacterial performance in 3D printing applications. *Polymers, 14*(3), 440. https://doi.org/10.3390/polym14030440.

Vidakis, N., Petousis, M., Velidakis, E., Spiridaki, M., & Kechagias, J. D. (2021). Mechanical performance of fused filament fabricated and 3D-printed polycarbonate polymer and polycarbonate/cellulose nanofiber nanocomposites. *Fibers, 9*(11), 74. https://doi.org/10.3390/fib9110074.

Vidakis, N., Petousis, M., Velidakis, E., Tzounis, L., Mountakis, N., Kechagias, J., & Grammatikos, S. (2021). Optimization of the filler concentration on fused filament fabrication 3D printed polypropylene with titanium dioxide nanocomposites. *Materials, 14*(11), 3076. https://doi.org/10.3390/ma14113076.

Vidakis, N., Petousis, M., Velidakis, E., Mountakis, N., Grammatikos, S., & Tzounis, L. (2023). Multi-functional medical grade Polyamide12/Carbon black nanocomposites in material extrusion 3D printing. *Composite Structures, 311*, 116788. https://doi.org/10.1016/j.compstruct.2023.116788.

Wang, H., Chen, Q., & Zhou, S. (2018). Carbon-based hybrid nanogels: A synergistic nanoplatform for combined biosensing, bioimaging, and responsive drug delivery. *Chemical Society Reviews, 47*(11), 4198–4232. https://doi.org/10.1039/c7cs00399d. http://pubs.rsc.org/en/journals/journal/cs.

Wang, L., Wang, Q., Slita, A., Backman, O., Gounani, Z., Rosqvist, E., Peltonen, J., Willför, S., Xu, C., Rosenholm, J. M., & Wang, X. (2022). Digital light processing (DLP) 3D-fabricated antimicrobial hydrogel with a sustainable resin of methacrylated woody polysaccharides and hybrid silver-lignin nanospheres. *Green Chemistry, 24*(5), 2129–2145. https://doi.org/10.1039/d1gc03841a.

Wen, J. (2007). *Heat capacities of polymers (pp. 145–154)*. Springer Nature. . 10.1007/978-0-387-69002-5_9.

Wu, C. S., & Liao, H. T. (2003). Modification of polyethylene-octene elastomer by silica through a sol-gel process. *Journal of Applied Polymer Science, 88*(4), 966–972. https://doi.org/10.1002/app.11725.

Xu, Y., Wang, X., Zhang, W. L., Lv, F., & Guo, S. (2018). Recent progress in two-dimensional inorganic quantum dots. *Chemical Society Reviews, 47*(2), 586–625. https://doi.org/10.1039/c7cs00500h.

Yamamoto, B. E., Trimble, A. Z., Minei, B., & Ghasemi Nejhad, M. N. (2019). Development of multifunctional nanocomposites with 3-D printing additive manufacturing and low graphene loading. *Journal of Thermoplastic Composite Materials, 32*(3), 383–408. https://doi.org/10.1177/0892705718759390. http://jtc.sagepub.com/content/by/year.

Yan, C., Wen, S., Shi, Y., Li, Z., & Wei, Q. (2021). Research on preparation and forming technologies of selective laser sintering polymer materials. In Selective laser sintering additive manufacturing technology (pp. 251–500, Chapter 3). Academic Press.

Zabidi, A. Z., Li, S., Felfel, R. M., Thomas, K. G., Grant, D. M., McNally, D., & Scotchford, C. (2019). Computational mechanical characterization of geometrically transformed Schwarz P lattice tissue scaffolds fabricated via two photon polymerization (2PP). *Additive Manufacturing, 25*, 399–411. https://doi.org/10.1016/j.addma.2018.11.021.

Zhang, D., Chi, B., Li, B., Gao, Z., Du, Y., Guo, J., & Wei, J. (2016). Fabrication of highly conductive graphene flexible circuits by 3D printing. *Synthetic Metals, 217*, 79–86. https://doi.org/10.1016/j.synthmet.2016.03.014.

Zhang, H., Li, X., Hou, J., Jiang, L., & Wang, H. (2022). Angstrom-scale ion channels towards single-ion selectivity. *Chemical Society Reviews, 51*(6), 2224–2254. https://doi.org/10.1039/d1cs00582k.

Zhang, J., Hu, Q., Wang, S., Tao, J., & Gou, M. (2019). Digital light processing based three-dimensional printing for medical applications. *International Journal of Bioprinting, 6*(1), 242. https://doi.org/10.18063/ijb.v6i1.242.

Zhu, W., Ma, X., Gou, M., Mei, D., Zhang, K., & Chen, S. (2016). 3D printing of functional biomaterials for tissue engineering. *Current Opinion in Biotechnology, 40*, 103–112. https://doi.org/10.1016/j.copbio.2016.03.014.

Material science insights into sustainable additive manufacturing with composite plastics

7

Kautilya S. Patel, Dhaval B. Shah, Mustafa Gandhi, Shashikant J. Joshi, and Kaushik M. Patel
Mechanical Engineering Department, Institute of Technology, Nirma University, Ahmedabad, Gujarat, India

7.1 Introduction

Plastic, celebrated for its versatility, lightweight properties, flexibility, moisture resistance, strength, and affordability, has become an integral part of modern life (Nayanathara Thathsarani Pilapitiya & Ratnayake, 2024). Its applications span diverse sectors, including construction, medical instruments, and food packaging (Czuba, 2014; Yadav et al., 2022; Yates et al., 2019). However, this widespread use has spawned a global crisis: plastic waste (PW) pollution. When plastic waste is improperly managed, failing to undergo recycling, incineration, or containment in secure landfills, it becomes a significant environmental pollutant. The consequences are severe, affecting both marine ecosystems and terrestrial environments (Asuquo, 2018; Mukheed & Khan, 2021). Marine life grapples with entanglement and ingestion of plastic debris, while chemicals leaching from plastic waste pose long-term health risks for humans, ranging from developmental issues to carcinogenic effects (Rustagi et al., 2011). Despite recycling efforts, the majority of plastic waste persists in the environment, necessitating the advancement of sophisticated recycling methods.

According to the organization for economic cooperation and development (OECD), global plastic waste generation has more than doubled from 156 million tons in 2000 to 353 million tons in 2019. Shockingly, only a mere 9% of plastic waste is recycled globally (Plastic pollution is growing relentlessly as waste management and recycling fall short, 2022). Even in countries like India, where efforts at recycling are significant, a substantial amount of plastic waste, approximately 3.8 million tons per annum remains unaddressed (Mohanty, 2017). This escalating crisis demands innovative and sustainable solutions for waste management.

This research delves into the realm of additive manufacturing (AM), also known as 3D printing, as a promising avenue for upcycling plastic waste into valuable components. Unlike traditional subtractive manufacturing methods, AM constructs

Additive Manufacturing of Functional Polymers and Nanocomposites. DOI: https://doi.org/10.1016/B978-0-443-26516-7.00007-9
© 2025 Elsevier Ltd. All rights are reserved, including those for text and data mining, AI training, and similar technologies.

objects layer-by-layer, thereby minimizing material wastage and enabling on-demand production (Javaid et al., 2021). The AM process is classified into seven groups based on the layering mechanism: material extrusion, vat photopolymerization, material jetting, binder jetting, powder bed fusion, direct energy deposition, and sheet lamination. Among the various AM techniques, fused deposition modeling (FDM) stands out due to its accessibility, user-friendliness, and cost-effectiveness (Hassanien et al., 2023). FDM printers build objects by melting and depositing thermoplastic filament layer by layer as shown in Fig. 7.1 (Patel et al., 2023).

While traditional methods of recycling plastics often face challenges such as complex sorting processes and limited markets for recycled materials, AM offers a promising solution. Moreover, the incorporation of carbon fiber reinforcement enhances the mechanical and thermal properties of recycled plastic products, paving the way for sustainable and economically viable waste management solutions. Various research studies are being conducted that focus on the production, mechanical, and thermal properties of 3D-printed carbon fiber reinforced polymer (CFRP) composites.

Pandit et al. (2023) conducted research on carbon fiber reinforced high-density polyethylene (CF-HDPE) composites using FDM. Their findings revealed that an appropriate carbon fiber (CF) content significantly improved the surface quality, tensile properties, thermal resistance, and stiffness of the composites. However, when CF concentrations were higher, issues such as microvoids and delamination emerged, highlighting a delicate balance between reinforcement and printability. In a separate study, Borkar et al. (2022) compared the mechanical properties of recycled HDPE (r-HDPE) filled with waste CF produced via injection molding and 3D printing. Both methods led to enhanced mechanical properties, with the most significant improvement observed at a fiber volume fraction of 29.5%. Interestingly, due to variations in fiber alignment, fragment length, and crystallinity caused by different cooling cycles, 3D printed samples showed slight performance differences compared to injection molded ones. In FDM printing, when fibers in composites like carbon fiber reinforced plastic waste (CFPW) are misaligned, it significantly compromises their strength. To illustrate, consider poorly placed beams in a bridge, they would be unable to bear the

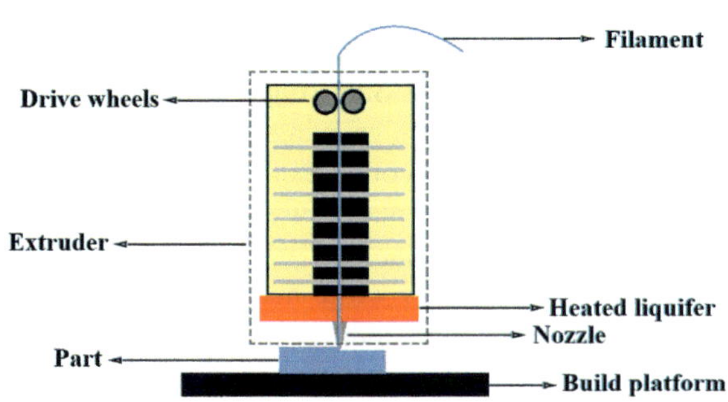

Figure 7.1 Working of FDM 3D printing.

load effectively. Similarly, misaligned fibers lead to uneven stress distribution, concentrating pressure on weaker regions. Consequently, this can result in cracks or delamination when subjected to stress.

However, research by Alarifi et al. (2023) suggests there's room for improvement. The authors discovered a 23% increase in yield strength for solid 3D-printed polyethylene glycol CF (PETG/CF) composites using finite element modeling and tensile testing, emphasizing the potential for design optimization. However, Ferreira et al. (2019) encountered an intriguing trade-off. Their unique testing setup revealed that the printing process reduced stress resistance by 28.21%, yet it significantly increased stiffness by 70.10%. This underscores the critical role of the printing process in determining the final properties of PETG/CF composites, as confirmed by thermogravimetric analysis and microscopic examination of fiber content and voids. In FDM printing, short carbon fiber (SCF) composites in nylon benefit from triangular infill patterns. The printing speed, ideally set at 50 mm/s, and the layer thickness significantly influence the resulting strength (Naranjo-Lozada et al., 2019). While SCF enhances tensile strength, it also reduces elongation and weakens the bonds between fibers and the polymer matrix (Liao et al., 2018). Overall, incorporating varying amounts of SCF improves the mechanical properties of 3D printed composite parts.

Almeshari et al. (2023) achieved a remarkable 150% increase in tensile strength for 22% short carbon fiber/polypropylene SCF/PP composites compared to neat PP, demonstrating the potential of CF reinforcement. However, their study also identified limitations such as microvoids and fiber pull-out. Similarly, Polline et al. (2021) explored challenges in processing recycled PP filaments with CF, encountering cavity formation and a trade-off between strength and interlayer bonding. Shifting the focus to a broader perspective, Oyinlola et al. (2023) highlighted the potential of using plastic waste for 3D printing in Sub-Saharan Africa, acknowledging the limitations of traditional recycling and identifying challenges like inconsistent filament diameter, material degradation, and potential health hazards during extrusion. While carbon fiber reinforcement holds great promise for enhancing the mechanical properties and thermal stability of 3D-printed components, several challenges persist. These include optimizing fiber content and distribution, minimizing voids, and ensuring consistent filament quality, especially when working with recycled materials.

This research explores the potential of 3D printed parts made from recycled plastic waste reinforced with carbon fiber. The study investigates the effects of the inclusion of 20% carbon fiber in plastic waste impacting the mechanical strength of the printed parts. Also, scanning electron microscopy (SEM) analysis has been employed to gain deeper insights into fiber alignment, distribution within the matrix, and the presence of voids that could impact performance. Lastly, the thermal behavior has been analyzed with thermogravimetric analysis (TGA), differential scanning calorimetry (DSC) analysis, and flame testing. TGA measured degradation behavior, while DSC measured the heat flow of material as it varies with temperature, and flame testing determined flammability characteristics according to a standardized method.

7.2 Materials and methodology

7.2.1 Raw material

This study utilizes a recycled plastic matrix obtained from the southern coastal region of Gujarat and rural Ahmedabad, India. The plastic waste comprises a mix of high-density and low-density plastics. Common items such as discarded packaging, textile waste, water bottles, edible oil cans, and polycarbonate roofing sheets contribute to this source material. Additionally, to enhance the mechanical and thermal properties of the recycled plastic, 20% chopped carbon fiber (with a 6 mm fiber length), procured from ZOLTEK Corporation, a subsidiary of the Toray Group, is incorporated into the matrix. This combination of recycled plastic and high-performance carbon fiber offers an eco-friendly and potentially high-performing composite material.

7.2.2 Filament fabrication

The filament has been fabricated using a multistep process through recycled plastic waste as depicted in Fig. 7.2. Initially, shredding reduces the plastic into manageable pieces, followed by washing with surfactants like sodium carbonate to eliminate impurities. The washed plastic is then dried for optimal processing. The filament creation process begins with blending chopped carbon fiber and the prepared plastic pellets to achieve a uniform mixture. The glycidyl methacrylate (GMA) is added to strengthen the bond between fibers and plastic. This composite mixture is fed into a single-screw extruder with controlled heating zones that progressively melt the plastic pellets. The combined heat and pressure ensure thorough melting and mixing, while the rotating screw continuously blends the molten plastic for consistent properties. Finally, the molten composite is forced through a 1.75 mm die, resulting in the formation of the final filament.

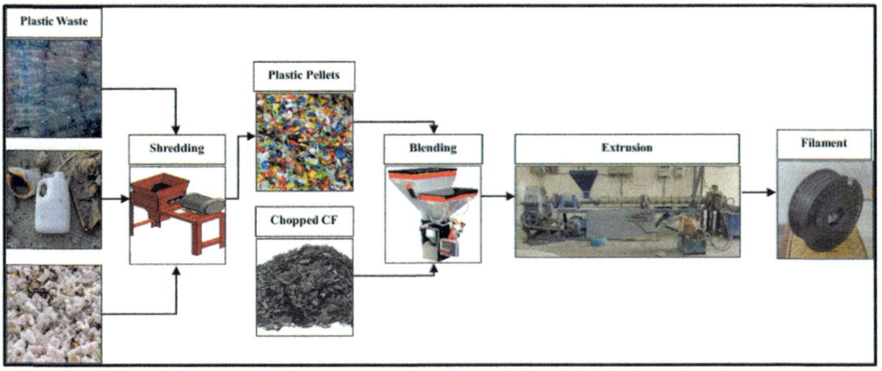

Figure 7.2 Filament manufacturing process.

Table 7.1 Process parameters.

Process parameters					
	Nozzle temperature (°C)	**Layer thickness (mm)**	**Bed temperature (°C)**	**Raster angle (°)**	**Infill density (%)**
Tensile test	250, 260, 270	0.1, 0.2, 0.3	110	0, 45, 90	100
Flame test	220	0.1	100	0	20, 40, 60, 80, 100

7.2.3 Process parameters

The 3D printed specimens for tensile and flame testing have been manufactured using a Creality Ender-3 V2 3D printer. The specific process parameters are mentioned in Table 7.1.

Commencing the printing process, the 3D CAD model of the testing samples for tensile and flame tests, as illustrated in Fig. 7.3, is uploaded into the slicing software. The testing samples are printed in a horizontal orientation on the printer bed to optimize material deposition and minimize printing challenges, as depicted in Fig. 7.4. Additionally, magigo glue adhesive is applied to the bed to enhance adhesion and prevent warping of the specimens during printing.

7.3 Results and discussion

7.3.1 Tensile testing

Tensile testing has been used to evaluate the strength and deformation behavior of 3D printed materials under a uniaxial tensile load. The tensile specimens have been fabricated according to ASTM D-638 type 1. For each combination of neat plastic and carbon fiber reinforced plastic, three specimens have been produced. The printing parameters including a nozzle temperature of 270°C and a layer thickness of 0.1 mm, remained consistent across these sets. However, to investigate the influence of raster angle on tensile properties the orientation of the printed filament relative to the tensile load direction has been varied at 0°, 45°, and 90°.

The tensile testing results demonstrated a positive impact of carbon fiber incorporation on the tensile strength of the 3D printed specimens. When compared to neat plastic specimens the addition of 20% carbon fiber resulted in an average increase of 1.637% strength, as outlined in Fig. 7.5. This improvement highlights the effectiveness of carbon fiber reinforcement in enhancing the mechanical performance of the composite material. However, it's important to note that this increase is not as significant as the results reported by Almeshari et al. (2023). This discrepancy could

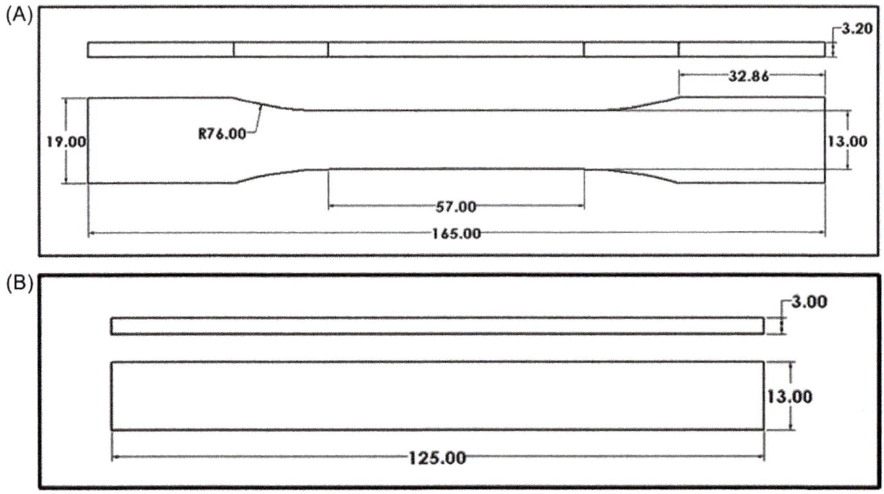

Figure 7.3 Testing specimens according to ASTM standards: (A) tensile test and (B) flame test (all dimensions are in mm).

Figure 7.4 FDM printing of tensile test specimen.

be attributed to the presence of numerous microvoids in the specimens caused by the high carbon fiber content (Pandit et al., 2023).

The tensile strength of both neat plastic and carbon fiber reinforced specimens has been significantly influenced by the raster angle during printing. In both materials, a 0° raster angle, where the printed filaments align with the tensile load, yielded the highest tensile strength. Specifically, for neat plastic, this alignment led to a 22.93% increase in strength when transitioning from 90° to 45°, and an additional 4.22% improvement from 45° to 0°. Similarly, carbon fiber reinforced plastic exhibited a 17.49% increase from 90° to 45°, followed by a 9.44% gain from 45° to 0°. These trends underscore the importance of a 0° raster angle, as it optimally aligns the fibers and filaments with the direction of the applied stress (Srinivasan Ganesh Iyer & Keles, 2022).

7.3.2 Scanning electron microscopy

Scanning electron microscopy (SEM) analysis played a crucial role in investigating the microstructure of the 3D printed specimens. The high-resolution micrographs obtained through SEM allowed for a detailed examination of several critical factors as shown in Fig. 7.6. Fiber orientation is one of the key features which can be analyzed by the SEM process. It can reveal whether the fibers are aligned along the printing direction or randomly scattered throughout the material. Also, it can help to assess the efficiency of the blending process during filament manufacturing. An even distribution of fibers throughout the polymer matrix is crucial. Uneven distribution can lead to inconsistencies in material properties across different regions of the printed part. Lastly, SEM analysis aids in identifying voids and defects within the printed parts. These imperfections significantly impact mechanical strength and overall quality.

The SEM analysis revealed that the carbon fibers have been predominantly aligned parallel to each other, following the direction of material deposition during the printing process. In line with the tensile test results, where a 0° raster angle (aligned fibers) resulted in the highest strength SEM analysis revealed good alignment of carbon fibers along the printing direction. While Fig. 7.5 confirms adequate polymer

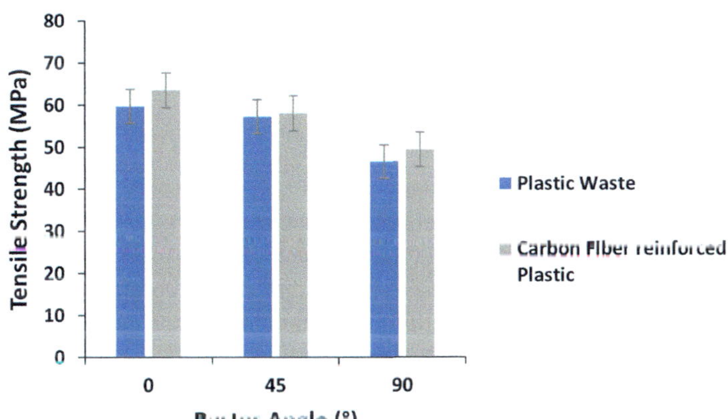

Figure 7.5 Comparison of tensile test of plastic waste and carbon fiber reinforced plastic waste.

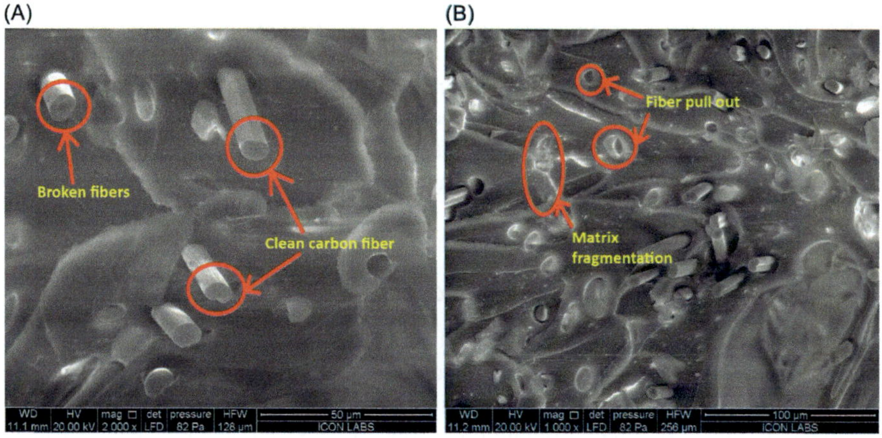

Figure 7.6 SEM images of (A) plastic waste and (B) carbon fiber reinforced plastic waste.

matrix coverage, some voids have also been observed. These voids can act as stress concentrators and contribute to early failure under tension. Therefore, optimizing printing parameters to minimize voids is crucial for achieving stronger composites.

SEM analysis also provides valuable insights beyond fiber alignment. Interestingly, it revealed fiber pull-out rather than breakage during tensile testing (Cao et al., 2022). While strong fiber alignment is crucial, weak fiber-matrix bonding allows for pull-out which hinders the composite's full potential. Additionally, SEM micrographs showed uneven fiber distribution potentially causing inconsistencies in mechanical properties. Techniques such as double extrusion or optimizing the mixing process during filament manufacturing could address this issue and achieve a more uniform fiber distribution. By addressing these areas identified through SEM analysis, high-quality 3D printed composites with superior strength and consistent mechanics can be produced.

7.3.3 Flame test

Material flammability plays a crucial role in determining potential applications as during the filament fabrication process lots of chemicals are used that can be harmful to the environment. The specimens have been fabricated according to ASTM D-635 as shown in Fig. 7.7. The impact of different infill densities and carbon fiber inclusion in plastic waste has been evaluated to check the fire resistance as shown in Fig. 7.8. Infill density represents the proportion of a specimen's volume occupied by solid material, which significantly influences overall fire behavior.

The results revealed that all plastic waste specimens displayed self-extinguishing behavior, regardless of infill density (20% or 100%). The flame never extended beyond the designated 25 mm mark. However, infill density significantly impacted the burn length. Specimens with a 20% infill density burned on average 93.8% longer

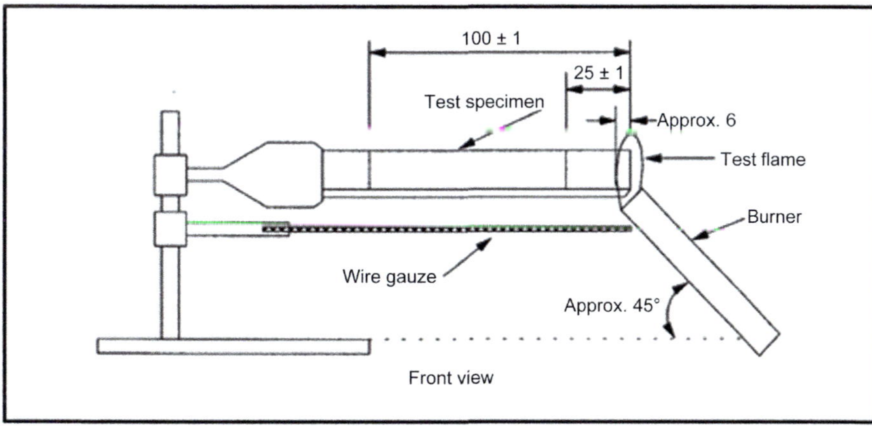

Figure 7.7 Schematic representation of flame test setup.

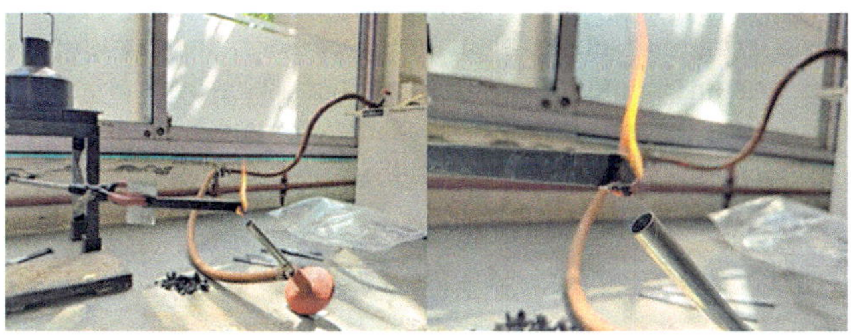

Figure 7.8 Flame test setup.

than those with a 100% infill density. This demonstrates that a denser infill acts as a stronger physical barrier against flame spread, resulting in a shorter burn length.

Similarly, carbon fiber reinforced plastic waste specimens are also self-extinguished, with the flame front never exceeding the 25 mm mark at any infill density. This suggests that incorporating carbon fiber has minimal to no impact on the self-extinguishing property of the material. However, just like the plastic waste infill density variations affected burn length. In this case, a decrease in infill density from 100% to 20% resulted in an average increase in burn length of 123.6%. Interestingly, when comparing the burn lengths of PW and CFPW specimens with the same infill density CFPW specimens exhibited slightly higher burn lengths, as shown in Fig. 7.9. For instance, at 100% infill density CFPW specimens burned 22.5% longer than plastic waste specimens. The reason behind this is the presence of carbon fiber reinforcement increases the material's thermal conductivity, allowing for more efficient heat transfer and flame propagation.

In conclusion, incorporating carbon fiber into the 3D printed filament offers a significant advantage in terms of material quality and performance. While both PW and

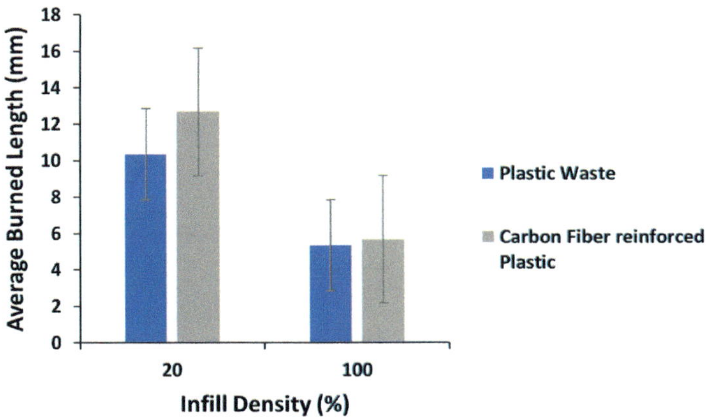

Figure 7.9 Comparison of flame test between plastic waste and carbon fiber reinforced plastic waste.

CFPW specimens exhibited self-extinguishing behavior, the addition of CF doesn't compromise the existing flame-retardant properties. More importantly, carbon fiber reinforcement significantly enhances the mechanical strength of the composite, making it suitable for applications requiring both high strength and flame resistance.

7.3.4 Thermogravimetric analysis (TGA)

TGA has been employed to assess the effectiveness of the filament manufacturing process and understand the thermal behavior of the composite material. TGA involves monitoring weight changes in a sample as its temperature is steadily raised in a controlled environment. In this study, a Perkin Elmer Pyris 1 TGA analyzer has been used which served two key purposes: first, it precisely determined the carbon fiber content within the 3D printed specimens, verifying the accuracy of fiber incorporation. Second, as the filament material is a complex blend of plastics (PP, LDPE, HDPE), TGA provided valuable insight into its thermal degradation profile, as each plastic degrades at a distinct temperature.

The TGA analysis revealed a carbon fiber content of 19.88% in the analyzed CFPW sample. This value has been obtained by comparing the remaining weight percentage of the CFPW sample at the point where the PW sample completely degraded (0% remaining weight). The slight discrepancy between the targeted 20% and the measured value can be attributed to uneven fiber distribution within the specimen and potential fiber loss during filament manufacturing. The TGA results also showcased a significant improvement in thermal stability for the CFPW samples. While neat PC samples reached 50% weight degradation at 427 °C, the CFPW samples achieved the same level of degradation at a much higher temperature of 524°C. This demonstrates the positive influence of CF addition on the thermal behavior of the composite filament as shown in Fig. 7.10.

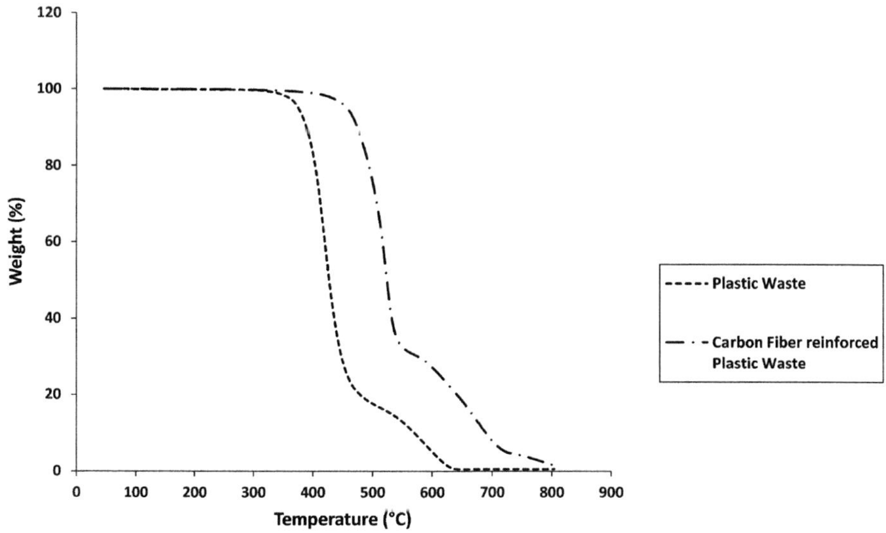

Figure 7.10 TGA plot to analyze mass degradation.

The TGA analysis also yielded valuable insights into the thermal decomposition behavior of the recycled plastic material. The TGA curve for the PW specimen exhibited a two-stage degradation pattern. In the first phase, which began around 300°C, the degradation of components like LDPE and PP within the plastic blend occurred. These polymers typically degrade at lower temperatures. The second phase starting around 450°C, most likely corresponds to the degradation of HDPE and CF, which is for its higher thermal stability. This result provides valuable insights for future endeavors in upcycling plastic waste, allowing for tailored processing techniques based on the specific composition of the recycled material.

7.3.5 Differential scanning calorimetry

Differential scanning calorimetry (DSC) is a technique used to measure the heat flow of a material as it varies with temperature. In the current research, DSC analysis on CFPW and PW has been conducted to compare their thermal behavior, with a specific focus on their melting points. During the DSC analysis, both CFPW and PW samples have been subjected to controlled temperature increases. The instrument then measured the heat absorbed or released by the samples at each temperature point.

The results of the DSC analysis revealed a distinct difference in the melting points of CFPW and PW. On average, CFPW melts at 151.225°C, whereas PW melts at 98.65°C. This difference can be attributed to the presence of carbon fiber in CFPW. Carbon fiber acts as a reinforcement agent, significantly enhancing the overall strength and thermal stability of the composite material (Magri et al., 2021). Consequently, CFPW exhibits a higher melting point compared to PW.

The observed difference in melting points between CFPW and PW has significant implications for the thermal behavior and quality of parts made from these materials. CFPW's higher melting point indicates greater thermal stability. Parts manufactured from CFPW are less likely to deform or melt at elevated temperatures compared to those made from PW. This enhanced thermal performance positions CFPW as a more suitable candidate for applications where parts experience high thermal stress, potentially leading to the production of longer-lasting and more reliable components.

7.4 Conclusions

The continuous buildup of plastic waste presents a grave danger to our environment. Recycling emerges as a promising remedy, converting plastic waste into valuable assets. In this research, the feasibility of employing recycled plastic waste combined with carbon fiber reinforcement for fabricating high-performance 3D printed components has been explored. The findings reveal a significant enhancement in material quality and performance when compared to plastic waste.

- CF reinforcement effectively improved the mechanical properties of the composite. The tensile testing revealed an average 4.637% increase in tensile strength for specimens with 20% carbon fiber compared to neat plastic. Also, the specimens with 0° raster angle displayed superior tensile strength due to better alignment between the filament and loading direction.
- SEM analysis highlighted positive fiber contribution to strength in certain regions. Areas with uneven fiber distribution emphasized the need for further optimization in blending or filament manufacturing processes.
- Both PW and CFPW specimens have been self-extinguished during flame testing. The infill density influenced burn length, but CF addition did not significantly alter self-extinguishing behavior.
- TGA analysis confirmed 19.88% CF content within the composite (slightly lower than the intended 20% due to manufacturing factors). The CF reinforced specimens exhibited improved thermal stability where PW degraded at 427°C while the composite reached the same level at 524°C, highlighting its ability to withstand higher temperatures and expanding its potential applications.
- The DSC analysis demonstrated that CFPW melts at a significantly higher temperature (151.225°C) compared to PW (98.65°C). This difference is attributed to the reinforcing effect of CF in CFPW, leading to enhanced thermal stability and potentially improving the quality and durability of parts made from CFPW.

In conclusion, this study successfully demonstrated the potential of using recycled PW reinforced with CF for 3D printing applications. The incorporation of CF resulted in improved mechanical properties, thermal stability, and flame retardancy, highlighting a significant improvement in overall material quality and performance compared to PW. Further research into optimizing fiber distribution and exploring different fiber content percentages can lead to more innovation in AM while addressing the PW challenge.

List of abbreviations

3D	Three Dimensional
AM	Additive manufacturing
CAD	Computer-aided design
CF	Carbon fiber
CFPW	Carbon fiber reinforced plastic waste
CFRP	Carbon fiber reinforced polymer
DSC	Differential Scanning Calorimetry
FDM	Fused deposition modeling
GMA	Glycidyl Methacrylate
HDPE	High density polyethylene
LDPE	Low-density polyethylene
OECD	Organization for economic cooperation and development
PC	Polycarbonate
PETG	Polyethylene terephthalate glycol
PET	Polyethylene terephthalate
PP	Polypropylene
PW	Plastic waste
rHDPE	Recycled high density polyethylene
SCF	Short carbon fiber
SEM	Scanning electron microscopy
TGA	Thermogravimetric analysis

References

Alarifi, I. M. (2023). PETG/carbon fiber composites with different structures produced by 3D printing. *Polymer Testing, 120*, 107949. https://doi.org/10.1016/j.polymertesting.2023.107949.

Almeshari, B., Junaedi, H., Baig, M., & Almajid, A. (2023). Development of 3D printing short carbon fiber reinforced polypropylene composite filaments. *Journal of Materials Research and Technology, 24*, 16–26. https://doi.org/10.1016/j.jmrt.2023.02.198. http://www.elsevier.com/journals/journal-of-materials-research-and-technology/2238-7854.

Asuquo, I. (2018). Plastic waste in the aquatic environment: Impacts and management. *Environment, 2*(1), 1. https://doi.org/10.31058/j.envi.2018.21001.

Borkar, A., Hendlmeier, A., Simon, Z., Randall, J. D., Stojcevski, F., & Henderson, L. C. (2022). A comparison of mechanical properties of recycled high-density polyethylene/waste carbon fiber via injection molding and 3D printing. *Polymer Composites, 43*(4), 2408–2418. https://doi.org/10.1002/pc.26550. http://onlinelibrary.wiley.com/journal/10.1002/(ISSN)1548-0569.

Cao, M., Cui, T., Yue, Y., Li, C., Guo, X., Jia, X., & Wang, B. (2022). Investigation of carbon fiber on the tensile property of FDM-produced PLA specimen. *Polymers, 14*(23), 5230. https://doi.org/10.3390/polym14235230.

Czuba, L. (2014). *Application of plastics in medical devices and equipment. Handbook of polymer applications in medicine and medical devices*. United States: Elsevier Inc. 9–19.

http://www.sciencedirect.com/science/book/9780323228053. https://doi.org/10.1016/B978-0-323-22805-3.00002-5.

Ferreira, I., Vale, D., Machado, M., & Lino, J. (2019). Additive manufacturing of polyethylene terephthalate glycol /carbon fiber composites: An experimental study from filament to printed parts. *Proceedings of the Institution of Mechanical Engineers, Part L: Journal of Materials: Design and Applications, 233*(9), 1866–1878. https://doi.org/10.1177/1464420718795197. http://pil.sagepub.com/content/by/year.

Hassanien, M., Alkhader, M., Abu-Nabah, B. A., & Abuzaid, W. (2023). A low-cost process for fabricating reinforced 3d printing thermoplastic filaments. *Polymers, 15*(2), 315. https://doi.org/10.3390/polym15020315.

Javaid, M., Haleem, A., Singh, R. P., Suman, R., & Rab, S. (2021). Role of additive manufacturing applications towards environmental sustainability. *Advanced Industrial and Engineering Polymer Research, 4*(4), 312–322. https://doi.org/10.1016/j.aiepr.2021.07.005. http://www.keaipublishing.com/en/journals/advanced-industrial-and-engineering-polymer-research/.

Liao, G., Li, Z., Cheng, Y., Xu, D., Zhu, D., Jiang, S., Guo, J., Chen, X., Xu, G., & Zhu, Y. (2018). Properties of oriented carbon fiber/polyamide 12 composite parts fabricated by fused deposition modeling. *Materials and Design, 139*, 283–292. https://doi.org/10.1016/j.matdes.2017.11.027.

Magri, A. E., El Mabrouk, K., Vaudreuil, S., & Touhami, M. E. (2021). Mechanical properties of CF-reinforced PLA parts manufactured by fused deposition modeling. *Journal of Thermoplastic Composite Materials, 34*(5), 581–595. https://doi.org/10.1177/0892705719847244. http://jtc.sagepub.com/content/by/year.

Mohanty, S. (2017). Recycling of plastics in Indian perspective.

Mukheed, M., & Khan, A. (2021). Journal of Pollution Effects & Control, 8, https://doi.org/10.35248/2375-4397.20.8.251.

Naranjo-Lozada, J., Ahuett-Garza, H., Orta-Castañón, P., Verbeeten, W. M. H., & Sáiz-González, D. (2019). Tensile properties and failure behavior of chopped and continuous carbon fiber composites produced by additive manufacturing. *Additive Manufacturing, 26*, 227–241. https://doi.org/10.1016/j.addma.2018.12.020.

Nayanathara Thathsarani Pilapitiya, P. G. C., & Ratnayake, A. S. (2024). The world of plastic waste: A review. *Cleaner Materials, 11*, 100220. https://doi.org/10.1016/j.clema.2024.100220.

Oyinlola, M., Okoya, S. A., & Whitehead, T. (2023). *Utilising plastic waste to create 3D-printed products in Sub-Saharan Africa. Digital innovations for a circular plastic economy in Africa.* United Kingdom: Taylor and Francis91–105. https://www.routledge.com/Digital-Innovations-for-a-Circular-Plastic-Economy-in-Africa/Oyinlola-Kolade/p/book/9781032244099. https://doi.org/10.4324/9781003278443-8.

Pandit, P. P., Liu, C., Iacono, S., Corti, G., & Hu, Y. (2023). Microstructural characterization and property of carbon fiber reinforced high-density polyethylene composites fabricated by fused deposition modeling. *Materials, 16*(1), 180. https://doi.org/10.3390/ma16010180.

Patel, K. S., Shah, D. B., Joshi, S. J., & Patel, K. M. (2023). Developments in 3D printing of carbon fiber reinforced polymer containing recycled plastic waste: A review. *Cleaner Materials, 9*, 100207. https://doi.org/10.1016/j.clema.2023.100207.

Plastic pollution is growing relentlessly as waste management and recycling fall short. https://www.oecd.org/en/about/news/press-releases/2022/02/plastic-pollution-is-growing-relentlessly-as-waste-management-and-recycling-fall-short.html (2022).

Polline, M., Mutua, J. M., Mbuya, T. O., & Ernest, K. (2021). Recipe development and mechanical characterization of carbon fibre reinforced recycled polypropylene 3D printing filament. *Open Journal of Composite Materials, 11*(03), 47–61. https://doi.org/10.4236/ojcm.2021.113005.

Rustagi, N., Pradhan, S. K., & Singh, R. (2011). Public health impact of plastics: An overview. *Indian Journal of Occupational and Environmental Medicine, 15*(3), 100–103. https://doi.org/10.4103/0019-5278.93198.

Srinivasan Ganesh Iyer, S., & Keles, O. (2022). Effect of raster angle on mechanical properties of 3D printed short carbon fiber reinforced acrylonitrile butadiene styrene. *Composites Communications, 32*, 101163. https://doi.org/10.1016/j.coco.2022.101163.

Yadav, A., Chandra, A., & Singh, S. (2022). Study on application of waste plastic in the construction industry. *Materials Today: Proceedings, 64*, 1455–1458. https://doi.org/10.1016/j.matpr.2022.04.743. https://www.sciencedirect.com/journal/materials-today-proceedings.

Yates, J., Deeney, M., White, H., Joy, E., Kalamatianou, S., & Kadiyala, S. (2019). PROTOCOL: Plastics in the food system: Human health, economic and environmental impacts. A scoping review. *Campbell Systematic Reviews, 15*(1-2)https://doi.org/10.1002/cl2.1033https://onlinelibrary.wiley.com/journal/18911803.

Synthesis and functionalization methods of polymer nanocomposites for additive manufacturing

8

Waleed Ahmed[1], Ameera F. Mohammad[1], Essam Zaneldin[2], Ali H. Al-Marzouqi[3], and Noura AlMazrouei[4]

[1]Engineering Requirements Unit, UAE University, Al-Ain, United Arab Emirates, [2]Department of Civil and Environmental Engineering, College of Engineering, United Arab Emirates University, Al Ain, United Arab Emirates, [3]Chemical and Petroleum Engineering Department, UAE University, Al-Ain, United Arab Emirates, [4]Department of Chemical and Biological Engineering, Monash University, Clayton, Victoria, Australia

8.1 Introduction

3D printing technology has revolutionized various industries by offering unprecedented capabilities in manufacturing complex structures with high precision and customization. Among the myriad materials used in 3D printing, as illustrated in Fig. 8.1, the 3D printing process goes from a design to the final model. Functional polymers have emerged as critical components due to their valuable properties and adaptability to distinct applications. Functional polymers, characterized by their ability to perform specific functions beyond mere structural roles, such as biocompatibility, conductivity, and responsiveness to external stimuli, have significantly enhanced the scope and efficacy of 3D printing technologies.

The start of 3D printing technology has revolutionized various industries by offering unprecedented capabilities in manufacturing complex structures with high precision and customization. Fig. 8.2 represents the 3D printer parts and illustrates the high precision and customization of 3D printing. Among the myriad materials used in 3D printing, functional polymers have emerged as critical components due to their versatile properties and adaptability to diverse applications. Functional polymers, characterized by their ability to perform specific functions beyond mere structural roles, such as biocompatibility, conductivity, and responsiveness to external stimuli, have significantly enhanced the scope and efficacy of 3D printing technologies.

We explore the transformative role of 27 distinct functional polymers in additive manufacturing, specifically within the context of 3D printing. Each polymer offers unique properties that cater to the dynamic needs of various industries, including biomedical engineering, aerospace, electronics, and more. For instance, polyglycolic acid (PGA) and poly(lactic-co-glycolic acid) (PLGA) are extensively used for their

Additive Manufacturing of Functional Polymers and Nanocomposites. DOI: https://doi.org/10.1016/B978-0-443-26516-7.00008-0
© 2025 Elsevier Ltd. All rights are reserved, including those for text and data mining, AI training, and similar technologies.

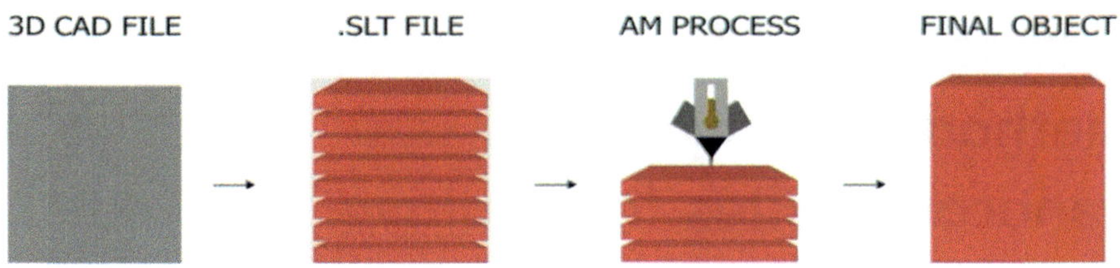

Figure 8.1 3D printering process.

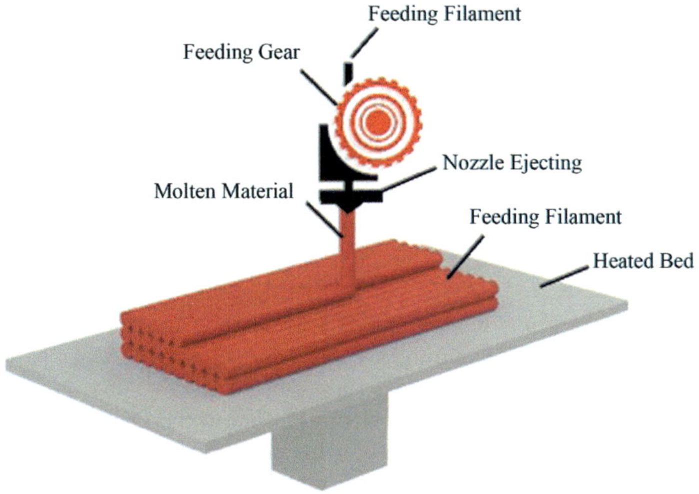

Figure 8.2 3D printer parts.

biodegradability and biocompatibility, making them ideal for medical implants and tissue engineering. Polylactic acid (PLA), a well-known biodegradable polymer, is widely utilized to create environmentally friendly products and prototypes. Polycarbonate (PC) and polymethylmethacrylate (PMMA) are valued for their optical clarity and impact resistance, making them suitable for applications requiring durable and transparent materials. Polyvinylidene fluoride (PVDF) and polyether ether ketone (PEEK) exhibit excellent chemical resistance and mechanical properties, which are essential for high-performance industrial components. Polyimide (PI) and polyacrylonitrile (PAN) are known for their thermal stability and strength, making them suitable for aerospace and automotive applications. In the realm of flexible and elastic materials, polyurethane (PU) and nylon stand out due to their durability and flexibility, which are crucial for wearable devices and dynamic mechanical parts. Polytetrafluoroethylene (PTFE) and polypropylene (PP) offer exceptional chemical resistance and low friction, making them ideal for nonstick coatings and fluid handling systems. Polyvinyl chloride (PVC) and polystyrene (PS) are versatile polymers used in various applications, from construction to packaging, due to their ease of processing and cost-effectiveness. Hydrophilic polymers such as polyethylene glycol (PEG), polyethylene Oxide (PEO), and polyvinyl alcohol (PVA) are employed in biomedical applications for their solubility and biocompatibility, enabling the fabrication of drug delivery systems and hydrogels. Polyethylene (PE) and polyethylene terephthalate (PET) are commonly used for their excellent mechanical properties and recyclability, essential for sustainable manufacturing practices. Specialty polymers like polyacrylamide (PAM), poly(N-isopropylacrylamide) (PNIPAM), and polydimethylsiloxane (PDMS) are used for their unique responsive properties, such as temperature sensitivity and elasticity, which are leveraged in smart materials and soft robotics. Conductive polymers like polyaniline (PANI) and polyethyleneimine (PEI) are utilized in electronic applications, including sensors and conductive coatings. Poly(3-hydroxybutyrate-co-3-hydroxyvalerate) (PHBV) and poly(N-vinyl pyrrolidone) (PVP) are notable for their biocompatibility and degradability, making them suitable for medical and environmental applications. By exploring the synthesis, processing, and application of these functional polymers, this chapter aims to understand how they reshape the landscape of additive manufacturing comprehensively. The continuous development and integration of these materials underscore the potential for groundbreaking advancements across multiple sectors, paving the way for future innovations in 3D printing technology. In fact, functional polymers' continuous development and integration into 3D printing processes highlight the potential for revolutionary advancements across multiple sectors. By exploring recent research and innovations, we aim to comprehensively understand how functional polymers reshape additive manufacturing landscely.

In this chapter, we explore the significant contributions of functional polymers to 3D printing, emphasizing their transformative potential across various applications. Our work highlights developing and utilizing these polymers to create advanced, biocompatible, and degradable materials for medical applications and sustainable and high-performance solutions for industrial and environmental uses. We demonstrate how functional polymers enhance the versatility, efficiency, and customization capabilities of 3D printing, paving the way for innovative advancements in manufacturing technologies. This comprehensive exploration underscores the pivotal role of functional polymers in driving future progress and expanding the horizons of additive manufacturing.

8.2 Polyglycolic acid

The exploration of fused granular fabrication (FGF) as a sustainable alternative to filament-based additive manufacturing (AM) has been a central theme in recent research. One study involved 3D printing PLA pellets with varying speeds and infill percentages, emphasizing the energy efficiency of higher printing speeds without compromising material strength. An interesting revelation emerged from porosity analysis, establishing a correlation between infill percentage and total pore volume. Mechanical characterization further indicated that 75% infill achieves the highest ultimate tensile strength, providing valuable insights for future investigations (Fontana et al., 2022). Shifting focus, another study investigates the realm of Digital Light Processing (DLP)-based 3D printing, centering on poly(glycerol sebacate) acrylate (PGSA). This degradable, biocompatible elastomer is a promising material for producing porous tubular structures. The research explores various ink formulations, optimizing printing efficiency and showcasing tunability in scaffold degradation kinetics and mechanical properties through postfabrication treatments. The creation of a user-friendly 3D model-generation tool enhances customization possibilities.

The resulting DLP-printed PGSA scaffolds exhibit high mimicry to complex 3D models, establishing PGSA's tunable properties for tissue regeneration applications (Wu, 2023, Wu et al., 2020). In the realm of rapid prototyping using desktop 3D printing, the study focuses on PLA and bronze-filled PLA. Through mechanical testing, it becomes evident that PLA exhibits superior tensile and flexural properties when contrasted with bronze-filled PLA. The emphasis on understanding these mechanical properties is crucial for material selection in engineering applications, providing valuable insights for 3D printing processes (Aveen et al., 2018). Transitioning to orthopedic implants, an extensive review investigates fabrication methodologies and property characterization of PLA, PGA, and their copolymeric variants. This encompasses various physical and mechanical characteristics, fabrication methods, and applications in the medical field. The study aims to contribute to standardizing 3D fabrication processes in tissue engineering, offering a detailed exploration of scaffold design, image analysis, and evaluation of different PLGA compositions and printing conditions (Guo et al., 2017). Shifting focus to drug delivery, a review explores the utilization of γ-PGA nanoparticles for anticancer drug delivery. The discussion covers production techniques controlled-release strategies, and provides insights into tumor physiology for targeted drug delivery in cancer chemotherapy (Manocha & Margaritis, 2008). A comprehensive review of PLA/cellulose/PEG composites provides insights into each material and recent trends. The study emphasizes the need for further exploration to enhance PLA/cellulose filaments' performance in 3D printing and identifies gaps in the literature regarding the use of PEG as an additive (Benini et al., 2023). Examining PGA, a water-soluble, biodegradable biopolymer, the review discusses applications in various fields. It covers fermentative production, downstream processing, and applications, contributing to the ongoing development of this commercially significant biopolymer (Bajaj & Singhal, 2011). Introducing an environmentally friendly approach, another study explores the creation of conductive PLA composites with graphene nanoplatelets, enhancing mechanical properties and electrical conductivity for advanced 3D printing technology (Kim et al., 2019). The application of PLA in biomedical fields through various additive manufacturing routes takes center stage in a review. The study emphasizes processing conditions and their effects, providing an insightful overview (Van den Eynde and Van Puyvelde, 2017). Summarizing the evolution of 3D printing technology, another overview provides insights into various applications of 3D-printed PLA and its composites in orthopedics. The discussion includes current limitations and prospects (Chen et al., 2019). Examining the degradation rate of PGA, an investigation focuses on the influence of the surface-to-volume ratio. Results reveal a dependency of mass loss on sample thickness, while properties like crystallinity, melting point, and molecular weight remain independent (Shawe et al., 2006). In the industrial domain, Kureha Corporation's groundbreaking technology for cost-effective mass production of high-molecular-weight PGA is discussed, along with applications in shale gas and oil exploration (Yamane et al., 2014). The appraisal concludes with an overview of PGA biosynthesis, microorganisms used, and applications. The importance of selecting suitable Bacillus species for fermentation is emphasized, along with potential modifications to enhance PGA quality (Dahiya et al., 2020).

8.3 Polylactic acid

The advancement of polylactic acid (PLA) as a versatile material is explored through diverse studies, each contributing unique insights. In 3D-printed structures, metal-reinforced PLA exhibits multifunctionality, albeit with a slight compromise in tensile properties. Challenges in metal-polymer adhesion and polishing necessitate improved adhesion or advanced 3D printing methods for optimal performance (Vakharia et al., 2021). Another avenue for PLA enhancement involves its combination with multi-wall carbon nanotubes (MWCNT). The resulting 3D-printed nanocomposites demonstrate significant improvements in mechanical, thermal, electrical, thermo-mechanical, and antibacterial properties. This study highlights the potential of these nanocomposites for applications such as sensor fabrication and health monitoring devices (Vidakis et al., 2020). A different approach involves the production of antimicrobial objects using PLA-silver nanoparticle nanocomposite filaments. This study showcases a cost-effective and environmentally friendly method, emphasizing the antimicrobial efficiency against bacteria. The developed 3D-printed objects hold promise as alternatives to metallic objects in biomedical applications (Vidakis et al., 2020). Biodegradable poly(lactic acid) nanocomposites for Fused Deposition Modeling 3D Printing addresses PLA's limitations by incorporating additives like cellulose, metals, and carbon-based materials. This study emphasizes ongoing research into novel additives and their combinations to replace nondegradable materials and foster environmental responsibility (Bardot and Schulz, 2020). Furthermore, preparing PLA carbon nanotube nanocomposites through melt blending proves effective in improving mechanical properties, thermal stability, crystallinity, and morphology. The study suggests the optimal composite with a favorable combination of strength and toughness (Zhou et al., 2018). 3D printing of poly (lactic acid) has significantly expanded PLA applications, particularly in the biomedical field. This review focuses on additive manufacturing methods such as extrusion-based 3D printing, powder-based laser sintering, and stereolithography, exploring their influence on the utilization of PLA (Van den Eynde and Van Puyvelde, 2018). Poly(lactides) synthesis and characterization methods provide a comprehensive overview of lactide polymerization mechanisms and the advantages and disadvantages of various synthesis methods. The paper stresses the importance of understanding the synthetic mechanism to control PLA's stereoregularity for improved properties (Pretula et al., 2016). The emphasis of "PLA Composites:

From Production to Properties" lies in delineating the significance of reinforcing fibers, micro- and/or nanofillers, and additives in surmounting PLA's limitations. The review encapsulates recent advancements and emerging trends, underscoring the transition from conventional applications to more technically demanding uses (Murariu & Dubois, 2016). PLA carbon nanotube nanocomposites garner attention due to CNTs' exceptional properties. This chapter explores recent developments in preparing and characterizing PLA/CNT nanocomposites, shedding light on factors influencing CNT dispersion in the PLA matrix (Akbari et al., 2015). New polylactide/layered silicate nanocomposites highlight successful production through melt extrusion. The study explores the impact of a compatibilizer on morphological analysis and mechanical properties, showcasing significant enhancements in nanocomposite properties (Ray et al., 2002). PLA biocomposite processing, additive manufacturing, and advanced applications delve into the potential of natural fiber reinforcement in PLA composites. The study highlights improvements in wear properties and offers sustainable alternatives for various applications, especially in the context of 3D and 4D printing (Ilyas et al., 2021). Polylactic acid (PLA) polymers, from properties to biomedical applications, provide a comprehensive overview of PLA properties, applications, and recent developments. It emphasizes the potential of PLA in biomedical applications, including recent advancements in PLA-based biomedical implants using additive manufacturing (Ebrahimi et al., 2021). Exploration of plasticized PLA/cellulose nanocomposites, fabricated via melt-extrusion and liquid feeding, focuses on assessing their mechanical, thermal, and optical properties. The study reveals improvements in crystallinity, transparency, elongation at break, and toughness, showcasing the potential of these nanocomposites (Herrera et al., 2015). Recent Advances in the Investigation of poly(lactic acid) (PLA) nanocomposites focus on enhancing PLA properties using various nanofillers. The review explores nanocomposites for applications such as active packaging, oxygen scavenging, UV shielding, sensors, and organic photovoltaic cells. Challenges include optimal dispersion and green synthesis methods (Bikiaris et al., 2023).

8.4 Polycarbonate

Polycarbonate (PC) has become a promising substitute for inorganic glass, boasting high optical transparency, lightweight characteristics, and exceptional toughness. Surface coating has become essential for broadening potential applications, especially in outdoor environments, and introducing novel functionalities. Recent coating technologies have focused on achieving enhanced properties, such as electrical conductivity, antireflection, and self-cleaning capabilities, aligning with the demands of developments in electrical and optical devices and energy-efficient automobiles (Kim & Research, 2016). Nanocomposites have played a key role in improving PC properties. PC/multiwalled carbon nanotube (MWCNT) nanocomposites demonstrated enhanced thermal stability and potential for varied applications (Larosa et al., 2017). Likewise, nanocomposites derived from polystyrene-cobutyl methacrylate copolymers and organo-modified clay demonstrated enhanced thermal stability, underscoring their potential across various applications (Siddiqui et al., 2013). Polyurethane (PUR) nanoclay composites showcased improvements in mechanical and thermal properties, emphasizing the significant impact of nanoclay on enhancing overall performance (Nayani et al., 2013). Direct melt intercalation produced PC/acrylonitrile butadiene styrene (ABS) polymer alloy montmorillonite (MMT) nanocomposites with enhanced thermal stability, indicating potential material property enhancement (Wang & Li, 2020). The mechanical properties of ABS matrix composites enhanced with nanosilica dioxide particles are investigated in this work. ABS is an ideal material for such applications because ABS is an amorphous, thermoplastic polymer with exceptional thermal characteristics that can also be employed with the fused filament fabrication (FFF) process. The composite material's enhanced mechanical characteristics facilitate the recycling of ABS from 3D printing while setting up the prospect of its reuse in related applications (Al-Mazrouei et al., 2022). Polymer composites for Melted and Extruded Manufacturing (MEM) technology exhibited tailored properties with improved flowability, hardness, and impact strength (Bulanda et al., 2021). Functional polycarbonates synthesized through ring-opening polymerization showcased potential applications in various fields. The study explored catalysts, reaction parameters, and deprotection, offering insights into tailoring polycarbonate properties for specific applications (Zhi, 2003). Recycled PC shards demonstrated effectiveness as feedstock for additive manufacturing, positioning them as an economical and practical material for sustainable manufacturing (Reich et al., 2019). Bio-based polycarbonate (bio PC) emerged as a sustainable alternative to fossil-fuel-based plastics. With superior tensile strength and low hazardous emissions during 3D printing, bio PC is suitable for ecofriendly indoor manufacturing of consumer products (Park et al., 2020). A comparative study on 3D-printed PC-reinforced ABS composites revealed improved hardness and strength with increased PC content, contributing to better 3D-printed structures (Kumar et al., 2019). Polycarbonate nanocomposite filaments infused with nanosilica for 3D printing showed enhanced thermal stability and mechanical properties. Even when present in minute amounts, nanosilica demonstrated its capacity to enhance both mechanical and optical characteristics, thereby expanding its prospective utilization across thin film interfaces and within the automotive sector (Kodali et al., 2021). Aliphatic polycarbonates synthesized from CO_2 and epoxides, especially functionalized polycarbonates, presented a resource-efficient, biocompatible, and biodegradable polymer with potential applications in various fields (Scharfenberg et al., 2018). The development of a novel class of photopolymerizable Bisphenol F polycarbonates (BFPs) showcased their potential in optical 3D printing for micro- and nanoscale fabrication. The BFPs exhibited tunable mechanical properties, high cell viability, and significant promise for various biomedical applications (Zhu et al., 2018). In summary, the versatility and potential of polycarbonate in multiple applications, coupled with advancements in materials and technologies, contribute to ongoing exploration for sustainable and tailored manufacturing solutions.

8.5 Polymethylmethacrylate

Hybrid materials, integrating organic and inorganic components, have garnered significant interest for their unique and synergistically derived properties, displacing pristine polymers in various applications due to their enhanced strength and stiffness. This chapter explores the synthesis and characterization of PMMA (polymethylmethacrylate) nanocomposites, including PMMA–SiO_2, PMMA–TiO_2, PMMA–clay, PMMA–silica–alumina, and PMMA–cellulose. The manufacturing process entailed the initiation of methyl methacrylate (MMA) polymerization through free radicals using benzoyl peroxide within an aqueous medium, along with the incorporation of inorganic phases through a "sol–gel" transformation, encompassing hydrolysis and condensation of Ti and Si alkoxides during MMA polymerization (Ahmad et al., 2007). Nanocomposites layered by PMMA silicate were effectively synthesized through both in-situ suspension polymerization and emulsion-polymerization methodologies, resulting in uniform exfoliated structures validated by wide-angle X-ray diffraction and atomic force microscopy (AFM) analyses. Structural characterization highlighted the coexistence of intercalated and exfoliated structures upon melt pressing, with the exfoliated structure preserved even with organic modifiers, leading to enhanced glass transition temperatures. This comprehensive exploration contributes insights into the structural dynamics and performance enhancements of PMMA-layered silicate nanocomposites (Huang & Brittain, 2001). Employing emulsifier-free emulsion polymerization, a novel approach yielded PMMA/clay nanocomposites with exfoliated structures, confirmed by XRD and TEM (transmission electron microscopy) analyses. These nanocomposites showed improved glass transition temperatures and storage moduli without a significant increase in refractive index, with transparency exceeding 70%. The study underscores the innovative strategy employed and the potential significance of PMMA/clay nanocomposites in optical applications (Yeom et al., 2004). Nanostructured PMMA surfaces were explored by incorporating silica, alumina, and silica-alumina nanocomposites synthesized via the sol–gel procedure. Structural analysis via SEM unveiled discernible modifications, while FT-IR spectra indicated filler incorporation and altered thermal stability, which were confirmed by TGA and DSC studies. This study offers insights into the interplay between PMMA and various nanostructured additives, elucidating resulting modifications in structural, chemical, and thermal characteristics (Orellana et al., 2011). The intercalative polymerization yielded by PMMA montmorillonite (MMT) nanocomposites with distinctive structures featuring partially exfoliated and partially intercalated configurations. Characterization revealed improved thermal stability, glass transition temperature, and mechanical properties compared to immaculate PMMA, emphasizing the potential of these nanocomposites in advanced materials research (Li, 2024).

Furthermore, PMMA and MMT nanocomposites synthesized by bulk polymerization demonstrated enhanced properties with a silicone coupling agent, highlighting the importance of surface modification. This research contributes to understanding structure–property relationships in PMMA/MMT nanocomposites (Xie et al., 2003). Moreover, the impact of nanofiller amount and sort of organic converter in PMMA-OMMT organomodified montmorillonite nanocomposites synthesized via in situ bulk polymerization was investigated, revealing enhancements in thermal stability, glass transition temperature, and mechanical properties (Nikolaidis et al., 2011). Additionally, well-defined PMMA–silica nanocomposites were synthesized using reversible addition-fragmentation chain transfer (RAFT) polymerization, demonstrating enhanced mechanical strength and control over molecular weight (Chinthamanipeta et al., 2008).

8.6 Polyvinylidene fluoride

Polyvinylidene fluoride (PVDF) is renowned for its exceptional mechanical, optical, thermal, piezoelectric, pyroelectric, and ferroelectric properties, rendering it highly desirable for various industrial applications (Batra, 2019). Researchers in one study focused on fabricating thick films of PVDF, both pristine and embedded with neodymium oxide (Nd_2O_3) nanoparticles, utilizing the solution casting method. These nanoparticles were synthesized through hydrothermal techniques, and their presence was confirmed via infrared vibrational spectroscopy. Furthermore, the photosensitive constants of the nanocomposite films were found by UV-VIS transmission spectroscopy, marking the first such analysis for films in the ferroelectric phase (Batra, 2019). Another investigation delved into synthesizing PVDF nanocomposites with few-layer black phosphorus nanosheets (BP) through liquid exfoliation. The process involved preparing a masterbatch via solution combination, tracked by a melting compounds procedure to produce various nanocomposites, which were then molded using compression molding. Characterization methods such as X-ray diffraction, infrared spectroscopy, and SEM analysis (scanning electron microscopy) validated the successful preparation of BP nanosheets, with notable enhancements observed in thermal stability, electrical conductivity, and mechanical strength upon the addition of BP nanofillers (Tiouitchi et al., 2019). Expanding on PVDF's multifunctional properties, another study explored the challenges and opportunities associated with material extrusion additive manufacturing (MEAM) of PVDF structures. The researchers investigated the effects of deposition parameters on mechanical and electromechanical characteristics and explored posttreatment methods such as annealing and corona poling to enhance the piezoelectric performance.

Additionally, microscale zirconium tungstate particulates were incorporated into the PVDF matrix to improve printability and reduce residual stress in addition to component warping in the printed substances, demonstrating the potential for rapid manufacturing of multifunctional products (Momenzadeh, 2020). A novel approach to fabricating thin PVDF films via 3D printing was proposed in another study, where optimal extrusion parameters were established, ensuring stable local polarization of the thin films for up to 20 hours (Dubkov et al., 2020). Furthermore, the synthesis of polymeric blends comprising (PVDF) and TPU (Thermoplastic polyurethane) occupied with CB-PPy (carbon black-polypyrrole) was explored to develop flexible and highly conductive materials suitable for 3D printing. Systematic optimization of blend composition and CB-PPy content yielded promising results in electrical conductivity, mechanical properties, and printability, underscoring the potential of PVDF/TPU/CB-PPy composites for various technological applications (Bertolini et al., 2020). These diverse studies collectively contribute to advancing the understanding and application of PVDF in multifunctional materials, spanning from nanocomposites to additive manufacturing techniques, paving the way for innovative solutions across various industries.

8.7 Polyether ether ketone

Polyether ether ketone (PEEK) is a versatile organic polymer synthesized using hydroxyl-containing oligoketones and dichloranhydrides of isophthalic and terephthalic acids via acceptor-catalytic polycondensation, yielding enhanced solubility, higher molecular weight, and superior thermal and mechanical properties. Investigations into the relationship between certain properties of polyethers and the condensation degree of oligoketones were conducted (Ozden et al., 1999). Fabrication of Sr-doped hydroxyapatite-dispersed PEEK composites through hot isostatic pressing with dispersoids content extending weight percentage of 5% to 20% exhibited uniform dispersal of dispersoids within the PEEK matrix, leading to improved mechanical properties confirmed through micro indentation measurements and finite element analysis. In vitro cytotoxicity assessment validated their potential as bioactive materials (Pal et al., 2020). The achievement of varying molecular masses in PEEK synthesis was realized through high-temperature polycondensation using nucleophilic substitution techniques. This was further optimized for superior 3D printing quality using selective laser sintering (SLS) technology, providing insights into their thermal and physicomechanical properties (Zhansitov et al., 2018). Nitration of PEEK and subsequent reduction to NH_2-PEEK produced versatile compounds soluble in dimethylformamide (DMF) with distinct thermal stability and fundamental nature, paving the way for further modifications and derivative synthesis (Conceição et al., 2008). Investigation into PEEK as a structural material for space applications involved thermomechanical design optimization and experimental characterization of 3D-printed PEEK structures, highlighting challenges such as thermal management and porosity while demonstrating the feasibility of using PEEK in additive layer manufacturing (ALM) for space component production (Rinaldi et al., 2020). Characterization of PEEK-hydroxyapatite nanocomposites revealed uniform dispersion of hydroxyapatite particles within the PEEK matrix, leading to enhanced mechanical properties and mitigating debonding issues, thereby offering promising solutions for biomedical applications (Wang & Li, 2020). The incorporation of carbon nanofibers (CNFs) into PEEK matrices via polymer crystallization technique improved properties. At the same time, the synthesis and characterization of h-BN-reinforced PEEK nanocomposites exhibited significant enhancements in wear and roughness resistance, hardness, thermal conductivity, storage modulus (Bartolucci et al., 2009; Liu et al., 2015).

Furthermore, the development of CNS-reinforced PEEK filament feedstocks and subsequent 3D printing for biomedical structural applications involved an assessment of in vitro bioactivity and mechanical characteristics, highlighting the synergistic effects of CNS and sulfonation treatment in enhancing osseointegration potential (Alam et al., 2020). Additionally, a review encompassing synthetic methodologies, modification procedures, and characterization techniques provided insights into the diverse range of properties and structural modifications of PEEK polymers, while a systematic review outlined the current applications of 3D-printed PEEK across various industries (Shukla et al., 2012; Dua et al., 2021). Finally, a comparative analysis of mechanical properties between 3D-printed PEEK and polyamide specimens elucidated the superior strength of PEEK, offering valuable guidance for their healthcare applications (Shetty et al., 2022). These studies collectively contribute to advancing the understanding and utilization of PEEK in diverse fields, from biomedical to aerospace applications, showcasing its versatility and potential for various industrial applications.

8.8 Polyimide

Polyimide (PI) has garnered significant attention as a matrix polymer for hybrid composites due to its commendable thermomechanical properties. Zhang et al. (2019) synthesized polyimide nanocomposite films by crafting a polyimide/silica nanoparticle medium resin over the poly-condensation and hydrolysis of varying ratios of TEOS (Tetraethoxysilane) and DMES (Dimethyldiethoxysilane) in a mixture of PPA (Polyamic acid) melted in DMAc (N,N-dimethylacetamide), tracked by high-temperature heating. The inclusion of 3-aminopropyltriethoxysilane (APrTEOS) served to improve compatibility among the PI (Polyimide) and SiO_2 (Silica). Thermal stability was examined via thermogravimetric analysis (TGA), revealing that well-dispersed inorganic components improved thermal stability by up to 16 wt%. Devaraju et al. (2012) adopted a two-step approach to synthesize polyhedral oligomeric silsesquioxane (POSS)–polyimide (PI) hybrid nanocomposites, initially dissolving BAED (Bisphenol-A ether diamine) and PMDA (pyromellitic dianhydride) in NMP (1-methyl-2-pyrrolidone) to produce PAA (Polyamic acid). Changing weight ratios of OAPS (octa(aminophenyl)silsesquioxane) were added to the PAA solution, followed by thermal imidization to effect polycondensation. Characterization using FTIR spectra analysis and thermal studies indicated enhanced properties, including a reduced dielectric constant and improved thermal stability, with the incorporation of POSS into polyimide. Feyzi et al. (2013) successfully synthesized a sequence of nanocomposites comprising organo-modified clay and organic polyimide via in situ polymerization, followed by comprehensive characterization using FT-IR spectroscopy, XRD, SEM, and TGA. Meanwhile, Huang et al. (2003) reported on synthesizing functional polyimide/polyhedral oligomeric silsesquioxane nanocomposites through a two-step approach involving thermal imidization after blending an OAPS/NMP solution with polyamic acid. Additionally Bergman et al. (1999) demonstrated the formation of an exfoliated polyolefin-silicate nanocomposite through the intercalation of a fluorohectorite modified organically by a complex of cationic palladium, while Ahmad et al.

(2007) synthesized poly(methyl methacrylate)-based nanocomposites incorporating silica and titanium dioxide nanoparticles via sol–gel transformation during in situ free radical polymerization. Moreover, Heo and Chang (2013) investigated polyimide films containing 4-phenylbutylamine-functionalized graphene sheets, showcasing thermal and electrical properties enhancements. Lastly, Chen et al. (2019) explored 3D printing of TPU/PLA/GO nanocomposites, demonstrating improved mechanical properties and biocompatibility, thus expanding their potential applications. These diverse studies underscore the versatility and promising applications of polyimide-based nanocomposites across various fields.

8.9 Polyacrylonitrile

Polyacrylonitrile (PAN) has been the subject of various synthesis approaches and applications, showcasing its versatility in nanocomposite fabrication. Lee et al. (2009) successfully conducted aqueous dispersal polymerization of itaconic acid, acrylonitrile, and methyl acrylate using PVA (polyvinyl alcohol) as a steric stabilizer in a medium of water and dimethylformamide (DMF). Their investigation systematically explored the synthesis and characterization of poly(acrylonitrile-co-itaconic acid-co-methyl acrylate) (PAN-co-IA-co-MA) nanoparticles by varying polymerization time and PVA concentration. The chemical structure was verified through 1 H NMR, FTIR, and GPC spectra analysis, with synthesized particles exhibiting nonsmooth surfaces and relatively spherical shapes. Moreover, Sun et al. (2022) manufactured a novel intercalation nanocomposite of PAN and kaolinite over straightforward in situ emulsion polymerization, employing organically modified layered silicates (kaolinite). Their process involved the modification of crude kaolinite followed by intercalation of acrylonitrile monomer, resulting in well-distributed and delaminated clay layers within the PAN matrix.

Similarly, Choi et al. (2002) synthesized nanocomposites montmorillonite of PAN/sodium (Na-MMT) by emulsion polymerization of acrylonitrile, achieving exfoliated morphologies and enhanced storage modulus compared to fresh PAN. Ahmad et al. (2007) explored the synthesis of PMMA nanocomposites through free radical polymerization of methyl methacrylate (MMA) in an aqueous medium, incorporating SiO_2 and TiO_2 via sol–gel transformation. Furthermore, Kuznetsova et al. (2013) synthesized organic-inorganic nanocomposites by reducing silver ions from silver acetate in DMSO and DMF mixtures, yielding stable polymer nanocomposites with uniformly dispersed silver nanoparticles. Siddiqui et al. (2013) investigated nanocomposites created on organomodified clay and poly(styrene-co-butyl methacrylate) copolymers, highlighting improved thermal stability with the introduction of clay. Riberi et al. (2018) proposed a method for incorporating nanoparticles into thermoplastic polyurethane filaments, enhancing the functional properties of printed parts. Finally, Jia et al. (2002) synthesized polyaniline–dodecylbenzenesulfonic acid (DBSA)/organophilic clay nanocomposites through aqueous medium polymerization, demonstrating enhanced thermal stability compared to bulk polymers. Nie et al. (2013) synthesized PAN/CuS compound nanofibers through monodispersed CuS nanoparticles, showcasing their potential as recyclable Fenton-like reagents for dye wastewater treatment. Şen and Çaylı (2010) successfully synthesized polymeric nanocomposites through the free radical polymerization of acylated epoxidized soybean oil and montmorillonite clay, achieving enhanced thermal and mechanical properties compared to the neat polymer matrix. Feng (2023) presented a novel method for synthesizing PAN nanofibers through an anodic aluminum oxide membrane as a model, offering simplicity in the fabrication process. Pérez et al. (2010) synthesized polypropylene nanocomposites using a molten-state mixing method, demonstrating enhanced thermal stability and heat deflection temperature by incorporating nanoclay and compatibilizer agent. These studies underscore PAN's versatility in nanocomposite fabrication, offering various approaches to enhance material properties across diverse applications.

8.10 Polyurethane

This chapter comprehensively explores the synthesis and characterization of polyurethane (PU) nanocomposites, elucidating various formulations and their resultant properties. Sin et al. (2015) initiated the investigation by synthesizing PU/clay nanocomposites utilizing ED (Ethylenediamine) and extenders chain of DAP (1,3-diamino-2-propanol), revealing a significant augmentation in both inter- and intrachain interactions, along with enhanced clay exfoliation facilitated by DAP. Nayani et al. (2013) extended the research by systematically varying hectorite clay percentages in PUR-nanoclay composites, achieving notable enhancements in foam properties such as tensile strength and thermal stability. Wong (2017) further diversified PU synthesis methodologies by incorporating nanoclay and castor oil, aiming to improve the mechanical properties and replace traditional polyol with a more environmentally friendly option. Fiayyaz et al. (2014) delved into the synthesis of PU/bentonite nanoclay nanocomposites through solution polymerization, successfully achieving homogeneous dispersion as confirmed by SEM analysis. Ch et al. (2012) explored PU nanocomposites derived from castor oil-HMDI, synthesizing PU-NPs with enhanced thermal stability and mechanical properties through innovative synthesis approaches. Rihayat et al. (2006) contributed insights into PU/clay nanocomposites based on palm oil polyol, demonstrating significantly improved thermal stability compared to virgin PU formulations. Zubir et al. (2016) investigated nanoclay/Palm Oil Polyol-based PU composites, highlighting the enhancements in thermal stability and tensile properties attributed to nanoclay incorporation. Satriananda et al. (2018) further advanced the field by synthesizing PU/clay nanocomposites extracted from palm oil for different applications, such as surface coating, showcasing superior thermal stability and corrosion resistance. Aziz et al. (2018) provided valuable insights into synthesizing bio-based PU nanocomposites from castor oil and PHB-diol, revealing substantial improvements in thermal and mechanical properties upon C25A incorporation. Moreover Riberi et al. (2018) proposed an innovative PU nanocomposite filament synthesis method, enabling functional 3D printing with modified filaments. Finally, Nikje and Farazfar (2012) investigated PET-based PU nanocomposites, unveiling the synergistic outcomes of recycled polyol and nanoclays on thermal and mechanical properties, while Das et al. (2019) offered a comprehensive overview of bio-based PU nanocomposites, underscoring their vast potential across various applications. Through diverse synthesis methodologies and meticulous characterization techniques, these studies collectively contribute to the advancement of PU nanocomposites, offering tailored solutions for a wide array of industrial applications.

8.11 Nylon

This chapter offers a comprehensive overview of the utilization of nanoparticles to enhance the design and performance of nylon-based materials, encompassing both foundational principles and recent advancements (Lim & Dasari, 2012). Various methodologies for preparing these nanocomposites and their physico-chemical characterization, mechanical properties, and functional attributes are discussed. Furthermore, the diverse applications of nylon-based nanocomposites and potential prospects in this field are explored. One innovative approach involves the preparation of nylon 6/Na-montmorillonite nanocomposites through a novel compounding technique utilizing Na-montmorillonite water slurry mixed with molten nylon 6 using an extruder, followed by water removal (Hasegawa et al., 2003). Electron transmission microscopy revealed exfoliated Na-montmorillonite silicate layers uniformly dispersed within the nanocomposites, offering enhancements in modulus, heat distortion temperature, strength, and gas permeability compared to pristine nylon/6. This method showcases notable improvements comparable to conventional nylon 6/clay nanocomposites. Another area of interest lies in exploring the potential of nylon 66 (PA66)/clay nanocomposites, with research primarily focusing on crystallization behavior and phase transition, yet lacking comprehensive studies on morphology, mechanical properties, and thermal stability (Phang et al., 2004). Similarly, investigations into nanocomposites of nylon-6 incorporating two sorts of marketable organoclays via liquify blending reveal the significance of clay dispersion on thermal and mechanical properties (Awad et al., 2008). Synthesis of nylon 66/montmorillonite nanocomposites utilizing co-treated montmorillonites demonstrates enhanced strength and modulus, highlighting the efficacy of the developed approach (Han et al., 2003).

Further advancements involve the synthesis of nylon 6 nanocomposites incorporating copper nanoparticles via ultrasound-assisted extrusion, exhibiting enhancements in mechanical properties and thermal stability attributed to efficient dispersion and crystal nucleation (Sierra-Ávila et al., 2018). Additionally, fabrication of nylon 11/organoclay nanocomposites through melt-compounding reveals superior properties with optimized nanoclay dispersion (Liu et al., 2003). The chapter also discusses the methods of preparing nylon 6/SiO_2 nanocomposites by anionic polymerization, elucidating the intricate interplay between silica content and mechanical properties (Rusu & Rusu, 2006). Lastly, a melt mixing masterbatch process for nylon 6 nanocomposites is explored, offering a balance between exfoliation and melt viscosity, showcasing a trade-off between modulus and melt processability (Shah & Paul, 2004). These studies contribute to advancing the understanding and potential applications of nylon-based polymer nanocomposites in various industries.

8.12 Polytetrafluoroethylene

Polytetrafluoroethylene (PTFE), a fluoropolymer renowned for its exceptional chemical stability, low friction coefficients, and extreme hydrophobicity, has garnered significant interest across various fields. In the study by Vidakis et al. (2020), PTFE was investigated as a strengthening factor in MEX (Material extrusion) additive industrial to enhance the mechanical performance of nanocomposites. Incorporating PTFE as fillers into polylactic acid (PLA) matrices and polyamide 12 (PA12) at loadings of up to 4.0 wt.%, 3D-printed samples were fabricated, showcasing improved mechanical responses within a filler concentration range of 2.0 to 3.0 wt.%. Meanwhile Cui et al. (2007) synthesized core–shell of PTFE-polyacrylate nanoparticles via seeded emulsion polymerization, encapsulating PTFE micropowder with acrylate in the presence of a fluorosurfactant. This core–shell structure enhanced the wettability of PTFE, expanding its compatibility with other materials. Yari and Moradi (2009) explored a novel method for synthesizing PTFE nanoparticles by adsorbing thermally synthesized tetrafluoroethylene (TFE) monomer onto different solid substrates, revealing variations in nanoparticle formation influenced by substrate properties. Jiang et al. (2019) introduced a multipurpose approach for 3D printing PTFE assemblies using DIW (direct ink writing), employing a specially formulated PTFE nanoparticle ink and a tailored thermal treatment process. Zhang et al., (2018, 2020) presented innovative optical and digital ultraviolet (UV) lithography-based μ-printing methods for crafting PTFE microstructures, showcasing their applications in medical implants and microfluidic control systems. Cheng et al. (2019) investigated surface functionalization of PTFE substrates through plasma treatment and chemical reactions, enhancing biocompatibility for medical device applications. Grewal and Yabu (2020) synthesized adhesive polymers based on biomimetic catechol for dispersing PTFE nanoparticles in aqueous media and generating silver nanoparticles, highlighting the versatility of these materials.

Additionally, Borkar et al. (2006) developed a solvent-free jet-blowing technique to produce PTFE micro and nanofibers, demonstrating their potential for surface modification applications. Lastly, Bapu et al. (1995) electrodeposited a nickel-fluoropolymer composite from a Watts nickel bath containing suspended PTFE, optimizing parameters to achieve an optimal PTFE content in the deposit. These studies collectively underscore PTFE's diverse approaches and applications in various technological domains, ranging from additive manufacturing to surface modification and nanoparticle synthesis.

8.13 Polypropylene

Polypropylene (PP) is a widely used commercial polymer, prized for its versatility and cost-effectiveness across various industries. However, its inherent hydrophobicity has posed challenges in developing advanced polymer nanocomposites. Researchers have explored innovative methods to enhance PP properties by incorporating nanofillers. One such approach involves the utilization of polymer-layered silicate nanocomposites, which typically consist of a PP matrix reinforced with solid mineral fillers like montmorillonite clay. Employing Fourier transform infrared spectroscopy (FT-IR) and X-ray photoelectron spectroscopy (XPS), Korakianiti et al. (2004) characterized these nanocomposites, elucidating their structural properties and interfacial interactions. Pérez et al. (2010) synthesized PP nanocomposites using melt-state mixing and twin-screw co-rotating extruders, investigating the effects of nanoclay content and compatibilizer agents like PP-g-MA on thermal, morphological, and mechanical properties through techniques such as DSC, TGA, XRD, TEM, and DMA. Sevil and Zuhal (2010) explored the synthesis of conducting polypyrrole (PPy) nanoparticles and their incorporation into PP matrices via melt-mixing and injection molding, highlighting improvements in mechanical and thermal properties and enhanced conductivity. Fina et al. (2005) studied the impact of functionalized polyhedral oligomeric silsesquioxanes (POSS) on PP nanocomposites, observing enhancements in thermal and morphological properties through DSC, TGA, SEM, and XRD analyses. Solomon and Somwangthanaroj (2004) reviewed recent advancements in PP nanocomposite synthesis, characterization, and application, emphasizing the clay intercalation structure and future research challenges. Varela et al. (2006) investigated the compatibilization and dispersion of clay nanocomposites in PP matrices using modified PP and organophilic clay, observing improvements in mechanical properties through TEM, DSC, and mechanical testing. Moodley (2007) explored the synthesis and characterization of nano-phased PP structures with montmorillonite nanoclays, analyzing mechanical properties and morphological changes through SEM and XRD studies. Gianelli et al. (2005) investigated the effect of matrix features on organoclay-PP nanocomposites, assessing mechanical properties and dispersion using XRD, TEM, TGA, MFR, and mechanical testing. Sun et al. (2022) synthesized PP nanocomposites using synthetic nanoplatelets derived from α-zirconium phosphate, highlighting improvements in compatibility and exfoliation through in situ propylene polymerization. Thakur et al. (2014) developed a green surface modification method for PP using plasma treatment and polydopamine functionalization, demonstrating enhanced properties in SiO2-reinforced nanocomposites. Overall, these studies underscore the potential of PP nanocomposites in various industrial applications, providing valuable insights into their synthesis, characterization, and performance enhancements.

8.14 Polyvinyl chloride

Polyvinyl chloride (PVC) is a significant thermoplastic polymer, distinct from other commodity polymers like polyethylene and polystyrene due to its specific structure, processing, and formulation requirements. Unlike its counterparts, PVC formulations involve various additives and exist in two primary forms: unbending PVC or uPVC and elastic PVC or pPVC. One notable avenue for enhancing PVC properties is integrating nanofillers, which are introduced and examined individually in subsequent sections. These nanofillers, ranging from organically modified montmorillonite (MMT) clays to hybrid silicone/acrylic core–shell type nanofillers, are vital in augmenting PVC properties. For instance, P4VP nanocomposites synthesized via in situ polymerization with MMT clays exhibited improved thermal stability and dynamic mechanical performance compared to other nanocomposites, demonstrating the effectiveness of this approach (Şen et al., 2006).

Similarly, PVC nanocomposites prepared through suspension polymerization with hybrid nanofillers displayed enhanced transparency and mechanical properties, with blends comprising 0.5–1.0 wt.%/VCM of nanofiller exhibiting the highest impact strength (Abramowicz et al., 2014). Moreover, incorporating modified montmorillonite into PVC matrices via melt blending resulted in improved nanocomposite properties, as evidenced by SEM, XRD, and TGA analyses (Hosseinloo & Sedaghat, 2014). These studies underscore the potential of nanofillers in enhancing PVC properties across various domains. Furthermore, advances in PVC-based blend nanocomposites offer promising opportunities for developing novel PVC products with superior properties compared to existing ones (Fariha et al., 2023). Notably, the utilization of nanoclays in PVC has been revitalized with novel nanoclays engineered for enhanced compatibility, leading to significant improvements in heat stability, mechanical properties, and barrier performance (Benderly et al., 2008).

Moreover, the comprehensive characterization of PVC nanocomposites, including thermal and morphological analyses, sheds light on their structure–property relationships and potential applications (Peprníček et al., 2006). Collectively, these studies pave the method for the advancement of PVC nanocomposites with tailored characteristics, offering new avenues for innovation in polymer materials. As researchers continue to explore novel synthesis methods and characterization techniques, the potential for PVC nanocomposites to address a wide range of industrial and technological challenges becomes increasingly evident. With ongoing advancements in materials science and engineering, PVC nanocomposites are poised to play a pivotal character in shaping the future of polymer materials and applications.

8.15 Polystyrene

Polystyrene (PS) is a widely utilized thermoplastic polymer due to its versatility in various applications, including abrasion, chemical resistance, and thermal stability. Researchers have explored the synthesis of PS-based nanocomposites to enhance its properties with different approaches. Giannakas et al. (2012) synthesized PS/laponite nanocomposites via a solution blending method, incorporating laponite clay modified with various surfactants like OCTA, OCTYL, and 2HT. While X-ray diffraction revealed an exfoliated structure with 2HT-modified laponite, conventional composites were obtained with OCTA and OCTYL. These nanocomposites exhibited improved thermal stability and mechanical properties, with exfoliated ones showing enhanced water barrier properties. Olad et al. (2011) investigated PS/clay nanocomposites synthesized through in-situ bulk polymerization and emulsion methods, incorporating Na-MMT and naturally improved clays. Characterization via SEM,

TGA, and optical transparency tests demonstrated improved thermal stability and transparency in nanocomposites compared to pure PS. Kostas et al. (2020) introduced a novel approach by synthesizing carbon nanowires from MCM-41 and incorporating them into PS via the solution–precipitation method. The resulting nanocomposites exhibited enhanced thermal and mechanical properties, with 3 wt% loading showing the most favorable outcomes. Siddiqui et al. (2013) explored the synthesis of nanocomposites using a poly(styrene-co-butyl methacrylate) copolymer matrix with various organomodified MMT clays. Their study revealed improved thermal stability in nanocomposites without significant changes in glass transition temperature or molecular weight. Awad and Engineering (2006) developed PS/clay nanocomposites using different imidazolium-modified montmorillonite clays, showing an intercalated structure with DMDIM-MMT and an exfoliated structure with DMHDIM-MMT. Neves et al. (2019) investigated expanded nanocomposites reinforced with cellulose nanofibers and nanocrystals, showing enhanced mechanical properties and reduced cell size in CNFs-reinforced samples. Blanco et al. (2014) discussed the potential of POSS fillers in improving the properties of polymer nanocomposites. Okamoto et al. (2000) synthesized clay/PMMA and clay/PS nanocomposites via in situ intercalative polymerization, demonstrating enhanced mechanical properties and thermal stability. He and Pan (2004) synthesized PS nanoparticles via differential microemulsion polymerization, achieving particle sizes below 20 nm under milder conditions than prior reports. These diverse approaches highlight the potential of PS-based nanocomposites in various applications, offering enhanced properties and functionalities.

8.16 Polyacrylamide

Polyacrylamide (PAM) is a synthetic polymer used in various industrial applications such as water treatment and soil conditioning. While direct 3D printing of PAM is challenging due to its chemical properties, efforts focus on blending PAM with other materials for compatibility with existing 3D printing technologies. Research explores creating PAM-based filaments or resins that could enable its indirect use in additive manufacturing, potentially opening new applications in the field. The study utilizes PAM and DLP 3D printing to create flexible hydrogel polymer electrolytes (HPEs) for Zinc ion batteries (ZIBs), demonstrating enhanced ionic conductivity and Zn2+ ions diffusion due to higher porosity, highlighting its potential for wearable applications and addressing limitations of traditional casted HPEs (Poompiew, 2023). It presents a 3D printing scaffold combining regenerated silk fibroin (RSF), polyacrylamide (PAM), and oxidized bacterial cellulose nanofibers (OBC), which enables effective strain sensing and promotes ectopic chondrogenic activity, indicating its promise for flexible bioelectronics in cartilage tissue engineering (Geng, 2024). The study employs PAM hydrogels PAM for simulating brain tissue properties in a human traumatic brain injury (TBI) surrogate model, aiming for minimal swelling and optical clarity adjustments.

Additionally, it employs 3D-printed polylactic acid (PLA) polymers for biomimetic skull structures, testing their mechanical strength, particularly with 100% grid infill patterns, to match human skull properties (Baker et al., 2023). The study develops a photocurable bioink using DLP 3D printing, integrating algin, acrylic amide, poly(ethylene glycol) diacrylate, phenyl(2,4,6-trimethylbenzoyl)phosphinic acid, and 3,4-dihydroxybenzoic acid, to address limitations in radiotherapy boluses, resulting in enhanced properties suitable for superficial tumor radiotherapy (Lu, 2024). The communication highlights PAM versatile properties and introduces a simple 3D printing method for polyacrylamide hydrogels, achieving a resolution of 100–150 μm, enabling stiffness-guided collective cell segregation in a lab-on-a-chip, with potential applications in biocompatible elastic medical devices and artificial organs (Wang & Li, 2020). The communication introduces biocompatible shape-memory double network (DN) hydrogels, combining PAM and gelatin, via a one-pot 3D printing method, enhancing mechanical integrity and shape transformation for potential use in medical robots and self-deployable devices (Chen et al., 2020). The study introduces a double-network (DN) hydrogel by combining κ-carrageenan and PAM. This hydrogel exhibits excellent recoverability, self-healing properties, and mechanical strength for 3D printing. It also demonstrates high strain sensitivity, making it suitable for sensitive strain sensors in robotics and human motion detection (Liu & Li, 2017). The study integrates sodium alginate (SA) with amino-graphene oxide (aGO) to form SA-functionalized GO (A-aGO), incorporated into a double-network (DN) hydrogel with alginate and PAM, enabling thixotropic properties for 3D printing before UV curing, resulting in intricate structures with high stackability and fidelity, and a UV-cured hydrogel with robust mechanical strength and toughness from physical and chemical cross-linking (Liu & Bastola, & Li, 2017). The study introduces a soft hydrogel-based actuator with autonomic perspiration and thermoregulation capabilities, featuring a N-isopropylacrylamidebody and PAM microporous postern layer, exhibiting a 600% cooling rate enhancement over nonsweating devices and efficient manipulation of heated objects mechanically and thermally (Mishra et al., 2020).

8.17 Polyaniline

Polyaniline (PANI) displays versatile properties in 3D printing, enabling the creation of conductive composites, flexible electrodes, and tissue engineering scaffolds, expanding additive manufacturing applications in flexible electronics, energy storage, and biomedical implants. A novel synthesis method enhances the gas sensitivity of PANI doped with phytic acid (Ph) while preserving its quality, enabling the development of wearable 3D-printed sensors by combining PANI with gelatin in a unified 3D network, showcasing multifunctional capabilities and potential for sustainable, biocompatible materials (Sun et al., 2022). The study investigates 3D-printed composites of PANI and titanium dioxide (TiO_2) to enhance photocatalytic efficiency for wastewater treatment and hydrogen gas production, focusing on optimizing TiO_2 crystal structure and particle size to develop more efficient photocatalysts with environmental and energy production advantages (Saengchai, 2023). The study explores the use of PANI in Digital Light Processing (DLP) 3D printing of conductive polymers, employing photo rheology and Jacobs working curves to predict printability. PANI loading up to $\simeq$3 wt% enhances conductivity significantly, reaching 10^{-5} S/cm, while maintaining the mechanical properties of the acrylic matrix, suggesting the potential of these composites for flexible electronic device applications (Arias-Ferreiro et al., 2021). Polyaniline nanofibers and graphene sheets are integrated into a polyacrylate resin solution for improved conductivity in DLP 3D printing, achieving significant enhancement in electrical conductivity (4.00 $\times$ 10^{-9} S/cm) and demonstrating the potential for this printing technology (Han & Cho, 2018). Polyurethane (PU) composites incorporating PANI and graphene sheets (GS) are optimized for DLP-type 3D printing, demonstrating significantly improved sheet resistance and higher tensile strengths, highlighting their potential for highly conductive PU composites in DLP-type 3D printing applications (Joo & Cho, 2020). The study investigates the modification of polyacrylonitrile (PAN) fibers using PANI through two methods: polyacrylonitrile/polyaniline blended and polyacrylonitrile/ polyaniline in situ, revealing solid interactions in the latter. While PAN/PANI mixture fibers showed improved mechanical parameters, they were weaker than unmodified PAN fibers, and PANI-synthesized in situ fibers exhibited dielectric properties and a mass-specific resistance of 5.47 kΩg/cm^2 (Karbownik et al., 2019). This study employs direct ink writing to fabricate PANI/RGO (reduced graphene oxide) composites using PANI/ GO (graphene oxide) gel inks, enabling the construction of planar solid-state supercapacitors with high performance, indicating a promising method for creating PANI-based 3D architectures for future micro-nanoscale energy and electric devices (Wang et al., 2018). The study employs PANI in 3D printing technology to manufacture conductive polymeric composite scaffolds for tissue engineering. 3D printing to produce conductive polymeric composite scaffolds for tissue engineering, demonstrating that 0.1% wt. PANI loading in polycaprolactone scaffolds enhances compressive strength (6.45 $\pm$ 0.16 MPa) and conductivity (2.46 $\pm$ 0.65 $\times$ 10^{-4} S/cm), showcasing potential suitability for bone tissue engineering due to observed favorable cytocompatibility with human adipose-derived stem cells. Polyaniline is pivotal in developing memristive elements via 3D printing, examining its impact on planar conductive materials and 3D fibrous structures, considering factors like molar mass, solvent, subphase nature, and PANI-PEO ratio (Malakhova et al., 2017). Polyaniline enables the fabrication of flexible nanohybrid paper electrodes for energy storage devices through inkjet printing, exhibiting enhanced electrochemical properties suitable for flexible supercapacitor applications (Chi et al., 2014).

8.18 Polydimethylsiloxane

Polydimethylsiloxane (PDMS) offers favorable properties like flexibility, biocompatibility, and optical transparency, making it valuable in 3D printing for diverse applications such as microfluidics and tissue engineering. Its use allows for the fabrication of complex, precise structures, enhancing prototyping and research in various scientific and technological fields, paving the way for advancements in science and technology. Manufacturing of intricate PDMS techniques has become more accessible as a result of recent developments in 3D printing, with enhanced mechanical properties up to threefold compared to traditional methods, enabling superior cell adhesion qualities for medical applications such as lab-on-a-chip devices and tissue/organ-on-a-chip systems (Özbolat et al., 2018). PDMS is employed in Digital Light Processing (DLP) to creating integrated microfluidic devices with 3D printing, with a focus on determining optimal design parameters and assessing printing accuracy for complex taper structures, ultimately demonstrating fluidic control capabilities, highlighting its practical functionality in microfluidic applications (Cheon & Kim, 2021). PDMS devices from 3D-printed molds support physiological cultures of HeLa cells and ovarian tissues, outperforming static conventional cultures postwashing and conditioning despite leaching concerns, suggesting potential for viable microfluidic applications (Ferraz et al., 2020). Polydimethylsiloxane is explored for 3D printing porous structures, showcasing tunability in mechanical properties through print patterns and infill densities, offering insights for designing adaptive soft robots and actuators across diverse fields like biomedical engineering and oil/water separation (Woo et al., 2021). The study introduces a novel method called freeform reversible embedding (FRE) for 3D printing PDMS within a hydrophilic Carbopol gel support, overcoming challenges of low elastic modulus and need for support, thus broadening biomaterial options in additive manufacturing (Hinton et al., 2016). Polydimethylsiloxane is crucial in 3D printing for miniaturized analytical devices and soft lithography molds; this study investigates 16 commercial stereolithography (SLA) resins to identify efficient treatments, revealing phosphine oxide-based photo-initiators leaching from 3D prints and demonstrates that UV and thermal treatments effectively eliminate inhibitors, enabling reliable PDMS replication and sequestration of high molecular weight species in the molds (Venzac et al., 2021). The study presents a method for producing flexible circuits on PDMS surfaces through inkjet printing semi-wrapped structures, enabling the creation of continuous and uniform conductive lines with high transparency and bendability, offering a straightforward approach without complex modifications or specialized equipment (Sun et al., 2016). The study explores PDMS molds fabricated using polylactic acid (PLA) filament-based 3D printing as a cost-effective alternative, showing excellent adhesion to glass coverslips for leak-proof well formation, with cell studies demonstrating efficacy and versatility in research applications (Borg et al., 2023). The study investigates the use of 3D-printed PDMS barriers to regulate the flow in paper microfluidics flexibly, demonstrating effective confinement of nonpolar solvents and enabling sequential fluid delivery, highlighting their practical implications in environmental and medical detection applications (Chen et al., 2020). The study presents a novel method for high fill-factor cylindrical microlens arrays (CMLAs) fabrication using electric-field-driven microscale 3D printing of PDMS, demonstrating favorable optical properties, simplicity, and flexibility in creating dual-microlens arrays with enhanced image-projection and light-diffraction capabilities, suggesting its potential for large-scale production at minimal cost and complexity (Zhang et al., 2021).

8.19 Polyethylene

Polyethylene (PE), including both low and high-density forms, emerges as a promising material for 3D printing, offering exceptional flexibility, durability, and chemical resistance, making low-density polyethylene (LDPE) ideal for flexible components while high-density polyethylene (HDPE) suits durable prototypes and functional parts across diverse industries, owing to its affordability and resistance to moisture and chemicals. The study addresses challenges in 3D printing with HDPE by incorporating yellow birch fibers and a coupling agent, significantly reducing deformation and improving Young's modulus, offering solutions to traditional HDPE printing limitations (Koffi et al., 2021). The study overcomes challenges in FFF 3D printing with high-density HDPE by blending waste-derived HDPE with additives and using a thin "brim" adhered with polyvinyl acetate-based glue, showcasing its potential for recycling plastic waste into 3D printing filament (Gudadhe et al., 2019). This study introduces composite filaments containing 50% HDPE/LDPE as a shell and a PC/ABS blend core for standard fused filament fabrication, showing improved impact resistance and elongation at break attributed to LDPE's characteristics, presenting cost-effective and mechanically enhanced options for additive manufacturing (Peng et al., 2019). High-density polyethylene emerges as pivotal in 3D printing for construction, offering advantages such as mass customization, reduced waste, and recycled material use, demonstrated through Finite Element Analysis for structurally feasible, steel-reinforcement-free arched roofs, promising affordable housing solutions (Tahmasebinia et al., 2018). Combining recycled high-density polyethylene with waste carbon fiber enhances mechanical properties, particularly tensile and flexural strength, and modulus, with variations in performance observed between injection molded and 3D-printed samples, indicating methodological influences on material outcomes (Borkar et al., 2022). The study developed HDPE nanocomposites with TiO_2 fillers to enhance mechanical properties in FFF materials, demonstrating significant improvements in tensile strength, particularly at 10 wt.% TiO_2 concentration, supported by morphological and thermal characterization (Vidakis et al., 2020). The study investigates the impact of printing values and degrading variables on the mechanical characteristics of FDM-printed PE fibers, revealing that smaller diameter fibers and higher collecting speeds enhance tensile strength and Young's modulus, higher printer tip temperatures and exposure to UV radiation and thermal deterioration reduce the mechanical features, emphasizing the need for optimization and durability assessment in practical applications (Xiao et al., 2023). The study investigates PE and EVA blends for 3D printing, finding that increasing EVA concentration enhances the pseudoplastic effect and amorphization of PE, leading to reduced crystallization rates and improved mechanical properties, showcasing the potential for enhanced printability with optimized PE/EVA ratios (Slonov, 2023). The study explores the feasibility of using reused HDPE filament for

3D printing. It entails the development of a material extrusion machine to produce a high-quality filament with similar characteristics to virgin HDPE, offering a sustainable option for additive manufacturing feedstock with the potential for further optimization and market competitiveness (Shiferaw et al., 2023). The study investigates recycling HDPE parts from defunct boats into marine-friendly materials using screw-assisted extrusion 3D printing, adding glass fibers to mitigate mechanical performance loss, and compares properties with compression molding, emphasizing extended thermos-mechanical and UV stability assessments for potential marine applications (Daniele, 2023). The study examines the addition of local natural sand silica to HDPE to create composite sheets for thin film and recycling industries, observing significant improvements in toughness and elastic modulus with 25 μm filler size but a general decrease in these properties with smaller particle sizes, impacting tensile strength and composite density (Siraj et al., 2022).

8.20 Polyethylene glycol

Polyethylene glycol (PEG) stands out in 3D printing for biomedical uses due to its biocompatibility, solubility, and tunable properties, offering promise in creating precise implants, scaffolds, and drug delivery systems, elevating its significance in regenerative medicine and biomedical research. A revolutionary 3D bioprinting device using extrusion technology such as PEG-Fibrinogen biomaterial is developed for skeletal muscle tissue engineering. It produces highly organized 3D constructs with aligned muscle fibers, demonstrating the potential for regenerating muscle tissue in a murine model of volumetric muscle loss (VML) (Fornetti et al., 2023). Modified PBS-DLS copolymers incorporating PEG exhibit enzymatic degradation, reduced mass loss, and bioactivity, enabling the fabrication of hierarchical tubular structures via electrospinning and 3D printing, ideal for tissue engineering applications (Zarei et al., 2023). The study investigates vat photopolymerization printing of hydrogels utilizing stable and hydrolysable crosslinkers such as PEG-dimethacrylate and PEO-related derivatives, enabling the fabrication of initially highly crosslinked networks that transition to low density, offering potential for medical and other applications necessitating 3D-shaped hydrogel materials (Liz-Basteiro et al., 2023). A novel 3D printing composite comprising animal bone powder, PEG, and PLA was investigated. It exhibited improved mechanical properties and biocompatibility, particularly with 14% PEG content, making it promising for bone tissue engineering applications (Sun et al., 2022). The study explores PBS-DLS copolymers modified with PEG, revealing their structure–properties correlation and suitability for electrospinning and 3D printing, demonstrating promising printability and nanofiber formation capabilities for diverse applications (Zarei et al., 2023). A 3D-printed porous PCL/PEG composite scaffold was implanted in beagle dogs, stimulating bile duct regeneration with epithelial cell proliferation, collagen deposition, and biliary muscular formation, showcasing its efficacy for extrahepatic bile duct regeneration compared to nonimplanted areas (Cai et al., 2023). The study presents self-healing elastomeric networks based on hydrophobic PEG-PLA star-shaped copolymers, showing high self-healing efficiencies, suitable mechanical properties for soft tissues, linear degradation profiles, and cytocompatibility, offering promise for self-healable 3D-printed biomedical devices (Grosjean et al., 2023). The research presents degradable elastomeric networks from hydrophobic PEG-PLA copolymers, offering high self-healing efficiency, suitable mechanical properties for soft tissues, linear degradation profiles, and cytocompatibility, suggesting potential for self-healable 3D-printed biomedical devices (Lage-Rivera et al., 2023). The incorporation of placenta powder into synthetic Hydrogels based on polyethylene glycol (PEG) is formed via a process called UV-initiated thiol-ene coupling (TEC), resulting in MoDPEG+ hydrogels with tunable storage moduli, enhanced biocompatibility, and improved mechanical properties, suitable for 3D printing in tissue regeneration applications (Fan et al., 2023). The study explores the structure–properties relationship of biodegradable PBS-DLS copolymers modified with hydrophilic PEG for 3D printing and electro-spinning, revealing direct fabrication of 1.75 mm filaments, biphasic morphology, improved processability, and enhanced cell viability, suggesting potential biomedical applications for these materials (Zarei et al., 2023). The study proposes 3D printing of personalized hybrid scaffolds for critical-size bone defects, featuring a gradient cell-laden PEG hydrogel within a porous PCL cage structure. Optimized designs, such as zigzag/spiral, offer mechanical strength and nutrient diffusion, facilitating hBMSC viability, differentiation, and mineralization, promising personalized solutions for bone tissue engineering (Altunbek et al., 2023).

8.21 Polyethyleneimine

Polyethyleneimine (PEI) demonstrates versatile potential in 3D printing, enhancing cell adhesion and proliferation in tissue engineering scaffolds and effectively capturing heavy metals in 3D-printed hydrogels for environmental remediation, highlighting its multifunctional role in creating stable structures for diverse applications. Polyethyleneimine facilitates direct-ink writing 3D printing at room temperature; a study on SiO_2–PEI composites explores extrusion flow, rheological behavior, and mechanical properties, revealing increased tensile strength with greater density and reduced nozzle diameter, and enhanced compressive strength with thinner layers (Feng, 2023). Polyethyleneimine enhances polycaprolactone (PCL) scaffold properties in 3D printing for bone tissue engineering by improving hydrophilicity, enzymatic degradation, and cellular attachment, differentiation, and proliferation, signifying promise in biomedical applications (Khoshnood & Zamanian, 2022). Polyethyleneimine enhances the mechanical properties of 3D-printed polyetherimide parts, expanding PEI's application potential in high-performance contexts through optimized high thermal processing parameters and carbon fiber incorporation, showcasing a novel approach for developing improved CF/PEI composites (Zhang et al., 2019). Polyethyleneimine enhances the formulation of a novel bioink alongside alginate for 3D printing applications, where increasing PEI content improves printability, mechanical strength, and cellular functions, facilitating wound healing and tissue regeneration with promising in vivo outcomes (Khoshnood & Zamanian,

2022). Polyethyleneimine is essential in a novel 3D printable ecofriendly hydrogel for heavy metal removal from aqueous environments, optimizing ink formulation to bind metals effectively enabling scalable water remediation through 3D printing technology (Finny et al., 2022). Polyethyleneimine is instrumental in a 3D-printed sorbent for CO_2 capture, achieving high purity and recovery values in biomass-fired power plant flue gas, showcasing its potential for efficient, low-energy CO_2 capture (Krishnamurthy et al., 2021). Polyethyleneimine improves the properties of 3D-printed alginate scaffolds by enhancing fibroblast cell functions, promoting adhesion, proliferation, and spreading, offering promise for tissue engineering and wound dressing applications by reducing degradation and forming a solid barrier layer (Khoshnood et al., 2021).

8.22 Polyethylene oxide

Polyethylene oxide (PEO) stands out in 3D printing for its water solubility, facilitating easy processing and blending with other materials, offering versatility for creating complex structures, bioinks for tissue engineering, and scaffolds for regenerative medicine, promising advancements in diverse fields and industries. A GelMA-PEO 3D hydrogel microspheroid system called 3D-HyGTIC mimics the tumor microenvironment, showing stepwise upregulation of immune checkpoint molecules in CD8+ T cells. The system models CD8+ T cell exhaustion and response to anti-PD-1 and anti-LAG-3 therapies, aiding in immunotherapy evaluation and combination therapy design (Meng et al., 2023). The study investigates vat photopolymerization for printing hydrogels with stable and hydrolysable crosslinkers, such as PEO/PEG-related formulations, enabling adjustable water content and showing promise in medicine and other fields requiring 3D-shaped hydrogel materials (Liz-Basteiro et al., 2023). A plasticizer-free filament of PEO and gabapentin, created through hot melt extrusion, facilitates FDM 3D printing for intragastric floating tablets, showcasing improved drug absorption and potential for enhancing gabapentin delivery in the management of overactive bladder (Khizer et al., 2023). The study investigates ethosuximide-loaded PEO scaffolds through 3D printing, demonstrating sustained drug release and offering the potential for personalized epilepsy treatment with enhanced efficacy and reduced cytotoxicity (Karabulut et al., 2023). Poly(ethylene glycol) diacrylate/poly(ethylene oxide) (PEG-DA/PEO) interpenetrating polymer network hydrogels (IPNH) were 3D printed into filaments and grids, validated for enzyme immobilization and CO2 capture, exhibiting favorable mechanical properties and enzyme activity retention, suggesting versatile applications in biocatalytic reactors and biosensors (Shen et al., 2023). Recent advancements in conductive polymer-based hydrogels (CPHs) for wearable sensor designs are explored, highlighting challenges such as limited strain sensing range and signal instability, with insights into future research directions and prospects in the field (Liu et al., 2023). The review emphasizes the synergistic potential of integrating electrospinning with 3D printing, enabling digitally controlled shape-morphing fabrication and the creation of porous structures for diverse biomedical applications, including tactile sensors, filtration systems, tissue engineering, and biomedical scaffolds (Ghosh et al., 2023). The study introduces a functional bioink for cartilage bioprinting, utilizing PEO in gelatin methacrylate (GelMA) for microporous hydrogels, combining chondrocyte spheroids (CSs) to print predefined lattice-shape constructs and enhance cell interactions, presenting a promising strategy for patient-specific cartilage fabrication (Li, 2024). The study explores the impact of PEO addition on the combustion properties of Aluminium/ cuprous oxide-based nanoscale energetic materials (nEMs), revealing desensitization to high electrical spark discharges with increased PEO content but compromised explosion-induced pressure beyond 15 wt.%, suggesting potential use in composite solid propellants (CSPs) for enhanced propulsion in small rocket motors (Kim & Kim, 2023).

8.23 Polyethylene terephthalate

Polyethylene terephthalate (PET) is a commonly used thermoplastic in 3D printing because of its remarkable mechanical properties and resistance to chemicals, compatible with Fused Deposition Modeling (FDM) and some Stereolithography (SLA) techniques, offering good strength, stiffness, and toughness for functional prototypes and end-use parts, with shrinkage management through heated print bed and nozzle, making it a versatile choice for diverse 3D printing applications. The study investigates gamma radiation effects on 3D-printed PETG and ASA polymer components, finding PETG's superior strength and ductility compared to ASA, with both showing cross-linking and decreased mechanical properties under exposure, suggesting potential applications in demanding environments like nuclear facilities and space exploration (Demir et al., 2024). The study explores using PET and HDPE in local distributed recycling and additive manufacturing processes, investigating the feasibility of using shredded plastic waste directly as 3D printing feedstock, optimizing parameters for multimaterial recycling printability, and demonstrating successful large-scale 3D printing of objects from rPET/rHDPE flakes, potentially reducing production costs by up to 88% (Gonzalez, 2024). The study compares synthetic temporal bone models for surgical training, highlighting photopolymer resin's superior anatomical accuracy and dissection experience compared to PETG and Simubone™, suggesting resin and PETG as cost-effective alternatives to cadaveric dissection for surgical practice (Cafino, 2024). The study investigates the influence of 3D printing parameters on the vibrational behavior and mechanical properties of PET-G tapered beams, revealing relationships between surface roughness, hardness, tensile strength, elasticity modulus, and printing parameters like infill rates and layer heights, suggesting potential enhancements with fiber reinforcements in future research (Ergene et al., 2023). The study investigates the impact of printing speed on the dimensional stability of 3D-printed items made from recycled container filament. It specifically focuses on how different printing parameters might affect the quality of recycled PET materials used in additive manufacturing applications (Budiono et al., 2023). The study evaluates the mechanical strength of wrist-hand orthoses (WHO) made from PET-G filament using material extrusion (MEX) additive manufacturing (AM), employing experimental testing and simulation to assess individualized orthoses for mechanical properties and comparing results between experimental tests and finite element simulation models (Łukaszewski, 2023). The study investigates the impact of 3D printing values on composites made of carbon fiber-reinforced polylactide and polyethylene terephthalate glycol, particularly PET-G, finding that raster angle, printing speed, and extrusion temperature significantly affect mechanical properties, highlighting the need for parameter optimization to enhance mechanical performance in 3D-printed PET-G composites (Ammar, 2024). The study investigates the effects of a chain extender on the properties of recycled PET (rPET) for 3D printing filament, finding that while it reduces melt flow index and increases viscosity, it enhances rigidity and thermal resistance, supporting sustainable manufacturing through recycled PET utilization in additive manufacturing (Pongmuksuwan, 2023). The study explores the development of a recycled composite using PET waste and carbon fibers, with samples tested for mechanical properties, revealing that samples containing 20% CF exhibit the highest elastic modulus, emphasizing the reinforcing effect of fibers and suggesting potential applications in recycling and composite materials (Almahri et al., 2023). The study investigates the impact of varying percentages of ionic liquid (IL) solutions on the recycling process of poly(ethylene terephthalate) (PET) recycled bottles, highlighting improvements in thermal stability and mechanical properties while acknowledging challenges in cost and effective recycling (Ahmed et al., 2021).

8.24 Poly(3-hydroxybutyrate-co-3-hydroxyvalerate)

Poly(3-hydroxybutyrate-co-3-hydroxyvalerate) (PHBV), prized for its biodegradability and biocompatibility, is employed in 3D printing for scaffold fabrication, offering customizable structures for tissue engineering with additives like CaSH and CS to enhance osteogenesis and tissue regeneration, highlighting its potential in biomedical implants, drug delivery systems, and tissue constructs. PHBV's role in tissue engineering is highlighted as it's combined with 45S5 Bioglass for 3D scaffold printing via FDM, enhanced by CTMS-treated BG particles for improved printability, offering scaffolds with mechanical and biological properties akin to trabecular bone ECM, showing potential for biomedical applications with cell-friendly features (Aráoz et al. 2021). In order to overcome the difficulties with compatibility and heat resistance in PHBV:PLA mixes for 3D printing, a styrene-acrylate copolymer containing oxirane moieties was used as a chain extender, enhancing printability through optimized parameters like temperature and speed, leading to improved mechanical properties and offering insights for broader biopolymer blend applications (Fuentes et al., 2020). Blending PHBV with PLA and PBAT enhances 3D printing capabilities, as evidenced by improved crystallinity, microstructure, and mechanical properties of PLA/PHBV/PBAT-printed specimens, showcasing the potential of these blends for broader applications in 3D printing (Yang et al., 2019). PHBV, combined with ZrO2·nH2O particles in composite filaments, shows altered matrix behavior, enhanced amorphous phase, and improved relative density, leading to increased microhardness and potential applications in regenerative medicine via FDM-printed scaffolds (Carvalho et al., 2021). The study evaluates chondrogenic differentiation of hMSCs and macrophage polarization on PC/PH blended scaffolds with dual PP and SP pores, revealing that the 60B_PC/PH/50SC scaffold supports hMSC attachment and promotes chondrogenesis, while macrophages exhibit transiently elevated IL-1β levels, suggesting potential for tissue engineering applications (Kaewkong et al., 2022). Researchers investigate 3D architectures using regenerated silk (RS) and PHBV with embedded functionalities, designing sutureless thermoresponsive clips for anastomized porcine intestines with programmable shape-changing behavior, demonstrating potential clinical applications in vital sign monitoring and sutureless sealant patches (Bon et al., 2021). The study demonstrates the fabrication of 3D-printed PHBV/CaSH scaffolds coated with chitosan hydrogel, which enhances osteogenesis by promoting cell adhesion, proliferation, and upregulating osteogenic gene expression, proposing an optimistic strategy for the restoration of bone deformities and the generation of new bone tissue. PHBV in 3D printing scaffold creation, combined with CaSH and coated with CS hydrogel, enhances adhesion, proliferation, and osteogenesis of rBMSCs compared to PHBV and PHBV/CaSH scaffolds alone, showing promise for bone defect repair with in vivo studies confirming efficacy in promoting new bone formation (Ye et al., 2018).

8.25 Poly(lactic-co-glycolic acid)

Poly(lactic-co-glycolic acid)'s (PLGA) integration into 3D printing enables tailored scaffold fabrication for tissue engineering, offering biocompatibility, tunable degradation kinetics, and controlled drug release, fostering applications in regenerative medicine and advanced biomedical constructs. PLGA facilitates the process of binder-free 3D graphene structures via a combination of 3D printing and atmospheric pressure chemical vapor deposition (CVD), offering the potential for tailored graphene structures with excellent thermal stability and diverse applications in energy storage electrodes (Kondapalli et al., 2021). PLGA integrated into 3D printing enables the fabrication of custom-designed, mechanically enhanced PLGA/CaSO4 scaffolds for bone tissue engineering, demonstrating improved osteogenesis properties and suitability for repairing irregular load-bearing bone defects (Liu et al., 2022). PLGA in 3D printing creates porous scaffolds for bone tissue engineering, incorporating TiO_2 nanoparticles to enhance mechanical properties and cellular activity, demonstrating improved osteoblast proliferation and calcium secretion, making it promising for bone tissue regeneration (Rasoulianboroujeni et al., 2019). PLGA in 3D printing with dimethyl sulfoxide (DMSO) facilitates the incorporation of bioactive molecules, improving scaffold stretchability and enhancing expression of genetic markers for chondrogenesis, osteogenesis, and adipogenesis in vitro, advancing therapeutic applications in regenerative medicine (Guo et al., 2018). PLGA in 3D printing for bone tissue engineering offers customizable scaffolds with intricate structures, as detailed in a comprehensive review outlining research progress, modified component properties, structural variations, and future directions for bone tissue repair applications (Sun et al., 2022). 3D-printed PLGA scaffolds intended for preserving alveolar sockets and bone regeneration exhibited nondegraded characteristics but impeded bone regeneration in rabbit tibia defects, suggesting caution regarding their use in noncritical bone defects to avoid slowing fracture healing (Lim et al., 2021). PLGA's vital role in 3D printing for cartilage tissue engineering is highlighted, showing stable degradation over 56 days with no adverse effects on scaffold properties or human bone marrow mesenchymal stem cells (HBMSC) viability, suggesting promising potential for cartilage tissue regeneration (Dastidar, 2023). PLGA in 3D printing creates scaffolds via diffusion and antisolvent phase separation, offering efficient fabrication of biocompatible scaffolds with microporous and hollow fiber architectures influenced by initial polymer solution concentrations and printing parameters (Mironov et al., 2019). PLGA in 3D printing facilitates the creation of osteogenesis-promoting scaffolds, exemplified by PLGA/hydroxyapatite (PLGA/HA) scaffolds which significantly enhance bone strength, serum calcium, and alkaline phosphatase expressions, reduce tartrate-resistant acid phosphatase levels, and elevate bone morphogenetic protein 2 (BMP2) levels, suggesting their potential in bone tissue engineering applications (Wang & Li, 2020). 3D-printed PLGA/hydroxyapatite (HA) scaffolds demonstrate excellent biocompatibility and osteogenesis, facilitating bone repair in rat full-thickness bone defects both in vivo and in vitro, supported by improved bone healing, upregulated osteoblast-related gene expression, complete scaffold degradation, and well-distributed porous latticework promoting osteoblast attachment and growth, suggesting their potential for orthopedic applications in bone grafting (Gao et al., 2019). PLGA's suitability for patient-specific scaffold fabrication via direct melt extrusion printing was investigated, correlating variations in lactic acid:glycolic acid (LA:GA) ratios and end caps with scaffold precision using micro-computed tomography, revealing linear solid relationships between actual and predicted precision and paving the way for standardized tissue engineering applications (Guo et al., 2017).

8.26 Poly(N-isopropylacrylamide)

Poly(N-isopropylacrylamide)'s (PNIPAM) temperature-responsive properties enable the creation of dynamic structures and functional materials in 3D printing, offering precise control over deformation and response, facilitating the development of stimuli-responsive constructs for applications such as smart actuators, sensors, biomedical devices, tissue engineering, and drug delivery systems. PNIPAM in 3D printing enables the synthesis of a novel PNIPAM/graphene oxide nanocomposite hydrogel with an improved molding method, response mode, shear thinning properties optimized by nanoclay, enhanced photothermal responsiveness with graphite oxide, and reversible behavior modulated by GO content and illumination time, showcasing potential in actuators and functional programs (Zhang et al., 2019). It is used in the development of copolymer scaffolds for dentoalveolar regeneration via 3D printing, showcasing successful copolymerization, printability, degradation up to 35% within 7 days, 1200% swelling, and initial cell viability, suggesting potential for dentoalveolar tissue engineering pending further research (Amoli, 2024). PNIPAM's thermoresponsive properties enable precise control over 3D printing processes, facilitating the development of scaffolds with tunable mechanical properties, biocompatibility, and degradation rates, offering versatility for biomedical applications like tissue engineering and drug delivery (Gialouri, 2023). Zhou et al. (2018) employed PNIPAM in 3D printing through a printable hybrid hydrogel, combining microscopic particles made of Poly(N-isopropylacrylamide) with a silica-alumina composition that has a high water content (Si/Al) gel matrix, responsive to heat and electrical stimuli, facilitating direct layer-by-layer printing without curing, allowing the fabrication of adaptable equipment like intelligent windows and 3D visual waveguide controls, showcasing PNIPAM's versatility in 3D printing technologies. PNIPAM is pivotal in 3D printing soft hydrogel-based actuators, enabling thermoregulation via multimaterial stereolithography, with a chemomechanical response facilitating pressurization at low temperatures (< 30°C) and perspiration at elevated temperatures (> 30°C), demonstrating superior thermoregulatory performance akin to animal systems with a temporary actuation efficiency decrease (Mishra et al., 2020). PNIPAM enables 4D printing using photoactive Prussian blue nanoparticles dispersed within thermo-responsive hydrogels, allowing for light-triggered shape-morphing effects and diverse 3D structure fabrication, demonstrating PNIPAM's integral role in responsive 3D printing (Pelluau, 2023). PNIPAM hydrogel's biocompatibility and thermoresponsive properties enable tissue engineering applications, facilitating cell proliferation and encapsulation/release of therapeutic agents in 3D printing, showcasing its potential as a customizable scaffold for tissue regeneration in additive manufacturing (Wu, 2023). PNIPAM enables the development of hydrogel-based actuators with programmable deformation, as demonstrated by a temperature-responsive GelMA-PNIPAM hydrogel network, exhibiting 3D printability, tunable mechanical properties, and significant volume reduction at physiological temperature. In contrast, myoblasts laden in the hydrogel showcase its potential for biomedical applications in biomimetic devices, artificial muscles, and soft robotics (Huang et al., 2023). Also, it facilitates the fabrication of a hybrid PNIPAM/cellulose nanofibrils (CNFs) hydrogel composite via inverted stereolithography, enabling the control of parameters related to the lower critical solution temperature (LCST), optical and bioadhesive properties tuning, and switchable bioadhesivity to bacteria, highlighting its potential for durable temperature-sensitive sensors and functional biomedical devices (Sun et al., 2022). PNIPAM hydrogels are essential in 3D printing for soft actuators and soft robotics, particularly in bilayer hydrogels combined with polyacrylamide (PAM) hydrogels, enabling different curvatures during bending and deformation, thus expanding the application potential of bilayer hydrogels in various fields (Li, 2024). PNIPAM's pivotal role in 3D printing is showcased through projection microstereolithography (PμSL), allowing high-resolution digital additive manufacturing with temperature-responsive properties, enabling control over deformation and sequential deformation of 3D-printed structures, paving the way for versatile applications in flexible sensors, actuators, biomedical devices, and tissue engineering (Han et at., 2018).

8.27 Polyvinyl alcohol

Polyvinyl alcohol (PVA) in 3D printing demonstrates versatility across pharmaceuticals, tissue engineering, biomedical devices, drug delivery systems, and dental applications, offering controlled release profiles, scaffolds for tissue regeneration, prosthetics, implants, and precise drug release kinetics, presenting a promising avenue for tailored healthcare solutions. Polyvinyl alcohol-based magnetic hydrogel microrobots, fabricated using 3D-printed sacrificial micromolds and a salting-out process, offer high mobility, tunable stability, and excellent biocompatibility, promising platforms for targeted drug and cell delivery in biomedical applications (Sanchis-Gual et al., 2023). A modified surgical technique combining PVA hydrogel hemiarthroplasty, a 3D-printed baseplate, and Akin osteotomy aims to prevent potential implant failure in addressing hallux rigidus, offering a solution for pain and reduced motion in the first metatarsophalangeal joint (Madi et al., 2022). The study investigates polyvinyl alcohol (PVA) blending with sodium alginate (SA) to optimize composite hydrogels for 3D printing cartilage scaffolds, determining that the 8SA/2PVA blend hydrogel exhibits the most suitable mechanical properties and printability for this application (Chai et al., 2023). The study introduces a novel support bath for embedded extrusion printing, featuring chemically crosslinked cationic polyvinyl alcohol (PVA) microgels, enhancing stability, long-term storage, and high-resolution printing of hydrogel-based biological structures, facilitating commercial development of embedded bioprinting (Zeng et al., 2023). Polyvinyl alcohol (PVA)/β-tricalcium phosphate (β-TCP) composite scaffolds containing icariin (ICA), fabricated via direct-ink 3D printing, exhibit slow-release behavior, enhanced cell compatibility, and accelerated in situ bone regeneration, presenting promise for skull defect treatment (Xu et al., 2022). The study introduces chitosan (CS) hydrogels prepared under novel alkaline conditions, featuring superior mechanical properties through a solid cross-linked network with polyvinyl alcohol (PVA), graphene oxide (GO) incorporation, and excellent printability, paving the way for fast 3D printing hydrogels with ultrastrength properties and flexible sensor applications (Hao & Maimaitiyiming, 2022). The review highlights polyvinyl alcohol (PVA) as a preferred polymer in pharmaceutical 3D printing, particularly in fused deposition modeling (FDM) technology, showcasing its versatility in overcoming formulation and processing challenges to create diverse dosage forms with suitable designs and release profiles, emphasizing PVA's adaptability and efficacy in pharmaceutical dosage form manufacturing (Couți, 2024). The study presents a novel tomographic volumetric bioprinting (VBP) approach utilizing a dynamic resin of thiol–ene photo-clickable PVA and thermo-sensitive sacrificial gelatin, enabling rapid fabrication of cell-laden hydrogel constructs in one step, supporting fast cell spreading and 3D osteogenic differentiation, and offering site-specific immobilization of molecules for spatiotemporal control of cell–material interactions, showcasing high efficiency in creating functional ultrasoft hydrogel constructs (Qiu et al., 2023). The study examines the impact of silica filler addition on the physical, thermal, and overall density properties of composite materials derived from polymeric waste used in 3D printing technology, showing increased tensile strength in polymers such as polypropylene, PVA, and PLA, with notable effects on crystallinity and morphology, thereby suggesting potential for enhancing recycling and reuse processes within 3D printing technology (Ahmed et al., 2021).

8.28 Poly(N-vinyl pyrrolidone)

Poly(N-vinyl pyrrolidone) (PVP) is a versatile polymer with high solubility, biocompatibility, and film-forming ability, finding extensive applications in pharmaceuticals, cosmetics, materials science, and biomedical fields, enhancing drug solubility, serving as a thickening agent, and enabling various formulations in tablets, suspensions, coatings, adhesives, textiles, wound dressings, tissue engineering, and drug delivery systems, driving innovation across diverse industries. PVP plays a crucial role in 3D printing, particularly in personalized drug delivery devices, enabling tailored medication release profiles via fused deposition modeling (FDM) alongside PVA, highlighting the significance of PVP in personalized medicine and its influence on drug customization and release kinetics in FDM-based drug fabrication processes (Bianchi et al., 2023). PVP significantly enhances the biological and physical performance of 3D-printed cartilage-like constructs, serving as a matrix polymer alongside poly(ε-caprolactone) (PCL), low molecular mass poly(D-glucosamine known as chitosan (CS), hyaluronate (HA), and sodium alginate (SA), with the PCL/3wt.%PVP/1wt.%HA composite demonstrating promising compressive strength and higher cell viability for potential cartilage tissue engineering applications (Izgordu et al., 2020). PVP is explored in fused deposition modeling for tablet manufacturing, aiming to confer immediate-release properties to drugs; the study investigates PVP's combination with polylactic acid (PLA) in hot-melt extrusion, with theophylline as the model drug, resulting in optimized formulations, particularly FP2, which achieved a 10.35% drug load and over 95% drug release in an hour, highlighting the potential of PVP-PLA-based filaments for immediate drug release in pharmaceutical 3D printing formulations (Surini, 2023). PVP is pivotal in synthesizing polyvinylpyrrolidone-based halide perovskite nanocrystals (CsPbBr3/PVP) powders via microfluidic electrostatic spinning, showcasing excellent fluorescent properties and stability, which, coupled with their suitability for various applications including 3D printing, enable the fabrication of fluorescent patterns, opening new avenues for fluorescent 3D printable materials and offering efficient production methods alongside exploring novel application directions (Dong et al., 2021). PVP is integral in an innovative electrohydrodynamic jet 3D printing technology facilitating the fabrication of controllable filament composite scaffolds with optimized ink viscosity, jetting morphology, and solidification, resulting in stable jets and high aspect ratio structures, thus yielding PCL/PVP composite scaffolds with controllable filament diameter and superior cell biocompatibility, promising for tissue engineering (Li, 2024). PVP is pivotal in drop-on-demand (DOD) bioprinting, ensuring precise material deposition and volume control, with optimized PVP-based bio-inks promoting enhanced cell viability and homogeneity, as validated by a Z value

≤9.30 threshold for consistent cell output and minimal impairment during printing (Ng et al., 2017). PVP contributes significantly to a composite 3D model for cartilage cell culture, featuring a gelatin film and a printed PCL/PVP scaffold, created via electrospray and electrohydrodynamic jet printing methods, which promotes cartilage cell adhesion and ingrowth, holding potential for tissue engineering endeavors (Li et al., 2021).

References

Abramowicz, A., Obloj-Muzaj, M., & Kozakiewicz, J. (2014). Poly(vinyl chloride) nanocomposites prepared in the suspension polymerization process. Part I. PVC filled with hybrid nanofiller.. Journal of Materials Science and Engineering B, *4*.

Ahmad, S., Ahmad, S., Ahmad, S., & Agnihotry, S. A. J. B.o M. S. (2007). *Synthesis and characterization of in situ prepared poly (methyl methacrylate) nanocomposites, 30*, 31–35.

Ahmed, W., Siraj, S., & Al-Marzouqi, A. H. (2021). Comprehensive characterization of polymeric composites reinforced with silica microparticles using leftover materials of fused filament fabrication 3D printing. *Polymers, 13*(15), 2423 Chicago.

Akbari, A., Majumder, M., & Tehrani, A. (2015). Polylactic acid (PLA) carbon nanotube nanocomposites. In Processing, performance and application. Berlin, Heidelberg: Springer.

Alam, F., Varadarajan, K. M., Koo, J. H., Wardle, B. L., & Kumar, S. (2020). Additively manufactured polyetheretherketone (PEEK) with carbon nanostructure reinforcement for biomedical structural applications. *Advanced Engineering Materials, 22*(10), 2000483.

Almahri, G., Madi, K., Alkaabi, F., Badran, Y., Shehadeh, K., ElHassan, A., & Alzahmi, S. (2023). Characterization of Hybrid FRP Composite Produced from Recycled PET and CFRP. *Polymers, 15*(13), 2946.

Al-Mazrouei, N., Al-Marzouqi, A. H., & Ahmed, W. (2022). Characterization and sustainability potential of recycling 3D-printed nylon composite wastes. *Sustainability, 14*(17), 10458.

Altunbek, M., Afghah, S. F., Fallah, A., Acar, A. A., & Koc, B. (2023). Design and 3D printing of personalized hybrid and gradient structures for critical size bone defects. *ACS Applied Bio Materials, 6*(5), 1873–1885.

Ammar, S. (2024). Mechanical performances of printed carbon fiber-reinforced PLA and PETG composites. *Proceedings of the Institution of mechanical Engineers Part L Journal of Materials Design and Applications, 238*(8), 1488–1499. 10.1177/14644207231225761.

Amoli, M. (2024). Development of 3d printed pnipam-chitosan scaffolds for dentoalveolar tissue engineering. *Gels, 10*(2), 140. https://doi.org/10.3390/gels10020140.

Arias-Ferreiro, G., Ares-Pernas, A., Lasagabáster-Latorre, A., Aranburu, N., Guerrica-Echevarría, G., Dopico-García, M., & Abad, M. (2021). Printability study of a conductive polyaniline/acrylic formulation for 3d printing. *Polymers, 13*(13), 2068.

Aráoz, B., Karakaya, E., Wusener, A., Detsch, R., Bizzotto, J., Gueron, G., & Hermida, É. (2021). 3d printed poly(hydroxybutyrate-co-hydroxyvalerate)—45s5 bioactive glass composite resorbable scaffolds suitable for bone regeneration. *Journal of Materials Research, 36*(19), 4000–4012. https://doi.org/10.1557/s43578-021-00272-9.

Aveen, K., Bhajathari, F. V., & Jambagi, S. C. (2018). 3D printing & mechanical characteristion of polylactic acid and bronze filled polylactic acid components. *Paper presented at the IOP Conference Series: Materials Science and Engineering*.

Awad, W.H., Esawi, A.M.K., & Ramadan, A.R. (2008). Fabrication and properties of Nylon-6/ layered silicate nanocomposites by melt blending. In *Proceedings of the ASME 2008 2nd Multifunctional Nanocomposites and Nanomaterials International Conference. ASME 2008 2nd Multifunctional Nanocomposites and Nanomaterials.* Sharm El Sheikh, Egypt.

Awad, W. H. J. P.-P. T. Engineering. (2006). *High-throughput method for the synthesis of high performance polystyrene nanocomposites, 45*, 1117–1122.

Aziz, M.A., Elsoholy, M.G., Saad, G.R. (2018). (3-hydroxybutyrate) and their nanocomposites. Preparation and characterization of bio-based polyurethanes obtained from castor oil and poly(3-hydroxybutyrate) and their nanocomposites. *Polymer-Plastics Technology and Engineering, 53*(16).

Bapu, G. N. K. R., Mohan, S. J. P., & Finishing, S. (1995). *Electrodeposition of nickel-polytetrafluoroethylene (PTFE) polymer composites, 82*, 86–88.

Bajaj, I., & Singhal, R. (2011). Poly (glutamic acid)–an emerging biopolymer of commercial interest. *Bioresource Technology, 102*(10), 5551–5561. https://doi.org/10.1016/j.biortech.2011.02.047. Epub 2011 Feb 15. PMID: 21377358.

Baker, A., Galindo, E., Angelos, J., Salazar, D., Sterritt, S., Willis, A., & Tartis, M. (2023). Mechanical characterization data of polyacrylamide hydrogel formulations and 3d printed pla for application in human head phantoms. *Data in Brief, 48*, 109114. https://doi.org/10.1016/j.dib.2023.109114.

Bardot, M., & Schulz, M. D. (2020). Biodegradable poly (Lactic acid) nanocomposites for fused deposition modeling 3d printing. *Nanomaterials, 10*(12), 2567.

Bartolucci, S.F., Mago, G., Gevgilili, H., Vural, S., Dikovics, K.B., Kalyon, D.M., & Fisher, F. T. (2009). Investigation of the properties of PEEK-nanotube composites prepared by solution methods. In *Proceedings of the ASME 2009 International Mechanical Engineering Congress and Exposition. Volume 12: Micro and Nano Systems, Parts A and B.* Lake Buena Vista, Florida, USA. November 13–19, 2009. pp. 279-282. ASME

Batra, A. $ Materials Science Group. (2019). Fabrication and optical characterization of polyvinylidene fluoride/neodymium oxide nanocomposite films. *Open Access Library Journal, 6*(12), 1.

Benderly, D., Osorio, F., Ijdo, W. L. J. J.o V., & Technology, A. (2008). *PVC nanocomposites—Nanoclay chemistry and performance, 14.*

Benini, K. C. C.d C., Bomfim, A. S. C., & Voorwald, H. J. C. J. P. (2023). *Cellulose-reinforced polylactic acid composites for three-dimensional printing using polyethylene glycol as an additive: A comprehensive review, 15.*

Bergman, J., Coates, G., Giannelis, E., & Thomas, M. (1999). Synthesis and characterization of polyolefin–silicate nanocomposites: a catalyst intercalation and in situ polymerization approach. *Chemical Communications,* (21), 2179–2180.

Bertolini, M.C., Dul, S., Barra, G.M.O., & Pegoretti, A. (2020). Poly(vinylidene fluoride)/ thermoplastic polyurethane flexible and 3D printable conductive composites. Journal of Applied Polymer Science, 138, e50305.

Bianchi, M., Pegoretti, A., & Fredi, G. (2023). An overview of poly(vinyl alcohol) and poly (vinyl pyrrolidone) in pharmaceutical additive manufacturing. *Journal of Vinyl and Additive Technology, 29*(2), 223–239. https://doi.org/10.1002/vnl.21982.

Bikiaris, N. D., Koumentakou, I., Samiotaki, C., Meimaroglou, D., Varytimidou, D., Karatza, A., & Papageorgiou, G. Z. (2023). Recent advances in the investigation of poly (lactic acid)(PLA) nanocomposites: incorporation of various nanofillers and their properties and applications. *Polymers, 15*(5), 1196.

Blanco, I., Bottino, F. A., Cicala, G., Latteri, A., & Recca, A. (2014). Synthesis and characterization of differently substituted phenyl hepta isobutyl-polyhedral oligomeric silsesquioxane/polystyrene nanocomposites. *Polymer Composites, 35*(1), 151–157.

Bon, S., Chiesa, I., Morselli, D., Esposti, M., Fabbri, P., Maria, C., & Valentini, L. (2021). Printable smart 3d architectures of regenerated silk on poly(3-hydroxybutyrate-co-3-hydroxyvalerate). *SSRN Electronic Journal*. https://doi.org/10.2139/ssrn.3805241.

Borg, G., Warner, H., Ioannidis, M., Bogaart, G., & Roos, W. (2023). PLA 3D printing as a straightforward and versatile fabrication method for PDMS molds. *Polymers, 15*(6), 1498. https://doi.org/10.3390/polym15061498.

Borkar, A., Hendlmeier, A., Simon, Ž., Randall, J., Stojcevski, F., & Henderson, L. (2022). A comparison of mechanical properties of recycled high-density polyethylene/waste carbon fiber via injection molding and 3d printing. *Polymer Composites, 43*(4), 2408–2418. https://doi.org/10.1002/pc.26550.

Borkar, S., Gu, B., Dirmyer, M., Delicado, R., Sen, A., Jackson, B. R., & Badding, J. V. (2006). Polytetrafluoroethylene nano/microfibers by jet blowing. *Polymer, 47*(25), 8337–8343.

Budiono, H., Hilmy, F., & Taufik, I. (2023). The effect of printing speed variations on dimensional stability of 3D printing results made from waste bottle filament. *Jurnal E-Komtek (Elektro-Komputer-Teknik), 7*(1), 187–194. https://doi.org/10.37339/e-komtek.v7i1.1114.

Bulanda, K., Oleksy, M., Oliwa, R., Budzik, G., Przeszłowski, Ł., Fal, J., & Jesionowski, T. (2021). Polymer composites based on polycarbonate (PC) applied to additive manufacturing using melted and extruded manufacturing (MEM) technology. *Polymers, 13*(15), 2455.

Cafino, R. (2024). Evaluation of polyethylene terephthalate glycol (petg), simubone ™, and photopolymer resin as 3d-printed temporal bone models for surgical simulation. *Asian Journal of Surgery, 47*(1), 237–244. https://doi.org/10.1016/j.asjsur.2023.08.095.

Cai, Z., Li, M., Zhu, Z., Wang, X., Huang, Y., Li, T., & Yan, M. (2023). Biological degradation of plastics and microplastics: a recent perspective on associated mechanisms and influencing factors. *Microorganisms, 11*(7), 1661.

Carvalho, J., Zanini, N., Claro, A., Amaral, N., & Barud, H. (2021). Composite filaments of phbv reinforced with zro2·nh2o particles for 3D printing. *Polymer Bulletin, 79*(4), 2113–2132. https://doi.org/10.1007/s00289-021-03610-3.

Ch, K. P. S., Pradhan, R., & Nayak, P. L. J. A.i A. S. R. (2012). *Synthesis and characterization of polyurethane nanocomposite from castor oil-hexamethylene diisocyanate (HMDI). 3*, 3045–3052.

Chai, W., Yang, R., Zhou, J., & Wei, Q. (2023). Performance comparison of pva/sa composite hydrogels for 3d printing of cartilage scaffolds with different compositions. *Journal of Physics Conference Series, 2437*(1), 012042. https://doi.org/10.1088/1742-6596/2437/1/012042.

Chen, J., Huang, J., & Hu, Y. (2020). 3d printing of biocompatible shape-memory double network hydrogels. *Acs Applied Materials & Interfaces, 13*(11), 12726–12734. https://doi.org/10.1021/acsami.0c17622.

Chen, Q., Mangadlao, J. D., Wallat, J., De Leon, A., Pokorski, J. K., & Advincula, R. C. (2017). 3D printing biocompatible polyurethane/poly (lactic acid)/graphene oxide nanocomposites: anisotropic properties. *ACS Applied Materials & Interfaces, 9*(4), 4015–4023.

Chen, X., Chen, G., Wang, G., Zhu, P., & Gao, C. J. A. E. M. (2019). Recent progress on 3D-printed polylactic acid and its applications in bone repair, *22*.

Cheng, B., Inoue, Y., Ishihara, K. J. C., & Surfaces B, B. (2019). Surface functionalization of polytetrafluoroethylene substrate with hybrid processes comprising plasma treatment and chemical reactions, *173*, 77–84.

Cheon, J., & Kim, S. (2021). Fabrication and demonstration of a 3D-printing/pdms integrated microfluidic device. *Recent Progress in Materials, 4*(1), 1. https://doi.org/10.21926/rpm.2201002 -1.

Chi, K., Zhang, Z., Xi, J., Huang, Y., Xiao, F., Wang, S., & Liu, Y. (2014). Freestanding graphene paper supported three-dimensional porous graphene–polyaniline nanocomposite synthesized by inkjet printing and in flexible all-solid-state supercapacitor. *Acs Applied Materials & Interfaces, 6*(18), 16312–16319. https://doi.org/10.1021/am504539k.

Chinthamanipeta, P. S., Kobukata, S., Nakata, H., & Shipp, D. A. (2008). Synthesis of poly (methyl methacrylate)–silica nanocomposites using methacrylate-functionalized silica nanoparticles and RAFT polymerization. *Polymer, 49*(26), 5636–5642.

Choi, Y. S., Wang, K. H., Xu, M., & Chung, I. J. (2002). Synthesis of exfoliated polyacrylonitrile/Na– MMT nanocomposites via emulsion polymerization. *Chemistry of Materials, 14*(7), 2936–2939.

Conceição, T. F., Bertolino, J. R., Barra, G. M., Mireski, S. L., Joussef, A. C., & Pires, A. T. (2008). Preparation and characterization of poly (ether ether ketone) derivatives. *Journal of the Brazilian Chemical Society, 19*, 111–116.

Couţi, N. (2024). Polyvinyl alcohol, a versatile excipient for pharmaceutical 3D printing. *Polymers, 16*(4), 517. https://doi.org/10.3390/polym16040517.

Cui, X., Zhong, S., Zhang, H., Gu, Q., Li, J., & Wang, H. (2007). Preparation and characterization of polytetrafluoroethylene-polyacrylate core–shell nanoparticles. *Polymers for Advanced Technologies, 18*(7), 544–548.

Dahiya, D., Chettri, R., & Nigam, P.S. (2020). Biosynthesis of polyglutamic acid (γ-PGA), a biodegradable and economical polyamide biopolymer for industrial applications. In Microbial and natural macromolecules. Academic Press.

Daniele, R. (2023). From nautical waste to additive manufacturing: Sustainable recycling of high-density polyethylene for 3d printing applications. *Journal of Composites Science, 7*(8), 320. https://doi.org/10.3390/jcs7080320.

Das, S., Kumar, S., Mohanty, S., & Nayak, S. K. (2019). Synthesis, characterization and application of bio-based polyurethane nanocomposites. *Sustainable Polymer Composites and Nanocomposites,* 1121–1158.

Dastidar, A. (2023). In vitro degradation of 3d-printed poly(l-lactide-co-glycolic acid) scaffolds for tissue engineering applications. *Polymers, 15*(18), 3714. https://doi.org/10.3390/polym15183714.

Demir, E., Duygun, İ., & Bedeloğlu, A. (2024). The mechanical properties of 3d-printed polylactic acid/polyethylene terephthalate glycol multi-material structures manufactured by material extrusion. *3D Printing and Additive Manufacturing, 11*(1), 197–206. https://doi.org/10.1089/3dp.2021.0321.

Van den Eynde, M., & Van Puyvelde, P. (2017). 3D Printing of Poly(lactic acid). In: M. Di Lorenzo, & R. Androsch (Eds.), *Industrial Applications of Poly(lactic acid). Advances in Polymer Science: vol 282*. Cham: Springer. https://doi.org/10.1007/12_2017_28

Van den Eynde, M., & Van Puyvelde, P. (2018). 3D Printing of Poly (lactic acid). *Industrial Applications of Poly (lactic acid),* 139–158.

Devaraju, S., Vengatesan, M. R., Selvi, M., Kumar, A. A., & Alagar, M. (2012). Synthesis and characterization of bisphenol-A ether diamine-based polyimide POSS nanocomposites for low K dielectric and flame-retardant applications. *High Performance Polymers, 24*(2), 85–96.

Dong, T., Zhao, J., Li, G., Li, F., Li, Q., & Chen, S. (2021). In situ synthesis of robust polyvinylpyrrolidone-based perovskite nanocrystal powders by the fiber-spinning chemistry method and their versatile 3d printing patterns. *Acs Applied Materials & Interfaces, 13*(33), 39748–39754. https://doi.org/10.1021/acsami.1c10806.

Dua, R., Rashad, Z., Spears, J., Dunn, G., & Maxwell, M. (2021). Applications of 3d printed peek via fused filament fabrication: A systematic review. *Polymers, 13*(22), 4046.

Dubkov, S.V., Silibin, M.V., Lebedev, S.V., Ryazanov, R.I., & Shvartsman, V.V. (2020). Development of a 3D Printing Technique for PVDF Thin Films for Sensor Elements of Electronic Devices. In *2020 IEEE Conference of Russian Young Researchers in Electrical and Electronic Engineering (EIConRus)* (pp. 2587-2590). IEEE.

Ebrahimi, F., Ramezani Dana, H. J. I. J.o P. M., & Biomaterials, P. (2021). *Poly lactic acid (PLA) polymers: from properties to biomedical applications, 71*, 1117–1130.

Ergene, B., Atlıhan, G., & Pinar, A. (2023). Experimental and finite element analyses on the vibration behavior of 3d-printed pet-g tapered beams with fused filament fabrication. *Multidiscipline Modeling in Materials and Structures, 19*(4), 634–651. https://doi.org/10.1108/mmms-11-2022-0265.

Fan, Y., Luchow, M., Badria, A., Hutchinson, D. J., & Malkoch, M. (2023). Placenta powder-infused thiol-ene PEG hydrogels as potential tissue engineering scaffolds. *Biomacromolecules, 24*(4), 1617–1626.

Fariha, S., Islam, M. S., Rockshat, M., Hani, S. U., Islam, J., Zabed, H. M., & Chowdhury, F. I. (2023). Advances in PVC-based blend nanocomposites. *ECS Journal of Solid State Science and Technology, 12*(12), 121005.

Fina, A., Tabuani, D., Frache, A., & Camino, G. (2005). Polypropylene–polyhedral oligomeric silsesquioxanes (POSS) nanocomposites. *Polymer, 46*(19), 7855–7866.

Feng, H. (2023). Effect of 3d printing process parameters on the mechanical properties of silica/polyethyleneimine composites using direct-ink writing. *Polymer Composites, 44*(11), 7687–7700. https://doi.org/10.1002/pc.27656.

Feng, L., Li, S. H., Zhai, J., Song, Y. L., Jiang, L., & Zhu, D. B. (2003). Template based synthesis of aligned polyacrylonitrile nanofibers using a novel extrusion method. *Synthetic Metals, 135*, 817–818.

Ferraz, M., Nagashima, J., Venzac, B., Gac, S., & Songsasen, N. (2020). 3D printed mold leachates in pdms microfluidic devices. *Scientific Reports, 10*(1). https://doi.org/10.1038/s41598-020-57816-y.

Feyzi, A., Faghihi, K., & Zolanvari, A.A. (2013). Synthesis and characterization of new polyimide/organoclay nanocomposites derived from 3,3′,4,4′-biphenyltetracarboxylic dianhydride and 1,2-bis(4-aminophenoxy)ethane. High Temperature Materials and Processes, 32.

Fiayyaz, M., Zia, K. M., Zuber, M., Jamil, T., Khosa, M. K., & Jamal, M. A. (2014). Synthesis and characterization of polyurethane/bentonite nanoclay based nanocomposites using toluene diisocyanate. *Korean Journal of Chemical Engineering, 31*, 644–649.

Finny, A., Cheng, N., Popoola, O., & Andreescu, S. (2022). 3d printable polyethyleneimine based hydrogel adsorbents for heavy metal ions removal. *Environmental Science Advances, 1*(4), 443–455. https://doi.org/10.1039/d2va00064d.

Fontana, L., Giubilini, A., Arrigo, R., Malucelli, G., & Minetola, P. (2022). Characterization of 3D printed polylactic acid by fused granular fabrication through printing accuracy, porosity, thermal and mechanical analyses. *Polymers, 14*(17), 3530.

Fornetti, E., De Paolis, F., Fuoco, C., Bernardini, S., Giannitelli, S. M., Rainer, A., & Gargioli, C. (2023). A novel extrusion-based 3D bioprinting system for skeletal muscle tissue engineering. *Biofabrication, 15*(2), 025009.

Fuentes, M., Thakur, S., Wu, F., Misra, M., Gregori, S., & Mohanty, A. (2020). Study on the 3d printability of poly(3-hydroxybutyrate-co-3-hydroxyvalerate)/poly(lactic acid) blends with chain extender using fused filament fabrication. *Scientific Reports, 10*(1). https://doi.org/10.1038/s41598-020-68331-5.

Gao, C., Chi, J., Wang, S., Wu, Z., Wang, C., Li, J., & Gu, F. (2019). Evaluation of the osteogenesis of 3d printing poly(lactic-co-glycolic acid)-hydroxyapatite scaffolds in a rat femur model and its biocompatibility. *Science of Advanced Materials, 11*(10), 1449–1457. https://doi.org/10.1166/sam.2019.3551.

Geng, Y. (2024). Silk fibroin/polyacrylamide-based tough 3d printing scaffold with strain sensing ability and chondrogenic activity. *Composites Part B Engineering, 271*, 111173. https://doi.org/10.1016/j.compositesb.2023.111173.

Ghosh, A., Orasugh, J. T., Ray, S. S., & Chattopadhyay, D. (2023). Integration of 3D printing–coelectrospinning: Concept shifting in biomedical applications. *ACS Omega, 8*(31), 28002–28025.

Gialouri, A. (2023). Thermoresponsive alginate-graft-pnipam/methyl cellulose 3D-printed scaffolds promote osteogenesis in vitro. *Gels, 9*(12), 984. https://doi.org/10.3390/gels9120984.

Gianelli, W., Ferrara, G., Camino, G., Pellegatti, G., Rosenthal, J., & Trombini, R. C. (2005). Effect of matrix features on polypropylene layered silicate nanocomposites. *Polymer, 46*(18), 7037–7046.

Giannakas, A., Giannakas, A. E., & Ladavos, A. K. (2012). *Preparation and characterization of polystyrene/organolaponite nanocomposites, 51*, 1411–1415.

Gonzalez, C. (2024). Multi-material distributed recycling via material extrusion: Recycled high density polyethylene and poly (ethylene terephthalate) mixture. *Polymer Engineering & Science*. https://doi.org/10.1002/pen.26643.

Grewal, M. S., & Yabu, H. (2020). Biomimetic catechol-based adhesive polymers for dispersion of polytetrafluoroethylene (PTFE) nanoparticles in an aqueous medium. *RSC Advances, 10*(7), 4058–4063.

Grosjean, M., Guth, L., Déjean, S., Paniagua, C., & Nottelet, B. (2023). Dynamic and Degradable Imine-Based Networks for 3D-Printing of Soft Elastomeric Self-Healable Devices. *Advanced Materials Interfaces, 10*(17), 2300066.

Gudadhe, A., Bachhar, N., Kumar, A., Andrade, P., & Kumaraswamy, G. (2019). Three-dimensional printing with waste high-density polyethylene. *Acs Applied Polymer Materials, 1*(11), 3157–3164. https://doi.org/10.1021/acsapm.9b00813.

Guo, T., Lim, C., Noshin, M., Ringel, J., & Fisher, J. (2018). 3d printing bioactive plga scaffolds using dmso as a removable solvent. Bioprinting, 10, e00038. https://doi.org/10.1016/j.bprint.2018.e0003.

Guo, T., Holzberg, T., Lim, C., Gao, F., Gargava, A., Trachtenberg, J., & Fisher, J. (2017). 3D printing PLGA: A quantitative examination of the effects of polymer composition and printing parameters on print resolution. *Biofabrication, 9*(2), 024101. https://doi.org/10.1088/1758-5090/aa6370.

Han, B., Ji, G., Wu, S., & Shen, J. (2003). Preparation and characterization of nylon 66/montmorillonite nanocomposites with co-treated montmorillonites. *European Polymer Journal, 39*(8), 1641–1646.

Han, H., & Cho, S. (2018). Fabrication of conducting polyacrylate resin solution with polyaniline nanofiber and graphene for conductive 3d printing application. *Polymers, 10*(9), 1003. https://doi.org/10.3390/polym10091003.

Hao, F., & Maimaitiyiming, X. (2022). Fast 3d printing with chitosan/polyvinyl alcohol/graphene oxide multifunctional hydrogel ink that has ultrastretch properity. *Chemistryselect, 7*(10). https://doi.org/10.1002/slct.202200201.

Hasegawa, N., Okamoto, H., Kato, M., Usuki, A., & Sato, N. (2003). Nylon 6/ Na–montmorillonite nanocomposites prepared by compounding Nylon 6 with Na–montmorillonite slurry. *Polymer, 44*(10), 2933–2937.

He, G., & Pan, Q. (2004). Synthesis of polystyrene and polystyrene/poly (methyl methacrylate) nanoparticles. *Macromolecular Rapid Communications, 25*(17), 1515–1518.

Heo, C., & Chang, J. H. (2013). Polyimide nanocomposites based on functionalized graphene sheets: Morphologies, thermal properties, and electrical and thermal conductivities. *Solid State Sciences, 24*, 6–14.

Herrera, N., Mathew, A. P., Oksman, K. J. C. S., & Technology (2015). *Plasticized polylactic acid/cellulose nanocomposites prepared using melt-extrusion and liquid feeding: Mechanical, thermal and optical properties, 106*, 149–155.

Hinton, T., Hudson, A., Pusch, K., Lee, A., & Feinberg, A. (2016). 3d printing pdms elastomer in a hydrophilic support bath via freeform reversible embedding. *Acs Biomaterials Science & Engineering, 2*(10), 1781–1786. https://doi.org/10.1021/acsbiomaterials.6b00170.

Hosseinloo, S., & Sedaghat, S. (2014). Modification of nanoclay for preparation of polyvinyl chloride nanocomposites and evaluation of their morphological and thermal properties.

Huang, J. C., He, C. B., Xiao, Y., Mya, K. Y., Dai, J., & Siow, Y. P. (2003). Polyimide/POSS nanocomposites: interfacial interaction, thermal properties and mechanical properties. *Polymer, 44*(16), 4491–4499.

Huang, X., & Brittain, W. J. (2001). Synthesis and characterization of PMMA nanocomposites by suspension and emulsion polymerization. *Macromolecules, 34*(10), 3255–3260.

Huang, Y., Cheng, Q., Jeng, U., & Hsu, S. (2023). A biomimetic bilayer hydrogel actuator based on thermoresponsive gelatin methacryloyl–poly(n-isopropylacrylamide) hydrogel with three-dimensional printability. *Acs Applied Materials & Interfaces, 15*(4), 5798–5810. https://doi.org/10.1021/acsami.2c18961.

Ilyas, R. A., Sapuan, S. M., Harussani, M. M., Hakimi, M. Y. A. Y., Haziq, M. Z. M., Atikah, M. S. N., & Asrofi, M. (2021). Polylactic acid (PLA) biocomposite: Processing, additive manufacturing and advanced applications. *Polymers, 13*(8), 1326.

Izgordu, M., Uzgur, E., Ulag, S., Sahin, A., Karademir, B., Kilic, B., & Gunduz, O. (2020). Investigation of 3d-printed polycaprolactone-/polyvinylpyrrolidone-based constructs. *Cartilage, 13*(2_suppl), 626S–635S. https://doi.org/10.1177/1947603519897302.

Jia, W., Segal, E., Kornemandel, D., Lamhot, Y., Narkis, M., & Siegmann, A. A. (2002). Polyaniline–DBSA/organophilic clay nanocomposites: synthesis and characterization. *Synthetic Metals, 128*(1), 115–120.

Jiang, Z., Erol, O., Chatterjee, D., Xu, W., Hibino, N., Romer, L. H., & Gracias, D. H. (2019). Direct ink writing of poly (tetrafluoroethylene)(PTFE) with tunable mechanical properties. *ACS Applied Materials & Interfaces, 11*(31), 28289–28295.

Joo, H., & Cho, S. (2020). Comparative studies on polyurethane composites filled with polyaniline and graphene for dlp-type 3d printing. *Polymers, 12*(1), 67. https://doi.org/10.3390/polym12010067.

Kaewkong, P., Kosorn, W., Sonthithai, P., Lertwimol, T., Thavornyutikarn, B., Chantaweroad, S., & Janvikul, W. (2022). Chondrogenic differentiation of human mesenchymal stem cells and macrophage polarization on 3d-printed poly(ε-caprolactone)/poly(3-hydroxybutyrate-co-3-hydroxyvalerate) blended scaffolds with different secondary porous structures. *Acs Applied Bio Materials, 5*(6), 2689–2702. https://doi.org/10.1021/acsabm.2c00161.

Karabulut, H., Dutta, A., Moukbil, Y., Cisen Akyol, A., Ulag, S., Aydin, B., & Gunduz, O. (2023). Fabrication of ethosuximide loaded alginate/polyethylene oxide scaffolds for epilepsy research using 3D-printing method. *Frontiers in Bioengineering and Biotechnology, 11*, 1244323.

Karbownik, I., Rac-Rumijowska, O., Fiedot-Toboła, M., Rybicki, T., & Teterycz, H. (2019). The preparation and characterization of polyacrylonitrile-polyaniline (pan/pani) fibers. *Materials, 12*(4), 664. https://doi.org/10.3390/ma12040664.

Khizer, Z., Akram, M. R., Tahir, M. A., Liu, W., Lou, S., Conway, B. R., & Ghori, M. U. (2023). Personalised 3D-Printed Mucoadhesive Gastroretentive Hydrophilic Matrices for Managing Overactive Bladder (OAB). *Pharmaceuticals, 16*(3), 372.

Khoshnood, N., & Zamanian, A. (2022). Development of novel alginate-polyethyleneimine cell-laden bioink designed for 3d bioprinting of cutaneous wound healing scaffolds. *Journal of Applied Polymer Science, 139*(21). https://doi.org/10.1002/app.52227.

Khoshnood, N., Zamanian, A., & Abbasi, M. (2021). The potential impact of polyethylenimine on biological behavior of 3d-printed alginate scaffolds. *International Journal of Biological Macromolecules, 178*, 19–28. https://doi.org/10.1016/j.ijbiomac.2021.02.152.

Kim, H. S., & Kim, S. H. (2023). Additive manufacturing and combustion characteristics of polyethylene oxide/aluminum/copper oxide-based energetic nanocomposites for enhancing the propulsion of small projectiles. *Nanomaterials, 13*(6), 1052.

Kim, N. (2016). Recent progress of functional coating materials and technologies for polycarbonate. *Journal of Coatings Technology and Research, 14*, 21–34.

Kim, M., Jeong, J. H., Lee, J. Y., Capasso, A., Bonaccorso, F., Kang, S. H., & Lee, G. H. (2019). Electrically conducting and mechanically strong graphene–polylactic acid composites for 3D printing. *ACS Applied Materials & Interfaces, 11*(12), 11841–11848.

Kodali, D., Umerah, C. O., Idrees, M., Jeelani, S. A. K., & Rangari, V. K. J. J.o C. M. (2021). *Fabrication and characterization of polycarbonate-silica filaments for 3D printing applications, 55*, 4575–4584.

Koffi, A., Toubal, L., Jin, M., Koffi, D., Döpper, F., Schmidt, H., & Neuber, C. (2021). Extrusion-based 3d printing with high-density polyethylene birch-fiber composites. *Journal of Applied Polymer Science, 139*(15). https://doi.org/10.1002/app.51937.

Kondapalli, V., He, X., Khosravifar, M., Khodabakhsh, S., Collins, B., Yarmolenko, S., & Shanov, V. (2021). cvd synthesis of 3d-shaped 3d graphene using a 3d-printed nickel–plga catalyst precursor. *Acs Omega, 6*(43), 29009–29021. https://doi.org/10.1021/acsomega.1c04072.

Korakianiti, A., Papaefthimiou, V., Daflou, T., Kennou, S., & Gregoriou, V. G. (2004). Characterization of polypropylene (PP) nanocomposites for industrial applications. *Macromolecular Symposia, Vol. 205*, Weinheim: WILEY-VCH Verlag, 71–84, No. 1.

Kostas, V., Baikousi, M., Barkoula, N. M., Giannakas, A., Kouloumpis, A., Avgeropoulos, A., & Karakassides, M. A. (2020). Synthesis, characterization and mechanical properties of nanocomposites based on novel carbon nanowires and polystyrene. *Applied Sciences, 10*(17), 5737.

Krishnamurthy, S., Blom, R., Andreassen, K., Rombouts, M., & Borrás, A. (2021). 3d printed pei containing adsorbents supported by carbon nanostructures for post-combustion carbon capture from biomass fired power plants. *Frontiers in Climate, 3*. https://doi.org/10.3389/fclim.2021.733499.

Kumar, M. N. S., Ramakrishnan, R., & Omarbekova, A. J. J.o M. S. Technology. (2019). 3D printed polycarbonate reinforced acrylonitrile–butadiene–styrene composites: Composition effects on mechanical properties, micro-structure and void formation study, *33*, 5219–5226.

Kuznetsova, N. P., Ermakova, T. G., Pozdnyakov, A. S., Emel'Yanov, A. I., & Prozorova, G. F. (2013). Synthesis and characterization of silver polymer nanocomposites of 1-vinyl-1, 2, 4-triazole with acrylonitrile. *Russian Chemical Bulletin, 62*, 2509–2513.

Lage-Rivera, S., Ares-Pernas, A., Becerra Permuy, J. C., Gosset, A., & Abad, M. J. (2023). Enhancement of 3D printability by FDM and electrical conductivity of PLA/MWCNT filaments using lignin as bio-dispersant. *Polymers, 15*(4), 999.

Larosa, C., Patra, N., Salerno, M., Mikac, L., Meri, R. M., & Ivanda, M. (2017). Preparation and characterization of polycarbonate/multiwalled carbon nanotube nanocomposites. *Beilstein Journal of Nanotechnology, 8*(1), 2026–2031.

Lee, J.-M., Kang, S.-J., & Park, S. J. M. R. (2009). *Synthesis of polyacrylonitrile based nanoparticles via aqueous dispersion polymerization, 17*, 817–820.

Li, K., Liu, M., Liu, T., Han, X., Wang, D., Zhang, F., & Yu, A. (2021). Sandwich-like gelatin/polycaprolactone/polyvinyl pyrrolidone 3d model with significantly improved cartilage cells adhesion and regeneration. *Macromolecular Materials and Engineering, 306*(10). https://doi.org/10.1002/mame.202100338.

Li, Y., Zhao, B., Xie, S., & Zhang, S. (2003). Synthesis and properties of poly (methyl methacrylate)/montmorillonite (PMMA/MMT) nanocomposites. *Polymer International, 52*(6), 892–898.

Li, Z. (2024). 3d printed heterogeneous bilayer temperature-responsive hydrogel with multi-curvature deformation. *Polymer, 296*, 126701. https://doi.org/10.1016/j.polymer.2024.126701.

Lim, J., He, M., & Chong, A. (2021). 3d-printed poly-lactic co-glycolic acid (plga) scaffolds in non-critical bone defects impede bone regeneration in rabbit tibia bone. *Bio-medical Materials and Engineering, 32*(6), 375–381. https://doi.org/10.3233/bme-216017.

Lim, S. H., & Dasari, A. (2012). Nylon-based polymer nanocomposites. In Advances in Polymer Nanocomposites. Woodhead Publishing, pp. 238–276.

Liu, D., Huyan, C., Wang, Z., Guo, Z., Zhang, X., Torun, H., & Chen, F. (2023). Conductive polymer based hydrogels and their application in wearable sensors: a review. *Materials Horizons, 10*(8), 2800–2823.

Liu, L., Xiao, L., Li, M., Zhang, X., Chang, Y., Shang, L., & Science, P. (2015). Effect of hexagonal boron nitride on high-performance polyether ether ketone composites, *294*, 127–133.

Liu, R., Jia, L., Chen, J., Long, Y., Zeng, J., Liu, S., & Jiang, H. (2023). Chondrocyte spheroid-laden microporous hydrogel-based 3D bioprinting for cartilage regeneration. *International Journal of Bioprinting, 10*(1), 0161.

Liu, S., Bastola, A., & Li, L. (2017). A 3d printable and mechanically robust hydrogel based on alginate and graphene oxide. *Acs Applied Materials & Interfaces, 9*(47), 41473–41481. https://doi.org/10.1021/acsami.7b13534.

Liu, S., & Li, L. (2017). Ultra stretchable and self-healing double-network hydrogel for 3d printing and strain sensor. *Acs Applied Materials & Interfaces, 9*(31), 26429–26437. https://doi.org/10.1021/acsami.7b07445.

Liu, T., Lim, K. P., Tjiu, W. C., Pramoda, K. P., & Chen, Z. K. (2003). Preparation and characterization of nylon 11/organoclay nanocomposites. *Polymer, 44*(12), 3529–3535.

Liz-Basteiro, P., Reviriego, F., Martínez-Campos, E., Reinecke, H., Elvira, C., Rodríguez-Hernández, J., & Gallardo, A. (2023). Vat photopolymerization 3D printing of hydrogels with re-adjustable swelling. *Gels, 9*(8), 600.

Łukaszewski, K. (2023). Mechanical evaluation of pet-g 3d-printed wrist-hand orthosis: An integrated experimental and numerical approach. *Materials, 16*(18), 6132. https://doi.org/10.3390/ma16186132.

Lu, Y. (2024). 3d bioprinting of adhesive, anti-bacterial alginate/polyacrylamide-based customized boluses using digital light processing for radiotherapy applications. *International Journal of Bioprinting, 10*(2), 1589. https://doi.org/10.36922/ijb.1589.

Madi, N., Chopra, A., & Parekh, S. (2022). Polyvinyl alcohol hydrogel hemiarthroplasty of the first metatarsophalangeal joint augmented with 3d-printed baseplate. *Techniques in Foot & Ankle Surgery, 22*(1), 27–34. https://doi.org/10.1097/btf.0000000000000351.

Malakhova, Y., Коровин, A, Lapkin, D., Малахов, C, Shcherban, V., Pichkur, E., & Erokhin, V. (2017). Planar and 3d fibrous polyaniline-based materials for memristive elements. *Soft Matter, 13*(40), 7300–7306. https://doi.org/10.1039/c7sm01773a.

Manocha, B., & Margaritis, A. (2008). Production and characterization of γ-polyglutamic acid nanoparticles for controlled anticancer drug release. *Critical Reviews in Biotechnology, 28*(2), 83–99.

Meng, Z., Niu, X., Xia, L., Chen, Y., Wang, Z., Wang, H., & Lu, S. (2023). A 3D Ex Vivo Tumor-Immune Coculture System Mimicking In Vivo Tumor Environmental Stress on CD8+ T Cells Exhaustion. *Advanced Biology, 7*(7), 2200264.

Mironov, A., Mironova, O., Syachina, M., & Popov, V. (2019). 3d printing of polylactic-co-glycolic acid fiber scaffolds using an antisolvent phase separation process. *Polymer, 182*, 121845. https://doi.org/10.1016/j.polymer.2019.121845.

Mishra, A., Wallin, T., Pan, W., Xu, P., Wang, K., Giannelis, E., & Shepherd, R. (2020). Autonomic perspiration in 3d-printed hydrogel actuators. *Science Robotics, 5*(38). https://doi.org/10.1126/scirobotics.aaz3918.

Momenzadeh, N. (2020). *Influences of material extrusion additive manufacturing (MEAM) parameters and additives on polyvinylidene fluoride (PVDF) properties.*

Moodley, V.K. (2007). *The synthesis, structure and properties of polypropylene nanocomposites* (Doctoral dissertation).

Murariu, M., & Dubois, P. (2016). PLA composites: From production to properties. *Advanced Drug Delivery Reviews, 107*, 17–46.

Nayani, M., Gunashekar, S., & Abu-Zahra, N. H. J. I. J.o P. S. (2013). *Synthesis and characterization of polyurethane-nanoclay composites. 2013,* 1–5.

Neves, R. M., Lopes, K. S., Zimmermann, M. V. G., Poletto, M., & Zattera, A. J. (2019). Characterization of polystyrene nanocomposites and expanded nanocomposites reinforced with cellulose nanofibers and nanocrystals. *Cellulose, 26*, 4417–4429.

Ng, W., Yeong, W., & Naing, M. (2017). Polyvinylpyrrolidone-based bio-ink improves cell viability and homogeneity during drop-on-demand printing. *Materials, 10*(2), 190. https://doi.org/10.3390/ma10020190.

Nie, G., Li, Z., Lu, X., Lei, J., Zhang, C., & Wang, C. (2013). Fabrication of polyacrylonitrile/CuS composite nanofibers and their recycled application in catalysis for dye degradation. *Applied Surface Science, 284*, 595–600.

Nikje, M. A., & Farazfar, M. (2012). Synthesis and characterization of polyurethane nanocomposites formulated by incorporation of polyols from PET waste. *Cellular polymers, 31*(2), 69–82.

Nikolaidis, A. K., Achilias, D. S., Karayannidis, G. P. J. I., & Research, E. C. (2011). Synthesis and characterization of PMMA/organomodified montmorillonite nanocomposites prepared by in situ bulk polymerization, *50*, 571–579.

Okamoto, M., Morita, S., Taguchi, H., Kim, Y. H., Kotaka, T., & Tateyama, H. (2000). Synthesis and structure of smectic clay/poly (methyl methacrylate) and clay/polystyrene nanocomposites via in situ intercalative polymerization. *Polymer, 41*(10), 3887–3890.

Olad, A., & Hayasi, M. J. P.-P. T. Engineering. (2011). A comparative study of polystyrene/layered silicate nanocomposites, synthesized by emulsion and bulk polymerization methods, *50*, 1487–1495.

Orellana, F., Lisperguer, J., & Perez-Guerrero, A. J. J.o T. C. C. S. (2011). Synthesis and characterization of poly(methyl methacrylate)-silica, -alumina and -silica-alumina sol-gel nanocomposites, *56*, 668–671.

Özbolat, V., Dey, M., Ayan, B., Povilianskas, A., Demirel, M., & Özbolat, İ. (2018). 3d printing of pdms improves its mechanical and cell adhesion properties. *Acs Biomaterials Science & Engineering, 4*(2), 682–693. https://doi.org/10.1021/acsbiomaterials.7b00646.

Ozden, S., Charayev, A. M., & Shaov, A. H. J. J.o M. S. (1999). The synthesis of polyetheretherketones and investigations of their properties, *34*, 2741–2744.

Pal, A., Biswas, B., Das, A., Chakraborty, A., Datta, P., Roy Chowdhury, A., & Sinha, A. (2020). Synthesis and characterization of Sr-doped HAp-incorporated polyether ether ketone composite. *Journal of Composite Materials, 54*(3), 287–298.

Park, S. J., Lee, J. E., Lee, H. B., Park, J., Lee, N. K., Son, Y., & Park, S. H. (2020). 3D printing of bio-based polycarbonate and its potential applications in ecofriendly indoor manufacturing. *Additive Manufacturing, 31*, 100974.

Pelluau, T. (2023). 4d printing nanocomposite hydrogel based on pnipam and prussian blue nanoparticles using stereolithography. *Macromolecular Materials and Engineering*. . https://doi.org/10.1002/mame.202300305.

Peng, F., Jiang, H., Woods, A., Joo, P., Amis, E., Zacharia, N., & Vogt, B. (2019). 3d printing with core–shell filaments containing high or low density polyethylene shells. *Acs Applied Polymer Materials, 1*(2), 275–285. https://doi.org/10.1021/acsapm.8b00186.

Peprníček, T., Kalendová, A., Pavlova, E., Šimoník, J., Duchet, J., & Gérard, J. F. J. P. D. (2006). Stability. *Poly(vinyl chloride)-paste/clay nanocomposites: Investigation of thermal and morphological characteristics, 91*, 3322–3329.

Phang, I. Y., Chen, L., Tjiu, W. C., Pisharath, S., & Liu, T. J. M. R. I. (2004). Nylon 66/organoclay nanocomposites: I. *Preparation, Morphology, Thermal And Mechanical Properties, 8*, 159–160.

Pongmuksuwan, P. (2023). Recycling pet bottles for 3d-printing filament: Effect of chain extender. *Materials Science Forum, 1103*, 103–108. https://doi.org/10.4028/p-ck8evp.

Poompiew, N. (2023). 3d-printed polyacrylamide-based hydrogel polymer electrolytes for flexible zinc-ion battery. *Electrochimica Acta, 466*, 143076. https://doi.org/10.1016/j.electacta.2023.143076.

Pretula, J., Slomkowski, S., & Penczek, S. J. Addr (2016). *Polylactides—Methods of synthesis and characterization, 107*, 3–16.

Pérez, M. A., Rivas, B. L., Rodriguez, S., Maldonado, A., & Venegas, C. (2010). *Polypropylene/clay Nanocomposites: Synthesis and characterization, 55*, 440–444.

Qiu, W., Gehlen, J., Bernero, M., Gehre, C., Schädli, G., Müller, R., & Qin, X. (2023). A synthetic dynamic polyvinyl alcohol photoresin for fast volumetric bioprinting of functional ultrasoft hydrogel constructs. *Advanced Functional Materials, 33*(20). https://doi.org/10.1002/adfm.202214393.

Rasoulianboroujeni, M., Fahimipour, F., Shah, P., Khoshroo, K., Tahriri, M., Eslami, H., & Tayebi, L. (2019). Development of 3d-printed plga/tio2 nanocomposite scaffolds for bone tissue engineering applications. *Materials Science and Engineering C, 96*, 105–113. https://doi.org/10.1016/j.msec.2018.10.077.

Ray, S. S., Maiti, P., Okamoto, M., Yamada, K., & Ueda, K. J. M. (2002). New polylactide/layered silicate nanocomposites. 1. *Preparation, Characterization, and Properties, 35*, 3104–3110.

Reich, M., Woern, A. L., Tanikella, N. G., & Pearce, J. M. J. M. (2019). Mechanical properties and applications of recycled polycarbonate particle material extrusion-based additive manufacturing, 12.

Riberi, K., Pereyra, J. Y., Planes, G. A., Abel, S. B., Acevedo, D. F., & Barbero, C. A. (2018). Synthesis of Polymeric Nancocomposites by Infiltration. Applications in 3D (Fused Filament Molding) Printing. *MRS Advances, 3*, 3793–3798.

Rihayat, T., Saari, M., Suraya, A. R., Mahmood, M. H., Dahlan, K. Z. H. M., Yunus, W. W., & Sapuan, S. M. (2006). Synthesis and thermal characterization of polyurethane/clay nanocomposites based on palm oil polyol. *Polymer-Plastics Technology and Engineering, 45*(12), 1323–1326.

Rinaldi, M., Cecchini, F., Pigliaru, L., Ghidini, T., Lumaca, F., & Nanni, F. J. P. (2020). Additive manufacturing of polyether ether ketone (PEEK) for space applications: A nanosat polymeric. *Structure (London, England: 1993)*, 13.

Rusu, G., & Rusu, E. (2006). Nylon 6/sio2nanocomposites synthesized by in situanionic polymerization. *High Performance Polymers, 18*(3), 355–375.

Saengchai, N. (2023). 3d printing of titanium dioxide and polyaniline-based photocatalytic composites. *Journal of Physics Conference Series, 2653*(1), 012014. https://doi.org/10.1088/1742-6596/2653/1/012014.

Sanchis-Gual, R., Hao, Y., Ueno, T., Landers, F., Hertle, L., Deng, S., & Chen, X. (2023). 3d printed template-assisted casting of biocompatible polyvinyl alcohol-based soft microswimmers with tunable stability. *Advanced Functional Materials, 33*(39). https://doi.org/10.1002/adfm.202212952.

Satriananda, Riza, M., Mulyati, S., & Mulana, F. (2018). Polyurethane/clay nanocomposites from palm oil for surface-coating applications. *Polymers from Renewable Resources, 9*(3-4), 103–110.

Scharfenberg, M., Hilf, J., & Frey, H. J. A. F. M. (2018). *Functional polycarbonates from carbon dioxide and tailored epoxide monomers: Degradable materials and their application potential,* 28.

Şen, S., Memeşa, M., Nugay, N., & Nugay, T. (2006). Synthesis of effective poly (4-vinylpyridine) nanocomposites: in situ polymerization from edges/surfaces and interlayer galleries of clay. *Polymer international, 55*(2), 216–221.

Şen, S., & Çaylı, G. (2010). Synthesis of bio-based polymeric nanocomposites from acrylated epoxidized soybean oil and montmorillonite clay in the presence of a bio-based intercalant. *Polymer International, 59*(8), 1122–1129.

Sevil, B., & Zuhal, K. (2010). *Synthesis and characterization of polypyrrole nanoparticles and their nanocomposites with poly (propylene). Macromolecular symposia, Vol. 295*, Weinheim: WILEY-VCH Verlag, 59–64, No. 1.

Shah, R. K., & Paul, D. R. (2004). Nylon 6 nanocomposites prepared by a melt mixing masterbatch process. *Polymer, 45*(9), 2991–3000.

Shawe, S., Buchanan, F., Harkin-Jones, E., & Farrar, D. (2006). A study on the rate of degradation of the bioabsorbable polymer polyglycolic acid (PGA). *Journal of Materials Science, 41*(15), 4832–4838.

Shen, J., Zhang, S., Fang, X., & Salmon, S. (2023). Carbonic anhydrase enhanced UV-crosslinked PEG-DA/PEO extruded hydrogel flexible filaments and durable grids for CO2 capture. *Gels, 9*(4), 341.

Shetty, S., Nandish, B. T., Amin, V., Jayaprakash, K., Shahira, & Stanly Selva Kumar, G. S. (2022). Engineering. *3D printed polyether ether ketone (PEEK), Polyamide (PA) and its evaluation of mechanical properties and its uses in healthcare applications,* 1224.

Shiferaw, M. Z., Gebremedhen, H. S., & Mengistie, A. K. (2023). Developing filament extruder and characterization of recycled high-density polyethylene for 3d printing filament material. *EOA, 1*(1). https://doi.org/10.33140/eoa.01.01.02.

Shukla, D., Negi, Y. S., Uppadhyaya, J. S., & Kumar, V. (2012). Synthesis and modification of poly (ether ether ketone) and their properties: a review. *Polymer Reviews, 52*(2), 189–228.

Siddiqui, M. N., Redhwi, H. H., Gkinis, K., & Achilias, D. S. (2013). Synthesis and characterization of novel nanocomposite materials based on poly (styrene-co-butyl methacrylate) copolymers and organomodified clay. *European Polymer Journal, 49*(2), 353–365.

Sierra-Ávila, R., Pérez-Alvarez, M., Valdez-Garza, J., Avila-Orta, C. A., Jiménez-Regalado, E. J., Mata-Padilla, J. M., & Cadenas-Pliego, G. (2018). Synthesis and Thermomechanical Characterization of Nylon 6/Cu Nanocomposites Produced by an Ultrasound-Assisted Extrusion Method. *Advances in Materials Science and Engineering, 2018*(1), 4792735.

Sin, S. L., Yan, H., & Xu, J. W. J. A. M. R. (2015). *Synthesis and characterization of polyurethane/clay nanocomposites, 1110*, 73–76.

Siraj, S., Al-Marzouqi, A. H., Iqbal, M. Z., & Ahmed, W. (2022). Impact of micro silica filler particle size on mechanical properties of polymeric based composite material. *Polymers, 14*(22), 4830.

Slonov, A. (2023). Investigation of the properties of polyethylene and ethylene-vinyl acetate copolymer blends for 3d printing applications. *Polymers, 15*(20), 4129. https://doi.org/10.3390/polym15204129.

Solomon, M. J., & Somwangthanaroj, A. (2004). Intercalated polypropylene nanocomposites. *Dekker Encyclopedia of Nanoscience and Nanotechnology, 2*, 1483–1488.

Sun, D., Li, Y., Zhang, B., & Pan, X. (2010). Preparation and characterization of novel nanocomposites based on polyacrylonitrile/kaolinite. *Composites Science and Technology, 70*(6), 981–988.

Sun, J., Jiang, J., Bao, B., Wang, S., He, M., Zhang, X., & Song, Y. (2016). Fabrication of bendable circuits on a polydimethylsiloxane (pdms) surface by inkjet printing semi-wrapped structures. *Materials, 9*(4), 253. https://doi.org/10.3390/ma9040253.

Sun, S., Hao, F., & Maimaitiyiming, X. (2022). 3d print polyaniline/gelatin hydrogels as wearable multifunctional sensors. *Chemistryselect, 7*(45). https://doi.org/10.1002/slct.202203286.

Sun, X., Tyagi, P., Agate, S., McCord, M., Lucia, L., & Pal, L. (2020). Highly tunable bioadhesion and optics of 3d printable pnipam/cellulose nanofibrils hydrogels. *Carbohydrate Polymers, 234*, 115898. https://doi.org/10.1016/j.carbpol.2020.115898.

Surini, S. (2023). Utilization of pvp-pla combination in fabricating theophylline-loaded filament for 3d printing with immediate release behavior. *Pharmaceutical Sciences and Research, 10*(2). https://doi.org/10.7454/psr.v10i2.1307.

Tahmasebinia, F., Niemelä, M., Sepasgozar, S., Lai, T., Su, W., Reddy, K., & Marroquin, F. (2018). Three-dimensional printing using recycled high-density polyethylene: technological challenges and future directions for construction. *Buildings, 8*(11), 165. https://doi.org/10.3390/buildings8110165.

Thakur, V. K., Vennerberg, D., & Kessler, M. R. J. Aam interfaces. (2014). *Green aqueous surface modification of polypropylene for novel polymer nanocomposites, 6*(12), 9349–9356.

Tiouitchi, G., Raji, M., Mounkachi, O., Ali, M. A., Mahmoud, A., Boschini, F., & Bouhfid, R. (2019). Black phosphorus-based polyvinylidene fluoride nanocomposites: Synthesis, processing and characterization. *Composites Part B: Engineering, 175*, 107165.

Vakharia, V. S., Kuentz, L., Salem, A., Halbig, M. C., Salem, J. A., & Singh, M. (2021). Additive manufacturing and characterization of metal particulate reinforced polylactic acid (PLA) polymer composites. *Polymers, 13*(20), 3545.

Varela, C., Rosales, C., Perera, R., Matos, M., Poirier, T., Blunda, J., & Rojas, H. (2006). Functionalized polypropylenes in the compatibilization and dispersion of clay nanocomposites. *Polymer Composites, 27*(4), 451–460.

Venzac, B., Deng, S., Mahmoud, Z., Lenferink, A., Costa, A., Bray, F., & Gac, S. (2021). pdms curing inhibition on 3d-printed molds: why? also, how to avoid it? *Analytical Chemistry, 93*(19), 7180–7187. https://doi.org/10.1021/acs.analchem.0c04944.

Vidakis, N., Petousis, M., Moutsopoulou, A., Papadakis, V., Spiridaki, M., Mountakis, N., & Maravelakis, E. (2023). Nanocomposites with optimized polytetrafluoroethylene content as a reinforcement agent in PA12 and PLA for material extrusion additive manufacturing. *Polymers, 15*(13), 2786.

Vidakis, N., Petousis, M. A., Velidakis, E., Liebscher, M., & Tzounis, L. J. B. (2020). *Three-dimensional printed antimicrobial objects of polylactic acid (PLA)-silver nanoparticle nanocomposite filaments produced by an in-situ reduction reactive melt mixing process, 5*.

Wang, C., Yang, J., Gao, C., Wu, Z., Yang, Y., Liu, R., & Gu, F. (2019). Effects of 3d printing poly(lactic-co-glycolic acid)-hydroxyapatite scaffolds on femoral defect restoration and serum factors. *Journal of Biomaterials and Tissue Engineering, 9*(11), 1499–1506. https://doi.org/10.1166/jbt.2019.2169.

Wang, G., Wang, L., Mark, L. H., Shaayegan, V., Wang, G., Li, H., & Park, C. B. (2018). Ultralow-threshold and lightweight biodegradable porous PLA/MWCNT with segregated conductive networks for high-performance thermal insulation and electromagnetic interference shielding applications. *ACS Applied Materials & Interfaces, 10*(1), 1195–1203.

Wang, L., Weng, L., Song, S., Zhang, Z., Tian, S., & Ma, R. (2011). Characterization of polyetheretherketone–hydroxyapatite nanocomposite materials. *Materials Science and Engineering: A, 528*(10-11), 3689–3696.

Wang, Y., & Li, D. (2020). Creating complex polyacrylamide hydrogel structures using 3d printing with applications to mechanobiology. *Macromolecular Bioscience, 20*(7). https://doi.org/10.1002/mabi.202000082.

Wong, K.Y. (2017). *Synthesis and Characterization of Polyurethane (PU) Nanocomposites* (Doctoral dissertation, Tunku Abdul Rahman University College).

Woo, R., Chen, G., Zhao, J., & Bae, J. (2021). Structure–mechanical property relationships of 3d-printed porous polydimethylsiloxane. *Acs Applied Polymer Materials, 3*(7), 3496–3503. https://doi.org/10.1021/acsapm.1c00417.

Wu, Y. L., D'Amato, A. R., Yan, A. M., Wang, R. Q., Ding, X., & Wang, Y. (2020). Three-dimensional printing of poly (glycerol sebacate) acrylate scaffolds via digital light processing. *ACS Applied Bio Materials, 3*(11), 7575–7588.

Wu, K. (2023). Investigation of 3d-printed pnipam-based constructs for tissue engineering applications: A review. *Journal of Materials Science, 58*(47), 17727–17750. https://doi.org/10.1007/s10853-023-09125-8.

Xiao, Y., Zhang, S., Chen, J., Guo, B., & Chen, D. (2023). Mechanical performance of 3d-printed polyethylene fibers and their durability against degradation. *Materials, 16*(14), 5182. https://doi.org/10.3390/ma16145182.

Xie, T., Yang, G., Fang, X., & Ou, Y. (2003). Synthesis and characterization of poly (methyl methacrylate)/montmorillonite nanocomposites by in situ bulk polymerization. *Journal of Applied Polymer Science, 89*(8), 2256–2260.

Yamane, K., Sato, H., Ichikawa, Y., et al. (2014). Development of an industrial production technology for high-molecular-weight polyglycolic acid. *Polymer Journal, 46*, 769–775. https://doi.org/10.1038/pj.2014.69.

Yang, J., Ching, Y. C., & Chuah, C. H. (2019). Applications of lignocellulosic fibers and lignin in bioplastics: A review. *Polymers, 11*(5), 751.

Yari, M., & Moradi, O. (2009). Synthesis polytetrafluoroethylene (PTFE) nanoparticles via adsorption of TFE monomer onto different surfaces.

Ye, X., Li, L., Lin, Z., Yang, W., Duan, M., Xia, Y., & Zhang, Y. (2018). Integrating 3d-printed phbv/calcium sulfate hemihydrate scaffold and chitosan hydrogel for enhanced osteogenic property. *Carbohydrate Polymers, 202*, 106–114. https://doi.org/10.1016/j.carbpol.2018.08.117.

Yeom, E., Kim, W., Kim, J., Lee, S. S., & Park, M. (2004). Synthesis and characterization of poly (methyl methacrylate)/montmorillonite nanocomposites by emulsifier-free emulsion polymerization. *Molecular Crystals and Liquid Crystals, 425*(1), 85–91.

Xu, Z., Sun, Y., Dai, H., Ma, Y., & Han, B. (2022). Engineered 3d-printed polyvinyl alcohol scaffolds incorporating β-tricalcium phosphate and icariin induce bone regeneration in rat skull defect model. *Molecules (Basel, Switzerland), 27*(14), 4535. https://doi.org/10.3390/molecules27144535.

Zarei, M., Michalkiewicz, B., & El Fray, M. (2023). *The effect of segmental composition of bio-based PBS-DLS copolymers containing PEG on their properties, 3D printing and electrospinning*.

Zeng, J., Xie, Z., Dekishima, Y., Kuwagaki, S., Sakai, N., & Matsusaki, M. (2023). "Out-of-the-box" granular gel bath based on cationic polyvinyl alcohol microgels for embedded extrusion printing. *Macromolecular Rapid Communications, 44*(8). https://doi.org/10.1002/marc.202300025.

Zhang, L., Zhang, X., Li, L., Liu, Y., Wang, D., Xu, L., & Zhang, A. (2019). Fabrication of photothermally responsive nanocomposite hydrogel through 3d printing. *Macromolecular Materials and Engineering, 305*(2). https://doi.org/10.1002/mame.201900718.

Zhang, H., Qi, T., Zhu, X., Zhou, L., Li, Z., Yang, W., & Lan, H. (2021). 3d printing of a pdms cylindrical microlens array with 100% fill-factor. *Acs Applied Materials & Interfaces, 13*(30), 36295–36306. https://doi.org/10.1021/acsami.1c08652.

Zhang, Y., Yin, M., Xia, O., Zhang, A.P., & Tam, H.Y. (2018). Optical 3D μ-printing of polytetrafluoroethylene (PTFE) microstructures. In *2018 IEEE Micro Electro Mechanical Systems (MEMS)* (pp. 37-40). IEEE.

Zhang, Y., Yin, M. J., Ouyang, X., Zhang, A. P., & Tam, H. Y. (2020). 3D μ-printing of polytetrafluoroethylene microstructures: A route to superhydrophobic surfaces and devices. *Applied Materials Today, 19*, 100580.

Zhansitov, A. A., Slonov, A. L., Shetov, R. A., Baikaziev, A. E., Shakhmurzova, K. T., Kurdanova, Z. I., & Khashirova, S. Y. (2018). Synthesis and properties of polyether-etherketones for 3d printing. *Fibre Chemistry, 49*, 414–419.

Zhou, Y., Lei, L., Yang, B., Li, J., & Ren, J. (2018). Preparation and characterization of polylactic acid (PLA) carbon nanotube nanocomposites. *Polymer Testing, 68*, 34–38.

Zhu, W., Pyo, S. H., Wang, P., You, S., Yu, C., Alido, J., & Chen, S. (2018). Three-dimensional printing of bisphenol A-free polycarbonates. *ACS Applied Materials & Interfaces, 10*(6), 5331–5339.

Zubir, S. A., Ernie, S. A., & Ahmad, S. A. H. R. I. M. (2016). Thermal and Mechanical Characterizations of Nanoclay/Palm Oil Polyol based Polyurethane Composites. *Key Engineering Materials, 673*, 127–130.

Further Reading

Agrawal, C. M., Niederauer, G. G., & Athanasiou, K. J. Te (1995). *Fabrication and characterization of PLA-PGA orthopedic implants, 1*(3), 241–252.

Alimohammadi, F., Parvinzadeh Gashti, M., Mozaffari, A. J. F., & Polymers (2018). *Polyvinylpyrrolidone/carbon nanotube/cotton functional nanocomposite: Preparation and characterization of properties, 19*, 1940–1947.

Bao, Y., Huang, Z.-m, & Weng, Z.-x J. J.o A. P. S. (2006). *Preparation and characterization of poly(vinyl chloride)/layered double hydroxides nanocomposite via in situ suspension polymerization, 102*, 1471–1477.

Bergius, W. N. A., Hutchings, L. R., Sarih, N. M., Thompson, R. L., Jeschke, M., & Fisher, R. J. P. C. (2013). *Synthesis and characterisation of end-functionalised poly(N-vinylpyrrolidone) additives by reversible addition–fragmentation transfer polymerisation, 4*, 2815–2827.

Cho, J., & Paul, D. R. J. P. (2001). *Nylon 6 nanocomposites by melt compounding, 42*, 1083–1094.

Feldman, D. J. J.o M. S. (2014). Poly(vinyl chloride) nanocomposites. *Part A, 51*, 659–667.

Gilbert, M. (2012). *Poly(vinyl chloride)(PVC)-based nanocomposites.*

Gowd, E. B., & Ramesh, C. (2014). *Crystallization and polymorphic behavior of nylon-clay nanocomposites.*

Han, D., Lu, Z., Chester, S., & Lee, H. (2018). Micro 3d printing of a temperature-responsive hydrogel using projection micro-stereolithography. *Scientific Reports, 8*(1). https://doi.org/10.1038/s41598-018-20385-2.

Kim, M., Park, C. I., Choi, W. M., Lee, J., Lim, J. G., Park, O. O., & Kim, J. M. J. J.o A. P. S. (2004). *Synthesis and material properties of Syndiotactic polystyrene/organophilic clay nanocomposites, 92*, 2144–2150.

Leon, Ad, Chen, Q., Palaganas, N. B., Palaganas, J. O., Manapat, J. Z., Advíncula, R. C. J. R., & Polymers, F. (2016). *High performance polymer nanocomposites for additive manufacturing applications, 103*, 141–155.

Liu, D., Huyan, C., Wang, Z., Guo, Z., Zhang, X., Torun, H., & Chen, F. (2023). Conductive polymer based hydrogels and their application in wearable sensors: A review. *Materials Horizons, 10*(8), 2800–2823.

Maity, N. K., Mandal, A., & Nandi, A. K. (2018). *Enhancement of mechanical, electrical and dielectric properties in poly(vinylidene fluoride) composites with two dimensional nanomaterials.*

Mohamed, M. G., & Kuo, S. W. J. P. (2018). *Functional polyimide/polyhedral oligomeric silsesquioxane nanocomposites, 11.*

Momenzadeh, N., Stewart, C. M., Berfield, T. A. J. M.o A., & Advanced Manufacturing, V. (2018). *Mechanical and Thermal Characterization of Fused Filament Fabrication Polyvinylidene Fluoride (PVDF) Printed Composites.*

Panda, A., Dyadyura, K., Valícek, J., Harničárová, M., Zajac, J., Modrák, V., & Pavelek, Z. J. M. (2017). *Manufacturing technology of composite materials—Principles of modification of polymer composite materials technology based on polytetrafluoroethylene, 10.*

Qu, X., Guan, T., Liu, G., She, Q., & Zhang, L. J. J.o A. P. S. (2005). *Preparation, structural characterization, and properties of poly(methyl methacrylate)/montmorillonite nanocomposites by bulk polymerization, 97*, 348–357.

Sain, S., Ray, D., Mukhopadhyay, A., Sengupta, S., Kar, T., Ennis, C. J., & Rahman, P. K. S. M. J. J.o A. P. S. (2012). *Synthesis and characterization of PMMA-cellulose nanocomposites by in situ polymerization technique,* 126.

Slonov, A. (2023). Investigation of the properties of polyethylene and ethylene-vinyl acetate copolymer blends for 3d printing applications. *Polymers, 15*(20), 4129. https://doi.org/10.3390/polym15204129.

Tian, X., Xin, B., Gao, W., Jin, S., & Chen, Z. J. T. J.o T. T. I. (2018). *Preparation and characterization of polyvinylidene fluoride/polysulfone amide composite nanofiber mats, 110*, 815–821.

Tseng, C.-R., Wu, J.-Y., Lee, H.-y, & Chang, F.-C. J. J.o A. P. S. (2002). *Preparation and characterization of polystyrene–clay nanocomposites by free-radical polymerization, 85*, 1370–1377.

Vaezi, M. R., Yang, S. J. V., & Prototyping, P. (2015). *Extrusion-based additive manufacturing of PEEK for biomedical applications, 10*, 123–135.

Wahab, M. A., Kim, I. K., Cho, W. J., Ha∗, C.-S. J. M. C., & Crystals, L. (2004). *Polyimide based hybrid nanocomposites, 417*, 127–134.

Zarei, M., Michalkiewicz, B., & El Fray, M. (2023). *Hierarchical tubular structures from biodegradable and bioactive PEG-modified PBS-DLS copolymers.*

Thermal and mechanical properties of additively manufactured functional polymers and nanocomposites

9

Santosh Kumar[1], and Rakesh Kumar[2]
[1]Department of Mechanical Engineering, Chandigarh Group of Colleges, Mohali, Punjab, India, [2]Department of Regulatory Affairs & Quality Assurance, Kaushik Orthopaedic Private Limited, Sonipat, Haryana, India

9.1 Introduction

The current state of additive manufacturing (AM), also known as 3D printing, is one of the ongoing advancements and increasing adoption across distinct industries. As technology continues to improve, materials become more diverse, and costs decrease. It uses a wide range of technologies that are capable of translating virtual solid model data into physical parts (Zhakeyev et al., 2017). The process of manufacturing the 3D printed parts is shown in Fig. 9.1.

After intensive research and development in the areas of materials, processes, software, equipment, and integration, AM has been used directly and indirectly to produce prototype parts with suitable material properties for evaluation and testing, as well as to make tools, dies, and molds. Currently, the direct fabrication of functional endues products is becoming the main trend of AM technologies, in particular, for polymeric and metallic products, and it has been increasingly implemented to manufacture parts in small or medium quantities.

The history development of AM methods started in 1983 for SLA, 1988 for FDM, 1995 for SLS, 2002 for human kidney development, in 2014 for 3D-printed house development, 2016 for human bone development, and 2018 for software development. Several 3D printer experts revealed that in the study this technology will be used for full-body organs in 2030 (Chia & Wu, 2015; Kumar et al., 2021a, 2021b). According to the ASTM F42 committee, AM processes are classified into seven categories such as jetting, binder jetting, vat photopolymerization, powder bed fusion, material extrusion, energy deposition, and sheet lamination. These processes differ from each other in terms of the techniques used to deposit layers and the ways in which the deposited layers are bonded together. There are different ways to categorize AM processes. For example, one approach is to classify them into three broad categories based on whether the starting materials exist as a liquid, filament/paste, or powder (Kumar et al., 2021a, 2021b; Ligon et al., 2017; Shahrubudin et al., 2019) as shown in Fig. 9.2.

Additive Manufacturing of Functional Polymers and Nanocomposites. DOI: https://doi.org/10.1016/B978-0-443-26516-7.00009-2
© 2025 Elsevier Ltd. All rights are reserved, including those for text and data mining, AI training, and similar technologies.

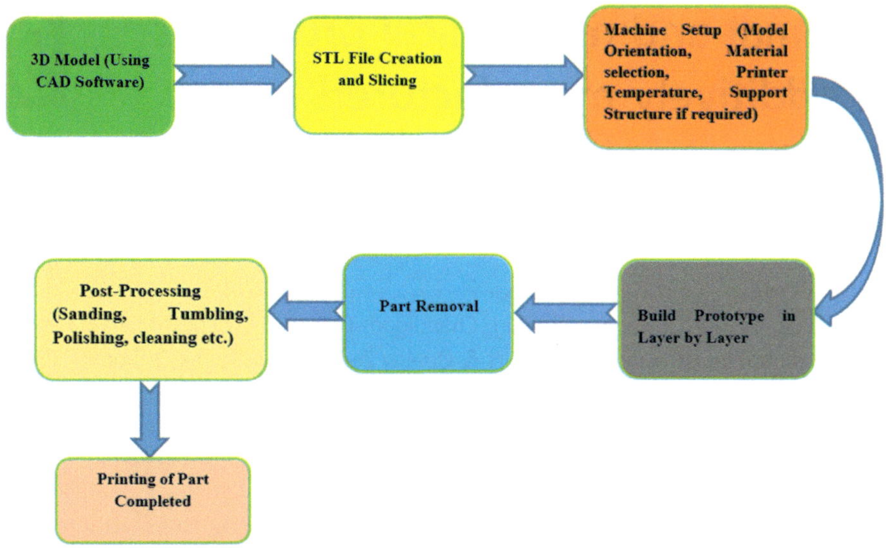

Figure 9.1 Step-by-step part manufacturing process using 3D printing.

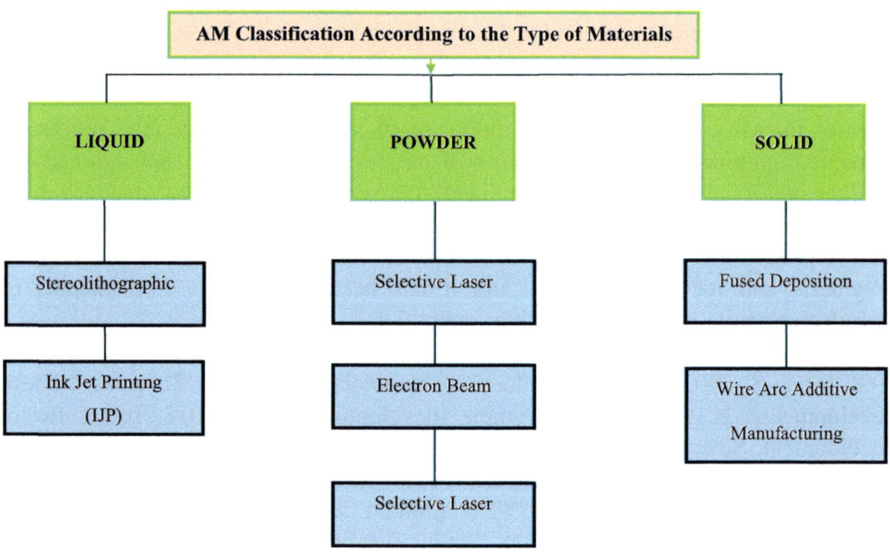

Figure 9.2 Three main categories of AM materials based on starting material.

9.2 Printing of different polymeric and nanocomposite materials

3D printing offers exciting possibilities for printing polymers and nanocomposite materials. AM used for printing polymers, for example, ABS, PLA, PC, nylon, etc.

Apart from this, the different AM processes are suitable for polymers depending upon the material characteristics and desired outcomes, for example, FDM, SLA, and SLS. The printing of nanocomposite materials improved properties such as strength, stiffness, and heat resistance. In addition, several 3D printing processes can be used for nanocomposites, but due to potential threats, careful selection is crucial. Challenges are nozzle clogging, requiring proper dispersion within the polymer matrix for consistent printing. For nanocomposites such as FDM and SLA are suitable processes.

Polymers: Acrylonitrile butadiene styrene (ABS) is used in the FDM process because of its good toughness, ease of printing, affordability, and balance of strength. Similarly polylactic acid (PLA) is a biodegradable material derived from corn starch, that is, from natural sources; it is easy to use and has low printing temperature and smooth finish. But the drawback is lesser strength and heat resistance than ABS. Polycarbonate (PC) has high strength, superior durability, and heat resistance but the limitation is that it requires high printing temperature as compared to PLA and ABS is more issue to print. Another polymer material is polyamide (nylon). It has good strength, resistance against chemicals, and toughness and is SLS process. Polypropylene (PP): polymer of lightweight, flexible, with good chemical resistance but challenging to print owing to warping. Hence the choice of polymers depends upon the specific requirement of the part such as strength, heat resistance, biocompatibility flexibility, etc. Composite materials: polymer matrix composites (PMCs) are the combined polymers as the matrix with reinforcing materials such as fibers or particles such as fibre-reinforced composite (FRP), particle-reinforced polymers, metal matrix composites, ceramic matrix composites (CMC), functional graded materials (FGMs), etc. (Attal et al., 2024; Kumar et al., 2021a,b; Ligon et al., 2017).

Apart from polymers and nanocomposite materials, the different materials used in AM techniques are metals, stainless steel, that is 316 L, which is a popular choice owing to good corrosion resistance, high strength and less cost. Similarly, aluminum alloy has a good strength-to-weight ratio and is lightweight. Titanium alloy (e.g., Ti6Al4V) provides excellent corrosion resistance, light weight as compared to SS, excellent strength-to-weight ratio, but high cost is the major limitation of this material. Cobalt chrome has good wear resistance and is biocompatible and resistance to body fluid. Inconel (e.g., Inconel 718 and 625) is known for high-temperature strength, resistance to oxidation, as well as corrosion (Bhatia & Sehgal, 2021; Kumar et al., 2024; Kumar et al., 2023; Lakhdar et al., 2021).

Special material: It includes the following applications such as food materials printing (meat, pizza, sauce, and chocolate), textile application, coal dust, etc. Smart Materials: it includes shape memory polymer, alloy, etc. In addition, ceramic material is also an example of AM materials. 3D printing methods like fused deposition modeling (FDM), selective laser sintering (SLS), direct ink writing (DIW), and stereolithography (SLA) are frequently employed in the production of polymeric functional devices. Polymer materials, being the most substantial building block, hold the key to determining the performance of functional devices. However, the restricted varieties and problems in the large-scale manufacture of 3D printable materials impede the advancement of 3D printing technology. As a result, developing innovative raw materials is critical for large-scale 3D printing manufacturing (Duda & Raghavan, 2016; Kumar, 2024a,b).

9.2.1 Selection of materials

It is now unable to market items manufactured with 3D printing technology since 3D printing has significant limitations in terms of strength and usefulness. 3D-printed products are generally delicate, limiting their applicability in a variety of sectors. To circumvent these constraints, reinforcements can be achieved by combining the matrix with composites to make a more usable product that would not be possible with only the matrix. Thermoplastic polymers are favored over metals because they are lightweight, low-cost, do not require lubrication, and are silent and efficient. As a result, it is recommended that these thermoplastics be combined to create a composite material suitable for 3D printing. For example, when nylon is blended with chopped carbon fibers, it exhibits chemical and heat resistance as well as high strength. Tensile strength rose to approximately 6 times greater than nonreinforced polymer (Blok et al., 2018; Dickson et al., 2017; Trimble & Ghasemi, 2020; Zhang et al., 2020).

9.2.2 Distinct AM printers and their application

The different printers have different fields of application and their merits and demerits as shown in Table 9.1.

The different types of composite materials used in 3D printing and printing methods (Borrello et al., 2018; Cheng & Chou, 2015; Fan & Feng, 2018; Murr et al., 2012; Scharff et al., 2016; Shiwpursad & Xue, 2018; Stansbury & Idacavage, 2016; Tofail et al., 2018; Zaleski et al., 2018) are shown in Table 9.2.

Importance of polymer nanocomposite: Improvements in mechanical characteristics such as impact strength, tensile and bending modulus and strength, at much lower concentrations of the nanofillers, abrasion resistance improvement, reduction in shrinkage, electrical, electronics, and optical, thermal properties improvement, greater thermal stability, thermal endurance, and higher HDT, excellent flame resistance and fire retardant characteristics, significant improvement in moisture and gas barrier properties. As a result, it may be treated without loss of quality. In addition, the synthesis of nanocomposites includes mainly three different techniques to develop polymer nanocomposites such as solution processing, in situ polymerization and melt processing. Thus pharmacy, metallurgy, computer science, environmental chemistry, and polymers are the major applications of nanotechnology in distinct fields. Polymer materials have anisotropic behaviors and metals have isotropic behaviors, therefore the properties of polymer composite depend on the direction and vary from point to point which depends upon homogeneous mixing of polymer and fibers, Production process, Injection molding or compression (Anajafi et al., 2020; Pant et al., 2021; Piyush et al., 2020; Sasmal et al., 2020). Composites have several advantages, including ease of manufacture, cheap cost, less secondary operations than other composites, long life, high strength-to-weight ratio, excellent

Table 9.1 Merits and demerits of AM processes.

AM method	Printing application	Merits	Demerits	References
FDM	Models, prototyping, etc.	Less cost as compared to SLA	Required more processing time for a few geometrical shapes. In addition the material utilized is limited to thermoplastic	Scharff et al. (2016); Tofail et al. (2018)
SLA	Patterns, prototypes, models, etc.	Less time is required to print the part, Provides an excellent smooth surface and is cost-effective also	In case of complicated shape requirements, the support structure is needed	Borrello et al., (2018); Stansbury and Idacavage (2016)
SLM	Aerospace parts, Medical devices (Implant and instrument), tooling and automobile parts, etc.	(a) As this SLM process does not require tooling the lead times is reduced (b) The number of parts can be printed at the same time	This process required post-processing thus increasing the cost	Shiwpursad and Xue (2018)
EBM	Custom-made skull, etc.	Rapid part printing process and more density of metal powder used	High machine cost and material used also print the limited height of the part	Cheng and Chou (2015); Murr et al. (2012); Zaleski et al. (2018)
SLS	Hearing aids, prosthetics, etc.	No need for structure support and versatile materials (plastic, ceramic metals, and glass can be printed)	As it has small holes and large flat surfaces hence accurate part cannot print	Fan and Feng (2018)

Table 9.2 AM methods, materials and their characteristics.

Materials	AM printing method	Characteristics
Metal/plastic filament	FDM	Metallic
Alumide	SLS	Heat resistant
Wood/plastic filament	FDM	Fragile
Sandstone	BJ	Unique sandstone appearance

fatigue characteristics, product design flexibility, and so on (Khan et al., 2017; Lee et al., 2009; Umarfarooq & Singh, 2023).

9.3 Mechanical and thermal properties of distinct AM functional polymers and nanocomposites materials

The melting point, glass transition temperature (Tg), and thermal conductivity are the three properties of functional polymers (AM) and nanocomposites. The functional polymers may increase due to crystallinity control. While noncomposite (AM) may increase or decrease depending upon nanoparticle type. The Tm in the case of functional polymer may increase because of crystallinity control. But Nanocomposite (AM) may increase/decrease depending upon nanoparticle type. However, the Tg may increase owing to the crystallinity control in the case of functional polymer. But for nanocomposite polymer may increase/decrease depending upon nanoparticle types. AM creates functional polymers and nanocomposite materials with tailored thermal properties in the case of functional polymers the AM can be influenced by crystallinity and porosity. However, for nanoparticle materials the nanoparticle type and dispersion, interface between polymer and nanoparticles (De Leon et al., 2016).

The literature of various researchers that described the thermal and mechanical properties is given in Table 9.3.

Polymeric nanocomposites are widely used in industry because of their cost-effectiveness. Nylon-6 clay hybrids are constructed of nanoclay and have been used in the automobile industry since 1989. Many metal oxide-based nanomaterials, including TiO_2, ZnO, and SiO_2, have been used in recent years due to their self-cleaning properties. Graphene oxide (GO) and carbon nanotubes are rapidly being used due to their selective properties and different forms, which have created new opportunities in highly industrialized nations. Graphene and its derivatives have a large surface area and antibacterial characteristics, making them ideal candidates for the production of biopolymer nanocomposites for biomedical purposes. To focus on the scientific development of micro/nanocomposites/bio-nanocomposites and their electrical properties about power engineering, as well as the electrical properties of insulating materials, particularly cross-linked polyethene (XLPE) and epoxy resins, both unfilled and filled with various filler materials. The effect of filler size, type, and

Table 9.3 Mechanical and thermal properties of distinct AM materials.

Authors/references	Material	Process	Printing parameters used			Mechanical testing		Conclusion
						Tensile strengths	Flexure strength	
Kumar et al. (2023)	PLA	FDM	Raster angle: 30°, 45°, and 60°	Number of top and bottom layers: 2, 3, and 4	Coating thickness: 50, 100, and 150 µm	29.51 MPa	98 MPa	Results revealed that the optimum parameters are, when raster angle 30°, coating thicknes100 µm, number of top and bottom layers 3, maximum tensile strength 29.51 MPa and Bending strength 98 MPa were achieved.
Tamaşag et al. (2022)	PLA, PLA–, and PETG	FDM	Infill densities 15%, 30%, and 50%, respectively	Color distribution was 33%, 66%, and 99%, respectively	Color pigments (blue, green, and red)	NA	Max.13.24 MPa	The infill density had the greatest impact (54.98%), followed by material nature (29.71%), colour pigment (12.85%), and colouring percentage (2.46%). In terms

(Continued)

Table 9.3 (Continued)

Authors/references	Material	Process	Printing parameters used			Mechanical testing		Conclusion
						Tensile strengths	Flexure strength	
								of infill density, tensile strength varied from 5.94 MPa at 15% to 13.28 MPa at 50%.
Leite et al. (2018)	PLA	FDM	Infill density (%) 20%, 40%, and 60%, respectively	Extrusion temperature:200°C and 220°C	Raster angle: 0/90°, –45°/+45° and layer thickness 0.1 mm and 0.2 mm	Max. 26.71 MPa	NA	In water absorption testing, the configuration with the lowest water absorption was 200°C extrusion and 0.2 mm layer thickness. To reduce water absorption, two protective coatings were used: acrylic and polyurethane. The latter proved to be the most effective, consistent with earlier findings.
Khosravani et al. (2022)	PLA	FDM	Raster angle: 0°, 30°, 45°, 60°,	Printing speed: 20 mm/second	Infill percentage 100%; bed	Max. 90 MPa	NA	The raster direction has a substantial impact on the

			and 90°, respectively.	and 80 mm/second	temperature 55°C			strength of the specimen. 0° specimens have the highest elastic modulus. The specimen's maximum tensile strength was lowered by 53.76% when printed in a 90° raster direction at 80 mm/s.
Letcher and Waytashek (2015)	PLA	FDM	Raster orientation	Bed temperature: NA	45° raster orientation	64 MPa	102 MPa	The mechanical characteristics of PLA filament and printed specimens were evaluated. Tensile tests showed that specimens with a 45° raster orientation had the highest strength. In fatigue testing, 90° specimens had the lowest resistance to fatigue loadings. The fatigue lifetimes of 45°

(*Continued*)

Table 9.3 (Continued)

Authors/references	Material	Process	Printing parameters used			Mechanical testing		Conclusion
						Tensile strengths	Flexure strength	
								and 0° specimens were similar and require more investigation. However, 45° specimens had the greatest fatigue endurance limit.
Moradi et al. (2021)	PLA	FDM	Bed temperature: 60°C	Printing speed: 20 mm/s	Raster angle: 45° and layer thickness (mm): 0.2	15.397 MPa	NA	The triangular design exhibited the highest ultimate tensile strength values. In terms of absorbed energy as a toughness indicator, the wiggling pattern and rapid honeycomb got the highest and lowest values, respectively.
Yu et al. (2019)	PLA	FDM	Printing speed	Filling density	Layer height	NA	58.25 MPa	When layer height is 0.15 mm, filling density is 100%, and printing speed

								is 30 mm/s, samples achieve the greatest tensile and compressive strength of 58.25 MPa and 86.11 MPa, respectively. The optimized technique can improve the mechanical qualities of 3D-printed items.
Atakok et al (2022)	PLA	FDM	Temperature of bed: 60◦C	Printing speed: 3000 mm/min	Layer thickness: 0.15e0.20–0-.25	56 MPa	NA	In this study, PLA and Re-PLA filaments and FDM from additive manufacturing techniques were used. The produced test parts were subjected to tensile, izod impact and three-point bending tests. The best results were 0.25 mm, 70%, and PLA for all test

(Continued)

Table 9.3 (Continued)

Authors/ references	Material	Process	Printing parameters used			Mechanical testing		Conclusion
						Tensile strengths	Flexure strength	
								outcomes, including layer thickness, occupancy rates, and filament materials. hey were calculated to be 60.006 MPa, 16.961 kJ/m2 for Izod impact strength, and 125.423 MPa for three-point bending strength.
Patil et al. (2024)	PLA	FDM	Speed of printing: 50 mm/s	Surface bed temperature: 60◦C	The temperature of the nozzle is 210°C and the diameter of the nozzle is 0.4 mm	14.161 MPa	38.850 MPa	In this experimental study, the objective is to analyze the influence of three printing variables (specifically, part orientation, layer height, and infill) on the tensile strength and flexural strength of the FDM process for PLA material. The tensile

								strength of 14.16 MPa was achieved.
Torres et al. (2015)	PLA	FDM	Layer thickness: 0.10–0.40 mm	Relative density/ infill: 0%–100%	Temperature (Tm): 280°C	73 MPa	106 MPa	Torsion tests were conducted on FDM PLA components, and the impact of processing factors such as layer thickness, % infill, and post-processing via heat treatment on the resulting shear characteristics was assessed. Authors suggested Low-level heat treatment can enhance strength while maintaining ductility and dependability.
Camargo et al. (2019)	PLA	FDM	Infill (%): 22	Layer thickness (mm): 0.13	Heated bed temperature: 60°C, printing speed: 3000 mm/ min	37.9 MPa	68.5 MPa	In this study, the mechanical properties of tensile strength, flexural strength, and impact energy of 3D printed parts

(*Continued*)

Table 9.3 (Continued)

Authors/ references	Material	Process	Printing parameters used			Mechanical testing		Conclusion
						Tensile strengths	Flexure strength	
								manufactured with FDM technology and PLA-graphene raw material were investigated by varying the infill and layer thickness parameters using a statistical technique CCD—central composite design. The findings revealed that the mechanical qualities improve as the linear layer thickness parameter rises. Mechanical qualities, such as tensile and flexural strength, improved as infill increased, whereas impact energy dropped.

Zisopol et al. (2021)	PLA	FDM	Print speed (Ps): 80 mm/s	Deposition temperature (Dt): 200°C	Infill model: lines, 45° orientated, bed temperature: 25°C–60°C	Max. 40.07 MPa	NA	The article aimed to investigate how the mechanical properties of PLA material placed by 3D printing differed based on process factors. The ultimate tensile strength (Rm) of the specimens ranged from 21.0 to 40.07 MPa. The experimental data analysis revealed that the greatest Rm value for the test piece was 40.07 MPa.

particle distribution in the polymer matrix of polymeric micro/nanocomposites/bio-nanocomposite on electrical, mechanical, and thermal characteristics is often investigated in comparison to plain polymers and conventionally utilized materials. Most nanoparticles are suitable for a wide range of applications due to their matrix compatibility, dispersion, specific surface area, and other properties. The use of nanoparticles in the materials/membranes industry to produce nanocomposite materials yields highly promising results in a variety of industrial applications (Karki et al., 2021; Kumar et al., 2023). Functionalized 3D-printed polymer devices play an important part in the next industrial revolution. According to Web of Science data (2010–2020), the number of articles on 3D-printed functional polymeric materials increases year after year, from 5 in 2010 to 193 in 2020.

9.3.1 Polymer nanocomposites material and application

The different polymer nanocomposites manufactured by processes and their enhanced characteristics as given in Table 9.4.

The application of polymer nanomaterial includes drug delivery, bioimaging, electronic device parts, catalysis, biomaterials magnetic application, etc. (Bhardwaj & Kaushik, 2017; Blum et al., 2015; Goto, 2023; Taylor-Pashow et al., 2010).

9.3.2 Limitations of nanocomposites in the case of AM printer

The manufacture of 3D-printed composites frequently encounters issues such as solution or mixture inhomogeneity, viscosity control challenges, necking after printing, and so on. All printing processes are optimal, as certain printing techniques' resolutions are insufficient to fulfill the minimal requirements of printed products. Some 3D printers print at a low pace, which can lead to nozzle blockage. The nozzle may also incur wear and tear after repeating the printing process several times, lowering the resolution of the produced object. The optimal concentration of additive is wanted to determine which is compatible with the speed and temperature of the 3D printers before the printing process.

According to Shrestha (2018), one disadvantage of the research is that the Ag nanoparticles would become stuck at the 3D printer nozzle once the amount reaches 5 wt% or more. In theory, the greater the number of nanoparticles, the better the performance of the specific functionality. Nonetheless, the fraction of nanoparticles must be kept at an optimal level to reduce nozzle blockage. The concentration of nanoparticles inserted must be controlled to minimize agglomeration formation during the printing process while allowing the composite to demonstrate its increased capabilities (Hart et al., 2019).

9.4 Challenges and future scope

The AM of functional polymer nanocomposites opens up interesting opportunities for manufacturing high-performance bespoke components. However, important risks include processing problems (i.e., maintaining constant printing parameters), such as

Table 9.4 Polymer nanocomposites material and application.

Authors/references	Year	Material	Reinforcement (wt %)	Synthesis method	Applications
Malas et al. (2019)	2019	Photopolymer	TO or CCT (5–20%)	Mechanical mixing	Electromagnetic elements
Gnanasekaran et al. (2017)	2017	PBT	CNT, Graphene	Solution mixing	Electrical
Kim et al. (2017)	2017	PVDF	MWCNT/BT	Melt blending	Pressure sensors
Dul et al. (2016)	2016	ABS	Graphene (2–8%)	Melt blending	Mechanical
Carrasco et al. (2010)	2010	ABS	Montmorillonite	Melt blending	Thermal
Mora et al. (2020)	2020	PLA	CNT	Melt blending	Electroconductive
Petrovskaya et al. (2018)	2018	PLA	HA (5–15%)	Sol-gel	Medical devices (implants)

flow rate and temperature, which make it difficult to achieve the appropriate properties in the finished components. In addition, certain 3D printing processes require postprocessing, which is challenging for nanocomposites. Aside from that, the nanocomposite is costly for AM process development and deployment, and there are health risks associated with nanoparticles if breathed.

The number of functional materials suitable for 3D printing into functional devices with high performance is currently somewhat restricted. Future efforts will be made to increase the range with the use of sophisticated material modification methods. Aside from that, methodologies for effective mass preparation of functional materials for 3D printing with high economic efficiency should be prioritized to align with the large-scale manufacturing plan of 3D printing (Hiremath et al., 2021). In addition, the studies of various researchers focused on how CNTs alter FRP properties in composite constructions. However, more research is needed in the future to investigate the impacts of a variety of nanoparticles (Kumar & Kumar, 2020; Kumar & Kumar, 2022a, 2022b; Kumar, 2024a, 2024b; Kumar, Kumar, Kumar, 2022; Kumar, Kumar, Kumar & Luthra, 2022; Rakesh et al., 2021; Selvan & Vedaraj, 2023).

9.5 Conclusions

Polymer nanocomposites AM provide a one-of-a-kind method for producing objects with specific thermal and mechanical characteristics. While the use of nanoparticles can considerably increase qualities like strength and heat resistance, inhomogeneous nanoparticle dispersion and limited compatibility with AM techniques might make it difficult to accomplish these improvements. As material research, processing, methods, and safety protocols increase, AM holds enormous potential for the future of high-performance, functionally graded polymers, and nanocomposite components.

References

Anajafi, Z., Naseri, M., & Neri, G. (2020). Acetone sensing behavior of p-SmFeO3/n-ZnO nanocomposite synthesized by thermal treatment method. *Sensors and Actuators B: Chemical, 304*, 127252. https://doi.org/10.1016/j.snb.2019.127252.

Atakok, G., Kam, M., & Koc, H. B. (2022). Tensile, three-point bending and impact strength of 3D printed parts using PLA and recycled PLA filaments: A statistical investigation. *Journal of Materials Research and Technology, 18*, 1542–1554. https://doi.org/10.1016/j.jmrt.2022.03.013.

Attal, A., Kumar, S., & Khanna, V. (2024). Comprehensive study on production methods and applications of functionally graded coatings. Thermal Spray Coatings: Materials, Techniques & Applications, 1, 25. https://doi.org/10.2174/9789815223552124010004.

Bhardwaj, V., & Kaushik, A. (2017). Biomedical applications of nanotechnology and nanomaterials. *Micromachines, 8*(10), 298. https://doi.org/10.3390/mi8100298.

Bhatia, A., & Sehgal, A. K. (2021). Additive manufacturing materials, methods and applications: A review. *Materials Today: Proceedings*. https://doi.org/10.1016/j.matpr.2021.04.379.

Blok, L. G., Longana, M. L., Yu, H., & Woods, B. K. S. (2018). An investigation into 3D printing of fibre reinforced thermoplastic composites. *Additive Manufacturing, 22*, 176–186. https://doi.org/10.1016/j.addma.2018.04.039.

Blum, A.P., Kammeyer, J.K., Rush, A.M., Callmann, C.E., Sei Kwang Hahn, & Gianneschi, N. C. (2015). Stimuli-responsive nanomaterials for biomedical applications. Journal of the American Chemical Society, 137(6), 2140–2154. https://doi.org/10.1021/ja510147n.

Borrello, J., Nasser, P., Iatridis, J. C., & Costa, K. D. (2018). 3D printing a mechanically-tunable acrylate resin on a commercial DLP-SLA printer. *Additive Manufacturing, 23*, 374–380. https://doi.org/10.1016/j.addma.2018.08.019.

Camargo, J. C., Machado, Á. R., Almeida, E. C., & Silva, E. F. M. S. (2019). Mechanical properties of PLA-graphene filament for FDM 3D printing. *The International Journal of Advanced Manufacturing Technology, 103*(5-8), 2423–2443. https://doi.org/10.1007/s00170-019-03532-5.

Carrasco, F., Pagès, P., Gámez-Pérez, J., Santana, O. O., & Maspoch, M. L. (2010). Processing of poly(lactic acid): Characterization of chemical structure, thermal stability and mechanical properties. *Polymer Degradation and Stability, 95*(2), 116–125. https://doi.org/10.1016/j.polymdegradstab.2009.11.045.

Cheng, B., & Chou, K. (2015). Geometric consideration of support structures in part overhang fabrications by electron beam additive manufacturing. *Computer-Aided Design, 69*, 102–111. https://doi.org/10.1016/j.cad.2015.06.007.

Chia, H. N., & Wu, B. M. (2015). Recent advances in 3D printing of biomaterials. *Journal of Biological Engineering, 9*(1). https://doi.org/10.1186/s13036-015-0001-4.

Dickson, A. N., Barry, J. N., McDonnell, K. A., & Dowling, D. P. (2017). Fabrication of continuous carbon, glass and Kevlar fibre reinforced polymer composites using additive manufacturing. *Additive Manufacturing, 16*, 146–152. https://doi.org/10.1016/j.addma.2017.06.004.

Duda, T., & Raghavan, L. V. (2016). Durrës. *3D Metal Printing Technology, 49*(29), 103–110. https://doi.org/10.1016/j.ifacol.2016.11.111.

Dul, S., Fambri, L., & Pegoretti, A. (2016). Fused deposition modelling with ABS–graphene nanocomposites. *Composites Part A: Applied Science and Manufacturing, 85*, 181–191. https://doi.org/10.1016/j.compositesa.2016.03.013.

Fan, Z., & Feng, H. (2018). Study on selective laser melting and heat treatment of Ti-6Al-4V alloy. *Results in Physics, 10*, 660–664. https://doi.org/10.1016/j.rinp.2018.07.008.

Gnanasekaran, K., Heijmans, T., van Bennekom, S., Woldhuis, H., Wijnia, S., de With, G., & Friedrich, H. (2017). 3D printing of CNT- and graphene-based conductive polymer nanocomposites by fused deposition modeling. *Applied Materials Today, 9*, 21–28. https://doi.org/10.1016/j.apmt.2017.04.003.

Goto, H. (2023). Polymerization in centrifugally oriented liquid crystal reaction field. *Molecular Crystals and Liquid Crystals, 760*(1), 27–33. https://doi.org/10.1080/15421406.2023.2165328.

Hart, L.R., He, Y., Ruiz-Cantu, L., Zhou, Z., Irvine, D., Wildman, R., & Hayes, W. (2019). 3D and 4D printing of biomaterials and biocomposites, bioinspired composites, and related transformers. In 3D and 4D printing of polymer nanocomposite materials (pp. 467–504). Elsevier EBooks. https://doi.org/10.1016/b978-0-12-816805-9.00015-6.

Hiremath, A., Murthy, A. A., Thipperudrappa, S., & Bharath, K. N. (2021). Nanoparticles filled polymer nanocomposites: A technological review. *Cogent Engineering, 8*(1). https://doi.org/10.1080/23311916.2021.1991229.

Karki, S., Gohain, M. B., Yadav, D., & Ingole, P. G. (2021). Nanocomposite and bio-nanocomposite polymeric materials/membranes development in energy and medical sector: A review. *International Journal of Biological Macromolecules, 193*, 2121–2139. https://doi.org/10.1016/j.ijbiomac.2021.11.044.

Khan, I., Saeed, K., & Khan, I. (2017). Nanoparticles: Properties, applications and toxicities. *Arabian Journal of Chemistry, 12*(7), 908–931. https://doi.org/10.1016/j.arabjc.2017.05.011.

Khosravani, M. R., Berto, F., Ayatollahi, M. R., & Reinicke, T. (2022). Characterization of 3D-printed PLA parts with different raster orientations and printing speeds. *Scientific Reports, 12*(1). https://doi.org/10.1038/s41598-022-05005-4.

Kim, K., Park, J., Suh, J., Kim, M., Jeong, Y., & Park, I. (2017). 3D printing of multiaxial force sensors using carbon nanotube (CNT)/thermoplastic polyurethane (TPU) filaments. *Sensors and Actuators A: Physical, 263*, 493–500. https://doi.org/10.1016/j.sna.2017.07.020.

Kumar, R., Kumar, M., & Chohan, J. S. (2021a). The role of additive manufacturing for biomedical applications: A critical review. *Journal of Manufacturing Processes, 64*, 828–850. https://doi.org/10.1016/j.jmapro.2021.02.022.

Kumar, R., Kumar, M., & Chohan, J. S. (2021b). Material-specific properties and applications of additive manufacturing techniques: A comprehensive review. *Bulletin of Materials Science, 44*(3). https://doi.org/10.1007/s12034-021-02364-y.

Kumar, R., Kumar, M., & Chohan, J. S. (2023). Investigation of tensile and flexural property of 316L stainless steel-coated polylactic acid parts. *Journal of Materials Engineering and Performance, 33*(6), 3087–3100. https://doi.org/10.1007/s11665-023-08200-0.

Kumar, R., & Kumar, S. (2020). Trending applications of 3D printing: A study. *Asian Journal of Engineering and Applied Technology, 9*(1), 1–12. https://doi.org/10.51983/ajeat-2020.9.1.1085.

Kumar, R., & Kumar, S. (2022a). *Overview of 3D-printing technology.* CRC Press EBooks265–289. https://doi.org/10.1201/9781003301066-15.

Kumar, R., & Kumar, S. (2022b). Overview on 3D and 4D Printing techniques and their emerging applications in medical sectors. *Current Materials Science, 16*. https://doi.org/10.2174/2666145416666221019105748.

Kumar, R., Kumar, S., Kumar, M., & Luthra, G. (2022). Conventional and 3D printing technology for the manufacturing of metal-matrix composite (pp. 197–223). *CRC Press eBooks*. https://doi.org/10.1201/9781003194897-9.

Kumar, S., Kumar, K. M., & Kumar, R. (2022). Comprehensive study on additive manufacturing process methods, cost comparision, challenges and industrial applications. I-Manager's. *Journal on Future Engineering and Technology, 17*(4), 31. https://doi.org/10.26634/jfet.17.4.18909.

Kumar, S. (2024a). *Role of thermal spray additive manufacturing in combating climate changes (pp. 235–259)*. BENTHAM SCIENCE PUBLISHERS eBooks. https://doi.org/10.2174/9789815223613124010013.

Kumar, S. (2024b). Comprehensive review on high entropy alloy-based coating. *Surface and Coatings Technology, 477*, 130327. https://doi.org/10.1016/j.surfcoat.2023.130327.

Kumar, S., Kumar, S., Singh, H., & Mehra, R. (2024). Cold spray coating of nano crystallization material, method, properties and challenges: A critical review. BENTHAM SCIENCE PUBLISHERS eBooks, 250–274.

Lakhdar, Y., Tuck, C., Binner, J., Terry, A., & Goodridge, R. (2021). Additive manufacturing of advanced ceramic materials. *Progress in Materials Science, 116*, 100736. https://doi.org/10.1016/j.pmatsci.2020.100736.

Lee, K., Hwang, Y., Cheong, S., Choi, Y., Kwon, L., Lee, J., & Kim, S. H. (2009). Understanding the role of nanoparticles in nano-oil lubrication. *Tribology Letters, 35*(2), 127–131. https://doi.org/10.1007/s11249-009-9441-7.

Leite, M., Fernandes, J., Deus, A.M., Reis, L., & Vaz, M.F. (2018). Study of the influence of 3D printing parameters on the mechanical properties of PLA. Conference: 3rd International Conference on Progress in Additive Manufacturing (Pro-AM 2018). Singapore. Dr.ntu.edu.sg, https://doi.org/10.25341/D4988C.

De Leon, A. C., Chen, Q., Palaganas, N. B., Palaganas, J. O., Manapat, J., & Advincula, R. C. (2016). High performance polymer nanocomposites for additive manufacturing applications. *Reactive and Functional Polymers, 103*, 141–155. https://doi.org/10.1016/j.reactfunctpolym.2016.04.010.

Letcher, T., & Waytashek, M. (2015). Material property testing of 3D-printed specimen in PLA on an entry-level 3D printer. In: *Proceedings of the ASME 2014 International Mechanical Engineering Congress and Exposition. Volume 2A: Advanced Manufacturing. Montreal, Quebec, Canada*, https://doi.org/10.1115/IMECE2014-39379.

Ligon, S. C., Liska, R., Stampfl, J., Gurr, M., & Mülhaupt, R. (2017). Polymers for 3D printing and customized additive manufacturing. *Chemical Reviews, 117*(15), 10212–10290. https://doi.org/10.1021/acs.chemrev.7b00074.

Malas, Isakov, Couling, & Gibbons (2019). Fabrication of high permittivity resin composite for vat photopolymerization 3D Printing: Morphology, thermal, dynamic mechanical and dielectric properties. *Materials, 12*(23), 3818. https://doi.org/10.3390/ma12233818.

Mora, A., Verma, P., & Kumar, S. (2020). Electrical conductivity of CNT/polymer composites: 3D printing, measurements and modeling. *Composites Part B: Engineering, 183*, 107600. https://doi.org/10.1016/j.compositesb.2019.107600.

Moradi, M., Aminzadeh, A., Rahmatabadi, D., & Hakimi, A. (2021). Experimental investigation on mechanical characterization of 3D printed PLA produced by fused deposition modeling (FDM). *Materials Research Express, 8*(3), 035304. https://doi.org/10.1088/2053-1591/abe8f3.

Murr, L. E., Gaytan, S. M., Ramirez, D. A., Martinez, E., Hernandez, J., Amato, K. N., Shindo, P. W., Medina, F. R., & Wicker, R. B. (2012). Metal fabrication by additive manufacturing using laser and electron beam melting technologies. *Journal of Materials Science & Technology, 28*(1), 1–14. https://doi.org/10.1016/s1005-0302(12)60016-4.

Pant, A., Lee, A. Y., Karyappa, R., Lee, C. P., An, J., Hashimoto, M., Tan, U.-X., Wong, G., Chua, C. K., & Zhang, Y. (2021). 3D food printing of fresh vegetables using food hydrocolloids for dysphagic patients. *Food Hydrocolloids, 114*, 106546. https://doi.org/10.1016/j.foodhyd.2020.106546.

Patil, S., Sathish, T., Makki, E., & Giri, J. (2024). Experimental study on mechanical properties of FDM 3D printed polylactic acid fabricated parts using response surface methodology. *AIP Advances, 14*(3). https://doi.org/10.1063/5.0191017.

Petrovskaya, T. S., Toropkov, N. E., Mironov, E. G., & Azarmi, F. (2018). 3D printed biocompatible polylactide-hydroxyapatite based material for bone implants. *Materials and Manufacturing Processes, 33*(16), 1899–1904. https://doi.org/10.1080/10426914.2018.1476764.

Piyush, Kumar, R., & Kumar, R. (2020). 3D printing of food materials: A state of art review and future applications. *Materials Today: Proceedings, 33*. https://doi.org/10.1016/j.matpr.2020.02.005.

Rakesh, K., Rajesh, S., Santosh, K., & Naman, G. (2021). Trending applications and mechanical properties of 3d printing: A review. *I-Manager's Journal on Mechanical Engineering, 11*(1), 22. https://doi.org/10.26634/jme.11.1.17627.

Sasmal, A., Sen, S., & Devi, P. S. (2020). Synthesis and characterization of SmFeO3 and its effect on the electrical and energy storage properties of PVDF. *Materials Research Bulletin, 130*, 110941. https://doi.org/10.1016/j.materresbull.2020.110941.

Scharff, R. B. N., Doubrovski, E. L., Poelman, W. A., Jonker, P. P., Wang, C. C. L., & Geraedts, J. M. P. (2016). Towards behavior design of a 3D-printed soft robotic hand. In C. Laschi, J. Rossiter, & F. Lida (Eds.). *Soft robotics: Trends, applications and challenges: Proceedings of the soft robotics week*Springer Nature. Advance online publication. https://doi.org/10.1007/978-3-319-46460-2_4. April 25-30, 2016, Livorno, Italy (pp. 23-29). (Biosystems & Biorobotics; Vol. 17).

Selvan, S., & Vedaraj, I. S. R. (2023). Effects of nanoparticles on the mechanical and thermal behavior of fiber reinforced polymer composites—A review. *Materials Today Proceedings*. https://doi.org/10.1016/j.matpr.2023.07.008.

Shahrubudin, N., Lee, T. C., & Ramlan, R. (2019). An overview on 3D printing technology: Technological, materials, and applications. *Procedia Manufacturing, 35*(35), 1286–1296. https://doi.org/10.1016/j.promfg.2019.06.089.

Shrestha, C. (2018). A Study of 3D Printed Silver-Polymer Composite Structures.

Shiwpursad, J., & Xue, J. (2018). Comparison of different types of 3D printing technologies. *Int J Sci Res Publ, 8*(4), (ISSN: 2250-3153). http://www.ijsrp.org/research-paper-0418.php?rp=P767298.

Stansbury, J. W., & Idacavage, M. J. (2016). 3D printing with polymers: Challenges among expanding options and opportunities. *Dental Materials, 32*(1), 54–64. https://doi.org/10.1016/j.dental.2015.09.018.

Tamaşag, I., Suciu, C., Beşliu-Băncescu, I., Dulucheanu, C., & Cerlincă, D.-A. (2022). Experimental study on the possibilities of FDM direct colour printing and its implications on mechanical properties and surface quality of the resulting parts. *Polymers, 14*(23), 5173. https://doi.org/10.3390/polym14235173.

Taylor-Pashow, K. M. L., Della Rocca, J., Huxford, R. C., & Lin, W. (2010). Hybrid nanomaterials for biomedical applications. *Chemical Communications, 46*(32), 5832. https://doi.org/10.1039/c002073g.

Tofail, S. A. M., Koumoulos, E. P., Bandyopadhyay, A., Bose, S., O'Donoghue, L., & Charitidis, C. (2018). Additive manufacturing: Scientific and technological challenges, market uptake and opportunities. *Materials Today, 21*(1), 22–37. https://doi.org/10.1016/j.mattod.2017.07.001.

Torres, J., Cole, M., Owji, A., DeMastry, Z., & Gordon, A. (2016). An approach for mechanical property optimization of fused deposition modeling with polylactic acid via design of experiments. *Semantic Scholar*. https://doi.org/10.1108/RPJ-07-2014-0083.

Trimble, A. Z., & Ghasemi, N. (2020). *Additive manufacturing/3D printing of polymer nanocomposites: Structure-related multifunctional properties*. Elsevier EBooks87–113. https://doi.org/10.1016/b978-0-12-819535-2.00004-1.

Umarfarooq, M.A., & Singh, B. (2023). Composite Nanomaterials and Their Development. 131–142. https://doi.org/10.1007/978-981-99-4149-0_7.

Yu, Z., Gao, Y., Jiang, J., Gu, H., Lv, S., Ni, H., Wang, X., & Jia, C. (2019). Study on effects of FDM 3D printing parameters on mechanical properties of polylactic acid. *IOP Conference Series: Materials Science and Engineering, 688*, 033026. https://doi.org/10.1088/1757-899x/688/3/033026.

Zaleski, V. G., Pobol, I. L., Bakinouski, A. A., & Gubko, A. D. (2018). Metal parts manufacturing by electron beam additive technologies. *Proceedings of the National Academy of Sciences of Belarus, Physical-Technical Series, 63*(2), 169–180. https://doi.org/10.29235/1561-8358-2018-63-2-169-180.

Zhakeyev, A., Wang, P., Zhang, L., Shu, W., Wang, H., & Xuan, J. (2017). Additive manufacturing: Unlocking the evolution of energy materials. *Advanced Science, 4*(10), 1700187. https://doi.org/10.1002/advs.201700187.

Zhang, Y., Purssell, C., Mao, K., & Leigh, S. (2020). A physical investigation of wear and thermal characteristics of 3D printed nylon spur gears. *Tribology International, 141*, 105953. https://doi.org/10.1016/j.triboint.2019.105953.

Zisopol, D. G., Nae, I., Portoaca, A. I., & Ramadan, I. (2021). A theoretical and experimental research on the influence of FDM parameters on tensile strength and hardness of parts made of polylactic acid. *Engineering, Technology & Applied Science Research, 11*(4), 7458–7463. https://doi.org/10.48084/etasr.4311.

Section 3

Applications of additively manufactured functional polymers and nanocomposites

Additive manufacturing of 3D printed polymers and nanocomposites

10

Laldinthari Suamte, T. Stanis Mangsianlun, and Punuri Jayasekhar Babu
Biomaterials and Bioengineering Research Laboratory, Department of Biotechnology, Pachhunga University College, Mizoram University, Aizawl, Mizoram, India

10.1 Introduction

Additive manufacturing (AM) techniques are a class of technologies exercised for the modification and conversion of a three-dimensional (3D) digital model into a tangible object. The transformation is performed through a layer-by-layer technique in a computer-controlled process, and this layer-adding methodology works differently compared to other subtractive technologies such as traditional milling procedures. AM is also part of the "Industry 4.0" movement, also referred to as 3D printing. The superiority and advantages of AM are highly involved in the degradation of long-established conventional manufacturing methods like injection molding, casting, and machining (Colorado et al., 2024). Since the early initiation of 3D printing in the 1980s, it has quickly gained recognition as a proficient technology and was instantly labelled as a superior method for fabricating hierarchically complex architecture, while traditional production methods may not always be efficient (Hofmann, 2014; Oropallo & Piegl, 2016). The competency of 3D printed objects has expanded due to vigorous modifications in the structural and functional properties of 3D printed objects. Such changes reach out to the fourth dimension of time and have designed the modernistic idea of four-dimensional printing (4DP) (Bodaghi et al., 2016; Khoo et al., 2015). 4D printed materials, being a model of printed operative components, extract the advantageous utility of functional shape modification that can be acquired by various techniques. Shape morphing by swelling and shape memory effect are consistently the most investigated transformation mechanisms (Ge et al., 2016; Sundaram et al., 2017). From its roots in sampling new designs, AM has already acquired to improve the manufacturer's flexibility, reduce the duration of the notion cycle, and extend recently developed design prototypes by supplying a mechanism to immediately attain physical outcomes from digital inputs with no requiring traditional tooling (Thompson et al., 2016). There are a handful of implementations set up by AM, which include customized musical instruments and sports materials, lightweight prosthetic implants, dental splints, moulds for injection moulding and composite layup, etc. AM has diverse beneficial applications in industrial products, such as sensors, flexible electronics, and antennas (Quinlan et al., 2017). Production of certain crucial materials by aerospace manufacturers could be greatly facilitated by AM, such as the creation of multiplex parts with lighter structures, which contributes significantly to the reduction of fuel

Additive Manufacturing of Functional Polymers and Nanocomposites. DOI: https://doi.org/10.1016/B978-0-443-26516-7.00010-9
© 2025 Elsevier Ltd. All rights are reserved, including those for text and data mining, AI training, and similar technologies.

costs. An appreciable employment of AM in the medical industry is the customization of medical devices personalized for the individual patient that is both rapid and profitable, as in the case of customization for hearing aids, surgical guides and models, orthopedics, and prosthetics (Gantenbein et al., 2018, Suamte et al., 2023). AM also plays an important function in assisting the automotive industry; it is economically unwasteful and could hasten manufacturing small batch customized parts. Worth mentioning is the ongoing development of novel better-performing polymers. Many AM-specific materials are designed and manufactured by various chemical companies and research institutes. For example, Gantenbein et al. (2018) recently produced printable hierarchical liquid-crystal-polymer structures with top-notch mechanical characteristics. AM equipment companies are known to dispense investments in technological innovations. One such breakthrough is the utilization of several printing heads to print multiple materials simultaneously (Gnanasekaran et al., 2017; Kim et al., 2017). The usage of 3D printing ultimately results in the possibility of tailoring smart objects into modern and highly pliable designs in well-controlled methods. This rules out the requirement of slow assembly of parts by manufacturing monolithic objects (Weller et al., 2015). Due to the establishment of smart printable materials, there will be a rapid amalgamation of 3D printing technologies into the soaring as well as advancing field of technologies (Nadgorny et al., 2016). There is an increase in investigating active materials and their 3D printing to efficiently match the rising demand of manufacturing requirements of complex functional platforms (Nadgorny & Ameli, 2018).

Due to the significant advantages over conventional materials, polymer nanocomposites have gained popularity in research and development for various industrial applications (Al Rashid & Khan, Al-Ghamdi, et al., 2021). Nanocomposite materials are composite materials categorized in such a way that at least one component has dimensions of less than 100 nm; this nanometric size allows the materials to have enhanced properties allowing their use in several areas of sensors and bioimagining (Babu et al., & Doble, 2018, 2021). Considering mechanical, electrical, and thermal characteristics, these nanocomposite materials are better performing in comparison to their respective micro or macro materials; this superiority is due to the unique nanoscale properties, such as large surface area and almost perfect crystal order. Polymer matrix, ceramic matrix nanocomposites, and metal matrix are some common nanocomposite materials (El-Eskandarany, 2020), which are extensively explored and studied for numerous applications, such as aerospace, electronics, energy storage, biomedicine and even in food packaging (Babu & Tingirikari, 2023).

Polymer-based nanocomposites are seen to show a wide range of upper-hand features over conventional materials. Such supremacy is because of their greater strength-to-weight ratio, extreme pliability and flexibility in the manufacturing process, effortless customizable product properties, and strong corrosion resistance. Besides, the present growth of AM procedures has thrown a fresh perspective on the manufacturing systems of various industry sectors, including electronics, construction, telecommunication, defence, and biomedical. AM techniques and methods offer considerable advantages, such as the liberty to create endless designs and modify customised parts, reduced waste, speedy product manufacturing, significantly fewer human-caused errors, high accuracy, sharp precision, and budget-friendliness (Ahangar et al., 2019; Nachal et al., 2019; Rane & Strano, 2019).

Even though there are a significant number of achievements and successes, there are numerous limitations and obstacles that must be addressed to institutionalized AM as large-scale manufacturing machinery. A substantial number of these issues are due to a shortage of satisfying properties (porosity, thermo-mechanical properties, creep, anisotropy, cost, corrosion properties, and long-term stability, etc.) of the building materials employed presently (Ligon et al., 2017). However, AM is slower than traditional production techniques like injection molding. Limitations of AM may also include its high-priced raw materials. Such high expense is often required because satisfactory print products demand top-quality filaments or powders. Furthermore, when modeling techniques are used, the interrelationships between design-material-process parameters and the ultimate qualities of the AM part are often unpredictable (Simpson et al., 2017). To this date, the advancement of AM and desktop modification are revolutionizing customized manufacturing elating innovation, ideas, and creativity, and hastily converting ideas into thrilling products. AM extend to simplify strategic coordination, particularly for intricate multiplex (Ligon et al., 2017). This paper focuses on the properties of polymers and nanomaterials used in additive manufacturing and in the development of various techniques, namely VAT photopolymerization, powder bed fusion (PBF), direct energy deposition (DED), material jetting (MJ), material extrusion (ME), binder jetting, and sheet lamination. Thus, it is projected that AM would continue to improve and outperform in the coming years.

10.2 Properties of polymers and nanomaterials used in additive manufacturing

AM adopts many types of polymer materials, which can be used effectively in forming various compositions and shapes. Some of the polymers used in additive manufacturing can be classified as thermoplastics, nanocomposites, polymer matrix composites, thermoresponsive polymers, thermoplastic elastomers, fibre-reinforced polymers, and particle-reinforced polymer composites, as shown in Fig. 10.1 (Saleh Alghamdi et al., 2021). Nylon 12, polycarbonate (PC), epoxy resin, polylactide (PLA), acrylonitrile butadiene styrene (ABS), glass-filled nylon, and polyamide (Nylon) are some of the common thermoplastic polymer materials utilized in fused deposition modeling (FDM). Commercial polymers materials that the FDM process includes are polycarbonate (PC), ABS_M30, which are widely famous (Jasiuk et al., 2018).

Thermoplastics are melt-processable plastics that can be softened on heating and solidified on cooling while keeping their natural characteristics (Muzaffar et al., 2018). They are employed due to various qualities such as their high strength, stability, rigidity, and high-temperature tolerance (Gladman et al., 2019). The polymers were found to have poor mechanical characteristics, rendering them unsuitable for a wide range of load-bearing applications. Polymers were reinforced with fibers and particles to improve their mechanical qualities and enable them to be

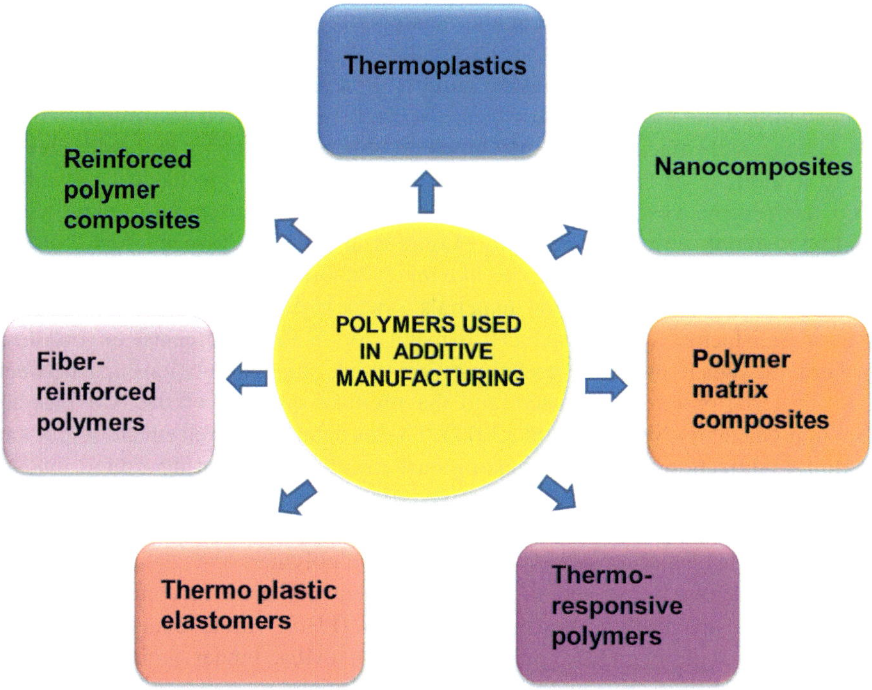

Figure 10.1 Types of polymers commonly used in different AM technologies.

employed in load-bearing applications. These polymers can comprise both synthetic and biomaterial resins (El Moumen et al., 2019).

Nanotechnology combined with AM allows the creation of 3D objects with optimal characteristics and multifunctionality. The synthesis and incorporation of nanomaterials, such as gold nanoparticles and carbon nanotubes, improves biocompatibility, electrical conductivity, mechanical strength, electromechanical and chemical sensitivity (Farahani et al., 2016; Sharma et al., 2012). To create a vast range of 3D nanocomposites, AM technologies are employed for applications in a variety of sectors, namely, tissue engineering, microfluidics, biosystems, engineered materials and composites, lab-on-a-chip, microelectromechanical systems (MEMS), and microelectronics. Nanocomposites are commonly manufactured using AM processes such as inkjet printing, microstereolithography, extrusion-based technologies, and powder bed technologies. They are also frequently employed in the disciplines of biosensors, biostructure, electronics, and electromagnetics (Babu et al., 2020; Babu & Doble, 2018; Manapat et al., 2017).

The use of carbon-based nanomaterials in AM has received widespread attention due to its ability to improve the characteristics and performance of printed materials. For example, carbon nanotubes and graphene of carbon-based nanomaterials preserve unique structural, mechanical, and electrical properties and contain desired features that make them the perfect well-qualified candidate for increasing the functioning and robustness of AM components (Acquah et al., 2016). It was discovered that incorporation of carbon nanotubes into AM procedures enhanced the conductivity, mechanical strength, and thermal stability of printed materials. Ma Q and team conducted a scientific study in which the importance of inserting multiwalled carbon nanotube (MWCNT) into 3D printed bio-inspired spherical-roof cubic cores was explored in terms of energy-absorbing capabilities, stability, and peak force. The study demonstrated the augmentation of compressive performance and highlighted their potential as lightweight energy-absorbing components due to the coating of the cores with multiwalled carbon nanotubes (MWCNTs) into 3D-printed bio-inspired spherical-roof cubic cores (Ma et al., 2022). Dul et al. discovered the importance of carbon nanotubes in 3D printing for creating electrically conductive polymer nanocomposites with enhanced mechanical properties. Multiwalled carbon nanotube (MWCN) and single-walled carbon nanotube (SWCN) combinations elevated the modulus, tensile strength, and electrical conductivity of the printed composites. Findings from different studies have emphasized on the promising use of these nanocomposites in sensors and thermoelectric devices, profiting from their diverse functional properties and capacity to generate complicated geometries via 3D printing. Thus, single-walled carbon nanotubes (SWCNs) boost thermal conductivity, making them well-suited for applications requiring heat dissipation (Dul et al., 2022). Another carbon-based nanomaterial, known as graphene, has been discovered to be efficient in AM. As a result, adding graphene increased the electrical conductivity of printed objects, making them ideal for applications in electronics and sensors (Htwe & Mariatti, 2022). Thus, graphene-based nanomaterials can be used in the 3D printing process as an ink formulation method. Tran et al. (2020) highlighted the potential of graphene nano-inks to allow the direct layer-by-layer synthesis of high-quality graphene nanosheets in three-dimensional graphene freeze 3D printing (Tran et al., 2020).

The latest study conducted by Cao et al. has indicated that combining nanomaterials with AM can boost the final product value. Nanomaterials added as a filler can have a positive impact on mechanical and magnetic characteristics. For example, Cao et al. used laser power bed fusion (PBF) AM to create a nickel-based alloy (Haynes 230) with potential aerospace and defense applications. To boost performance, multiwalled carbon nanotubes (MWCNT) was used as reinforcement filler. The nickel nanocomposite containing 2.5% MWCNT showed higher relative density (99.36%), yield strength (164 ± 2 ksi), ultimate strength (197 ± 5.4 ksi), and modulus of elasticity (34 ± 2.1 msi) compared to the control nickel (Cao et al., 2023).

10.3 Types of additive manufacturing techniques

According to the standard followed by the American Society of Testing and Materials, AM techniques are classified into different categories (Fig. 10.2) namely, vat photopolymerization, powder bed fusion, direct energy deposition, material extrusion, material jetting, binder jetting, and sheet lamination (Zhakeyev et al., 2017). Further classification can be done depending on the base material that is used

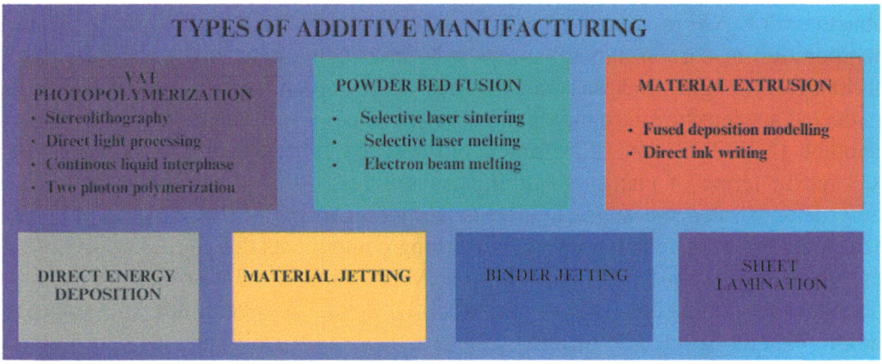

Figure 10.2 Different types of AM techniques classified as per the American Society of Testing and Materials.

and the medium for processing the base material to form the final product (Saleh Alghamdi et al., 2021). Each one of these techniques requires a unique set of materials and principles that are utilized in the fabrication of the desired end product (Tan et al., 2020).

10.3.1 Vat photopolymerization

Vat photopolymerization relies on the use of radiations like ultraviolet and visible lights for layer-wise solidification of liquid resins to form a desired three-dimensional structure as displayed in Fig. 10.3 (Bachmann et al., 2021). This technique is known for its high resolution, accuracy, and outstanding surface finish in comparison to other AM techniques (Garcia et al., 2020). Different vat photopolymerization techniques include stereolithography (SLA), direct light processing (DLP), and continuous liquid interphase production (CLIP) (Davoudinejad, 2021).

10.3.1.1 Stereolithography (SLA)

SLA is the first technique developed under AM. It allows the fabrication of a three-dimensional object by selective solidification of a photosensitive resin using a photopolymerization reaction (Huang et al., 2020). SLA is widely employed for a diverse range of materials like complex alumina, hydroxyapatite and zirconia for fabrication of the desired end product (Liu et al., 2020). Composites of boron nitride nanoplatelets and polymer were fabricated using this technique. The presence of nanoplatelets improved the compressive strength, damping and microhardness (Bustillos et al., 2019). Xu et al. (2021) developed a multilayered 3D-printed oral dosage polyprintlet using SLA, which was incorporated with antihypertensive drugs like atenolol, amlodipine, hydrochlorothiazide and irbesartan, allowing the delivery of low-dose combination therapy. Bioresorbable scaffolds consisting of bioactive glass-b-tricalcium phosphate and triply-periodic minimal surfaces (TPMS) were fabricated by SLA. This structure had an accurate imitation of the environment found in a natural bone tissue and provided a valuable insight in the design of bioactive

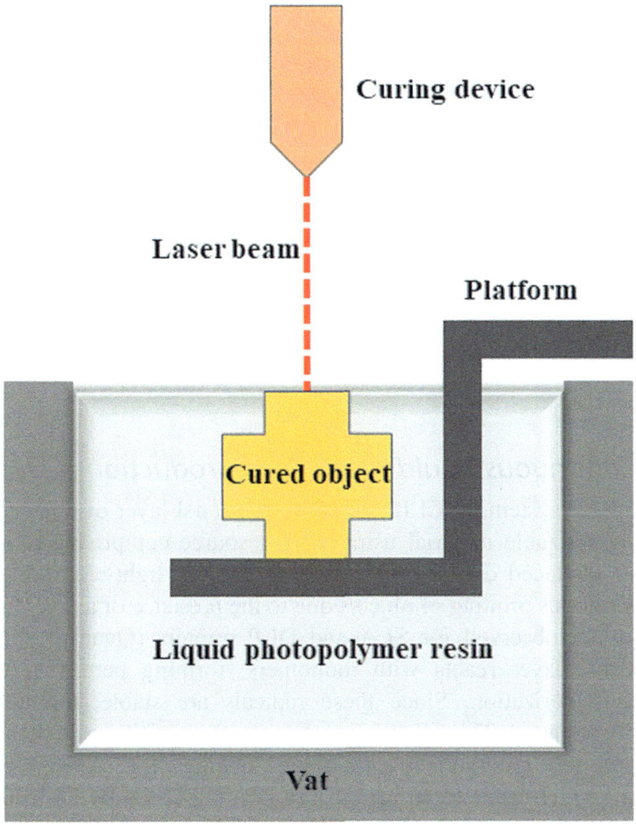

Figure 10.3 Diagrammatic illustration of the working principle of vat photopolymerization technology.

components and bionic structures for use in bone defect treatment. In SLA, the printing resolution although dependent on the speed of printing, volume and direction can be improved upon lowering the laser spot size (Li et al., 2023).

10.3.1.2 Direct light processing (DLP)

In DLP, a digital projector screen is used for curing onto the photopolymerizable material. As polymerization occurs all at once, the printing rate is faster on comparison to SLA (Zhao et al., 2020). Functional nanoparticle-enhanced conduits were developed using this technique for effective peripheral nerve regeneration. The conduit consisted of gelatin-methacryloyl (GelMA) hydrogels loaded with poly (ethylene glycol)-poly(3-caprolactone) (MPEG-PCL) nanoparticles, which provided sustained drug release and a physical microenvironment required for elongation and regeneration of axons (Tao et al., 2019). A photocurable bioink of chitosan (CHI-MA) was developed for use in DLP. The CHI-MA bioink through DLP printing was processed further to form complex structures of 3D hydrogels having high biocompatibility, good resolution and fidelity (Shen et al., 2020). Dexamethasone-loaded punctal plugs were developed from polyethylene glycol diacrylate (PEGDA) and polyethylene glycol 400 (PEG 400) using DLP to aid in controlled drug delivery to the eye. In vitro release kinetics of the punctal plugs revealed the sustained release of the drugs for both plugs made with PEGDA and PEG 400. This study demonstrated the potential use of DLP for the fabrication of personalized punctal plugs for ocular administration (Xu et al., 2021). A novel transdermal drug system comprising gelatin methacryloyl (GelMA) microneedles containing amoxicillin (AMX) was fabricated through DLP-based 3D printing technology. The result from the in vitro antibacterial assays displayed significant inhibition on the growth of *Escherichia coli* and *Staphylococcus aureus* suggesting the use of these GelMA microneedles as a potential delivery agent for transdermal therapeutic treatments (Erkus et al., 2023).

10.3.1.3 Continuous liquid interphase production (CLIP)

Similar to the DLP system, in CLIP, the cross-sectional layer of objects is projected on a photopolymerizable material using a light source comprising of small pixels. CLIP employs advanced digital projectors like laser or light-emitting diode (LED) and allows continuous printing of objects due to the presence of an oxygen permeable layer which is not observed for SLA and DLP printers (Quan et al., 2020). The oxygen-permeable layer reacts with monomers, forming peroxy radicals, which inhibit photopolymerization. Since these radicals are stable, the polymerization cannot be easily restarted. This allows a continuous flow of the resin between the base of the vat and the polymerizable material making it feasible for continuous polymerization, which leads to an increase in printing speed by 25–100 times more than SLA and DLP (Hahn et al., 2020; O'Brien & Bowman, 2006). An implantable device loaded with cisplatin and paclitaxel was printed using a Carbon3D S1 prototype CLIP printer for intraoperative treatment of cancer (Hagan et al., 2022). In a similar work, Caudill and the team fabricated microneedles using a Carbon3D S1 prototype CLIP printer for transdermal vaccine delivery. Microneedle patches coated with vaccine antigens and adjuvants were developed and on comparison with a subcutaneous bolus injection, the microneedle displayed improved retention of cargo in the skin and enhanced immune cell activation. It also elicited T cell response and induced humoral immune response achieving dose sparing. These findings suggested the potential of CLIP-printed microneedles for use as a self-applicable and noninvasive vaccination (Caudill et al., 2021).

10.3.1.4 Two-photon polymerization (TPP)

TPP employs a near-infrared femtosecond laser beam for the photopolymerization process. Against a solidified resin, two laser photon beams are deposited simultaneously to obtain a predefined 3D object. This technique offers a very high resolution which is capable of achieving 100 nm and is very efficient for printing micro-structures with absolute precision (Lee et al., 2008; Wang et al., 2021). Round tip polymer microneedle arrays were prepared by polydimethylsiloxane micromolding and TPP using a photoreactive acrylate polymer. These microneedle array sustained their elastic deformation displaying compressive strength for use in transdermal drug delivery. In vitro studies also demonstrated successful penetration of the epidermis and human stratum corneum (Gittard et al., 2009). In a similar work, using TPP printers, microneedles for vaccine and drug administration was developed by Plamadeala and team. Ex vivo tests conducted on human skin revealed good delivery of the of the vaccine/drug loaded on the surface without any sign of breakage in the microneedle (Plamadeala et al., 2020). A light-driven microscale 3D hydrogel actuator embedded with photothermal surface-modified Fe_3O_4 nanoparticles was developed successfully by TPP microfabrication. The small size and biocompatible nature displayed its probable use in the field of biomedical microelectromechanical systems MEMS (Zheng et al., 2020).

10.3.2 Powder bed fusion (PBF)

PBF is an AM process in which thermal energy is used for fusing powdered particles to design a 3D object using computer-aided design (CAD) as shown in Fig. 10.4 (Singh et al., 2021). This technique is widely used because of its fast production speed, low cost, reusability of powder material, and expansive range of compatible materials (Chin et al., 2020). Several kinds of this PBF process are present on the

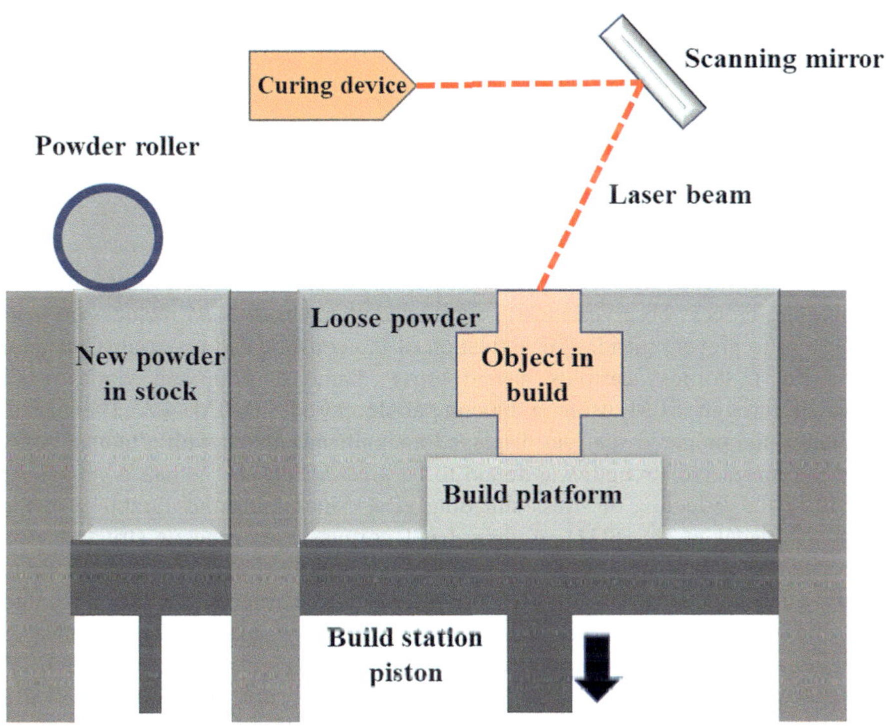

Figure 10.4 Illustrative figure of working principle of powder bed fusion technique.

basis of the source of heat and the materials used. Selective laser melting (SLM), selective laser sintering (SLS), and electron beam melting (EBM) are different categories under PBF (Singh et al., 2020). Since the melting process occurs in a vacuum, it can be used for the production of parts having high densities (Zhang et al., 2018).

10.3.2.1 Selective laser sintering (SLS)

SLS is a widely used AM technique because of its high capability and versatility in processing any kind of polymer that is available in a sintered powder form. It uses a laser for successive layer scanning of powdered polymers to produce a 3D object (Shirazi et al., 2015). This technique was utilized by Fina and the team to construct a bilayered cylindrical gyroid lattice that had the team to construct a bilayered cylindrical gyroid lattice with the team to construct a bilayered cylindrical gyroid lattice that adjustable release kinetics. Drug modulation by the gyroid lattice was possible for all four polymers loaded with paracetamol and this revealed the probable use of SLS for customized drug release of different polymers in a fast and sustainable manner without the need for alteration of any formulation in the composition (Fina et al., 2018). A dexamethasone eluting porous scaffold was constructed via SLA for bone regeneration. Drug release studies showed that dexamethasone was released in a continuous manner for a month and ex vivo studies revealed new formation of blood vessels and bone upon implantation of the scaffold in cranium defected rats (Sun et al., 2020). In a similar work, an SLS bio-inspired silicon-wollastonite scaffold having a porous structure was developed by Kamboj and team. The scaffold displayed improved viability and provided aid in the guidance of the human mesenchymal cells (hMSC) on the scaffold with promotion of osteogenic differentiation in vitro (Kamboj et al., 2020).

10.3.2.2 Selective laser melting (SLM)

In SLM, a laser beam provides heat which allows the powder to melt fully instead of the sintering process observed in SLS. The build chamber in SLM includes an inert atmosphere to prevent nitriding or oxidation of the consolidated material (Olakanmi et al., 2015). Porous scaffolds named cross, diamond, grid, and vinties were fabricated through SLM using a biocompatible porous Ti-6AL-4V. The elastic modulus of the prepared scaffold displayed a significant match with a human bone, and the compressive strength was found to be greater than the human bone, which will aid in the reduction of stress shielding effect, increasing the durability of the implant (Mondal et al., 2022). A three-level porous heterogeneous structure was developed using a combined process of SLM, a hydrothermal process and finite element analysis (FEA). Using CAD, the microporous structure was designed, and the hydroxyapatite crystals were coated using the HT technique. The presence of nanohydroxyapatite and micropores were observed to enhance proliferation of preosteoblasts and osteogenic behaviors of rBMSCs in vitro. This study opened a new pathway for researchers in building orthopedic implants possessing enhanced osteointegration (Pei et al., 2021).

10.3.2.3 Electron beam melting (EBM)

EBM is a technique used for melting metallic powders using an electron beam (Galati, 2021). Commonly used metal alloys in EBM for the production of implants in the biomedical industry are chromium cobalt alloys, 316L steel and titanium alloys (Tamayo et al., 2021). Co–29Cr–6Mo femoral prototypes comprising of a mesh structure developed by Murr and team using EBM displayed significant bone stiffness (Murr et al., 2011). Ozan and team fabricated different TI-Nb-Zr alloys having good mechanical properties for use in the production of orthopedic implants. The implants had relatively low stiffness, higher mechanical strength, excellent plastic deformation, good cytocompatibility, and high elastic admissible strain, which are qualities needed for application in biomedicine (Ozan et al., 2015). EBM is also used in the successful fabrication of Ti6Al4V ankle implants, hip stents, and cervical and sacral vertebral fusion cages, which were implanted into patients successfully. Till date, adequate functioning of the implants in the body were observed after implantation (Tang et al., 2018).

10.3.3 Direct energy deposition (DED)

DED is an AM technique in which a focused thermal energy source (laser or an electron beam) is used for the fusion of materials upon melting as they are deposited (Ahn, 2021). Powdered materials like aluminum, copper, cobalt, tin, titanium, nickel, and stainless steel are commonly used materials in the process (Gibson et al., 2015). Depending on the method or application, DED is also referred to as direct metal deposition (DMD), electron beam additive manufacturing (EBAM), and laser-engineered net shaping (LENS) (Saboori et al., 2019). In a novel work, Direct Energy Deposition-Laser Additive Manufacturing (DED-LAM) was utilized in the fabrication of a biomedical alloy, β-type Titanium Molybdenum (Ti-15Mo). This study revealed a new insight into the methodology of DED-LAM technique for its potential use in the fabrication of biomedical implants (Bhardwaj et al., 2019). Biomedical alloys of 13Nb-13Zr fabricated using DED were investigated for their anisotropic microstructure, corrosion and nanomechanical behaviour. The observations from the study revealed that the alloy had an outstanding quality with minimal pores. Because of the unique microstructure, improved corrosion and nanomechanical properties of the x–y plane which is normal to the building direction on comparison to the x–z plane lying parallel to the building direction was observed in the anisotropic investigation. This study also opened a new path for biomedical alloy fabrication as it was the first work that used DED for the production of 13Nb-13Zr biomedical alloys (Ansari et al., 2023).

10.3.4 Material jetting (MJ)

Among all the 3D printing processes, MJ is one of the most accurate and fastest processes. It is a technique in which liquid material droplets from the inkjet print heads of the support and build material are jetted onto a build platform (Dilag et al., 2019). A major advantage of MJ is multimaterial printing making it feasible in the designing of haptic and visual prototypes resembling molded parts with homogenous thermal and mechanical features. A Few disadvantages of this technique would be the poor mechanical strength, high cost, degradation of mechanical properties with time and photosensitivity (Saleh Alghamdi et al., 2021). Since it produces parts based on the ink jetting technology, it is also called as direct inkjet printing or 3D inkjet printing (Elkaseer et al., 2022). It is also the most suitable technique for the fabrication of complex 2D or 3D biological constructs (Kumar et al., 2022). Lo and team reported the use of inkjet printing for the production of a soft resistive pressure sensor for use in wearable electronics. This was achieved by printing a layer of silver nanoparticles directly on a polydimethylsiloxane substrate and encapsulated using a VHB tape. The sensor could measure arterial pulse waveforms and allow the detection of acoustic vibrations under different levels of sound pressure (Lo et al., 2020). A heating element was developed through inkjet printing using a nanoparticle silver ink on a flexible polymer film accompanied by thermal sintering. The capability of the printed heaters was revealed by a long term test for several days in which negligible fluctuations and good resistance stability were observed (Mitra et al., 2021). Using inkjet printing, a novel method for controlled deposition of self-assembled peptides-diphenylalanine (FF), was achieved by Safaryan and team, which can be employed for the production of peptide-based nanostructures for use in nanotechnological applications (Safaryan et al., 2018).

10.3.5 Material extrusion (ME)

ME is a process in which continuous extrusion of polymer pellets, filaments, and viscous inks occurs through a heated nozzle at a constant pressure onto a build platform where they eventually solidify as displayed in Fig. 10.5 (Tan et al., 2020). The most common technique employed under this process is FDM or fused filament fabrication (FFF) (Al Rashid & Koç, 2021). Compared to other AM techniques, they provide affordability and compatibility with a vast range of materials.

10.3.5.1 Fused deposition modeling (FDM)

FDM is an ME process that involves the use of polymers as the raw material filaments. The filament is heated into a molten state, which is extruded out from a nozzle and solidifies upon cooling (Mwema et al., 2020). Polylactic acid (PLA)/ hydroxyapatite microsphere porous scaffolds were developed using FDM for evaluation of their potential use as substitutes in bone lesions. Morphological analysis confirmed a smooth surface and displayed stiffness values that were similar with bone tissue (Corcione et al., 2019). A multifunctional polymer nanocomposite scaffold of halloysite nanotube reinforced with polylactic acid filaments and loaded with metformin (MET) was developed using FDM for its potential use in biomedical applications. In vitro drug release kinetics and cell viability study revealed good cell proliferation with no signs of cytotoxicity and 50% of the total drug was released at the end of 120 hours (Kempin et al., 2017). Sakthivel and team optimized FDM parameters using response surface methodology (RSM) for obtaining a high surface finish and dimensional accuracy for PLA stainless-steel composites. Biocompatibility

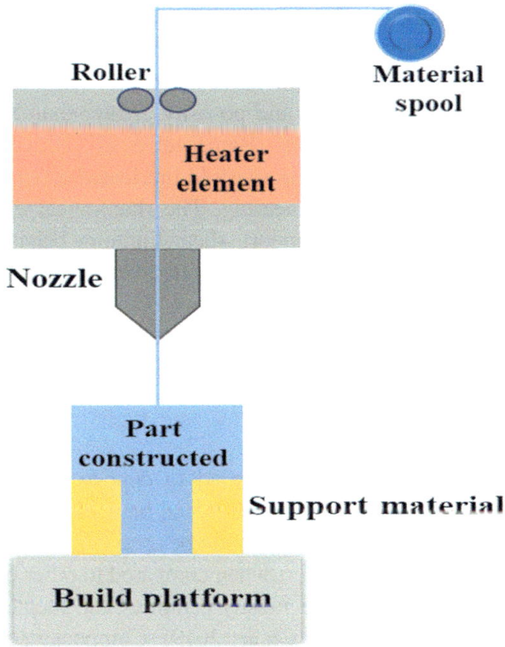

Figure 10.5 Diagrammatic representation of material extrusion technique.

study on the scaffold assessed using preosteoblasts cell confirmed good compatibility similar to PLA and the toughness was measured at 18 kJ/m^2 with an ultimate tensile strength of 23.1828 MPa at 90° raster orientation of the print and 69.0288 MPa at 45° raster orientation of the print. This study proved that numerical optimization of FDM can facilitate in the development of next-generation biomaterials to be used as orthodontic and orthopedic implants (Sakthivel et al., 2020).

10.3.5.2 Direct ink writing (DIW)

This technique is also referred to as liquid deposition modeling (LDM) (Revelo & Colorado, 2018). It is a process where viscous inks of concentrated polymer solutions and pastes are printed and solidified through drying, cooling and chemical reactions. One of the advantages of DIW is its malleable materials requirements and ambient operation temperature, offering a broad range of printable materials (Truby & Lewis, 2016). Ajdary reported the use of DIW for synthesizing a layer-by-layer mesh of renewable polysaccharides (chitosan and nanocellulose). The DIW technique is employed because it can fulfil all the geometric details required for mesh dimensions, opening size, shape, and infill density. Biological studies revealed noncytotoxicity and good biocompatibility against human macrophages/monocytes indicating the promising use of these biocomposites for implant administration (Ajdary et al., 2022). A hydrogel composed of cellulose nanofibril reinforced with aloe vera was employed for the development of 3D geometries using the DIW technique. The hydrogels demonstrated excellent viscoelastic qualities and printing of the lattice structures was done with a fitting resolution. The printed 3D structures displayed excellent wet stability and provided protection to the shape from extensive shrinkage. All printed structures had a high capacity of water uptake (46 g/g) and porosity greater than 80% signifying their possible use for biomedical applications (Baniasadi et al., 2021). A graphene oxide (GO)-reinforced 13–93B1 bioactive glass scaffold fabricated using DIW was evaluated for its application in bone tissue engineering. The rheological characteristics of the scaffold was assessed for structural support, shear thinning behaviour, and recovery. Morphological studies revealed good pore size, suitable porosity and pore throat size with excellent interconnectivity. Live dead assay confirmed no cytotoxicity of the scaffold, and adequate cell viability and growth were observed, confirming its suitable use for bone tissue engineering (Dixit et al., 2024).

10.3.6 Binder jetting

Binder jetting employs the use of a print head to deposit droplets of liquid agents (binding agent solution) onto powder particles placed on a powder bed, followed by the joining of the particles to form a 3D project. It can be used for processing a wide range of materials and is has a relatively lower cost when compared to other printing process (Liu et al., 2020). The powder materials are mostly ceramics and metals, and polymers are often used as binding agents (Ziaee & Crane, 2019). Composite scaffold of tricalcium phosphate (TCP)-45S5 Bioactive glass were developed using this binder jetting technique and conventional processing. In vitro studies revealed good interaction of bone cell materials and this study demonstrated the probable use of this technique in developing porous scaffolds for bone grafting (Bose et al., 2021). Stochastic porous structures of Voronoi tessellation were fabricated through a binder jetting technique using a 316 L SS powder, and further modifications were carried out using DLC coating to determine its potential use in biomedical fields (Laskowska et al., 2022). Zhang and team reported the first use of this technique for printing silk powders (SP) to form complex parts in which polyvinyl alcohol (PVA) was utilized as a solid binder. The SP/PVA formulation was able to successfully print an intricate structure having a high resolution of 200 μm. This printed construct possessed good compressive modulus comparable to printed parts obtained from ceramics demonstrating its potential for use in load-bearing biomedical applications (Zhang et al., 2021). In another work by Mostafaei and team, the binder jetting technique was used for the production of a removable partial denture framework from a metallic powder (Mostafaei et al., 2018).

10.3.7 Sheet lamination

Sheet lamination is an AM technology that involves the production of a 3D project by stacking and lamination of thin sheets of materials. The method for lamination is carried out through brazing, bonding, or ultrasonic welding and the final form of the object is executed by computer numerical control machining or laser cutting. Different materials like plastics, paper, and metal sheets can be used for fabrication and this technique is further classified on the basis of the lamination technique that is applied for bonding the sheets together. Laminated object manufacturing (LOM) is one of the first sheet lamination processes that employs the use of both additive and subtractive methods that bond sheets of materials to produce a 3D object using this principle, other processes like ultrasonic additive manufacturing (UAM) and plastic sheet lamination (PSL) are also developed with time (Saleh Alghamdi et al., 2021; Tan et al., 2020). Based on LOM, a novel graphene (LIG) foam printing process induced by a 3D laser was developed by Luong and team. Good mechanical strength and electrical conductivity were observed on the developed LIG graphene foams demonstrating its probable use in electronic sensor applications and energy storage (Luong et al., 2018). A novel process for the production of porous ceramics was developed using an aqueous slurry-based laminated object manufacturing by Zhang and team. This new method allowed the flexible design of pore structures through the use of different organic mesh templates (Zhang et al., 2021).

10.4 Conclusion

AM holds a wide range of applications in industries, sports and musical instruments, as well as electronics. It provides aid in aerospace manufacturing, medical industry, and the automotive industry through the use of high-performance polymers, which are undergoing constant development. Polymer nanocomposites are becoming increasingly popular in research for the production of a variety of smart materials due to their advantages. Polymer matrix, metal matrix, and ceramic matrix nanocomposites are examples of common nanocomposite materials that have been extensively researched and examined for a variety of applications. AM, in cooperation with nanotechnology, enables the creation of 3D objects with excellent properties having multifunctionality thereby boosting the value of a finished product.

Each of these additive manufacturing (AM) techniques is accompanied by its own advantages. Vat photopolymerization is characterized by high resolution, precision, and excellent surface quality. Powder Bed Fusion (PBF) is renowned for its rapid production speed, low cost, reusability of powder material, and a wide range of compatible materials. Material Jetting (MJ) facilitates multimaterial printing, while Fused Deposition Modeling (FDM) allows for compatibility with a vast array of materials. However, AM technology should not be perceived as a complete replacement for conventional techniques; rather, the integration of these two approaches should serve to complement each other, yielding synergistic outcomes. With the ongoing research and advancements observed in this field, AM is expected to further expand its capabilities to produce novel, high-functioning smart materials.

References

Acquah, S. F., Leonhardt, B. E., Nowotarski, M. S., Magi, J. M., Chambliss, K. A., Venzel, T. E., Delekar, S. D., & Al-Hariri, L. A. (2016). Carbon nanotubes and graphene as additives in 3D printing. *Carbon nanotubes-current progress of their polymer composites,* 227–251. InTech.

Ahangar, P., Cooke, M. E., Weber, M. H., & Rosenzweig, D. H. (2019). Current biomedical applications of 3D printing and additive manufacturing. *Applied Sciences, 9*(8), 1713.

Ajdary, R., Reyes, G., Kuula, J., Raussi-Lehto, E., Mikkola, T. S., Kankuri, E., & Rojas, O. J. (2022). Direct ink writing of biocompatible nanocellulose and chitosan hydrogels for implant mesh matrices. *ACS Polymers Au, 2*(2), 97–107. https://doi.org/10.1021/acspolymersau.1c00045.

Al Rashid, A., Khan, S. A., Al-Ghamdi, S. G., & Koç, M. (2021). Additive manufacturing of polymer nanocomposites: Needs and challenges in materials, processes, and applications. *Journal of Materials Research and Technology, 14*, 910–941.

Al Rashid, A., & Koç, M. (2021). Fused filament fabrication process: A review of numerical simulation techniques. Polymers, 13, 3534.

Ansari, N., Lee, D.-H., Huang, E.-W., Jain, J., & Lee, S. Y. (2023). Anisotropic microstructure, nanomechanical and corrosion behavior of direct energy deposited Ti–13Nb–13Zr biomedical alloy. *Journal of Materials Research and Technology, 26*, 2682–2694.

Babu, P. J., & Doble, M. (2018). Albumin capped carbon-gold (C-Au) nanocomposite as an optical sensor for the detection of Arsenic (III). *Optical Materials, 84*, 339–344.

Babu, P. J., Raichur, A. M., & Doble, M. (2018). Synthesis and characterization of biocompatible carbon-gold (C-Au) nanocomposites and their biomedical applications as an optical sensor for creatinine detection and cellular imaging. *Sensors and Actuators B: Chemical, 258*, 1267–1278.

Babu, P. J., Saranya, S., Raichur, A. M., & Doble, M. (2020). Design of photoluminescence point-of-care membrane strip for the detection of dopamine. *Materials Letters, 277*, 128316. https://doi.org/10.1016/j.matlet.2020.128316.

Babu, P. J., Saranya, S., Singh, Y. D., Venkataswamy, M., Raichur, A. M., & Doble, M. (2021). Photoluminescence carbon nano dots for the conductivity based optical sensing of dopamine and bioimaging applications. *Optical Materials, 117*, 111120.

Babu, P. J., & Tingirikari, J. M. R. (2023). A review on polymeric nanomaterials intervention in food industry. *Polymer Bulletin, 80*(1), 137–164. https://doi.org/10.1007/s00289-022-04104-6.

Bachmann, J., Obst, P., Knorr, L., Schmölzer, S., Fruhmann, G., Witt, G., Osswald, T., Wudy, K., & Hinrichsen, O. (2021). Cavity vat photopolymerisation for additive manufacturing of polymer-composite 3D objects. *Communications Materials, 2*(1), 107.

Bhardwaj, T., Shukla, M., Paul, C. P., & Bindra, K. S. (2019). Direct energy deposition-laser additive manufacturing of titanium-molybdenum alloy: Parametric studies, microstructure and mechanical properties. *Journal of Alloys and Compounds, 787*, 1238–1248.

Bodaghi, M., Damanpack, A. R., & Liao, W. H. (2016). Self-expanding/shrinking structures by 4D printing. *Smart Materials and Structures, 25*(10), 105034.

Bose, S., Bhattacharjee, A., Banerjee, D., Boccaccini, A. R., & Bandyopadhyay, A. (2021). Influence of random and designed porosities on 3D printed tricalcium phosphate-bioactive glass scaffolds. *Additive Manufacturing, 40*, 101895.

Bustillos, J., Montero-Zambrano, D., Loganathan, A., Boesl, B., & Agarwal, A. (2019). Stereolithography-based 3D printed photosensitive polymer/boron nitride nanoplatelets composites. *Polymer Composites, 40*(1), 379–388. https://doi.org/10.1002/pc.24662.

Cao, L., Bradford-Vialva, R. L., Eckerle, R. M., Herman, F., Jones, C., & Klosterman, D. A. (2023). Development of carbon nanotube-reinforced nickel-based nanocomposites using laser powder bed fusion. *Advanced Engineering Materials, 25*(4), 2201197. https://doi.org/10.1002/adem.202201197.

Caudill, C., Perry, J. L., Iliadis, K., Tessema, A. T., Lee, B. J., Mecham, B. S., Tian, S., & DeSimone, J. M. (2021). Transdermal vaccination via 3D-printed microneedles induces potent humoral and cellular immunity. *Proceedings of the National Academy of Sciences, 118*(39), e2102595118. https://doi.org/10.1073/pnas.2102595118.

Chin, S. Y., Dikshit, V., Meera Priyadarshini, B., & Zhang, Y. (2020). Powder-based 3D printing for the fabrication of device with micro and mesoscale features. *Micromachines, 11*(7), 658.

Colorado, H. A., Gutierrez-Velasquez, E. I., Gil, L. D., & De Camargo, I. L. (2024). Exploring the advantages and applications of nanocomposites produced via vat photopolymerization in additive manufacturing: A review. *Advanced Composites and Hybrid Materials, 7*(1), 1. https://doi.org/10.1007/s42114-023-00808-z.

Corcione, C. E., Gervaso, F., Scalera, F., Padmanabhan, S. K., Madaghiele, M., Montagna, F., Sannino, A., Licciulli, A., & Maffezzoli, A. (2019). Highly loaded hydroxyapatite microsphere/PLA porous scaffolds obtained by fused deposition modelling. *Ceramics International, 45*(2), 2803–2810.

Davoudinejad, A. (2021). *Vat photopolymerization methods in additive manufacturing. in* additive manufacturing. Elsevier, 159–181. https://www.sciencedirect.com/science/article/pii/B9780128184110000070.

Dilag, J., Chen, T., Li, S., & Bateman, S. A. (2019). Design and direct additive manufacturing of three-dimensional surface micro-structures using material jetting technologies. *Additive Manufacturing, 27*, 167–174.

Dixit, K., Vishwakarma, A., Kumar, H., Kim, K., & Sinha, N. (2024). Direct ink writing of graphene oxide reinforced 13-93B1 bioactive glass scaffolds for bone tissue engineering applications. bioRxiv, 2024–03.

Dul, S., Gutierrez, B. J. A., Pegoretti, A., Alvarez-Quintana, J., & Fambri, L. (2022). 3D printing of ABS nanocomposites. Comparison of processing and effects of multi-wall and single-wall carbon nanotubes on thermal, mechanical and electrical properties. *Journal of Materials Science & Technology, 121*, 52–66.

El-Eskandarany, M. S. (2020). *Fabrication of and consolidation of hard nanocomposite materials.* Mechanical alloying. Elsevier, 193–228. https://doi.org/10.1016/B978-0-12-818180-5.00008-X.

Elkaseer, A., Chen, K. J., Janhsen, J. C., Refle, O., Hagenmeyer, V., & Scholz, S. G. (2022). Material jetting for advanced applications: A state-of-the-art review, gaps and future directions. *Additive Manufacturing, 60*, 103270.

Erkus, H., Bedir, T., Kaya, E., Tinaz, G. B., Gunduz, O., Chifiriuc, M.-C., & Ustundag, C. B. (2023). Innovative transdermal drug delivery system based on amoxicillin-loaded gelatin methacryloyl microneedles obtained by 3D printing. *Materialia, 27*, 101700.

Farahani, R. D., Dubé, M., & Therriault, D. (2016). Three-dimensional printing of multifunctional nanocomposites: Manufacturing techniques and applications. *Advanced Materials, 28*(28), 5794–5821. https://doi.org/10.1002/adma.201506215.

Fina, F., Goyanes, A., Madla, C. M., Awad, A., Trenfield, S. J., Kuek, J. M., Patel, P., Gaisford, S., & Basit, A. W. (2018). 3D printing of drug-loaded gyroid lattices using selective laser sintering. *International Journal of Pharmaceutics, 547*(1–2), 44–52.

Galati, M. (2021). Electron beam melting process: A general overview. *Additive Manufacturing*, 277–301.

Gantenbein, S., Masania, K., Woigk, W., Sesseg, J. P., Tervoort, T. A., & Studart, A. R. (2018). Three-dimensional printing of hierarchical liquid-crystal-polymer structures. *Nature, 561*(7722), 226–230.

Garcia, E. A., Ayranci, C., & Qureshi, A. J. (2020). Material property-manufacturing process optimization for form 2 vat photo polymerization 3D Printers. *Journal of Manufacturing and Materials Processing, 4*(1), 12.

Ge, Q., Sakhaei, A. H., Lee, H., Dunn, C. K., Fang, N. X., & Dunn, M. L. (2016). Multimaterial 4D printing with tailorable shape memory polymers. *Scientific Reports, 6*(1), 31110.

Gibson, I., Rosen, D., & Stucker, B. (2015). *Directed energy deposition processes. Additive manufacturing technologies*. Springer New York, 245–268. https://doi.org/10.1007/978-1-4939-2113-3_10.

Gittard, S. D., Ovsianikov, A., Monteiro-Riviere, N. A., Lusk, J., Morel, P., Minghetti, P., Lenardi, C., Chichkov, B. N., & Narayan, R. J. (2009). Fabrication of polymer microneedles using a two-photon polymerization and micromolding process. *Journal of Diabetes Science and Technology, 3*(2), 304–311. https://doi.org/10.1177/193229680900300211.

Gladman, A., Garcia-Leiner, M., & Sauer-Budge, A. F. (2019). Emerging polymeric materials in additive manufacturing for use in biomedical applications. *AIMS Bioengineering, 6*(1) http://www.aimspress.com/fileOther/PDF/Bioengineering/bioeng-06-01-001.pdf.

Gnanasekaran, K., Heijmans, T., Van Bennekom, S., Woldhuis, H., Wijnia, S., De With, G., & Friedrich, H. (2017). 3D printing of CNT-and graphene-based conductive polymer nanocomposites by fused deposition modeling. *Applied Materials Today, 9*, 21–28.

Hagan, C. T., IV, Bloomquist, C., Warner, S., Knape, N. M., Kim, I., Foley, H., Wagner, K. T., Mecham, S., DeSimone, J., & Wang, A. Z. (2022). 3D printed drug-loaded implantable devices for intraoperative treatment of cancer. *Journal of Controlled Release, 344*, 147–156.

Hahn, V., Kiefer, P., Frenzel, T., Qu, J., Blasco, E., Barner-Kowollik, C., & Wegener, M. (2020). Rapid assembly of small materials building blocks (voxels) into large functional 3D metamaterials. *Advanced Functional Materials, 30*(26), 1907795. https://doi.org/10.1002/adfm.201907795.

Hofmann, M. (2014). 3D printing gets a boost and opportunities with polymer materials. *ACS Macro Letters, 3*(4), 382–386. https://doi.org/10.1021/mz4006556.

Htwe, Y. Z. N., & Mariatti, M. (2022). Printed graphene and hybrid conductive inks for flexible, stretchable, and wearable electronics: Progress, opportunities, and challenges. *Journal of Science: Advanced Materials and Devices, 7*(2), 100435.

Huang, J., Qin, Q., & Wang, J. (2020). A review of stereolithography: Processes and systems. *Processes, 8*(9), 1138.

Jasiuk, I., Abueidda, D. W., Kozuch, C., Pang, S., Su, F. Y., & McKittrick, J. (2018). An overview on additive manufacturing of polymers. *JOM, 70*(3), 275–283. https://doi.org/10.1007/s11837-017-2730-y.

Kamboj, N., Kazantseva, J., Rahmani, R., Rodríguez, M. A., & Hussainova, I. (2020). Selective laser sintered bio-inspired silicon-wollastonite scaffolds for bone tissue engineering. *Materials Science and Engineering: C, 116*, 111223.

Kempin, W., Franz, C., Koster, L.-C., Schneider, F., Bogdahn, M., Weitschies, W., & Seidlitz, A. (2017). Assessment of different polymers and drug loads for fused deposition modeling of drug loaded implants. *European Journal of Pharmaceutics and Biopharmaceutics, 115*, 84–93.

Khoo, Z. X., Teoh, J. E. M., Liu, Y., Chua, C. K., Yang, S., An, J., Leong, K. F., & Yeong, W. Y. (2015). 3D printing of smart materials: A review on recent progresses in 4D printing. *Virtual and Physical Prototyping, 10*(3), 103–122. https://doi.org/10.1080/17452759.2015.1097054.

Kim, K., Park, J., Suh, J., Kim, M., Jeong, Y., & Park, I. (2017). 3D printing of multiaxial force sensors using carbon nanotube (CNT)/thermoplastic polyurethane (TPU) filaments. *Sensors and Actuators A: Physical, 263*, 493–500.

Laskowska, D., Bałasz, B., Kaczorowski, W., Grabarczyk, J., Svobodova, L., Szatkiewicz, T., & Mitura, K. (2022). The DLC coating on 316L stainless steel stochastic voronoi tessellation structures obtained by binder jetting additive manufacturing for potential biomedical applications. *Coatings, 12*(10), 1373.

Lee, K.-S., Kim, R. H., Yang, D.-Y., & Park, S. H. (2008). Advances in 3D nano/ microfabrication using two-photon initiated polymerization. *Progress in Polymer Science, 33*(6), 631–681.

Li, M., Jiang, J., Liu, W., Huang, X., Wu, X., Wei, W., Zhu, H., Zhang, J., Xiao, J., & Dai, H. (2023). Bioadaptable bioactive glass-β-tricalcium phosphate scaffolds with TPMS gyroid structure by stereolithography for bone regeneration. *Journal of Materials Science & Technology, 155*, 54–65.

Ligon, S. C., Liska, R., Stampfl, J., Gurr, M., & Mülhaupt, R. (2017). Polymers for 3D printing and customized additive manufacturing. *Chemical Reviews, 117*(15), 10212–10290. https://doi.org/10.1021/acs.chemrev.7b00074.

Liu, X., Zou, B., Xing, H., & Huang, C. (2020). The preparation of ZrO2-Al2O3 composite ceramic by SLA-3D printing and sintering processing. *Ceramics International, 46*(1), 937–944.

Lo, L., Shi, H., Wan, H., Xu, Z., Tan, X., & Wang, C. (2020). Inkjet-printed soft resistive pressure sensor patch for wearable electronics applications. *Advanced Materials Technologies, 5*(1), 1900717. https://doi.org/10.1002/admt.201900717.

Luong, D. X., Subramanian, A. K., Silva, G. A. L., Yoon, J., Cofer, S., Yang, K., Owuor, P. S., Wang, T., Wang, Z., Lou, J., Ajayan, P. M., & Tour, J. M. (2018). Laminated object manufacturing of 3D-printed laser-induced graphene foams. *Advanced Materials, 30*(28), 1707416. https://doi.org/10.1002/adma.201707416.

Ma, Q., Rejab, M. R. M., Hassan, S. A., Hu, H., & Kumar, A. P. (2022). Potentiality of MWCNT on 3D-printed bio-inspired spherical-roof cubic core under quasi-static loading. *Journal of the Mechanical Behavior of Biomedical Materials, 136*, 105514.

Manapat, J. Z., Chen, Q., Ye, P., & Advincula, R. C. (2017). 3D printing of polymer nanocomposites via stereolithography. *Macromolecular Materials and Engineering, 302*(9), 1600553. https://doi.org/10.1002/mame.201600553.

Mitra, D., Thalheim, R., & Zichner, R. (2021). Inkjet printed heating elements based on nanoparticle silver ink with adjustable temperature distribution for flexible applications. *Physica Status Solidi (a), 218*(17), 2100257. https://doi.org/10.1002/pssa.202100257.

Mondal, P., Das, A., Mondal, A., Chowdhury, A. R., & Karmakar, A. (2022). Fabrication of Ti-6Al-4V porous scaffolds using selective laser melting (SLM) and mechanical compression test for biomedical applications. *Journal of The Institution of Engineers (India): Series D, 103*(1), 181–190. https://doi.org/10.1007/s40033-022-00333-1.

Mostafaei, A., Stevens, E. L., Ference, J. J., Schmidt, D. E., & Chmielus, M. (2018). Binder jetting of a complex-shaped metal partial denture framework. *Additive Manufacturing, 21*, 63–68.

El Moumen, A., Tarfaoui, M., & Lafdi, K. (2019). Additive manufacturing of polymer composites: Processing and modeling approaches. *Composites Part B: Engineering, 171*, 166–182.

Murr, L. E., Amato, K. N., Li, S. J., Tian, Y. X., Cheng, X. Y., Gaytan, S. M., Martinez, E., Shindo, P. W., Medina, F., & Wicker, R. B. (2011). Microstructure and mechanical properties of open-cellular biomaterials prototypes for total knee replacement implants fabricated by electron beam melting. *Journal of the Mechanical Behavior of Biomedical Materials, 4*(7), 1396–1411.

Muzaffar, A., Ahamed, M. B., Deshmukh, K., Faisal, M., & Pasha, S. K. (2018). Enhanced electromagnetic absorption in NiO and BaTiO3 based polyvinylidenefluorlde nanocomposites. *Materials Letters, 218*, 217–220.

Mwema, F. M., Akinlabi, E. T., & Fatoba, O. S. (2020). Visual assessment of 3D printed elements: A practical quality assessment for home-made FDM products. *Materials Today: Proceedings, 26*, 1520–1525.

Nachal, N., Moses, J. A., Karthik, P., & Anandharamakrishnan, C. (2019). Applications of 3D printing in food processing. *Food Engineering Reviews, 11*(3), 123–141. https://doi.org/10.1007/s12393-019-09199-8.

Nadgorny, M., & Ameli, A. (2018). Functional polymers and nanocomposites for 3D printing of smart structures and devices. *ACS Applied Materials & Interfaces, 10*(21), 17489–17507. https://doi.org/10.1021/acsami.8b01786.

Nadgorny, M., Xiao, Z., Chen, C., & Connal, L. A. (2016). Three-dimensional printing of pH-responsive and functional polymers on an affordable desktop printer. *ACS Applied Materials & Interfaces, 8*(42), 28946–28954. https://doi.org/10.1021/acsami.6b07388.

O'Brien, A. K., & Bowman, C. N. (2006). Impact of oxygen on photopolymerization kinetics and polymer structure. *Macromolecules, 39*(7), 2501–2506. https://doi.org/10.1021/ma051863l.

Olakanmi, E. O., Cochrane, R. F., & Dalgarno, K. W. (2015). A review on selective laser sintering/melting (SLS/SLM) of aluminium alloy powders: Processing, microstructure, and properties. *Progress in Materials Science, 74*, 401–477.

Oropallo, W., & Piegl, L. A. (2016). Ten challenges in 3D printing. *Engineering with Computers, 32*(1), 135–148. https://doi.org/10.1007/s00366-015-0407-0.

Ozan, S., Lin, J., Li, Y., Ipek, R., & Wen, C. (2015). Development of Ti–Nb–Zr alloys with high elastic admissible strain for temporary orthopedic devices. *Acta Biomaterialia, 20*, 176–187.

Pei, X., Wu, L., Lei, H., Zhou, C., Fan, H., Li, Z., Zhang, B., Sun, H., Gui, X., & Jiang, Q. (2021). Fabrication of customized Ti6AI4V heterogeneous scaffolds with selective laser melting: Optimization of the architecture for orthopedic implant applications. *Acta Biomaterialia, 126*, 485–495.

Plamadeala, C., Gosain, S. R., Hischen, F., Buchroithner, B., Puthukodan, S., Jacak, J., Bocchino, A., Whelan, D., O'Mahony, C., Baumgartner, W., & Heitz, J. (2020). Bio-inspired microneedle design for efficient drug/vaccine coating. *Biomedical Microdevices, 22*(1), 8. https://doi.org/10.1007/s10544-019-0456-z.

Quan, H., Zhang, T., Xu, H., Luo, S., Nie, J., & Zhu, X. (2020). Photo-curing 3D printing technique and its challenges. *Bioactive Materials, 5*(1), 110–115.

Quinlan, H. E., Hasan, T., Jaddou, J., & Hart, A. J. (2017). Industrial and consumer uses of additive manufacturing: A discussion of capabilities, trajectories, and challenges. *Journal of Industrial Ecology, 21*(S1). https://doi.org/10.1111/jiec.12609.

Rane, K., & Strano, M. (2019). A comprehensive review of extrusion-based additive manufacturing processes for rapid production of metallic and ceramic parts. *Advances in Manufacturing, 7*(2), 155–173. https://doi.org/10.1007/s40436-019-00253-6.

Saboori, A., Aversa, A., Marchese, G., Biamino, S., Lombardi, M., & Fino, P. (2019). Application of directed energy deposition-based additive manufacturing in repair. *Applied Sciences, 9*(16), 3316.

Safaryan, S., Slabov, V., Kopyl, S., Romanyuk, K., Bdikin, I., Vasilev, S., Zelenovskiy, P., Shur, V. Y., Uslamin, E. A., Pidko, E. A., Vinogradov, A. V., & Kholkin, A. L. (2018). Diphenylalanine-based microribbons for piezoelectric applications via inkjet printing. *ACS Applied Materials & Interfaces, 10*(12), 10543–10551. https://doi.org/10.1021/acsami.7b19668.

Sakthivel, N., Bramsch, J., Voung, P., Swink, I., Averick, S., & Vora, H. D. (2020). Investigation of 3D-printed PLA–stainless-steel polymeric composite through fused deposition modelling-based additive manufacturing process for biomedical applications. *Medical Devices & Sensors, 3*(6), e10080. https://doi.org/10.1002/mds3.10080.

Saleh Alghamdi, S., John, S., Roy Choudhury, N., & Dutta, N. K. (2021). Additive manufacturing of polymer materials: Progress, promise and challenges. *Polymers, 13*(5), 753.

Sharma, P., Babu, P. J., & Bora, U. (2012). Sapindus mukorossi aqueous fruit extract as reducing, capping and dispersing agents in synthesis of gold nanoparticles. *Micro & Nano Letters, 7*(12). https://doi.org/10.1049/mnl.2012.0684

Shen, Y., Tang, H., Huang, X., Hang, R., Zhang, X., Wang, Y., & Yao, X. (2020). DLP printing photocurable chitosan to build bio-constructs for tissue engineering. *Carbohydrate Polymers, 235*, 115970.

Shirazi, S. F. S., Gharehkhani, S., Mehrali, M., Yarmand, H., Metselaar, H. S. C., Kadri, N. A., & Osman, N. A. A. (2015). A review on powder-based additive manufacturing for tissue engineering: Selective laser sintering and inkjet 3D printing. *Science and Technology of Advanced Materials, 16*(3), 033502.

Simpson, T. W., Williams, C. B., & Hripko, M. (2017). Preparing industry for additive manufacturing and its applications: Summary & recommendations from a National Science Foundation workshop. *Additive Manufacturing, 13*, 166–178.

Singh, D. D., Mahender, T., & Reddy, A. R. (2021). Powder bed fusion process: A brief review. *Materials Today: Proceedings, 46*, 350–355.

Singh, R., Gupta, A., Tripathi, O., Srivastava, S., Singh, B., Awasthi, A., Rajput, S. K., Sonia, P., Singhal, P., & Saxena, K. K. (2020). Powder bed fusion process in additive manufacturing: An overview. *Materials Today: Proceedings, 26*, 3058–3070.

Suamte, L., Tirkey, A., Barman, J., & Babu, P. J. (2023). Various manufacturing methods and ideal properties of scaffolds for tissue engineering applications. *Smart Materials in Manufacturing, 1*, 100011.

Sun, Z., Wu, F., Gao, H., Cui, K., Xian, M., Zhong, J., Tian, Y., Fan, S., & Wu, G. (2020). A dexamethasone-eluting porous scaffold for bone regeneration fabricated by selective laser sintering. *ACS Applied Bio Materials, 3*(12), 8739–8747. https://doi.org/10.1021/acsabm.0c01126.

Sundaram, S., Kim, D. S., Baldo, M. A., Hayward, R. C., & Matusik, W. (2017). 3D-printed self-folding electronics. *ACS Applied Materials & Interfaces, 9*(37), 32290–32298. https://doi.org/10.1021/acsami.7b10443.

Tan, L. J., Zhu, W., & Zhou, K. (2020). Recent progress on polymer materials for additive manufacturing. *Advanced Functional Materials, 30*(43), 2003062. https://doi.org/10.1002/adfm.202003062.

Tang, H. P., Zhao, P., Xiang, C. S., Liu, N., & Jia, L. (2018). *Ti-6Al-4V orthopedic implants made by selective electron beam melting. Titanium in medical and dental applications*. Elsevier. 239–249. https://www.sciencedirect.com/science/article/pii/B9780128124567000111.

Tao, J., Zhang, J., Du, T., Xu, X., Deng, X., Chen, S., Liu, J., Chen, Y., Liu, X., & Xiong, M. (2019). Rapid 3D printing of functional nanoparticle enhanced conduits for effective nerve repair. *Acta Biomaterialia, 90*, 49–59.

Thompson, M. K., Moroni, G., Vaneker, T., Fadel, G., Campbell, R. I., Gibson, I., Bernard, A., Schulz, J., Graf, P., & Ahuja, B. (2016). Design for Additive Manufacturing: Trends, opportunities, considerations, and constraints. *CIRP Annals, 65*(2), 737–760.

Tran, T. S., Dutta, N. K., & Choudhury, N. R. (2020). Poly(ionic liquid)-stabilized graphene nanoinks for scalable 3d printing of graphene aerogels. *ACS Applied Nano Materials, 3*(11), 11608–11619. https://doi.org/10.1021/acsanm.0c02781.

Truby, R. L., & Lewis, J. A. (2016). Printing soft matter in three dimensions. *Nature, 540*(7633), 371–378.

Wang, J., Zhang, Y., Aghda, N. H., Pillai, A. R., Thakkar, R., Nokhodchi, A., & Maniruzzaman, M. (2021). Emerging 3D printing technologies for drug delivery devices: Current status and future perspective. *Advanced Drug Delivery Reviews, 174*, 294–316.

Weller, C., Kleer, R., & Piller, F. T. (2015). Economic implications of 3D printing: Market structure models in light of additive manufacturing revisited. *International Journal of Production Economics, 164*, 43–56.

Xu, X., Awwad, S., Diaz-Gomez, L., Alvarez-Lorenzo, C., Brocchini, S., Gaisford, S., Goyanes, A., & Basit, A. W. (2021). 3D printed punctal plugs for controlled ocular drug delivery. *Pharmaceutics, 13*(9), 1421.

Zhakeyev, A., Wang, P., Zhang, L., Shu, W., Wang, H., & Xuan, J. (2017). Additive manufacturing: Unlocking the evolution of energy materials. *Advanced Science, 4*(10), 1700187. https://doi.org/10.1002/advs.201700187.

Zhang, J., Allardyce, B. J., Rajkhowa, R., Wang, X., & Liu, X. (2021). 3D printing of silk powder by Binder Jetting technique. *Additive Manufacturing, 38*, 101820.

Zhang, L., Liu, Y., Li, S., & Hao, Y. (2018). Additive manufacturing of titanium alloys by electron beam melting: A review. *Advanced Engineering Materials, 20*(5), 1700842. https://doi.org/10.1002/adem.201700842.

Zhao, Z., Tian, X., & Song, X. (2020). Engineering materials with light: Recent progress in digital light processing based 3D printing. *Journal of Materials Chemistry C, 8*(40), 13896–13917.

Zheng, C., Jin, F., Zhao, Y., Zheng, M., Liu, J., Dong, X., Xiong, Z., Xia, Y., & Duan, X. (2020). Light-driven micron-scale 3D hydrogel actuator produced by two-photon polymerization microfabrication. *Sensors and Actuators B: Chemical, 304*, 127345.

Ziaee, M., & Crane, N. B. (2019). Binder jetting: A review of process, materials, and methods. *Additive Manufacturing, 28*, 781–801.

Environmental advancements through additive manufacturing of functional polymers and nanocomposites

11

Viswapriya Viswalingam[1], *and Dileep Kumar*[2]
[1]Department of Pharmaceutical Chemistry, Poona College of Pharmacy, Bharati Vidyapeeth (Deemed to be University), Pune, Maharashtra, India, [2]Department of Pharm Chemistry, Manipal College of Pharmaceutical Sciences, Manipal Academy of Higher Education, Manipal, Karnataka, India

11.1 Introduction

The growing demand has spurred research interest in employing polymer composites (Al Rashid, Khalid, et al., 2020). These composites offer versatile final product properties, adaptable design, various raw material options, and continuously evolving synthesis methods (Khalid et al., 2021). They are commercially accessible and cost-effective, especially for low-cost applications compared to materials like metals and ceramics (Al Rashid & Koç, 2021). They are increasingly preferred for critical applications such as aerospace, automotive, and medical devices due to their enhanced and tailored properties (Yilmaz et al., 2021). Polymer-based nanocomposites outshine traditional materials with their superior strength, weight ratio, customizable product traits, adaptable manufacturing processes, and corrosion resistance. Additionally, the recent advancements in AM have ushered in a new production era across multiple industrial sectors, including biomedical, construction, electronics, telecommunication, mechanical, and defense (Salmi, 2021). AM processes offer increased design freedom, customized part fabrication, swift production, reduced waste (sometimes elimination), minimized human error, high precision, and accuracy at low costs (Pal et al., 2011).

The emergence of 3DP processes have revolutionized the creation of polymeric materials, surpassing basic structures and typical resins. Advances in 3DP machinery, printing precision, and production speeds persist, alongside efforts to refine existing printing methods and broaden their applications (Karakurt & Lin, 2020, Patel et al., 2023; Tu et al., 2023). Although 3DP allows the advanced production of large objects, its scalability for swift mass production remains restricted. Most research in this domain focuses on crafting specific functional prototypes through 3DP (Adel et al., 2020; Berman et al., 2023; Zeng et al., 2023).

Ongoing engineering efforts aim to enhance mechanical properties, prolong the lifespan of printed objects, and eliminate structural flaws. Fresh 3DP methods are

Additive Manufacturing of Functional Polymers and Nanocomposites. DOI: https://doi.org/10.1016/B978-0-443-26516-7.00011-0
© 2025 Elsevier Ltd. All rights are reserved, including those for text and data mining, AI training, and similar technologies.

under development to address issues such as weak mechanical strength, interlayer adhesion, and filler alignment (Sahoo, 2017; Wickramasinghe et al., 2020; Wolfs et al., 2019). Nevertheless, many of these methods still necessitate additional postprocessing, leading to increased complexity, time, and expenses (Sarker, 2021).

The rapid advancement in AM technologies calls for the development of versatile materials. Although smart polymers and nanocomposites have shown promise for 3D printing, this field is still in its infancy, holding untapped potentials and novel applications (Jagadeesh et al., 2022; Li et al., 2023; Revilla-León et al., 2020). The future of this domain involves the creation of 3D printable materials that integrate multiple functionalities to produce sensors and actuators suitable for intricate environments (Nazir et al., 2023; Xu et al., 2017). The development of self-healing polymers and gels in 3DP aims to bolster mechanical properties and shelf life. However, further exploration of cross-linked covalent structures of self-healing inks is vital for expanding functional versatility and stability (Arif et al., 2022; Gasperini et al., 2014; Nadgorny et al., 2017; Wang et al., 2023).

Opportunities exist for the development of 3D printable light- and pH-responsive polymers, mechanochromic materials, and sustainable polymers. While materials from sustainable sources, biocompatible materials, and high-performance polymers offer potential in various fields, their exploration in 3DP remains limited (Hofmann, 2014; Nadgorny et al., 2016; Vogt, 2020). PNCs offer a promising solution to real-world issues and daily challenges, showing significant potential for the future. These composites are designed on the premise that size and surface area correlate with increased reactivity (Huseien, 2023; Lin et al., 2023; Vilatela & Eder, 2012). They combine polymers as the matrix and nanomaterials as nanofillers, resulting in diverse functionalities suitable for electronic, magnetic, and optical applications due to their integrated and compatible structure (Bustamante-Torres et al., 2021; Jamróz et al., 2019; Li et al., 2010).

There exist two kinds of polymers: natural and synthetic. Natural polymers, derived from sources like silk, wool, DNA, cellulose, and proteins, are often water-based (Vroman & Tighzert, 2009). In contrast, synthetic polymers like nylon, polyethylene, polyester, Teflon, and epoxy are artificially produced. A variety of inorganic nanofillers such as nanoclays, metal-oxide nanoparticles, carbon nanomaterials, and metal nanoparticles can be blended with a polymer matrix to enhance specific properties for tailored applications (Simionescu & Ivanov, 2014).

The improved properties of resulting PNCs stem from the uniform distribution of nanofillers within the polymer matrix. Yet, issues arise when nanofillers aggregate due to Van der Waals forces, leading to a decline in desired properties (Guler et al., 2024; Oliveira et al., 2018). This challenge can be mitigated by using nanofillers with modified surfaces, improving their dispersion within the polymer matrix by enhancing reactions and compatibility at the interface (Chanda et al., 2021; Pourhashem et al., 2020). Surface functionalization can be achieved through physical or chemical reactions, enabling advanced applications of PNCs (Fernández & Orozco, 2021; Niu et al., 2022).

A wide range of applications across various fields has been identified for these PNCs. Among them are stimuli-responsive polymers or smart polymers, capable of significant property changes in response to minor environmental alterations. These polymers are sensitive to triggers like temperature, light, electrical and magnetic fields, and chemicals (De, Mahata and Kim, 2022; Jingcheng et al., 2021; Wells et al., 2019).

The advantages of 3DP technologies over traditional manufacturing methods are anticipated to attract more scientists and engineers, prompting the exploration of innovative functional materials for integration with 3DP technologies, leading to endless possibilities.

11.2 What, how, and when AM

During the late 1700s, the invention of machinery like the flying shuttle and the steam engine, which replaced manual labor, brought about immense societal changes. This marked the onset of the industrial revolution, impacting nearly all the aspects of daily life (Mohajan, 2019). As this revolution progressed, society underwent all-encompassing transformations. Over the next 200 years, there was an unprecedented surge in both average income and population growth. Global per capita income increased by more than tenfold, and the world's population expanded sixfold. Such extensive changes had never been witnessed before (Perera, 2018).

Modern AM originated in the mid-1900s with Otto John Munz's 1951 patent, which essentially marked the beginnings of contemporary stereolithograph (Mancilla-De-la-Cruz et al., 2022). Munz devised a method involving selective exposure of a transparent photo emulsion in layers, gradually forming an object's cross-sectional layers. However, the resulting 3D object required manual carving or etching from the solid transparent cylinder, posing a limitation to this process (Zocca et al., 2021).

Key contributions by individuals like Hideo Kodama and Herbert also played a part in stereolithography-related techniques. These innoVations involved directing a UV laser beam to scan a layer within a photopolymer. The significant leap in commercially available AM systems occurred with Charles W. Hull's 1986 stereolithography patent, leading to the establishment of 3D Systems in 1988 and the introduction of the first commercial SLA machine. Since then, there has been a consistent and exponential increase in available systems, technologies, and materials.

AM commences with the creation of a virtual 3D CAD model, eventually converted into an STL file format. This file is then prepared for printing by configuring print orientation, setting parameters like resolution and material, and slicing it into layers. The actual printing process varies according to the technology used. Postprocessing typically involves cleaning and, at times, additional treatments such as machining or coloring.

Despite industry acceptance of the term "AM," the popular press often refers to it as "3D printing" for its wider public understanding. Recent advances have introduced new file formats for printing that include color and material information to enhance model quality. Furthermore, researchers are exploring direct printing from native CAD formats, expanding the potential applications of AM.

AM has the potential to create a similar radical shift in society, influencing every aspect of our lives. Substantial research has been dedicated to the various technical aspects of AM, covering materials, processes, applications, and management. This book provides an overview of many of these technologies. This chapter introduces AM, displays current applications, and explores upcoming and future areas of its use.

In the last three decades, AM has significantly expanded its applications. Notably, while slightly more than half of its applications involve rapid prototyping (encompassing fit and assembly, functional models, presentation models, and visual aids), close to 39% of AM usage is now dedicated to creating actual components, whether directly or indirectly. This broader scope involves generating patterns for prototype tooling, patterns for metal casting, tooling components, and directly manufacturing parts.

For instance, in the technique of powder bed fusion, finely powdered materials are layered and solidified using a heat source to create successive layers. Once each layer is fused, a new layer of powdered material is added for the subsequent layer (Soundararajan et al., 2021). Different heat sources are employed to either melt or sinter the powder. Whereas, Selective Laser Melting (SLM) and Selective Laser Sintering (SLS) methods use laser energy, Electron Beam Melting (EBM) employs an electron beam, and Selective Heat Sintering (SHS) utilizes a thermal printhead (Gonzalez-Gutierrez, 2022; Keshavamurthy et al., 2021; Leary, 2020; Nugroho et al., 2021). These powder bed fusion processes offer superior resolution compared to extrusion-based methods and can handle various materials such as polymer powders, metal alloys, and ceramics (Goodridge & Ziegelmeier, 2017).

Among these, SLS stands as the most extensively used powder bed fusion process, where 3D objects are crafted layer by layer using powdered material. Thermal energy in the form of focused laser radiation selectively bonds the material according to the object's design (Lupone et al., 2022). In contrast, SLM differs from SLS in the way material solidifies and the resulting bonding mechanisms. SLM involves melting the powdered material entirely into a liquid phase, whereas SLS incorporates partial liquefaction (Gong et al., 2021). The choice of laser sources depends on the absorbent nature of the powder material, with CO_2, Nd:YAG, short-pulse Cu-vapor, and fiber lasers being commonly employed sources of thermal energy for SLS (Baldassarre, 1982).

One distinctive advantage of powder bed fusion processes is the absence of a need for substrate, as the unsintered powder serves as support for complex or overhanging structures (Awad et al., 2021; Chadha et al., 2023; Yao et al., 2023)Additionally, this all unused material can be reused for further use in part fabrication without major alterations in its properties. However, a major downside of these processes is the necessity for postprocessing treatments to achieve improved surface finishing and heat treatment to enhance mechanical properties (Jiang et al., 2018; Liu & Shin, 2019).

On the other hand, Vat photopolymerization (VP) represents another category of AM that involves exposing photosensitive materials to controlled radiation to form layers of polymerized material. Successive layers combine to produce a complete 3D object, although the utility of this process is restricted to materials that polymerize when exposed to light (Salonitis, 2014; Shah et al., 2023; Mauriello et al., 2023). This technology works well with photopolymers and resins. Depending on the variation in the curing source, VP processes can be further categorized into Stereolithography, digital light processing, two-photon polymerization, and volumetric 3D printing (Al Rashid et al., 2021; Salas et al., 2023; Schwartz, 2022).

For example, SLA photopolymer resins primarily consist of monomers, oligomers/binders, photoinitiators, and a few additives. Monomers and oligomers play a significant role in solidifying photopolymer resin through crosslinking. Photoinitiators generate radicals that react with oligomers and monomers, facilitating crosslinks between polymer chains and cure upon exposure light. Additives might also include nanoparticles and pigmentation for enhanced resin properties and desired coloring respectively (Bagheri & Jin, 2019; Zakeri et al., 2020).

Digital light processing (DLP) similarly relies on UV rays induced polymerization of resin. The key distinguishment between these technologies lie in DLP's ability to instantly polymerize entire layers using digital micromirror devices (DMD), offering a faster printing speed compared to SLA and other competing technologies (Amini et al., 2023; Kadry et al., 2019; Wang et al., 2022). The light source used has evolved from standard lamps to light-emitting diodes, offering a broader range of wavelengths at a substantial cost. DLP's hardware setup provides higher resolution for lateral printing, while z-resolution depends on light penetration and scattering, which can be improved by incorporating light absorbers and facilitators of polymerization (Ahn et al., 2020). DLP achieves a resolution of approximately 1 μm, regardless of the layer's lateral area and complexity, while printing speed averages about 30 mm^3/s. Two-photon polymerization (2PP) is another Vat polymerization-based printing system like SLA and DLP, offering better control and higher print quality (Dias Resende et al., 2023). Utilizing near-infrared femtosecond (Fs) laser pulses, 2PP enables the direct writing of any desired 3D model into the volume of photosensitive materials transparent in the NIR and highly absorptive in the UV spectral range (Chergui, 2012).

Volumetric 3D printing directly constructs structures in three dimensional form using a rotating Vat and an index-matching resin (Madrid-Wolff et al., 2023). By projecting six distinctive images of the object taken from specific angles and alternately during rotation, the method enables the creation of objects directly in 3D, unlike other methods that build structures through 2D layers (Al Rashid et al., 2021; Bargmann et al., 2018).

VP processes provide the highest printing resolution among the discussed methods. Despite its high resolution, many other factors influence the properties of 3D-printed parts. One limitation is the printing time, where even though one layer can be quickly produced, curing that layer takes time. Furthermore, continuous refilling of the resin tank often delays the process. Finally, the mechanical and structural properties of 3D-printed parts are impacted by the degree of polymerization and the post print curing process (Agrawaal & Thompson, 2021; Park et al., 2022). In contrast, extrusion-based processes involve the extrusion of semisolid liquid onto a build surface through a nozzle, where it solidifies layer by layer to create a solid part. Most commonly used technique in this category is fused filament fabrication (FFF), also known as fused deposition modeling (FDM) (Fico et al., 2022; Tao et al., 2021). In FFF, material is provided in filamentous form, which is heated in a chamber to the required viscosity and then extruded through a nozzle onto a platform (Dey et al., 2021; Sola, 2022). FFF dominated 96% of the global AM market sales in 2016, making it the most widely used AM technology (Meyer et al., 2021). Materials such as thermoplastics, metal particle-reinforced polymers, and hydrogels have been processed using extrusion methods (Leary, 2020). Generally, FFF feedstock material comes in the form of filament spools, and a mechanism of rollers and gears feeds the filament material into a heating chamber for deposition onto the building platform. Some FFF printers offer dual extruders to enable multimaterial fabrication or 3D fabrication of support structures for overhanging designs (Bankapalli et al., 2023). Strong bonding between adjacent extruded filaments is crucial for achieving the desired strength by promoting intermolecular interaction between polymer chains. Turner et al. conducted research on this phenomenon, concluding that several parameters, including viscosity, thermal conductivity, heat capacity, and cooling rate, govern polymer fusion (Soundararajan et al., 2021). Interlaminar adhesion is another critical issue associated with FFF processes, as fabricated parts may experience delamination due to insufficient adhesion between layers under shear forces (Fallahi et al., 2023). The primary advantage of this AM technology over others is its lower cost and widespread commercial use (Attaran, 2017). Although the printing resolution is lower, FFF can handle different amorphous and crystalline materials. However, its main limitation lies in the inferior mechanical properties compared to injection molded parts due often owing to anisotropy and increased porosity (Zhuo et al., 2021; Gao et al., 2020). Optimizing the process parameters involved in FFF is an active area of research, given that various parameters affect the performance of final product.

11.3 AM in producing polymer nanocomposites

Commercially available polymers used in AM lack the necessary performance for high-demand applications. Thus, it's crucial to create functional polymer materials compatible with the aforementioned AM processes (Govender et al., 2021; Srivastava et al., 2022). Generally, these polymers lack significant mechanical, thermal, and electrical properties, such as strength, heat transfer capabilities, and conductivity. To address this, alterations involve introducing filler particles or reinforcements to attain the desired properties for functional applications.

Polymer nanocomposites are achieved by integrating nanofillers dispersed within polymer matrices. These nanoparticles, typically at a nanometer scale, serve as reinforcements, enhancing various properties. These materials have garnered attention in research due to their remarkable improvements in mechanical, electrical, thermal, or chemical characteristics, even with minimal reinforcement content (Sahoo, 2017). Compared to fiber-reinforced composites, polymer nanocomposites offer the advantage of greater homogeneity, minimizing anisotropic effects associated with unidirectional reinforcement (Vlasvelda et al., 2005; Yuvaraj et al., 2023). Several nanoparticle-based composites have been developed for AM, primarily

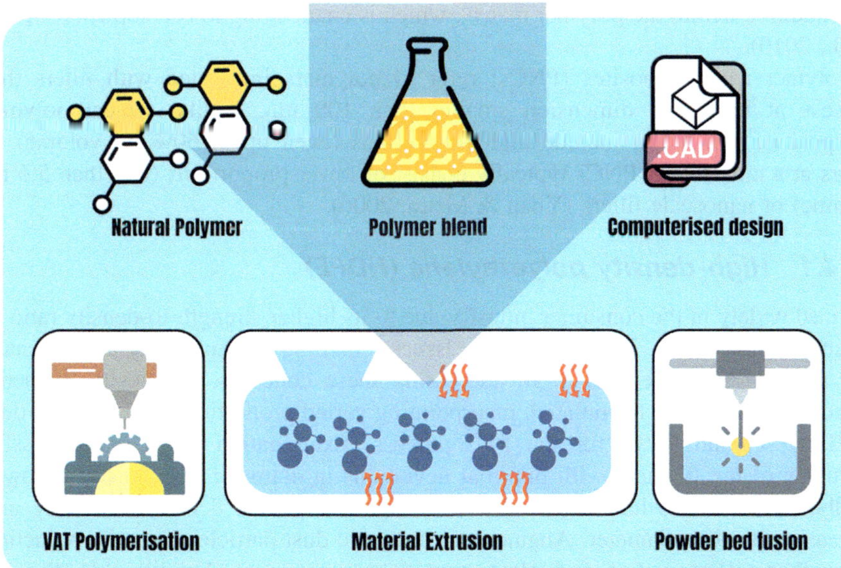

Figure 11.1 AM for nanocomposites are often carried put using VAT polymerization, material extrusion and powder bed fusion.

focusing on bolstering mechanical and electrical properties. For instance, there are reports in the literature of nanoclay for improved mechanical aspects and carbon nanotubes and graph electrical conductivity (Guadagno et al., 2023; Nurazzi et al., 2021) (Fig. 11.1).

The process of creating polymer nanocomposites (PNCs) includes blending various nanofillers into the polymer matrix (Darwish et al., 2022). PNCs can be manufactured through diverse methods, including in situ synthesis, solution blending, melt processing, electrospinning, and more (Pathak et al., 2023; Pathak et al., 2019).

11.4 Materials that make polymer nanocomposites using AM

The concept of polymer nanocomposites (PNCs) is theoretically rooted in the creation of an extensive interface between nanoscale heterogeneities and the macromolecules within a pure polymer (Huang et al., 2022). The aim is to generate exceptional properties compared to standard microfilled polymer composites by establishing a significant interface between nanoparticles and macromolecules (Hanemann & Szabó, 2010; Mishra et al., 2022). In nanocomposites, significant reinforcement and improvements in properties, such as reduced flammability and increased conductivity, are often achieved at low concentrations of nanofillers. However, these effects are greatly influenced by the uniform dispersion of the nanoadditive within the polymer matrix, which is challenging to accomplish (Šupová et al., 2010).

Polymer nanocomposites (PNCs) refer to polymers integrated with fillers that possess at least one dimension smaller than 100 nm. Unlike typical polymer composites, which can include high proportions (even up to 60% by volume) of fillers at a microscale, PNCs typically consist of lower proportions (less than 5% by volume) of nanoscale fillers (Yuan & Misra, 2006).

11.4.1 High-density polyethylene (HDPE)

Utilized widely in the consumer industry due to its higher strength-to-density ratio in comparison to LDPE, HDPE confronts issues regarding its reusability and mechanical strength (Grigore, 2017). In addressing these concerns, composite filaments based on HDPE were formulated, incorporating varied proportions of cardboard dust (20%, 50%, and 75%) for the FFF process. Examination indicated successful extrusion of the filament with minimal deviations in diameter and density at lower cardboard content, while higher filler contents resulted in decreased density and increased filament diameter. Augmented cardboard dust particles caused a reduction in mechanical properties and glass transition temperature. Despite this, the 3D printing process remained viable at these levels.

The research involved creating uniform silver nanoparticles (Ag NPs) and hybrid Fe3O4-Ag nanoparticles (FeAg NPs) through a seeding growth process using specific surfactants (Nguyen-Tri, 2019). Investigating their antibacterial properties, the study observed enhanced biological effects through electron transfer between the nanoparticles. Analyzing via nanoscale techniques revealed the significant role of the lamellar structure in the spherulite of FeAg NPs/HDPE nanocomposites in their antibacterial activity (Aguiar et al., 2021). The presence of FeAg NPs in polyethylene (PE) nanocomposites displayed stronger antibacterial effects compared to those reinforced solely with Ag NPs. This superiority was attributed to the faster release of Ag+ ions from the hybrid nanoparticles and the ionization of Ag NPs within the hybrid nanostructure.

11.4.2 Polycaprolactone (PCL)

Pierantozzi et al. presented an innovative extrusion-based method for 3D printing of composite scaffolds, aiming to investigate the use of PCL in biomedical applications. This process involved directly printing a PCL matrix embedded with a bioactive reinforcing phase, comprising HA and strontium replaced HA (SrHA), thus eliminating the need for filament production and streamlining the manufacturing process. The resulting composites, varying in reinforcing content, underwent comprehensive characterization of their physical, mechanical, and biological properties (Pierantozzi et al., 2020). Various structural designs were implemented to replicate bone tissue structures, and printing parameters were fine-tuned for diverse material configurations. Micro-CT analysis validated that the porosities of the 3D-printed samples closely matched the 3D CAD models, affirming the reliability of this introduced technique (Pei et al., 2017). The study did not establish any correlation between the mechanical properties of the nanocomposites and their specific formulations.

Developing biodegradable-bioactive nanocomposites, the research focused on utilizing baghdadite nanoparticles as a reinforcement phase. Produced via the sol-gel method, baghdadite was incorporated into PCL-based nanocomposites ranging from 0 to 30 wt.% using the solvent-casting technique. The nanosize of baghdadite particles was determined through TEM images, while crystallite size was assessed using a modified Scherrer equation (Emadi et al., 2023). Adding baghdadite to PCL proved beneficial, enhancing mechanical and thermal properties, noted through various characterizations like tensile strength, DSC, DTA, and TG. Prediction models were explored for the composite system's effective modulus, showing these analytical models as reliable for approximating mechanical properties. Additionally, the baghdadite/PCL composites demonstrated improved bioactivity and apatite formation, while degradation tests revealed increased decomposition and degradation rate compared to pure PCL films with higher baghdadite proportions. Cell culture results displayed baghdadite's positive impact on cell growth, favoring a 20 wt.% baghdadite composition due to its mechanical properties. This study suggests that baghdadite could replace existing bioactive fillers, enhancing the fabrication of improved biodegradable-bioactive bio-/nanocomposites, particularly for bone regeneration applications.

11.4.3 Polylactic acid (PLA)

Poly(lactic acid) (PLA) is highly regarded as a promising alternative to petroleum-based polymers owing to its biodegradability, biocompatibility, renewable nature, and favorable thermomechanical properties. Nevertheless, PLA faces several limitations including low heat distortion temperature, limited thermal resistance, and slow crystallization rates. Furthermore, specific properties such as flame retardancy, UV resistance, antibacterial features, barrier properties, and electrical conductivity, among others, are required for diverse applications. Addressing these drawbacks, the inclusion of various nanofillers offers an appealing avenue to improve and augment the inherent traits of pure PLA. Extensive research into different nanofillers with various structures and properties has resulted in successful advancements in the development of PLA nanocomposites.

The need to enhance PLA matrix properties, including thermal resistance and specific characteristics like flame retardancy, electrical conductivity, and antibacterial features, to meet diverse end-user demands, has led to extensive investigation of various nanofillers for PLA-based nanocomposites. These nanofillers, from natural compounds like lignin and tannin nanoparticles to materials such as MgO, ZnO, TiO_2, carbon-based fillers, and fullerenes, offer a wide array of benefits, from breakdown acceleration in packaging films to extending food shelf-life, UV protection, water purification, mechanical reinforcement, and potential use in photovoltaic cells. Despite significant research progress, challenges remain in achieving uniform dispersion within the PLA matrix and employing entirely sustainable production methods, posing obstacles for academic and industrial sectors.

Nonetheless, PLA nanocomposites are actively researched and increasingly approaching commercial viability.

Microstructural, mechanical, and compositional analyses were conducted on 3D-printed models replicating trabecular bone structures and standardized compression test samples. Backes et al.'s research highlighted improved thermomechanical properties in these composites. Increase in cell proliferation was noted with a 10% HA content. The study synthesized PLA/HA nanocomposites via melt processing, examining their microstructure, mechanical strength, and thermal characteristics. Additionally, the study prepared FFF filaments, demonstrating the potential use of PLA/HA nanocomposites in AM with high print quality and accuracy in 3D-printed models.

11.4.4 Acrylonitrile butadiene styrene (ABS)

Enhancing the mechanical properties of pure form of acrylonitrile butadiene styrene (ABS) polymer by approximately 41% was observed through the incorporation of graphene as a nanofiller (Sharma et al., 2020). Nanocomposites of ABS-graphene were produced using a solution blending method, facilitating the uniform dispersion and integration of graphene within the ABS polymer. The impact of varying weight percentages (wt%) of the nanofiller on the properties of the ABS polymer was studied. The research revealed that an increase in the weight content of Graphene (ranging from 1–6 wt%) improved dispersion, reaching an optimal limit at 5 wt%, beyond which it started to agglomerate at 6 wt%, subsequently diminishing the material's mechanical properties (Wu et al., 2019). Remarkably, the findings of field-emission scanning, electron, microscopy, Raman spectroscopy, and mechanical test results exhibited remarkable consistency (Tables 11.1 and 11.2).

11.5 Polymer nanocomposites for environmental causes

Polymeric nanocomposites (PNC) possess significant potential for diverse applications due to their integrated structure, which combines unique properties not found in any individual component materials. Additionally, their ability to be shaped and sized according to specific requirements allows for control over properties like surface area, magnetic behavior, optical characteristics, and catalytic activity. The cost-effectiveness and lightweight nature of PNCs have further enhanced their potential across a range of environmental and industrial uses. Stimuli-responsive nanocomposites, a subset of PNCs, exhibit at least one remarkable chemical or physical property that can be regulated or triggered by an external stimulus. These exceptional properties and adaptable behaviors broaden the potential applications of these nanocomposites. Over the period of past few decades, there has been significant progress in the advancement of polymeric biodegradable nanoadsorbents. These materials are increasingly seen as a potential substitute for activated carbon due to their large surface area, robust structure, varied pore sizes, favorable surface chemistry, and the

Table 11.1 Additive manufacturing processes for printed material and their functions.

S. No.	AM process	Material	Printed material	Applied functions	References
1.	Selective laser sintering	Polyamide-12	Membrane	Superhydrophobicity and superoleophobicity	Yuan et al. (2019)
2.	Immersed surface accumulation	Photocurable resin with multiwalled carbon nanotubes	Microscale artificial hairs with eggbeater heads	Biomimetic solution for oil/water separation	Yang et al. (2018)
3.	Selective laser sintering	PSU	Membrane	Superhydrophobicity	Yuan et al. (2017)
4.	Stereolithography	UV-curable resin and surface modified with octadecyltri-chlorosilane	Mesh	skimmer for oil collection during spills	Al-Shimmery et al. (2019)
5.	Multijet	Urethane acrylate oligomers and paraffin wax	Porous contactors	O/W demulsification	Pellejero et al. (2020)
6.	Fused deposition modeling	Polylactic Acid	Packing	Superhydrophobicity and superoleophobicity	Xing et al. (2019)
7.	Direct inkjet writing	Cellulose acetate polyvinyl alcohol SiO2 nanoparticles	Composite membrane	Superhydrophobicity and superoleophobicity	Chen et al. (2019)

Table 11.2 AM manufacturing processes and their applied functions.

S. No.	AM method	Advantages	Disadvantages	References
1.	VAT polymerization	Distinguished resolution	Curing causes delayed process Incompatible mechanical properties	Cullen and Price. (2019)
2.	Material extrusion	Useful for both amorphous and crystalline substance	Lowered resolution High porosity	Torrado et al. (2015)
3.	Powder bed fusion	Structural compatibility Useful for complex structures	Post heating and curing might be required	Al Rashid, Khan, et al. (2020)

ability to regenerate under mild conditions (Anjum et al., 2019). Generally, these biodegradable polymeric nanoadsorbents have demonstrated effective capabilities in adsorbing a wide array of common organic pollutants such as organic acids, phenolic compounds, aromatic or polyaromatic hydrocarbons, alkanes, and their derivatives. To further improve their ability to absorb different contaminants, like highly water soluble compounds such as sulfonated toxins and heavy metal particles, modifying or functionalizing their surfaces has proven to be an effective strategy. This alteration allows the functional groups attached to the polymeric structure to interact more efficiently with the specific pollutants.

Initially developed in the 1960s using gel permeation chromatography, polymeric adsorbents have gained recognition for their distinctive physical properties, making them a respected material for processes involving adsorption and filtration (Pan et al., 2009). Currently, they are widely employed in the removal of organic pollutants from industrial and natural waters, leveraging their vast surface areas, strong structural integrity, and adaptable physicochemical properties that can be adjusted by changing the conditions during the polymerization process.

Addressing environmental concerns, the primary polymeric nanoadsorbents rely on porous resins, polyaniline that are biodegradable, polyacrylamides, chitosan, cellulose acetate, cellulose or carboxymethyl cellulose, and alginate polymers.

11.5.1 Air purification

Air pollution is a pressing concern for developing countries, with air filtration being a primary and effective method to combat this issue. While the respiratory system can block most particulate matter during breathing, tiny PM 2.5 particles manage to pass through, posing health risks due to their size and potential to carry other harmful compounds (Xing et al., 2016). Black carbon, a common atmospheric pollutant from combustion processes, and the recent pandemic, releasing virus particles into the air, contribute to air pollution, emphasizing the need for more efficient filters (Rathod & Beig, 2021). Conventional filters with low porosity and small pore sizes can be enhanced by using nanofibers, significantly improving pollutant trapping efficiency.

To combat air pollution, industries like automotive and aerospace are seeking solutions. New materials, such as nanofibers produced by electrospinning, show promise for efficient air filtration. These nanofibers can be made from various natural and synthetic polymers and integrated into filtration systems (Rakhi et al., 2023). Efforts are directed toward materials with advanced characteristics suitable for removing pollutants and correlating their removal mechanisms with particle sizes to enhance filtration efficiency. Air filtration operates through several key mechanisms, including impaction, interception, diffusion, and electrostatic attraction.

An essential consideration lies in the sustainability and reusability of filters concerning their impact on the environment. While the creation of new materials poses challenges, the structures of nanofibers and nanocomposites hold great promise for effective filtration.

Traditional air filtration membranes, typically comprising larger-scale fibers from materials like glass, allow the passage of ultrafine particles (PM 2.5) and bacteria due to their larger pores, impacting air quality (Liu, Huang, et al., 2019). In contrast, electrospun nanofiber membranes exhibit an interconnected pore structure, resulting in manageable pores with extensive surface areas and high porosity.

The synthesis of air filtration membranes presents challenges, often involving the use of toxic organic solvents during production, which could affect the environment and pose safety risks. Another hurdle is ensuring multifunctionality to simultaneously filter various materials. Various synthetic polymers, such as polyimide, polyurethane, polyacrylonitrile, polyamide, and polysulfone, have successfully been utilized to produce nanofibrous membranes for filtration purposes.

Efforts are directed not only toward improving air filtration efficiency but also toward reducing the environmental impact of membrane production. Consequently, researchers are exploring ecofriendly materials derived from green synthesis methods. They are experimenting with green electrospun materials created from natural and biocompatible polymers dissolved in nontoxic solvents to produce fibrous membranes.

In line with ecofriendly practices, techniques like electrospinning with bio-based chitosan/poly(vinyl alcohol) nanofibers, integrating superhydrophobic silica nanoparticles, have been developed to enhance filtration efficiency (Ahmadijokani et al., 2023). Additionally, the incorporation of Ag nanoparticles into electrospun antibacterial nanofibrous membranes has demonstrated high performance, biological compatibility, and antibacterial properties against particulate matter 2.5 (Bortolassi et al., 2019; Sabarees et al., 2022). These advancements suggest potential applications for ecofriendly air filtration materials, particularly in the domain of personal air filtration.

11.5.2 Water and soil restoration

Researchers consistently seek innovative and cost-effective solutions to address heavy metal ion contamination in water. Various techniques, such as different sorting methods based on properties like size, volatility, solubility, or chemical reactivity, have been explored for efficient ion removal (Kumar et al., 2021). Polymers have emerged as a crucial class of materials for water treatment, particularly for retaining metal ions, owing to their adaptable functional features that cater to a broad range of wastewaters (Berber, 2020). Polymeric nanocomposites, integrating properties of both polymers and nanomaterials, offer enhanced performance in comparison to traditional treatment technologies, prompting a shift toward polymer use for water remediation.s

Water not only harbors heavy metals but also contains significant amounts of impactful organic pollutants like dyes, pharmaceuticals, endocrine-disrupting chemicals, personal care products, and flame retardants, influencing environmental factors. However, while environmental quality and aquatic ecosystem improvement remain primary focuses, official pollutant limits are yet to be established, leading to continuous research in this area (Fuller et al., 2022; Geissen et al., 2015; Manisalidis et al., 2020).

Additionally, polycyclic aromatic hydrocarbons (PAHs), persistent organic pollutants (POPs), such as dioxins, furans, chlorinated pesticides (OCP), and polychlorinated biphenyls (PCBs), are closely monitored in water and soil due to their widespread presence in ecosystems from human activities (Alawi et al., 2018; Guo et al., 2019). Adsorption, particularly employing nanomaterials, nanocomposites, nanoparticles, clays, biopolymers, metal-organic frameworks (MOFs), and zeolites, is one of the commonly utilized methods for organic pollutant removal from water, with a focus on their effectiveness and cost-efficiency (Dehghani et al., 2023).

Polymer materials, including natural and nanocomposite polymers, have been widely used in water decontamination (Khodakarami & Bagheri, 2021; Pandey et al., 2017). Their applications span from flocculants or coagulants to membrane systems, ensuring effective water treatment, while methods such as polymer grafting on membrane surfaces and the utilization of extracellular polymers containing polysaccharides and proteins play a significant role in metal binding (Jiang et al., 2021). Bacterial polymers derived from various species, like *Pseudomonas, Halomonas, Paenibacillus, Bacillus*, and *Herbaspirillum*, display potential for removing heavy metals, especially from industrial effluents. These polymers, which can be sourced from activated sludge, wastewater, and other origins, exhibit high efficiencies in immobilizing heavy metals (Balíková et al., 2022; Igiri et al., 2018; Matei et al., 2022).

Soil, also affected by heavy metal and radionuclide pollution, experiences severe consequences on ecosystems and human health. Remediation techniques like stabilization, excavation, bioremediation, landfill, and soil rinsing utilize materials such as biosolid compost and biochar as organic amendments for restoring soil quality (Chibuike & Obiora, 2014; Palansooriya et al., 2020). Biochar, combined with chitosan, shows promise in accelerating the degradation of polyhydroxybutyrate-co-valerate (PHBV) and its composites, indicating its potential in soil restoration (Kong et al., 2018; Mingke et al., 2020). Moreover, natural polymers like chitosan, when used for soil stabilization, prove effective due to their particle sizes, large specific surface areas, and varying surface charges (Yanat & Schroën, 2021).

Polymer usage extends to improving soil erosion, impacting soil quality, and immobilizing heavy metals through chelation and complex stabilization processes. Bacterial polymers contribute to reducing the bioavailability of Pb and Cd in contaminated soils by forming stable complexes and reducing leaching concentrations (Zulfiqar et al., 2023; Tao et al., 2016). These polymers enhance soil quality, aid in the passivation process for lead immobilization, and show potential as coagulants for replacing traditional inorganic coagulants (Oladoja, 2019).

Furthermore, natural polymer-based stabilizers, such as lignin, carboxymethyl cellulose (CMC), and succinic acid (SA), display effective chelation and immobilization abilities for reducing leaching concentrations of heavy metals like Pb and Cd (Liu et al., 2020; Rabiee et al., 2023). Moreover, microbial polymers, characterized by their high molecular weight, contribute significantly to soil quality and heavy metal immobilization (Ojuederie & Babalola, 2017; Li et al., 2022). Their applications encompass binding, cohesion, adhesion, and enhancement of the soil's mechanical properties. These polymers exhibit high efficiency in soil bioremediation, emphasizing their importance in restoring and protecting ecosystems.

In summary, polymers, particularly natural and microbial variants, play pivotal roles in water and soil remediation, from efficient heavy metal removal to enhancing soil quality and mitigating the impact of environmental pollutants. The study encompasses various techniques and materials used in addressing heavy metal ion contamination, organic pollutant removal, soil stabilization, and erosion control, highlighting their significance in environmental protection and ecosystem restoration.

11.5.3 Biocatalysis and nanocomposites

Nanoparticles possess significant potential as biocatalysts and redox active agents due to their extensive specific surface area, high reactivity, and shape-dependent optical, electronic, and catalytic properties. Researchers are focusing on designing highly efficient photo-/chem-catalytic structure to purify contaminated water and gases using various nanoparticles, including nanosized semiconductor materials such as nano-TiO_2 [97Mil, 04Kon], ZnO [04Dan, 05Lin], and CdS [98Tor, 08Dat, 09Zhu], zero-valence metals, such as Cu0 [05Lin, 09Wu], and Zn0 [08Son, 08Wan], and bimetallic nanoparticles including Fe/Pd [98Zha, 01Ell, 05Guc, 08Wan] (Armaković et al., 2023; Tamiolakis et al., 2016; Yang et al., 2022).

However, the application of aqueous suspensions poses challenges related to particle separation and catalyst recycling. To address these concerns, researchers have been immobilizing nanoparticles on polymer matrices like porous resins [01Pon, 05Lin, 09Lin], ion exchangers [08Kim], and polymeric membranes [03Ike] to reduce particle loss, prevent agglomeration, and enhance the potential for convective flow (Gómez & Hernandez, 2023; Khdary et al., 2023).

Bio- and synthetic polymer-nanoparticle composites have been extensively studied for heterogeneous catalysis. These materials have been investigated in a wide range of reactions, such as oxidation, hydrogenation, cyclic polymerization, Suzuki cross-coupling reaction, and Tsuji–Trost reaction (Michela Dell'Anna et al., 2013; Ohtaka, 2021). The choice of the polymer matrix is crucial, as it must provide sufficient permeability for target ions, molecules, or reactants to access the nanoparticles, where catalytic reactions take place (Heinz et al., 2017; Sarkar et al., 2012).

For instance, titanium dioxide nanoparticles (nano-TiO_2) have been extensively researched for degrading organic pollutants. Under illumination, nano-TiO_2 generates an electron movement to the conduction band and creates a hole in the valence band, resulting in a significant reducing and oxidizing activity (Al-Nuaim et al., 2023; Anucha et al., 2022; Li et al., 2020). Given their high photocatalytic activity, the polymeric substrates of these nanocomposite catalysts are anticipated to act as antioxidants under UV and visible light. The reported polymeric substrates usually include saturated carbon polymers or fluoropolymers, such as poly(dimethylsiloxane) (PDMS), polyvinyl pyrrolidone (PVP), polyethylene (PE), polypropylene (PP), poly(3-hexylthiophene) (P3HT), polyaniline (PANI), poly(tetrafluoroethylene), and Nafion (Liu, Ye, et al., 2019; Tennakone & Kottegoda, 1996). For instance,

Ameen et al. created a poly 1-naphthylamine (PNA)/TiO_2 nanocomposite through in situ polymerization, exhibiting enhanced photocatalytic activity in degrading methylene blue (MB) dye under visible light.

Research has demonstrated the effectiveness of nanoscale metals such as Cu0, Fe0, Zn0, Fe/Pd, etc., in destroying various organic contaminants (Kumar et al., 2022; Xu et al., 2022; Zhu et al., 2009). However, the high reactivity of these metal nanoparticles often necessitates support to prevent their oxidation and preserve their chemical nature until they can interact with the targeted pollutants (Dikshit et al., 2021; Naseem & Durrani, 2021; Yaqoob et al., 2020). Stabilizing these nanoparticles prevents aggregation, enhances their chemical reactivity, and supports their soil transport. For instance, nanoscale zero valent copper (nZVC) has been synthesized supported on a cation exchange resin, effectively preventing surface area reduction due to particle agglomeration and simultaneously recycling the copper ions produced from the reaction between CCl4 and Cu0 (Kumar et al., 2022). Additionally, composite of iron (Fe/Pd) nanoparticles with sodium carboxymethyl cellulose (CMC) as a stabilizer has been reported, demonstrating improved stability against aggregation, chemical reactivity, and soil transport (Zinchenko et al., 2022; He et al., 2006). Batch dechlorination tests revealed that the CMC-stabilized nanoparticles degraded trichloroethene (TCE) significantly faster than their nonstabilized counterparts (Rajajayavel & Ghoshal, 2015) (Fig. 11.2).

Figure 11.2 Various environmental applications of nanocomposites.

11.6 Hurdles and future prospects

The ongoing exploration of novel materials and the synthesis of advanced nanocomposites, influenced by both material science advancements and the evolution of AM (AM) processes, have substantially broadened the landscape of design possibilities (Rahman et al., 2023). This pursuit is driven by the quest for materials that can meet the demands of highly functional applications, pushing the boundaries of what is achievable (Purvis et al., 2019). While the rise of AM has been transformative, it's crucial to understand that it is not intended to entirely replace traditional manufacturing methods. Instead, it serves as a complementary force, working in tandem to enhance and diversify production capabilities (Duchek, 2020).

Embracing a design for AM (DfAM) approach is essential. It's a strategic mindset that interconnects the intricacies of the manufacturing process, the choice of materials, and the intended design objectives. By aligning these elements, industries such as aerospace, automotive, biomedical, and electrical sectors can benefit immensely. This holistic approach paves the way for the creation of parts and products that are not only technologically advanced but also efficiently manufactured (Kumke et al., 2016; Verma et al., 2023).

Looking toward the future, the key areas for further exploration and development are critical. The continuous development of novel materials remains a high priority. Particularly, there is a need to delve deeper into polymers and biocomposites, expanding the existing repertoire of materials suitable for AM (Chafiq et al., 2023; Criado-Gonzalez et al., 2021; Razzaq et al., 2023). Understanding the relationships between material properties, manufacturing processes, and the final product's characteristics is pivotal in expanding the scope and compatibility of AM processes.

One of the most substantial hurdles lies in the optimization and modeling of AM-produced parts. The complexity of multiple printing parameters influencing the quality of the final product necessitates advanced tools and methodologies for precise prediction and optimization of properties (Badini et al., 2023; Bouzaglou et al., 2023; Pepelnjak et al., 2023; Rouf et al., 2022). This step is essential in tailoring functional parts for specific applications, ensuring their efficiency and reliability.

In essence, the synergy between materials, AM processes, and design methodologies holds immense potential for driving innovation across industries. It not only expands the horizons of what can be created but also enables the production of highly functional and tailored solutions for various applications, setting the stage for a new era of manufacturing and design excellence (Berggren et al., 2020; Fornasiero et al., 2021).

On a commercial scale, the manufacturing of polymer nanocomposites poses a significant challenge due to the intricate fabrication procedures involved (Pfaendner, 2010). Besides the complexity of fabrication, the selection of appropriate techniques and operational conditions is crucial. Numerous parameters must be meticulously defined, including the stability and dispersion of nanofillers within the polymer matrix, interactions between the nanofillers and the polymer matrix, surface charge, polymer chain flexibility, surface chemistry, and the crystallization of nanofillers (Goh et al., 2020; Tanahashi, 2010). Additionally, maintaining particle sizes at the nanoscale is challenging due to the tendency for aggregation over time. This aggregation issue is compounded by gravitational forces, which are difficult to counteract. While the addition of surfactants or stabilizers can help sustain nanofiller dispersion stability, their inclusion may compromise the inherent properties of the nanomaterials (Miyazawa et al., 2021, Yasser, 2016). Techniques such as coagulation or precipitation can also modify the characteristics of nanocomposites, diminishing their exceptional features.

Regarding the healthcare applications of polymer nanocomposites, their toxicity profile emerges as a primary concern (Agboola et al., 2021; Ghazzy et al., 2023). The size of nanoparticles presents a significant challenge in clinical and in vivo use. The ability of nanoparticles to easily traverse the blood-brain barrier raises concerns about their potential to access sensitive organs within the human body (De Jong & Borm, 2008; Mitchell et al., 2021; Patra et al., 2018). There is a pressing need for toxicological assessments, such as quantifying the release of free nanoparticles from polymer nanocomposites, which has been the focus of several research endeavors. For instance, Schlagenhauf et al. (2014) introduced a measurement technique to quantify the amount of carbon nanotubes (CNTs) released from CNT/epoxy nanocomposites. This quantification method involves various steps, including prelabeling CNTs with lead ions, fabricating CNT/epoxy nanocomposites, abrasion to collect freely available nanoparticles, and quantifying these nanoparticles by measuring the lead ion concentration (Schlagenhauf et al., 2015). In vitro toxicity studies using THP-1 monocyte-derived macrophages and A549 human alveolar epithelial cells revealed that CNT release was lower in cases where agglomerated CNTs were dispersed in the polymer matrix. While polymer nanocomposites show promising potential for healthcare applications, thorough research to understand their toxicity and adversity is essential before advancing to clinical trials.

11.7 Conclusion

As discussed, these applications necessitate the production of stable nanoparticles capable of being functionalized and demonstrating biocompatible characteristics. The primary challenge faced by the research community in this domain is the successful creation of particles with specific functionality suitable for industrial and environmental purposes. Polymer nanocomposites play a pivotal role in the advancement of material science. The impressive physical and chemical properties of these PNC components offer the potential to expand intelligent functionalities in material applications. Nevertheless, there remains a considerable scope for exploration within these systems. Effectively managing the intricate and pivotal variables would further drive the development of these materials and ensure their potential for scalability.

While various preparation methods have been documented for polymer nanocomposites, further investigation in this area is warranted. This is because the inherent properties of nanofillers and achieving their uniform dispersion within the polymer matrix continue to present significant challenges. The preparation technique and the types of nanofillers used significantly influence the suitability of a nanocomposite for a specific application. Polymer nanocomposites find applicability in diverse areas such as water treatment, sensors, electromagnetic shielding, and food packaging. This is due to the enhanced properties displayed by polymer nanocomposites as compared to the natural form of polymer.

References

Adel, N., Mahmoud, A., & Maged, A. (2020). 3D printing: An appealing route for customized drug delivery systems. *International Journal of Pharmaceutics, 588*, 119732. https://doi.org/10.1016/j.ijpharm.2020.119732.

Agboola, O., Fayomi, O. S. I., Ayodeji, A., Ayeni, A. O., Alagbe, E. E., Sanni, S. E., Okoro, E. E., Moropeng, L., Sadiku, R., Kupolati, K. W., & Oni, B. A. (2021). A review on polymer nanocomposites and their effective applications in membranes and adsorbents for water treatment and gas separation. *Membranes, 11*(2), 1–33. https://doi.org/10.3390/membranes11020139. https://www.mdpi.com/2077-0375/11/2/139/pdf.

Agrawaal, H., & Thompson, J. E. (2021). Additive manufacturing (3D printing) for analytical chemistry. *Talanta Open, 3*, 100036. https://doi.org/10.1016/j.talo.2021.100036.

Aguiar, R., Miller, R. E., & Petel, O. E. (2021). Microstructural evidence of the toughening mechanisms of polyurethane reinforced with halloysite nanotubes under high strain-rate tensile loading. *Scientific Reports, 11*(1). https://doi.org/10.1038/s41598-021-92663-5. http://www.nature.com/srep/index.html.

Ahmadijokani, F., Molavi, H., Bahi, A., Wuttke, S., Kamkar, M., Rojas, O. J., Ko, F., & Arjmand, M. (2023). Electrospun nanofibers of chitosan/polyvinyl alcohol/UiO-66/nanodiamond: Versatile adsorbents for wastewater remediation and organic dye removal. *Chemical Engineering Journal, 457*, 141176. https://doi.org/10.1016/j.cej.2022.141176.

Ahn, D., Stevens, L. M., Zhou, K., & Page, Z. A. (2020). Rapid high-resolution visible light 3D printing. *ACS Central Science, 6*(9), 1555–1563. https://doi.org/10.1021/acscentsci.0c00929. http://pubs.acs.org/journal/acscii.

Al Rashid, A., & Koç, M. (2021). Creep and recovery behavior of continuous fiber-reinforced 3DP composites. *Polymers, 13*(10). https://doi.org/10.3390/polym13101644 Article 10.

Al Rashid, A., Khalid, M. Y., Imran, R., Ali, U., & Koc, M. (2020). Utilization of banana fiber-reinforced hybrid composites in the sports industry. *Materials, 13*(14)https://doi.org/10.3390/ma13143167 Article 14.

Al Rashid, A., Khan, S. A., G. Al-Ghamdi, S., & Koç, M. (2020). Additive manufacturing: Technology, applications, markets, and opportunities for the built environment. *Automation in Construction, 118*, 103268. https://doi.org/10.1016/j.autcon.2020.103268.

Al Rashid, A., Ahmed, W., Khalid, M. Y., & Koç, M. (2021). Vat photopolymerization of polymers and polymer composites: Processes and applications. *Additive Manufacturing, 47*, 102279. https://doi.org/10.1016/j.addma.2021.102279.

Al-Nuaim, M. A., Alwasiti, A. A., & Shnain, Z. Y. (2023). The photocatalytic process in the treatment of polluted water. *Chemical Papers, 77*(2), 677–701. https://doi.org/10.1007/s11696-022-02468-7. https://www.springer.com/journal/11696.

Al-Shimmery, A., Mazinani, S., Flynn, J., Chew, J., & Mattia, D. (2019). 3D printed porous contactors for enhanced oil droplet coalescence. *Journal of Membrane Science, 590*, 117274. https://doi.org/10.1016/j.memsci.2019.117274.

Alawi, M., Masaad, M., & Al-Hussaini, M. (2018). Comparative study of persistent organic pollutant (POP) (chlorinated pesticides, PCBs, and dioxins/furans) concentrations in cancer-affected human organs with those of healthy organs. *Environmental Monitoring and Assessment, 190*(8). https://doi.org/10.1007/s10661-018-6856-3.

Amini, A., Guijt, R. M., Themelis, T., De Vos, J., & Eeltink, S. (2023). Recent developments in digital light processing 3D-printing techniques for microfluidic analytical devices. *Journal of Chromatography, A, 1692*, 463842. https://doi.org/10.1016/j.chroma.2023.463842.

Anjum, M., Miandad, R., Waqas, M., Gehany, F., & Barakat, M. A. (2019). Remediation of wastewater using various nano-materials. *Arabian Journal of Chemistry, 12*(8), 4897–4919. https://doi.org/10.1016/j.arabjc.2016.10.004.

Anucha, C. B., Altin, I., Bacaksiz, E., & Stathopoulos, V. N. (2022). Titanium dioxide (TiO_2)-based photocatalyst materials activity enhancement for contaminants of emerging concern (CECs) degradation: In the light of modification strategies. *Chemical Engineering Journal Advances, 10.* https://doi.org/10.1016/j.ceja.2022.100262. https://www.journals.elsevier.com/chemical-engineering-journal-advances.

Arif, Z. U., Khalid, M. Y., Zolfagharian, A., & Bodaghi, M. (2022). 4D bioprinting of smart polymers for biomedical applications: Recent progress, challenges, and future perspectives. *Reactive and Functional Polymers, 179.* https://doi.org/10.1016/j.reactfunctpolym.2022.105374. https://www.journals.elsevier.com/reactive-and-functional-polymers.

Armaković, S. J., Savanović, M. M., & Armaković, S. (2023). Titanium dioxide as the most used photocatalyst for water purification: An overview. *Catalysts, 13*(1). https://doi.org/10.3390/catal13010026. https://www.mdpi.com/journal/catalysts.

Attaran, M. (2017). The rise of 3-D printing: The advantages of additive manufacturing over traditional manufacturing. *Business Horizons, 60*(5), 677–688. https://doi.org/10.1016/j.bushor.2017.05.011.

Awad, A., Fina, F., Goyanes, A., Gaisford, S., & Basit, A. W. (2021). Advances in powder bed fusion 3D printing in drug delivery and healthcare. *Advanced Drug Delivery Reviews, 174*, 406–424. https://doi.org/10.1016/j.addr.2021.04.025. http://www.elsevier.com/locate/drugdeliv.

Badini, S., Regondi, S., Frontoni, E., & Pugliese, R. (2023). Assessing the capabilities of ChatGPT to improve additive manufacturing troubleshooting. *Advanced Industrial and Engineering Polymer Research, 6*(3), 278–287. https://doi.org/10.1016/j.aiepr.2023.03.003.

Bagheri, A., & Jin, J. (2019). Photopolymerization in 3D Printing. *ACS Applied Polymer Materials, 1*(4), 593–611. https://doi.org/10.1021/acsapm.8b00165.

Baldassarre, L. (1982). Effetti termici della radiazione dei laser Nd:YAG e CO2 sui tessuti biologici. *Bollettino Della Societa Italiana di Biologia Sperimentale, 58*(6), 320–326.

Balíková, K., Vojtková, H., Duborská, E., Kim, H., Matúš, P., & Urík, M. (2022). Role of exopolysaccharides of Pseudomonas in heavy metal removal and other remediation strategies. *Polymers, 14*(20), 4253. https://doi.org/10.3390/polym14204253.

Bankapalli, N. K., Gupta, V., Saxena, P., Bajpai, A., Lahoda, C., & Polte, J. (2023). Filament fabrication and subsequent additive manufacturing, debinding, and sintering for extrusion-based metal additive manufacturing and their applications: A review. *Composites Part B: Engineering, 264*, 110915. https://doi.org/10.1016/j.compositesb.2023.110915.

Bargmann, S., Klusemann, B., Markmann, J., Schnabel, J. E., Schneider, K., Soyarslan, C., & Wilmers, J. (2018). Generation of 3D representative volume elements for heterogeneous materials: A review. *Progress in Materials Science, 96*, 322–384. https://doi.org/10.1016/j.pmatsci.2018.02.003.

Berber, M. R. (2020). Current advances of polymer composites for water treatment and desalination. *Journal of Chemistry, 2020.* https://doi.org/10.1155/2020/7608423. http://www.hindawi.com/journals/chem/contents/.

Berggren, K., Xia, Q., Likharev, K. K., Strukov, D. B., Jiang, H., Mikolajick, T., Querlioz, D., Salinga, M., Erickson, J. R., Pi, S., Xiong, F., Lin, P., Li, C., Chen, Y., Xiong, S., Hoskins, B. D., Daniels, M. W., Madhavan, A., Liddle, J. A., ... Raychowdhury, A. (2020). Roadmap on emerging hardware and technology for machine learning. *Nanotechnology, 32*(1). https://doi.org/10.1088/1361-6528/aba70f. https://iopscience.iop.org/article/10.1088/1361-6528/aba70f.

Berman, O., Weizman, M., Oren, A., Neri, R., Parnas, H., Shashar, N., & Tarazi, E. (2023). Design and application of a novel 3D printing method for bio-inspired artificial reefs. *Ecological Engineering, 188*, 106892. https://doi.org/10.1016/j.ecoleng.2023.106892.

Bortolassi, A. C. C., Nagarajan, S., de Araújo Lima, B., Guerra, V. G., Aguiar, M. L., Huon, V., Soussan, L., Cornu, D., Miele, P., & Bechelany, M. (2019). Efficient nanoparticles removal and bactericidal action of electrospun nanofibers membranes for air filtration. *Materials Science and Engineering C, 102*, 718–729. https://doi.org/10.1016/j.msec.2019.04.094.

Bouzaglou, O., Golan, O., & Lachman, N. (2023). Process design and parameters interaction in material extrusion 3d printing: A review. *Polymers, 15*(10), 2280. https://doi.org/10.3390/polym15102280.

Bustamante-Torres, M., Romero-Fierro, D., Arcentales-Vera, B., Pardo, S., & Bucio, E. (2021). Interaction between filler and polymeric matrix in nanocomposites: Magnetic approach and applications. *Polymers, 13*(17), 2998. https://doi.org/10.3390/polym13172998.

Chadha, U., Selvaraj, S. K., Abraham, A. S., Khanna, M., Mishra, A., Sachdeva, I., Kashyap, S., Dev, S. J., Swatish, R. S., Joshi, A., Anand, S. K., Adefris, A., Lokesh Kumar, R., Kaliappan, J., & Dhanalakshmi, S. (2023). Powder bed fusion via machine learning-enabled approaches. *Complexity, 2023*. https://doi.org/10.1155/2023/9481790. https://www.hindawi.com/journals/complexity/.

Chafiq, M., Chaouiki, A., & Ko, Y. G. (2023). Recent advances in multifunctional reticular framework nanoparticles: A paradigm shift in materials science road to a structured future. *Nano-Micro Letters, 15*(1). https://doi.org/10.1007/s40820-023-01180-9. https://www.springer.com/journal/40820.

Chanda, S., Bajwa, D. S., Holt, G. A., Stark, N., Bajwa, S. G., & Quadir, M. (2021). Silane compatibilzation to improve the dispersion, thermal and mechancial properties of cellulose nanocrystals in poly (ethylene oxide). *Nanocomposites, 7*(1), 87–96. https://doi.org/10.1080/20550324.2021.1942641. http://www.tandfonline.com/ynan.

Chen, Q., Zhao, J., Ren, J., Rong, L., Cao, P. F., & Advincula, R. C. (2019). 3D printed multifunctional, hyperelastic silicone rubber foam. *Advanced Functional Materials, 29*(23). https://doi.org/10.1002/adfm.201900469. http://onlinelibrary.wiley.com/journal/10.1002/(ISSN)1616-3028.

Chergui, M. (2012). Ultrafast structural dynamics of biological systems. *Comprehensive Biophysics, 1*, 398–424. https://doi.org/10.1016/B978-0-12-374920-8.00123-5. http://www.sciencedirect.com/science/referenceworks/9780080957180.

Chibuike, G. U., & Obiora, S. C. (2014). Heavy metal polluted soils: Effect on plants and bioremediation methods. *Applied and Environmental Soil Science, 2014*, 1–12. https://doi.org/10.1155/2014/752708.

Criado-Gonzalez, M., Dominguez-Alfaro, A., Lopez-Larrea, N., Alegret, N., & Mecerreyes, D. (2021). Additive manufacturing of conducting polymers: Recent advances, challenges, and opportunities. *ACS Applied Polymer Materials, 3*(6), 2865–2883. https://doi.org/10.1021/acsapm.1c00252.

Cullen, A. T., & Price, A. D. (2019). Fabrication of 3D conjugated polymer structures via vat polymerization additive manufacturing. *Smart Materials and Structures, 28*(10). https://

doi.org/10.1088/1361-665X/ab35f1. https://iopscience.iop.org/article/10.1088/1361-665X/ab35f1/pdf.

Darwish, M. S. A., Mostafa, M. H., & Al-Harbi, L. M. (2022). Polymeric nanocomposites for environmental and industrial applications. *MDPI, Egypt International Journal of Molecular Sciences, 23*(3). https://doi.org/10.3390/ijms23031023. https://www.mdpi.com/1422-0067/23/3/1023/pdf.

Dehghani, M. H., Ahmadi, S., Ghosh, S., Othmani, A., Osagie, C., Meskini, M., AlKafaas, S. S., Malloum, A., Khanday, W. A., Jacob, A. O., Gökkuş, Ö., Oroke, A., Martins Chineme, O., Karri, R. R., & Lima, E. C. (2023). Recent advances on sustainable adsorbents for the remediation of noxious pollutants from water and wastewater: A critical review. *Arabian Journal of Chemistry, 16*(12). https://doi.org/10.1016/j.arabjc.2023.105303. http://colleges.ksu.edu.sa/Arabic%20Colleges/CollegeOfScience/ChemicalDept/AJC/default.aspx (ScienceDirect http://www.sciencedirect.com/science/journal/18785352).

Dey, A., Nanjin Roan Eagle, I., & Yodo, N. (2021). A Review on filament materials for fused filament fabrication. *Journal of Manufacturing and Materials Processing, 5*(3), 69. https://doi.org/10.3390/jmmp5030069.

Dias Resende, C. C., Quirino Barbosa, T. A., Moura, G. F., Piola Rizzante, F. A., Mendonça, G., Zancopé, K., & Domingues das Neves, F. (2023). Cost and effectiveness of 3-dimensionally printed model using three different printing layer parameters and two resins. *Journal of Prosthetic Dentistry, 129*(2), 350–353. https://doi.org/10.1016/j.prosdent.2021.06.006. http://www.sciencedirect.com/science/journal/00223913.

Dikshit, P. K., Kumar, J., Das, A. K., Sadhu, S., Sharma, S., Singh, S., Gupta, P. K., & Kim, B. S. (2021). Green synthesis of metallic nanoparticles: Applications and limitations. *Catalysts, 11*(8). https://doi.org/10.3390/catal11080902. https://www.mdpi.com/2073-4344/11/8/902/pdf.

Duchek, S. (2020). Organizational resilience: A capability-based conceptualization. *Business Research, 13*(1), 215–246. https://doi.org/10.1007/s40685-019-0085-7.

Emadi, H., Karevan, M., Masoudi Rad, M., Sadeghzade, S., Pahlevanzadeh, F., Khodaei, M., Khayatzadeh, S., & Lotfian, S. (2023). Bioactive and biodegradable polycaprolactone-based nanocomposite for bone repair applications. *Polymers, 15*(17), 3617. https://doi.org/10.3390/polym15173617.

Fallahi, H., Kaynan, O., & Asadi, A. (2023). Insights into the effect of fiber–matrix interphase physiochemical- mechanical properties on delamination resistance and fracture toughness of hybrid composites. *Composites Part A: Applied Science and Manufacturing, 166*, 107390. https://doi.org/10.1016/j.compositesa.2022.107390.

Fernández, M., & Orozco, J. (2021). Advances in functionalized photosensitive polymeric nanocarriers. *Polymers, 13*(15), 2464. https://doi.org/10.3390/polym13152464.

Fico, D., Rizzo, D., Casciaro, R., & Esposito Corcione, C. (2022). A Review of polymer-based materials for fused filament fabrication (FFF): Focus on sustainability and recycled materials. *Polymers, 14*(3), 465. https://doi.org/10.3390/polym14030465.

Fornasiero, R., Marchiori, I., Pessot, E., Zangiacomi, A., Sardesai, S., Barros, A. C., Thanous, E., Weerdmeester, R., & Muerza, V. (2021). *Paths to innovation in supply chains: The landscape of future research*. Springer Science and Business Media LLC169–233. https://doi.org/10.1007/978-3-030-63505-3_8.

Fuller, R., Landrigan, P. J., Balakrishnan, K., Bathan, G., Bose-O'Reilly, S., Brauer, M., Caravanos, J., Chiles, T., Cohen, A., Corra, L., Cropper, M., Ferraro, G., Hanna, J., Hanrahan, D., Hu, H., Hunter, D., Janata, G., Kupka, R., Lanphear, B., ... Yan, C. (2022). Pollution and health: A progress update. *The Lancet Planetary Health, 6*(6), e535. https://

doi.org/10.1016/S2542-5196(22)00090-0. https://www.journals.elsevier.com/the-lancet-planetary-health.

Gao, X., Qi, S., Kuang, X., Su, Y., Li, J., & Wang, D. (2020). Fused filament fabrication of polymer materials: A review of interlayer bond. *Additive Manufacturing, 37*, 101658. https://doi.org/10.1016/j.addma.2020.101658.

Gasperini, L., Mano, J. F., & Reis, R. L. (2014). Natural polymers for the microencapsulation of cells. *Journal of the Royal Society Interface, 11*(100). https://doi.org/10.1098/rsif.2014.0817. http://rsif.royalsocietypublishing.org/content/11/100/20140817.full.pdf+html.

Geissen, V., Mol, H., Klumpp, E., Umlauf, G., Nadal, M., van der Ploeg, M., van de Zee, S. E. A. T. M., & Ritsema, C. J. (2015). Emerging pollutants in the environment: A challenge for water resource management. *Netherlands International Soil and Water Conservation Research, 3*(1), 57–65. https://doi.org/10.1016/j.iswcr.2015.03.002. http://www.journals.elsevier.com/international-soil-and-water-conservation-research/.

Ghazzy, A., Naik, R. R., & Shakya, A. K. (2023). Metal–polymer nanocomposites: A promising approach to antibacterial materials. *Polymers, 15*(9)https://doi.org/10.3390/polym15092167http://www.mdpi.com/journal/polymers.

Goh, P. S., Wong, K. C., Yogarathinam, L. T., Ismail, A. F., Abdullah, M. S., & Ng, B. C. (2020). Surface modifications of nanofillers for carbon dioxide separation nanocomposite membrane. *Symmetry, 12*(7). https://doi.org/10.3390/sym12071102. https://res.mdpi.com/d_attachment/symmetry/symmetry-12-01102/article_deploy/symmetry-12-01102.pdf.

Gong, G., Ye, J., Chi, Y., Zhao, Z., Wang, Z., Xia, G., Du, X., Tian, H., Yu, H., & Chen, C. (2021). Research status of laser additive manufacturing for metal: Aa review. *Journal of Materials Research and Technology, 15*, 855–884. https://doi.org/10.1016/j.jmrt.2021.08.050.

Gonzalez-Gutierrez, J. (2022). Indirect additive manufacturing techniques for metal parts: Binder-based additive manufacturing techniques. In F. G. Caballero (Ed.). *Encyclopedia of materials: Metals and alloys* (pp. 319–329). Elsevier. https://doi.org/10.1016/B978-0-12-819726-4.00100-9.

Goodridge, R., & Ziegelmeier, S. (2017). *Powder bed fusion of polymers*. Elsevier BV181–204. https://doi.org/10.1016/b978-0-08-100433-3.00007-5.

Govender, R., Ofosu Kissi, E., Larsson, A., & Tho, I. (2021). Polymers in pharmaceutical additive manufacturing: A balancing act between printability and product performance. *Advanced Drug Delivery Reviews, 177*, 113923. https://doi.org/10.1016/j.addr.2021.113923.

Grigore, M. E. (2017). Methods of recycling, properties and applications of recycled thermoplastic polymers. *Romania Recycling, 2*(4). https://doi.org/10.3390/recycling2040024. https://www.mdpi.com/2313-4321/2/4/24/pdf.

Guadagno, L., Naddeo, C., Sorrentino, A., & Raimondo, M. (2023). Thermo-mechanical performance of epoxy hybrid system based on carbon nanotubes and graphene nanoparticles. *Nanomaterials, 13*(17), 2427. https://doi.org/10.3390/nano13172427.

Guler, S. H., Simsek, T., Guler, O., & Dikici, B. (2024). Possible interaction of PVC with micro-and nano-fillers. *Engineering Materials, 1657*, 335–360. https://doi.org/10.1007/978-3-031-45375-5_16. https://www.springer.com/series/4288.

Guo, W., Pan, B., Sakkiah, S., Yavas, G., Ge, W., Zou, W., Tong, W., & Hong, H. (2019). Persistent organic pollutants in food: contamination sources, health effects and detection methods. *International Journal of Environmental Research and Public Health, 16*(22), 4361. https://doi.org/10.3390/ijerph16224361.

Gómez, E. E. R., & Hernandez, J. H. M. (2023). Obtaining electrospun membranes of chitosan/PVA and TiO_2 as a solid polymer electrolyte with potential application in ion exchange membranes. *Membranes, 13*(11). https://doi.org/10.3390/membranes13110862. http://www.mdpi.com/journal/membranes.

Hanemann, T., & Szabó, D. V. (2010). Polymer-nanoparticle composites: From synthesis to modern applications. *Materials, 3*(6), 3468–3517. https://doi.org/10.3390/ma3063468Germany. http://www.mdpi.com/1996-1944/3/6/3468/pdf.

He, F., Zhao, D., Liu, J., & Roberts, C. (2006). Stabilization of Fe–Pd nanoparticles with sodium carboxymethyl cellulose for enhanced transport and dechlorination of trichloroethylene in soil and groundwater. *Industrial & Engineering Chemistry Research, 46*. https://doi.org/10.1021/ie0610896.

Heinz, H., Pramanik, C., Heinz, O., Ding, Y., Mishra, R. K., Marchon, D., Flatt, R. J., Estrela-Lopis, I., Llop, J., Moya, S., & Ziolo, R. F. (2017). Nanoparticle decoration with surfactants: Molecular interactions, assembly, and applications. *Surface Science Reports, 72*(1), 1–58. https://doi.org/10.1016/j.surfrep.2017.02.001. http://www.sciencedirect.com/science/journal/01675729.

Hofmann, M. (2014). 3D printing gets a boost and opportunities with polymer materials. *ACS Macro Letters, 3*(4), 382–386. https://doi.org/10.1021/mz4006556 HTTP.

Huang, J., Zhou, J., & Liu, M. (2022). Interphase in polymer nanocomposites. *JACS Au, 2*(2), 280–291. https://doi.org/10.1021/jacsau.1c00430.

Huseien, G. F. (2023). Potential applications of core-shell nanoparticles in construction industry revisited. *Applied Nano, 4*(2), 75–114. https://doi.org/10.3390/applnano4020006.

Igiri, B. E., Okoduwa, S. I. R., Idoko, G. O., Akabuogu, E. P., Adeyi, A. O., & Ejiogu, I. K. (2018). Toxicity and bioremediation of heavy metals contaminated ecosystem from tannery wastewater: A review. *Journal of Toxicology, 2018*. https://doi.org/10.1155/2018/2568038. http://www.hindawi.com/journals/jt/.

Jagadeesh, P., Puttegowda, M., Rangappa, S. M., Alexey, K., Gorbatyuk, S., Khan, A., Doddamani, M., & Siengchin, S. (2022). A comprehensive review on 3D printing advancements in polymer composites: Technologies, materials, and applications. *International Journal of Advanced Manufacturing Technology, 121*(1-2), 127–169. https://doi.org/10.1007/s00170-022-09406-7. https://www.springer.com/journal/170.

Jamróz, E., Kulawik, P., & Kopel, P. (2019). The effect of nanofillers on the functional properties of biopolymer-based films: A review. *Polymers, 11*(4), 675. https://doi.org/10.3390/polym11040675.

Jiang, J., Xu, X., & Stringer, J. (2018). Support structures for additive manufacturing: A review. *Journal of Manufacturing and Materials Processing, 2*(4). https://doi.org/10.3390/jmmp2040064 Article 4.

Jiang, X., Li, Y., Tang, X., Jiang, J., He, Q., Xiong, Z., & Zheng, H. (2021). Biopolymer-based flocculants: A review of recent technologies. *Environmental Science and Pollution Research, 28*(34), 46934–46963. https://doi.org/10.1007/s11356-021-15299-y.

Jingcheng, L., Reddy, V. S., Jayathilaka, W. A. D. M., Chinnappan, A., Ramakrishna, S., & Ghosh, R. (2021). Intelligent polymers, fibers and applications. *Polymers, 13*(9). https://doi.org/10.3390/polym13091427. https://www.mdpi.com/2073-4360/13/9/1427/pdf.

Kadry, H., Wadnap, S., Xu, C., & Ahsan, F. (2019). Digital light processing (DLP) 3D-printing technology and photoreactive polymers in fabrication of modified-release tablets. *European Journal of Pharmaceutical Sciences, 135*, 60–67. https://doi.org/10.1016/j.ejps.2019.05.008.

Karakurt, I., & Lin, L. (2020). 3D printing technologies: Techniques, materials, and post-processing. *Current Opinion in Chemical Engineering, 28*, 134–143. https://doi.org/10.1016/j.coche.2020.04.001.

Keshavamurthy, R., Tambrallimath, V., & Saravanabavan, D. (2021). Development of polymer composites by additive manufacturing process. *Encyclopedia of Materials: Composites, 2*, 804–814. https://doi.org/10.1016/B978-0-12-803581-8.11885-5. https://www.sciencedirect.com/book/9780128197318.

Khalid, M. Y., Rashid, A. A., Arif, Z. U., Akram, N., Arshad, H., & Márquez, F. P. G. (2021). Characterization of failure strain in fiber reinforced composites: Under on-axis and off-axis loading. *Crystals, 11*(2), 1–11. https://doi.org/10.3390/cryst11020216. https://www.mdpi.com/2073-4352/11/2/216/pdf.

Khdary, N. H., Almuarqab, B. T., & El Enany, G. (2023). Nanoparticle-embedded polymers and their applications: A review. *Membranes, 13*(5). https://doi.org/10.3390/membranes13050537http://www.mdpi.com/journal/membranes.

Khodakarami, M., & Bagheri, M. (2021). Recent advances in synthesis and application of polymer nanocomposites for water and wastewater treatment. *Journal of Cleaner Production, 296*, 126404. https://doi.org/10.1016/j.jclepro.2021.126404.

Kong, L., Gao, Y., Zhou, Q., Zhao, X., & Sun, Z. (2018). Biochar accelerates PAHs biodegradation in petroleum-polluted soil by biostimulation strategy. *Journal of Hazardous Materials, 343*, 276–284. https://doi.org/10.1016/j.jhazmat.2017.09.040.

Kumar, M., Seth, A., Singh, A. K., Rajput, M. S., & Sikandar, M. (2021). Remediation strategies for heavy metals contaminated ecosystem: A review. *Environmental and Sustainability Indicators, 12*. https://doi.org/10.1016/j.indic.2021.100155. http://www.journals.elsevier.com/environmental-and-sustainability-indicators.

Kumar, S., Kaur, P., Brar, R. S., & Babu, J. N. (2022). Nanoscale zerovalent copper (nZVC) catalyzed environmental remediation of organic and inorganic contaminants: A review. *Heliyon, 8*(8). https://doi.org/10.1016/j.heliyon.2022.e10140. http://www.journals.elsevier.com/heliyon/.

Kumke, M., Watschke, H., & Vietor, T. (2016). A new methodological framework for design for additive manufacturing. *Virtual and Physical Prototyping, 11*(1), 3–19. https://doi.org/10.1080/17452759.2016.1139377. http://www.tandf.co.uk/journals/titles/17452759.asp.

Leary, M. (2020). *Material extrusion*. Elsevier BV223–268. https://doi.org/10.1016/b978-0-12-816721-2.00008-7.

Leary, M. (2020). *Powder bed fusion*. Elsevier BV295–319. https://doi.org/10.1016/b978-0-12-816721-2.00011-7.

Li, N., Khan, S. B., Chen, S., Aiyiti, W., Zhou, J., & Lu, B. (2023). Promising new horizons in medicine: medical advancements with nanocomposite manufacturing via 3D printing. *Polymers, 15*(20). https://doi.org/10.3390/polym15204122. http://www.mdpi.com/journal/polymers.

Li, Q., Wang, Y., Li, Y., Li, L., Tang, M., Hu, W., Chen, L., & Ai, S. (2022). Speciation of heavy metals in soils and their immobilization at micro-scale interfaces among diverse soil components. *Science of The Total Environment, 825*, 153862. https://doi.org/10.1016/j.scitotenv.2022.153862.

Li, R., Li, T., & Zhou, Q. (2020). Impact of titanium dioxide (TiO2) modification on its application to pollution treatment—A review. *Catalysts, 10*(7). https://doi.org/10.3390/catal10070804 Article 7.

Li, S., Meng Lin, M., Toprak, M. S., Kim, D. K., & Muhammed, M. (2010). Nanocomposites of polymer and inorganic nanoparticles for optical and magnetic applications. *Nano Reviews, 1*(1), 5214. https://doi.org/10.3402/nano.v1i0.5214.

Lin, F., Li, M., Zeng, L., Luo, M., & Guo, S. (2023). Intermetallic nanocrystals for fuel-cells-based electrocatalysis. *Chemical Reviews, 123*(22), 12507–12593. https://doi.org/10.1021/acs.chemrev.3c00382.

Liu, G., Liao, L., Dai, Z., Qi, Q., Wu, J., Ma, L. Q., Tang, C., & Xu, J. (2020). , Organic adsorbents modified with citric acid and Fe3O4 enhance the removal of Cd and Pb in contaminated solutions. *Chemical Engineering Journal, 395*. https://doi.org/10.1016/j.cej.2020.125108. http://www.elsevier.com/inca/publications/store/6/0/1/2/7/3/index.htt.

Liu, H., Huang, J., Mao, J., Chen, Z., Chen, G., & Lai, Y. (2019). Transparent antibacterial nanofiber air filters with highly efficient moisture resistance for sustainable particulate matter capture. *iScience, 19*, 214–223. https://doi.org/10.1016/j.isci.2019.07.020.

Liu, J., Ye, L., Wooh, S., Kappl, M., Steffen, W., & Butt, H. J. (2019). Optimizing hydrophobicity and photocatalytic activity of PDMS-coated titanium dioxide. *ACS Applied Materials and Interfaces, 11*(30), 27422–27425. https://doi.org/10.1021/acsami.9b07490. http://pubs.acs.org/journal/aamick.

Liu, S., & Shin, Y. C. (2019). Additive manufacturing of Ti6Al4V alloy: A review. *Materials & Design, 164*, 107552. https://doi.org/10.1016/j.matdes.2018.107552.

Lupone, F., Padovano, E., Casamento, F., & Badini, C. (2022). Process phenomena and material properties in selective laser sintering of polymers: A review. *Materials, 15*(1), 183. https://doi.org/10.3390/ma15010183.

Madrid-Wolff, J., Toombs, J., Rizzo, R., Bernal, P. N., Porcincula, D., Walton, R., Wang, B., Kotz Helmer, F., Yang, Y., Kaplan, D., Zhang, Y. S., Zenobi-Wong, M., McLeod, R. R., Rapp, B., Schwartz, J., Shusteff, M., Talyor, H., Levato, R., & Moser, C. (2023). A review of materials used in tomographic volumetric additive manufacturing. *MRS Communications, 13*(5), 764–785. https://doi.org/10.1557/s43579-023-00447-x. https://www.springer.com/journal/43579.

Mancilla-De-la-Cruz, J., Rodriguez-Salvador, M., An, J., & Kai Chua, C. (2022). Three-dimensional printing technologies for drug delivery applications: Processes, materials, and effects. *International Journal of Bioprinting, 8*(4), 622. https://doi.org/10.18063/ijb.v8i4.622.

Manisalidis, I., Stavropoulou, E., Stavropoulos, A., & Bezirtzoglou, E. (2020). Environmental and health impacts of air pollution: A review. *Frontiers in Public Health, 8*. https://doi.org/10.3389/fpubh.2020.00014.

Matei, E., Predescu, A. M., Râpă, M., Ţurcanu, A. A., Mateş, I., Constantin, N., & Predescu, C. (2022). Natural polymers and their nanocomposites used for environmental applications. *MDPI, Romania Nanomaterials, 12*(10). https://doi.org/10.3390/nano12101707. https://www.mdpi.com/2079-4991/12/10/1707/pdf?version=1652792577.

Mauriello, J., Maury, R., Guillaneuf, Y., & Gigmes, D. (2023). 3D/4D printing of polyurethanes by vat photopolymerization. advanced materials technologies. Advanced Materials Technologies, 8, 2300366. https://doi.org/10.1002/admt.202300366.

Meyer, M. M., Glas, A. H., & Eßig, M. (2021). Systematic review of sourcing and 3D printing: make-or-buy decisions in industrial buyer–supplier relationships. *Management Review Quarterly, 71*(4), 723 752. https://doi.org/10.1007/s11301-020-00198-2. http://link.springer.com/journal/volumesAndIssues/11301.

Michela Dell'Anna, M., Mali, M., Mastrorilli, P., Rizzuti, A., Ponzoni, C., & Leonelli, C. (2013). Suzuki–Miyaura coupling under air in water promoted by polymer supported palladium nanoparticles. *Journal of Molecular Catalysis A: Chemical, 366*, 186–194. https://doi.org/10.1016/j.molcata.2012.09.022.

Mingke, L., Lin, H., He, Y., & Zhang, Y. (2020). The influence of corncob-based biochar on remediation of arsenic and cadmium in yellow soil and cinnamon soil. *Science of The Total Environment, 717*, 137014. https://doi.org/10.1016/j.scitotenv.2020.137014

Mishra, T., Mandal, P., Rout, A. K., & Sahoo, D. (2022). A state-of-the-art review on potential applications of natural fiber-reinforced polymer composite filled with inorganic nanoparticle. *Composites Part C: Open Access, 9*, 100298. https://doi.org/10.1016/j.jcomc.2022.100298.

Mitchell, M. J., Billingsley, M. M., Haley, R. M., Wechsler, M. E., Peppas, N. A., & Langer, R. (2021). Engineering precision nanoparticles for drug delivery. *Nature Reviews. Drug Discovery, 20*(2), 101–124. https://doi.org/10.1038/s41573-020-0090-8. http://www.nature.com/nrd/index.html.

Miyazawa, T., Itaya, M., Burdeos, G. C., Nakagawa, K., & Miyazawa, T. (2021). A critical review of the use of surfactant-coated nanoparticles in nanomedicine and food nanotechnology. *International Journal of Nanomedicine, 16*, 3937–3999. https://doi.org/10.2147/IJN.S298606. https://www.dovepress.com/getfile.php?fileID=70364.

Mohajan, H. (2019). The first industrial revolution: Creation of a new global human era. *Journal of Social Sciences and Humanities, 5*, 377–387 2019.

Nadgorny, M., Xiao, Z., Chen, C., & Connal, L. A. (2016). Three-dimensional printing of pH-responsive and functional polymers on an affordable desktop printer. *ACS Applied Materials & Interfaces, 8*(42), 28946–28954. https://doi.org/10.1021/acsami.6b07388.

Nadgorny, M., Xiao, Z., & Connal, L. A. (2017). 2D and 3D-printing of self-healing gels: Design and extrusion of self-rolling objects. *Molecular Systems Design & Engineering, 2*(3), 283–292. https://doi.org/10.1039/C7ME00023E.

Naseem, T., & Durrani, T. (2021). The role of some important metal oxide nanoparticles for wastewater and antibacterial applications: A review. *Environmental Chemistry and Ecotoxicology, 3*, 59–75. https://doi.org/10.1016/j.enceco.2020.12.001.

Nazir, A., Gokcekaya, O., Md Masum Billah, K., Ertugrul, O., Jiang, J., Sun, J., & Hussain, S. (2023). Multi-material additive manufacturing: A systematic review of design, properties, applications, challenges, and 3D printing of materials and cellular metamaterials. *Materials & Design, 226*, 111661. https://doi.org/10.1016/j.matdes.2023.111661.

Nguyen-Tri, P. (2019). Butyl rubber-based composite: Thermal degradation and prediction of service lifetime. *Journal of Composites Science, 3*, 2019.

Niu, Y. Q., Liu, J. H., Aymonier, C., Fermani, S., Kralj, D., Falini, G., & Zhou, C. H. (2022). Calcium carbonate: Controlled synthesis, surface functionalization, and nanostructured materials. *Chemical Society Reviews, 51*(18), 7883–7943. https://doi.org/10.1039/d1cs00519g. http://pubs.rsc.org/en/journals/journal/cs.

Nugroho, W. T., Dong, Y., & Pramanik, A. (2021). *3D printing composite materials: A comprehensive review. Composite materials: Manufacturing, properties and applications*. Australia: Elsevier65–115. https://www.sciencedirect.com/book/9780128205129. https://doi.org/10.1016/B978-0-12-820512-9.00013-7.

Nurazzi, N. M., Asyraf, M. R. M., Khalina, A., Abdullah, N., Sabaruddin, F. A., Kamarudin, S. H., Ahmad, S., Mahat, A. M., Lee, C. L., Aisyah, H. A., Norrahim, M. N. F., Ilyas, R. A., Harussani, M. M., Ishak, M. R., & Sapuan, S. M. (2021). Fabrication, functionalization, and application of carbon nanotube-reinforced polymer composite: An overview. *Polymers, 13*(7). https://doi.org/10.3390/polym13071047. https://www.mdpi.com/2073-4360/13/7/1047/pdf.

Ohtaka, A. (2021). Recent progress of metal nanoparticle catalysts for C–C bond forming reactions. *Catalysts, 11*(11), 1266. https://doi.org/10.3390/catal11111266.

Ojuederie, O. B., & Babalola, O. O. (2017). Microbial and plant-assisted bioremediation of heavy metal polluted environments: A review. *International Journal of Environmental Research and Public Health, 14*(12). https://doi.org/10.3390/ijerph14121504. http://www.mdpi.com/1660-4601/14/12/1504/pdf.

Oladoja, N. (2019). Polymers for coagulation and flocculation in water treatment. In R. Das (Ed.). *Polymeric Materials for Clean Water* (pp. 77–92). Springer Nature. https://doi.org/10.1007/978-3-030-00743-0_4.

Oliveira, A. D.de, Beatrice, C. A. G., Oliveira, A. D.de, & Beatrice, C. A. G. (2018). *Polymer nanocomposites with different types of nanofiller. Nanocomposites—Recent Evolutions.* IntechOpen. https://doi.org/10.5772/intechopen.81329.

Pal, A.K., Mohanty, A.K., & Misra, M. (2011). Additive manufacturing technology of polymeric materials for customized products: Recent developments and future prospective. RSC Advances, 11(58), 36398–36438. https://doi.org/10.1039/d1ra04060j.

Palansooriya, K. N., Shaheen, S. M., Chen, S. S., Tsang, D. C. W., Hashimoto, Y., Hou, D., Bolan, N. S., Rinklebe, J., & Ok, Y. S. (2020). Soil amendments for immobilization of potentially toxic elements in contaminated soils: A critical review. *Environment International, 134.* https://doi.org/10.1016/j.envint.2019.105046. http://www.elsevier.com/locate/envint.

Pan, B., Pan, B., Zhang, W., Lv, L., Zhang, Q., & Zheng, S. (2009). Development of polymeric and polymer-based hybrid adsorbents for pollutants removal from waters. *Chemical Engineering Journal, 151*(1-3), 19–29. https://doi.org/10.1016/j.cej.2009.02.036.

Pandey, N., Shukla, S. K., & Singh, N. B. (2017). Water purification by polymer nanocomposites: An overview. *Nanocomposites, 3*(2), 47–66. https://doi.org/10.1080/20550324.2017.1329983.

Park, S., Shou, W., Makatura, L., Matusik, W., & Fu, K. (2022). 3D printing of polymer composites: Materials, processes, and applications. *Matter, 5*(1), 43–76. https://doi.org/10.1016/j.matt.2021.10.018. http://www.cell.com/matter.

Pathak, C., Vaidya, F. U., & Pandey, S. M. (2019). Chapter 3—Mechanism for development of nanobased drug delivery system. In S. S. Mohapatra, S. Ranjan, N. Dasgupta, R. K. Mishra, & S. Thomas (Eds.). *Applications of targeted nano drugs and delivery systems* (pp. 35–67). Elsevier. https://doi.org/10.1016/B978-0-12-814029-1.00003-X.

Patel, K. S., Shah, D. B., Joshi, S. J., & Patel, K. M. (2023). Developments in 3D printing of carbon fiber reinforced polymer containing recycled plastic waste: A review. *Cleaner Materials, 9.* https://doi.org/10.1016/j.clema.2023.100207. http://www.elsevier.com/locate/issn/2772-3976.

Pathak, R., Punetha, V. D., Bhatt, S., & Punetha, M. (2023). Multifunctional role of carbon dot-based polymer nanocomposites in biomedical applications: A review. *Journal of Materials Science, 58*(15), 6419–6443. https://doi.org/10.1007/s10853-023-08408-4. https://www.springer.com/journal/10853.

Patra, J. K., Das, G., Fraceto, L. F., Campos, E. V. R., Rodriguez-Torres, M. D. P., Acosta-Torres, L. S., Diaz-Torres, L. A., Grillo, R., Swamy, M. K., Sharma, S., Habtemariam, S., & Shin, H. S. (2018). Nano based drug delivery systems: Recent developments and future prospects. *Journal of Nanobiotechnology, 16*(1). https://doi.org/10.1186/s12951-018-0392-8. https://jnanobiotechnology.biomedcentral.com.

Pei, X., Ma, L., Zhang, B., Sun, J., Sun, Y., Fan, Y., Gou, Z., Zhou, C., & Zhang, X. (2017). Creating hierarchical porosity hydroxyapatite scaffolds with osteoinduction by three-dimensional printing and microwave sintering. *Biofabrication, 9*(4), 045008. https://doi.org/10.1088/1758-5090/aa90ed.

Pellejero, I., Almazán, F., Lafuente, M., Urbiztondo, M. A., Drobek, M., Bechelany, M., Julbe, A., & Gandía, L. M. (2020). Functionalization of 3D printed ABS filters with MOF for toxic gas removal. *Journal of Industrial and Engineering Chemistry, 89*, 194–203. https://doi.org/10.1016/j.jiec.2020.05.013. http:www.sciencedirect.com/science/journal/1226086X.

Pepelnjak, T., Stojšić, J., Sevšek, L., Movrin, D., & Milutinović, M. (2023). Influence of process parameters on the characteristics of additively manufactured parts made from advanced biopolymers. *Polymers, 15*(3), 716. https://doi.org/10.3390/polym15030716.

Perera, F. (2018). Pollution from fossil-fuel combustion is the leading environmental threat to global pediatric health and equity: Solutions exist. *International Journal of Environmental Research and Public Health, 15*(1), 16. https://doi.org/10.3390/ijerph15010016.

Pfaendner, R. (2010). Nanocomposites: Industrial opportunity or challenge. *Polymer Degradation and Stability, 95*(3), 369–373. https://doi.org/10.1016/j.polymdegradstab.2009.11.019.

Pierantozzi, D., Scalzone, A., Jindal, S., Stīpniece, L., Šalma-Ancāne, K., Dalgarno, K., Gentile, P., & Mancuso, E. (2020). 3D printed Sr-containing composite scaffolds: Effect of structural design and material formulation towards new strategies for bone tissue engineering. *Composites Science and Technology, 191*, 108069. https://doi.org/10.1016/j.compscitech.2020.108069.

Pourhashem, S., Saba, F., Duan, J., Rashidi, A., Guan, F., Nezhad, E. G., & Hou, B. (2020). Polymer/Inorganic nanocomposite coatings with superior corrosion protection performance: A review. *Journal of Industrial and Engineering Chemistry, 88*, 29–57. https://doi.org/10.1016/j.jiec.2020.04.029. http:www.sciencedirect.com/science/journal/1226086X.

Purvis, B., Mao, Y., & Robinson, D. (2019). Three pillars of sustainability: In search of conceptual origins. *Sustainability Science, 14*(3), 681–695. https://doi.org/10.1007/s11625-018-0627-5.

Rabiee, N., Sharma, R., Foorginezhad, S., Jouyandeh, M., Asadnia, M., Rabiee, M., Akhavan, O., Lima, E. C., Formela, K., Ashrafizadeh, M., Fallah, Z., Hassanpour, M., Mohammadi, A., & Saeb, M. R. (2023). Green and sustainable membranes: A review. *Environmental Research, 231*. https://doi.org/10.1016/j.envres.2023.116133. https://www.sciencedirect.com/journal/environmental-research.

Rahman, M., Islam, K. S., Dip, T. M., Chowdhury, M. F. M., Debnath, S. R., Hasan, S. M. M., Sakib, M. S., Saha, T., Padhye, R., & Houshyar, S. (2023). A review on nanomaterial-based additive manufacturing: dynamics in properties, prospects, and challenges. *Progress in Additive Manufacturing*. https://doi.org/10.1007/s40964-023-00514-8. https://www.springer.com/journal/40964.

Rajajayavel, S. R. C., & Ghoshal, S. (2015). Enhanced reductive dechlorination of trichloroethylene by sulfidated nanoscale zerovalent iron. *Water Research, 78*, 144–153. https://doi.org/10.1016/j.watres.2015.04.009. http://www.elsevier.com/locate/watres.

Rakhi, T., Pragati, M., Prerna, B., & Jyoti, S. (2023). Chapter 8—Multifaceted approach for nanofiber fabrication. In N. Sharma, & B. S. Butola (Eds.). *Fiber and Textile engineering in drug delivery systems* (pp. 253–283). Woodhead Publishing. https://doi.org/10.1016/B978-0-323-96117-2.00012-1.

Rathod, A., & Beig, G. (2021). Impact of biomass induced black carbon particles in cascading COVID-19. *Urban Climate, 38*, 100913. https://doi.org/10.1016/j.uclim.2021.100913.

Razzaq, M. Y., Balk, M., Mazurek-Budzyńska, M., & Schadewald, A. (2023). From nature to technology: Exploring bioinspired polymer actuators via electrospinning. *Polymers, 15*(19), 4029. https://doi.org/10.3390/polym15194029.

Revilla-León, M., Sadeghpour, M., & Özcan, M. (2020). An update on applications of 3D printing technologies used for processing polymers used in implant dentistry. *Odontology / the Society of the Nippon Dental University, 108*(3), 331–338. https://doi.org/10.1007/s10266-019-00441-7.

Rouf, S., Raina, A., Ul Haq, M. I., Naveed, N., Jeganmohan, S., & Farzana Kichloo, A. (2022). 3D printed parts and mechanical properties: Influencing parameters, sustainability aspects, global market scenario, challenges and applications. *Advanced Industrial and Engineering Polymer Research, 5*(3), 143–158. https://doi.org/10.1016/j.aiepr.2022.02.001.

Sabarees, G., Velmurugan, V., Padmini Tamilarasi, G., Alagarsamy, V., & Raja Solomon, V. (2022). Recent advances in silver nanoparticles containing nanofibers for chronic wound management. *Polymers, 14*(19), 3994. https://doi.org/10.3390/polym14193994.

Sahoo, T. R. (2017). *Polymer nanocomposites for environmental applications.* In *D. K. Tripathy & B.P. Sahoo (Eds.), Properties and applications of polymer nanocomposites: clay and carbon based polymer nanocomposites.* India: Springer Berlin Heidelberg77–106. http://www.springer.com/in/book/9783662535158. https://doi.org/10.1007/978-3-662-53517-2_5.

Salas, A., Zanatta, M., Sans, V., & Roppolo, I. (2023). Chemistry in light-induced 3D printing. *ChemTexts, 9*(1). https://doi.org/10.1007/s40828-022-00176-z.

Salmi, M. (2021). Additive manufacturing processes in medical applications. *Materials, 14*(1), 191. https://doi.org/10.3390/ma14010191.

Salonitis, K. (2014). Stereolithography. *Comprehensive Materials Processing, 10*, 19–67. https://doi.org/10.1016/B978-0-08-096532-1.01001-3. http://www.sciencedirect.com/science/referenceworks/9780080965338.

Sarkar, S., Guibal, E., Quignard, F., & SenGupta, A. K. (2012). Polymer-supported metals and metal oxide nanoparticles: Synthesis, characterization, and applications. *Journal of Nanoparticle Research, 14*(2). https://doi.org/10.1007/s11051-011-0715-2.

Sarker, I. H. (2021). Machinelearning: Algorithms, real-world applications and research directions. *SN Computer Science, 2*(3). https://doi.org/10.1007/s42979-021-00592-xhttps://www.springer.com/journal/42979.

Schlagenhauf, L., Nüesch, F., & Wang, J. (2014). Release of carbon nanotubes from polymer nanocomposites. *Fibers, 2*(2). https://doi.org/10.3390/fib2020108 Article 2.

Schlagenhauf, L., Buerki-Thurnherr, T., Kuo, Y.-Y., Wichser, A., Nüesch, F., Wick, P., & Wang, J. (2015). Carbon nanotubes released from an epoxy-based nanocomposite: Quantification and particle toxicity. *Environmental Science & Technology, 49*(17), 10616–10623. https://doi.org/10.1021/acs.est.5b02750.

Schwartz, J. J. (2022). Additive manufacturing: Frameworks for chemical understanding and advancement in vat photopolymerization. *MRS Bulletin, 47*(6), 628–641. https://doi.org/10.1557/s43577-022-00343-0. https://www.springer.com/journal/43577.

Shah, M., Ullah, A., Azher, K., Rehman, A. U., Juan, W., Aktürk, N., Tüfekci, C. S., & Salamci, M. U. (2023). Vat photopolymerization-based 3D printing of polymer nanocomposites: Current trends and applications. *RSC Advances, 13*(2), 1456–1496. https://doi.org/10.1039/d2ra06522c. http://pubs.rsc.org/en/journals/journal/ra.

Sharma, V., Kapoor, S., Goyal, M., & Jindal, P. (2020). Enhancement of the mechanical properties of graphene based acrylonitrile butadiene styrene (ABS) nanocomposites. *Materials Today: Proceedings, 28*, 1744–1747. https://doi.org/10.1016/j.matpr.2020.05.155.

Simionescu, B. C., & Ivanov, D. (2014). Natural and synthetic polymers for designing composite materials. In I. V. Antoniac (Vol. Ed.), *Handbook of bioceramics and biocomposites: 1–54*Springer International Publishinghttps://doi.org/10.1007/978-3-319-09230-0_11-1.

Sola, A. (2022). Materials requirements in fused filament fabrication: A framework for the design of next-generation 3D printable thermoplastics and composites. *Macromolecular Materials and Engineering, 307*(10). https://doi.org/10.1002/mame.202200197.

Soundararajan, B., Sofia, D., Barletta, D., & Poletto, M. (2021). Review on modeling techniques for powder bed fusion processes based on physical principles. *Additive Manufacturing, 47*, 102336. https://doi.org/10.1016/j.addma.2021.102336.

Srivastava, M., Rathee, S., Patel, V., Kumar, A., & Koppad, P. G. (2022). A review of various materials for additive manufacturing: Recent trends and processing issues. *Journal of*

Materials Research and Technology, 21, 2612–2641. https://doi.org/10.1016/j.jmrt.2022.10.015. http://www.elsevier.com/journals/journal-of-materials-research-and-technology/2238-7854.

Šupová, M., Simha Martynková, G., & Čech Barabaszová, K. (2010). Effect of nanofillers dispersion in polymer matrices: A review. *Science of Advanced Materials, 3*, 1–25. https://doi.org/10.1166/sam.2011.1136.

Tamiolakis, I., Papadas, I. T., Spyridopoulos, K. C., & Armatas, G. S. (2016). Mesoporous assembled structures of Cu_2O and TiO_2 nanoparticles for highly efficient photocatalytic hydrogen generation from water. *RSC Advances, 6*(60), 54848–54855. https://doi.org/10.1039/c6ra08546f.

Tanahashi, M. (2010). Development of fabrication methods of filler/polymer nanocomposites: With focus on simple melt-compounding-based approach without surface modification of nanofillers. *Materials, 3*(3), 1593–1619. https://doi.org/10.3390/ma3031593.

Tao, X., Li, A., & Yang, H. (2016). Immobilization of metals in contaminated soils using natural polymer-based stabilizers. *Environmental Pollution (Barking, Essex : 1987), 222*. https://doi.org/10.1016/j.envpol.2016.12.028.

Tao, Y., Kong, F., Li, Z., Zhang, J., Zhao, X., Yin, Q., Xing, D., & Li, P. (2021). A review on voids of 3D printed parts by fused filament fabrication. *Journal of Materials Research and Technology, 15*, 4860–4879. https://doi.org/10.1016/j.jmrt.2021.10.108.

Tennakone, K., & Kottegoda, I. R. M. (1996). Photocatalytic mineralization of paraquat dissolved in water by TiO_2 supported on polythene and polypropylene films. *Journal of Photochemistry and Photobiology A: Chemistry, 93*(1), 79–81. https://doi.org/10.1016/1010-6030(95)04141-9.

Torrado, A. R., Shemelya, C. M., English, J. D., Lin, Y., Wicker, R. B., & Roberson, D. A. (2015). Characterizing the effect of additives to ABS on the mechanical property anisotropy of specimens fabricated by material extrusion 3D printing. *Additive Manufacturing, 6*, 16–29. https://doi.org/10.1016/j.addma.2015.02.001.

Tu, H., Wei, Z., Bahrami, A., Ben Kahla, N., Ahmad, A., & Onuralp Özkılıç, Y. (2023). Recent advancements and future trends in 3D concrete printing using waste materials. *Developments in the Built Environment, 16*, 100187. https://doi.org/10.1016/j.dibe.2023.100187.

Verma, A., Kapil, A., Klobčar, D., & Sharma, A. (2023). A review on multiplicity in multi-material additive manufacturing: Process, capability, scale, and structure. *Materials, 16*(15), 5246. https://doi.org/10.3390/ma16155246.

Vilatela, J. J., & Eder, D. (2012). Nanocarbon composites and hybrids in sustainability: A review. *ChemSusChem, 5*(3), 456–478. https://doi.org/10.1002/cssc.201100536.

Vlasvelda, D. P. N., Parlevliet, P. P., Bersee, H. E. N., & Pickena, S. J. (2005). Fibre–matrix adhesion in glass-fibre reinforced polyamide-6 silicate nanocomposites Composites Part A. *Applied Science and Manufacturing, 36*, 1–11. https://doi.org/10.1016/j.compositesa.2004.06.035.

Vogt, B. D. (2020). A virtual issue of applied polymer materials: 3D printing of polymers. *ACS Applied Polymer Materials, 2*(6), 2102–2104. https://doi.org/10.1021/acsapm.0c00517 Available from: pubs.acs.org/journal/aapmcd.

Vroman, I., & Tighzert, L. (2009). Biodegradable polymers. *Materials, 2*(2), 307–344. https://doi.org/10.3390/ma2020307.

Wang, L., Wang, Q., Slita, A., Backman, O., Gounani, Z., Rosqvist, E., Peltonen, J., Willför, S., Xu, C., Rosenholm, J. M., & Wang, X. (2022). Digital light processing (DLP) 3D-fabricated antimicrobial hydrogel with a sustainable resin of methacrylated woody polysaccharides and hybrid silver-lignin nanospheres. *Green Chemistry, 24*(5), 2129–2145. https://doi.org/10.1039/D1GC03841A.

Wang, X., Liu, J., Zhang, Y., Kristiansen, P. M., Islam, A., Gilchrist, M., & Zhang, N. (2023). Advances in precision microfabrication through digital light processing: System development, material and applications. *Virtual and Physical Prototyping, 18*(1). https://doi.org/10.1080/17452759.2023.2248101. http://www.tandf.co.uk/journals/titles/17452759.asp.

Wells, C. M., Harris, M., Choi, L., Murali, V. P., Guerra, F. D., & Jennings, J. A. (2019). Stimuli-responsive drug release from smart polymers. *Journal of Functional Biomaterials, 10*(3). https://doi.org/10.3390/jfb10030034http://www.mdpi.com/journal/jfb.

Wickramasinghe, S., Do, T., & Tran, P. (2020). FDM-Based 3D printing of polymer and associated composite: A review on mechanical properties, defects and treatments. *Polymers, 12*(7), 1–42. https://doi.org/10.3390/polym12071529. https://res.mdpi.com/d_attachment/polymers/polymers-12-01529/article_deploy/polymers-12-01529.pdf.

Wolfs, R. J. M., Bos, F. P., & Salet, T. A. M. (2019). Hardened properties of 3D printed concrete: The influence of process parameters on interlayer adhesion. *Cement and Concrete Research, 119*, 132–140. https://doi.org/10.1016/j.cemconres.2019.02.017.

Wu, Y., Zhan, K., Yang, Z., Sun, W., Zhao, B., Yan, Y., & Yang, J. (2019). Graphene oxide/Al composites with enhanced mechanical properties fabricated by simple electrostatic interaction and powder metallurgy. *Journal of Alloys and Compounds, 775*, 233–240. https://doi.org/10.1016/j.jallcom.2018.10.158.

Xing, R., Yang, B., Huang, R., Qi, W., Su, R., Binks, B. P., & He, Z. (2019). Three-dimensionally printed bioinspired superhydrophobic packings for oil-in-water emulsion separation. *Langmuir. the ACS Journal of Surfaces and Colloids, 35*(39), 12799–12806. https://doi.org/10.1021/acs.langmuir.9b02131. http://pubs.acs.org/journal/langd5.

Xing, Y.-F., Xu, Y.-H., Shi, M.-H., & Lian, Y.-X. (2016). The impact of PM2.5 on the human respiratory system. *Journal of Thoracic Disease, 8*(1), E69–E74. https://doi.org/10.3978/j.issn.2072-1439.2016.01.19.

Xu, W., Yang, T., Liu, S., Du, L., Chen, Q., Li, X., Dong, J., Zhang, Z., Lu, S., Gong, Y., Zhou, L., Liu, Y., & Tan, X. (2022). Insights into the Synthesis, types and application of iron nanoparticles: The overlooked significance of environmental effects. *Environment International, 158*, 106980. https://doi.org/10.1016/j.envint.2021.106980.

Xu, Y., Wu, X., Guo, X., Kong, B., Zhang, M., Qian, X., Mi, S., & Sun, W. (2017). The boom in 3D-printed sensor technology. *Sensors, 17*(5). https://doi.org/10.3390/s17051166 Article 5.

Yanat, M., & Schroën, K. (2021). Preparation methods and applications of chitosan nanoparticles; with an outlook toward reinforcement of biodegradable packaging. *Reactive and Functional Polymers, 161*, 104849. https://doi.org/10.1016/j.reactfunctpolym.2021.104849.

Yang, H., Yang, B., Chen, W., & Yang, J. (2022). Preparation and photocatalytic activities of TiO2-based composite catalysts. *Catalysts, 12*(10). https://doi.org/10.3390/catal12101263 Article 10.

Yang, Y., Li, X., Zheng, X., Chen, Z., Zhou, Q., & Chen, Y. (2018). 3D-printed biomimetic super-hydrophobic structure for microdroplet manipulation and oil/water separation. *Advanced Materials, 30*(9). https://doi.org/10.1002/adma.201704912http://www3.interscience.wiley.com/journal/119030556/issue.

Yao, L., Ramesh, A., Xiao, Z., Chen, Y., & Zhuang, Q. (2023). Multimetal research in powder bed fusion: A review. *Materials, 16*(12), 4287. https://doi.org/10.3390/ma16124287.

Yaqoob, A. A., Ahmad, H., Parveen, T., Ahmad, A., Oves, M., Ismail, I. M. I., Qari, H. A., Umar, K., & Mohamad Ibrahim, M. N. (2020). Recent advances in metal decorated nanomaterials and their various biological applications: A review. *Frontiers in Chemistry, 8*. https://doi.org/10.3389/fchem.2020.00341. http://journal.frontiersin.org/journal/chemistry.

Yasser, Z. (2016). Study of nanoparticles aggregation/agglomeration in polymer particulate nanocomposites by mechanical properties. *Composites Part A: Applied Science and Manufacturing, 84*, 158–164. https://doi.org/10.1016/j.compositesa.2016.01.020.

Yilmaz, B., Rashid, A. A., Ait Mou, Y., Evis, Z., & Koç, M. (2021). Bioprinting: A review of processes, materials and applications. *Bioprinting, 23*, e00148. https://doi.org/10.1016/j.bprint.2021.e00148.

Yuan, Q., & Misra, R. D. K. (2006). Polymer nanocomposites: Current understanding and issues. *Materials Science and Technology, 22*(7), 742–755. https://doi.org/10.1179/174328406X101292.

Yuan, S., Strobbe, D., Kruth, J. P., Van Puyvelde, P., & Van Der Bruggen, B. (2017). Superhydrophobic 3D printed polysulfone membranes with a switchable wettability by self-assembled candle soot for efficient gravity-driven oil/water separation. *Journal of Materials Chemistry A, 5*(48), 25401–25409. https://doi.org/10.1039/c7ta08836a. http://pubs.rsc.org/en/journals/journalissues/ta.

Yuan, S., Zhu, J., Li, Y., Zhao, Y., Li, J., Van Puyvelde, P., & Van Der Bruggen, B. (2019). Structure architecture of micro/nanoscale ZIF-L on a 3D printed membrane for a superhydrophobic and underwater superoleophobic surface. *Journal of Materials Chemistry A, 7*(6), 2723–2729. https://doi.org/10.1039/c8ta10249j. http://pubs.rsc.org/en/journals/journal/ta.

Yuvaraj, G., Ramesh, M., & Rajeshkumar, L. (2023). Carbon and cellulose-based nanoparticle-reinforced polymer nanocomposites: A critical review. *Nanomaterials, 13*(11), 1803. https://doi.org/10.3390/nano13111803.

Zakeri, S., Vippola, M., & Levänen, E. (2020). A comprehensive review of the photopolymerization of ceramic resins used in stereolithography. *Additive Manufacturing, 35*, 101177. https://doi.org/10.1016/j.addma.2020.101177.

Zeng, L., Ling, S., Du, D., He, H., Li, X., & Zhang, C. (2023). Direct ink writing 3D printing for high-performance electrochemical energy storage devices: A minireview. *Advanced Science, 10*, 2303716. https://doi.org/10.1002/advs.202303716.

Zhu, N., Luan, H., Yuan, S., Chen, J., Wu, X., & Wang, L. (2009). Effective dechlorination of HCB by nanoscale Cu/Fe particles. *Journal of Hazardous Materials, 176*, 1101–1105. https://doi.org/10.1016/j.jhazmat.2009.11.092.

Zhuo, P., Li, S., Ashcroft, I. A., & Jones, A. I. (2021). Material extrusion additive manufacturing of continuous fibre reinforced polymer matrix composites: A review and outlook. *Composites Part B: Engineering, 224*. https://doi.org/10.1016/j.compositesb.2021.109143. https://www.journals.elsevier.com/composites-part-b-engineering.

Zinchenko, A., Sakai, T., Morikawa, K., & Nakano, M. (2022). Efficient stabilization of soil, sand, and clay by a polymer network of biomass-derived chitosan and carboxymethyl cellulose. *Journal of Environmental Chemical Engineering, 10*(1), 107084. https://doi.org/10.1016/j.jece.2021.107084.

Zocca, A., Franchin, G., Colombo, P., & Günster, J. (2021). Additive manufacturing. *Encyclopedia of Materials: Technical Ceramics and Glasses, 1-3*, 203–221. https://doi.org/10.1016/B978-0-12-803581-8.12081-8. https://www.sciencedirect.com/referencework/9780128222331/encyclopedia-of-materials-technical-ceramics-and-glasses.

Zulfiqar, U., Haider, F. U., Maqsood, M. F., Mohy-Ud-Din, W., Shabaan, M., Ahmad, M., Kaleem, M., Ishfaq, M., Aslam, Z., & Shahzad, B. (2023). Recent advances in microbial-assisted remediation of cadmium-contaminated soil. *Plants, 12*(17). https://doi.org/10.3390/plants12173147. http://www.mdpi.com/journal/plants.

Additive manufacturing of functional polymers and nanocomposites for soft robotics

12

Syed Khawar Hussain Shah, Ramisha Sajjad, Muzamil Hussain, Muhammad Tuoqeer Anwar, Muhammad Salman Nawaz, and Muhammad Rafi Raza
Department of Mechanical Engineering, COMSATS University Islamabad, Sahiwal Campus, Sahiwal, Punjab, Pakistan

12.1 Introduction

Additive manufacturing is a process in which layers of a material are placed over other to build a structure; the process is also known as 3D printing (3DP). In this technique, a solid 3D model is built through AM from a virtual model file. 3DP is gaining popularity as a manufacturing technique, as it enables to printing of complex geometries that were not possible using conventionally available techniques. Due to this, there is an increasing demand for novel materials for 3DP. As a result, smart materials came into existence that can undergo a control structural change upon application of an external stimulus.

3DP methods have expanded to include multipurpose and varied shapes throughout their life cycle rather than being restricted to producing permanent 3D things with a single purpose. The process known as 4D printing (4DP) provides additional adaptability and diversity in the structures. The same methods that AM used are also employed in 4DP, although additional transformations or changes in material properties over time are included. The ability of the final product to adapt over time to different environmental conditions is therefore indicated by the fourth dimension, which is time. These materials are referred to as smart materials, application of these materials is in complicated parts, sensors, robots, and biomedical field.

Smart materials are programmable materials whose structure and form are established using CAD modeling software, and which are then cut into layers with the necessary thickness for 3DP/4DP. The advanced form of products is made from another type of smart materials that could be triggered with the aid of external stimuli which results in different material properties and morphologies. These externally applied stimuli may include light, humidity, pH, temperature, and electromagnetic fields (Holman et al., 2021). Functional polymers, their nanocomposites, and additive manufacturing trio is crucial for making a robot or their different components as presented in Fig. 12.1.

Additive Manufacturing of Functional Polymers and Nanocomposites. DOI: https://doi.org/10.1016/B978-0-443-26516-7.00012-2
© 2025 Elsevier Ltd. All rights are reserved, including those for text and data mining, AI training, and similar technologies.

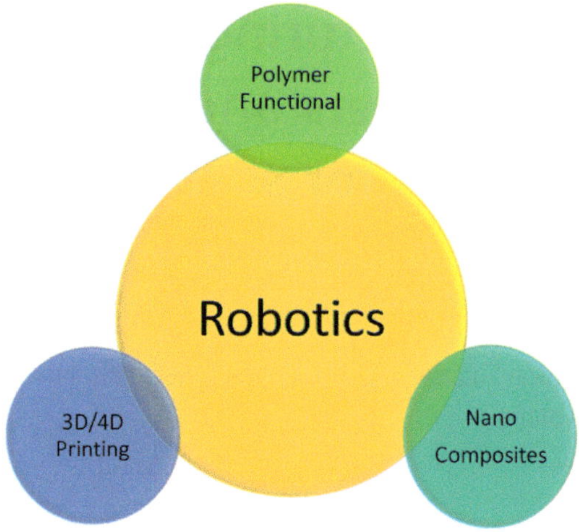

Figure 12.1 Functional polymers and nanocomposites are used to make robotics parts through 3D/4D printing.

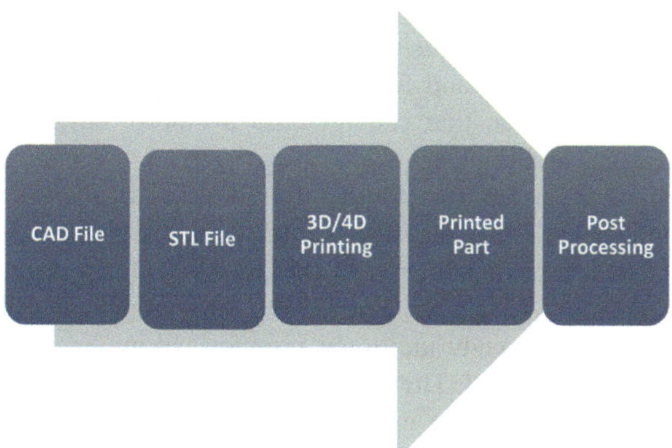

Figure 12.2 The process of making a 3D-printed part from a Virtual Solid model.

12.2 3D printing methods

Additive manufacturing is a process that allows to making of a real-world 3D model from a computer-aided design (CAD) file. The CAD file is converted into a format that is readable by a 3D printer. Additive manufacturing models build up by placing layer over layer using G code; the process is also presented through process diagram (Fig. 12.2). Even though the 3D printing field has been growing for the last few decades, recent research in optimized 3D printing resolved various long-standing issues associated with this field, for example, slow production rates, low

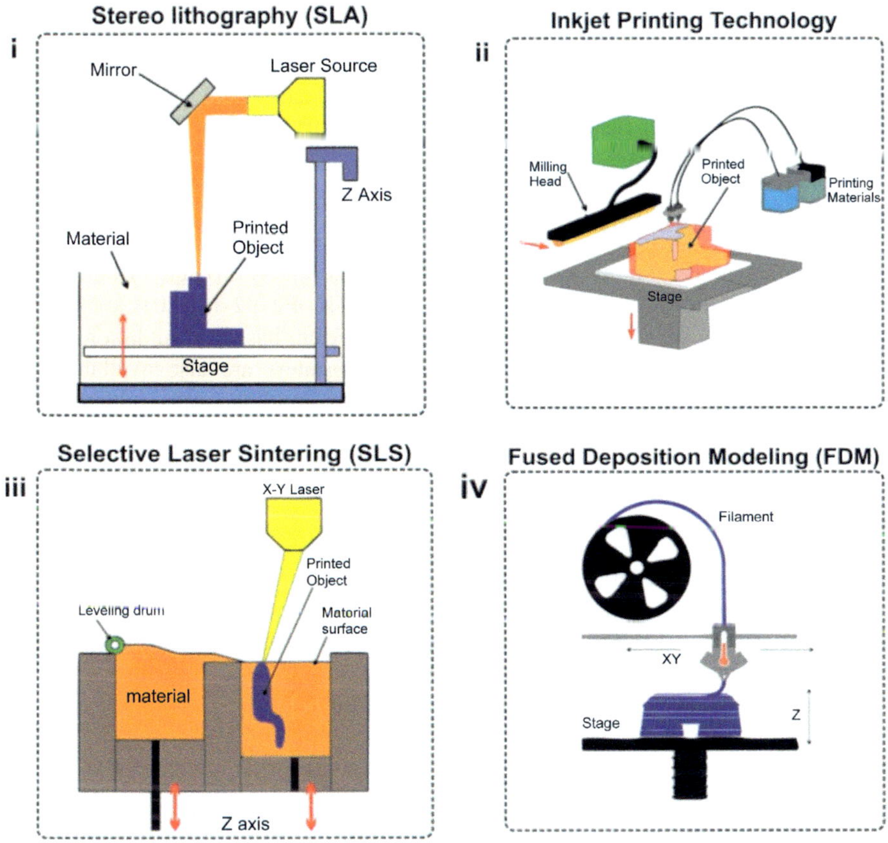

Figure 12.3 Different 3DP techniques: (i) stereo lithography (SLA), (ii) inkjet printing technology, (iii) selective laser sintering (SLS), and (iv) fused deposition modeling (FDM).

resolution, and structural issues associated with additive manufacturing (Chen et al., 2011) (Fig. 12.3).

The most common types of additive manufacturing techniques are discussed below.

12.2.1 Stereolithography (SLA)

Stereolithography (SLA) is often known as vat polymerization; in this technique, one or multiple laser beams are used to cure layers of liquid photopolymer resins (Hofmann, 2014). After the solidification of a layer, the model again sinks by a specified increment for another photo-curing printing cycle over the previously built layer.

12.2.2 Selective laser sintering (SLS)

Selective laser sintering (SLS), commonly known as powder bed fusion, utilizes laser beams to melt and fuse powders to manufacture 3D objects. There are several binding mechanisms for powder fusion, for example, solid-state sintering, and partial or full melting (Kruth et al., 2005).

12.2.3 Inkjet printing

Inkjet printing is a technique in which a printer head ejects droplets of photo resin when pressure and voltage are applied to it (ASTM F2792-12, 2012). The fluid normally consists of ultraviolet (UV)-curable acrylate-based formulation, which solidifies when exposed to a UV source. Inkjet printers are also available with multiple heads to support printing with multiple inks simultaneously. Inkjet printers are highly accurate and can print with a spatial resolution of a magnitude tenth of a micrometer (Barclift & Williams, 2012).

12.2.4 Extrusion-based printing

It is comparatively the most commonly used printing process among all; in this process, a filament is forced through a small nozzle, the process is known as extrusion (ASTM F2792-12, 2012). Stepper motors are used for accurate placement of the jet of the material with the help of G codes.

Common types of extrusion printing are discussed below.

12.2.4.1 Fused deposition modeling (FDM)

Fused deposition modeling is a technique in which a filament is extruded from a heated nozzle, which converts the filament into a jet, then stepper motors place the jet on defined coordinates with the help of G codes and place layer on layer to build the 3D structure (Aliheidari et al., 2017).

12.2.4.2 Liquid deposition modeling (LDM)

Liquid deposition modeling (LDM), also known as direct ink-writing (DIW), is a 3DP technique that is widely used for micro-scale structures. This is an extrusion-based technique in which small size nozzles are used, which pour ink on the defined paths and ink solidifies forming a 3D structure. The quality of a printer and its extrusion efficacy are reliant on the viscoelastic properties of the ink, which should be in line with the capabilities of the printing system used.

The 3D printing techniques are graphically presented in Fig. 12.1.

12.3 An overview of robotics

Robotics is a field in which such machines/devices are made to perform a task without or with very little involvement of a human beings; there are multiple stages involved in making a robot that are designing, developing structures, and utilizing different kinds of sensors and actuators with microcontroller as per the requirement of the task.

There are two major classes of robots discussed in the following sections:

12.3.1 Hard robot

Hard robots are manufactured using inflexible materials like metals, plastic, and acrylic sheets. They comprise a stiff body with a controller attached to it, which is further interfaced with sensors and actuators to fulfill the requirement of automation (Sadik Batcha, 2017).

12.3.2 Soft robot

Soft robots are built using flexible material and used in places where typically hard robots cannot work such as in robotic surgeries and prostheses. Stretchable and elastic materials are used to fabricate soft robots such as silicones and hydrogels. The actuators and sensors used in soft robots are also made up of flexible materials (Schmitt et al., 2018).

12.4 Functional polymers and their nanocomposites

Functional polymers are a class of polymers that possess special properties, for example, magnetic, biomedical, electrical, and biodegradable properties (Wegner, 2000). They can be designed on the molecular level by incorporating functional fillers in a polymer.

Nanocomposites belong to the class of composites in which one of the phases of the material has nanoscale morphology like in the form of nanoparticles and nanotubes. Taking dimensions to the nanolevel has a significant impact in inducing properties in the material (Okpala, 2013), that is, enhancing various aspects of the material such as electrical, mechanical, and chemical ones (Paul & Robeson, 2008). The rapidly expanding field is developing many new nanocomposites with novel properties.

Functional polymers and their nanocomposites are popular in additive manufacturing, more specifically in 4D printing; the fourth dimension refers to an extra dimension added to a printed structure to alter their properties with time. The properties of functional polymers and their nanocomposites can be controlled as per the requirement by using an external stimulus like temperature, light, voltage, pH, and mechanical loading (Liu & Urban, 2010); this opened various new possibilities which previously looked impossible, for example, 4D-printed structures have the ability to self-assemble and self-repair them.

The process of converting a 3D printed structure into a 4D one upon the application of external stimuli is presented in Fig. 12.4.

The latest advancements in functional polymers and nanocomposites' 3DP and their use in robotics are outlined in the following section.

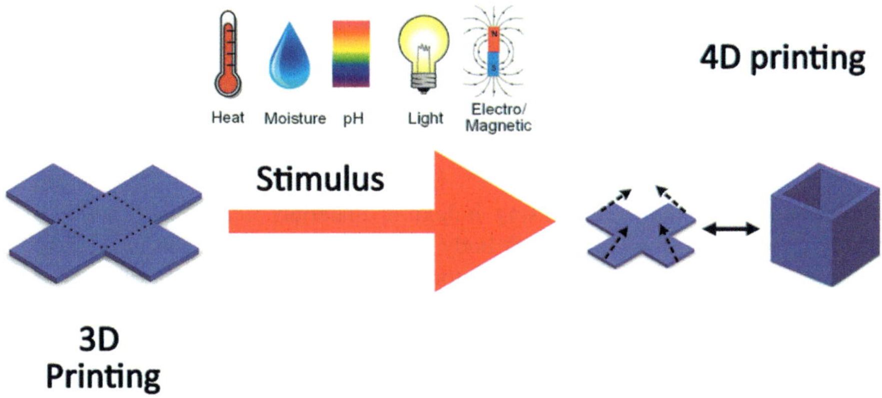

Figure 12.4 The process of making a 4D-printed part from a 3D structure through the application of external stimuli such as temperature, light, electromagnetic or other environmental stimulus.

12.4.1 Temperature responsive polymers

The polymers that are receptive to change in temperature lie in the category of temperature responsive polymers. This class of polymers includes a variety of polymers; the additive manufacturing of these polymers is covered in the section that follows.

12.4.1.1 Shape memory polymers

Shape memory polymers (SMPs) belong to the class of polymers that possess the capability to regain their actual shape by recovering from temporarily induced deformations. SMPs possesses many desirable properties such as low density, good ability to recover with in a wide range of stimuli, recyclability at low cost, biocompatibility and easier processing (Sun et al., 2012). SMPs are extensively utilized in making actuators for robotic systems.

Researchers reported one-way, two-way, and multiple-way SMPs. Conventional cross-linked polymers exhibit a one-way shape memory effect (Berg et al., 2014). Two-way shape memory polymers also known as reversible SMPs, possess the capability to deform autonomously at high temperature, and revert at low temperature between two desired shapes. The polymers that can remember more than one shape are known as multiple-shape memory polymers.

There are different types of SMPs triggered by different types of stimuli, and among all of them, thermally triggered SMPs are the most studied ones.

SMPs also known as smart polymers are widely used in 3D printing, various materials lie in this category including polycaprolactone (PCL), polyurethane (PU) (Estelle et al., 2017), methacrylic, and epoxy-based resins. Their capability is very useful for developing structures that can evolve by themselves (Ge et al., 2016), for example, DiAPLEX MM-4520, a shape memory polymer; its pellets were mixed with FDM filament and utilized for the manufacturing of a robotic gripper (Yang et al., 2016).

Various printing techniques are used for printing of soft robotics. Nowadays, shape memory polymers are widely used for printing along with different materials simultaneously by utilizing direct ink and material jetting techniques, which results in robots with variable stiffness depending on the application. Numerous printing techniques with multiple material printing capabilities have been introduced and they have shown potential in printing of soft robotics (Yap et al., 2020). So, the additive manufacturing of thermosetting SMPs is an important development that increases the applications of SMPs in a wide range and resolves its drawbacks.

12.4.1.2 Hydrogel-based polymers

Hydrogels are a potential candidate for bio-inspired robotics and soft robotics. Hydrogels show superior biocompatibility, high water content, and responsiveness to an external stimulus. These gels are made by chemical cross-linking. Polymer matrices in these gels are tied up using cross-linked points, which results in a stable bulk structure. However, as gels are normally porous, their integrity is destroyed when deformation occurs. Dimensional changes occur at the micro level in stimuli-sensitive gels; this was observed in poly(N-isopropylacrylamide) or PNIPAm gels by using laser scanning confocal microscopy (LSCM) (Hirokawa et al., 1999). Hydrogels have large amounts of water-filled interstitial sites. Hydrogels with large amount of water show the ability to quickly change their morphology and shape through swelling/deswelling with change in water content, which can be triggered by different stimulus such as pH, light, temperature, and biochemicals (Sano et al., 2018). By utilizing different molding and fabrication techniques, hydrogels can be modified in various shapes and morphologies; this property makes these gels a good candidate to design sensors and actuators (Banerjee et al., 2018) for mimicking function of different biological counterparts. For this reason, hydrogels gather attention as their application in soft robotics.

A study has been conducted on using shape responsive hydrogels (SRH) for manufacturing small scale soft grippers, for which a predesigned four-arm shaped well is prepared along with an Ecoflex mold. It was demonstrated that gripper can automatically trigger under various stimulus because of the captured logical output information. It is believed that the proposed method is a very good approach for controlling motion of small soft robots (Zhang et al., 2022).

Environment interactive robots are normally made up of elastomeric materials such as silicon that have pneumatic pathways that cause actuation in the robot. A very flexible skin that can combine with pneumatic actuators is printed by using direct ink writing technique (Robinson et al., 2015). The fabricated skin allows the robot to gather the force and shape information from interaction with environment, a hydrogel and a silicon elastomer precursor were developed, the fluids were cross-linked using in situ photo polymerization into conductive hydrogels, and were insulated with a large ultimate strain.

Although several researchers investigated the behavior of hydrogels in additive manufacturing, studies are limited to some specific hydrogels, and many experiments are anticipated to study their true potential in additive manufacturing.

12.4.2 Electromagnetically responsive polymers

These polymers exhibit a change in the shape due to an electrical stimulus. 3DP of metallic structures got great attention due to their application in sensor and actuators for robotics and smart structures. The cost attached with it and the use of specialized alloy made it a challenging task (Wang et al., 2014). In order to cope with this issue, conductive polymers composites have been developed; they have the capability to make light weight elastic electroconductive materials with easy processing (Ameli et al., 2014).

Ferro magnetically sensitive composite resins were formulated through adding magnetite particles within a photo curable resin. This process enables direct manufacturing of high-resolution cantilevers in a single step. The fabricated composite cantilever beam was triggered by an external magnetic field and displayed a high ferromagnetic sensitivity that was measured using the printed cantilever beam. Furthermore, as compared to conventionally molded cantilever, these cantilever beams exhibited a higher magnetic sensitivity (Credi et al., 2016).

Electromagnetically responsive polymers are widely being used in electronics and tissue engineering, and they hold a great potential for application in robotics.

12.4.3 Piezoelectric materials

Piezoelectric materials are a special type of material that possesses the capability to generate an equivalent electric charge when a pressure, force, or vibration is applied to them. On the other hand, when an electric charge is supplied to them, they deform accordingly. Due to their unique capability, piezoelectric materials are widely used in making sensors, transducers, actuators, and energy-harvesting devices (Ramadan et al., 2014). Ceramics are one of the materials that possess the strongest piezoelectric properties. Although they possess good piezoelectric properties, they are brittle, and it is difficult to process them. To mitigate this shortcoming of ceramics, and for easy fabrication of sensors and actuators, 3DP is a good option that can convert 2D ceramic fibers into thin 3D structures (Chen et al., 2016).

Piezoelectric polymers are also used for making piezoelectric devices, especially polyvinylidene fluoride (PVDF) (Kim et al., 2017). Although its piezoelectric properties are inferior as compared to ceramics (Kim et al., 2014), however, the biocompatibility, flexibility (Kennedy et al., 2017), and easy processability of PVDF makes it a good candidate for 3D printing. PVDF possesses four phases (α, β, γ, δ), among all these phases. The β phase have the strongest piezoelectric properties. To enhance the β phase and keep its alignment, electrical poling is used in conjunction with extrusion-based 3DP (Lee & S, 2014).

Piezoelectric properties of PVDF with nanoparticles of barium titanate and multiwalled carbon nanotubes (MWCNTs) were studied for their application in pressure sensing and energy harvesting (Kim, Islam, et al., 2017). In the manufactured nanocomposite, the main constituents for piezoelectric properties were PVDF and barium titanate, whereas, in order to support stress and facilitate transmission of electrons, MWCNTs were added that resulted in enhanced concentration of B phase in PVDF, ultimately leading to the increase in the efficiency. The highest piezoelectric coefficient was recorded at 129 pC/N at 0.4 wt%-MWCNTs and 18 wt %-BT loading.

A 3D multiaxial force sensor was printed using thermoplastic polyurethane (TPU) and carbon nanotubes (CNTs); TPU offers high flexibility in structure whereas CNTs increase the electrical conductivity of the sensor. The filament was composed of 4 wt % of MWCNTs, which were melt compounded for this application, and the average electrical resistivity of the filament was noted to be 0.143 ± 0.036 Ω·m (Kim, Park, et al., 2017).

Piezoelectric material 3D/4D printing offers the following advantages over traditional methods: ability to create intricate structures and accurately regulate and adjust parameters like layer thickness and random piezoelectric coefficient can be combined through 3D printing. Piezoelectric smart materials work well in embedded applications (Bodkhe et al., 2017) and can be combined with other printed materials to provide efficient electrical conversion. A huge number of cases are reported in which piezoelectric materials are used in additive manufacturing, whereas further studies can be carried out to enhance their efficiency.

12.4.4 Peizoresistive materials

Peizoresistive materials are a special kind of materials, which show a change in resistivity whenever they experience a stress or strain; the change in resistivity depends on the magnitude of applied force. This property made it a good choice for sensing stimulus. Metals and semiconductors have been used as peizoresistive elements, but sensitivity to high temperature, hardness, and expensiveness limits their applications (Liu et al., 2017). In order to overcome these issues, peizoresistive materials have been reinforced into the flexible polymers or matrices (Stassi et al., 2014). In additive manufacturing of stress and strain sensors, these materials have been used.

3DP is being utilized to deposit conductive carbon black ink on a reservoir of silicon elastomer and a filler fluid. A highly flexible strain sensor was produced from the postcuring. The sensor produced from this 3DP technique exhibited an enhanced printing speed (4 mm/s), decreased area of resistor, and resulted in an enhanced conductivity and strain sensitivity. A minute change in the resistance, sensor could sustain cycled straining (Muth et al., 2014).

A multiaxial force sensor was fabricated using 3DP from TPU composites and MWCNTs. The sensor was printed using FDM printer by depositing an active composite material on flexible passive TPU composite. This technique is used for fabricating monolithic sensor, which do not need any postprocessing operations. The fabricated device could easily survive 1000 bending cycles with a minimal decrease in resistivity of magnitude ~0.65% (Kim, Park, et al., 2017).

A flexible photo resin matrix was manufactured using an inkjet 3D printer. By removing the supporting material, microchannels were formed, which were filled using silver ink. This resulted in a force sensor, capable of detecting forces across two axes. In order to maintain mechanical shape and sensitivity of the sensor, the strain on the sensor must be below 5% (Agarwala et al., 2017).

Different-shaped strain sensors are manufactured and integrated into Ecoflex. The CAD/CAM program that manages the tool path or nozzle translation during printing first renders the sensor design. Each printed sensor's cross-sectional dimensions can be altered by varying the printing speed, applied pressure, and nozzle size (Muth et al., 2014).

The peizoresistive composite can be utilized to detect mechanical flexing when it is incorporated into an already-existing object or attached to an interface device called an exo-glove that detects hand flexion. Additionally, it has been shown how the material may be utilized to make integrated capacitive sensors for smart vessels that can feel the amount of liquid placed inside them as well as capacitive sensing devices for customized 3D-printed human interface devices (HIDs). The printed sensors can actually be monitored with open-source programming libraries and open-source electronics that are already in existence. They are also easy to interface and don't require complex electronic circuits or amplification. There is no need to change the printer's settings; that is, standard print settings are employed. The ability to print sockets for connecting to common equipment like interface boards and multimeters as part of the printed structure is a major benefit of using 3DP to create electronic components like these, as opposed to a 2D printed electronics approach that uses technology like inkjet printing and would require the use of conductive glues and paints (Leigh et al., 2012).

For strain sensing applications, a flexible, conductive, 3D-printable thermoplastic-based material was successfully created. After compounding thermoplastic polyurethane and multiwalled carbon nanotubes (TPU/MWCNT) in the form of filaments, sensors were manufactured by employing FDM. The behaviors of mechanical, electrical, and piezoresistivity were examined under cyclic and monotonous loadings. MWNCTs increased the stiffness of TPU, improving its printing ability. The elasticity modulus of printed nanocomposites showed very slight decreases (~14%, relative to bulk equivalents), suggesting higher performance and outstanding interlayer adhesion over those documented in the literature. As a result, both the through-layer and cross-layer conductivity was substantially maintained post printing. Under applied stresses as high as 100%, piezoresistivity gauge factors as high as 176 was attained. Based on the findings, TPU/MWCNT is a great peizoresistive material for 3DP and has potential uses in prosthetics, wearable electronics, and soft robotics—all fields that need complicated design, multidirectionality, and customizability (Christ et al., 2017).

A study was carried out in order to investigate the change in sensitivity of 3D printed ethanol sensors by altering infill density and geometrical infill patterns. With the decrease in infill percentage from 75% to 50 %, an immense increase in relative resistance was recorded and a decrease in time from ~600 to ~200 seconds was observed to attain resistance saturation. In addition, different infill patterns have been investigated in the study, which resulted in different resistance. The reason behind the behavior is change in surface to volume ratio. It is found from this study that 3D-printed sensors can perform better as compared to their nonprinted analogues (Aliheidari et al., 2017).

12.4.5 Light responsive materials

Light sensitive materials show transformation on macroscopic level when they are subjected to irradiations (Liu & Urban, 2010). A printing method was developed to manufacture core and shell capsule to govern the discharge of a bioactive compounds. 3DP started with the printing of a core made of aqueous media with an active biomolecule, a polymeric emulsifier and a humectant; after that a water immiscible shell is printed. The shell was made using polylactic-co-glycolic (PLGA) loaded with nanorods of gold to promote a photo thermal effect, which initiated the shell rupture to release the bioactive compound from the core. The printing technique resulted in a stable shell with minimal defects (Gupta et al., 2015).

A 3D light sensitive shape memory device was fabricated by using polyurethane (PU) and carbon black (CB)-based light sensitive shape memory composite and FDM. The fabricated part depicted a good response to light due to addition of CB with an efficient photo thermal conversion. It was also noticed that fabricated part can also be triggered using the sunlight. It is believed by the author that this strategy will open up new avenues in stimuli responsive devices, which could be used in making soft robotics and smart devices (Yang et al., 2017).

There are a number of a cases in additive manufacturing where light responsive materials are used but their potential for robotics application is still deficient.

12.4.6 pH-sensitive materials

The pH-sensitive polymers belong to a class of polymers, which show a transition above and below a specific pH. These polymers have ionizable functional groups and they can be classified as polyacids and polybases.

A magnetic actuated pH-responsive hydrogel-based soft microrobot was fabricated for targeting and delivery of drug. The biocompatible 2-hydroxyethyl methacrylate (PHEMA) and poly(ethylene glycol) acrylate (PEGDA) bilayer structure of the suggested soft microrobot comprised eight radial arms and contained Fe_3O_4 nanoparticles. Initially, as a basic description of the hydrogels, the circular hydrogel patterns' ability was investigated to swell and deswell. The PEGDA-based hydrogel remained unaltered while the diameter of the PHEMA hydrogel pattern rose from 300 μm to 1800 μm when the pH value changed from 2.0 to 10.0. Secondly, the soft microrobot's form using a bilayer structure made of two hydrogels was examined at various pH levels. In the low-pH solution (pH=2.6), the soft microrobot had an unfolded shape, while in the high-pH solution (pH=9.58), it was folded. Thirdly, it was investigated that the soft microrobot containing the trapped microbead could be controlled by an electromagnetic actuation (EMA) system's magnetic field and that the trapped microbead's release action was indicated by the way it unfolded at a low pH in an H-shaped chamber (Li et al., 2016).

There is comparatively a smaller number of cases reported where pH-sensitive polymers are used in additive manufacturing, but they still have a great scope in additive manufacturing.

12.4.7 Self-healing materials

These are the materials, which have the capability to recover from damages either partially or fully. Multiple studies reported the self-healing materials (Collins et al., 2017). While 3DP of self-healing materials is rarely documented, it has been used recently to surpass the design constraints of self-healing materials beyond conventional casting (Zhang et al., 2015) and to initiate a self-healing mechanism to improve the mechanical structures of 3D printed components (Davidson et al., 2016).

Among all the self-healing materials, which are being used for 3D printing, physical gels are the most explored self-healing material. Self-healing gels that are prepared from 1-vinyl-1,2,4-triazole and Poly(N-acryloyl glycinamide) (PNAGA) are used for 3DP. The prepared gels are antiinflammatory, biocompatible and highly flexible. The gel possess excellent elasticity and have a potential to reach a magnitude of ~1300% elongation (Wang et al., 2017).

A versatile method was proposed for creating conductive hydrogels with high strength, which is achieved by merely doping the supramolecular copolymer hydrogels in situ, which are cross-linked with poly(3,4-ethylenedioxythiophene)-poly(styrenesulfonate) (PEDOT/PSS) through numerous hydrogen bonds during the polymerization process. The fabricated hydrogel possesses excellent mechanical strength, flexibility and conductivity up to 2.2 S/m. The most important characteristic of the fabricated hydrogel is self-healing property that assists in the recovery of the conductivity and mechanical characteristics. Due to the thermoplastic nature, proposed hydrogel could be used to fabricate 3D models using 3D printer. The proposed hydrogel have great properties for fabricating flexible electrode, electroactive scaffold material, and biosensors (Wu et al., 2017).

Even though self-healing materials got a significant attention recently regarding their application in 3D printing, but mostly self-healing material used for 3DP are noncovalently cross-linked gels. There is still a lot of potential for self-healing materials in additive manufacturing, particularly in robots.

12.5 PLA-based polymers and nanocomposites

Polylactic acid (PLA) is a thermoplastic-based polyester; the basic source of PLA is from biodegradable resources and PLA possesses various excellent properties including good processability, outstanding thermal and mechanical properties, and ecofriendly. Due to its origin from plants, PLA is acknowledged as a biodegradable material and used in tissue engineering and biomedical application specifically in biomedical implants (Tian et al., 2017).

A nanocomposite of PLA was manufactured in which PLA was used along with a filler, that is, reduced graphene oxide (rGO), for making flexible circuits. The synthetization of the manufactured nanocomposites (PLA/rGO) is carried out using melt compounding method. It was noticed that a 4–6 wt% increase in the concentration of rGO resulted in a substantial enhancement in electrical conductivity; the optimum range of rGO was found to be in between 4 and 8 wt% (Zhang et al., 2016).

Soft actuators are used in different robotics application such as in biomedical, for instance, catheters, and soft pipes are made up of soft robotics. An omnidirectional soft modular actuator (Gul et al., 2017) was fabricated using multilayer polylactic acid-based composites consisting of SMP, shape memory alloy (SMA), polyurethane, and graphene filament. Several printing techniques were used to fabricate the actuator including FDM printing, dip coating, and ultraviolet (UV) 3D printing. Experimental setup consisted of a transparent human artery made up of natural PLA. The actuator was tested under two different loading conditions: 0 and 10 g. Under these conditions, actuator showed an excellent control. It was believed by the authors that the alignment of nanoparticle is the reason behind the enhanced electrical conductivity (Zhang et al., 2016).

A flexure pressure sensor was fabricated using Makerboat PLA filament (Shi et al., 2015). A peizoresistive tactile sensor was fabricated using FDM by incorporating polylactic acid graphene (PLA-G) composite and thermoplastic polyurethane (TPU); the sensor demonstrated an excellent sensitivity and recovery to bending induced strains (Mousavi et al., 2018).

A novel type of robot was fabricated that had the ability to grow their own structure; the concept was taken from the plant roots and their ability to grow and move. A plant root is just like a robot that is designed and fabricated and develops its body through 3D printing. Each root of the robot is made up of a tubular structure and a growing head with sensory tip that governs robot behavior. The head is a 3D printer like system that builds the root by depositing PLA filament and robot moves by growing the root. A symmetry in the structure is created by differential deposition of filament, which results in curvature of the structure that helps the root to bend in order to reach the desired position. Experiments were carried out in order to validate efficacy of innovative robot, and results showed that a 50 mm diameter root grows with a speed of 4 mm/min and bending with a minimum radius of 100 mm (Sadeghi et al., 2017).

12.6 ABS-based nanocomposites

Acrylonitrile butadiene styrene (ABS) is one of the polymer filaments, which is widely used in 3D printing. Due to its solubility in acetone, ABS is easily extrudable and has the ability to hold higher proportion of filler due to that it is considered as a good material for fabrication of polymer composites (Palmero et al., 2019). ABS possesses good stability and flexural and impact strength and offers resistance against chemical environments.

A robotic arm has been made using ABS and PLA material, and a strength comparison is performed; the strength test was carried out on two different loading conditions, that is 5 N and 10 N. During both tests, it was noticed that less Von Misses stresses are generated in the robotic arm which was made up of ABS material (Demir & Coşgun, 2019).

12.7 Conclusion

This work aims to provide a thorough overview of the use of functional polymers and their nanocomposites fabricated through additive manufacturing, with a focus on their applications specifically in soft robotics. There has been substantial advancement in the field of printable materials in recent times, with smart polymer nanocomposites emerging as an exciting new class. However, the discipline is still in its infancy and is developing at a rapid pace. Different types of functional polymer nanocomposites are discussed; some have frequent application in electrical and robotics such as temperature-responsive polymers, electrically responsive polymers, and piezoelectric and peizoresistive-based polymers, whereas other polymers have comparatively less application at present in robotics, which includes light-responsive polymers, pH-sensitive polymers, and self-healing polymers. In the manufacturing of sensors and actuators, piezoelectric and piezoresistive materials are extensively employed. Self-healing polymers and gels are becoming popular as a way to enhance the mechanical properties and shelf-life of 3D-printed materials. Despite their popularity for years, pH and light responsive polymers have less applications and still have much more to offer. In short, functional polymers and nanocomposites have been used in additive manufacturing of various robotics applications; however, many nanocomposites exist whose potential needs to be explored. Functional polymers and nanocomposites have a promising future in robotics, and further research could result in more fruitful utilization of them in robotics.

References

Agarwala, S., Goh, G. L., Yap, Y. L., Goh, G. D., Yu, H., Yeong, W. Y., & Tran, T. (2017). Development of bendable strain sensor with embedded microchannels using 3D printing. *Sensors and Actuators A: Physical, 263*, 593–599. https://doi.org/10.1016/j.sna.2017.07.025.

Aliheidari, N., Hohimer, C., & Ameli, A. (2017). 3D-printed conductive nanocomposites for liquid sensing applications. Conference on Smart Materials, Adaptive Structures and Intelligent Systems, SMASIS 2017. United States: American Society of Mechanical Engineers.

Ameli, A., Nofar, M., Wang, S., & Park, C. B. (2014). Lightweight polypropylene/stainless-steel fiber composite foams with low percolation for efficient electromagnetic interference shielding. *ACS Applied Materials & Interfaces, 6*(14), 11091–11100. https://doi.org/10.1021/am500445g.

ASTM F2792-12 (2012). *Standard terminology for additive manufacturing technologies, ASTM International.*

Banerjee, H., Tse, Z. T. H., & Ren, H. (2018). Soft robotics with compliance and adaptation for biomedical applications and forthcoming challenges. *International Journal of Robotics and Automation, 33*(1), 69–80. https://doi.org/10.2316/Journal.206.2018.1.206-4981. www.actapress.com.

Barclift, M. W. & Williams, C. B. (2012). Examining variability in the mechanical properties of parts manufactured via polyjet direct 3D printing. 23rd Annual International Solid Freeform Fabrication Symposium—An Additive Manufacturing Conference, SFF 2012 876 890. Denmark: University of Texas at Austin (freeform).

Berg, G. J., McBride, M. K., Wang, C., & Bowman, C. N. (2014). New directions in the chemistry of shape memory polymers. *Polymer, 55*(23), 5849–5872. https://doi.org/10.1016/j.polymer.2014.07.052. http://www.journals.elsevier.com/polymer/.

Bodkhe, S., Turcot, G., Gosselin, F. P., & Therriault, D. (2017). One-step solvent evaporation-assisted 3D printing of piezoelectric PVDF nanocomposite structures. *ACS Applied Materials & Interfaces, 9*(24), 20833–20842. https://doi.org/10.1021/acsami.7b04095.

Chen, Y., Zhou, C., & Lao, J. (2011). A layerless additive manufacturing process based on CNC accumulation. *Rapid Prototyping Journal, 17*(3), 218–227. https://doi.org/10.1108/13552541111124806.

Chen, Z., Song, X., Lei, L., Chen, X., Fei, C., Chiu, C. T., Qian, X., Ma, T., Yang, Y., Shung, K., Chen, Y., & Zhou, Q. (2016). 3D printing of piezoelectric element for energy focusing and ultrasonic sensing. *Nano Energy, 27*, 78–86. https://doi.org/10.1016/j.nanoen.2016.06.048.

Christ, J. F., Aliheidari, N., Ameli, A., & Pötschke, P. (2017). 3D printed highly elastic strain sensors of multiwalled carbon nanotube/thermoplastic polyurethane nanocomposites. *Materials and Design, 131*, 394–401. https://doi.org/10.1016/j.matdes.2017.06.011.

Collins, J., Nadgorny, M., Xiao, Z., & Connal, L. A. (2017). Doubly dynamic self-healing materials based on oxime click chemistry and boronic acids. *Macromolecular Rapid Communications, 38*(6.).

Credi, C., Fiorese, A., Tironi, M., Bernasconi, R., Magagnin, L., Levi, M., & Turri, S. (2016). 3D printing of cantilever-type microstructures by stereolithography of ferromagnetic photopolymers. *ACS Applied Materials & Interfaces, 8*(39), 26332–26342. https://doi.org/10.1021/acsami.6b08880.

Davidson, J. R., Appuhamillage, G. A., Thompson, C. M., Voit, W., & Smaldone, R. A. (2016). Design paradigm utilizing reversible Diels-Alder reactions to enhance the mechanical properties of 3D printed materials. *ACS Applied Materials and Interfaces, 8*(26), 16961–16966. https://doi.org/10.1021/acsami.6b05118. http://pubs.acs.org/journal/aamick.

Demir, H., & Coşgun, A. E. (2019). *Comparison of PLA and ABS on robot arm model and 3D technology. Internet Archive, 6*, 38–44.

Estelle, K., Blair, D., Evans, K., & Gozen, B. A. (2017). Manufacturing of smart composites with hyperelastic property gradients and shape memory using fused deposition. *Journal of Manufacturing Processes, 28*, 500–507. https://doi.org/10.1016/j.jmapro.2017.04.018.

Ge, Q., Sakhaei, A. H., Lee, H., Dunn, C. K., Fang, N. X., & Dunn, M. L. (2016). Multimaterial 4D Printing with tailorable shape memory polymers. *Scientific Reports, 6*(1). https://doi.org/10.1038/srep31110.

Gul, J. Z., Yang, Y. J., Su, K. Y., & Choi, K. H. (2017). Omni directional multimaterial soft cylindrical actuator and its application as a steerable catheter. *Soft Robotics, 4*(3), 224–240. https://doi.org/10.1089/soro.2016.0042. http://www.liebertpub.com/overview/soft-robotics/616/.

Gupta, M. K., Meng, F., Johnson, B. N., Kong, Y. L., Tian, L., Yeh, Y. W., Masters, N., Singamaneni, S., & McAlpine, M. C. (2015). 3D printed programmable release capsules. *Nano Letters, 15*(8), 5321–5329. https://doi.org/10.1021/acs.nanolett.5b01688. http://pubs.acs.org/journal/nalefd.

Hirokawa, Y., Jinnai, H., Nishikawa, Y., Okamoto, T., & Hashimoto, T. (1999). Direct Observation of internal structures in poly(N-isopropylacrylamide) chemical gels. *Macromolecules, 32*(21), 7093–7099. https://doi.org/10.1021/ma990437v.

Hofmann, M. (2014). 3D Printing gets a boost and opportunities with polymer materials. *ACS Macro Letters, 3*(4), 382–386. https://doi.org/10.1021/mz4006556.

Holman, H., Kavarana, M. N., & Rajab, T. K. (2021). Smart materials in cardiovascular implants: Shape memory alloys and shape memory polymers. *Artificial Organs, 45*(5), 454–463. https://doi.org/10.1111/aor.13851. http://onlinelibrary.wiley.com/journal/10.1111/(ISSN)1525-1594.

Kennedy, Z. C., Christ, J. F., Evans, K. A., Arey, B. W., Sweet, L. E., Warner, M. G., Erikson, R. L., & Barrett, C. A. (2017). 3D-printed poly(vinylidene fluoride)/carbon nanotube composites as a tunable, low-cost chemical vapour sensing platform. *Nanoscale, 9*(17), 5458–5466. https://doi.org/10.1039/c7nr00617a. http://www.rsc.org/publishing/journals/NR/Index.asp.

Kim, H., Torres, F., Villagran, D., Stewart, C., Lin, Y., & Tseng, T. L. B. (2017). 3D Printing of BaTiO3/PVDF composites with electric in situ poling for pressure sensor applications. *Macromolecular Materials and Engineering, 302*(11). https://doi.org/10.1002/mame.201700229. http://www3.interscience.wiley.com/journal/117930403/grouphome/home.html.

Kim, K., Zhu, W., Qu, X., Aaronson, C., McCall, W. R., Chen, S., & Sirbuly, D. J. (2014). 3D optical printing of piezoelectric nanoparticle–polymer composite materials. *ACS Nano, 8*(10), 9799–9806. https://doi.org/10.1021/nn503268f.

Kim, K., Park, J., Suh, J.-hoon, Kim, M., Jeong, Y., & Park, I. (2017). 3D printing of multiaxial force sensors using carbon nanotube (CNT)/thermoplastic polyurethane (TPU) filaments. *Sensors and Actuators A: Physical, 263*, 493–500. https://doi.org/10.1016/j.sna.2017.07.020.

Kim, H., Islam, M. T., Md Didarul, I., Chavez, L. A., Garcia Rosales, C. A., Wilburn, B. R., Stewart, C. M., Noveron, J. C., Tseng, T.-L. B., & Lin, Y. (2017). Increased piezoelectric response in functional nanocomposites through multiwall carbon nanotube interface and fused-deposition modeling three-dimensional printing. *MRS Communications, 7*(4), 960–966. https://doi.org/10.1557/mrc.2017.126.

Kruth, J.-P., Mercelis, P., Van Vaerenbergh, J., Froyen, L., & Rombouts, M. (2005). Binding mechanisms in selective laser sintering and selective laser melting. *Rapid Prototyping Journal, 11*(1), 26–36. https://doi.org/10.1108/13552540510573365.

Lee, C., & Tarbutton, J. (2014). Tarbutton, and structures, Electric poling-assisted additive manufacturing process for PVDF polymer-based piezoelectric device applications. *Smart Materials and Structures, 23*, 2014.

Leigh, S. J., Bradley, R. J., Purssell, C. P., Billson, D. R., Hutchins, D. A., & Hong, J. (2012). A simple, low-cost conductive composite material for 3d printing of electronic sensors. *PLoS ONE, 7*(11), e49365. https://doi.org/10.1371/journal.pone.0049365.

Li, H., Go, G., Ko, S. Y., Park, J.-O., & Park, S. (2016). Magnetic actuated pH-responsive hydrogel-based soft micro-robot for targeted drug delivery. *Smart Materials and Structures, 25*(2), 027001. https://doi.org/10.1088/0964-1726/25/2/027001.

Liu, F., & Urban, M. W. (2010). Recent advances and challenges in designing stimuli-responsive polymers. *Progress in Polymer Science (Oxford), 35*(1-2), 3–23. https://doi.org/10.1016/j.progpolymsci.2009.10.002.

Liu, H., Dong, M., Huang, W., Gao, J., Dai, K., Guo, J., Zheng, G., Liu, C., Shen, C., & Guo, Z. (2017). Lightweight conductive graphene/thermoplastic polyurethane foams with ultrahigh compressibility for piezoresistive sensing. *Journal of Materials Chemistry C, 5*(1), 73–83. https://doi.org/10.1039/C6TC03713E.

Mousavi, S., Howard, D., Wu, S., & Wang, C. (2018). An ultrasensitive 3D printed tactile sensor for soft robotics. arXiv, Australia arXiv. https://arxiv.org. https://doi.org/10.48550/arxiv.1810.09236.

Muth, J. T., Vogt, D. M., Truby, R. L., Mengüç, Y., Kolesky, D. B., Wood, R. J., & Lewis, J. A. (2014). Embedded 3D printing of strain sensors within highly stretchable elastomers. *Advanced Materials, 26*(36), 6307–6312. https://doi.org/10.1002/adma.201400334. http://onlinelibrary.wiley.com/journal/10.1002/(ISSN)1521-4095.

Okpala, C. C. (2013). Nanocomposites—An overview. *International Engineering Journal for Research & Development, 8*, 17–23.

Palmero, E. M., Casaleiz, D., de Vicente, J., Hernández-Vicen, J., López-Vidal, S., Ramiro, E., & Bollero, A. (2019). Composites based on metallic particles and tuned filling factor for 3D-printing by Fused Deposition Modeling. *Composites Part A: Applied Science and Manufacturing, 124*, 105497. https://doi.org/10.1016/j.compositesa.2019.105497.

Paul, D. R., & Robeson, L. M. (2008). Polymer nanotechnology: Nanocomposites. *Polymer, 49*(15), 3187–3204. https://doi.org/10.1016/j.polymer.2008.04.017.

Ramadan, K. S., Sameoto, D., & Evoy, S. (2014). A review of piezoelectric polymers as functional materials for electromechanical transducers. *Smart Materials and Structures, 23*(3), 033001. https://doi.org/10.1088/0964-1726/23/3/033001.

Robinson, S. S., O'Brien, K. W., Zhao, H., Peele, B. N., Larson, C. M., Mac Murray, B. C., Van Meerbeek, I. M., Dunham, S. N., & Shepherd, R. F. (2015). Integrated soft sensors and elastomeric actuators for tactile machines with kinesthetic sense. *Extreme Mechanics Letters, 5*, 47–53. https://doi.org/10.1016/j.eml.2015.09.005. http://www.journals.elsevier.com/extreme-mechanics-letters/.

Sadeghi, A., Mondini, A., & Mazzolai, B. (2017). Toward self-growing soft robots inspired by plant roots and based on additive manufacturing technologies. *Soft Robotics, 4*(3), 211–223. https://doi.org/10.1089/soro.2016.0080. http://www.liebertpub.com/overview/soft-robotics/616/.

Sadik Batcha, M. (2017). Research output analysis on robotic technology: A scientometric study. *Indian Journal of Information Sources and Services, 7*(1), 25–31. https://doi.org/10.51983/ijiss.2017.7.1.461.

Sano, K., Ishida, Y., & Aida, T. (2018). Synthesis of anisotropic hydrogels and their applications. *Angewandte Chemie—International Edition, 57*(10), 2532–2543. https://doi.org/10.1002/anie.201708196. http://onlinelibrary.wiley.com/journal/10.1002/(ISSN)1521-3773.

Schmitt, F., Piccin, O., Barbé, L., & Bayle, B. (2018). Soft robots manufacturing: A review. *Frontiers in Robotics and AI, 5*. https://doi.org/10.3389/frobt.2018.00084.

Shi, H. L., Kim, Y., Dagalakis, N., & Duan X. C. (2015). A 3D printing flexure pressure sensor for robot impact safety testing. IET Conference Publications. United States: Institution of Engineering and Technology. www.ietdl.org/CP.

Stassi, S., Cauda, V., Canavese, G., & Pirri, C. (2014). Flexible tactile sensing based on piezoresistive composites: A review. *Sensors (Basel, Switzerland), 14*(3), 5296–5332. https://doi.org/10.3390/s140305296.

Sun, L., Huang, W. M., Ding, Z., Zhao, Y., Wang, C. C., Purnawali, H., & Tang, C. (2012). Stimulus-responsive shape memory materials: A review. *Materials & Design, 33*(1), 577–640. https://doi.org/10.1016/j.matdes.2011.04.065.

Tian, X., Liu, T., Wang, Q., Dilmurat, A., Li, D., & Ziegmann, G. (2017). Recycling and remanufacturing of 3D printed continuous carbon fiber reinforced PLA composites. *Journal of Cleaner Production, 142*, 1609–1618. https://doi.org/10.1016/j.jclepro.2016.11.139.

Wang, H., Zhu, H., Fu, W., Zhang, Y., Xu, B., Gao, F., Cao, Z., & Liu, W. (2017). A high strength self-healable antibacterial and anti-inflammatory supramolecular polymer hydrogel. *Macromolecular Rapid Communications, 38*(9), 1600695. https://doi.org/10.1002/marc.201600695.

Wang, X., Guo, Q., Cai, X., Zhou, S., Kobe, B., & Yang, J. (2014). Initiator-integrated 3D printing enables the formation of complex metallic architectures. *ACS Applied Materials & Interfaces, 6*(4), 2583–2587. https://doi.org/10.1021/am4050822.

Wegner, G. (2000). Functional polymers. *Acta Materialia, 48*(1), 253–262. https://doi.org/10.1016/S1359-6454(99)00298-0.

Wu, Q., Wei, J., Xu, B., Liu, X., Wang, H., Wang, W., Wang, Q., & Liu, W. (2017). A robust, highly stretchable supramolecular polymer conductive hydrogel with self-healability and thermo-processability. *Scientific Reports, 7*(1). https://doi.org/10.1038/srep41566.

Yang, H., Leow, Wan R., Wang, T., Wang, J., Yu, J., He, K., Qi, D., Wan, C., & Chen, X. (2017). 3D printed photoresponsive devices based on shape memory composites. *Advanced Materials, 29*(33). https://doi.org/10.1002/adma.201701627.

Yang, Y., Chen, Y., Wei, Y., & Li, Y. (2016). 3D printing of shape memory polymer for functional part fabrication. Springer London, Hong Kong. *International Journal of Advanced Manufacturing Technology, 84*(9-12), 2079–2095. https://doi.org/10.1007/s00170-015-7843-2. http://www.springerlink.com/content/0268-3768.

Yap, Y. L., Sing, S. L., & Yeong, W. Y. (2020). A review of 3D printing processes and materials for soft robotics. *Rapid Prototyping Journal, 26*(8), 1345–1361. https://doi.org/10.1108/RPJ-11-2019-0302. http://www.emeraldinsight.com/info/journals/rpj/rpj.jsp.

Zhang, D., Chi, B., Li, B., Gao, Z., Du, Y., Guo, J., & Wei, J. (2016). Fabrication of highly conductive graphene flexible circuits by 3D printing. *Synthetic Metals, 217*, 79–86. https://doi.org/10.1016/j.synthmet.2016.03.014.

Zhang, M., Vora, A., Han, W., Wojtecki, R. J., Maune, H., Le, A. B. A., Thompson, L. E., McClelland, G. M., Ribet, F., Engler, A. C., & Nelson, A. (2015). Dual-responsive hydrogels for direct-write 3D printing. *Macromolecules, 48*(18), 6482–6488. https://doi.org/10.1021/acs.macromol.5b01550.

Zhang, X., Wu, Y., Li, Y., Jiang, H., Yang, Q., Wang, Z., Liu, J., Wang, Y., Fan, X., & Kong, J. (2022). Small-scale soft grippers with environmentally responsive logic gates. *Materials Horizons, 9*(5), 1431–1439. https://doi.org/10.1039/d2mh00097k.

Additive manufacturing of functional polymers and nanocomposites for automotive applications

13

Mohammad Harun-Ur-Rashid
Department of Chemistry, International University of Business Agriculture and Technology (IUBAT), Uttara Model Town, Uttara, Dhaka, Bangladesh

13.1 Introduction

Additive manufacturing (AM), known broadly as 3D printing, marks a transformative shift in the manufacturing landscape, particularly within the automotive industry. This technology transcends traditional manufacturing barriers, introducing unprecedented levels of flexibility, customization, and efficiency. The application of AM in automotive manufacturing is not just a technical evolution but a strategic revolution, enabling manufacturers to overcome design constraints, reduce lead time, and achieve significant cost efficiencies. The automotive industry is under constant pressure to innovate, driven by consumer demands for higher performance and stricter environmental regulations. AM emerges as a key enabler in this context, offering the ability to produce lighter, stronger, and more complex components that traditional methods cannot. Unlike conventional subtractive manufacturing processes that carve out materials, AM builds objects layer by layer according to digital 3D models, drastically reducing material waste. The precision and versatility of AM allow for the integration of design and function in component fabrication, promoting a more agile manufacturing process adaptable to changing market needs. AM encompasses several technologies, each suited to different materials and applications. Fused Deposition Modeling (FDM), the most widely used AM technique, extrudes thermoplastic polymers to create high-strength parts (Kristiawan et al., 2021). Selective Laser Sintering (SLS), another prevalent method, uses a laser to sinter powdered material, making it ideal for creating complex and durable components (Duan et al., 2010). Stereolithography (SLA) employs an ultraviolet laser to cure photosensitive resins, offering high-resolution finishes ideal for detailed esthetic models (Lakkala et al., 2023). Direct Metal Laser Sintering (DMLS) (Kumar Maurya et al., 2021), a technique similar to SLS but for metals, is particularly transformative in automotive manufacturing, enabling the production of robust metal parts directly from digital files. These technologies collectively broaden the spectrum of materials and applications in the automotive sector, from prototyping to full-scale production.

The adoption of AM in automotive manufacturing introduces several benefits. The most significant is the reduction in part weight, crucial for enhancing fuel efficiency

Additive Manufacturing of Functional Polymers and Nanocomposites. DOI: https://doi.org/10.1016/B978-0-443-26516-7.00013-4
© 2025 Elsevier Ltd. All rights are reserved, including those for text and data mining, AI training, and similar technologies.

and reducing emissions in line with global environmental standards. AM's ability to consolidate multiple components into a single, complex structure without the need for assembly further streamlines production and increases the structural integrity of parts. Additionally, the technology's capacity for rapid prototyping accelerates the design process, allowing for faster iteration and innovation cycles. This agility is vital in a competitive industry where speed to market can be a critical advantage. Moreover, the customization potential of AM is particularly valuable in high-performance and luxury automotive segments, where bespoke designs and features can provide a competitive edge. The integration of AM into automotive manufacturing dates back to the 1980s when it was initially employed for rapid prototyping (Pei et al., 2023). The ability to quickly fabricate models and parts significantly shortened the development cycles for new vehicles. As AM technologies evolved, their application expanded beyond prototyping to include tooling and direct part manufacturing. Today, leading automotive manufacturers and suppliers utilize AM for a broad range of applications, from creating intricate lightweight structures to fabricating end-use parts. Examples include the production of complex fuel nozzles, lightweight chassis components, and customized interior features. These applications highlight the maturation of AM from a tool for visual and functional prototypes to an integral part of the manufacturing process capable of producing durable, operational components directly from digital data (Bacciaglia et al., 2022; Vasco, 2021). Table 13.1 provides an at-a-glance comparison of key aspects of traditional and AM methods relevant to the automotive industry. Each method offers unique advantages and faces specific challenges, making the choice between them dependent on the particular needs of a production scenario.

As explored in the subsequent sections, the adoption of functional polymers and nanocomposites further enhances the capabilities of AM. These materials are engineered to meet specific performance criteria such as high temperature resistance, mechanical strength, and flexibility, essential for the demanding environments of automotive applications. The synergy between AM technologies and these advanced materials not only fuels innovation in automotive design and production but also aligns with broader industry objectives such as sustainability and efficiency optimization.

13.2 Overview of AM technologies

13.2.1 Types of AM technologies

13.2.1.1 Fused deposition modeling

FDM is a popular 3D printing method within the AM spectrum, notable for its ability to efficiently produce customized items without the additional costs associated with traditional manufacturing methods like mold and tool creation (Kristiawan et al., 2021). This technique allows for the production of complex shapes with minimal material waste, leveraging its capability for mass customization at reduced costs. The FDM process involves melting a thermoplastic filament, which is then extruded layer by layer to form structures as dictated by CAD software inputs. The material

Table 13.1 Comparison of traditional versus AM techniques in automotive production.

Feature	Traditional manufacturing	AM
Flexibility in design	Limited by tooling and machinery	High, allows complex geometries
Speed of production	High for large volumes	Lower, best for small batches
Cost of production	Economical at scale	Higher for small runs, decreases with scale
Material waste	Generally higher, excess material often trimmed away	Minimal, materials added layer by layer
Customization	Costly and time-consuming	Easily accommodates customization
Tooling requirements	Significant, molds and dies required	Minimal to none, reduces upfront investment
Lead time	Long, due to tooling and setup	Short, direct from digital files
Material diversity	Wide variety available	Growing, but more limited than traditional methods
Part consolidation	Limited, often requires assembly	High, can produce complex integrated parts
Energy consumption	High, due to scale and machinery	Generally lower per piece, but dependent on technology
Prototype development	Slow and expensive	Rapid and cost-effective
Strength of final product	Generally strong, dependent on manufacturing process	Varies, can be optimized with design and material choice
Surface finish	Typically smooth, dependent on process	May require postprocessing for smooth finish
Suitability forlarge-scale production	High	Developing, currently more suited to low-volume or specialized applications

properties of the filament, such as thermoplasticity, are crucial as they influence the bonding between layers and the overall structural integrity of the final product. Currently, polymers like polylactic acid (PLA) and acrylonitrile butadiene styrene (ABS) are extensively used, with materials like polypropylene (PP) gaining attention due to their commonality in daily life and superior properties like heat and chemical resistance. Research indicates variability in mechanical properties of FDM products based on filament composition and printing parameters, highlighting the complex interplay of factors that affect the FDM process. This study reviews the critical parameters and their interactions within the FDM process, from filament production to the final printing stages, aiming to optimize these factors to enhance the mechanical properties of printed products suitable for specific applications.

The filament, a crucial component in the FDM process, is typically made from low-melting-point pure polymers. However, to enhance the strength and functional properties of these filaments, polymer composites are developed by mixing a matrix with reinforcing agents. These composite filaments require special treatments due to their varied characteristics compared to pure polymers. For pure polymer filaments, the production process involves extruding polymer pellets through dies using extruders to form the filament. In contrast, producing composite filaments involves premixing materials, which can be done via solution mixing followed by drying or by dry mixing using a stirring machine with spinning roller blades at the polymer's melting point. After adding compatibilizers and additives, the mixture is allowed to cool to ensure uniformity, then processed into small pellets or fabric pieces for extrusion. During extrusion, parameters like die temperature, roller speed, and spindle speed are critical as they influence the viscosity and the eventual diameter of the filament. Additionally, in composite filaments, the thread shape impacts the filler orientation, affecting the filament's regularity. The entire process involves setting up the machine, defining the filament diameter, preparing the material, and extruding it to wrap around the roller, ensuring consistent filament production. Fig. 13.1 illustrates the filament making process.

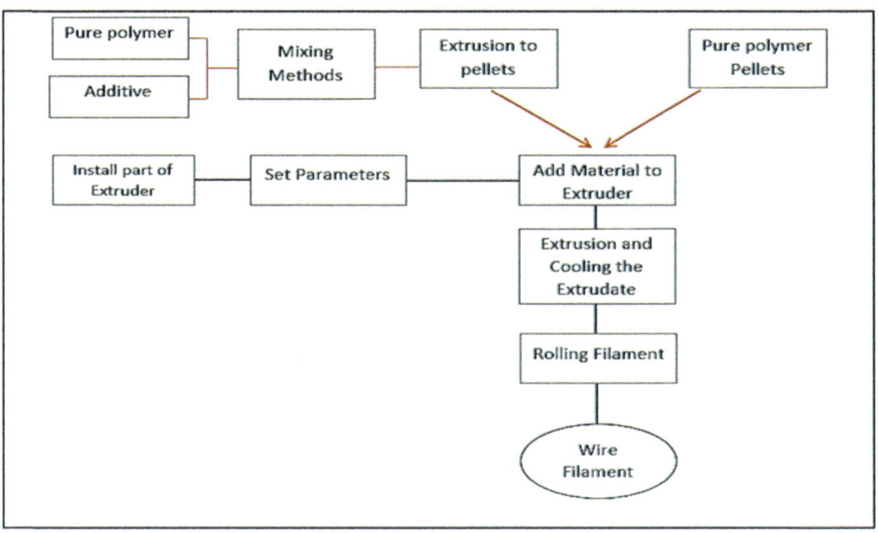

Figure 13.1 Filament fabrication process for FDM.
The figure has been reproduced with the permission from Kristiawan et al. (2021). Copyright 2021 by the authors. Published by De Gruyter. From Kristiawan, R.B., Imaduddin, F., Ariawan, D., Ubaidillah, & Arifin, Z. (2021). A review on the fused deposition modeling (FDM) 3D printing: Filament processing, materials, and printing parameters. Open Engineering, 11(1), 639–649. https://doi.org/10.1515/eng-2021-0063.

In FDM, filament production involves both pure polymers and polymer composites. Pure polymers are extruded from pellets through dies, whereas composites require premixing with compatibilizers and additives to enhance their properties before extrusion. Parameters like die temperature and roller speeds are critical during extrusion as they affect the filament's diameter and viscosity. For printing parameters, raster angle and width are crucial for part strength and material efficiency. Raster angle influences load distribution, with certain angles enhancing or weakening a specimen's strength. Raster width affects the build's internal strength and production speed. Contour width and the number of contours also play roles in defining part structure, while the air gap impacts the final product's surface quality and dimensional accuracy. Various build styles in FDM allow customization of the part's density and appearance, influencing manufacturing time and material usage. These styles range from solid fills to sparse configurations using different raster patterns, affecting the part's surface finish and shrinkage, thus impacting overall quality and accuracy.

13.2.1.2 Selective laser sintering (SLS)

SLS is a sophisticated AM technique vital for producing high-strength, complex parts for the automotive industry. Using a high-power laser, SLS fuses polymer powder into solid structures layer by layer, starting with a thin layer of powder on the build platform and selectively sintering it according to the part's design, without needing support structures. This allows for more intricate designs and a cleaner manufacturing process. SLS uses functional polymers like polyamide (Nylon 12), known for their durability and heat resistance. The integration of nanocomposites, such as carbon nanotubes or silica, further enhances the mechanical and thermal properties of these materials, making them ideal for essential automotive components like underhood parts and hinges, thereby improving component performance, customization, and production efficiency.

Meng et al. (2022) reported on the preparation of graphene platelets (GnPs)-coated polyamide (PA) nanocomposites using an electrostatic self-assembly method, followed by SLS. The study verified that a low concentration (0.2 wt%) of GnPs is sufficient to improve the mechanical properties significantly, increasing the Young's modulus by 81%. This enhancement is attributed to the high structural integrity of GnPs, which also improves electrical and thermal conductivity when integrated into the PA matrix. The authors developed a numerical model to simulate the mechanical behavior of these 3D printed composites, showing good agreement with experimental results. This work presents a promising approach to fabricate high-performance materials for engineering applications, highlighting the role of GnPs in enhancing the functionalities of polymer nanocomposites through a cost-effective and industrially scalable process. Fig. 13.2 illustrates the process of fabricating graphene platelets (GnPs) coated polyamide (PA) nanocomposites for SLS. Initially, GnPs are mixed with polyamide in a stirring process to ensure even distribution. The mixture undergoes filtration and drying to prepare the composite material in powder form. Subsequently, the powder is screened to ensure uniformity and then used in an SLS

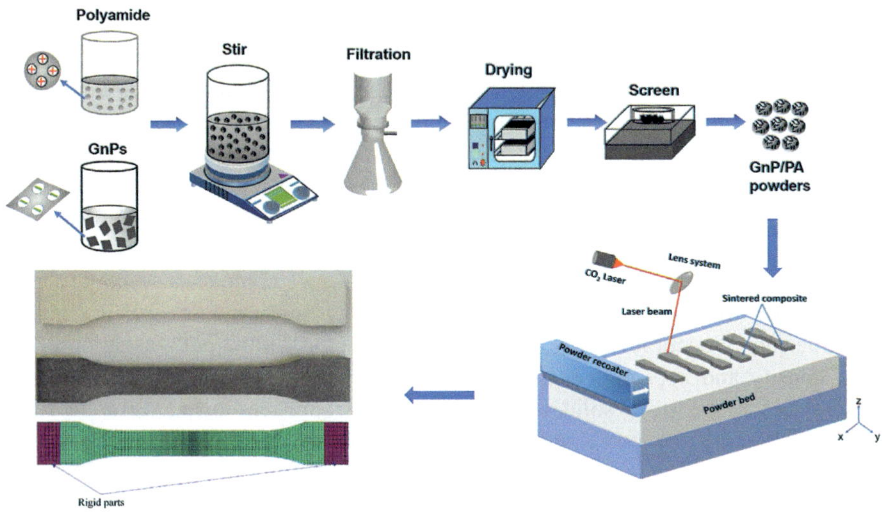

Figure 13.2 Process flow diagram illustrating the preparation of polyamide/graphene nanocomposites and their application in SLS to produce mechanically enhanced 3D-printed components.
The figure has been reproduced with the permission from Meng et al. (2022). Copyright 2022 by Elsevier B.V. From Meng, Qingshi, Song, Xuyang, Han, Sensen, Abbassi, Fethi, Zhou, Zhiqiang, Wu, Bin, Wang, Xiangming & Araby, Sherif. (2022). Mechanical and functional properties of polyamide/graphene nanocomposite prepared by chemicals free-approach and selective laser sintering. Composites Communications, 36. https://doi.org/10.1016/j.coco.2022.101396.

printer where a CO_2 laser selectively sinters the powder to form rigid composite structures.

The SLS significantly benefits automotive manufacturing by enabling economical production of complex, custom parts without additional tooling. This makes it ideal for luxury vehicles and replacement parts where traditional methods are cost-prohibitive. SLS parts exhibit isotropic mechanical properties, ensuring consistent performance under dynamic loads. However, SLS faces challenges such as a limited range of suitable materials and the need for postprocessing to improve the esthetics of visible parts, increasing costs and production time. Addressing thermal stress management and dimensional accuracy in larger parts remains crucial. Future improvements in SLS aim to enhance its application in automotive manufacturing by introducing faster sintering processes, larger build volumes, and advanced powder management techniques. The development of tailored polymer-nanocomposite blends is expected to enhance performance, supporting the industry's shift toward lighter, more efficient vehicles. SLS continues to evolve, promising greater material efficiency and design flexibility in automotive production.

13.2.1.3 Stereolithography (SLA)

SLA is a sophisticated AM technology particularly well-suited for creating detailed and high-quality parts using photo-reactive resin materials. In the automotive industry, SLA is primarily utilized for prototypes and complex components that require precise dimensions and smooth surface finishes. This technology harnesses the power of ultraviolet (UV) light to cure liquid resins into solid plastics, layer by layer, enabling the production of parts with intricate details and superior esthetic qualities. SLA operates by using an ultraviolet laser to trace and cure the surface of a liquid photopolymer resin contained in a vat. The laser beam selectively cures the resin according to the 3D model's specifications, hardening it into a solid layer. Once a layer is complete, the build platform moves up or down, depending on the machine configuration, allowing a new layer of resin to cover the surface and be cured by the laser. This process is repeated until the entire part is built. The ability of SLA to produce fine details and complex geometries makes it especially valuable for high-fidelity prototypes and parts that require tight tolerances. A typical SLA AM technology apparatus includes five main components: the build platform, optics, vat, recoater, and control system, as shown in Fig. 13.3 A (Lakkala et al., 2023). The vat holds viscous liquid resin that is smoothed by a recoater to ensure an even surface for printing. The build platform, which moves vertically, facilitates the layer-by-layer fabrication process. The optics system employs a light source, optical modulators, a

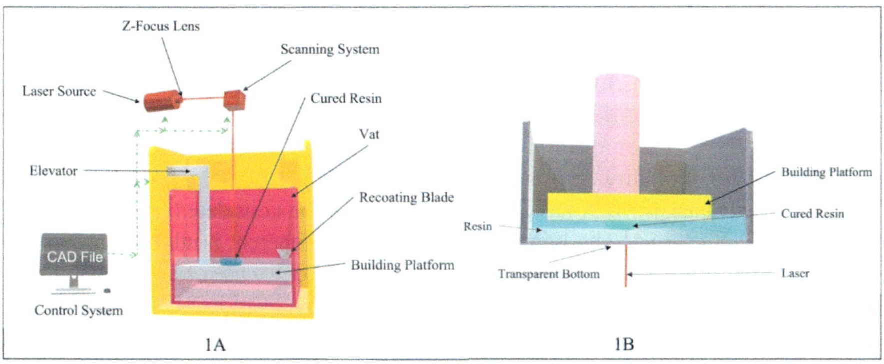

Figure 13.3 Diagram of the apparatus configurations for SLA in AM process. Panel A (left) depicts a top-down SLA setup, highlighting components such as the laser source, scanning system, recoating blade, and elevator within the vat. Panel B (right) shows a bottom-up SLA setup with a transparent bottom vat and a building platform positioned above it, illustrating the path of the laser beam and the positioning of the cured resin layers. Both setups are controlled by a system connected to a CAD file for precise design execution.
The figure has been reproduced with the permission from Lakkala et al. (2023). Copyright 2023 by the authors. Published by Elsevier B.V. From Lakkala, Preethi, Munnangi, Siva Ram, Bandari, Suresh & Repka, Michael. (2023). Additive manufacturing technologies with emphasis on stereolithography 3D printing in pharmaceutical and medical applications: A review. International Journal of Pharmaceutics: X, 5. https://doi.org/10.1016/j.ijpx.2023.100159.

Z-focus lens, and galvanometers to precisely focus and scan the laser beam across the resin. The control system manages machine operations, scanning speeds, and environmental conditions within the vat chamber. SLA printers are classified into "Top-Down" and "Bottom-Up" based on the light source's orientation and projection direction. In the top-down setup (Fig. 13.3A), the light source is above the vat, curing the resin as the build platform descends into it layer by layer. This method can be inefficient due to surface disturbances and oxygen inhibition affecting curing. Conversely, the bottom-up approach (Fig. 13.3B) features a light source below the vat, projecting upward. This setup requires less resin and uses a nonsticky, transparent film at the vat's bottom to facilitate layer detachment. It avoids oxygen inhibition, enhancing curing efficiency but is limited to smaller prints due to vat capacity and the need for supports to counteract gravity during printing.

In SLA, the choice of material is pivotal to achieving the desired part characteristics. The resins used are typically composed of acrylic or epoxy photopolymers that can be formulated with various additives to enhance mechanical properties, thermal resistance, and optical clarity. Recently, the incorporation of nanocomposites into SLA resins has been explored to further enhance these properties. For example, adding silica or titanium dioxide nanoparticles can improve the material's stiffness and heat deflection temperature, making SLA-produced parts more suitable for demanding automotive applications such as lens covers, light fixtures, and fluid flow systems. One of the most significant advantages of SLA is its ability to produce parts with very high resolution and excellent surface finish, which are essential for esthetic and functional components in automotive applications. The precision of SLA is particularly useful for creating complex, thin-walled structures and intricate designs that are difficult to achieve with other AM technologies. Additionally, the wide range of resin formulations available allows engineers to tailor material properties to specific application requirements, enhancing the functionality of prototypes and end-use parts.

Despite its advantages, SLA has limitations that must be considered in the context of automotive manufacturing. The primary challenge is the inherent brittleness and limited durability of standard photopolymer resins, which can restrict the use of SLA parts in load-bearing or high-impact applications. Moreover, the materials used in SLA generally have lower thermal and chemical resistance compared to those used in other AM processes like FDM or SLS, potentially limiting their use in under-hood components or areas exposed to harsh environmental conditions. Additionally, the postprocessing required for cleaning and curing SLA parts can be labor-intensive and time-consuming. Advancements in resin chemistry and SLA technology are continuously expanding the range of applications for this method in the automotive industry. Ongoing research aims to develop new resin formulations with enhanced properties such as increased toughness, higher thermal stability, and improved environmental resistance. These developments could potentially open up new avenues for using SLA in more structurally demanding and functional applications. Moreover, innovations in machine technology that allow for larger build volumes and faster printing speeds are poised to enhance the efficiency and cost-effectiveness of SLA for automotive manufacturing. SLA offers distinct benefits for the production of high-precision and esthetically superior automotive parts. As resin technologies evolve and SLA equipment improves, the potential applications of this technology in the automotive sector are expected to grow, making it an increasingly vital component of the industry's AM toolkit.

13.2.1.4 Direct metal laser sintering (DMLS)

DMLS is a pivotal AM technology particularly beneficial for the automotive industry, due to its capacity to fabricate high-strength, durable metal parts directly from digital data. As a derivative of the powder bed fusion process, DMLS has transformed the production of complex metal components, enabling innovations in design and functionality that traditional manufacturing methods cannot match. This technology is especially crucial in producing critical automotive components that require superior mechanical properties and thermal resistance. DMLS utilizes a high-power laser to sinter fine metal powder, layer by layer, to build parts that are nearly fully dense and comparable in strength to forged parts. The process begins with the application of a thin layer of metal powder to a build platform. A laser then selectively sinters the powder, melting it so that it solidifies and bonds to the layer below. After each layer is sintered, the build platform descends slightly, and a new layer of powder is spread on top. This sequence repeats until the part is complete. The capability of DMLS to produce geometrically complex parts without the need for additional tooling or molds offers substantial benefits in terms of design flexibility and speed of production. DMLS can process a wide variety of metals and alloys, including stainless steel, titanium, aluminum, and nickel-based superalloys, making it highly versatile for automotive applications. The introduction of metal matrix composites (MMCs) into DMLS—where metal powders are mixed with ceramic particles such as carbides or oxides—further enhances the mechanical properties of parts, such as wear resistance, hardness, and operational stability at high temperatures. These materials are particularly useful in manufacturing components like turbine blades, engine parts, and transmission components, which require high performance under extreme conditions. Fig. 13.4 illustrates the working procedure of DMLS (Kumar Maurya et al., 2021).

The primary advantage of DMLS in the automotive sector is its ability to consolidate multiple parts into a single, more complex component, reducing assembly time and increasing the overall strength and reliability of the part. This consolidation is crucial for lightweighting efforts, as it helps reduce the overall weight of the vehicle, thereby enhancing fuel efficiency and reducing emissions. Additionally, DMLS can significantly shorten the development cycle of new parts, allowing for rapid prototyping and testing, which accelerates innovation and reduces time to market. Despite its advantages, DMLS faces challenges, particularly in terms of cost and production speed. The initial setup and material costs can be high, and the sintering process itself is relatively slow compared to other AM technologies. Moreover, parts produced by DMLS often require significant postprocessing, including heat treatment to relieve residual stresses and surface finishing to improve their esthetic and functional qualities. Managing these aspects is crucial for ensuring the economic viability of DMLS for large-scale automotive applications. The future of DMLS in automotive manufacturing looks promising with ongoing advancements in laser technology, machine capabilities, and powder metallurgy.

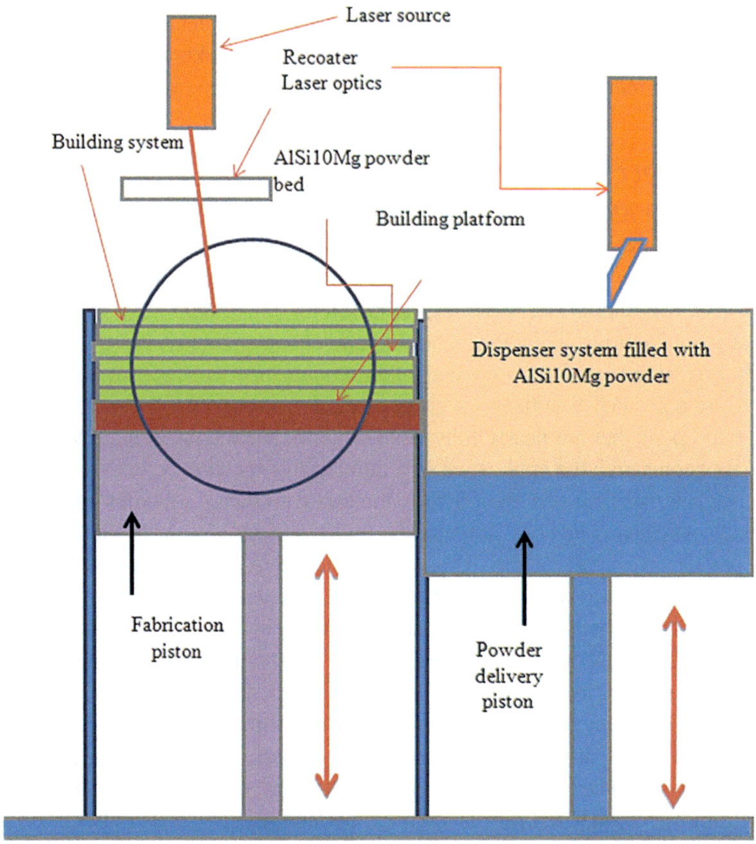

Figure 13.4 Schematic representation of the working procedure of DMLS.
From Kumar Maurya, Nagendra, Sharma, Rohit, Kumar, Nikhil, Kumar, Anubhav, Anand, Piyush, Rai, Prakhar & Singh, Harjeet. (2021). An overview of investigation of Fatigue, tensile strength and hardness of the components fabricated through direct metal laser sintering (DMLS) process. Materials Today: Proceedings, 47, 3979–3984. https://doi.org/10.1016/j.matpr.2021.04.131. Copyright 2021 by Elsevier B.V.

Research is focused on developing faster, more efficient DMLS machines and introducing new powder materials that can further expand the application range of this technology. Optimizations in process control and postprocessing techniques are also expected to reduce costs and improve the surface quality of DMLS parts, making this technology even more attractive for automotive manufacturers. DMLS stands as a transformative technology in the field of AM, offering significant potential to drive innovation in automotive design and production. With its ability to handle complex geometries and high-performance materials, DMLS is poised to play an increasingly critical role in the future of automotive manufacturing, aligning with industry trends toward greater efficiency, customization, and performance.

13.2.1.5 Electron beam melting (EBM)

EBM is an advanced AM technique that uses a high-power electron beam to fuse metal powder into solid parts. EBM is distinguished by its ability to process high-melting-point materials such as titanium alloys, which are extensively used in high-performance automotive applications. This technique is especially valuable for producing complex, lightweight, and strong components that can withstand extreme operating conditions typically encountered in automotive environments. EBM works under vacuum conditions, where a powerful electron beam generates energy that selectively melts metal powder layer by layer according to CAD data. The process takes place at elevated temperatures, which typically results in parts with minimal residual stresses and enhanced mechanical properties. This high-energy beam can be precisely controlled to achieve high densities and superior properties, making EBM ideal for components that require high strength and durability, such as engine components and structural elements of vehicles (Zhang et al., 2018). The primary materials used in EBM are high-performance metals such as titanium, cobalt chrome, and nickel-based superalloys. These materials are chosen for their exceptional strength-to-weight ratios, corrosion resistance, and high-temperature performance. The ability of EBM to handle such advanced materials makes it uniquely suited for automotive applications where such traits are mandatory, such as in the manufacturing of parts that are exposed to high thermal and mechanical stress.

EBM offers several specific advantages for automotive manufacturing. The ability to produce parts with excellent mechanical properties and high complexity without the need for tooling is a significant benefit. Furthermore, the process's high energy efficiency and the ability to fully melt the material lead to parts with very low porosity and high fatigue resistance. These characteristics are critical for safety-critical components in automotive applications. Another advantage is the geometric freedom that EBM provides, allowing for the design and production of optimized structures like lightweight lattice frames that are not feasible with traditional manufacturing methods. Despite its advantages, EBM faces several challenges. The range of materials compatible with EBM is currently more limited compared to other AM technologies, focusing mainly on metals and not extending to polymers or composites. The cost of EBM equipment and materials can also be high, potentially limiting its use to high-value applications. Additionally, the surface finish of EBM parts often requires postprocessing, which can add to production times and costs. Managing these aspects is crucial for ensuring that EBM can be economically viable in the competitive automotive market. The future of EBM in automotive manufacturing looks promising with ongoing advancements aimed at enhancing the process's capabilities and economic viability. Research is focused on expanding the range of materials that can be used, improving the surface quality of finished parts, and increasing the production rate. Innovations in machine and software technology that allow for larger build volumes and faster printing speeds are expected to make EBM even more attractive for automotive applications. Furthermore, the development of new alloy powders designed specifically for EBM could lead to parts with tailored properties, enhancing vehicle performance and efficiency.

13.2.1.6 Digital light processing (DLP)

DLP is a type of AM technology that uses light to cure photosensitive resins into solid objects, layer by layer. This method is highly valued in the automotive industry for its speed, precision, and the ability to produce high-quality parts with excellent surface finish. DLP is particularly well-suited for creating complex parts that require intricate details and smooth surfaces, such as prototypes, interior components, and elements that necessitate fine esthetic qualities. DLP technology utilizes a digital projector screen to flash a single image of each layer all at once across the entire platform. This process allows the light to cure a whole layer of photo-reactive resin in a few seconds, significantly speeding up the production time compared to other methods like SLA that use a point-to-point approach with a laser. The DLP process involves building parts from the bottom up in a vat of liquid resin, where the build platform lifts the model upward, creating space for uncured resin to flow beneath the part for the next layer. This technique provides high accuracy and good repeatability, crucial for automotive applications where precision is necessary (Chaudhary et al., 2023). DLP primarily uses photopolymer resins that can be formulated with various additives to enhance their properties. Recent advancements have seen the development of resins infused with ceramic or glass particles to improve their thermal stability and structural integrity. These innovations make DLP an even more compelling option for producing components that require not only fine detail but also enhanced mechanical properties, such as light covers, intricate connectors, and decorative elements.

DLP offers several advantages for automotive manufacturing. The capability to achieve high resolution and excellent surface quality makes DLP ideal for parts that are visible or require minimal postprocessing. Additionally, the speed of DLP allows for rapid iteration and development of design concepts, which is invaluable during the prototyping phase of automotive development. Moreover, DLP's efficient use of materials, requiring only enough resin to fill the vat and minimal waste, aligns with the industry's increasing push toward sustainability and cost-effectiveness. However, DLP is not without its challenges. The primary limitation is the durability of the resins used, which typically do not match the strength and temperature resistance of metals or more robust thermoplastics. This can limit the application of DLP-produced parts to nonstructural, esthetic, or low-load components within the vehicle. Additionally, the size of parts that can be produced is generally smaller than those made with other AM techniques, restricted by the size of the projector and the vat. Scaling up production for larger parts remains a technical and economic challenge. Looking forward, significant efforts are being made to expand the capabilities and applications of DLP in the automotive sector. Innovations in resin chemistry are aimed at producing new materials that offer better performance characteristics, such as higher impact resistance, flexibility, or heat resistance. Advances in DLP machine design are also anticipated, which could allow for larger build volumes and even faster production speeds. Such advancements could broaden the scope of DLP's applications to include more functional parts and larger components. DLP stands out as a valuable AM technology in the automotive industry, particularly for applications requiring high detail and surface quality. As the technology and materials continue to evolve, DLP is expected to play an increasingly significant role in the production of automotive components, enhancing both the efficiency and creativity of automotive design and manufacturing processes.

13.2.1.7 Multi jet fusion (MJF)

MJF is an innovative AM technology that has rapidly gained traction in the automotive industry due to its ability to produce high-quality, functional parts at a relatively fast pace. Utilizing a fine-grained material that allows for ultrathin layers and high resolution, MJF is particularly suitable for creating both prototypes and end-use components in the automotive sector. MJF operates by applying a fusing agent to a bed of nylon powder, which is then selectively fused by heating elements. A detailing agent is applied around the contours of each layer to ensure fine details and smooth surfaces. The process does not require support structures because the unused powder in the bed supports the part during production. This ability to produce complex geometries without supports significantly reduces postprocessing time and waste. The precision and consistency of MJF make it an excellent choice for manufacturing intricate automotive components such as connectors, hinges, and ductwork (Sillani et al., 2019). The primary materials used in MJF are various types of nylons, including PA 12, which offers excellent mechanical properties and durability. These polymers are ideal for the demands of the automotive industry, providing the necessary strength, flexibility, and thermal stability. Moreover, the recent development of incorporating nanocomposites into these nylons enhances their functionalities. For instance, the addition of glass beads or carbon fibers can improve dimensional stability and heat resistance, qualities critical for under-hood applications and parts exposed to high operational stress.

MJF offers several significant advantages for automotive applications. The technology enables high throughput, making it suitable for mid-to-high volume production, a rare capability in AM. Additionally, MJF produces parts with isotropic mechanical properties, meaning the created components perform consistently in all directions. This isotropic nature is crucial for parts that undergo variable stresses in an automotive environment. The ability to integrate complex features and internal channels into single components also allows for weight reduction and part consolidation, leading to more efficient designs and less material usage. Despite its advantages, MJF faces certain challenges. The limited range of materials, primarily focused on nylon-based polymers, can restrict application diversity. Additionally, while MJF offers excellent dimensional accuracy and detail resolution, the surface finish of the parts may still require postprocessing, particularly when a smooth or esthetically pleasing appearance is required. The cost of MJF systems and their operation, including the powders used, can also be a barrier for smaller manufacturers or those new to AM. The future development of MJF is focused on expanding the range of materials available and improving the scalability of the technology. Efforts are underway to develop new polymer blends and introduce metals and ceramics that are compatible with MJF processes, broadening its application scope. Advancements in machine design that increase build volumes and printing speeds are anticipated to make MJF even more competitive with traditional manufacturing methods. These improvements could see MJF becoming a cornerstone technology in automotive manufacturing, capable of producing a wider variety of parts efficiently and cost-effectively. MJF represents a dynamic and rapidly evolving technology in the field of AM, offering substantial benefits for the automotive industry. Its ability to produce durable, complex parts quickly and in volume aligns well with the industry's needs, positioning MJF as a key player in the future of automotive manufacturing.

13.2.1.8 Continuous liquid interface production (CLIP)

CLIP is a cutting-edge AM technology that significantly enhances the speed and quality of 3D printing. Developed by Carbon3D, CLIP uses a photochemical process to rapidly produce parts from a pool of resin. This technology stands out for its ability to create objects continuously without the layer-by-layer approach typical of traditional 3D printing methods, making it particularly advantageous for applications in the automotive industry where both production speed and part quality are critical. CLIP operates on the principle of using ultraviolet (UV) light to cure a photosensitive resin, combined with a special transparent window that allows UV light to pass through while also permitting oxygen to permeate. The presence of oxygen creates a "dead zone" at the bottom of the resin pool, preventing the resin from curing against the window and allowing for continuous drawing of the part upward. This mechanism enables incredibly fast production speeds, as the printing process does not need to pause between layers. The result is a smooth, high-quality finish with excellent mechanical properties, making CLIP ideal for producing parts with intricate details and complex geometries (Lipkowitz et al., 2022).

CLIP is designed to work with a range of specially formulated resins that can be tuned to exhibit specific properties required by automotive manufacturers, such as flexibility, strength, and temperature resistance. These resins can be modified with various additives, including nanocomposites, to enhance their performance under different conditions. For instance, incorporating carbon nanotubes or fiberglass can improve the material's stiffness and durability, making it suitable for functional components in vehicles that require high structural integrity and resilience (Balli et al., 2017). The primary advantage of CLIP in automotive manufacturing is its speed, which far surpasses that of conventional layer-by-layer 3D printing technologies. This rapid production capability is particularly beneficial for prototyping and the manufacturing of custom or limited-run parts. Additionally, the continuous nature of the printing process eliminates the layer lines that can weaken parts and require postprocessing in other AM methods. This feature ensures that parts produced by CLIP have not only better esthetics but also superior mechanical properties and a smoother surface finish, reducing the need for secondary operations (Pazhamannil & Govindan, 2021).

Despite its benefits, CLIP technology does come with challenges. The range of materials compatible with CLIP is currently somewhat limited to specific resins developed for the process, potentially restricting its application range. Moreover, the cost of CLIP technology and its resins can be high, making it less accessible for some manufacturers. Ensuring the consistency of mechanical properties across larger parts is also a critical issue, as variations in curing intensity and resin behavior can affect performance. The future of CLIP in automotive manufacturing looks promising with ongoing advancements aimed at expanding the range of compatible materials and enhancing the scalability of the technology. Research is focused on developing new resin formulations that can withstand the demanding conditions of automotive applications, including higher temperatures and more variable loads. Innovations in machine design are also expected to increase the size and reduce the costs of CLIP systems, making this technology more viable for larger production runs and more widespread use within the industry. CLIP represents a revolutionary step in AM, offering remarkable speed and quality that are highly beneficial for the automotive industry. As the technology and materials continue to evolve, CLIP is expected to play an increasingly significant role in the production of automotive components, contributing to faster development cycles and more innovative vehicle designs.

13.2.2 Advantages over traditional methods

AM technologies offer a suite of advantages over traditional manufacturing methods that are increasingly relevant in a world prioritizing customization, efficiency, and sustainability. These advantages are particularly pronounced in industries like automotive, where the demand for innovation and environmental responsibility is high.

13.2.2.1 Design flexibility and complexity

One of the most significant advantages of AM over traditional methods is the unparalleled flexibility in design. AM allows for the creation of complex geometries that are often impossible to achieve with conventional manufacturing techniques such as casting or machining. This capability enables designers to optimize part geometries for performance rather than manufacturability, integrating complex features like internal channels for lightweighting or thermal management without additional cost in tooling (Rosen, 2014).

13.2.2.2 Reduced material waste

AM processes typically build parts layer by layer, using only the material necessary to form the part, significantly reducing waste compared to subtractive manufacturing methods where material is removed from a larger block. This not only makes AM a more sustainable choice but also can lead to cost savings on expensive materials, such as titanium or high-performance alloys used in automotive applications (Peng et al., 2018).

13.2.2.3 Rapid prototyping

Rapid prototyping is another area where AM technologies shine. Traditional methods often require extensive time and cost to produce molds, dies, or specialized tooling.

In contrast, AM can produce parts directly from digital models in a matter of hours or days, accelerating the design cycle and allowing for quicker iteration and refinement. This speed in prototyping is invaluable for staying competitive in dynamic markets by enabling faster product development and market testing (Touri et al., 2019).

13.2.2.4 Customization and mass customization

AM excels in producing customized or personalized products at no additional production cost. Traditional manufacturing typically relies on economies of scale, where producing different versions of a product can be prohibitively expensive. AM allows for individual customization in production runs, even in small quantities, which is especially beneficial in industries such as automotive, where there is a trend toward personalization in luxury or high-performance vehicle markets (Lacroix et al., 2021).

13.2.2.5 Cost-effective small production runs

For small production runs or bespoke items, AM can be more cost-effective than traditional manufacturing, which often involves significant setup costs. Since AM requires no special tooling, the fixed costs are much lower, making it economically viable to produce small quantities of parts as needed. This advantage is critical not only for niche markets but also for reducing the inventory costs associated with large stocks of spare parts (Griffin et al., 2022).

13.2.2.6 Supply chain simplification

AM has the potential to dramatically simplify supply chains by reducing the number of steps between design and production. Parts can be manufactured at or near their point of use, reducing the need for extensive logistics and warehousing. This reduction in the supply chain complexity not only speeds up production but also decreases the carbon footprint associated with transportation of parts across multiple locations (Cui et al., 2021).

13.2.2.7 Enhanced mechanical properties

Through precise control over material microstructure and the possibility of creating topology-optimized designs, AM parts can exhibit superior mechanical properties compared to those made with traditional methods. Innovations in AM materials, including advanced polymers and metal matrix composites, have further enhanced the performance characteristics of AM-produced parts, making them suitable for critical applications in the automotive sector (Zhu et al., 2021).

AM technologies provide a myriad of advantages that address the key needs of modern manufacturing: efficiency, customization, and sustainability. As these technologies continue to mature and integrate into production environments, their role in industries like automotive is set to expand, offering new opportunities for innovation and improvement. Table 13.2 encapsulates the core attributes of each AM technique, highlighting their suitability for different types of automotive applications

Table 13.2 Comparison of various AM techniques based on their material compatibility, resolution, speed, and cost, specifically in the context of automotive applications.

AM technique	Material compatibility	Resolution	Speed	Cost
FDM	Thermoplastics, composites	Moderate to low	Moderate	Low to moderate
SLA	Photopolymer resins	High	Moderate to slow	Moderate to high
SLS	Polymers, metal powders, ceramics	High	Moderate	High
DMLS	Metals and alloys	High	Slow	Very high
EBM	Metals and alloys	Moderate to high	Moderate	Very high
DLP	Photopolymer resins	High	Fast	Moderate
MJF	Polymers	High	Fast	Moderate to high
CLIP	Photopolymer resins	Very high	Very fast	High

based on the required material properties, resolution, manufacturing speed, and cost considerations. Such a comparison is crucial for making informed decisions about which AM technique to employ for specific automotive parts or systems.

13.3 Functional polymers and their properties

Functional polymers play a crucial role in AM, especially in sectors like automotive where the performance demands on materials are stringent and varied. These polymers are engineered to provide specific properties that make them suitable for use in environments that demand durability, thermal stability, and mechanical strength. Commonly used functional polymers in automotive AM include ABS (Acrylonitrile Butadiene Styrene), which offers good heat resistance and toughness; Nylon, known for its flexibility and fatigue resistance; and PEEK (Polyether Ether Ketone), which stands out due to its exceptional thermal stability, chemical resistance, and mechanical properties suitable for high-performance parts. Each of these materials brings distinct advantages to automotive applications, such as reduced weight, increased design flexibility, and the ability to withstand harsh operating conditions. Furthermore, the development of nanocomposites has enhanced these polymers' properties by incorporating nanoparticles, such as carbon nanotubes or silica, into the polymer matrix, thus improving their mechanical strength, thermal resistance, and dimensional stability. This synergy between polymers and nanoparticles is opening new frontiers in the design and manufacturing of automotive components, leading to parts that are not only lighter and stronger but also more efficient to produce.

13.3.1 Polymer type

In the realm of AM for automotive applications, the choice of polymer significantly influences the properties and suitability of finished parts. Polymers such as ABS (Acrylonitrile Butadiene Styrene), Polyamide (often known as Nylon), and PEEK (Polyether Ether Ketone), along with their various blends, are particularly valued for their unique characteristics that cater to the stringent requirements of the automotive industry.

13.3.1.1 ABS (acrylonitrile butadiene styrene)

ABS is a common thermoplastic polymer used extensively in AM due to its good mechanical properties, excellent impact resistance, and ease of processing. It is particularly favored for applications that require durability and moderate resistance to temperature and chemicals. ABS is widely utilized in the automotive sector for components like dashboards, panels, and other interior parts that need to withstand impact and stress while offering a good esthetic finish. However, ABS can warp during printing, which requires careful control of the printing environment.

13.3.1.2 Polyamide (nylon)

Polyamide, particularly Nylon 12 and Nylon 6, is renowned for its strength, flexibility, and fatigue resistance. It absorbs less moisture compared to other grades of nylon, making it more stable and suitable for complex automotive components. Nylon's ability to resist wear and impact makes it ideal for under-the-hood applications as well as gears and bearings that experience regular mechanical stress. Furthermore, polyamides can be reinforced with glass fibers or carbon fibers to enhance their thermal and mechanical properties, making them even more robust for critical automotive applications.

13.3.1.3 PEEK (polyether ether ketone)

PEEK is a high-performance polymer in the arena of AM, prized for its exceptional thermal stability, high chemical resistance, and excellent mechanical properties. It can operate at higher temperatures and withstand harsher environments than most other polymers, which makes it suitable for applications such as engine parts, transmission components, and other areas exposed to extreme conditions. PEEK can also be blended with carbon fibers to improve its structural and thermal properties, further extending its use within the automotive industry where high performance is crucial.

13.3.1.4 Polymer blends and composites

Blending different polymaterials or reinforcing them with nanocomposites can significantly enhance their properties. For instance, blending ABS with polycarbonate (PC) can improve the thermal and impact resistance of the resulting polymer, making it suitable for more demanding applications. Similarly, incorporating nanofillers like carbon nanotubes or graphene can enhance the electrical conductivity, thermal stability, and mechanical strength of basic polymers. These advancements are particularly beneficial in automotive applications that require materials to perform reliably in challenging environments.

13.3.2 Properties relevant to automotive applications

In the automotive industry, the choice of materials is critical, especially when it involves AM technologies. The properties of thermal stability, durability, and lightweight nature are particularly vital for automotive applications, as they directly impact vehicle performance, safety, and efficiency. Polymers and nanocomposites selected for automotive components via AM must align with these requirements to ensure the functionality and longevity of the parts under extreme conditions.

13.3.2.1 Thermal stability

Thermal stability is a crucial property for automotive materials, especially for components exposed to high temperatures during operation, such as engine parts, exhaust systems, and lighting enclosures. Materials used in these applications must withstand prolonged exposure to heat without degrading, losing structural integrity, or emitting harmful byproducts. High-performance polymers like PEEK and polyimides are often chosen for their excellent heat resistance, withstanding temperatures up to 250°C and beyond. Moreover, the incorporation of nanocomposites can enhance this property further; for example, adding graphene or carbon nanotubes can improve thermal conductivity and stability, thereby increasing the heat dissipation capabilities of polymers and extending their operational life under high-temperature conditions.

13.3.2.2 Durability

Durability in automotive components refers to their ability to resist wear, impact, and fatigue over extended periods and under varied environmental conditions. This property is essential for structural components, chassis, and body parts that are subjected to continuous stress and must perform reliably throughout the vehicle's lifespan. Polymers like nylon and reinforced polypropylene are favored for their toughness and impact resistance. These materials can be further enhanced through the addition of fibers (e.g., glass, carbon), which significantly improve their mechanical strength and fatigue resistance, making them suitable for more demanding applications. Durability also involves resistance to chemicals and UV light, critical for exterior parts exposed to road chemicals and sunlight.

13.3.2.3 Lightweight nature

The drive toward lightweighting in the automotive industry is motivated by the need to improve fuel efficiency and reduce emissions. Lightweight materials are essential for electric vehicles, where weight directly affects range and performance. AM offers significant advantages in this area by allowing for the design of parts that use less material without compromising strength through techniques like lattice structures and topology optimization. Materials such as aluminum alloys and titanium used in DMLS or EBM are examples, but even lighter options include polymer-based composites. These materials not only reduce the weight but also maintain or enhance the performance compared to traditional metals. Using advanced polymers and nanocomposites also facilitates the integration of functionalities such as thermal management and electrical conductivity, further contributing to vehicle efficiency and performance.

13.3.2.4 Integration into automotive applications

To effectively integrate these materials into automotive applications, engineers and designers must consider the interaction of these properties with the part design and the manufacturing process. AM provides the flexibility to optimize material distribution within a part to enhance properties like durability and thermal stability while minimizing weight. This approach not only improves material efficiency but also opens up new possibilities for designing parts that were not feasible with traditional manufacturing methods.

The properties of thermal stability, durability, and lightweight nature are foundational to the selection of materials for AM in the automotive industry. By leveraging the capabilities of polymers and nanocomposites, manufacturers can meet the stringent demands of automotive applications, achieving a balance of performance, safety, and efficiency. Advances in material science and AM technologies continue to push the boundaries, enhancing the properties of materials to meet the evolving needs of the automotive sector. Table 13.3 encapsulates key information regarding the performance characteristics of various polymers when subjected to the stressors typical in automotive environments. Such properties influence the selection and optimization of materials for specific applications within the industry, ensuring that parts meet or exceed operational requirements.

Table 13.3 Properties of functional polymers used in automotive AM.

Polymer	Mechanical properties	Thermal properties	Chemical properties
ABS	High impact resistance, good toughness and strength	Heat deflection temperature: ~98°C	Resistant to many acids and bases, vulnerable to solvents
Polyamide (Nylon)	High strength, excellent fatigue resistance	Melting point: ~180°C (Nylon 6), ~220°C (Nylon 12)	Good chemical resistance, hygroscopic (absorbs water)
PEEK	Exceptional mechanical strength, stiffness	Continuous service temperature: up to 250°C	Excellent resistance to organic and inorganic chemicals
Polypropylene (PP)	Good flexibility, high impact resistance	Melting point: ~160°C	Resistant to acids and bases, poor solvent resistance
Polycarbonate (PC)	High impact strength, good dimensional stability	Glass transition temperature: ~147°C	Vulnerable to acids, good resistance to many oils and fats
PLA	Moderate strength and stiffness, low impact resistance	Heat deflection temperature: ~50°C–60°C	Good resistance to many household chemicals
TPU	High flexibility, excellent abrasion resistance	Melting point: ~200°C	Good resistance to oil, grease, and various solvents

13.4 Nanocomposites: enhancing material properties

Nanocomposites are engineered materials made by embedding nanoparticles, typically less than 100 nm in size, into a matrix of polymer, metal, or ceramic. These nanoparticles enhance the base material with superior properties due to their nanoscale behaviors, such as increased surface area to volume ratio and quantum effects. In AM, the use of nanocomposites has revolutionized applications that require high-performance characteristics, particularly in the automotive industry. Nanotechnology, which manipulates materials at the molecular or atomic level, is crucial for developing these nanocomposites. It produces nanomaterials that are stronger, lighter, and more reactive than larger-scale materials, enhancing composite materials without increasing weight or sacrificing integrity. The application of nanomaterials extends beyond structural enhancements. Innovative forms such as nanocomposite gels (Harun-Ur-Rashid, 2019; Harun-Ur-Rashid, Seki, et al., 2009; Karim et al., 2023; Rezaul Karim et al., 2020) and films (Harun-Ur-Rashid et al., 2021; Huda et al., 2014; Nurul Huda et al., 2010) have advanced the capabilities of intelligent materials across various industries. For example, structural colored nanomaterials (Harun-Ur-Rashid et al., 2010; Harun-Ur-Rashid, Imran, et al., 2009; Takeoka et al., 2013) and organo-metallic nanomaterials (Harun-Ur-Rashid & Imran, 2022) are pioneering new functional materials with enhanced optical and electronic properties. Similarly, in the medical field, groundbreaking developments like nanovaccines (Harun-Ur-Rashid, Jahan, & Imran, 2023) and nanobiomedicine (Harun-Ur-Rashid et al., 2023a) have leveraged nanotechnology to enhance treatment personalization and effectiveness. Furthermore, the diagnostics and sensing sectors have benefitted from the introduction of nanobiosensors (Harun-Ur-Rashid & Foyez, Jahan, et al., 2022), offering enhanced sensitivity and precision. These diverse applications of nanomaterials showcase their potential not only in enhancing the mechanical properties of polymers used in automotive AM but also in revolutionizing sectors like healthcare (Imran et al., 2019), environmental management (Harun-Ur-Rashid et al., 2024a, 2024b; Harun-Ur-Rashid, Pal, et al., 2024; Hassan, 2024), and energy (Islam et al., 2024). In industrial manufacturing, particularly, the integration of nanomaterials has led to more efficient processes and innovative products (Harun-Ur-Rashid et al., 2023b; Harun-Ur-Rashid et al., 2024; Harun-Ur-Rashid, 2023a, 2023b, 2023c; Harun-Ur-rashid, Imran & Susan, 2023), aiding in pollution control and contributing to cleaner energy solutions. This broad impact underscores the transformative power of nanotechnology and nanocomposites, indicating a promising future in which these materials drive advancements across multiple domains (Harun-Ur-Rashid, Krishna, et al., 2024; Khair et al., 2021).

13.4.1 Nanocomposites in AM

Nanocomposites in AM typically consist of a polymer base integrated with nanoparticles such as carbon nanotubes, graphene, nano-clays, or silica. These nanoparticles significantly enhance the polymer's mechanical, thermal, and electrical properties. For example, carbon nanotubes can transform a standard polymer into a conductive, robust material suitable for structural components or electronic interfaces in vehicles. The use of nanocomposites in AM offers multiple advantages, including enhanced mechanical properties such as increased tensile strength and durability, which are essential for automotive parts exposed to long-term mechanical stress. They also provide improved thermal stability, allowing components to withstand high temperatures near engines without degrading. Additionally, materials like graphene improve the electrical and thermal conductivity of polymers, beneficial for parts requiring efficient heat dissipation or electrical integration. Moreover, these nanoparticles make it possible to create materials that are both lighter and stronger than traditional composites, aligning with the automotive industry's goals of improving fuel efficiency and vehicle performance. However, integrating nanocomposites into AM processes presents challenges, including the even dispersion of nanoparticles and the high costs associated with their production and handling. Future developments are expected to address these issues through the creation of new nanoparticle types and improved compatibility with AM technologies. Research is ongoing to develop cost-effective production methods and enhance environmental safety and sustainability. Overall, nanocomposites are poised to significantly impact the development of next-generation automotive components, offering enhanced performance and design capabilities.

13.4.2 Role of nanofillers

Nanofillers such as carbon nanotubes, graphene, and nano-clays significantly enhance the properties of polymers used in AM for automotive applications. Carbon nanotubes improve mechanical strength, thermal conductivity, and electrical properties, making polymers suitable for high-stress and high-temperature environments. Graphene increases the stiffness and strength of polymers, enhances barrier properties to gases, and improves thermal stability, which is crucial for components like fuel tanks and electronic interfaces. Nano-clays enhance the modulus and heat resistance of polymers, improve fire retardancy, and increase dimensional stability, reducing shrinkage and warping during manufacturing. However, integrating these nanofillers into polymers presents challenges such as achieving uniform dispersion and maintaining consistent quality in mass production. Advances in processing technologies are crucial to maximize the benefits of nanocomposites, aiming to overcome these limitations and enhance the performance of automotive components, making them lighter, stronger, and more efficient.

13.5 Applications in automotive manufacturing

13.5.1 Component fabrication

AM has revolutionized the automotive industry by providing unprecedented flexibility and precision in the fabrication of components. From interior elements to critical engine parts and customized features, AM technologies facilitate the creation of complex designs that are both lightweight and durable, which is essential for modern automotive manufacturing.

13.5.1.1 Interior components

In the realm of interior design, AM allows for the production of intricate components that meet both functional and esthetic requirements. Items such as dashboard elements, air vents, and console fittings are commonly manufactured using technologies like SLS and FDM (Aslan & Yıldız, 2020; Dalpadulo et al., 2022; Mohanavel et al., 2021). These methods enable the integration of complex geometries and customized designs that would be difficult or impossible to achieve with traditional manufacturing processes. For example, AM can produce lightweight, ergonomic designs for dashboard components that incorporate complex, hollow structures to reduce weight without sacrificing strength. The use of advanced polymers and composites can also enhance the esthetic appeal and tactile quality of interior parts.

13.5.1.2 Engine parts

Engine components are particularly challenging to manufacture due to the demanding conditions they must withstand, including high temperatures and constant mechanical stress. AM, particularly DMLS, has been increasingly used to produce parts such as turbocharger wheels, fuel nozzles, and complex cooling ducts (Gray & Depcik, 2020; Zhao et al., 2023). These parts benefit from the superior thermal and mechanical properties of the metals and alloys used in DMLS, which can be optimized for high performance. For instance, DMLS allows for the creation of lightweight yet durable metal parts with integrated cooling channels that improve thermal management, a critical factor in engine performance and efficiency.

13.5.1.3 Custom elements

Customization is a significant advantage of AM, offering tailored solutions that cater to specific consumer preferences or design specifications. Custom elements such as personalized trim pieces, bespoke brackets, and even entire body panels can be manufactured using AM to meet individual customer demands. This capability is particularly valued in high-performance and luxury automotive sectors, where differentiation and personalization are crucial. For example, AM can be used to create custom-fitted components that perfectly match the contours of a specific vehicle model, adding both functional value and a unique esthetic touch. Table 13.4 serves as a reference for understanding how different AM technologies are applied in the automotive industry to meet various design and performance criteria. It highlights the flexibility of AM in producing complex and customized parts that traditional manufacturing methods cannot achieve, thereby enhancing the innovation and efficiency of automotive design and production.

Table 13.4 Overview of AM applications in automotive component fabrication.

Component type	Examples	AM technology used	Material properties	Advantages of AM
Interior components	Dashboard elements, air vents, console fittings	SLS, FDM	Advanced polymers and composites	Enables complex geometries, customization, lightweight designs, enhanced esthetic and tactile qualities
Engine parts	Turbocharger wheels, fuel nozzles, cooling ducts	DMLS, EBM	Metals and high-performance alloys	Produces parts that withstand high temperatures and mechanical stress, integrates cooling features
Custom elements	Personalized trim pieces, bespoke brackets, body panels	Various AM technologies	Polymers, metals, depending on application	Allows for high customization, tailored to individual specifications, unique designs

13.5.2 Performance and testing

In the automotive industry, AM technologies necessitate rigorous performance testing to ensure that components made from functional polymers and nanocomposites meet high standards. These tests encompass mechanical evaluations like stress and fatigue resistance, essential for assessing the suitability of structural components under real-world conditions. Thermal and environmental durability tests are also crucial, as they measure the ability of parts to withstand extreme temperatures and corrosive environments, simulating the challenging conditions of automotive settings. Additionally, real-world performance trials on actual vehicles assess how these parts affect overall vehicle functionality, including safety, fuel efficiency, and handling. Compliance with safety and regulatory standards through crash and emissions testing ensures these innovative components not only enhance performance but also align with global safety norms, thereby supporting the broader acceptance and integration of AM technologies in automotive manufacturing.

Borrelli et al. (2020) reported an innovative technique for manufacturing a plastic automotive component using AM to reduce weight and costs while maintaining safety. The study specifically focused on a dashboard glove box and evaluated its crash performance through simulated frontal impact sled tests. The materials analyzed included polyamide (PA) and polypropylene (PP), both reinforced with 5% basalt fibers (PAB and PPB, respectively). These materials were chosen to enhance the mechanical properties of the thermoplastics, as pure thermoplastics typically exhibit poor mechanical performance. The reinforcement with basalt fibers was selected due to its lower cost compared to carbon or glass fibers. The numerical simulations were carried out using the LS-DYNA solver, with the finite element method (FEM) modeling involving three phases: importing CAD models of the current and additive glove boxes, defining properties and materials, and meshing the geometries. The FEM model focused on the front interior of a Fiat 500, specifically the dashboard glove box, to evaluate knee impacts during a frontal crash. The dashboard body, including the glove box, was meshed using mainly shell linear elements (Fig. 13.5).

A subsystem test was initially conducted using a simplified model with an impactor representing the knee to analyze the mechanical behavior and stiffness of the glove box (Fig. 13.6). This test showed that the additive glove box generally performed better than the current production case, especially in terms of force reduction on the knee. The PP+5%B material consistently demonstrated lower force values compared to PA+5%B.

For a more comprehensive analysis, a rating test was performed, simulating a frontal impact against a deformable barrier as prescribed by Euro NCAP. This test involved using a Hybrid III 50th Percentile Male Dummy and aimed to evaluate biomechanical parameters such as femur compression and knee sliding. The simulation showed that the additive glove box achieved comparable biomechanical scores to the current production box while offering a significant weight reduction. The complex geometry of the dummy-dashboard system used in this test is illustrated in Fig. 13.7.

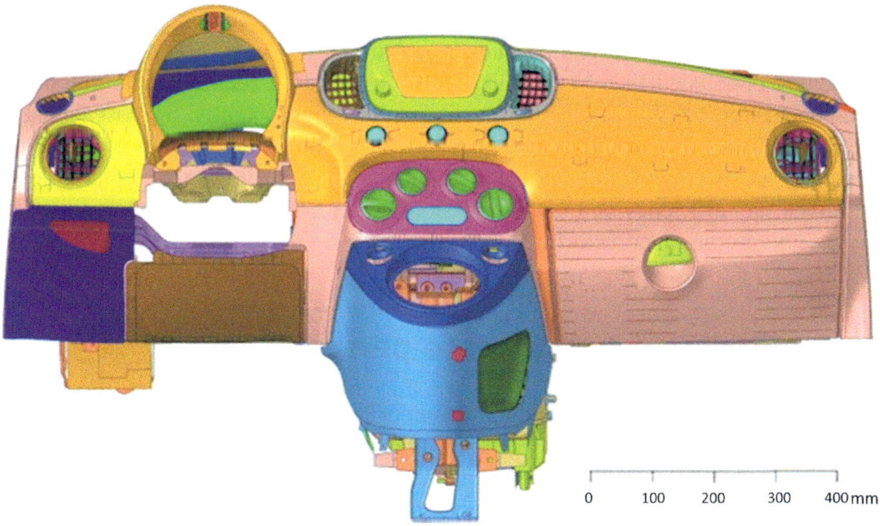

Figure 13.5 FEM model utilized in the described investigation.
From Borrelli, Alessandro, D'Errico, Giuseppe, Borrelli, Corrado & Citarella, Roberto. (2020). Assessment of Crash Performance of an Automotive Component Made through Additive Manufacturing. Applied Sciences, 10(24). https://doi.org/10.3390/app10249106. Copyright 2020 by the authors. Licensee MDPI, Basel, Switzerland.

Figure 13.6 Subassembly test demonstrating the impactor simulating the knee.
From Borrelli, Alessandro, D'Errico, Giuseppe, Borrelli, Corrado & Citarella, Roberto. (2020). Assessment of Crash Performance of an Automotive Component Made through Additive Manufacturing. Applied Sciences, 10(24). https://doi.org/10.3390/app10249106. Copyright 2020 by the authors. Licensee MDPI, Basel, Switzerland.

The results indicated that the adoption of AM for the glove box did not compromise safety performance and provided the advantage of reduced weight. The study concluded that AM could be effectively used for producing automotive components, especially for premium vehicles, due to its potential for cost savings in tooling and inventory, and its ability to create complex geometries that enhance structural integrity. Additionally, the reduction in vehicle weight contributes to improved fuel efficiency and lower emissions, making AM a promising approach for future automotive manufacturing.

13.6 Challenges and future perspectives

AM with functional polymers and nanocomposites presents significant opportunities and challenges in the automotive industry. One of the main hurdles is scaling up AM for mass production, which is complicated by the technology's current limitations in production speed and part size. Additionally, there are challenges related to the materials themselves, which must not only meet stringent automotive standards for durability and performance but also remain cost-effective. Barriers to technology adoption also include the high cost of AM equipment, the need for specialized operator training, and integration issues within traditional manufacturing lines. On the sustainability front, while AM reduces waste compared to subtractive manufacturing methods, concerns about the environmental impact of the polymers and nanocomposites used remain. Many of these materials are petroleum-based and nonbiodegradable, posing challenges for disposal and recycling. Efforts are underway

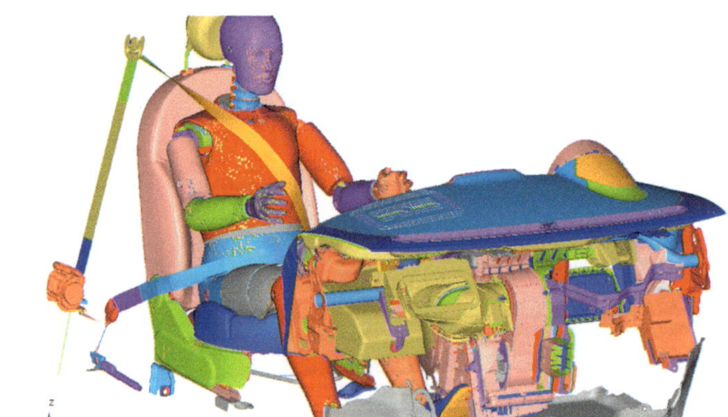

Figure 13.7 FEM analysis of the surrogate-dashboard configuration.
From Borrelli, Alessandro, D'Errico, Giuseppe, Borrelli, Corrado & Citarella, Roberto. (2020). Assessment of Crash Performance of an Automotive Component Made through Additive Manufacturing. Applied Sciences, 10(24). https://doi.org/10.3390/app10249106. Copyright 2020 by the authors. Licensee MDPI, Basel, Switzerland.

to develop more sustainable materials, such as biodegradable polymers and more energy-efficient AM processes, to mitigate these impacts.

Looking to the future, the AM sector is likely to see significant advancements in both materials and technology. Innovations are expected in the development of new composite materials that are lighter and stronger, tailored for the unique demands of the automotive industry. Technological improvements in AM machines will likely increase production speeds and enable the manufacturing of larger parts, which are crucial for scaling up to meet automotive production demands. Furthermore, as the automotive industry continues its shift toward electric and autonomous vehicles, AM will be essential in developing specialized components that facilitate this transition. The industry is expected to increasingly embrace customization and on-demand manufacturing, leveraging AM's capabilities to meet specific customer and market needs more efficiently. These developments will help address current limitations and expand the role of AM in automotive manufacturing, paving the way for more innovative and sustainable vehicle production.

13.7 Conclusion

AM profoundly impacts the automotive industry by enabling the production of components that are complex, lightweight, and durable, surpassing the capabilities of traditional manufacturing methods. Throughout this chapter, we've explored a variety of AM techniques including FDM, SLS, and DMLS, each bringing unique benefits in terms of material compatibility, resolution, and cost-effectiveness. The use of functional polymers and nanocomposites enhanced with materials like carbon nanotubes and graphene has proven crucial, significantly improving the mechanical, thermal, and environmental resilience of automotive parts. Despite the advantages, challenges such as scalability and environmental sustainability persist. However, ongoing advancements in AM materials and technologies continue to push the boundaries, indicating a promising future with broader applications in automotive manufacturing. This evolution is set to further integrate AM within the automotive sector, revolutionizing vehicle design and production, enhancing manufacturing sustainability, and leading to the development of more efficient and customizable automotive solutions. As AM continues to mature, it promises to catalyze significant innovations in automotive design and production, driving the industry toward more sustainable and efficient practices.

Acknowledgment

The author extends his deepest gratitude to the Department of Chemistry at IUBAT—International University of Business Agriculture and Technology for their unwavering support throughout the research project.

Statement of conflict of interest

The author declares no potential conflict of interest with respect to the research, authorship, and/or publication of this chapter. The research was conducted in the absence of any commercial or financial relationships that could be construed as a potential conflict of interest. Any opinions, findings, and conclusions or recommendations expressed in this material are those of the authors and do not necessarily reflect the views of the institutions or companies mentioned herein.

References

Aslan, B., & Yıldız, A. R. (2020). Optimum design of automobile components using lattice structures for additive manufacturing. *Materials Testing, 62*(6), 633–639. https://doi.org/10.3139/120.111527.

Bacciaglia, A., Ceruti, A., & Liverani, A. (2022). Towards large parts manufacturing in additive technologies for aerospace and automotive applications. *Procedia Computer Science, 200*, 1113–1124. https://doi.org/10.1016/j.procs.2022.01.311. http://www.sciencedirect.com/science/journal/18770509.

Balli J., Kumpaty S., & Anewenter, V. (2017). Continuous liquid interface production of 3D objects: An unconventional technology and its challenges and opportunities. ASME International Mechanical Engineering Congress and Exposition, Proceedings (IMECE), 5. https://doi.org/10.1115/IMECE2017-71802, http://www.asmedl.org/journals/doc/ASMEDL-home/proc/

Borrelli, A., D'Errico, G., Borrelli, C., & Citarella, R. (2020). Assessment of crash performance of an automotive component made through additive manufacturing. *Applied Sciences, 10*(24), 9106. https://doi.org/10.3390/app10249106.

Chaudhary, R., Fabbri, P., Leoni, E., Mazzanti, F., Akbari, R., & Antonini, C. (2023). Additive manufacturing by digital light processing: A review. *Progress in Additive Manufacturing, 8*(2), 331–351. https://doi.org/10.1007/s40964-022-00336-0.

Cui, W., Yang, Y., Di, L., & Dababneh, F. (2021). Additive manufacturing-enabled supply chain: Modeling and case studies on local, integrated production-inventory-transportation structure. *Additive Manufacturing, 48*, Article 102471. https://doi.org/10.1016/j.addma.2021.102471.

Dalpadulo, E., Petruccioli, A., Gherardini, F., & Leali, F. (2022). A review of automotive spare-part reconstruction based on additive manufacturing. *Journal of Manufacturing and Materials Processing, 6*(6), 133. https://doi.org/10.3390/jmmp6060133.

Duan, B., Wang, M., Zhou, W. Y., Cheung, W. L., Li, Z. Y., & Lu, W. W. (2010). Three-dimensional nanocomposite scaffolds fabricated via selective laser sintering for bone tissue engineering. *Acta Biomaterialia, 6*(12), 4495–4505. https://doi.org/10.1016/j.actbio.2010.06.024.

Gray, J., & Depcik, C. (2020). Review of additive manufacturing for internal combustion engine components. *SAE International Journal of Engines, 13*(5), 617–632. https://doi.org/10.4271/03-13-05-0039. https://www.sae.org/publications/collections/content/E-JOURNAL-03/.

Griffin, C., Hale, J., & Jin, M. (2022). A framework for assessing investment costs of additive manufacturing. *Progress in Additive Manufacturing, 7*(5), 903–915. https://doi.org/10.1007/s40964-022-00260-3. https://link.springer.com/journal/40964.

Harun-Ur-Rashid, A. B. I. (2019). *Superabsorbent hydrogels from carboxymethyl cellulose*Nova Science Publisher, Inc.

Harun-Ur-Rashid, M. (2023a). *Engineered s*. Materials Research Forum, LLC103–126. https://doi.org/10.21741/9781644902554-4.

Harun-Ur-Rashid, M. (2023b) *Nanomaterials in the automobile sector*. Materials Research Forum, LLC124–150. https://doi.org/10.21741/9781644902288-6.

Harun-Ur-Rashid, M. (2023c). Prospective nanomaterials for food packaging and safety. *Materials Research Forum,* 327–352. https://doi.org/10.21741/9781644902288-13.

Harun-Ur-Rashid, M., Imran, A. B., & Susan, M. A. B. H. (2023a). Fire-resistant polymeric foams and their applications. *ACS Symposium Series, 1440*, 97–121. https://doi.org/10.1021/bk-2023-1440.ch005. https://pubs.acs.org/series/symposium.

Harun-Ur-rashid, M., Imran, A. B., & Susan, M. A. B. H. (2023b). Green polymer nanocomposites in automotive and packaging industries. *Current Pharmaceutical Biotechnology, 24*(1), 145–163. https://doi.org/10.2174/1389201023666220506111027. https://www.eurekaselect.com/issue/11976.

Harun-Ur-Rashid, M., Jahan, I., & Imran, A. B. (2023). *Recent advancement of nanostructured materials for clinical challenges in vaccinology*. Springer Science and Business Media LLC135–160. https://doi.org/10.1007/978-3-031-35395-6_8.

Harun-Ur-Rashid, M., Krishna, S. B. N., Golla, N., & Imran, A. B. (2024). *Polyphenol oxidases: The future toward global sustainability polyphenol oxidases: Function, wastewater remediation, and biosensors*. Bangladesh: De Gruyter309–343. https://www.degruyter.com/document/doi/10.1515/9783111033525/html. https://doi.org/10.1515/9783111033525-012.

Harun-Ur-Rashid, M., Pal, K., & Imran, A. B. (2024). Hybrid nanocomposite fabrication of nanocatalyst with enhanced and stable photocatalytic activity. *Topics in Catalysis, 67*(1-4), 17–45. https://doi.org/10.1007/s11244-023-01809-4. https://www.springer.com/journal/11244.

Harun-Ur-Rashid, M., Imran, A. B., & Hasan Susan, M. A. B. (2024a). *Magnetic polymer nanocomposites for removal of dyes and metals from wastewater*. Elsevier BV. https://doi.org/10.1016/b978-0-323-95486-0.00008-9.

Harun-Ur-Rashid, M., Imran, A. B., & Susan, M. A. B. H. (2024b). *Emerging 3D printed polymers and composites for water quality preservation*. Elsevier BV10.1016/b978-0-323-95486-0.00010-7.

Harun-Ur-Rashid, M., Jahan, I., Islam, M. J., Kumer, A., Nurul Huda, M., Imran, A. B., Gouadria, S., & Alsalhi, S. A. (2024). Global advances and smart innovations in supramolecular polymers. *Journal of Molecular Structure, 1304*, 137665. https://doi.org/10.1016/j.molstruc.2024.137665.

Harun-Ur-Rashid, M., Seki, T., & Takeoka, Y. (2009). Structural colored gels for tunable soft photonic crystals. *Chemical Record, 9*(2), 87–105. https://doi.org/10.1002/tcr.20169Japan. http://www3.interscience.wiley.com/cgi-bin/fulltext/122268886/PDFSTART.

Harun-Ur-Rashid, M., Imran, A. B., Seki, T., Takeoka, Y., Ishii, M., & Nakamura, H. (2009). Template synthesis for stimuli-responsive angle independent structural colored smart materials. *Transactions of the Materials Research Society of Japan, 34*(2), 333–337. https://doi.org/10.14723/tmrsj.34.333.

Harun-Ur-Rashid, M., Foyez, T., & Imran, A. B. (2021). *Fabrication of stretchable composite thin film for superconductor applications*. Informa UK Limited63–78. https://doi.org/10.1201/9781003123781-5.

Harun-Ur-Rashid, M., Foyez, T., Jahan, I., Pal, K., & Bin Imran, A. (2022). Rapid diagnosis of COVID-19 via nano-biosensor-implemented biomedical utilization: A systematic review. *RSC Advances, 12*(15), 9445–9465. https://doi.org/10.1039/d2ra01293f.

Harun-Ur-Rashid, M., & Imran, A. B. (2022). *Organometallic nanomaterials synthesis and sustainable green nanotechnology applications*. Informa UK Limited249–270. 10.1201/9781003318606-11.

Harun-Ur-Rashid, M., Bin Imran, A., Seki, T., Ishii, M., Nakamura, H., & Takeoka, Y. (2010). Angle-Independent structural color in colloidal amorphous arrays. *Chemphyschem: A European Journal of Chemical Physics and Physical Chemistry, 11*(3), 579–583. https://doi.org/10.1002/cphc.200900869.

Hassan, M.K. (2024). Computational investigation for tetragonal crystals of Zn(GaS2)2, Zn (GaSe2)2, and Zn(GaTe2)2 photocatalysts for wastewater treatment: First principle approaches. *Advanced Journal of Chemistry, Section B: Natural Products and Medical Chemistry*, 6, 46–66. https://doi.org/10.48309/ajcb.2024.421434.12042024.

Huda, N., Rashid, M., & Harun-Ur-Rashid, M. (2014). A review on stimuli-responsive grafted membranes based on facile synthesis process and extensive applications. *International Journal of Innovation Applied Studies, 8*(3), 1296–1312 2014.

Imran, A. B., Harun-Ur-Rashid, M., & Takeoka, Y. (2019). *Polyrotaxane actuators. Soft actuators: Materials, modeling, applications, and future perspectives: second edition.* Bangladesh: Springer Singapore81–147. https://link.springer.com/book/10.1007/978-981-13-6850-9. 10.1007/978-981-13-6850-9_6.

Islam, M. J., Pal, K., Abdullah, M. M., & Harun-Ur-Rashid, M. (2024). Threshold bandgap of lead halide hybrid perovskites for making stable single crystal by NIR laser trapping. *Journal of Molecular Structure, 1301*, Article 137322. https://doi.org/10.1016/j.molstruc.2023.137322.

Karim, M. R., Harun-Ur-Rashid, M., & Imran, A. B. (2023). Effect of sizes of vinyl modified narrow-dispersed silica cross-linker on the mechanical properties of acrylamide based hydrogel. *Scientific Reports, 13*(1)https://doi.org/10.1038/s41598-023-32185-4https://www.nature.com/srep/.

Khair, A., Dey, N. K., Harun-Ur-Rashid, M., Alim, M. A., Bahadur, N. M., Mahamud, S., & Ahmed, S. (2021). Diffusimetry renounces Graham's law, achieves diffusive convection, concentration gradient induced diffusion, heat and mass transfer. *Defect and Diffusion Forum, 407*, 173–184. https://doi.org/10.4028/www.scientific.net/ddf.407.173.

Kristiawan, R. B., Imaduddin, F., Ariawan, D., Ubaidillah, & Arifin, Z. (2021). A review on the fused deposition modeling (FDM) 3D printing: Filament processing, materials, and printing parameters. *Open Engineering, 11*(1), 639–649. https://doi.org/10.1515/eng-2021-0063. http://www.degruyter.com/view/j/eng.

Kumar Maurya, N., Sharma, R., Kumar, N., Kumar, A., Anand, P., Rai, P., & Singh, H. (2021). An overview of investigation of Fatigue, tensile strength and hardness of the components fabricated through direct metal laser sintering (DMLS) process. *Materials Today: Proceedings, 47*, 3979–3984. https://doi.org/10.1016/j.matpr.2021.04.131.

Lacroix, R., Seifert, R. W., & Timonina-Farkas, A. (2021). Benefiting from additive manufacturing for mass customization across the product life cycle. *Operations Research Perspectives, 8.* https://doi.org/10.1016/j.orp.2021.100201, http://www.journals.elsevier.com/operations-research-perspectives/.

Lakkala, P., Munnangi, S. R., Bandari, S., & Repka, M. (2023). Additive manufacturing technologies with emphasis on stereolithography 3D printing in pharmaceutical and medical applications: A review. *International Journal of Pharmaceutics: X, 5*, Article 100159. https://doi.org/10.1016/j.ijpx.2023.100159.

Lipkowitz, G., Samuelsen, T., Hsiao, K., Lee, B., Dulay, M. T., Coates, I., Lin, H., Pan, W., Toth, G., Tate, L., Shaqfeh, E. S. G., & DeSimone, J. M. (2022). Injection continuous liquid interface production of 3D objects. *Science Advances, 8*(39). https://doi.org/10.1126/sciadv.abq3917.

Meng, Q., Song, X., Han, S., Abbassi, F., Zhou, Z., Wu, B., Wang, X., & Araby, S. (2022). Mechanical and functional properties of polyamide/graphene nanocomposite prepared by chemicals free approach and selective laser sintering. *Composites Communications, 36*, Article 101396. https://doi.org/10.1016/j.coco.2022.101396.

Mohanavel, V., Ashraff Ali, K. S., Ranganathan, K., Allen Jeffrey, J., Ravikumar, M. M., & Rajkumar, S. (2021). The roles and applications of additive manufacturing in the aerospace and automobile sector. *Materials Today: Proceedings, 47*, 405–409. https://doi.org/10.1016/j.matpr.2021.04.596.

Nurul Huda, M., Seki, T., Suzuki, H., Hamidul Kabir, A. N. M., Harunur-Rashid, M., & Takeoka, Y. (2010). Characteristics of high-density poly(N-isopropylacrylamide) (PNIPA) brushes on silicon surface by atom transfer radical polymerization. *Transactions of the Materials Research Society of Japan, 35*(4), 845–848. https://doi.org/10.14723/tmrsj.35.845.

Pazhamannil, R. V., & Govindan, P. (2021). Current state and future scope of additive manufacturing technologies via vat photopolymerization. *Materials Today: Proceedings, 43*, 130–136. https://doi.org/10.1016/j.matpr.2020.11.225.

Pei, E., Kabir, I. R., & Leutenecker-Twelsiek, B. (2023). History of AM. *Springer Handbooks, 1592*, 3–29. https://doi.org/10.1007/978-3-031-20752-5_1. https://www.springer.com/series/15558.

Peng, T., Kellens, K., Tang, R., Chen, C., & Chen, G. (2018). Sustainability of additive manufacturing: An overview on its energy demand and environmental impact. *Additive Manufacturing, 21*, 694–704. https://doi.org/10.1016/j.addma.2018.04.022.

Rezaul Karim, M., Harun-Ur-Rashid, M., & Imran, A. B. (2020). Highly stretchable hydrogel using vinyl modified narrow dispersed silica particles as cross-linker. *ChemistrySelect, 5*(34), 10556–10561. https://doi.org/10.1002/slct.202003044. http://onlinelibrary.wiley.com/journal/10.1002/(ISSN)2365-6549.

Rosen, D. W. (2014). Research supporting principles for design for additive manufacturing. *Virtual and Physical Prototyping, 9*(4), 225–232. https://doi.org/10.1080/17452759.2014.951530.

Sillani, F., Kleijnen, R. G., Vetterli, M., Schmid, M., & Wegener, K. (2019). Selective laser sintering and multi jet fusion: Process-induced modification of the raw materials and analyses of parts performance. *Additive Manufacturing, 27*, 32–41. https://doi.org/10.1016/j.addma.2019.02.004.

Takeoka, Y., Yoshioka, S., Teshima, M., Takano, A., Harun-Ur-Rashid, M., & Seki, T. (2013). Structurally coloured secondary particles composed of black and white colloidal particles. *Scientific Reports, 3*(1). https://doi.org/10.1038/srep02371.

Touri, M., Kabirian, F., Saadati, M., Ramakrishna, S., & Mozafari, M. (2019). Additive manufacturing of biomaterials—The evolution of rapid prototyping. *Advanced Engineering Materials, 21*(2). https://doi.org/10.1002/adem.201800511.

Vasco, J. C. (2021). *Additive manufacturing for the automotive industry. Additive manufacturing*. Portugal: Elsevier 505–530. https://www.sciencedirect.com/book/9780128184110. https://doi.org/10.1016/B978-0-12-818411-0.00010-0.

Zhang, L. C., Liu, Y., Li, S., & Hao, Y. (2018). Additive manufacturing of titanium alloys by electron beam melting: A review. *Advanced Engineering Materials, 20*(5). https://doi.org/10.1002/adem.201700842.

Zhao, N., Parthasarathy, M., Patil, S., Coates, D., Myers, K., Zhu, H., & Li, W. (2023). Direct additive manufacturing of metal parts for automotive applications. *Journal of Manufacturing Systems, 68*, 368–375. https://doi.org/10.1016/j.jmsy.2023.04.008.

Zhu, J., Zhou, H., Wang, C., Zhou, L., Yuan, S., & Zhang, W. (2021). A review of topology optimization for additive manufacturing: Status and challenges. *Chinese Journal of Aeronautics, 34*(1), 91–110. https://doi.org/10.1016/j.cja.2020.09.020.

Applications of additively manufactured functional polymers and nanocomposites in aerospace and aviation industry

14

Ravi Shankar Rai
Automation and Robotics Engineering Department, Sandip Institute of Technology and Research Centre, Nashik, Maharashtra, India

14.1 Introduction

14.1.1 Overview of additive manufacturing

Advancements in technology and the manufacturing industry require novel, sustainable and cost saving alternatives. Additive manufacturing (AM) is a technique that involves the creation of objects by building them layer by layer, where after particles from different layers are lined together through heat or chemicals, the process is replicated, and further layers are added (Abdulhameed et al., 2019; Blakey-Milner et al., 2021). The process of AM results in a 3D complex structure as the layers are linked together one at a time. As apparent from the term, additive manufacturing process integrates the addition of raw materials rather than removing them in order to create light and robust 3D design (Diegel et al., 2019; Huang et al., 2023). Generally, 3D printing is a method for manufacturing 3D models using Computer-Aided Modeling (CAD). Hence, depending on the creation process AM can be categorized as the following: easy implementation, automated manufacturing, layered manufacturing, and solid free foam (Mohan et al., 2017). Over the past decades, additive manufacturing has revolutionized the manufacturing industry as it is able to transform engineering ideas into completely functional and robust structures, regardless of the complicated geometry associated with them. Therefore, AM is able to create products easily, accurately, at a low cost and utilizing environmentally friendly material. Additive manufacturing and 3D printing has been utilized in many sectors (Favero et al., 2021; McDonough, 2020; Srinivasan et al., 2019), especially in aerospace and automotive applications, where 33% of the AM revenue is received from these industries (Froes et al., 2019). Not only that, but the aerospace and automotive industries are the fastest growing sectors, reporting an increase in the range of 12% annually (Froes et al., 2019). The material in the aerospace and automotive industry varies between ceramic, metal and its alloys and polymeric based materials. These materials ought to possess exceptional performance, strength, and heat resistance with high durability, long term reliability and fatigue resistance (Santhosh et al., 2023; Zopp et al., 2017). Moreover, in the aerospace field, international standards are in place to sustain the process of material manufacturing.

Additive Manufacturing of Functional Polymers and Nanocomposites. DOI: https://doi.org/10.1016/B978-0-443-26516-7.00014-6
© 2025 Elsevier Ltd. All rights are reserved, including those for text and data mining, AI training, and similar technologies.

Recently, standards such as AMS (70007004) are being developed to maintain the materials and their production through additive manufacturing, which highlights the important and developing role of AM in the aerospace industry (Froes et al., 2019). AM techniques can be employed for rapid prototyping, creation of tools, and creating or finishing components and parts. Utilization of 3D printing and AM reduces the waste and consumption of energy during the manufacturing process, as time and energy are conserved throughout the various stages of production, in turn lowering the production costs and contributing to the sustainable development of manufacturing processes (Mercado Rivera & Arciniegas, 2020). Various techniques are used in additive manufacturing, including powder bed fusion, material extrusion, VAT polymerization, binder jetting, and directed energy deposition as depicted in Fig. 14.1 (Alami et al., 2023). One of the key advantages of AM is its ability to produce highly complex geometries and intricate internal structures that would be difficult or impossible to achieve with conventional manufacturing methods. This capability opens up new possibilities for design optimization, light weighting, and functional integration (Fidan et al., 2023).

AM also offers rapid prototyping capabilities, allowing for the quick and cost-effective production of prototypes and proof-of-concept models. This accelerates the product development cycle and facilitates iterative design processes, leading to faster innovation and time-to-market (Roscoe et al., 2023). In addition to rapid prototyping, AM is increasingly being used for end-use production applications across various

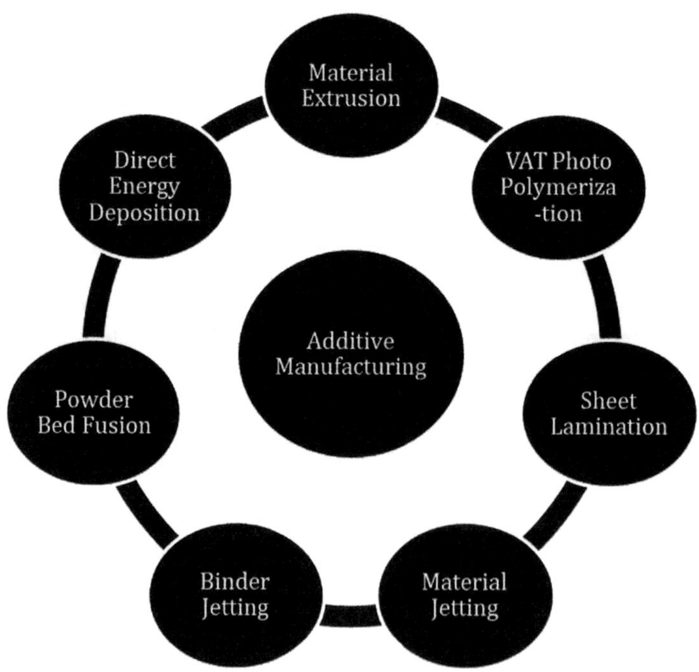

Figure 14.1 Additive manufacturing processes.

industries, including aerospace, automotive, healthcare, and consumer goods. In aerospace and aviation, AM enables the production of lightweight, high-performance components with tailored material properties, contributing to fuel efficiency, performance optimization, and supply chain resilience (Cicconi et al., 2021; Mueller, 2012). However, challenges such as material limitations, process repeatability, and postprocessing requirements still exist and need to be addressed for wider adoption of AM in industrial settings. Nevertheless, ongoing research and technological advancements continue to push the boundaries of additive manufacturing, unlocking new opportunities for innovation and efficiency in manufacturing processes across diverse industries (Neuenfeldt-Junior et al., 2021; Nys et al., 2023).

14.1.1.1 Importance of functional polymers and nanocomposites

The importance of functional polymers and nanocomposites lies in their ability to offer tailored material properties and enhanced functionalities, making them highly valuable across a wide range of industries, including aerospace and aviation (Nadgorny & Ameli, 2018). Here are several key reasons highlighting their significance in this section. Functional polymers and nanocomposites can be engineered to exhibit specific mechanical, thermal, electrical, and chemical properties. This tunability allows for the creation of materials with characteristics precisely matched to the requirements of a particular application. In aerospace, where performance and durability are paramount, the ability to customize material properties is crucial for ensuring safety and efficiency (Jalali et al., 2023). Polymers and nanocomposites are inherently lightweight materials, making them ideal for applications where weight reduction is critical, such as aerospace. By incorporating functional additives or nanomaterials, its possible to further enhance their strength-to-weight ratio, contributing to fuel efficiency and performance optimization in aircraft and spacecraft (Hore & Composto, 2014). Despite their lightweight nature, functional polymers and nanocomposites can possess remarkable strength and durability. Through advanced manufacturing techniques and material formulations, these materials can withstand harsh environmental conditions, including extreme temperatures, high pressures, and corrosive substances, making them well-suited for aerospace applications where reliability is paramount (Kallambadi Sadashivappa et al., 2023). Many polymers and nanocomposites exhibit excellent resistance to corrosion, which is particularly advantageous in aerospace environments where exposure to moisture, salt, and other corrosive agents is common. By using corrosion-resistant materials, aerospace components can maintain their structural integrity and longevity, reducing maintenance requirements and enhancing operational efficiency (Menge et al., 2022). Functional polymers and nanocomposites can be engineered to possess specific electrical and thermal conductivity properties. This is beneficial for applications requiring electromagnetic shielding, thermal management, or electrical insulation, all of which are critical considerations in aerospace electronics and avionics systems (Tamjid et al., 2024). Polymers and nanocomposites offer unparalleled design flexibility, allowing for the creation of complex geometries and intricate structures that would be challenging or impossible to achieve with traditional materials. This flexibility enables innovative design solutions and lightweight optimization strategies, leading to improved performance and efficiency in aerospace components and systems (Kausar, 2022). Functional polymers and nanocomposites often offer cost advantages over traditional materials such as metals and ceramics. They can be manufactured using relatively low-cost processes, such as injection molding or additive manufacturing, reducing production costs while still meeting stringent performance requirements. This cost-effectiveness is particularly appealing in aerospace, where cost efficiency is a significant consideration (Jalali et al., 2023). Overall, functional polymers and nanocomposites play a vital role in advancing technology and innovation in aerospace and aviation. Their customizable properties, lightweight design, strength, durability, corrosion resistance, electrical and thermal conductivity, design flexibility, and cost-effectiveness make them indispensable materials for a wide range of aerospace applications, from structural components to electronics and beyond (Dewada & Telang, 2021; Fico et al., 2022; Maalihan et al., 2022; Rinaldi et al., 2023).

14.1.1.2 Significance of aerospace and aviation industry

The aerospace and aviation industry holds immense significance on both a global economic scale and in advancing human technological capabilities. Here are several key reasons highlighting its importance: The aerospace and aviation industry facilitates global connectivity by enabling the movement of people, goods, and information across vast distances. Commercial airlines, cargo carriers, and satellite networks connect remote regions, facilitate international trade, and support tourism, fostering economic growth and cultural exchange (Kaur & Waraich, 2023). Aerospace technology plays a crucial role in national security and defense. Military aircraft, satellites, and unmanned aerial vehicles (UAVs) provide surveillance, reconnaissance, and strategic capabilities for defense forces worldwide. Additionally, space-based communication and navigation systems support military operations and enhance situational awareness (Dhakal, 2023). The aerospace and aviation industry is a catalyst for technological innovation, driving advancements in materials science, propulsion systems, aerodynamics, avionics, and space exploration technologies. Breakthroughs in these areas have far-reaching implications, influencing other industries and paving the way for future technological developments (Lade et al., 2023). The aerospace and aviation industry generates significant economic value through manufacturing, research and development, employment, and related services. Aircraft manufacturers, airlines, maintenance facilities, and aerospace suppliers contribute to GDP growth and create employment opportunities across diverse sectors of the economy (Gisario et al., 2019). Aerospace technology enables humanitys exploration of space, including missions to the Moon, Mars, and beyond. Space agencies, such as NASA, ESA, and private companies like SpaceX and Blue Origin, conduct research, launch satellites, and develop space infrastructure, expanding our understanding of the universe and paving the way for future space colonization and resource utilization. The aerospace industry is actively pursuing environmental sustainability initiatives to reduce emissions, improve fuel efficiency, and mitigate its environmental impact. Advancements in aircraft design, alternative fuels, and operational practices aim to minimize carbon footprint and address environmental concerns associated with air travel (Maleviti & Stamoulis, 2017). Aerospace technology supports humanitarian assistance and disaster relief efforts by providing airlift capabilities, aerial reconnaissance, and satellite communication during emergencies and natural disasters. Aircraft and UAVs play critical roles in search and rescue operations, medical evacuations, and delivering humanitarian aid to affected regions. The aerospace and aviation industry fosters international collaboration and partnerships among governments, organizations, and research institutions worldwide (Szabo & Koblen, 2017). Collaborative initiatives facilitate knowledge sharing, technology transfer, and joint research endeavors, addressing global challenges and promoting peaceful cooperation in space exploration and aviation safety. The aerospace and aviation industry plays a multifaceted role in shaping our world, from enabling global connectivity and national security to driving technological innovation and space exploration. Its significance extends beyond economic considerations, encompassing societal, environmental, and humanitarian dimensions, making it a cornerstone of modern civilization (Ceruti et al., 2019; Debnath et al., 2022; Dhakal, 2023).

14.2 Fundamentals of additively manufactured functional polymers and nanocomposites

14.2.1 Principles of additive manufacturing

The principles of additive manufacturing (AM) revolve around the layer-by-layer fabrication of objects based on digital 3D models (Valjak et al., 2022). Here are the key principles underlying the various AM processes. The process begins with the creation or acquisition of a digital 3D model using computer-aided design (CAD) software or 3D scanning technologies. This digital model serves as the blueprint for the physical object to be manufactured (Wu et al., 2022). The digital 3D model is sliced into thin horizontal layers using slicing software. Each layer represents a cross-section of the object to be printed. The thickness of these layers determines the resolution and surface finish of the final part (Pragana et al., 2021). Additive manufacturing techniques involve the deposition of material layer by layer to build up the final object. The material may be in the form of powders, filaments, resins, or liquids, depending on the specific AM process being used. Each layer of material is selectively deposited and bonded to the previous layer, typically through processes such as fusion, curing, or solidification. The bonding mechanism may involve heat, light, chemical reactions, or mechanical force, depending on the material and AM technique employed. In some cases, support structures may be required to prevent overhanging features from collapsing during printing. These support structures are typically generated automatically by slicing software and are removed after printing is complete. After printing, additional postprocessing steps may be necessary to improve the surface finish, accuracy, or mechanical properties of the printed part. This may include machining, sanding, polishing, or surface treatment processes. Quality control measures are essential to ensure that printed parts meet the desired specifications and performance requirements. This may involve inspection techniques such as dimensional measurement, surface analysis, and mechanical testing (Vayre et al., 2012). One of the key advantages of additive manufacturing is its ability to facilitate iterative design processes. Since AM allows for rapid prototyping and on-the-fly design modifications, designers and engineers can quickly iterate on designs, test prototypes, and refine designs based on real-world feedback. Additive manufacturing supports a wide range of materials, including polymers, metals, ceramics, composites, and biomaterials. The choice of material depends on factors such as mechanical properties, thermal characteristics, chemical compatibility, and intended application (Zhu et al., 2021). Additive manufacturing offers scalability, allowing for the production of one-off prototypes, small batches, or mass customization with minimal additional cost or lead time. This scalability makes AM suitable for a wide range of applications across various industries, from aerospace and automotive to healthcare and consumer goods (Javaid et al., 2021). Overall, the principles of additive manufacturing enable unprecedented design freedom, customization, and efficiency in the production of complex geometries and customized components, driving innovation and advancing manufacturing capabilities across diverse sectors (Hasanov et al., 2022).

14.3 Types of functional polymers and nanocomposites

Functional polymers and nanocomposites are materials engineered to exhibit specific properties and functionalities tailored to their intended applications. Here is an overview of the various types of functional polymers.

14.3.1 Conductive polymers

Conductive polymers are a class of polymers that exhibit electrical conductivity similar to metals or semiconductors. Unlike traditional insulating polymers, conductive polymers possess conjugated pi-electron systems along their backbone, which allows for the movement of charge carriers. This unique property enables their use in a wide range of applications, such as organic electronics, flexible displays, sensors, and antistatic coatings (Yan et al., 2024). Notable conductive polymers include polyaniline (PANI), polypyrrole (PPy), and poly(3,4-ethylenedioxythiophene) (PEDOT), each offering distinct electrical, mechanical, and environmental stability characteristics. In the aerospace industry, conductive polymers are utilized to create lightweight, flexible components for electromagnetic interference (EMI) shielding, lightweight antennas, and as parts of structural health monitoring systems (Blachowicz et al., 2023). Their ability to combine conductivity with the inherent advantages of polymerssuch as flexibility, lightweight, and process abilitymakes conductive polymers invaluable for modern technological applications where traditional metallic conductors may be less suitable (Tirado-Garcia et al., 2021).

14.3.2 Shape-memory polymers

Shape-memory polymers (SMPs) are advanced materials capable of returning to a predefined shape when exposed to an external stimulus such as heat, light, or an electric field. This remarkable ability stems from their unique molecular architecture, typically involving a network of hard and soft segments (Verma & Verma, 2022). At ambient temperatures, SMPs can be deformed and held in a temporary shape, but upon exposure to the stimulus, they revert to their original configuration. This shape-memory effect is leveraged in various high-tech applications, including aerospace, biomedical devices, and consumer products. In aerospace, SMPs are used to develop self-deploying structures, such as adaptive airfoils, morphing skins, and deployable space components, offering significant benefits in weight reduction, compact storage, and adaptability (Garces et al., 2019). Their versatility also extends to medical applications, such as stents and minimally invasive surgical tools, where the materials ability to change shape can facilitate insertion and deployment within the body. Shape-memory polymers represent a critical advancement in smart materials technology, providing innovative solutions to complex engineering challenges through their programmable and reversible transformation capabilities (Mahmood et al., 2024).

14.3.3 Biodegradable polymers

Biodegradable polymers are a category of polymers designed to break down into natural byproducts, such as water, carbon dioxide, and biomass, through the action of microorganisms, enzymes, or hydrolysis. These materials are engineered to address the environmental challenges associated with traditional, non-degradable plastics by offering a sustainable alternative that reduces long-term waste accumulation (Mukherjee et al., 2023). Common biodegradable polymers, which are primarily used for additive manufacturing, include polylactic acid (PLA), polycaprolactone (PCL) and polyethylene terephathalate (PET) as depicted in Fig. 14.2. Their applications span diverse fields, from packaging materials and agricultural films to biomedical devices like sutures, drug delivery systems, and tissue engineering scaffolds. In the aerospace and aviation industry, biodegradable polymers are explored for applications such as single-use components, environmentally friendly packaging for sensitive equipment, and support materials in additive manufacturing that dissolve after serving their purpose (Tian et al., 2023). The adoption of biodegradable polymers not only contributes to reducing environmental impact but also supports regulatory compliance and promotes ecofriendly practices across various industries.

14.3.4 Self-healing polymers

Self-healing polymers are an innovative class of materials designed to autonomously repair damage such as cracks or fractures, extending the lifespan and reliability of components without the need for external intervention. These polymers mimic natural

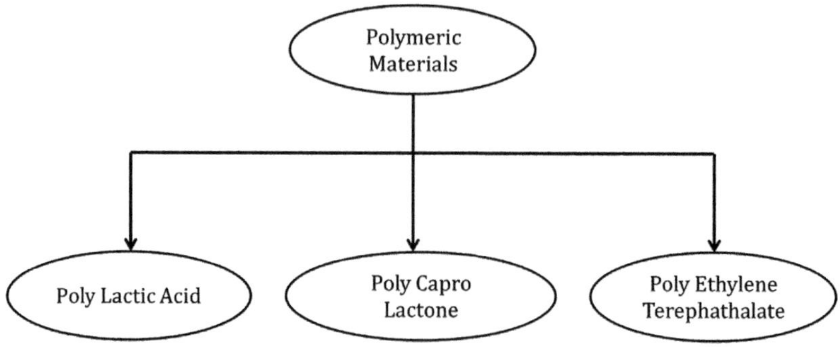

Figure 14.2 Primary polymeric materials for additive manufacturing.

healing processes found in biological systems, incorporating mechanisms such as microencapsulated healing agents, dynamic covalent bonds, or reversible cross-links that activate under specific conditions (e.g., heat, light, or moisture) (Ritzen et al., 2021). When damage occurs, the healing agents are released or the bonds re-form, effectively healing the material by restoring its structural integrity and functionality. This self-repair capability is particularly valuable in demanding environments such as aerospace, where material failure can have critical consequences. In aerospace applications, self-healing polymers are utilized in structural composites, coatings, and sealants to enhance safety and reduce maintenance costs by automatically addressing minor damages. They also find use in electronics, automotive components, and consumer goods, where durability and long-term performance are critical (Wang et al., 2022). The development of self-healing polymers represents a significant advancement in materials science, offering the potential for smarter, more resilient products across a wide range of industries.

14.3.5 Photopolymers

Photopolymers are a specialized class of polymers that undergo a transformation when exposed to light, typically ultraviolet (UV) or visible light. This transformation can include curing, hardening, or altering their chemical structure, allowing for precise control over material properties and functionality. The process is driven by photo initiators, which absorb light energy and initiate polymerization or cross-linking reactions (Voet et al., 2021). Photopolymers are extensively used in applications where rapid and localized material changes are required. In 3D printing, particularly in stereolithography (SLA) and digital light processing (DLP), photopolymers enable the creation of highly detailed and complex structures with fine resolution and smooth surfaces. They are also crucial in producing high-performance coatings, adhesives, and photoresists used in electronics manufacturing. In the aerospace industry, photopolymers are employed for rapid prototyping, creating lightweight and high-precision components, and for producing complex molds and tooling (Joseph et al., 2022). Additionally, their use in medical devices, dental applications, and microfabrication showcases their versatility and ability to meet stringent performance requirements. The unique responsiveness of photopolymers to light makes them indispensable for modern manufacturing processes, facilitating innovation and efficiency in producing complex, high-quality parts and products (Zhu et al., 2023).

14.3.6 High-temperature polymers

High-temperature polymers are a class of polymers engineered to maintain their mechanical and chemical properties under extreme heat conditions, typically ranging from 200C (392F) to over 400C (752F). These polymers are crucial in applications where conventional materials would degrade or lose performance at elevated temperatures (Tang et al., 2022). Common types include polyimides (PIs), polybenzimidazole (PBI), polyether ether ketone (PEEK), and polytetrafluoroethylene (PTFE). They offer exceptional thermal stability, resistance to oxidation, and chemical inertness, making them ideal for use in aerospace, automotive, and electronics industries. In aerospace, high-temperature polymers are used in components such as engine parts, insulation, and structural composites where they contribute to weight reduction while enduring the high temperatures and harsh environments of flight (Lacelle et al., 2023). Their ability to combine heat resistance with desirable mechanical propertiessuch as strength, toughness, and lightweightmakes them indispensable for applications requiring high-performance materials that can withstand severe thermal stresses. The development and application of high-temperature polymers continue to push the boundaries of what is possible in high-heat environments, enabling innovations in design and functionality across various advanced engineering fields (Deng et al., 2018).

14.3.7 Thermoplastic elastomers

Thermoplastic elastomers (TPEs) are a versatile class of materials that combine the elasticity of rubber with the process ability of thermoplastics. Unlike traditional vulcanized rubbers, TPEs can be repeatedly melted, reshaped, and solidified, allowing for efficient manufacturing through processes such as injection molding, extrusion, and 3D printing (Bu et al., 2023). This unique combination of properties results from their molecular structure, typically comprising hard and soft segments or blocks that enable both flexibility and mechanical strength. TPEs offer a range of benefits including excellent wear resistance, impact resistance, and the ability to stretch and recover their shape. They are widely used in applications requiring durable, flexible components, such as seals, gaskets, automotive parts, medical devices, and consumer goods. In the aerospace and aviation industry, TPEs are employed for vibration dampening, flexible seals, cable insulation, and ergonomic grips (Musa et al., 2022). Their ability to maintain performance over a broad temperature range and under dynamic mechanical stress makes them ideal for components exposed to varying environmental conditions. The ease of recycling and the potential for combining multiple functional properties in a single material make TPEs a valuable choice for both high-performance applications and sustainable manufacturing practices (Timoshenko et al., 2021).

14.4 Types of nanocomposites

14.4.1 Polymer matrix nanocomposites (PMNCs)

Polymer matrix nanocomposites (PMNCs) are advanced materials consisting of a polymer matrix embedded with nanoscale fillers or reinforcements, such as nanoparticles, nanoclays, carbon nanotubes (CNTs), or graphene. These nanofillers enhance the polymers mechanical, thermal, electrical, and barrier properties by leveraging their high surface area and unique interaction with the polymer matrix. The incorporation of nanomaterials leads to significant improvements in strength, stiffness, toughness, and thermal stability, while maintaining or even reducing the materials weight (Raina et al., 2023). PMNCs are employed across a broad spectrum of industries for applications that demand high performance and multifunctionality. In aerospace and aviation, they are used to create lightweight, high-strength components such as structural parts, skins, and interior fittings, contributing to fuel efficiency and performance optimization. Additionally, their improved thermal and electrical properties make them suitable for advanced electronic applications, such as sensors and conductive coatings. The ability to tailor the properties of PMNCs through the selection of specific nanofillers and polymer matrices makes them a powerful tool in material engineering, enabling the development of next-generation composites with customized properties for diverse and demanding applications (Prabhakaran, 2014).

14.4.2 Metal matrix nanocomposites (MMNCs)

Metal matrix nanocomposites (MMNCs) are materials composed of a metal matrix reinforced with nanoscale particles or fibers, which significantly enhance the mechanical, thermal, and electrical properties of the base metal. These nanoreinforcements, such as ceramics, carbides, nitrides, or carbon-based nanomaterials (e.g., carbon nanotubes or graphene), provide superior strength, hardness, wear resistance, and thermal stability while maintaining or even reducing the overall weight of the composite (Suresh et al., 2019). The nanoscale nature of the reinforcements allows for an extensive interface area with the metal matrix, leading to improved load transfer and the ability to fine-tune properties at the microstructural level. MMNCs are increasingly used in high-performance applications where traditional materials fall short, particularly in the aerospace, automotive, and electronics industries. In aerospace, MMNCs contribute to the development of lightweight yet robust components, such as turbine blades, structural parts, and thermal protection systems, which are crucial for enhancing fuel efficiency and overall performance. Their ability to withstand extreme conditions makes them ideal for parts exposed to high temperatures, pressures, and mechanical stress. The development of MMNCs represents a significant advancement in materials science, offering the potential to create engineered materials with tailored properties for specific, demanding applications (Hariharasakthisudhan et al., 2020).

14.4.3 Ceramic matrix nanocomposites (CMNCs)

Ceramic matrix nanocomposites (CMNCs) are advanced materials that integrate a ceramic matrix with nanoscale reinforcements, such as nanoparticles, nanowires, or nanotubes. These reinforcements significantly enhance the mechanical properties, toughness, and thermal stability of the ceramic matrix while often preserving its inherent hardness, wear resistance, and high-temperature performance (Rathod et al., 2017). The inclusion of nanomaterials, such as silicon carbide (SiC), alumina (Al_2O_3), or carbon nanotubes (CNTs), at the nanoscale level helps to improve crack resistance, fracture toughness, and overall strength by providing effective load transfer, crack deflection, and energy absorption mechanisms. CMNCs are highly valued in demanding applications where traditional ceramics may fail, including aerospace, defense, automotive, and energy sectors. In aerospace, they are used for components like turbine blades, thermal barriers, and lightweight armor, offering the ability to withstand extreme conditions such as high temperatures, aggressive chemical environments, and mechanical stresses. The advanced properties of CMNCs make them ideal for applications requiring high structural integrity and durability under harsh conditions (Gurnani & Mukhopadhyay, 2020). Their use in cutting tools, protective coatings, and advanced engineering components highlights their versatility and the significant improvements they bring to performance and longevity in critical applications.

14.4.4 Nanofiber-reinforced composites

Nanofiber-reinforced composites are materials in which a matrixtypically a polymer, metal, or ceramicis strengthened with nanofibers to enhance mechanical properties, such as tensile strength, toughness, and stiffness. These nanofibers, which can be made from materials like carbon, glass, alumina, or polymers, provide significant reinforcement due to their high aspect ratio, large surface area, and exceptional mechanical properties (Uddin et al., 2023). The integration of nanofibers improves load distribution within the composite and provides a mechanism for arresting crack propagation, resulting in superior durability and resilience. In aerospace and aviation, these composites are used for creating lightweight, high-strength structural components, aircraft skins, and impact-resistant materials, contributing to improved fuel efficiency and performance. Additionally, nanofiber-reinforced composites find applications in medical devices, automotive parts, and sports equipment, where their high-performance characteristics lead to advanced functionality and enhanced safety (Zhou et al., 2023). Their ability to combine the advantageous properties of nanofibers with the flexibility and processability of various matrices makes nanofiber-reinforced composites critical for developing next-generation materials with tailored properties for specific high-performance applications (Gou et al., 2010).

14.4.5 Graphene nanocomposites

Graphene nanocomposites are a cutting-edge class of materials where graphene, a single layer of carbon atoms with exceptional mechanical, electrical, and thermal properties, is embedded into a matrix material, such as polymers, metals, or ceramics.

The inclusion of graphene enhances the host materials strength, conductivity, thermal stability, and barrier properties due to its unique two-dimensional structure and high surface area. In graphene nanocomposites, graphene can serve as a reinforcement phase, providing significant improvements in tensile strength and stiffness, as well as electrical and thermal conductivity (Khalid, Kamal, et al., 2023). These composites are utilized across various industries, including aerospace, automotive, electronics, and energy storage. In the aerospace sector, graphene nanocomposites are used to develop lightweight, high-strength structural components, efficient thermal management systems, and advanced sensors. Their superior electrical properties make them ideal for applications in conductive coatings, electromagnetic interference (EMI) shielding, and next-generation electronic devices (Divakaran et al., 2020). Additionally, graphene nanocomposites are explored for use in high-performance batteries and supercapacitors, where they contribute to increased energy density and charge-discharge efficiency. The ability to tailor the properties of graphene nanocomposites by adjusting the graphene content and its dispersion within the matrix material enables the design of multifunctional materials that meet specific performance criteria, driving innovation and technological advancements across a broad spectrum of applications (Monetta et al., 2015; Prabhakaran, 2014).

14.4.6 Carbon nanotube (CNT) nanocomposites

Carbon nanotube (CNT) nanocomposites are materials that incorporate carbon nanotubescylindrical nanostructures with remarkable mechanical, electrical, and thermal propertiesinto various matrices, including polymers, metals, and ceramics. The integration of CNTs significantly enhances the host materials mechanical strength, electrical conductivity, thermal conductivity, and durability due to their high aspect ratio, intrinsic strength, and exceptional surface area (Iqbal et al., 2021). These nanocomposites leverage the unique attributes of CNTs, such as their ability to conduct electricity more efficiently than metals and their mechanical strength, which is an order of magnitude higher than that of traditional materials. In aerospace and aviation, CNT nanocomposites are used to develop lightweight, high-strength structural components, conductive materials for electromagnetic interference (EMI) shielding, and advanced thermal management solutions. They are also explored for applications in flexible electronics, high-performance batteries, and supercapacitors, where they contribute to enhanced energy storage capacity and efficiency (Paunikar & Kumar, 2014). In addition, CNT nanocomposites are utilized in the automotive industry, sports equipment, and medical devices, offering superior performance and longevity. The ability to customize the distribution and alignment of CNTs within the matrix allows for the design of materials with specific and enhanced properties, making CNT nanocomposites a pivotal innovation for creating next-generation high-performance materials (Rezzig et al., 2023).

14.4.7 Nanoclay composites

Nanoclay composites are materials in which nanoclay particles, typically montmorillonite or other layered silicates, are uniformly dispersed within a matrix such as a polymer, metal, or ceramic. These nanoclays, consisting of nanoscale platelets with high aspect ratios, significantly enhance the mechanical, thermal, and barrier properties of the host material by improving stiffness, strength, heat resistance, and gas impermeability (Mahesh et al., 2021). The nanoclay platelets create a tortuous path for gas molecules, dramatically reducing permeability, which makes these composites particularly effective for packaging applications requiring excellent barrier properties. In aerospace and automotive industries, nanoclay composites are used for lightweight structural components, coatings, and flame-retardant materials, contributing to weight reduction and improved safety (Kasraie & Pour Shahid Saeed Abadi, 2021). Additionally, their superior thermal stability and mechanical reinforcement are valuable for manufacturing under-the-hood automotive parts and high-performance plastics. In consumer goods, nanoclay composites offer enhanced durability and reduced material use in packaging and protective coatings. The ability to process these composites using conventional manufacturing techniques, combined with their improved performance characteristics, makes nanoclay composites a versatile and cost-effective solution for a wide range of applications, from high-tech engineering to everyday products (Mahesh et al., 2023).

14.4.8 Quantum dot nanocomposites

Quantum dot nanocomposites are materials that incorporate quantum dots (QDs) semiconductor nanoparticles with unique optical and electronic properties into various matrices such as polymers, glasses, or ceramics. These quantum dots, typically ranging from 2 to 10 nm in size, exhibit size-dependent photoluminescence, allowing them to emit light of specific wavelengths based on their size and composition (Jafarawy et al., 2023). When embedded into a matrix, QDs enhance the materials optical, electronic, and mechanical properties, enabling applications such as tunable photonic devices, displays, solar cells, and bio-imaging tools. In the aerospace and aviation industry, quantum dot nanocomposites are explored for advanced sensor technologies, high-efficiency photovoltaic cells, and enhanced display systems for instrumentation. Their ability to produce vivid, stable colors and high quantum yields makes them ideal for high-performance display technologies, including QD-LEDs and quantum dot-enhanced LCDs, which offer better color accuracy and efficiency than traditional display technologies. Additionally, in renewable energy, QD nanocomposites are used to develop next-generation solar cells with higher power conversion efficiencies (Madhi & Shirkavand Hadavand, 2022). Their versatility extends to medical diagnostics and imaging, where their distinct optical properties enable precise tracking and imaging of biological processes. The combination of quantum dots unique properties with the supportive matrix materials allows for the creation of advanced, multifunctional composites that drive innovations across various cutting-edge technologies (Bokare et al., 2021). Functional polymers and nanocomposites provide superior mechanical, thermal, electrical, and chemical properties tailored to specific applications. These materials often offer significant weight reductions while maintaining or enhancing performance, crucial for aerospace and automotive industries (Nadgorny & Ameli,

2018). They enable complex designs and multifunctional capabilities, promoting innovation in product development. Biodegradable and self-healing polymers contribute to environmental sustainability by reducing waste and extending product life cycles. They drive advancements in various fields, including electronics, healthcare, and energy, by offering improved or entirely new functionalities (Alo et al., 2022). By leveraging the diverse capabilities of functional polymers and nanocomposites, industries can achieve better performance, cost efficiency, and sustainability in a wide array of applications (Atuanya et al., 2019; Jalali et al., 2023; Kallambadi Sadashivappa et al., 2023).

14.4.9 Material properties and characteristics

Functional polymers and nanocomposites are pivotal in advancing material technology, offering specialized properties tailored for various high-performance applications. Functional polymers are engineered with specific chemical functionalities, enabling them to display unique characteristics beyond those of conventional polymers. These polymers exhibit a wide range of mechanical properties, including elasticity, tensile strength, and stiffness, making them suitable for diverse uses from flexible electronics to robust structural components. Their thermal properties, such as glass transition temperature and thermal stability, allow them to function effectively under varying temperature conditions (Buketov et al., 2023; Ikram et al., 2022). Additionally, functional polymers can be chemically resistant and tailored to possess specific electrical properties, such as conductivity or dielectric strength. Optical properties, such as transparency and refractive index, can also be customized, making them ideal for applications in optics and photonics (Verma et al., 2024). Nanocomposites, on the other hand, are materials composed of a polymer matrix embedded with nanoscale fillers, enhancing the base polymers performance. The incorporation of nanoparticles, nanotubes, nanofibers, or nanoclays can significantly improve mechanical properties like strength, toughness, and stiffness due to the high surface area and strong interfacial interactions of the nanofillers. These nanocomposites also exhibit enhanced thermal properties, including improved thermal stability and conductivity, which are beneficial for applications requiring heat dissipation or resistance to thermal degradation. Electrically, nanocomposites can be engineered to increase conductivity, making them suitable for applications in flexible electronics and sensors (Prabhakaran, 2014). Furthermore, nanocomposites offer superior barrier properties against gases and moisture, crucial for packaging and protective coatings, while maintaining or enhancing optical properties such as transparency and UV resistance (Morgan et al., 2011). The development of these advanced materials involves overcoming challenges such as achieving uniform dispersion of nanofillers and ensuring strong interfacial adhesion within the polymer matrix. Despite these challenges, the future of functional polymers and nanocomposites is promising, with ongoing research focused on creating sustainable, biodegradable materials and smart composites that can respond to environmental stimuli (Hassan et al., 2021). These advancements are expected to revolutionize industries ranging from electronics to biomedical engineering, contributing to the creation of next-generation materials that combine superior performance with environmental sustainability (Kausar, 2020).

14.4.10 Lightweight design and structural optimization

Lightweight design and structural optimization are crucial aspects of engineering that aim to reduce the weight of structures while maintaining or enhancing their performance, durability, and safety. These approaches are essential in various industries, including aerospace, automotive, civil engineering, and manufacturing, where weight reduction can lead to improved fuel efficiency, lower emissions, cost savings, and enhanced functionality (Xia et al., 2023).

14.4.10.1 Importance of lightweight components in aerospace

Lightweight components are critical in the aerospace industry due to their substantial impact on performance, efficiency, and cost-effectiveness. The reduction of aircraft weight directly correlates with decreased fuel consumption, leading to significant cost savings and enhanced fuel efficiency, which are vital in an industry where fuel expenses constitute a major portion of operational costs. Furthermore, lightweight materials such as advanced composites, aluminum-lithium alloys, and titanium offer superior strength-to-weight ratios, enhancing structural integrity and safety while also allowing for innovative design solutions like blended wing bodies and more aerodynamic shapes. These materials contribute to extended range and increased payload capacity, enabling aircraft to carry more passengers or cargo over longer distances without compromising performance (Plocher & Panesar, 2019). In addition to operational benefits, lightweight components also play a crucial role in sustainability efforts by reducing greenhouse gas emissions, thus helping the industry meet stringent environmental regulations. The integration of lightweight materials also facilitates advancements in manufacturing technologies, such as additive manufacturing, which allows for the production of complex, highly optimized structures that are both durable and easier to maintain. Overall, the adoption of lightweight components in aerospace not only drives economic and performance gains but also supports environmental and technological advancements, underscoring their paramount importance in the industry (Nguyen et al., 2022).

14.4.10.2 Structural optimization techniques

Structural optimization techniques are essential in engineering disciplines to enhance the performance, efficiency, and sustainability of structures. These techniques involve the systematic alteration of a structures design to achieve the best possible performance while minimizing material usage and weight. Topology optimization is a key method, enabling the identification of the most efficient material distribution within a given design space, often resulting in innovative, non-intuitive geometries that maximize stiffness and strength while reducing weight. Shape optimization further refines these geometries by adjusting the contours of structural components to enhance their performance under specific loading conditions. Size optimization focuses on determining the optimal dimensions of structural elements, balancing the trade-off between material usage and mechanical performance (Peto et al., 2024). Multidisciplinary optimization (MDO) integrates various engineering domains, such as structural, aerodynamic, and thermal considerations, to achieve a holistic, optimized design. Advances in computational methods, including finite element analysis (FEA) and evolutionary algorithms, have significantly improved the feasibility and accuracy of these optimization techniques. Additionally, modern manufacturing technologies like additive manufacturing (3D printing) have expanded the practical application of optimized designs, allowing for the production of complex, lightweight structures that were previously unachievable (Bhuwal et al., 2018). The implementation of structural optimization techniques not only leads to more efficient and cost-effective designs but also promotes sustainability by reducing material waste and energy consumption in the manufacturing process.

14.4.10.3 Case studies onlghtweight design

Case studies on lightweight design demonstrate the profound impact of advanced materials and innovative engineering on various industries, particularly aerospace and automotive. The Boeing 787 Dreamliner exemplifies this with its extensive use of carbon fiber-reinforced polymers (CFRP), comprising approximately 50% of the aircrafts primary structure, resulting in a 20% reduction in fuel consumption compared to its predecessors (Mele et al., 2023). Similarly, the Airbus A350 XWB utilizes over 53% composites and advanced aluminum-lithium alloys, achieving a 25% improvement in fuel efficiency and CO_2 emissions reduction. In the automotive sector, the Tesla Model S features an all-aluminum body and chassis, enhancing its range and performance while maintaining high safety standards. The BMW i3, with its carbon fiber-reinforced plastic (CFRP) passenger cell and thermoplastic exterior panels, showcases the potential for lightweight design to maximize electric vehicle efficiency and sustainability. Additionally, SpaceXs Falcon 9 rocket leverages aluminum-lithium alloys and composite materials, contributing to significant cost reductions and performance gains in space transport (Abbas et al., 2018). These case studies highlight how lightweight design not only advances fuel efficiency and operational performance but also supports sustainability and economic viability across different sectors.

14.5 Customization and complexity

Customization and complexity are pivotal in the application of additively manufactured functional polymers and nanocomposites within the aerospace and aviation industry. The advent of additive manufacturing (AM) has revolutionized the production of highly complex geometries that were previously unattainable through traditional manufacturing methods. This capability is particularly beneficial in aerospace, where components often require intricate designs to meet stringent performance criteria. Functional polymers and nanocomposites, when used in AM, allow for the creation of lightweight, high-strength components with tailored properties such as enhanced thermal stability, electrical conductivity, and improved mechanical performance. Customization through AM enables the production of bespoke parts tailored to specific applications, reducing material waste and production time (Reddy, 2011). For instance, the ability to directly manufacture parts with integrated functions, such as sensors or thermal management systems, enhances the overall efficiency and performance of aerospace components. Moreover, the use of nanocomposites in AM allows for the precise control of material properties at the nanoscale, enabling the design of parts with superior strength-to-weight ratios and resistance to extreme environmental conditions. These advancements not only improve the performance and safety of aerospace systems but also offer significant cost savings by streamlining the production process and reducing the need for assembly of multiple components. Thus, the customization and complexity enabled by additively manufactured functional polymers and nanocomposites represent a transformative leap in aerospace manufacturing, driving innovation and efficiency in the industry (Ilangovan et al., 2020; Vidakis et al., 2022).

14.5.1 Tailored designs for specific requirements

Tailored designs for specific requirements are a cornerstone in the application of additively manufactured functional polymers and nanocomposites in the aerospace and aviation industry. Additive manufacturing (AM) enables the precise customization of components to meet the unique demands of aerospace applications, where each part must adhere to stringent performance, weight, and safety criteria. Functional polymers and nanocomposites offer a versatile material palette that can be engineered to exhibit specific properties such as enhanced thermal resistance, electrical conductivity, and superior mechanical strength. This material versatility allows for the creation of highly specialized parts, such as lightweight yet robust airframe components, heat-resistant engine parts, and multifunctional elements with embedded sensors or conductive pathways. The ability to design and manufacture parts with complex geometries, optimized for weight reduction and load-bearing efficiency, further exemplifies the advantages of AM in aerospace. For example, lattice structures and other intricate designs that reduce weight without compromising strength can be readily produced using AM. The integration of nanocomposites enables further enhancement of material properties at the molecular level, offering tailored solutions for high-stress, high-temperature, and corrosive environments typically encountered in aerospace applications (Bettermann et al., 2019). These tailored designs not only improve the performance and durability of aerospace components but also contribute to significant reductions in production time and cost, as well as material waste. By leveraging the capabilities of additively manufactured functional polymers and nanocomposites, the aerospace industry can achieve unprecedented levels of customization, efficiency, and innovation, ultimately leading to more advanced and reliable aerospace systems.

14.5.2 Complex geometries and internal structures

The capability to create complex geometries and intricate internal structures is a significant advantage of using additively manufactured functional polymers and nanocomposites in the aerospace and aviation industry. Additive manufacturing (AM) facilitates the production of components with highly complex shapes that are difficult or impossible to achieve with traditional manufacturing techniques. This is particularly beneficial in aerospace, where optimizing structural efficiency and reducing weight are paramount. Functional polymers and nanocomposites offer the flexibility to design and manufacture parts with integrated internal features, such as lattice structures, which provide high strength-to-weight ratios while minimizing material usage. These advanced materials can be engineered at the molecular level to enhance specific properties, such as thermal stability, electrical conductivity, and mechanical resilience, which are crucial for aerospace applications (Kappe et al., 2022). For instance, internal cooling channels within turbine blades or customized heat exchangers can be directly manufactured, improving thermal management and overall efficiency. The use of nanocomposites further enhances these capabilities by allowing for precise control over material properties, enabling the creation of parts that can withstand extreme environmental conditions and high mechanical stresses. This ability to produce complex geometries and internal structures not only improves the performance and durability of aerospace components but also enables the integration of multiple functions into single parts, reducing assembly requirements and potential points of failure (Liu et al., 2017). Consequently, the application of additively manufactured functional polymers and nanocomposites represents a transformative approach in the aerospace industry, driving advancements in design, performance, and manufacturing efficiency.

14.5.3 Applications of customized components

The application of customized components produced through additive manufacturing (AM) of functional polymers and nanocomposites is transforming the aerospace and aviation industry. AM enables the creation of highly specialized components tailored to meet the exacting requirements of aerospace applications, which often involve complex geometries, stringent weight constraints, and specific performance criteria. Customized components, such as lightweight airframe parts, optimized engine components, and bespoke cabin interiors, can be precisely engineered using functional polymers and nanocomposites to enhance properties such as strength, thermal resistance, and durability. For instance, custom-designed brackets and fittings made from carbon fiber-reinforced polymers can reduce overall aircraft weight, leading to improved fuel efficiency and reduced emissions. Additionally, nanocomposites allow for the integration of unique functionalities, such as self-sensing capabilities or enhanced thermal conductivity, directly into the components. This level of customization not only improves the performance and reliability of individual parts but also facilitates the integration of multiple functions into single components, reducing the need for assembly and decreasing the risk of failure. Furthermore, the ability to rapidly prototype and produce customized components accelerates the development cycle, allowing for quicker iteration and optimization of designs. This agility is particularly advantageous in aerospace, where the ability to swiftly adapt to new requirements and incorporate advanced materials can provide a significant competitive edge. Overall, the use of additively manufactured customized components made from functional polymers and nanocomposites offers substantial benefits in terms of performance, efficiency, and innovation in the aerospace and aviation industry.

14.6 Reduced lead times and rapid prototyping

Reduced lead times and rapid prototyping are critical advantages of using additively manufactured functional polymers and nanocomposites in the aerospace and aviation industry. Additive manufacturing (AM) enables the swift production of prototypes and functional components directly from digital models, significantly shortening the design and development cycle. This capability allows engineers to quickly iterate on designs, test various configurations, and refine components without the lengthy delays associated with traditional manufacturing methods. Functional polymers and nanocomposites further enhance this process by offering versatile material properties that can be precisely tailored to meet specific performance requirements (Kim et al., 2023). For instance, rapid prototyping of lightweight structural components or complex geometries, such as internal lattice structures, can be achieved efficiently, facilitating immediate performance evaluation and optimization. The use of AM in producing nanocomposite-enhanced components allows for the integration of advanced functionalities, such as improved thermal management or electrical conductivity, directly into the prototypes. This rapid development process not only accelerates time-to-market but also reduces the costs associated with tooling and material waste. Additionally, the ability to produce final-use parts quickly means that aerospace companies can respond more rapidly to market demands and technological advancements, maintaining a competitive edge (Kozmel et al., 2019). Thus, the implementation of additively manufactured functional polymers and nanocomposites is revolutionizing the aerospace industry by drastically reducing lead times and enabling rapid prototyping, leading to more efficient, innovative, and responsive design and manufacturing processes.

14.6.1 Traditional manufacturing challenges

Traditional manufacturing methods in the aerospace and aviation industry face several challenges that are effectively addressed by the application of additively manufactured functional polymers and nanocomposites. Conventional techniques such as machining, casting, and forging often involve complex, multistep processes that are time-consuming and labor-intensive. These methods typically require significant material removal, leading to high levels of waste, especially when working with expensive and difficult-to-machine materials like titanium and high-performance alloys (Getachew et al., 2023). Additionally, traditional manufacturing struggles with producing complex geometries and internal structures, limiting the potential for optimizing component performance and weight. The integration of advanced functionalities, such as embedded sensors or intricate cooling channels, is also challenging and often necessitates assembly of multiple parts, increasing the risk of failure points and reducing overall reliability. Moreover, tooling and setup costs for traditional manufacturing can be prohibitively high, particularly for low-volume production runs typical in aerospace applications (Mishra & Jagadesh, 2023). Additive manufacturing (AM) overcomes these obstacles by enabling the direct production of parts from digital designs, which allows for the creation of highly complex and optimized geometries without the need for extensive tooling. Functional polymers and nanocomposites used in AM provide enhanced material properties, such as improved thermal resistance, electrical conductivity, and mechanical strength that can be precisely controlled to meet specific performance requirements. This results in reduced material waste, lower production costs, and faster turnaround times, ultimately driving innovation and efficiency in aerospace manufacturing (Islam et al., 2024; Shah et al., 2023).

14.6.2 Role of additive manufacturing in lead time reduction

Additive manufacturing (AM) plays a pivotal role in reducing lead times in the aerospace and aviation industry, particularly through the use of functional polymers and nanocomposites. Traditional manufacturing methods often involve lengthy processes such as tooling, machining, and assembly, which significantly extend production timelines. In contrast, AM enables the direct fabrication of components from digital models, bypassing the need for complex tooling and multiple production stages. This capability is particularly beneficial for producing highly customized and intricate parts, which are common in aerospace applications. The use of functional polymers and nanocomposites in AM further enhances this efficiency by allowing the production of parts with tailored material properties, such as improved thermal resistance, mechanical strength, and electrical conductivity, in a single manufacturing step. This not only speeds up the prototyping phase but also accelerates the transition from design to final production (Totin et al., 2019). Additionally, AM facilitates rapid iteration and optimization of designs, enabling engineers to quickly test and refine components to meet specific performance criteria. The ability to produce complex geometries and internal structures without the need for assembly reduces the time required to bring new designs to market. Moreover, AMs digital nature supports on-demand production, reducing the need for large inventories and further shortening lead times. In summary, additive manufacturing significantly reduces lead times in the aerospace and aviation industry by streamlining production processes, enabling rapid prototyping and iteration, and allowing for the efficient production of customized, high-performance components (Chen, Li, et al., 2019; Palanisamy et al., 2020).

14.6.3 Benefits of rapid prototyping in aerospace

Rapid prototyping, facilitated by additive manufacturing (AM) of functional polymers and nanocomposites, offers significant benefits in the aerospace and aviation industry. One of the primary advantages is the acceleration of the design and development cycle. Traditional prototyping methods are often slow and costly, requiring extensive tooling and setup. In contrast, AM allows for the quick and cost-effective production of prototypes directly from digital models, enabling engineers to swiftly iterate on designs and incorporate feedback (Najmon et al., 2019). This rapid iteration is crucial in aerospace, where design optimization for weight reduction, strength, and performance is vital. Functional polymers and nanocomposites used in AM offer versatile material properties that can be tailored to meet specific requirements, such as enhanced thermal resistance, mechanical strength, and electrical conductivity. This flexibility allows for the creation of prototypes that closely mimic the final product, providing more accurate performance testing and validation. Furthermore, rapid prototyping with AM enables the production of complex geometries and internal structures that are difficult or impossible to achieve with traditional manufacturing methods (Ly et al., 2021). This capability allows for more innovative design solutions and the integration of multiple functions into a single component, reducing assembly time and potential points of failure. Additionally, the reduced lead times and lower costs associated with rapid prototyping allow aerospace companies to bring new products to market faster, maintain a competitive edge, and respond more quickly to changing market demands and technological advancements (Voicu et al., 2021). Overall, the benefits of rapid prototyping in aerospace through the use of additively manufactured functional polymers and nanocomposites include accelerated development cycles, enhanced design flexibility, improved performance testing, and faster time-to-market.

14.7 Material properties and performance enhancement

The application of additively manufactured functional polymers and nanocomposites in the aerospace and aviation industry significantly enhances material properties and performance. These advanced materials provide unique opportunities for engineers to tailor properties such as strength, stiffness, thermal stability, and electrical conductivity to meet the rigorous demands of aerospace applications (Rai & Bajpai, 2023, 2024). Functional polymers, when used in additive manufacturing (AM), can be engineered to achieve specific mechanical properties, including high tensile strength and impact resistance, essential for structural components subjected to extreme loads and stresses (Md Ali et al., 2021). The incorporation of nanocomposites further enhances these properties by integrating nanoparticles, such as carbon nanotubes or graphene, which significantly improve the materials overall performance. These nanoparticles can enhance the thermal and electrical conductivity of the polymers, crucial for applications requiring efficient heat dissipation and electromagnetic interference shielding. Additionally, nanocomposites can be designed to provide superior resistance to harsh environmental conditions, such as high temperatures, UV radiation, and corrosive environments, thereby extending the lifespan and reliability of aerospace components (Rai & Bajpai, 2021). The ability to precisely control the material composition and microstructure during the AM process allows for the optimization of these properties at a level not achievable with traditional manufacturing methods. This results in components that are not only lighter and stronger but also more efficient and durable. Ultimately, the use of additively manufactured functional polymers and nanocomposites in aerospace leads to significant performance enhancements, contributing to the development of more advanced, reliable, and efficient aircraft and spacecraft (Basheer, 2020; Pessoa & Fraga, 2023).

14.7.1 Mechanical properties of functional polymers and nanocomposites

The mechanical properties of functional polymers and nanocomposites play a critical role in their application within the aerospace and aviation industry, where materials must withstand extreme conditions while maintaining high performance. Additively manufactured (AM) functional polymers are valued for their versatility and the ability to be engineered with specific mechanical properties, such as high tensile strength, impact resistance, and elasticity (Gnanasekaran et al., 2017). These properties make them suitable for various aerospace components, from structural parts to intricate assemblies. The inclusion of nanocomposites further enhances these mechanical properties by embedding nanoparticles like carbon nanotubes, graphene, or silica within the polymer matrix (Su et al., 2022). These nanoparticles provide significant improvements in stiffness, strength, and toughness due to their superior mechanical characteristics and high surface area, which facilitate better load transfer within the composite material. For instance, the addition of carbon nanotubes to polymer matrices can result in a dramatic increase in tensile strength and modulus, making the resultant nanocomposites ideal for high-stress applications such as fuselage panels, wing components, and other critical structural elements. The improved fracture toughness of these materials also enhances their durability and resistance to crack propagation, which is vital for the long-term reliability of aerospace components. Moreover, the lightweight nature of these nanocomposites contributes to overall weight reduction in aircraft, directly impacting fuel efficiency and payload capacity. The ability to tailor these mechanical properties through the AM process enables the production of components that are optimized for specific performance criteria, ensuring that they meet the stringent standards required in the aerospace industry. This capability not only enhances the structural integrity and performance of aerospace systems but also drives innovation in material design and application (Zaghloul et al., 2023).

14.7.2 Thermal and electrical characteristics

The thermal and electrical characteristics of additively manufactured functional polymers and nanocomposites are crucial for their applications in the aerospace and aviation industry, where materials must perform reliably under extreme conditions. Functional polymers can be engineered to exhibit specific thermal properties, such as high thermal stability and low thermal conductivity, making them ideal for insulation and thermal management applications. The integration of nanocomposites, such as those incorporating graphene or carbon nanotubes, further enhances these thermal properties by significantly increasing thermal conductivity while maintaining a lightweight structure. This is particularly beneficial for components that require efficient heat dissipation, such as electronic housings, heat exchangers, and engine parts, where excessive heat can lead to performance degradation or failure (Platnieks et al., 2020). In addition to thermal properties, the electrical characteristics of these materials are equally important. Nanocomposites embedded with conductive nanoparticles, such as silver nanowires or carbon-based nanomaterials, exhibit enhanced electrical conductivity, making them suitable for applications requiring electromagnetic interference (EMI) shielding, antistatic components, and conductive pathways. For example, the high electrical conductivity of graphene-based nanocomposites can be leveraged to create lightweight, flexible conductive materials that are crucial for modern avionics and communication systems. The ability to fabricate these materials using additive manufacturing (AM) allows for the creation of complex geometries and integrated functionalities that are not possible with traditional manufacturing methods (Zhang et al., 2021). The precise control over the distribution and orientation of nanoparticles within the polymer matrix provided by AM further optimizes the thermal and electrical properties of the resulting components. This leads to improved performance and reliability of aerospace systems, reducing the risk of thermal runaway in electronic components and ensuring consistent electrical performance in demanding environments. Overall, the advanced thermal and electrical characteristics of additively manufactured functional polymers and nanocomposites significantly enhance the capabilities and efficiency of aerospace and aviation applications (Jorge et al., 2023).

14.7.3 Performance optimization strategies

Performance optimization strategies are integral to the effective application of additively manufactured functional polymers and nanocomposites in the aerospace and aviation industry. These strategies involve a multidisciplinary approach to material design, process optimization, and structural analysis to achieve components with superior performance characteristics (Franz et al., 2022). One key strategy is the tailored design of material properties through the precise control of the composition and distribution of nanoparticles within the polymer matrix. By adjusting the type, size, and concentration of nanoparticles such as carbon nanotubes, graphene, or ceramic particles, engineers can enhance mechanical properties like strength, stiffness, and toughness, as well as thermal and electrical properties (Panesar et al., 2018). Another crucial strategy is the optimization of the additive manufacturing (AM) process parameters. Variables such as printing speed, layer thickness, temperature, and postprocessing conditions are meticulously controlled to ensure uniform material deposition, minimize defects, and enhance the bonding between layers. Advanced simulation tools and machine learning algorithms are increasingly employed to predict and optimize these parameters, leading to improved part quality and performance consistency. Topology optimization is also a vital strategy in performance enhancement. This computational technique involves designing the components geometry to achieve maximum performance with minimum material usage, resulting in lightweight structures that meet the stringent weight and strength requirements of aerospace applications (Plocher & Panesar, 2019). The capability of AM to produce complex geometries, including internal lattice structures and conformal cooling channels, further amplifies the benefits of topology optimization. Additionally, integrating multifunctional capabilities into single components is a strategy that leverages the unique potential of AM. By incorporating functions such as thermal management, EMI shielding, and structural support into a single part, the overall system performance is significantly enhanced while reducing the assembly time and potential points of failure. Continuous monitoring and quality control using in-situ sensors and real-time data analytics during the AM process ensure that the manufactured parts meet the desired specifications and performance standards (Aziz et al., 2021). These comprehensive performance optimization strategies not only improve the reliability and efficiency of aerospace components but also drive innovation, enabling the development of next-generation aerospace systems with unparalleled capabilities.

14.8 Supply chain optimization and cost savings

Supply chain optimization and cost savings are significant advantages gained through the application of additively manufactured functional polymers and nanocomposites in the aerospace and aviation industry. Traditional manufacturing processes often involve long, complex supply chains with multiple stages of production, extensive tooling, and numerous suppliers, each adding time and cost. Additive manufacturing (AM) streamlines these supply chains by enabling the direct production of components from digital models, reducing the need for intermediate steps and minimizing dependency on multiple suppliers. This simplification leads to shorter lead times and more agile production capabilities, allowing aerospace manufacturers to respond more quickly to market demands and technological advancements (Roozkhosh et al., 2024). Functional polymers and nanocomposites enhance these benefits by offering tailored material properties that can be precisely controlled during the AM process, eliminating the need for multiple materials and complex assembly processes. The ability to produce complex geometries and integrate multiple functions into single components further reduces the number of parts required, which simplifies inventory management and reduces the risk of part shortages. This consolidation of parts not only streamlines the manufacturing process but also lowers costs associated with storage, handling, and transportation. Moreover, the digital nature of AM facilitates on-demand production, reducing the need for large inventories and associated warehousing costs. Aerospace companies can produce parts as needed, minimizing waste and optimizing resource utilization. This capability is particularly beneficial for low-volume, high-complexity parts typical in aerospace applications, where traditional manufacturing would be cost-prohibitive due to high tooling and setup costs (Gradl et al., 2020). Additionally, AM allows for the rapid prototyping and iterative design testing of functional polymers and nanocomposites, leading to faster development cycles and reduced time-to-market for new products. The integration of AM into the supply chain also enhances sustainability by reducing material waste and energy consumption. Traditional subtractive manufacturing methods often result in significant material waste, particularly when working with expensive aerospace-grade materials. In contrast, AM builds components layer by layer, using only the necessary material and significantly reducing waste. This efficiency translates into lower raw material costs and a smaller environmental footprint, aligning with the industrys growing emphasis on sustainability (Reece et al., 2022).

Furthermore, the reduced need for physical inventory and warehousing supports a more decentralized manufacturing model, allowing for production closer to the point of use. This decentralization can lead to significant cost savings in logistics and transportation while also providing greater flexibility in managing production schedules and responding to changes in demand. In summary, the application of additively manufactured functional polymers and nanocomposites in the aerospace and aviation industry leads to substantial supply chain optimization and cost savings (Bhosekar & Ierapetritou, 2021). By streamlining production processes, reducing dependency on multiple suppliers, minimizing waste, and enabling on demand manufacturing, AM not only enhances efficiency and agility but also supports sustainable practices and reduces overall production costs. These benefits make AM an increasingly attractive solution for aerospace manufacturers seeking to optimize their supply chains and achieve significant cost savings (Roozkhosh et al., 2024).

14.8.1 Decentralized manufacturing and on-demand production

Decentralized manufacturing and on-demand production are transformative strategies in the aerospace and aviation industry, greatly enabled by the use of additively manufactured functional polymers and nanocomposites. Traditional manufacturing relies on centralized production facilities and extensive supply chains, which can lead to long lead times, high transportation costs, and increased vulnerability to supply chain disruptions (Harrison et al., 2018). In contrast, additive manufacturing (AM) facilitates a decentralized approach, where production can occur at multiple, strategically located facilities closer to the point of use. This flexibility allows aerospace companies to respond swiftly to regional demands, reduce logistical complexities, and lower transportation costs (Qin et al., 2022). Functional polymers and nanocomposites are particularly well-suited for decentralized manufacturing due to their versatility and high performance. These materials can be tailored to meet specific aerospace requirements, such as enhanced thermal resistance, mechanical strength, and electrical conductivity, without the need for extensive postprocessing. This capability ensures that high-quality components can be produced reliably at various decentralized locations, maintaining consistency and performance standards. On-demand production, enabled by AM, further enhances the benefits of decentralized manufacturing. Aerospace companies can produce parts as needed, reducing the necessity for large inventories and associated warehousing costs. This just-in-time production approach not only minimizes waste but also allows for rapid iteration and customization of components, facilitating the quick integration of new technologies and design improvements. The ability to produce complex geometries and integrated components in a single manufacturing step reduces assembly requirements and potential failure points, improving overall system reliability and performance (Rinaldi et al., 2022). Moreover, decentralized manufacturing and on-demand production enhance supply chain resilience. By reducing dependence on a single production facility and diversifying manufacturing locations, aerospace companies can better withstand disruptions such as natural disasters, geopolitical events, or supply shortages. This distributed approach to manufacturing also supports sustainability goals by reducing the carbon footprint associated with transporting parts over long distances. In summary, decentralized manufacturing and on-demand production, enabled by additively manufactured functional polymers and nanocomposites, offer significant advantages in the aerospace and aviation industry (Lupi et al., 2023). These strategies improve supply chain efficiency, reduce costs, enhance responsiveness to market demands, and increase resilience to disruptions, all while maintaining high standards of quality and performance. This paradigm shift in manufacturing is driving innovation and efficiency, positioning the aerospace industry to meet the evolving challenges of the future.

14.8.2 Cost analysis of additive manufacturing

A comprehensive cost analysis of additive manufacturing (AM) in the context of applications of additively manufactured functional polymers and nanocomposites in the aerospace and aviation industry reveals significant potential for cost savings and efficiency improvements. Traditional manufacturing methods, such as machining, casting, and forging, often involve high initial setup costs, extensive tooling, and substantial material waste, particularly when dealing with complex geometries and high-performance materials (Gouveia et al., 2022). In contrast, AM processes, which build components layer by layer directly from digital models, drastically reduce these initial costs by eliminating the need for specialized tooling and minimizing material waste. The ability to produce complex geometries in a single step further reduces the need for assembly and the associated labor costs. Functional polymers and nanocomposites used in AM offer tailored material properties, enabling the production of lightweight yet strong components that contribute to overall cost efficiency. These materials can be optimized for specific applications, resulting in parts that perform better and last longer, thereby reducing maintenance and replacement costs (Cunningham et al., 2017). Additionally, the rapid prototyping capability of AM allows for faster design iterations and testing, accelerating the development cycle and reducing the time-to-market for new products. This agility not only cuts development costs but also provides a competitive edge in the rapidly evolving aerospace sector (Cicconi et al., 2021).

Moreover, the decentralized and on-demand nature of AM supports a leaner supply chain, reducing inventory costs and minimizing the financial impact of supply chain disruptions. By producing parts closer to the point of use, transportation costs are significantly lowered, and lead times are shortened. This approach also allows for the efficient production of small batch sizes and customized components without the financial penalties typically associated with low-volume production in traditional manufacturing. While the initial investment in AM technology and the cost of advanced materials like nanocomposites can be high, the long-term savings from reduced material waste, lower labor costs, decreased tooling expenses, and enhanced supply chain efficiency often outweigh these upfront expenditures. Furthermore, the continuous advancements in AM technology and the increasing availability of functional polymers and nanocomposites are driving down costs, making AM an increasingly viable and cost-effective option for aerospace manufacturing. In conclusion, the cost analysis of additive manufacturing in the aerospace and aviation industry highlights its ability to deliver substantial cost savings and operational efficiencies (Franchetti & Kress, 2017). By leveraging the benefits of functional polymers and nanocomposites, aerospace companies can achieve significant reductions in material and labor costs, streamline their supply chains, and accelerate their product development cycles, all while maintaining high standards of performance and reliability.

14.8.3 Case studies on supply chain optimization

Case studies on supply chain optimization through the application of additively manufactured functional polymers and nanocomposites in the aerospace and aviation industry illustrate significant improvements in efficiency, cost savings, and operational flexibility. One notable example is the collaboration between Boeing and various AM technology providers to streamline the production of lightweight components for their commercial aircraft. By leveraging additive manufacturing, Boeing has been able to produce complex parts such as air ducts, brackets, and interior paneling directly from high-performance polymers and nanocomposites. This approach has reduced the number of parts required, minimized material waste, and shortened production times (Simons, 2018). The result is a more agile supply chain that can quickly adapt to changes in demand and design modifications. Another case study involves Airbus, which has integrated AM into its supply chain to manufacture customized, weight-optimized cabin components. Airbus utilizes functional polymers reinforced with nanocomposites to produce parts that are not only lighter but also stronger and more durable than their traditionally manufactured counterparts (Fargnoli et al., 2022). This shift has led to a significant reduction in inventory requirements and transportation costs, as parts can be produced closer to the point of assembly. Additionally, the ability to manufacture on-demand has allowed Airbus to respond more swiftly to customer-specific requirements and maintenance needs, enhancing overall customer satisfaction. GE Aviation provides a compelling example of supply chain optimization through AM with their development of the LEAP engine fuel nozzle. Traditionally composed of 20 separate parts, the nozzle is now produced as a single, additively manufactured component using advanced materials. This consolidation has drastically reduced the supply chain complexity, cutting down on the number of suppliers and streamlining the assembly process. The use of nanocomposite-enhanced materials has also improved the nozzles performance and durability, leading to lower maintenance costs and improved engine efficiency.

Lockheed Martins adoption of AM for producing satellite components showcases the benefits of supply chain optimization in the aerospace sector. By using additively manufactured nanocomposite structures, Lockheed Martin has achieved significant reductions in lead times and production costs (Kihel et al., 2022). The ability to produce highly specialized and complex parts on-demand has enabled the company to enhance the performance of their satellites while reducing the logistical challenges associated with traditional manufacturing methods. These case studies demonstrate how the integration of additively manufactured functional polymers and nanocomposites into the supply chain can lead to substantial improvements in efficiency, cost savings, and flexibility. By reducing the reliance on extensive tooling, minimizing material waste, and enabling on-demand production, aerospace companies can optimize their supply chains to better meet the demands of modern aviation. The successful implementation of these technologies not only enhances operational performance but also positions these companies at the forefront of innovation in the aerospace industry (Fargnoli et al., 2022; Zhang et al., 2014).

14.9 Material innovation and advancements

Material innovation and advancements in the context of additively manufactured functional polymers and nanocomposites are driving transformative changes in the aerospace and aviation industry. These advancements enable the development of materials with superior properties tailored to meet the rigorous demands of aerospace applications. The integration of nanocomposites, such as those incorporating carbon nanotubes, graphene, and other nanomaterials, into polymer matrices has resulted in significant enhancements in mechanical, thermal, and electrical properties. These materials offer unprecedented strength-to-weight ratios, improved thermal stability, and enhanced electrical conductivity, making them ideal for a variety of aerospace components (Ramezani & Mohd Ripin, 2023). One of the most significant innovations is the development of high-performance nanocomposite materials that can withstand the extreme conditions encountered in aerospace environments. For instance, polymers reinforced with carbon nanotubes exhibit exceptional tensile strength and modulus, which are critical for structural components subjected to high loads and stresses. These materials also offer enhanced resistance to fatigue and fracture, improving the longevity and reliability of aerospace parts. Moreover, the thermal conductivity of nanocomposites can be engineered to efficiently manage heat dissipation in high-temperature applications, such as engine components and electronic housings, thereby enhancing performance and safety (Hamrani et al., 2023).

The advancements in additive manufacturing (AM) technologies have further propelled material innovation. Techniques such as selective laser sintering (SLS), fused deposition modeling (FDM), and stereolithography (SLA) allow for precise control over the fabrication process, enabling the production of complex geometries and intricate internal structures that are not feasible with traditional manufacturing methods. This capability not only optimizes material usage but also allows for the creation of multifunctional components that integrate various roles, such as structural support, thermal management, and electrical conductivity, into a single part (Hajare & Gajbhiye, 2022). Additionally, the continuous development of new polymer blends and nanocomposite formulations tailored for AM processes has expanded the range of applications in aerospace. For example, flame retardant polymers and radiation-resistant nanocomposites are being developed to meet specific safety and performance requirements for aircraft interiors and space applications. These materials are also being engineered to be more environmentally friendly, with a focus on recyclability and reduced carbon footprint, aligning with the industrys sustainability goals (Sivarao et al., 2024). Research and development in material science are also focused on improving the scalability and reproducibility of these advanced materials. Efforts are underway to standardize the production processes and material properties, ensuring consistent quality and performance across different manufacturing batches and facilities. This standardization is crucial for the widespread adoption of additively manufactured functional polymers and nanocomposites in the aerospace industry (Khalid et al., 2022). In summary, material innovation and advancements in the field of additively manufactured functional polymers and nanocomposites are revolutionizing the aerospace and aviation industry. These advancements enable the creation of high-performance, multifunctional components that meet the stringent requirements of aerospace applications, driving efficiency, performance, and sustainability in aerospace manufacturing (Chatterjee et al., 2022; Dada et al., 2023).

14.9.1 Emerging materials for aerospace applications

Emerging materials for aerospace applications, particularly those used in additively manufactured functional polymers and nanocomposites, are significantly advancing the capabilities and performance of aerospace components. These novel materials are designed to address the demanding requirements of the aerospace and aviation industry, including lightweight structures, high strength, thermal stability, and resistance to harsh environmental conditions. One of the most promising emerging materials is graphene-reinforced polymers (Toozandehjani, 2018). Graphene, a single layer of carbon atoms arranged in a two-dimensional lattice, offers exceptional mechanical strength and electrical conductivity. When integrated into polymer matrices, graphene enhances the composites tensile strength, thermal conductivity, and electrical properties, making it ideal for applications such as structural components, thermal management systems, and electronic housings. Another significant advancement is the development of carbon nanotube (CNT) composites (Das & Sarkar, 2018). Carbon nanotubes, known for their extraordinary strength and stiffness, are being incorporated into various polymers to create nanocomposites with superior mechanical properties. These materials are particularly beneficial for components that require high strength-to-weight ratios and resistance to fatigue and fracture, such as aircraft wings and fuselage sections. The electrical conductivity of CNTs also makes these composites suitable for applications that require electromagnetic interference (EMI) shielding and advanced sensor systems

(Ramachandran et al., 2022). Additionally, ceramic-polymer nanocomposites are emerging as critical materials for high-temperature aerospace applications. These composites combine the thermal stability and wear resistance of ceramics with the flexibility and processability of polymers, resulting in materials that can withstand extreme temperatures and mechanical stress (Rathod et al., 2017). Applications include engine components, thermal protection systems, and high-performance coatings. Metal-organic frameworks (MOFs) are another class of emerging materials showing great potential. MOFs are porous materials that can be designed to have specific properties such as high surface area, tunable porosity, and chemical functionality. In the aerospace context, MOFs are being explored for applications in gas storage, separation, and catalysis, as well as lightweight structural materials that can offer high strength and thermal resistance (Khalid, Umer, et al., 2023).

The integration of smart materials, such as shape memory polymers (SMPs) and self-healing composites, is also gaining traction in aerospace applications. SMPs can change shape in response to external stimuli (e.g., temperature), which can be used for adaptive structures and morphing wings. Self-healing composites can autonomously repair damage, enhancing the durability and lifespan of aerospace components (Basheer, 2020). The continuous advancement in additive manufacturing technologies facilitates the use of these emerging materials by allowing precise control over the fabrication process and enabling the creation of complex, multifunctional structures. This capability is essential for optimizing the material properties and performance of aerospace components. In conclusion, the development and application of emerging materials such as graphene-reinforced polymers, carbon nanotube composites, ceramic-polymer nanocomposites, metal-organic frameworks, and smart materials are revolutionizing the aerospace and aviation industry (Boyer et al., 2015). These materials offer enhanced mechanical, thermal, and electrical properties, driving innovation and efficiency in aerospace design and manufacturing. The integration of these advanced materials through additive manufacturing is poised to significantly improve the performance, reliability, and sustainability of aerospace systems.

14.9.2 Advancements in material development

Advancements in material development have profoundly impacted the applications of additively manufactured functional polymers and nanocomposites in the aerospace and aviation industry, driving innovation and enhancing performance. Recent breakthroughs in nanotechnology and polymer science have led to the creation of materials with tailored properties that meet the stringent requirements of aerospace applications. One key advancement is the development of high-performance thermoplastic polymers such as polyether ether ketone (PEEK) and polyimide (PI), which exhibit exceptional thermal stability, mechanical strength, and chemical resistance. These materials are increasingly being used in additive manufacturing to produce lightweight, high-strength components capable of withstanding the harsh conditions of aerospace environments. Nanocomposites have seen significant progress with the integration of nanomaterials like carbon nanotubes (CNTs), graphene, and silica nanoparticles into polymer matrices. These nanocomposites demonstrate remarkable enhancements in mechanical properties, such as increased tensile strength, stiffness, and impact resistance, along with improved thermal conductivity and electrical performance (Parveez et al., 2022). For instance, graphene-reinforced polymers offer a combination of high strength and lightweight, which is crucial for reducing the overall weight of aircraft and improving fuel efficiency. Similarly, CNT-infused polymers provide superior structural integrity and durability, essential for critical aerospace components subjected to high loads and stresses. The advent of multifunctional materials that combine structural and functional capabilities represents another major advancement. These materials integrate properties such as thermal management, electromagnetic interference (EMI) shielding, and self-sensing capabilities into a single component, thereby reducing the need for multiple materials and simplifying the design and manufacturing processes. For example, the development of nanocomposites with embedded sensors allows for real-time health monitoring of aerospace structures, enhancing safety and maintenance efficiency (Ozturk et al., 2023).

Advancements in additive manufacturing technologies, such as selective laser sintering (SLS), fused deposition modeling (FDM), and digital light processing (DLP), have facilitated the use of these advanced materials. These technologies enable precise control over the microstructure and composition of the printed parts, ensuring optimal material properties and performance. The ability to produce complex geometries and integrate multiple functions into a single part further maximizes the benefits of advanced materials in aerospace applications. Furthermore, the continuous improvement in material characterization techniques, such as scanning electron microscopy (SEM) and atomic force microscopy (AFM), has provided deeper insights into the behavior of nanocomposites at the microscopic level. This understanding has led to better optimization of material formulations and processing conditions, ensuring consistent quality and performance of the manufactured components. In summary, the advancements in material development, particularly in the realm of functional polymers and nanocomposites, are revolutionizing the aerospace and aviation industry (Chinchanikar & Shaikh, 2022; Khalid et al., 2024). These innovations are enabling the production of lightweight, high-performance components with enhanced mechanical, thermal, and multifunctional properties. The integration of these advanced materials through additive manufacturing technologies is not only improving the performance and reliability of aerospace systems but also driving the industry toward greater efficiency and sustainability.

14.9.3 Future trends in material innovation

Future trends in material innovation, particularly within the realm of additively manufactured functional polymers and nanocomposites, are poised to revolutionize the aerospace and aviation industry further. One significant trend is the development of ultra-lightweight materials with enhanced mechanical properties. Researchers are exploring advanced nanocomposites that incorporate novel nanomaterials such as boron nitride nanotubes (BNNTs) and two-dimensional materials like molybdenum disulfide (MoS2). These materials offer superior strength-to-weight ratios, exceptional thermal stability, and enhanced electrical and thermal conductivities, making them ideal for next-generation aerospace components (Ye et al., 2024). Another promising trend is the creation of multifunctional materials that integrate various performance capabilities into a single composite. Future materials will not only provide structural support but also incorporate functionalities such as self-sensing, self-healing, and energy storage. For instance, the integration of piezoelectric nanomaterials within polymer matrices could enable the development of aerospace components capable of real-time structural health monitoring and damage detection, significantly improving safety and maintenance efficiency (Yang, 2023). The trend toward sustainability and environmental consciousness is also driving innovation in material development. Biodegradable polymers and recyclable nanocomposites are being researched to reduce the environmental impact of aerospace manufacturing and end-of-life disposal. Materials such as polylactic acid (PLA) composites reinforced with natural fibers or biodegradable nanoparticles are gaining attention for their potential to combine high performance with ecofriendliness. Advancements in smart materials are another critical trend. These materials can respond to environmental stimuli such as temperature, pressure, or humidity, enabling adaptive and responsive aerospace structures. Shape memory polymers (SMPs) and temperature-responsive nanocomposites are examples of smart materials that can change shape or properties in response to external conditions, providing new possibilities for adaptive wing designs and morphing structures that enhance aerodynamic efficiency (Kumar Patel & Purohit, 2018).

The ongoing improvements in additive manufacturing techniques, such as multimaterial printing and nanoscale 3D printing, are expected to play a significant role in the future of material innovation. These techniques allow for the precise control of material composition and microstructure, enabling the fabrication of complex, highly customized aerospace components with optimized properties. The ability to print multiple materials simultaneously will further enhance the integration of different functionalities within a single component. Furthermore, the use of artificial intelligence (AI) and machine learning (ML) in material design and optimization is an emerging trend that holds great promise. AI and ML algorithms can analyze vast amounts of data to identify optimal material compositions and processing conditions, accelerating the development of new materials with tailored properties. This data-driven approach to material innovation is expected to lead to significant breakthroughs in the performance and efficiency of aerospace materials. In summary, future trends in material innovation for additively manufactured functional polymers and nanocomposites in the aerospace and aviation industry include the development of ultra-lightweight and multifunctional materials, sustainable and biodegradable composites, smart materials with adaptive capabilities, and advanced manufacturing techniques enabled by AI and ML. These innovations are set to further enhance the performance, efficiency, and sustainability of aerospace systems, driving the industry toward new heights of technological advancement (Parveez et al., 2022).

14.10 Maintenance, Repair, and Overhaul Applications

Maintenance, Repair, and Overhaul (MRO) applications in the aerospace and aviation industry are experiencing significant advancements through the use of additively manufactured functional polymers and nanocomposites. These materials offer remarkable benefits, including reduced lead times, cost efficiency, and enhanced performance for critical components. Additive manufacturing (AM) enables the on-demand production of replacement parts, reducing the need for extensive inventories and minimizing aircraft downtime (Boretti, 2024). This capability is particularly valuable for producing complex geometries and custom parts that are often challenging and expensive to manufacture using traditional methods. Functional polymers and nanocomposites play a crucial role in MRO applications due to their superior mechanical properties, lightweight nature, and adaptability (Ayeni et al., 2011). For instance, carbon nanotube (CNT)-reinforced polymers provide exceptional strength and durability, making them ideal for structural repairs and component replacements. The high tensile strength and resistance to wear and tear of these materials ensure that repaired or replaced parts can withstand the demanding operational conditions of aerospace environments.

Additionally, the use of nanocomposites in AM allows for the incorporation of self-healing properties, enhancing the longevity and reliability of repaired components. Self-healing materials can autonomously repair minor damages, such as cracks and scratches, during operation, reducing the frequency and cost of maintenance. This innovation significantly improves the overall lifecycle management of aerospace components, ensuring continued performance and safety. The precision and customization capabilities of AM are also beneficial for producing specialized tools and fixtures used in MRO processes. Custom jigs, fixtures, and molds can be rapidly produced to facilitate complex repair tasks, improving efficiency and accuracy. This flexibility allows MRO providers to adapt quickly to various repair scenarios and accommodate the specific needs of different aircraft models (Mathaisel, 2005). Furthermore, AM technology supports the development of lightweight, high-performance components that contribute to fuel efficiency and reduce the environmental impact of aircraft operations. For example, the production of lightweight polymer and nanocomposite parts for interior cabin refurbishments not only enhances passenger comfort but also reduces the overall weight of the aircraft, leading to lower fuel consumption and emissions. Case studies have demonstrated the successful application of AM and advanced materials in MRO contexts. For instance, Lufthansa Technik has utilized AM to produce replacement parts for aircraft interiors, achieving significant reductions in lead times and costs compared to traditional manufacturing methods. Similarly, GE Aviation has employed AM to create complex engine components for repair, enhancing the efficiency and reliability of their MRO operations. In conclusion, the integration of additively manufactured functional polymers and nanocomposites in Maintenance, Repair, and Overhaul applications offers substantial advantages in terms of efficiency, cost savings, and performance. These advanced materials and AM technologies enable rapid, on-demand production of high-quality replacement parts, facilitate innovative repair solutions, and support the development of lightweight components, ultimately enhancing the overall maintenance and operational capabilities of the aerospace and aviation industry (Harun Babu et al., 2022; Zhou et al., 2020).

14.10.1 Challenges in MRO operations

Challenges in Maintenance, Repair, and Overhaul (MRO) operations within the aerospace and aviation industry, particularly concerning the applications of additively manufactured functional polymers and nanocomposites, are multifaceted and complex. One of the primary challenges is the certification and regulatory approval of additively manufactured parts. The aerospace industry is highly regulated to ensure safety and reliability, and obtaining certification for new materials and manufacturing processes can be a time-consuming and costly process. This is particularly true for novel nanocomposites and functional polymers, which may not have a long history of use in critical aerospace applications (Siponen et al., 2019). Another significant challenge is ensuring the consistency and quality of additively manufactured parts. Variations in the AM process, such as differences in printer calibration, material quality, and environmental conditions, can lead to inconsistencies in the mechanical properties and performance of the produced parts. This variability poses a risk to the reliability and safety of aerospace components, necessitating rigorous quality control and validation procedures to ensure that each part meets the stringent standards required for aviation applications. The integration of advanced materials like nanocomposites into the AM process also presents technical challenges. Nanomaterials, such as carbon nanotubes and graphene, need to be evenly dispersed within the polymer matrix to achieve the desired enhancement in material properties. However, achieving a uniform dispersion is technically challenging and requires advanced manufacturing techniques and equipment. Inadequate dispersion can lead to weak points within the material, compromising the integrity and performance of the final component. The cost of developing and implementing AM technologies for MRO operations is another barrier. High initial investments in AM equipment, along with the costs associated with developing and certifying new materials, can be prohibitive, particularly for smaller MRO providers (Chen, Gusikhin, et al., 2019). Additionally, the specialized nature of AM processes may require training and skill development for technicians and engineers, further increasing the operational costs and complexity. Supply chain issues also pose a challenge. The availability of high-quality, aerospace-grade polymers and nanocomposites suitable for AM is limited, and sourcing these materials can be difficult and expensive. The dependency on specific suppliers for advanced materials adds a layer of risk to MRO operations, as any disruption in the supply chain can impact the ability to perform timely repairs and maintenance (Ayeni et al., 2011).

Moreover, the adoption of AM technologies in MRO operations requires a cultural shift within organizations. Traditional MRO practices are deeply entrenched, and transitioning to new methods involves overcoming resistance to change, updating standard operating procedures, and ensuring that all stakeholders are aligned with the new approach. This cultural shift is essential for the successful implementation and integration of AM and advanced materials into MRO processes (Wang et al., 2023). In summary, while the applications of additively manufactured functional polymers and nanocomposites offer significant benefits for MRO operations in the aerospace and aviation industry, several challenges must be addressed. These include regulatory hurdles, quality control issues, technical difficulties in material processing, high costs, supply chain limitations, and the need for organizational change. Overcoming these challenges is crucial for leveraging the full potential of AM technologies and advanced materials in enhancing the efficiency, reliability, and sustainability of MRO operations.

14.10.2 Role of additive manufacturing in MRO

Additive manufacturing (AM) plays a pivotal role in Maintenance, Repair, and Overhaul (MRO) operations within the aerospace and aviation industry, particularly through the use of functional polymers and nanocomposites. AM allows for the rapid and cost-effective production of complex, custom-designed parts that are often required for the maintenance and repair of aircraft. This capability significantly reduces lead times and inventory costs, as parts can be manufactured on-demand rather than being stored in large quantities (Mallikarjuna & Reutzel, 2022). For example, Airbus has successfully utilized AM to produce spare parts for its A350 XWB aircraft, demonstrating the potential of this technology to streamline MRO processes and enhance operational efficiency. The flexibility of AM enables the production of parts with intricate geometries and internal structures that are difficult or impossible to achieve with traditional manufacturing methods. This is particularly advantageous for producing lightweight components, which are critical in aerospace applications. Nanocomposites, such as carbon nanotube-reinforced polymers, can be precisely integrated into these complex structures, enhancing their mechanical properties and durability. This integration leads to longer-lasting parts that require less frequent replacement, thereby reducing maintenance downtime and costs. AM also facilitates the creation of multifunctional components, combining various material properties into a single part. This multifunctionality can include enhanced thermal management, electromagnetic interference (EMI) shielding, and self-sensing capabilities (Sarayakupolu, no date). For instance, functional polymers embedded with sensors can monitor the health of components in real-time, providing valuable data that can predict when maintenance is needed and prevent unexpected failures. This proactive approach to maintenance improves the safety and reliability of aircraft operations.

Moreover, the use of AM in MRO allows for greater customization and adaptability. Each aircraft can have unique repair needs, and AM enables the production of tailored solutions that precisely meet these requirements. This customization extends to the repair tools and fixtures themselves, which can be quickly produced to support specific maintenance tasks, thereby improving efficiency and precision in repair operations. The role of AM is further exemplified by its contribution to sustainability in MRO. By producing parts locally and on-demand, AM reduces the need for transportation and storage, lowering the carbon footprint associated with traditional supply chains. Additionally, the ability to recycle and reuse materials in the AM process aligns with the industrys sustainability goals, promoting the development of greener maintenance practices (Wang et al., 2023). In summary, additive manufacturing significantly enhances MRO operations in the aerospace and aviation industry by enabling rapid, cost-effective production of customized, high-performance parts made from advanced functional polymers and nanocomposites. This technology improves efficiency, reduces downtime, and supports sustainability, all while maintaining the high standards of safety and reliability required in aerospace applications (Oyesola et al., 2020). As AM continues to evolve, its role in MRO will likely expand, driving further innovations and improvements in aircraft maintenance and repair processes.

14.10.3 Case studies on MRO applications

Case studies on Maintenance, Repair, and Overhaul (MRO) applications illustrate the transformative impact of additively manufactured functional polymers and nanocomposites in the aerospace and aviation industry. One notable example is Lufthansa Technik, which has pioneered the use of additive manufacturing (AM) to produce replacement parts for aircraft interiors. By utilizing high-performance polymers like polyether ether ketone (PEEK) reinforced with carbon nanotubes (CNTs), Lufthansa Technik has successfully created lightweight, durable parts that meet stringent safety and performance standards (Kloppenburg et al., 2018). This approach has significantly reduced lead times and costs associated with traditional manufacturing methods, demonstrating the efficiency and practicality of AM in MRO operations (Henriximon, 2021). Another significant case study involves GE Aviation, which has adopted AM for the repair of complex engine components. GE uses functional polymers and nanocomposites to fabricate intricate parts that are critical to engine performance and longevity. For instance, GE has developed a process to repair fuel nozzles using additive manufacturing, incorporating ceramic-polymer nanocomposites to enhance thermal resistance and durability. This innovation not only extends the life of the components but also improves the overall efficiency of the engines, highlighting the potential of AM to revolutionize component repair and maintenance in the aerospace sector (Analysis of the U.S. aerospace industry, 2009). The United States Air Force (USAF) has also explored the applications of AM in MRO, particularly in the production of on-demand spare parts for aging aircraft fleets. In one case, the USAF used AM to produce a replacement part for a C-5 Galaxy transport aircraft, which was otherwise obsolete and difficult to source. By employing advanced polymers and nanocomposites, the USAF was able to fabricate a part that met all necessary specifications and was ready for installation within a fraction of the time and cost required for traditional manufacturing (Poulsen &

Valencia, 2017). This case underscores the critical role of AM in maintaining the readiness and operational capability of military aircraft. Additionally, Rolls-Royce has integrated AM into its MRO strategy for the production and repair of turbine blades. Utilizing metal-polymer hybrid composites, Rolls-Royce has developed methods to repair blades that exhibit superior mechanical properties and thermal stability. These additively manufactured blades undergo rigorous testing to ensure they meet the high safety standards required for aerospace applications. The success of this initiative demonstrates the feasibility of using AM for high-stress, high-temperature components in the aerospace industry. In summary, case studies from Lufthansa Technik, GE Aviation, the USAF, and Rolls-Royce illustrate the significant advantages of using additively manufactured functional polymers and nanocomposites in MRO applications. These examples highlight how AM can reduce lead times, lower costs, and improve the performance and longevity of aerospace components (Luniya & Chimata, 2021). As the technology continues to evolve, the role of AM in MRO is expected to expand, offering even greater benefits to the aerospace and aviation industry.

14.11 Conclusion and future outlook

In conclusion, the integration of additively manufactured functional polymers and nanocomposites is revolutionizing the aerospace and aviation industry, particularly in the realm of Maintenance, Repair, and Overhaul (MRO). The ability to produce high-performance, lightweight, and customized components on-demand significantly enhances the efficiency and cost-effectiveness of MRO operations. Case studies from leading industry players such as Lufthansa Technik, GE Aviation, the USAF, and Rolls-Royce demonstrate the practical benefits and transformative potential of these advanced materials and manufacturing technologies. The use of functional polymers and nanocomposites in additive manufacturing not only extends the lifespan of critical components but also improves overall aircraft performance and safety. Looking to the future, continued advancements in material science and additive manufacturing techniques are expected to further expand the applications and capabilities of these technologies in the aerospace sector. Emerging trends such as the development of ultra-lightweight nanocomposites, multifunctional materials, and smart structures will drive further innovation. Additionally, the integration of artificial intelligence and machine learning in the design and optimization of materials and manufacturing processes will accelerate the development of tailored solutions for specific aerospace applications. Sustainability will also play a crucial role in the future outlook of aerospace MRO. The adoption of recyclable and biodegradable polymers, along with efficient use of resources in additive manufacturing, aligns with the industrys goals of reducing environmental impact. Furthermore, the ability to produce parts locally and on-demand will enhance supply chain resilience and reduce logistics costs and carbon footprints. In summary, the future of the aerospace and aviation industry looks promising with the continued adoption and advancement of additively manufactured functional polymers and nanocomposites. These innovations will not only improve MRO operations but also contribute to the overall efficiency, safety, and sustainability of aerospace systems. As technology evolves, the industry will witness even greater integration of these advanced materials, paving the way for next-generation aircraft and more robust maintenance practices.

14.11.1 Summary of key findings

In summary, the key findings of this research emphasize the significant advantages of employing additively manufactured functional polymers and nanocomposites in aerospace applications. These materials not only improve the performance and efficiency of aerospace components but also contribute to cost savings, sustainability, and innovation in MRO practices. The continued advancements in this field are poised to drive further technological breakthroughs, ensuring a robust and dynamic future for the aerospace and aviation industry. Following are the derived key finding on applications of additively manufactured functional polymers and nanocomposites in aerospace and aviation industry.

- The research on the applications of additively manufactured functional polymers and nanocomposites in the aerospace and aviation industry has yielded several key findings that underscore their transformative potential. Firstly, the integration of these advanced materials into additive manufacturing processes has proven to significantly enhance the mechanical properties, thermal stability, and overall performance of aerospace components. Functional polymers, reinforced with nanomaterials such as carbon nanotubes and graphene, offer superior strength-to-weight ratios, making them ideal for producing lightweight yet durable parts that are crucial for improving fuel efficiency and reducing operational costs.
- Secondly, the customization capabilities of additive manufacturing enable the production of complex geometries and highly specialized components that are difficult to achieve with traditional manufacturing methods. This flexibility allows for the rapid and cost-effective creation of bespoke parts tailored to specific aircraft needs, thereby reducing lead times and minimizing inventory costs. Notably, case studies from industry leaders such as Lufthansa Technik, GE Aviation, and the USAF illustrate successful implementations of these technologies in Maintenance, Repair, and Overhaul (MRO) operations, highlighting significant reductions in downtime and enhancements in maintenance efficiency.
- Furthermore, the research highlights the environmental benefits of using additively manufactured functional polymers and nanocomposites. The ability to produce parts on-demand and locally reduces the need for extensive transportation and storage, thereby lowering the carbon footprint associated with traditional supply chains. Additionally, advancements in sustainable materials, such as biodegradable polymers and recyclable nanocomposites, align with the industrys goals of reducing environmental impact and promoting ecofriendly practices.
- Lastly, the continued development and integration of smart materials, such as self-healing and self-sensing nanocomposites, present exciting opportunities for real-time health monitoring and adaptive responses in aerospace structures. These innovations promise to enhance the safety, reliability, and longevity of aerospace components, further solidifying the role of additive manufacturing in the future of the aerospace and aviation industry.

In summary, the key findings of this research emphasize the significant advantages of employing additively manufactured functional polymers and nanocomposites in aerospace applications. These materials not only improve the performance and efficiency of aerospace components but also contribute to cost savings, sustainability, and innovation in MRO practices. The continued advancements in this field are poised to drive further technological breakthroughs, ensuring a robust and dynamic future for the aerospace and aviation industry.

14.11.2 Implications for aerospace industry

The applications of additively manufactured functional polymers and nanocomposites carry profound implications for the aerospace industry, marking a significant shift in how components are designed, produced, and maintained. One of the most notable implications is the potential for substantial weight reduction in aerospace structures. The superior strength-to-weight ratios of nanocomposite materials allow for the creation of lighter yet equally robust components, directly contributing to improved fuel efficiency and reduced operational costs. This weight reduction also enables aircraft to carry more payload or travel longer distances, enhancing overall performance and economic viability. Another key implication is the enhanced ability to produce highly customized and complex parts. Additive manufacturing (AM) techniques enable the fabrication of components with intricate geometries and internal structures that traditional manufacturing methods cannot easily achieve. This capability is particularly beneficial for producing optimized aerodynamic surfaces, internal cooling channels, and other critical features that enhance the performance and efficiency of aerospace systems. The flexibility of AM also allows for rapid prototyping and on-demand production, significantly shortening development cycles and reducing time-to-market for new aircraft designs. The integration of smart materials into aerospace components further enhances the industrys capability to monitor and maintain aircraft health. Functional polymers and nanocomposites with embedded sensors can provide real-time data on structural integrity, thermal conditions, and other critical parameters. This capability facilitates predictive maintenance, reducing the risk of unexpected failures and extending the lifespan of aircraft components. Consequently, airlines can achieve higher reliability and safety standards while minimizing maintenance costs and downtime.

Moreover, the environmental benefits of additively manufactured functional polymers and nanocomposites align with the growing emphasis on sustainability within the aerospace industry. The ability to produce parts locally reduces the carbon footprint associated with transportation and logistics. Additionally, the use of recyclable and biodegradable materials supports the industrys efforts to reduce waste and promote ecofriendly practices. These advancements contribute to the development of greener aerospace technologies and help the industry meet regulatory and consumer demands for sustainability. The economic implications are equally significant. The reduction in material waste, lower inventory requirements, and minimized lead times translate into cost savings for manufacturers and operators. Small and medium-sized enterprises (SMEs) in the aerospace supply chain can also benefit from the democratization of manufacturing capabilities enabled by AM, allowing them to compete more effectively and innovate without the need for substantial capital investment in traditional manufacturing infrastructure. In summary, the implications of incorporating additively manufactured functional polymers and nanocomposites in the aerospace industry are far-reaching. These advancements promise to enhance aircraft performance, improve safety and reliability, promote sustainability, and drive economic efficiency. As the technology continues to evolve, it is set to redefine the boundaries of what is possible in aerospace engineering and manufacturing, leading to a new era of innovation and growth in the industry.

14.11.3 Future directions and areas for research

In summary, the future research in the field of additively manufactured functional polymers and nanocomposites for aerospace applications is poised to drive significant advancements. Key areas of focus include material development, multifunctional and smart materials, improvements in AM techniques, sustainability, standardization, and economic analysis. These efforts will collectively contribute to enhancing the performance, safety, and sustainability of aerospace systems, paving the way for the next generation of innovative aerospace technologies. Following are the derived directions of research as a future work in the field of additively manufactured functional polymers and nanocomposites in aerospace and aviation industry.

- Future directions and areas for research in the application of additively manufactured functional polymers and nanocomposites in the aerospace and aviation industry hold tremendous potential for innovation and advancement. One critical area of focus is the development of new nanocomposite materials with enhanced properties. Research efforts are directed toward optimizing the dispersion and alignment of nanomaterials such as carbon nanotubes, graphene, and nanoclays within polymer matrices to maximize mechanical strength, thermal stability, and electrical conductivity. Achieving uniform dispersion and strong interfacial bonding between the nanomaterials and polymer matrix is essential for realizing the full potential of these advanced materials.
- Another promising direction is the exploration of multifunctional materials that can integrate multiple properties into a single component. For instance, research is being conducted on self-healing nanocomposites that can autonomously repair damage, thereby extending the lifespan of aerospace components. Similarly, the development of materials with embedded sensing capabilities for real-time health monitoring is an exciting area of innovation. These smart materials can provide critical data on the structural integrity and performance of components, facilitating predictive maintenance and enhancing safety.
- Advancements in additive manufacturing techniques themselves are also a key area for future research. Improving the precision, speed, and scalability of AM processes will enable the production of larger and more complex aerospace components with higher reliability and consistency. Innovations in multimaterial printing, where different materials can be combined in a single manufacturing process, are particularly relevant for creating components with tailored properties for specific applications.
- Research into the environmental impact and sustainability of AM materials and processes is another important direction. Developing recyclable and biodegradable polymers and nanocomposites will align with the aerospace industrys sustainability goals and regulatory

requirements. Life cycle assessment studies to evaluate the environmental benefits of AM compared to traditional manufacturing methods will provide valuable insights for promoting greener practices.

- Collaboration between academia, industry, and regulatory bodies is crucial for advancing the adoption of these technologies. Establishing standardized testing and certification procedures for additively manufactured nanocomposite parts is essential to ensure their safety and reliability in aerospace applications. Research into the long-term performance and durability of these materials under various operational conditions will also support their broader implementation.
- Lastly, exploring the economic aspects of integrating AM and advanced materials into the aerospace supply chain is an important area for research. Studies on cost-benefit analysis, supply chain optimization, and the impact on manufacturing workflows will help stakeholders make informed decisions about adopting these technologies. The role of digital twins and simulation in optimizing design and manufacturing processes using AM and nanocomposites is another emerging area that warrants further exploration.

References

Abbas, S., Li, F., & Qiu, J. (2018). A review on SHM techniques and current challenges for characteristic investigation of damage in composite material components of aviation Industry. *Materials Performance and Characterization, 7*(1), 224–258. https://doi.org/10.1520/mpc20170167.

Abdulhameed, O., Al-Ahmari, A., Ameen, W., & Mian, S. H. (2019). Additive manufacturing: Challenges, trends, and applications. *Advances in Mechanical Engineering, 11*(2). https://doi.org/10.1177/1687814018822880. http://ade.sagepub.com/.

Alami, A. H., Ghani Olabi, A., Alashkar, A., Alasad, S., Aljaghoub, H., Rezk, H., & Abdelkareem, M. A. (2023). Additive manufacturing in the aerospace and automotive industries: Recent trends and role in achieving sustainable development goals. *Emirates Ain Shams Engineering Journal, 14*(11). https://doi.org/10.1016/j.asej.2023.102516. http://www.elsevier.com/wps/find/journaldescription.cws_home/724208/descriptiondescription.

Ali, A. M., Omar, M. Z., Hashim, H., Salleh, M. S., & Mohamed, I. F. (2021). Recent development in graphene-reinforced aluminium matrix composite: A review. *Reviews on Advanced Materials Science, 60*(1), 801–817. https://doi.org/10.1515/rams-2021-0062. https://www.degruyter.com/view/j/rams.

Alo, O. A., Mauchline, D., & Otunniyi, I. O. (2022). 3D-printed functional polymers and nanocomposites: Defects characterization and product quality improvement. *Advanced Engineering Materials, 24*(5). https://doi.org/10.1002/adem.202101219. http://onlinelibrary.wiley.com/journal/10.1002/(ISSN)1527-2648.

Analysis of the U.S. aerospace industry. U.S. Aerosp. Ind. Anal. Prospect. (2009).

Atuanya, C. U., Aigbodion, V. S., Agbo, C. O. A., & Anene, F. A. (2019). Thermo-mechanical behaviour of functional polymer nanocomposites. *International Journal of Advanced Manufacturing Technology, 103*(9-12), 4771–4775. https://doi.org/10.1007/s00170-019-03997-1. http://www.springerlink.com/content/0268-3768.

Ayeni, P., Baines, T., Lightfoot, H., & Ball, P. (2011). State-of-the-art of Lean in the aviation maintenance, repair, and overhaul industry. *Proceedings of the Institution of Mechanical Engineers, Part B: Journal of Engineering Manufacture, 225*(11), 2108–2123. https://doi.org/10.1177/0954405411407122.

Aziz, N. A., Adnan, N. A. A., Wahab, D. A., & Azman, A. H. (2021). Component design optimisation based on artificial intelligence in support of additive manufacturing repair and restoration: Current status and future outlook for remanufacturing. *Journal of Cleaner Production, 296*. https://doi.org/10.1016/j.jclepro.2021.126401. https://www.journals.elsevier.com/journal-of-cleaner-production.

Basheer, A. A. (2020). Advances in the smart materials applications in the aerospace industries. *Aircraft Engineering and Aerospace Technology, 92*(7), 1027–1035. https://doi.org/10.1108/AEAT-02-2020-0040. http://www.emeraldinsight.com/journals.htm?issn=0002-2667.

Bettermann, S., Kandelhard, F., Moritz, H. U., & Pauer, W. (2019). Digital and lean development method for 3D-printed reactors based on CAD modeling and CFD simulation. *Chemical Engineering Research and Design, 152*, 71–84. https://doi.org/10.1016/j.cherd.2019.09.024. http://www.elsevier.com/wps/find/journaldescription.cws_home/713871/descriptiondescription.

Bhosekar, A., & Ierapetritou, M. (2021). A framework for supply chain optimization for modular manufacturing with production feasibility analysis. *Computers & Chemical Engineering, 145*, 107175. https://doi.org/10.1016/j.compchemeng.2020.107175.

Bhuwal, A. S., Kapdi, S., & Jinto, M. D. (2018). Structural optimization techniques to design light weight and low radiated noise components. *SAE International Journal of Vehicle Dynamics, Stability, and NVH, 2*(3), 203–212. https://doi.org/10.4271/10-02-03-0013.

Blachowicz, T., Ehrmann, G., & Ehrmann, A. (2023). Recent developments in additive manufacturing of conductive polymer composites. *Macromolecular Materials and Engineering, 308*(7). https://doi.org/10.1002/mame.202200692.

Blakey-Milner, B., Gradl, P., Snedden, G., Brooks, M., Pitot, J., Lopez, E., Leary, M., Berto, F., & du Plessis, A. (2021). Metal additive manufacturing in aerospace: A review. *Materials & Design, 209*, 110008. https://doi.org/10.1016/j.matdes.2021.110008.

Bokare, A., Chinnusamy, S., & Erogbogbo, F. (2021). TiO2-graphene quantum dots nanocomposites for photocatalysis in energy and biomedical applications. *Catalysts, 11*(3), 319. https://doi.org/10.3390/catal11030319.

Boretti, A. (2024). A techno-economic perspective on 3D printing for aerospace propulsion. *Journal of Manufacturing Processes, 109*, 607–614. https://doi.org/10.1016/j.jmapro.2023.12.044.

Boyer, R. R., Cotton, J. D., Mohaghegh, M., & Schafrik, R. E. (2015). Materials considerations for aerospace applications. *MRS Bulletin, 40*(12), 1055–1066. https://doi.org/10.1557/mrs.2015.278.

Bu, J., Shen, N., Qin, Z., & Xu, W. (2023). Integration of low-melting-point alloys and thermoplastic elastomers for 3D printing of multifunctional composites. *Cell Reports Physical Science, 4*(10), 101604. https://doi.org/10.1016/j.xcrp.2023.101604.

Buketov, A., Sapronov, O., Klevtsov, K., & Kim, B. (2023). Functional Polymer nanocomposites with increased anticorrosion properties and wear resistance for water transport. *Polymers, 15*(16), 3449. https://doi.org/10.3390/polym15163449.

Ceruti, A., Marzocca, P., Liverani, A., & Bil, C. (2019). Maintenance in aeronautics in an Industry 4.0 context: The role of augmented reality and additive manufacturing. *Journal of Computational Design and Engineering, 6*(4), 516–526. https://doi.org/10.1016/j.jcde.2019.02.001.

Chatterjee, A., Mishra, A., Sharma, S., & Bhagchandani, R.K. (2022). Additive manufacturing techniques and design optimization of an airplane. AIP Conference Proceedings, 2653, 020004, http://scitation.aip.org/content/aip/proceeding/aipcp. https://doi.org/10.1063/5.0110507.

Chen, J., Gusikhin, O., Finkenstaedt, W., & Liu, Y. N. (2019). Maintenance, repair, and operations parts inventory management in the era of industry 4.0. *IFAC-PapersOnLine, 52*(13), 171–176. https://doi.org/10.1016/j.ifacol.2019.11.171. http://www.journals.elsevier.com/ifac-papersonline/.

Chen, Z., Li, J., Liu, C., Liu, Y., Zhu, J., & Lao, C. (2019). Preparation of high solid loading and low viscosity ceramic slurries for photopolymerization-based 3D printing. *Ceramics International, 45*(9), 11549–11557. https://doi.org/10.1016/j.ceramint.2019.03.024.

Chinchanikar, S., & Shaikh, A. A. (2022). A Review on machine learning, big data analytics, and design for additive manufacturing for aerospace applications. *Journal of Materials Engineering and Performance, 31*(8), 6112–6130. https://doi.org/10.1007/s11665-022-07125-4. https://www.springer.com/journal/11665.

Cicconi, P., Mandolini, M., Favi, C., Campi, F., & Germani, M. (2021). Metal additive manufacturing for the rapid prototyping of shaped parts: A case study. *Computer-Aided Design and Applications, 18*(5), 1061–1079. https://doi.org/10.14733/cadaps.2021.1061-1079. http://www.cad-journal.net/files/vol_18/CAD_18(5)_2021_1061-1079.pdf.

Cunningham, C. R., Wikshland, S., Xu, F., Kemakolam, N., Shokrani, A., Dhokia, V., & Newman, S. T. (2017). Cost modelling and sensitivity analysis of wire and arc additive manufacturing. *Procedia Manufacturing, 11*, 650–657. https://doi.org/10.1016/j.promfg.2017.07.163.

Dada, M., Popoola, P., & Mathe, N. (2023). Recent advances of high entropy alloys for aerospace applications: A review. *World Journal of Engineering, 20*(1), 43–74. https://doi.org/10.1108/WJE-01-2021-0040. http://www.emeraldinsight.com/journal/wje.

Das, D. K., & Sarkar, J. (2018). Graphene-magnesium nanocomposite: An advanced material for aerospace application. *Modern Physics Letters B, 32*(6). https://doi.org/10.1142/S0217984918500756http://www.worldscinet.com/mplb/mplb.shtml.

Debnath, B., Shakur, M. S., Tanjum, F., Rahman, M. A., & Adnan, Z. H. (2022). Impact of additive manufacturing on the supply chain of aerospace spare parts industryA review. *Logistics, 6*(2). https://doi.org/10.3390/logistics6020028. http://www.mdpi.com/journal/logistics.

Deng, X., Zeng, Z., Peng, B., Yan, S., & Ke, W. (2018). Mechanical properties optimization of poly-ether-ether-ketone via fused deposition modeling. *Materials, 11*(2), 216. https://doi.org/10.3390/ma11020216.

Dewada, S. S., & Telang, A. (2021). A review of recently developed polymer composite materials for fused deposition modeling 3D printing. *Materials Research Express, 8*(12). https://doi.org/10.1088/2053-1591/ac3b13. https://iopscience.iop.org/article/10.1088/1741-2552/ac3b13.

Dhakal, S. (2023). The roles and applications of additive manufacturing technology in the aerospace industry. *International Journal of Scientific Research in Multidisciplinary Studies, 9*, 42–49.

Diegel, O., Nordin, A., & Motte, D. (2019). *Additive manufacturing technologies*. Springer Science and Business Media LLC19–39. https://doi.org/10.1007/978-981-13-8281-9_2.

Divakaran, N., Zhang, X., Kale, M. B., Senthil, T., Mubarak, S., Dhamodharan, D., Wu, L., & Wang, J. (2020). Fabrication of surface modified graphene oxide/unsaturated polyester nanocomposites via in-situ polymerization: Comprehensive property enhancement. *Applied Surface Science, 502*. https://doi.org/10.1016/j.apsusc.2019.144164. http://www.journals.elsevier.com/applied-surface-science/.

Fargnoli, M., Haber, N., & Tronci, M. (2022). Case study research to foster the optimization of supply chain management through the PSS approach. *Sustainability, 14*(4), 2235. https://doi.org/10.3390/su14042235.

Favero, G., Bonesso, M., Rebesan, P., Dima, R., Pepato, A., & Mancin, S. (2021). Additive manufacturing for thermal management applications: From experimental results to numerical modeling. *International Journal of Thermofluids, 10*, 100091. https://doi.org/10.1016/j.ijft.2021.100091.

Fico, D., Rizzo, D., Casciaro, R., & Corcione, C. E. (2022). A review of polymer-based materials for fused filament fabrication (FFF): Focus on sustainability and recycled materials. *Polymers, 14*(3). https://doi.org/10.3390/polym14030465. https://www.mdpi.com/2073-4360/14/3/465/pdf.

Fidan, I., Huseynov, O., Ali, M. A., Alkunte, S., Rajeshirke, M., Gupta, A., Hasanov, S., Tantawi, K., Yasa, E., Yilmaz, O., Loy, J., Popov, V., & Sharma, A. (2023). Recent Inventions in additive manufacturing: Holistic review. *Inventions, 8*(4). https://doi.org/10.3390/inventions8040103. http://www.mdpi.com/journal/inventions.

Franchetti, M., & Kress, C. (2017). An economic analysis comparing the cost feasibility of replacing injection molding processes with emerging additive manufacturing techniques. *The International Journal of Advanced Manufacturing Technology, 88*(9-12), 2573–2579. https://doi.org/10.1007/s00170-016-8968-7.

Franz, G., Vantomme, P., & Hassan, M. H. (2022). A review on drilling of multilayer fiber-reinforced polymer composites and aluminum stacks: Optimization of strategies for improving the drilling performance of aerospace assemblies. *Fibers, 10*(9). https://doi.org/10.3390/fib10090078. http://www.mdpi.com/journal/fibers.

Froes, F., Boyer, R., & Dutta, B. (2019). *Introduction to aerospace materials requirements and the role of additive manufacturing*. Elsevier BV1–6. https://doi.org/10.1016/b978-0-12-814062-8.00001-7.

Garces, I. T., Aslanzadeh, S., Boluk, Y., & Ayranci, C. (2019). Effect of moisture on shape memory polyurethane polymers for extrusion-based additive manufacturing. *Materials, 12*(2), 244. https://doi.org/10.3390/ma12020244.

Getachew, M. T., Shiferaw, M. Z., & Ayele, B. S. (2023). The current state of the art and advancements, challenges, and future of additive manufacturing in aerospace applications. *Advances in Materials Science and Engineering, 2023*. https://doi.org/10.1155/2023/8817006http://www.hindawi.com/journals/amse/.

Gisario, A., Kazarian, M., Martina, F., & Mehrpouya, M. (2019). Metal additive manufacturing in the commercial aviation industry: A review. *Journal of Manufacturing Systems, 53*, 124–149. https://doi.org/10.1016/j.jmsy.2019.08.005.

Gnanasekaran, K., Heijmans, T., van Bennekom, S., Woldhuis, H., Wijnia, S., de With, G., & Friedrich, H. (2017). 3D printing of CNT- and graphene-based conductive polymer nanocomposites by fused deposition modeling. *Applied Materials Today, 9*, 21–28. https://doi.org/10.1016/j.apmt.2017.04.003.

Gou, J., Tang, Y., Liang, F., Zhao, Z., Firsich, D., & Fielding, J. (2010). Carbon nanofiber paper for lightning strike protection of composite materials. *Composites Part B: Engineering, 41*(2), 192–198. https://doi.org/10.1016/j.compositesb.2009.06.009.

Gouveia, J. R., Pinto, S. M., Campos, S., Matos, J. R., Sobral, J., Esteves, S., & Oliveira, L. (2022). Life cycle assessment and cost analysis of additive manufacturing repair processes in the mold industry. *Sustainability (Switzerland), 14*(4). https://doi.org/10.3390/su14042105. https://www.mdpi.com/2071-1050/14/4/2105/pdf.

Gradl, P. R., Protz, C., Fikes, J., Clark, A., Evans, L., Miller, S., Ellis, D., & Hudson, T. (2020). Lightweight thrust chamber assemblies using multi-alloy additive manufacturing and composite overwrap. *AIAA Propulsion and Energy 2020 Forum, 25*(1). https://doi.org/10.2514/6.2020-3787.

Gurnani, L., & Mukhopadhyay, A. (2020). *Development of carbon nanotube-reinforced ceramic matrix nanocomposites for advanced structural applications*. Springer Science and Business Media LLC929–974. https://doi.org/10.1007/978-3-030-16347-1_30.

Hajare, D. M., & Gajbhiye, T. S. (2022). Additive manufacturing (3D printing): Recent progress on advancement of materials and challenges. *Materials Today: Proceedings, 58*,

736–743. https://doi.org/10.1016/j.matpr.2022.02.391. https://www.sciencedirect.com/journal/materials-today-proceedings.

Hamrani, A., Bouarab, F. Z., Agarwal, A., Ju, K., & Akbarzadeh, H. (2023). Advancements and applications of multiple wire processes in additive manufacturing: A comprehensive systematic review. *Virtual and Physical Prototyping, 18*(1). https://doi.org/10.1080/17452759.2023.2273303. http://www.tandf.co.uk/journals/titles/17452759.asp.

Hariharasakthisudhan, P., Sathickbasha, K., Surya Rajan, B., & Jeyanthram, V. (2020). Steady-state wear behavior of multi-scale hybrid compositeAA6061/Al2O3 (m)/Si3N4 (nm)/GraphiteEN31 steel tribo system. *Journal of Tribology, 142*(1). https://doi.org/10.1115/1.4044853.

Harun Babu, R. M., Sivaprakash, P., Kumar, K. A., Shebana, M., & Sowndharya, S. G. (2022). *Deep PHM: IoT-based deep learning approach on prediction of prognostics and health management of an aircraft engine. In Big data analytics in smart manufacturing: principles and practices*. India: Taylor and Francis, India Taylor and Francis121–134. http://www.tandfebooks.com/doi/book/9781000815740. https://doi.org/10.1201/9781003202776-7.

Harrison, R. P., Ruck, S., Rafiq, Q. A., & Medcalf, N. (2018). Decentralised manufacturing of cell and gene therapy products: Learning from other healthcare sectors. *Biotechnology Advances, 36*(2), 345–357. https://doi.org/10.1016/j.biotechadv.2017.12.013. http://www.elsevier.com/inca/publications/store/5/2/5/4/5/5/index.htt.

Hasanov, S., Alkunte, S., Rajeshirke, M., Gupta, A., Huseynov, O., Fidan, I., Alifui-Segbaya, F., & Rennie, A. (2022). Review on additive manufacturing of multi-material parts: Progress and challenges. *Journal of Manufacturing and Materials Processing, 6*(1), 4. https://doi.org/10.3390/jmmp6010004.

Hassan, T., Salam, A., Khan, A., Khan, S. U., Khanzada, H., Wasim, M., Khan, M. Q., & Kim, I. S. (2021). Functional nanocomposites and their potential applications: A review. *Journal of Polymer Research, 28*(2). http://www.springer.com/journal/10965. https://doi.org/10.1007/s10965-021-02408-1.

Henriximon (2021). Enhancing the diagnostic performance of Condition Based Maintenance through the fusion of sensor with maintenance data. *Annual Conference of the PHM Society, 13*(1). https://doi.org/10.36001/phmconf.2021.v13i1.2981.

Hore, M. J. A., & Composto, R. J. (2014). Functional polymer nanocomposites enhanced by nanorods. *Macromolecules, 47*(3), 875–887. https://doi.org/10.1021/ma402179w.

Huang, S., Wei, H., & Li, D. (2023). Additive manufacturing technologies in the oral implant clinic: A review of current applications and progress. *Frontiers in Bioengineering and Biotechnology, 11*. https://doi.org/10.3389/fbioe.2023.1100155.

Ikram, H., Al Rashid, A., & Ko, M. (2022). Additive manufacturing of smart polymeric composites: Literature review and future perspectives. *Polymer Composites, 43*(9), 6355–6380. https://doi.org/10.1002/pc.26948.

Ilangovan, S., Senthil Kumaran, S., & Naresh, K. (2020). Effect of nanoparticles loading on free vibration response of epoxy and filament winding basalt/epoxy and E-glass/epoxy composite tubes: Experimental, analytical and numerical investigations. *Materials Research Express, 7*(2), 025007. https://doi.org/10.1088/2053-1591/ab6e36.

Iqbal, A., Saeed, A., & Ul-Hamid, A. (2021). A review featuring the fundamentals and advancements of polymer/CNT nanocomposite application in aerospace industry. *Polymer Bulletin, 78*(1), 539–557. https://doi.org/10.1007/s00289-019-03096-0

Islam, M. A., Mobarak, M. H., Rimon, M. I. H., Al Mahmud, M. Z., Ghosh, J., Ahmed, M. M. S., & Hossain, N. (2024). Additive manufacturing in polymer research: Advances, synthesis, and applications. *Polymer Testing, 132*. https://doi.org/10.1016/j.polymertesting.2024.108364. https://www.sciencedirect.com/science/journal/01429418.

Jafarawy, M. S. A., Thirumalai, D., Lee, J., Jung, H. S., Chang, S. C., Yoon, J. H., & Kim, D. H. (2023). Graphene quantum dot nanocomposites: Electroanalytical and optical sensor technology perspective. *Journal of Analytical Science and Technology, 14*(1). https://doi.org/10.1186/s40543-023-00393-2. https://jast-journal.springeropen.com.

Jalali, A., Zhang, R., Rahmati, R., Nofar, M., Sain, M., & Park, C. B. (2023). Recent progress and perspective in additive manufacturing of EMI shielding functional polymer nanocomposites. *Nano Research, 16*(1), 1–17. https://doi.org/10.1007/s12274-022-5053-4. https://www.springer.com/journal/12274.

Javaid, M., Haleem, A., Singh, R. P., Suman, R., & Rab, S. (2021). Role of additive manufacturing applications towards environmental sustainability. *Advanced Industrial and Engineering Polymer Research, 4*(4), 312–322. https://doi.org/10.1016/j.aiepr.2021.07.005. http://www.keaipublishing.com/en/journals/advanced-industrial-and-engineering-polymer-research/.

Jorge, F. E., Tienne, L. G. P., Marques, M. D. F. V., & Monteiro, S. N. (2023). Evaluation of thermoelectric properties of hybrid polyaniline nanocomposites incorporated with graphene oxide and zinc oxide with different morphologies. *Journal of Materials Research and Technology, 27*, 6822–6832. https://doi.org/10.1016/j.jmrt.2023.11.098. http://www.elsevier.com/journals/journal-of-materials-research-and-technology/2238-7854.

Joseph, B., Sam, R. M., Tharayil, A., Sagarika, V. K., Kalarikkal, N., & Thomas, S. (2022). *Photopolymers for 3D printing*. Elsevier BV, 145–154. https://doi.org/10.1016/b978-0-12-818311-3.00011-2.

Kallambadi Sadashivappa, P., Venkatachalam, R., Pothu, R., Boddula, R., Banerjee, P., Naik, R., Radwan, A. B., & Al-Qahtani, N. (2023). Progressive review of functional nanomaterials-based polymer nanocomposites for efficient EMI shielding. *Journal of Composites Science, 7*(2), 77. https://doi.org/10.3390/jcs7020077.

Kappe, K., Bihler, M., Morawietz, K., Hgenell, P. P. C., Pfaff, A., & Hoschke, K. (2022). Design concepts and performance characterization of heat pipe wick structures by LPBF additive manufacturing. *Materials, 15*(24). https://doi.org/10.3390/ma15248930. http://www.mdpi.com/journal/materials.

Kasraie, M., & Pour Shahid Saeed Abadi, P. (2021). Additive manufacturing of conductive and high-strength epoxy-nanoclay-carbon nanotube composites. *Additive Manufacturing, 46*, 102098. https://doi.org/10.1016/j.addma.2021.102098.

Kaur, M., & Waraich, R. K. (2023). Trending digital innovations in aerospace and aviation industry. *International Journal for Multidisciplinary Research, 5*(4). https://doi.org/10.36948/ijfmr.2023.v05i04.5688.

Kausar, A. (2020). Shape memory polyurethane/graphene nanocomposites: Structures, properties, and applications. *Journal of Plastic Film & Sheeting, 36*(2), 151–166. https://doi.org/10.1177/8756087919865296.

Kausar, A. (2022). Fullerene reinforced polymeric nanocomposites for energy storageStatus and prognoses. *Frontiers in Materials, 9*. https://doi.org/10.3389/fmats.2022.874169.

Khalid, M. Y., Arif, Z. U., Noroozi, R., Zolfagharian, A., & Bodaghi, M. (2022). 4D printing of shape memory polymer composites: A review on fabrication techniques, applications, and future perspectives. *Journal of Manufacturing Processes, 81*, 759–797. https://doi.org/10.1016/j.jmapro.2022.07.035. http://www.elsevier.com/wps/find/journaldescription.cws_home/620379/descriptiondescription.

Khalid, M. Y., Arif, Z. U., Tariq, A., Hossain, M., Ahmed Khan, K., & Umer, R. (2024). 3D printing of magneto-active smart materials for advanced actuators and soft robotics applications. *Emirates European Polymer Journal, 205*. https://doi.org/10.1016/j.eurpolymj.2023.112718. https://www.sciencedirect.com/science/journal/00143057.

Khalid, M. Y., Kamal, A., Otabil, A., Mamoun, O., & Liao, K. (2023). Graphene/epoxy nanocomposites for improved fracture toughness: A focused review on toughening mechanism. *Chemical Engineering Journal Advances, 16*. https://doi.org/10.1016/j.ceja.2023.100537. https://www.journals.elsevier.com/chemical-engineering-journal-advances.

Khalid, M. Y., Umer, R., & Khan, K. A. (2023). Review of recent trends and developments in aluminium 7075 alloy and its metal matrix composites (MMCs) for aircraft applications. *Emirates Results in Engineering, 20*. https://doi.org/10.1016/j.rineng.2023.101372. https://www.journals.elsevier.com/results-in-engineering.

Kihel, Y. E., Kihel, A. E., & Embarki, S. (2022). Optimization of the sustainable distribution supply chain using the lean value stream mapping 4.0 Tool: A case study of the automotive wiring industry. *Processes, 10*(9). https://doi.org/10.3390/pr10091671. http://www.mdpi.com/journal/processes.

Kim, J., Menichella, B., Lee, H., Dayton, P. A., & Pinton, G. F. (2023). A rapid prototyping method for sub-MHz single-element piezoelectric transducers by using 3D-printed components. *Sensors, 23*(1). https://doi.org/10.3390/s23010313. http://www.mdpi.com/journal/sensors.

Kloppenburg, M., Kettenbohrer, J., Beimborn, D., & Bgle, M. (2018). Leading 20,000+ employees with a process-oriented management system: Insights into process management at Lufthansa Technik Group. *Management for Professionals, 612*, 505–520. https://doi.org/10.1007/978-3-319-58307-5_27. http://www.springer.com/series/10101.

Kozmel, T., Grabowski, J., Kantner, C., Sebastian, J., Nez, B., & Fetty, J. (2019). Additive manufacturing of ferrium c64 for aerospace gear applications. The Vertical Flight SocietyForum 75: The future of vertical flight. In Proceedings of the 75th Annual Forum and Technology Display Vertical Flight Society, United States.

Kumar Patel, K., & Purohit, R. (2018). Future Prospects of shape memory polymer nano-composite and epoxy based shape memory polymerA review. *Materials Today: Proceedings, 5*(9), 20193–20200. https://doi.org/10.1016/j.matpr.2018.06.389.

Lacelle, T., Sampson, K. L., Yazdani Sarvestani, H., Rahimizadeh, A., Barroeta Robles, J., Mirkhalaf, M., Rafiee, M., Jakubinek, M. B., Paquet, C., & Ashrafi, B. (2023). Additive manufacturing of polymer derived ceramics: Materials, methods, and applications. *APL Materials, 11*(7). https://doi.org/10.1063/5.0151661http://scitation.aip.org/content/aip/journal/aplmater.

Lade, J., Mohammed, K. A., Singh, D., Prasad Verma, R., Math, P., Saraswat, M., & Raj Gupta, L. (2023). A critical review of fabrication routes and their effects on mechanical properties of AMMCs. *Materials Today: Proceedings*. https://doi.org/10.1016/j.matpr.2023.03.041. https://www.sciencedirect.com/journal/materials-today-proceedings.

Liu, R., Wang, Z., Sparks, T., Liou, F., & Newkirk, J. (2017). Aerospace component repair with additive manufacturing. Laser Additive Manufacturing.

Luniya T. Chimata G. 2021 Extrending The Life Of Classic Cars, The Additive Manufacturing Way. ASME International Mechanical Engineering Congress and Exposition, Proceedings (IMECE) 2. http://www.asmedl.org/journals/doc/ASMEDL-home/proc/ https://doi.org/10.1115/IMECE2021-70355.

Lupi, F., Cimino, M. G. C. A., Berlec, T., Galatolo, F. A., Corn, M., Roman, N., Rossi, A., & Lanzetta, M. (2023). Blockchain-based shared additive manufacturing. *Computers and Industrial Engineering, 183*. https://doi.org/10.1016/j.cie.2023.109497. https://www.journals.elsevier.com/computers-and-industrial-engineering.

Ly, T.-H., Chau, T.-T., Nguyen, H. T., Le, T.-L., & Nguyen, V.-A.-D. (2021). Optimum design and rapid prototyping of mechanicsm based on cad/cae system and fused deposition modeling 3d printing technology. *Journal of Industrial Engineering and Halal Industries, 2*(1). https://doi.org/10.14421/jiehis.3068.

Maalihan, R. D., Chen, Q., Tamura, H., Joseph Rey, J. R. H., Pajarito, B. B., Caldona, E. B., & Advincula, R. C. (2022). Mechanically and thermally enhanced 3D-printed photocurable polymer nanocomposites containing functionalized chitin nanowhiskers by stereolithography. *ACS Applied Polymer Materials, 4*(4), 2513–2526. https://doi.org/10.1021/acsapm.1c01816. . pubs.acs.org/journal/aapmcd.

Madhi, A., & Shirkavand Hadavand, B. (2022). Fluorescent epoxy-graphene quantum dots nanocomposites: Synthesis and study of properties. *Polymer-Plastics Technology and Materials, 61*(2), 117–130. https://doi.org/10.1080/25740881.2021.1959929.

Mahesh, V., Joseph, A. S., Mahesh, V., Harursampath, D., & Chethan, V. N. (2021). Investigation on the mechanical properties of additively manufactured PETG composites reinforced with OMMT nanoclay and carbon fibers. *Polymer Composites, 42*(5), 2380–2395. https://doi.org/10.1002/pc.25985. http://onlinelibrary.wiley.com/journal/10.1002/(ISSN)1548-0569.

Mahesh, V., Maladkar, P. G., Sadaram, G. S. S., Joseph, A., Mahesh, V., & Harursampath, D. (2023). Experimental investigation of the in-plane quasi-static mechanical behaviour of additively-manufactured polyethylene terephthalate/organically modified montmorillonite nanoclay composite auxetic structures. *Journal of Thermoplastic Composite Materials, 36*(10), 4021–4041. https://doi.org/10.1177/08927057221147826. https://journals.sagepub.com/home/JTC.

Mahmood, A., Perveen, F., Chen, S., Akram, T., & Irfan, A. (2024). Polymer composites in 3D/4D printing: Materials, advances, and prospects. *Molecules (Basel, Switzerland), 29*(2), 319. https://doi.org/10.3390/molecules29020319.

Maleviti, E., & Stamoulis, E. (2017). Environmental management in aviation and aerospace industries: A baseline analysis on employees perspectives. *International Journal of Sustainable Aviation, 3*(2), 151. https://doi.org/10.1504/IJSA.2017.086196.

Mallikarjuna, B., & Reutzel, E. W. (2022). Reclamation of intermetallic titanium aluminide aero-engine components using directed energy deposition technology. *Manufacturing Review, 9*. https://doi.org/10.1051/mfreview/2022024. https://mfr.edp-open.org/.

Mathaisel, D. F. X. (2005). A lean architecture for transforming the aerospace maintenance, repair and overhaul (MRO) enterprise. *International Journal of Productivity and Performance Management, 54*(8), 623–644. https://doi.org/10.1108/17410400510627499.

McDonough, J. R. (2020). A perspective on the current and future roles of additive manufacturing in process engineering, with an emphasis on heat transfer. *Thermal Science and Engineering Progress, 19*, 100594. https://doi.org/10.1016/j.tsep.2020.100594.

Mele, M., Pisaneschi, G., Ciotti, M., Campana, G., Zucchelli, A., & Fiorini, M. (2023). Environmental drawbacks of lightweight design algorithms in material extrusion additive manufacturing: A case study. *Journal of the Brazilian Society of Mechanical Sciences and Engineering, 45*(10). https://doi.org/10.1007/s40430-023-04456-8.

Menge, H. G., Huynh, N. D., Cho, C., Choi, D., & Park, Y. T. (2022). Designable functional polymer nanocomposites via layer-by-layer assembly for highly deformable power-boosted triboelectric nanogenerators. *Composites Part B: Engineering, 230*. https://doi.org/10.1016/j.compositesb.2021.109513. https://www.journals.elsevier.com/composites-part-b-engineering.

Mercado Rivera, F. J., & Arciniegas, A. J. R. (2020). Additive manufacturing methods: techniques, materials, and closed-loop control applications. *International Journal of Advanced Manufacturing Technology, 109*(1-2), 17–31. https://doi.org/10.1007/s00170-020-05663-6. http://www.springerlink.com/content/0268-3768.

Mishra, P. K., & Jagadesh, T. (2023). Applications and challenges of 3D printed polymer composites in the emerging domain of automotive and aerospace: A converged review.

Journal of The Institution of Engineers (India): Series D, 104(2), 849–866. https://doi.org/10.1007/s40033-022-00426-x. https://www.springer.com/journal/40033.

Mohan, N., Senthil, P., Vinodh, S., & Jayanth, N. (2017). A review on composite materials and process parameters optimisation for the fused deposition modelling process. *Virtual and Physical Prototyping, 12*(1), 47–59. https://doi.org/10.1080/17452759.2016.1274490.

Monetta, T., Acquesta, A., & Bellucci, F. (2015). Graphene/epoxy coating as multifunctional material for aircraft structures. *Aerospace, 2*(3), 423–434. https://doi.org/10.3390/aerospace2030423.

Morgan, B.A.B., Matters, M., & Issue, V. (2011). Polymer-clay nanocomposites: Design and application of multi-functional materials. Material Matte.

Mueller, B. (2012). Additive manufacturing technologiesRapid prototyping to direct digital manufacturing. *Assembly Automation, 32*(2). https://doi.org/10.1108/aa.2012.03332baa.010.

Mukherjee, C., Varghese, D., Krishna, J. S., Boominathan, T., Rakeshkumar, R., Dineshkumar, S., Brahmananda Rao, C. V. S., & Sivaramakrishna, A. (2023). Recent advances in biodegradable polymersProperties, applications and future prospects. *European Polymer Journal, 192*, 112068. https://doi.org/10.1016/j.eurpolymj.2023.112068.

Musa, L., Krishna Kumar, N., Abd Rahim, S. Z., Mohamad Rasidi, M. S., Watson Rennie, A. E., Rahman, R., Yousefi Kanani, A., & Azmi, A. A. (2022). A review on the potential of polylactic acid based thermoplastic elastomer as filament material for fused deposition modelling. *Journal of Materials Research and Technology, 20*, 2841–2858. https://doi.org/10.1016/j.jmrt.2022.08.057. http://www.elsevier.com/journals/journal-of-materials-research-and-technology/2238-7854.

Nadgorny, M., & Ameli, A. (2018). Functional polymers and nanocomposites for 3D printing of smart structures and devices. *ACS Applied Materials & Interfaces, 10*(21), 17489–17507. https://doi.org/10.1021/acsami.8b01786.

Najmon, J. C., Raeisi, S., & Tovar, A. (2019). *Review of additive manufacturing technologies and applications in the aerospace industry. In Additive manufacturing for the aerospace industry*. United States: Elsevier Inc, 7–31. http://www.sciencedirect.com/science/book/9780128140628. https://doi.org/10.1016/B978-0-12-814062-8.00002-9.

Neuenfeldt-Junior, A., Cheiram, M., Eckhardt, M., Scheuer, C., Siluk, J., & Francescatto, M. (2021). Additive and subtractive rapid prototyping techniques: a comparative analysis of FDM & CNC processes. *International Journal of Industrial Engineering and Management, 12*(4), 262–273. https://doi.org/10.24867/ijiem-2021-4-293.

Nguyen, P. C. H., Kim, Y., & Choi, Y. (2022). Lightweight design with metallic additively manufactured cellular structures. *Journal of Computational Design and Engineering, 9*(1), 155–167. https://doi.org/10.1093/jcde/qwab078. https://academic.oup.com/jcde/issue.

Nys, N., Knig, M., Neugebauer, P., Jones, M. J., & Gruber-Woelfler, H. (2023). Additive manufacturing as a rapid prototyping and fabrication tool for laboratory crystallizersA proof-of-concept study. *Organic Process Research and Development, 27*(8), 1455–1462. https://doi.org/10.1021/acs.oprd.3c00126. http://pubs.acs.org/journal/oprdfk.

Oyesola, M. O., Mpofu, K., Mathe, N. R., & Daniyan, I. A. (2020). Hybrid-additive manufacturing cost model: A sustainable through-life engineering support for maintenance repair overhaul in the aerospace. *Procedia Manufacturing, 49*, 199–205. https://doi.org/10.1016/j.promfg.2020.07.019. http://www.journals.elsevier.com/procedia-manufacturing.

Ozturk, F., Cobanoglu, M., & Ece, R. E. (2023). Recent advancements in thermoplastic composite materials in aerospace industry. *Journal of Thermoplastic Composite Materials, 37*, 3084–3116. https://doi.org/10.1177/08927057231222820. https://journals.sagepub.com/home/JTC.

Palanisamy, M., Pugalendhi, A., & Ranganathan, R. (2020). Selection of suitable additive manufacturing machine and materials through bestworst method (BWM). *The*

International Journal of Advanced Manufacturing Technology, 107(5-6), 2345–2362. https://doi.org/10.1007/s00170-020-05110-6.

Panesar, A., Abdi, M., Hickman, D., & Ashcroft, I. (2018). Strategies for functionally graded lattice structures derived using topology optimisation for Additive Manufacturing. *Additive Manufacturing, 19*, 81–94. https://doi.org/10.1016/j.addma.2017.11.008.

Parveez, B., Kittur, M. I., Badruddin, I. A., Kamangar, S., Hussien, M., & Umarfarooq, M. A. (2022). Scientific advancements in composite materials for aircraft applications: A review. *Polymers, 14*(22), 5007. https://doi.org/10.3390/polym14225007.

Paunikar, S., & Kumar, S. (2014). Effect of CNT waviness on the effective mechanical properties of long and short CNT reinforced composites. *Computational Materials Science, 95*, 21–28. https://doi.org/10.1016/j.commatsci.2014.06.034.

Pessoa, R. S., & Fraga, M. A. (2023). Recent advances in solar cells for aerospace applications: Materials and technologies. *Journal of Aerospace Technology and Management, 15*. https://doi.org/10.1590/jatm.v15.1296. https://jatm.com.br/jatm/article/view/1296.

Peto, M., Garca-vila, J., Rodriguez, C. A., Siller, H. R., da Silva, J. V. L., & Ramrez-Cedillo, E. (2024). Review on structural optimization techniques for additively manufactured implantable medical devices. *Frontiers in Mechanical Engineering, 10*. https://doi.org/10.3389/fmech.2024.1353108. http://www.frontiersin.org/journals/mechanical-engineering.

Platnieks, O., Gaidukovs, S., Neibolts, N., Barkane, A., Gaidukova, G., & Thakur, V. K. (2020). Poly(butylene succinate) and graphene nanoplateletbased sustainable functional nanocomposite materials: Structure-properties relationship. *Materials Today Chemistry, 18*, 100351. https://doi.org/10.1016/j.mtchem.2020.100351.

Plocher, J., & Panesar, A. (2019). Review on design and structural optimisation in additive manufacturing: Towards next-generation lightweight structures. *Materials & Design, 183*, 108164. https://doi.org/10.1016/j.matdes.2019.108164.

Poulsen, S. N., & Valencia, V. V. (2017). *Additive manufacturing applicability for United States Air Force Civil Engineer contingency operations. Additive manufacturing handbook: Product development for the defense industry*. United States: CRC Press, 759–776. http://www.tandfebooks.com/doi/book/10.1201/9781315119106. https://doi.org/10.1201/9781315119106.

Prabhakaran, R. (2014). Nanocomposites for aircraft applications. *Journal of Aerospace Sciences and Technologies*, 169–185. https://doi.org/10.61653/joast.v66i3.2014.452.

Pragana, J. P. M., Sampaio, R. F. V., Bragana, I. M. F., Silva, C. M. A., & Martins, P. A. F. (2021). Hybrid metal additive manufacturing: A stateofthe-art review. *Advances in Industrial and Manufacturing Engineering, 2*, 100032. https://doi.org/10.1016/j.aime.2021.100032.

Qin, J., Hu, F., Liu, Y., Witherell, P., Wang, C. C. L., Rosen, D. W., Simpson, T. W., Lu, Y., & Tang, Q. (2022). Research and application of machine learning for additive manufacturing. *Additive Manufacturing, 52*. https://doi.org/10.1016/j.addma.2022.102691. https://www.journals.elsevier.com/additive-manufacturing.

Rai, R. S., & Bajpai, V. (2021). Recent advances in ZnO nanostructures and their future perspective. *Advances in Nano Research, 11*(1), 37–54. https://doi.org/10.12989/anr.2021.11.1.037. http://www.techno-press.org/.

Rai, R. S., & Bajpai, V. (2023). Improvement of interfacial adhesion of CuO nanostructured carbon fiber reinforced polymer composites. *Polymer Composites, 44*(3), 1789–1804. https://doi.org/10.1002/pc.27205. http://onlinelibrary.wiley.com/journal/10.1002/(ISSN)1548-0569.

Rai, R. S., & Bajpai, V. (2024). Surface functionalization of carbon fiber by hydrothermally synthesized ZnO nanostructures for mechanical strengthening of polymer composites. *Polymer*

Composites, 45(3), 2012–2031. https://doi.org/10.1002/pc.27901. http://onlinelibrary.wiley.com/journal/10.1002/(ISSN)1548-0569.

Raina, N., Rani, R., Kumari, A., Bhardwaj, B. Y., & Gupta, M. (2023). *Polymer-matrix nanocomposites and its potential applications*. Elsevier BV, 567–583. https://doi.org/10.1016/b978-0-323-91248-8.00017-9

Ramachandran, K., Boopalan, V., Bear, J. C., & Subramani, R. (2022). Multi-walled carbon nanotubes (MWCNTs)-reinforced ceramic nanocomposites for aerospace applications: A review. *Journal of Materials Science, 57*(6), 3923–3953. https://doi.org/10.1007/s10853-021-06760-x. http://www.springer.com/journal/10853.

Ramezani, M., & Mohd Ripin, Z. (2023). 4D printing in biomedical engineering: advancements, challenges, and future directions. *Journal of Functional Biomaterials, 14*(7), 347. https://doi.org/10.3390/jfb14070347.

Rathod, V. T., Kumar, J. S., & Jain, A. (2017). Polymer and ceramic nanocomposites for aerospace applications. *Applied Nanoscience (Switzerland), 7*(8), 519–548. https://doi.org/10.1007/s13204-017-0592-9. https://www.springer.com/materials/nanotechnology/journal/13204.

Reddy, B. (Ed.). (2011). Advances in NanocompositesSynthesis, Characterization and Industrial Applications. InTech. https://doi.org/10.5772/604.

Reece, E. M., Davis, M. J., Abu-Ghname, A., Castanon, J. M., Voris, M., Hoxworth, R., Winocour, S., & Buchanan, E. P. (2022). Working smarter, not harder: Using data-driven strategies to generate front-end cost savings through price negotiation and supply chain optimization. *Plastic and Reconstructive Surgery, 149*(6), 1488–1497. https://doi.org/10.1097/PRS.0000000000009121. http://journals.lww.com/plasreconsurg/pages/issuelist.aspx.

Rezzig, D., Tarfaoui, M., & Abdeslam, S. (2023). The effect of interphase and agglomeration on the mechanical properties of aluminum matrix nanocomposites reinforced with carbon nanotubes. *Materials Today Communications, 37*, 107103. https://doi.org/10.1016/j.mtcomm.2023.107103.

Rinaldi, M., Ferrara, M., Pigliaru, L., Allegranza, C., & Nanni, F. (2023). Additive manufacturing of polyether ether ketone-based composites for space application: a mini-review. *CEAS Space Journal, 15*(1), 77–87. https://doi.org/10.1007/s12567-021-00401-4.

Rinaldi, M., Caterino, M., & Macchiaroli, R. (2022). Additive manufacturing and supply chain configuration: Modelling and performance evaluation. *Journal of Industrial Engineering and Management, 15*(1), 103. https://doi.org/10.3926/jiem.3590.

Ritzen, L., Montano, V., & Garcia, S. J. (2021). 3d printing of a self-healing thermoplastic polyurethane through fdm: From polymer slab to mechanical assessment. *Polymers, 13*(2), 1–21. https://doi.org/10.3390/polym13020305. https://www.mdpi.com/2073-4360/13/2/305/pdf.

Roozkhosh, P., Pooya, A., Soleimani Fard, O., & Bagheri, R. (2024). Revolutionizing supply chain sustainability: An additive manufacturing-enabled optimization model for minimizing waste and costs. *Process Integration and Optimization for Sustainability, 8*(1), 285–300. https://doi.org/10.1007/s41660-023-00368-1.

Roscoe, S., Cousins, P. D., & Handfield, R. (2023). Transitioning additive manufacturing from rapid prototyping to high-volume production: A case study of complex final products. *Journal of Product Innovation Management, 40*(4), 554–576. https://doi.org/10.1111/jpim.12673. . http://onlinelibrary.wiley.com/journal/10.1111/(ISSN)1540-5885.

Santhosh, N., Praveena, B. A., Jain, R., Abul Hasan, M., Islam, S., Amir Khan, M., Razak, A., & Daniyal, M. (2023). Analysis of friction and wear of aluminium AA 5083/ WC composites for building applications using advanced machine learning models. *Ain Shams Engineering Journal, 14*(9), 102090. https://doi.org/10.1016/j.asej.2022.102090.

Sarayakupolu, T. The Implementation of Additive Manufacturing Technologies in Aircraft Maintenance, Repair, And Overhaul (MRO) Operations. Istanbul Gelisim University.

Shah, A., Aliyev, R., Zeidler, H., & Krinke, S. (2023). A review of the recent developments and challenges in wire arc additive manufacturing (WAAM) process. *Journal of Manufacturing and Materials Processing, 7*(3), 97. https://doi.org/10.3390/jmmp7030097.

Simons, M. (2018). Additive manufacturingA revolution in progress? Insights from a multiple case study. *The International Journal of Advanced Manufacturing Technology, 96*(1-4), 735–749. https://doi.org/10.1007/s00170-018-1601-1.

Siponen, M., Haapasalo, H., & Harkonen, J. (2019). Maintenance, repair, and operations inventory reduction and operational development. *International Journal of Industrial and Systems Engineering, 32*(1), 1. https://doi.org/10.1504/ijise.2019.099780.

Sivarao, S., Lee, K. Y. S., Kumaran, K., Ramasamy, D., Kassim, A. M., Salleh, M. S., Vatesh, U. K., Pujari, S., & Dharsyanth, R. (2024). Advancements in additive manufacturing: Innovations in direct ink writing materials and their transformative practical applications. *Journal of Engineering Science and Technology, 19*(1), 246–275. https://jestec.taylors.edu.my/Vol%2019%20Issue%201%20February%20%202024/19_1_18.pdf.

Srinivasan, K. V., Manimaran, A., Arulprakasajothi, M., Revanth, M., & Arolkar, V. A. (2019). Design and development of porous regenerator for Stirling cryocooler using additive manufacturing. *Thermal Science and Engineering Progress, 11*, 195–203. https://doi.org/10.1016/j.tsep.2019.03.013.

Su, X., Wang, R., Li, X., Araby, S., Kuan, H. C., Naeem, M., & Ma, J. (2022). A comparative study of polymer nanocomposites containing multi-walled carbon nanotubes and graphene nanoplatelets. *Nano Materials Science, 4*(3), 185–204. https://doi.org/10.1016/j.nanoms.2021.08.003. http://www.keaipublishing.com/en/journals/nano-materials-science/.

Suresh, S., Gowd, G. H., & Kumar, M. L. S. D. (2019). Mechanical and wear behavior of Al 7075/Al2O3/SiC/mg metal matrix nanocomposite by liquid state process. *Advanced Composites and Hybrid Materials, 2*(3), 530–539. https://doi.org/10.1007/s42114-019-00101-y. springer.com/journal/42114.

Szabo, S., & Koblen, I. (2017). Aviation technology life cycle management: importance for aviation companies, aerospace industry organizations and relevant stakeholders. *MADMagazine of Aviation Development, 5*(2), 15. https://doi.org/10.14311/MAD.2017.02.03.

Tamjid, E., Najafi, P., Khalili, M. A., Shokouhnejad, N., Karimi, M., & Sepahdoost, N. (2024). Review of sustainable, eco-friendly, and conductive polymer nanocomposites for electronic and thermal applications: Current status and future prospects. *Discover Nano, 19*(1). https://doi.org/10.1186/s11671-024-03965-2. https://www.springer.com/journal/11671.

Tang, Z., Gong, J., Cao, P., Tao, L., Pei, X., Wang, T., Zhang, Y., Wang, Q., & Zhang, J. (2022). 3D printing of a versatile applicability shape memory polymer with high strength and high transition temperature. *Chemical Engineering Journal, 431*, 134211. https://doi.org/10.1016/j.cej.2021.134211.

Tian, J., Zheng, Y., Ouyang, Q., Xue, P., Guo, B., & Xu, J. (2023). *Structure and properties of biodegradable polymer materials for fused deposition modeling 3D printing*. IntechOpen. https://doi.org/10.5772/intechopen.110175.

Timoshenko, M. V., Balabanov, S. V., Sychev, M. M., & Nikiforov, D. I. (2021). Thermoplastic elastomer for 3Dprinting by fused deposition modeling. *Polymer ScienceSeries A, 63*(6), 652–656. https://doi.org/10.1134/S0965545X21060134. https://rd.springer.com/journal/11498.

Tirado-Garcia, I., Garcia-Gonzalez, D., Garzon-Hernandez, S., Rusinek, A., Robles, G., Martinez-Tarifa, J. M., & Arias, A. (2021). Conductive 3D printed PLA composites: On the interplay of mechanical, electrical and thermal behaviours. *Composite Structures, 265*, 113744. https://doi.org/10.1016/j.compstruct.2021.113744.

Toozandehjani, M. (2018). Conventional and advanced composites in aerospace industry: Technologies revisited. *American Journal of Aerospace Engineering, 5*(1), 9. https://doi.org/10.11648/j.ajae.20180501.12.

Totin, A., Macdonald, E., & Conner, B. (2019). Additive manufacturing for aerospace maintenance and sustainment. DSIAC Journal, 6.

Uddin, M.B., Jahan, I., Kelkar, A.D., & Mohan, R.V. (2023). Development of conductive lightweight nanofiber reinforced composite for aircraft lightning strike protection composites and advanced materials expo, CAMX. https://doi.org/10.33599/nasampe/c.23.0161.

Valjak, F., Kosori, D., Reetar, M., & Bojeti, N. (2022). Function-based design principles for additive manufacturing. *Applied Sciences, 12*(7), 3300. https://doi.org/10.3390/app12073300.

Vayre, B., Vignat, F., & Villeneuve, F. (2012). Designing for additive manufacturing. *Procedia CIRP, 3*(1), 632–637. https://doi.org/10.1016/j.procir.2012.07.108.

Verma, A., Mordina, B., Mukhopadhyay, K., Dwivedi, M., & Banerjee, S. (2024). Functional polymer nanocomposites as supercapacitors for health Care. *Materials Horizons: From Nature to Nanomaterials, 1329*, 505–529. https://doi.org/10.1007/978-981-99-3021-0_21. http://www.springer.com/series/16122.

Verma, S., & Verma, V. K. (2022). Shape memory polymers for additive manufacturing: An overview. *Materials Today: Proceedings, 57*, 2077–2081. https://doi.org/10.1016/j.matpr.2021.11.507.

Vidakis, N., Petousis, M., Mountakis, N., Grammatikos, S., Papadakis, V., Kechagias, J. D., & Das, S. C. (2022). On the thermal and mechanical performance of Polycarbonate/titanium nitride nanocomposites in material extrusion additive manufacturing. *Composites Part C: Open Access, 8*, 100291. https://doi.org/10.1016/j.jcomc.2022.100291.

Voet, V. S. D., Guit, J., & Loos, K. (2021). Sustainable photopolymers in 3D printing: A Review on biobased, biodegradable, and recyclable alternatives. *Macromolecular Rapid Communications, 42*(3). https://doi.org/10.1002/marc.202000475. http://onlinelibrary.wiley.com/journal/10.1002/(ISSN)1521-3927.

Voicu, A. D., Hadr, A., & Vlsceanu, D. (2021). Benefits of 3D printing technologies for aerospace lattice structures. *Scientific Bulletin of Naval Academy, 24*(1), 8–16. https://doi.org/10.21279/1454-864X-21-I1-001. http://www.anmb.ro/buletinstiintific/eng/home.html.

Wang, L., Li, Y., Zhou, L., Lou, Y., Liu, S., Zheng, D., & Yi, M. (2023). Progress in additive manufacturing, additive repair and fatigue evaluation of aviation titanium alloy blades. *Materials Research Letters, 11*(12), 973–1012. https://doi.org/10.1080/21663831.2023.2275599.

Wang, Z., Cui, H., Liu, M., Grage, S. L., Hoffmann, M., Sedghamiz, E., Wenzel, W., & Levkin, P. A. (2022). Tough, transparent, 3D-printable, and self-healing poly(ethylene glycol)-gel (PEGgel). *Advanced Materials, 34*(11) https://doi.org/10.1002/adma.202107791. http://onlinelibrary.wiley.com/journal/10.1002/(ISSN)1521-4095.

Wu, W., Li, X., Liu, Q., Hsi Fuh, J. Y., Zheng, A., Zhou, Y., Ren, L., & Li, G. (2022). Additive manufacturing of bulk metallic glass: Principles, materials and prospects. *Materials Today Advances, 16*. https://doi.org/10.1016/j.mtadv.2022.100319. https://www.journals.elsevier.com/materials-today-advances/.

Xia, L., Bi, M., Wu, J., Wang, F., Wang, L., Xie, Y. M., & Ma, G. (2023). Integrated lightweight design method via structural optimization and path planning for material extrusion. *Additive Manufacturing, 62*. https://doi.org/10.1016/j.addma.2022.103387. https://www.journals.elsevier.com/additive-manufacturing.

Yan, Y., Han, M., Jiang, Y., Ng, E. L. L., Zhang, Y., Owh, C., Song, Q., Li, P., Loh, X. J., Qi, B., Chan, Y., & Chan, S. Y. (2024). Electrically conductive polymers for additive manufacturing. *ACS Applied Materials & Interfaces, 16*(5), 5337–5354. https://doi.org/10.1021/acsami.3c13258.

Yang, P. (2023). Nanomaterials in aerospace: Advancements, applications, and the path forward. *Highlights in Science, Engineering and Technology, 73*, 116–121. https://doi.org/10.54097/hset.v73i.12847.

Ye, W., Shi, Y., Zhou, Q., Xie, M., Wang, H., Bou-Sad, B., & Liu, W. (2024). Recent advances in self-lubricating metal matrix nanocomposites reinforced by carbonous materials: A review. *Nano Materials Science*. https://doi.org/10.1016/j.nanoms.2024.02.007.

Zaghloul, M. M. Y., Zaghloul, M. M. Y., & Fuseini, M. (2023). Recent progress in epoxy nanocomposites: Corrosion, structural, flame retardancy and applicationsA comprehensive review. *Polymers for Advanced Technologies, 34*(11), 3438–3472. https://doi.org/10.1002/pat.6144. http://onlinelibrary.wiley.com/journal/10.1002/(ISSN)1099-1581.

Zhang, Q., Shah, N., Wassick, J., Helling, R., & van Egerschot, P. (2014). Sustainable supply chain optimisation: An industrial case study. *Computers & Industrial Engineering, 74*(1), 68–83. https://doi.org/10.1016/j.cie.2014.05.002.

Zhang, X., Pan, S., Song, H., Guo, W., Zhao, S., Chen, G., Zhang, Q., Jin, H., Zhang, L., Chen, Y., & Wang, S. (2021). Polymerinorganic thermoelectric nanomaterials: Electrical properties, interfacial chemistry engineering, and devices. *Frontiers in Chemistry, 9*. https://doi.org/10.3389/fchem.2021.677821.

Zhou, R., Zhang, Z., & Hong, M. (2020). The art of laser ablation in aeroengine: The crown jewel of modern industry. *Journal of Applied Physics, 127*(8). https://doi.org/10.1063/1.5134813.

Zhou, Y., Zhang, Y., Wang, J., He, Y., Yuan, J., & Men, X. (2023). Well-aligned MXene@MoS2/Carbon nanofiber aerogels reinforced epoxy composites with splendid thermal and tribological performance. *Carbon, 213*, 118222. https://doi.org/10.1016/j.carbon.2023.118222.

Zhu, G., Zhang, J., Huang, J., Qiu, Y., Liu, M., Yu, J., Liu, C., Shang, Q., Hu, Y., Hu, L., & Zhou, Y. (2023). Recyclable and reprintable biobased photopolymers for digital light processing 3D printing. *Chemical Engineering Journal, 452*, 139401. https://doi.org/10.1016/j.cej.2022.139401.

Zhu, J., Zhou, H., Wang, C., Zhou, L., Yuan, S., & Zhang, W. (2021). A review of topology optimization for additive manufacturing: Status and challenges. *Chinese Journal of Aeronautics, 34*(1), 91–110. https://doi.org/10.1016/j.cja.2020.09.020.

Zopp, C., Blmer, S., Schubert, F., & Kroll, L. (2017). Processing of a metastable titanium alloy (Ti-5553) by selective laser melting. *Ain Shams Engineering Journal, 8*(3), 475–479. https://doi.org/10.1016/j.asej.2016.11.004.

Additive manufacturing of functional polymers and nanocomposites for sport goods applications

15

Vinay Deep Punetha[1], Akhila Raman[2], and Saritha Appukuttan[2]
[1]2D Materials and LASER Actuation Laboratory, Centre of Excellence for Research, PP Savani University, Kosamba, Gujarat, India, [2]Department of Chemistry, Amrita Vishwa Vidyapeetham, Amritapuri, Kollam, Kerala, India

15.1 Introduction

Additive manufacturing (AM), also known as 3D printing, has fundamentally transformed the sports industry by enabling the production of highly customized and innovative sports equipment. Historically, polymers were introduced in sports to replace traditional materials such as wood, metal, and leather, owing to their lighter weight, greater flexibility, and environmental resistance. This shift significantly improved athletic performance, reduced injuries, and provided greater comfort and protection. The mid-20th century saw the integration of polymers into sports equipment, driven by advancements in materials science and manufacturing techniques. AM techniques, particularly, have revolutionized the use of polymers in sports, offering substantial advantages over traditional methods that involve cutting away material. AM constructs objects layer by layer from digital models, allowing for the creation of custom-fit sports equipment tailored to individual body measurements and preferences, thereby enhancing performance, comfort, and injury prevention. For instance, custom-fitted mouthguards and protective padding can be produced using AM techniques, ensuring a perfect fit for athletes.

The most intriguing aspect of AM is the fabrication of complex geometries that are difficult or impossible to achieve with traditional methods. It has provided an opportunity to design and deploy lightweight structures with optimized strength and flexibility, such as lattice structures in helmets and sports shoes. The layer-by-layer construction minimizes material waste, making AM a more sustainable manufacturing process, which is especially crucial when working with high-performance and efficient polymers. In addition, use of AM has accelerated the lab trials for sophisticated designs as it facilitates rapid prototyping, allowing designers to test and iterate designs quickly. Accelerated lab trials are reducing the time needed to market new sports equipment. These techniques are especially useful in the multimaterial sports system. Advanced AM techniques can combine multiple materials within a single print job, creating composite structures with enhanced properties crucial for developing high-performance sports equipment requiring different material properties in various regions of the same component.

Additive Manufacturing of Functional Polymers and Nanocomposites. DOI: https://doi.org/10.1016/B978-0-443-26516-7.00015-8
© 2025 Elsevier Ltd. All rights are reserved, including those for text and data mining, AI training, and similar technologies.

Some of the polymers used in the sports industry have proven particularly impactful over others due to their unique properties and versatile applications. The integration of these advanced polymers into sports equipment has improved performance and safety, leading to innovations and applications that enhance athletic performance. The unique properties of each polymer allow for the creation of specialized sports gear tailored to specific needs, enhancing the overall sports experience. As research and development continue, the sports industry will likely see more groundbreaking applications of these versatile materials. Integrating advanced reinforcement materials with polymers has significantly enhanced sports equipment's performance, durability, and functionality. Reinforcements such as carbon fibers, glass fibers, Kevlar, and nanoparticles improve the mechanical properties of polymer composites, making them suitable for high-stress applications. Carbon fiber, known for its high strength-to-weight ratio and stiffness, is extensively used in bicycles, tennis rackets, and protective gear, providing enhanced performance and durability (Mohammadizadeh et al., 2019). Glass fiber offers high tensile strength and chemical resistance, used in skis, snowboards, and helmets, ensuring performance in various conditions (Nugraha et al., 2023). Nanoparticles, such as multiwalled carbon nanotubes (MWCNT), enhance the strength, stiffness, and conductivity of polymer composites used in advanced sports equipment to improve durability and functionality (Fekiri et al., 2020). Continuous innovation in reinforcement technologies promises a future where sports equipment meets the highest standards of performance and durability, allowing athletes to push their boundaries. This chapter aims to explore these advancements and the future trends in additive manufacturing and polymer nanocomposites in sports, providing insights into the current state and future direction of this rapidly evolving field. It will cover the types of functional polymers and nanocomposites, the enhancements brought by advanced reinforcements, and the recent advancements and future trends in the sports equipment industry.

15.2 Polymers for additive manufacturing in sports

The use of polymers in the sports industry has a rich history that dates back several decades. Initially, polymers were introduced to replace traditional materials such as wood, metal, and leather, primarily due to their lighter weight, greater flexibility, and resistance to environmental factors. This shift allowed athletes to perform at higher levels, reduce injuries, and experience greater comfort and protection during activities. The integration of polymers into sports equipment began in earnest in the mid-20th century, coinciding with advancements in materials science and manufacturing techniques. Unlike traditional subtractive manufacturing methods, which involve cutting away material, AM builds polymeric objects layer by layer from a digital model. This process offers several significant advantages for the production of sports equipment. AM allows for the creation of custom-fit sports equipment tailored to individual athletes' unique body measurements and preferences. This leads to better performance, comfort, and injury prevention. For example, using AM techniques, custom-fitted mouthguards and protective padding can be quickly produced. AM enables the fabrication of complex geometries that are difficult or impossible to achieve with traditional manufacturing methods. This is particularly beneficial for designing lightweight structures with optimized strength and flexibility, such as lattice structures in helmets and sports shoes. The layer-by-layer construction of AM minimizes material waste, making it a more sustainable manufacturing process. This efficiency is critical when working with high-performance polymers, which can be expensive. AM allows for rapid prototyping, allowing designers to test and iterate designs quickly. This accelerates the development process and reduces the time to market for new sports equipment. Advanced AM techniques can combine multiple materials within a single print job, creating composite structures with enhanced properties. This capability is crucial for developing high-performance sports equipment that requires varying material properties in different regions of the same component.

This section briefly describes various polymer matrix systems, highlighting their benefits and specific utility in sports. Understanding these polymers will help readers appreciate the technological advancements and material innovations that have enhanced athletic performance, safety, and overall sports experience. A broad range of classes of materials is utilized in additive manufacturing for sports equipment, enhancing performance and customization. These materials include thermoplastics, which offer flexibility and durability, and thermosetting plastics, known for their strength and heat resistance. Biodegradable polymers provide ecofriendly options, while metals deliver high strength and resilience for critical components. Ceramics are often used for their lightweight and hard-wearing properties, whereas composites combine multiple materials to optimize performance. Additionally, elastomers offer excellent elasticity, and polymer blends enhance material properties. Liquid crystal polymers provide exceptional mechanical performance, and foams are utilized for cushioning and impact absorption. Together, these diverse material classes facilitate innovative designs and functionalities in sports equipment. Poly(lactic acid) (PLA) is a biopolymer that has gained significant attention in the sports industry due to its biodegradability and excellent mechanical properties. PLA is widely used in sports equipment such as protective gear, sports shoes, and accessories. The primary benefit of PLA in sports applications is its ecofriendliness; it is derived from renewable resources like corn starch and sugarcane, making it a sustainable alternative to petroleum-based plastics. Additionally, PLA's good tensile strength and stiffness ensure durability and reliability in high-impact sports. For example, PLA is used to fabricate custom-fit mouthguards and protective padding, offering athletes safety and comfort (Joseph et al., 2023). Polyamide 12 (PA12) is another polymer that is extensively used in the sports industry. Known for its excellent mechanical properties, including high toughness, flexibility, and resistance to wear and chemicals, PA12 is ideal for manufacturing components that require long-term durability and performance. It is particularly favored in producing sports footwear and equipment that endure high-stress levels and abrasion. PA12 lightweight nature and strength make it suitable for sports helmets, protective gear, and even bike components, where reducing weight without compromising safety is crucial (Saliakas, 2023). Thermoplastic polyurethane (TPU) has revolutionized the sports equipment market with its unique combination of flexibility, elasticity, and resilience. TPU is known for its high abrasion resistance, making it perfect for applications that experience frequent friction and impact. This polymer is commonly used in the production of sports shoes, where its flexibility and durability contribute to better performance and comfort. TPU's biocompatibility is also suitable for wearable sports devices and protective gear, enhancing athlete safety without sacrificing comfort (Xu et al., 2020). High-density polyethylene (HDPE) is widely utilized in the sports industry for its excellent impact resistance and durability. HDPE is often used to construct sports equipment such as helmets, protective padding, and outdoor sports facilities. Its ability to withstand harsh environmental conditions and heavy use makes it a preferred choice for products that require high performance over extended periods. HDPE is also used in creating playground and sports field equipment, ensuring safety and durability in public and professional sports arenas (Chaudhary et al., 2021). Polyacrylonitrile (PAN) has been extensively applied in sports equipment due to its high tensile strength and thermal stability. PAN is primarily used to produce carbon fiber precursors, which are then utilized in high-performance sports equipment. PAN-based materials' strength and lightweight properties make them ideal for manufacturing items such as tennis rackets, bicycles, and fishing rods. The polymer's ability to be processed into very fine fibers allows for creating composites that significantly enhance the performance and durability of sports equipment (Agarwal et al., 2020). Polydimethylsiloxane (PDMS) is known for its flexibility, durability, and biocompatibility, making it an excellent choice for sports and wearable applications. PDMS is often used in the fabrication of flexible sensors and protective gear, providing athletes with enhanced safety and performance monitoring capabilities. Its ability to withstand various environmental conditions without degrading makes it suitable for various sports applications, from wearable electronics to impact protection devices (Fekiri et al., 2020). Shape Memory Polymers (SMPs) are an innovative class of materials that have made significant strides in the sports industry. SMPs can recover their original shape after deformation when exposed to an external stimulus such as heat. This unique property makes SMPs ideal for applications that require adaptability and resilience. In the sports industry, SMPs create customizable protective gear, sportswear, and even sports equipment that can adapt to different conditions. The ability to tailor the mechanical properties of SMPs to specific needs enhances their utility in high-performance sports applications (Garces & Ayranci, 2018). Polytetrafluoroethylene (PTFE) is a polymer known for its low friction, high heat resistance, and excellent chemical stability. These properties make PTFE an ideal material for sports equipment requiring low-friction surfaces, such as bearings, bushings, and sliding mechanisms. PTFE's non-reactivity and durability also make it suitable for high-performance sports gear that must withstand extreme conditions and repetitive use (Chen et al., 2013). Polyvinyl alcohol (PVA) is utilized in the sports industry for its excellent film-forming properties and high tensile strength. PVA is commonly used in applications such as water-soluble films for sports gear packaging and protective films for electronic sports devices. Its biodegradability and non-toxicity also make it an environmentally friendly option for various sports applications, contributing to sustainability efforts in the industry (Agarwal et al., 2020). Liquid crystal polymers

(LCPs) are used in the sports industry for their exceptional mechanical properties and resistance to high temperatures. LCPs are often utilized to fabricate high-performance sports equipment such as tennis rackets, golf clubs, and bicycle components. The strength and rigidity of LCPs, combined with their lightweight nature, make them ideal for applications that require high durability and precision (Gantenbein, 2021). Polyethylene (PE) is a versatile polymer used extensively in the sports industry for high-impact resistance and durability applications. PE is commonly found in products such as protective gear, sports flooring, and outdoor sports equipment. Its ability to withstand harsh environmental conditions and high mechanical stress makes it a reliable choice for long-lasting sports applications (Gisario et al., 2017). Polycarbonate (PC) is known for its high impact resistance and optical clarity, making it an excellent material for sports eyewear, helmets, and visors. PC's toughness and lightweight properties provide athletes essential protection while ensuring comfort and performance. Additionally, the PC's resistance to UV light and weathering makes it suitable for outdoor sports applications where long-term exposure to the elements is a concern (Agarwal et al., 2020). Integrating these advanced polymers into the sports industry has not only improved the performance and safety of sports equipment but also paved the way for innovations and applications. The unique properties of each polymer allow for the creation of highly specialized sports gear tailored to the specific needs of athletes, enhancing their overall performance and experience. As research and development continue to advance, the sports industry will undoubtedly see even more groundbreaking applications of these versatile materials.

15.3 Additive manufacturing (AM) techniques in sports

There are numerous additive manufacturing (AM) methods available, each offering unique capabilities and advantages for different applications. While a comprehensive discussion of all these techniques is beyond the scope of this chapter, it is important to highlight a couple of key methods that are particularly relevant to the development of advanced sports goods (Nath & Nilufar, 2020). The image presents six different additive manufacturing (AM) techniques, each employing unique mechanisms for creating three-dimensional objects, which are particularly relevant for the fabrication of advanced sports goods. These techniques can be differentiated by their processes, materials used, and the nature of the final product. Fused Deposition Modeling (FDM) uses a thermoplastic filament, which is heated and extruded through a nozzle to build objects layer by layer on a printing bed. The motorized system controls movement along the x, y, and z axes, ensuring precision in the layer deposition. FDM is widely used due to its simplicity and cost-effectiveness, making it suitable for producing custom sports equipment with specific geometrical requirements (Fig. 15.1A). SLA utilizes a UV laser to cure and solidify liquid photopolymer resin in a tank, layer by layer. The build platform moves downward after each layer is solidified. This method is known for producing parts with high resolution and smooth surface finishes, ideal for creating complex, detailed components in sports gear that require fine structural features (Fig. 15.1B). Similar to SLA, digital light processing

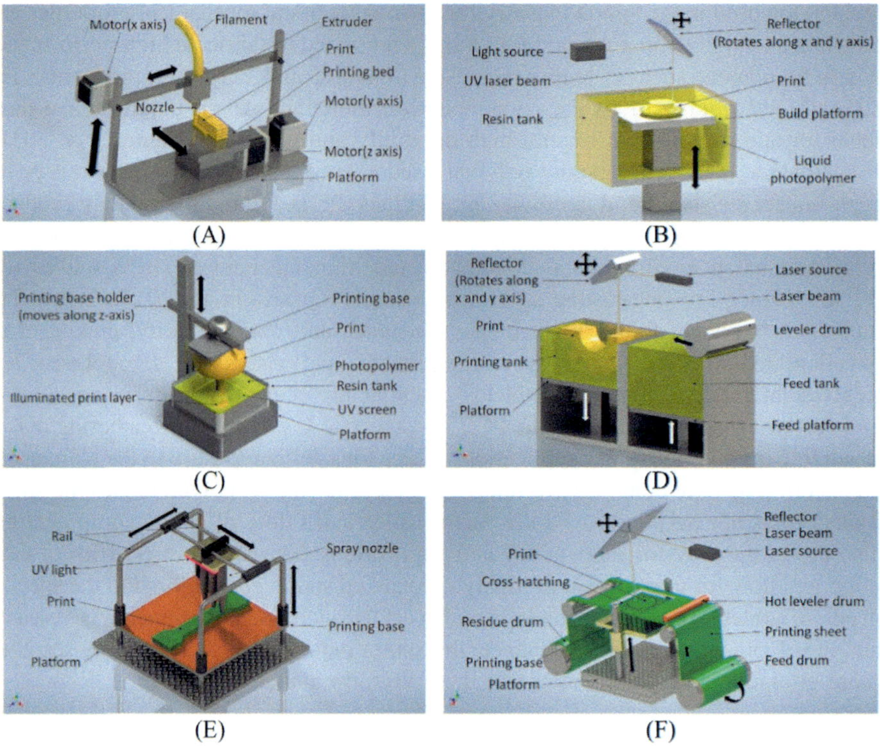

Figure 15.1 Selected polymer additive manufacturing (AM) techniques: (A) Fused deposition modeling (FDM) printers operate on the material extrusion principle to create desired parts. (B) Stereolithography (SLA) employs a UV laser beam to cure liquid UV-curable polymer, enabling high-precision printing. (C) Digital light processing (DLP) printers use UV screens for accurate and efficient printing, offering faster output than SLA. (D) Selective laser sintering (SLS) fuses fine polymer powders with a laser beam, although the resulting parts may have rough or grainy surfaces. (E) PolyJet printing involves spraying fine polymer droplets from multiple nozzles, which are then immediately cured by UV light, allowing for fast, multimaterial printing. (F) In laminated object manufacturing (LOM), polymer sheets are precisely cut and layered to form the final product, making it ideal for rapid, large-scale prints. Adapted from Nath and Nilufar (2020) under CC BY-NC-ND 4.0 license, © 2020, MDPI.

(DLP) uses a light source to cure photopolymer resin, but instead of a laser, it employs a digital light projector screen. The entire layer is exposed at once, making the process faster. DLP is beneficial for producing small, intricate parts in sports goods where precision is paramount, such as in protective gear or customized equipment (Fig. 15.1C) (Nath & Nilufar, 2020).

FDM has been utilized in the fabrication of protective gear, such as custom-fitted mouthguards (Hada et al., 2024), which improve user comfort and safety by precisely matching the wearer's anatomy. The flexibility in material selection also allows the development of functional sports prosthetics that enhance performance across various activities (Górski, 2023). Additionally, FDM has been applied to produce eco-friendly sports equipment, using composites that integrate sustainable materials like hemp fiber and ABS, offering high impact resistance and durability (Mahendran et al., 2024). FDM's versatility extends to creating equipment with varying mechanical properties tailored for specific sports needs. For instance, different infill patterns in FDM-printed PLA structures yield diverse levels of tensile and yield strength, useful for optimizing sports equipment based on flexibility and durability requirements (Rizwee & Kumar, 2024). Moreover, lightweight, durable, and stable components are essential in sports like biathlon, where FDM allows for the development of customized rifle supports that accommodate the unique shooting posture of athletes (Park et al., 2019). The impact of FDM is further exemplified in hybrid sports equipment, such as figure-eight models reinforced with glass fiber-reinforced polymers (GFRP). This integration results in enhanced tensile strength, making the equipment more resilient under dynamic loads, a requirement for intense sports environments (Nugraha et al., 2023). In water sports, FDM is leveraged to fabricate bio-inspired structures like honeycomb surfboards, which show superior bending performance under stress, ultimately contributing to increased safety and performance (Soltani et al., 2020). UAV propellers designed with FDM also demonstrate viable applications in sports, providing prototypes that match the performance of traditional materials at a fraction of the cost (Joseph et al., 2023; Toleos et al., 2020). Fig. 15.2 elaborates the manufacturing process of a figure-eight structure using additive manufacturing (AM), beginning with the design schematic (a), followed by 3D modeling (b), layout preparation in printing software (c), and the final 3D-printed product on the print bed (d).

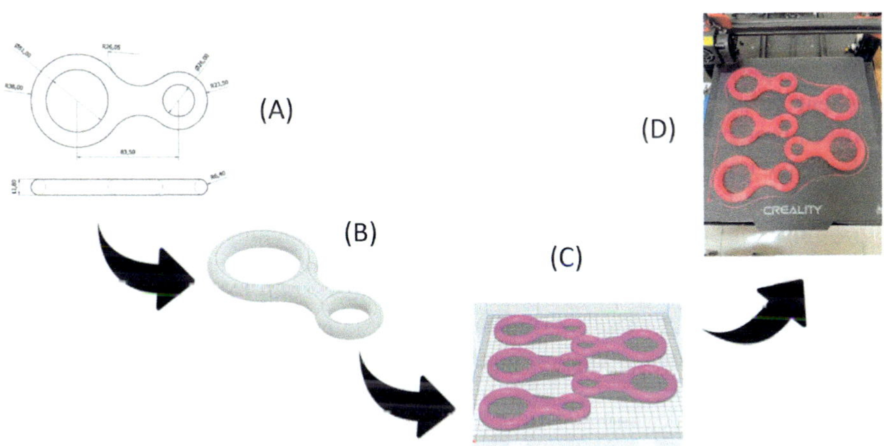

Figure 15.2 Manufacturing process of the figure-eight using AM. (A) Design, (B) modeling, (C) plotting and arranging in printing software, (D) printing process and finishing. Copyright 2023.
Reproduced with permission from Elsevier (Nugraha et al., 2023).

The flexibility of FDM in creating customized structures makes it highly suitable for sports gear, especially with auxetic structures that improve energy absorption, essential in protective equipment for high-impact sports (Joseph et al., 2023). Additionally, FDM is instrumental in manufacturing ankle braces with metamaterial structures that provide enhanced motion control without sacrificing the functional range of motion (Teng, 2022). FDM's role extends into functional applications beyond protection, as shown in studies where it aids in the efficient design of plaster cores for casting processes in sports equipment manufacturing (Choe et al., 2021; Tunay, 2024). In applications like small-scale UAVs and sports equipment, FDM provides structural components made of polymers like PLA and ABS. These materials offer high stiffness-to-weight ratios and energy absorption capabilities, though thermal aging can impact their durability over time (Tunay, 2024). Auxetic structures fabricated with ABS and PLA showcase the potential of FDM in sports applications requiring strength, stiffness, and specific energy absorption properties, while also enabling efficient production of these complex shapes (Vyavahare & Kumar, 2021). FDM's applicability in sports is also seen in flexible polymers, like PLA fabrics, structured for durability and tensile strength, thus opening new avenues in wearable sports technology (Fajardo et al., 2023). The potential for customized sports gear is further enhanced by combining FDM with carbon fiber-reinforced polymers, resulting in lightweight, high-performance equipment suited for activities that demand both strength and flexibility, such as custom sports car wheel rims (Filgueira et al., 2022). Continuous fiber reinforcement in FDM has allowed the development of sports equipment with enhanced mechanical properties, such as tennis rackets and surfboards. Carbon fiber and glass fibers integrated through FDM create high-stiffness, lightweight gear ideal for demanding sports environments (Goh, 2018). The technique's adaptability for multimaterial applications lets athletes achieve optimal comfort and performance by customizing the flexibility and hardness of equipment components (Lu et al., 2022). The automotive industry's interest in FDM for sports applications is illustrated in studies where car prototypes like the E-segment sedan were printed for structural and design testing, bridging the gap between prototyping and actual performance evaluations (Frizziero, 2021).

Selective laser sintering (SLS) process involves compacting and heating powdered material to just below its melting point, causing the particles to bond together into a solid mass. This process typically occurs without fully melting the material, allowing it to retain its shape while forming a denser and more cohesive structure. Sintering is commonly used in powder metallurgy, ceramics, and additive manufacturing techniques like SLS. The key advantage of sintering is that it allows for the production of complex shapes and components with controlled porosity and mechanical properties, which are essential in various applications, including sports equipment, aerospace, and electronics. In this process, a laser, guided by a scanner system, selectively fuses layers of powdered material to build a part layer by layer. The powder bed, consisting of either polymeric powders or nanocomposite mixtures, is spread uniformly across the build platform by a roller, ensuring consistent layering crucial for the final product's structural integrity. The laser beam, after passing through a heating system, targets specific areas of the powder bed to sinter the particles together, forming solid layers according to the digital design. The unsintered powder remains in place, providing support for subsequent layers and aiding in the creation of intricate geometries. This ability to precisely control the material properties and build complex shapes makes SLS a valuable tool in manufacturing high-performance, customized components for sports equipment, where both design and material strength are critical (Figs. 15.1D and 15.3).

In material jetting (MJ) process, droplets of photopolymer are deposited layer by layer onto a build platform, where they are cured by UV light. This method allows for the use of multiple materials in a single build, enabling the creation of complex parts with varying properties. Material jetting is ideal for creating prototypes or parts that require a combination of materials, such as different types of polymers in sports goods (Fig. 15.1E). Laminated object manufacturing works by layering sheets of material, which are bonded together and cut into shape by a laser or blade. The sheets could be paper, plastic, or metal. LOM is particularly effective for creating large, less complex parts, where material cost and build speed are crucial factors. This method might be used for making structural components of sports goods that do not require high precision but need to be robust and cost-effective (Fig. 15.1F). In sports footwear, for instance, LS has been used to create soles with variable stiffness tailored to specific performance needs. Adjusting mechanical properties like bending and flexural stiffness allows laser-sintered footwear to optimize athletes' sprinting and jumping dynamics. A study on sprint shoes with selective laser-sintered Nylon 12 soles demonstrated that shoes engineered with different stiffness levels helped athletes achieve improved performance in activities like squat and bounce drop jumps (Toon et al., 2011). LS has also proven essential in fabricating lightweight and durable helmets and padding. By using elastomeric materials, LS enables the production of energy-absorbing structures that can better protect athletes by dispersing impact forces effectively. Parameters such as laser power and powder bed temperature can be optimized to enhance material properties like toughness, durability, and flexibility, which are crucial for protective sports equipment (Soe et al., 2014). Additionally, LS has found applications in

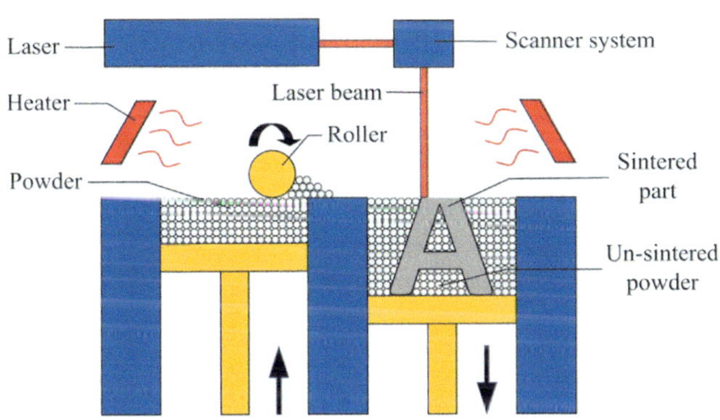

Figure 15.3 Selective laser sintering (SLS) process: layer-by-layer fabrication of complex parts using powdered materials.
Adapted from Xu et al. (2020) under CC BY-NC-ND 4.0 license, © 2020, MDPI.

motorsports, where it supports the direct manufacturing of structural motorcycle components. These components are designed with precise wall thicknesses and finely tuned positions, achieving high strength-to-weight ratios essential for speed and agility on the track. The LS process allows for control over crystallinity and porosity in parts, thereby enhancing performance while minimizing vehicle weight. Scientific evidences suggest that parts manufactured using LS systems can have varying mechanical properties based on the machine's setup, impacting the performance characteristics of motorsports components (Tasch et al., 2018). This precision and adaptability make laser sintering a transformative technology in sports, empowering manufacturers to deliver tailored, high-performance solutions that boost athletic capabilities and ensure safety. By allowing for material customization and intricate design capabilities, LS continues to reshape the future of sports manufacturing, bridging the gap between innovation and athlete performance (Srinivas et al., 2023; Toon et al., 2011; Vinet & Caine, 2011).

Fig. 15.4 illustrates an electrospinning setup where polyacrylonitrile (PAN) fibers are being generated and patterned using a syringe pump and high voltage DC (HVDC) supply, which creates an electrospun jet. As the voltage increases, electrostatic forces overcome the surface tension of the polymer solution at the tip of the syringe needle, leading to the formation of a Taylor cone. From the tip of this cone, a thin, continuous jet of the polymer solution is ejected. This jet undergoes rapid stretching and thinning due to the electrostatic repulsion of the charged polymer

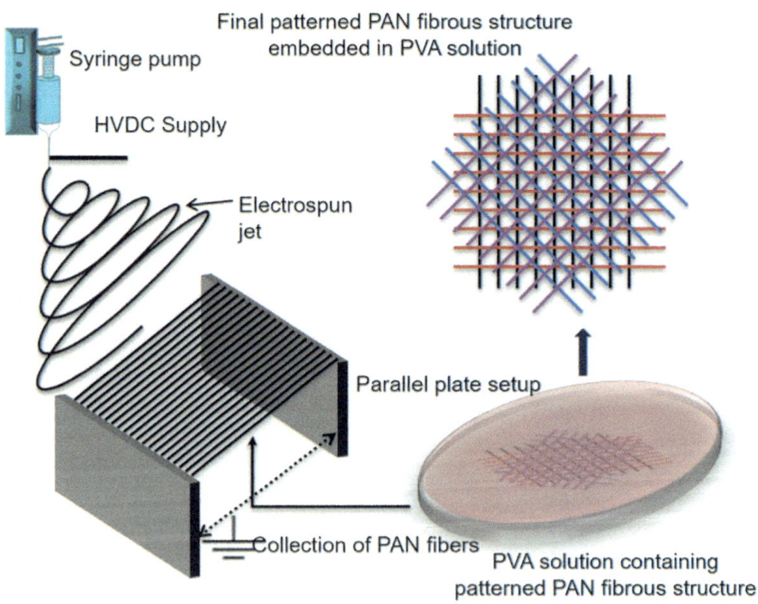

Figure 15.4 Electrospinning process for patterned PAN fibrous structure embedded in PVA matrix.
Adapted from Sahay et al. (2020) under CC BY-NC-ND 4.0 license, © 2020, MDPI.

molecules, resulting in the formation of fine nanofibers. These fibers are then directed toward a grounded collector plate, where they are deposited in a controlled, patterned manner, ultimately forming the fibrous structure shown in the image. The collected fibers form a structured, interwoven network, which is then embedded in a polyvinyl alcohol (PVA) solution. The outcome is a composite material where the patterned PAN fibrous structure is integrated within the PVA matrix, creating a material with potential applications where specific mechanical properties and fiber alignment are essential, such as in advanced sports goods.

One prominent example is the use of piezoelectric fibers made from materials like PVDF (polyvinylidene fluoride) and its composites. For instance, PVDF combined with SiC and FeCl3 through electrospinning has shown increased β-phase content, enhancing its piezoelectric response for applications like smart socks in gait monitoring (Gong, 2024). Another PVDF-based composite, incorporating CNTs, provides self-powered sensors for soccer shoes, allowing real-time monitoring of ball-shoe interactions (Zeng, 2023). Such applications help athletes track and improve their performance, offering personalized insights for training and injury prevention.

The PVDF combined with BaTiO3 forms a nanofiber composite with heightened piezoelectric sensitivity, suitable for wearable sensors to monitor physical activities such as heartbeat and respiration, crucial for tracking athletes' health and performance (He et al., 2024). Another innovative application includes a CNT/TPU composite nanofiber yarn, which, when sewn into garments, functions as a strain sensor, offering real-time feedback on body movements (Tang et al., 2022). In the realm of sports rehabilitation, electrospun nanofibers have contributed to advanced medical applications. For example, electrospun scaffolds made from ZrO_2/HAP composites mimic the mechanical properties of bone, supporting ligament and bone repair in sports injuries. The addition of nanoparticles like zirconia (ZrO_2) improves the composite's stiffness and compressive strength, essential for healing bone injuries and promoting recovery (Guo, 2023). Additionally, stem cell nanocomposite fiber scaffolds aid in ligament repair, enhancing the bonding strength at the tendon–bone interface, thereby accelerating recovery from ligament injuries (Xie & Tian, 2021). Beyond mechanical and piezoelectric applications, electrospinning is used in creating sports materials with enhanced comfort and functionality. For example, PVDF-based nanomesh structures exhibit radiative cooling properties, blocking sunlight while allowing heat dissipation, keeping athletes cooler during outdoor sports (Kim et al., 2021). Hydrophobic and antifouling properties are also achievable through electrospun nanofibers, such as PCL/TiO2, which adds water resistance and self-cleaning abilities to sports textiles, improving hygiene and durability (Pais et al., 2022). Electrospinning further supports energy generation in sports equipment. Nanogenerators using PVDF and biochar derived from rice straw create piezoelectric nanofibers capable of harvesting body energy, potentially powering wearable devices in a sustainable manner (Ali & Mostafa, 2024). Similarly, triboelectric nanogenerators composed of Ni3C/PDMS and electrospun nylon fibers convert mechanical energy into electrical power, suitable for heating applications like muscle relaxation patches in physiotherapy for sports recovery (Kulakarni et al., 2024).

15.4 Enhancing performance and functionality in sports equipment through advanced reinforcement

Since the rise of graphene, nanomaterials have transformed multiple fields, spurring advancements in synthesis techniques and applications (Dhapola, 2024; Jahangir et al., 2024; Kumar, 2023; Punetha 2017, 2022; Punetha et al., 2019; Punetha et al., 2020; Ranjan et al., 2024; Sahoo et al., 2016). One prominent area of growth is in the fabrication of polymer nanocomposites, where these materials are used to enhance the performance of various products. With unique mechanical, thermal, and electrical properties, nanomaterials have paved the way for advanced solutions in sectors such as healthcare, electronics, and most notably, sports. In sports applications, the integration of nanocomposites has opened up exciting avenues for developing equipment that is lighter, stronger, and more durable. Graphene, for instance, has been at the forefront of this evolution. Known for its remarkable strength and flexibility, graphene-enhanced composites have enabled the production of equipment that withstands rigorous use without compromising on weight or performance. Tennis rackets, golf clubs, and even racing bikes now utilize graphene nanocomposites, contributing to reduced energy loss during movement and enhancing overall control and stability for athletes. Beyond graphene, other nanomaterials, such as carbon nanotubes and nanosilica, have contributed to innovations in sports equipment. These materials improve shock absorption, impact resistance, and fatigue durability, all essential factors for high-performance sports gear. For example, polymer nanocomposites reinforced with carbon nanotubes provide increased resilience and flexibility, making them ideal for protective sports equipment like helmets and pads. One area where graphene-based nanocomposites excel is in wearable technology. A notable example is the creation of conductive, flexible, and breathable textile electrodes that are used for biomedical and health monitoring applications in sports. By using the pad-dry-cure method, graphene-coated textile electrodes showed significant conductivity improvements, from 3.5 MΩ to 400 ± 5.0 kΩ for 12 padding passes. The electrodes also demonstrated low impedance and high sensitivity for detecting electrocardiograms and pulse rates, making them valuable for athlete monitoring during activities such as walking or running (Shathi et al., 2020). Pressure sensors are another application where graphene has made an impact. The integration of poly (vinylidene fluoride-trifluoroethylene) (P(VDF-TrFE)/graphene composite films allows for high sensitivity and rapid response, crucial in sports training and injury prevention. For instance, these composite films achieved a sensitivity of 0.85 kPa^{-1} and response times as quick as 50 ms, helping monitor athletic biomechanics and performance optimization (Zhu et al., 2024). Furthermore, graphene's inclusion enhanced the film's mechanical properties, including a 65% increase in tensile strength, which is invaluable in creating durable sensors that withstand repeated physical impacts typical in sports training. Graphene has also enabled advancements in laser-induced graphene (LIG), a single-step technique that transforms various polymer substrates into wearable electronics. LIG can be patterned for use as strain and pressure sensors. In sports applications, these sensors, laminated onto polydimethylsiloxane (PDMS) and embedded in textiles, are used in volleyball sportswear for functions like spike force measurement and player movement tracking. This innovation in wearable sensors assists athletes by gathering real-time data that can inform training and competition strategies (Raza, 2022). Self-powered sensor systems, which are crucial as wearable technology advances, have found a solution in triboelectric nanogenerators (TENGs) doped with graphene. For instance, thermoplastic polyurethane (TPU) films combined with graphene in TENGs enable wearable sensors that generate power from movement, which can then monitor athletic performance. The graphene-enhanced TPU TENG devices power LEDs, sports timers, and movement sensors without external energy sources, demonstrating how graphene-doped TPU nanocomposites could revolutionize self-powered sports sensors (Yang et al., 2024). Graphene's versatility extends to athletic footwear, where ethylene vinyl acetate (EVA) foam, commonly used in midsoles, can be reinforced with graphene nanoplatelets to enhance its mechanical properties. This reinforcement leads to improved impact absorption and abrasion resistance, both of which are essential for athletic footwear used in high-impact activities. Such enhanced EVA foams offer higher fatigue resistance, energy return, and prolonged durability, supporting athletes by reducing stress on the lower extremities during training and competition (Hao et al., 2024). Beyond footwear, graphene-enhanced nanocomposites contribute to injury prevention. A study explored the clinical application of graphene in sports-related ankle fractures, utilizing graphene fiber and rubber composites to improve internal fixation strength and stability for orthopedic applications. Graphene's inclusion in joint fixation devices has been shown to promote recovery by enhancing joint stability and reducing the likelihood of reinjury, which is crucial in sports involving high-frequency joint movements (Retraction: Clinical Application of Graphene Composite in Internal Fixation of Ankle Fracture in Sports, 2023). In high-impact sports, such as basketball and soccer, wearable graphene-coated fiber sensors have proven effective in tracking players' movement. These sensors possess high sensitivity, fast response times, and exceptional reproducibility, enabling coaches to gather precise data on player motions. This information helps athletes improve their form and technique, contributing to better performance and reduced injury risk (Zhang, 2018). Graphene's role in energy storage and electronic applications is also being realized in the form of sensors for detecting doping substances in sports. For instance, electrochemical sensors modified with graphene and nickel oxide (NiO) can detect ephedrine, a compound commonly associated with sports doping. These sensors, when applied to sports doping control, show high sensitivity and selectivity, underscoring the potential of graphene-based sensors in ensuring fair competition (Li and Liu, 2024). Advances in molecularly imprinted polymer (MIP) sensors have also benefited from graphene technology. MIP sensors using graphene oxide for the selective detection of banned substances like sibutramine are a promising development for doping control. These sensors offer rapid detection times and high sensitivity, providing reliable data to prevent illicit drug use in sports, and their durability ensures consistent performance over time (Shi et al., 2024). Graphene's low-cost, ecofriendly production from biomass waste also adds value to sports equipment manufacturing. Due to its lightweight, strength, and superior electrical and thermal conductivity, biomass-derived graphene has been integrated into sports gear, including sensors in wearables, to monitor training parameters. This approach allows for sustainable production methods without compromising the quality or performance of sports equipment (Wu et al., 2024). Graphene's capacity for chemical modifications, as seen in molecularly imprinted electrochemical sensors, has enabled its use in glucose and lactate detection sensors in athletes' sweat or urine. Such sensors allow real-time monitoring of nutritional status, providing critical insights into an athlete's hydration and metabolic state, ultimately aiding in performance optimization and recovery (Li and Lin, 2024).

In addition to graphene, researchers have leveraged CNTs in polymer nanocomposites, developing materials that contribute to lighter, stronger, and more responsive sports equipment. These advancements are evident in applications across protective gear, performance wearables, and sports rehabilitation devices. The CNT-based polymer nanocomposites have demonstrated significant potential is in creating lightweight sports equipment. Carbon nanotubes embedded within polymer matrices provide a unique combination of strength and flexibility, making them ideal for high-impact sports. For instance, incorporating CNTs into thermoset polymer composites improved energy absorption in impact tests, a critical factor for sports safety gear. Low-velocity impact (LVI) testing showed that adding CNTs, along with other nanofillers, significantly enhanced the elastic modulus and energy absorption properties, which are crucial for reducing sports-related injuries (Gao et al., 2024). Carbon nanotubes also excel in improving the dynamic properties of sports structures, particularly where mechanical movements and support are critical, such as in spinning and bending components of sports equipment. By employing the nonlocal strain gradient theory, studies have examined the impact of CNTs on the behavior of cantilever nanobeams and nanotubes. Such configurations, when integrated into sports equipment, enable structures that can withstand and adapt to variable loads without compromising stability. The incorporation of functionally graded materials (FGMs) in these CNT-based nanostructures offers potential improvements in vibration damping and dynamic softening, thereby increasing durability and reducing energy loss during high-performance activities (Song et al., 2023). CNT-based nanocomposites have also proven effective in wearable sensors, where motion monitoring and injury prevention are crucial. Multiwalled carbon nanotubes (MWCNTs) have been incorporated into polyacrylonitrile (PAN) nanofiber membranes, resulting in enhanced tensile properties that withstand continuous flexing and stretching, ideal for wearable applications in sports. When combined with thermoplastic polyurethane (TPU), the tensile strength of these composites is further enhanced, ensuring durability under the demands of intense physical activity. Such wearable sensors can measure movements in real-time, providing athletes and trainers with precise biomechanical data to improve performance and prevent injuries (Zhao, 2024).

CNT-based nanocomposites have applications in sports rehabilitation, particularly in ethnic sports rehabilitation therapy, which integrates traditional exercise methods with nanotube-enhanced materials. By combining nanotechnology with cultural sports practices, therapeutic exercises for knee arthritis and other joint injuries have become more effective. Studies show that using nanotubes in rehabilitation aids enhances exercise effectiveness by supporting joint stability, allowing athletes to recover faster and resume training sooner. Comparative studies revealed that patients treated with CNT-enhanced aids recovered their mobility more quickly than those using traditional rehabilitation equipment (Li, 2022) Another essential application of CNTs is in sports nutrition and antidoping monitoring. For instance, molecularly imprinted polymer (MIP) sensors combined with MWCNTs have been developed for the rapid detection of artificial sweeteners like aspartame in sports drinks. These MWCNT-MIP sensors provide a cost-effective, portable, and sensitive solution for real-time testing, enabling accurate aspartame detection at nanomolar levels. This technology benefits athletes by ensuring they adhere to dietary guidelines, which can be vital for maintaining optimal health and performance during training and competition (Su & Su, 2024). Furthermore, CNT-based sensors can detect banned substances, an application particularly relevant in sports where antidoping measures are stringent. For example, a novel potentiometric sensor using MWCNTs was developed to detect acebutolol hydrochloride (AC), a prohibited β-adrenoreceptor blocker. This sensor displayed high selectivity, sensitivity, and stability, providing accurate results for AC in commercial products, making it a valuable tool for on-site antidoping testing (Al-Mohaimeed et al., 2023). Functionally graded CNT structures have also shown potential in high-performance sports equipment, such as bicycles. Thermal buckling studies of CNT-enhanced bidirectional functionally graded materials (FGMs) indicated their suitability for components subjected to temperature fluctuations and mechanical stresses. This attribute makes CNT-enhanced materials advantageous for applications where thermal stability and high strength-to-weight ratios are essential, as in the construction of high-performance bikes (Yan et al., 2024). As for sports footwear, CNT-based composites are being tested as potential replacements for traditional materials. MWCNT-reinforced ethylene vinyl acetate (EVA) foam has shown promise for use in shoe midsoles due to its increased durability, impact absorption, and energy return capabilities. The addition of CNTs to EVA composites not only improves mechanical properties but also extends the lifespan of footwear by reducing wear and maintaining cushioning properties, which is vital for athletic shoes subjected to constant high-impact forces (Lunchev et al., 2022). In terms of broader applications, the versatility of CNTs has even been explored in theoretical modeling of composite structures. For instance, classical and first-order beam theories have been applied to analyze the thermal behavior of CNT-based composite beams under various conditions. These models help inform the design of future sports structures, enabling more accurate predictions of how CNT composites respond to environmental stresses, which can directly impact the durability and performance of sports equipment (Yan et al., 2024) Moreover, the growing field of CNT nanocomposites for drug delivery has indirect implications for sports technology. Research into nonlocal strain gradient effects and functional grading in CNT-reinforced materials has revealed their potential for controlled drug release through vibration and mechanical response. This research may lead to therapeutic applications in sports medicine, where controlled drug delivery could aid in injury recovery and pain management, thereby supporting athletes' rehabilitation processes (Song et al., 2023).

Carbon fiber (CF) is one of the most widely used materials for its high strength-to-weight ratio, excellent stiffness, and low thermal expansion. These properties make it an ideal reinforcement for sports equipment that demands both lightweight and high strength. Carbon fiber-reinforced composites are extensively used in bicycles, tennis rackets, and protective gear, providing athletes with enhanced performance and durability. Continuous fiber-reinforced additively manufactured (CFRAM) components benefit significantly from carbon fiber reinforcement by achieving lightweight yet strong materials suitable for sports goods (Mohammadizadeh et al., 2019). Combining carbon fiber with polymers results in composites that can withstand high-impact forces while remaining light, which is critical in competitive sports where both speed and strength are essential. Moreover, reinforcements such as glass fiber (GF) offer high tensile strength, excellent chemical resistance, and good thermal stability. These attributes make glass fiber-reinforced polymers suitable for applications where durability and resistance to environmental factors are critical. Glass fiber is commonly used in sports equipment like skis, snowboards, and helmets, providing reliable performance in various conditions. Significant increases in tensile strength and hardness have been observed in figure-eight models manufactured via 3D printing and coated with GFRP (Nugraha et al., 2023). Glass fiber ensures that sports equipment can endure heavy use and harsh environmental conditions, making it a preferred material for winter sports and protective gear. Nanoparticles, including multiwalled carbon nanotubes (MWCNT), offer exceptional mechanical, electrical, and thermal properties. When integrated into polymer matrices, these nanoparticles enhance the overall performance of the composites, providing superior strength, stiffness, and conductivity. Nanoparticles are used in advanced sports equipment to improve durability, performance, and functionality. The fabrication of flexible piezoresistive pressure sensors using a 3D-printable carbon nanotube-based nanocomposite demonstrates the high sensitivity and mechanical strength of these materials, making them suitable for wearable electronics in sports (Fekiri et al., 2020). Using nanoparticles allows for the creation of intelligent sports equipment that can monitor performance and provide real-time feedback to athletes.

Polytetrafluoroethylene (PTFE) is characterized by its low friction, high heat resistance, and excellent chemical stability. These properties make PTFE a valuable reinforcement in sports equipment, requiring smooth surfaces and high wear resistance. PTFE is used in sports gear applications such as bearings, bushings, and sliding mechanisms. PTFE-reinforced composites exhibit improved wear resistance and reduced friction, enhancing the performance of components used in seawater lubrication conditions. The application of PTFE in sports equipment ensures smooth operation and longevity, particularly in components that are subject to repetitive motion and friction. Fibrous reinforcements like carbon fiber, glass fiber, and Kevlar are essential for enhancing the structural integrity and strength of sports equipment. Carbon fiber offers a superior strength-to-weight ratio, making it ideal for high-performance sports gear. Glass fiber provides durability and resistance to environmental factors, making it suitable for outdoor sports. Kevlar's impact resistance makes it perfect for protective gear. These fibrous reinforcements ensure that sports equipment can withstand the rigors of competitive sports while maintaining lightweight and agile properties. Particulate Reinforcements, such as nanoparticles, significantly enhance sports equipment's mechanical and functional properties. Integrating nanoparticles into polymer matrices results in composites with superior strength, stiffness, and conductivity. These properties are particularly beneficial in advanced sports equipment, such as flexible sensors and wearable electronics, which require high durability and real-time performance monitoring capabilities. In addition to nano reinforcements, polymers such as Kevlar, which is known for its high tensile strength, impact resistance, and lightweight nature, can be used to reinforce other polymer matrices. These properties make Kevlar an excellent reinforcement for protective sports equipment, such as helmets, gloves, and body armor. Kevlar-reinforced composites offer enhanced safety without compromising on comfort or mobility. In continuous fiber-reinforced additive manufacturing, Kevlar is used as a reinforcing agent to improve the mechanical properties of sports equipment, providing high tensile and flexural strengths (Goh, 2018). The incorporation of Kevlar into sports gear ensures that athletes are protected from high-impact injuries while maintaining the agility and flexibility needed for performance.

The synergy between polymers and reinforcements leads to better products and paves the way for continuous innovation and improved athletic performance. Each type of reinforcement offers unique benefits, enhancing the mechanical properties, durability, and functionality of sports gear. From carbon fiber's exceptional strength-to-weight ratio to the high impact resistance of Kevlar, these materials provide athletes with equipment that performs reliably under high-stress conditions. The advanced manufacturing techniques and the integration of innovative materials ensure that sports equipment continues to evolve, meeting the increasing demands of athletes for superior performance and safety. Integrating these advanced reinforcements into polymer composites has revolutionized the sports equipment industry. The unique properties of each reinforcement allow for the creation of highly specialized sports gear tailored to the specific needs of athletes, enhancing their overall performance and experience. As research and development continue to advance, the sports industry will undoubtedly see even more groundbreaking applications of these versatile materials. By leveraging the advanced properties of various reinforcements, the sports equipment industry can produce gear that enhances athletic performance and ensures athletes' safety and comfort. The continuous innovation in reinforcement technologies promises a future where sports equipment can meet the highest performance and durability standards, allowing athletes to push the boundaries of their capabilities

Fig. 15.5 illustrates a three-point bending test setup, a method commonly used to evaluate the flexural properties of materials. In this test, a specimen is supported at two points, while a load is applied at the midpoint using a loading nose. This setup measures the material's response to the applied force, providing critical data on its flexural strength, stiffness, and resistance to deformation. Such testing is essential in sports equipment development, where understanding how materials behave under stress is crucial for ensuring durability and reliability. For example, in the design of composite materials for skis, snowboards, or protective gear, this test helps optimize

Figure 15.5 Three-point bending test setup for evaluating flexural properties of composite materials.
Copyright 2023. Reproduced with permission from Elsevier (Alarifi & Alarifi, 2023).

material properties to enhance performance and safety. The results from this type of testing are integral to selecting and engineering materials that meet the rigorous demands of sports applications.

UTM analysis (Fig. 15.6) shows stress–strain curves of specimens printed in three different build orientations (flat, on-edge, and upright), is directly related to the investigation of how these orientations affect the mechanical properties of thermoplastic polyurethane (TPU), a material known for its high ductility, biocompatibility, and excellent abrasion resistance. These properties make TPU suitable for a range of applications, including aerospace engineering, medical devices, and sports equipment. The stress-strain curves clearly demonstrate that TPU parts printed in the flat and on-edge orientations exhibit superior tensile strength and greater deformability compared to those printed in the upright orientation. This is consistent with the study's findings, which indicate that the build orientation is a critical factor in determining the mechanical properties of printed TPU parts. In the flat and on-edge orientations, the stress-strain curves show higher stress values before failure, indicating better tensile strength. These orientations allow the material to leverage stronger intralayer bonds, which are aligned with the direction of the applied force. This alignment contributes to the material's ability to withstand higher loads and undergo more significant deformation before breaking. Conversely, the upright orientation results in lower tensile strength and reduced ductility, as seen in the stress–strain curve for this orientation. The lower performance in this configuration is attributed to the application of load perpendicular to the layer bonding, which exploits the weaker interlayer adhesion, leading to earlier failure.

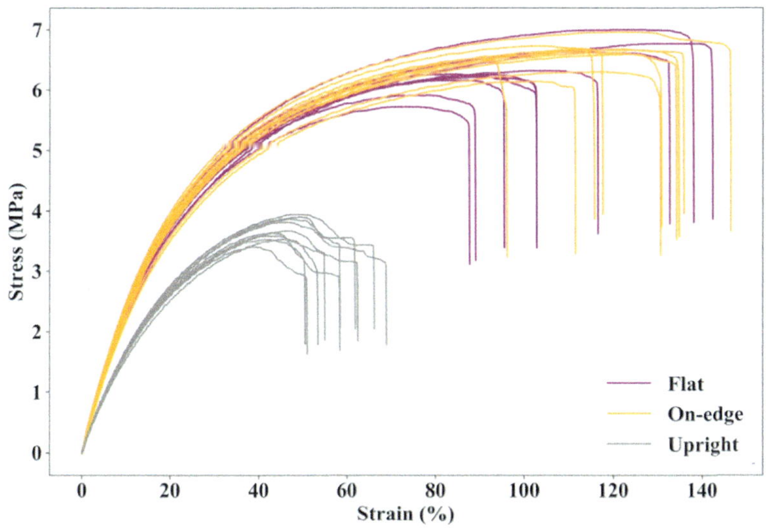

Figure 15.6 Stress–strain curves of thermoplastic polyurethane (TPU) specimens printed in flat, on-edge, and upright orientations.
Adapted from Xu et al. (2020) under CC BY-NC-ND 4.0 license, © 2020, MDPI.

Table 15.1 provides a comprehensive overview of various polymer matrices and their respective reinforcements, highlighting their diverse applications.

15.5 Sensors in sports

The advent of additive manufacturing has significantly transformed the landscape of sports equipment, particularly with the integration of advanced sensors. These sensors enhance performance monitoring, safety, and real-time data analysis, providing athletes with critical insights and feedback. The following segment explores various studies and developments in integrating sensors into sports equipment through additive manufacturing techniques. A significant innovation in sports helmets involves embedding optical fiber-based pressure sensors using additive manufacturing techniques. In a recent study, researchers utilized thermoplastic polyurethane (TPU) to embed these sensors into the lining of helmets. This integration allows for precise monitoring of head impacts, providing valuable data to enhance safety in contact sports. The sensors demonstrated sensitive shock-monitoring capabilities, significantly reducing inaccuracies compared to existing systems. This embedded system is unobtrusive to the wearer and can be powered by small, medical grade batteries, requiring minimal maintenance (Xhameni et al., 2022). The intersection of nanoscience and additive manufacturing has led to the development of printable, flexible electronics, including piezoresistive pressure sensors. One study demonstrated the fabrication of such a sensor using a 3D-printable carbon nanotube-based nanocomposite. This composite, made of

Table 15.1 Overview of polymer matrices and their reinforcements for sports and other applications.

S. No.	Polymer matrix	Reinforcement	Application	References
1.	Polylactic acid (PLA)	Recycled carbon fiber (rCF)	Sports, medical, automotive	Huanbo et al. (2023)
2.	Nylon	Carbon fiber (CF), fiberglass (FG), Kevlar	Auto industry, aerospace, sports goods, medical tools	Mohammadizadeh et al. (2019)
3.	Polylactic acid (PLA)	Glass fiber-reinforced polymer (GFRP)	Improved tensile strength, hardness, and density in figure-eight designs	Nugraha et al. (2023)
4.	Polydimethylsiloxane (PDMS)	Multiwalled carbon nanotubes (MWCNT)	Flexible sensors for sport, automotive, and biomedical fields	Fekiri et al. (2020)
5.	Liquid crystal polymers	Reinforced fibers	Lightweight materials for aircraft, automobiles, wind turbine blades, sports products	Gantenbein (2021)
6.	Poly(ethylene terephthalate) (PET)	Organic additives promoting biodegradation	Fabrication of medical and electronic devices, sports equipment, food and drug packaging solutions	Gisario et al. (2017)
7.	Nanocrystalline Al-Mn alloy	Polymer templates	Lightweight sandwich structures for aerospace, automotive, and sports industries	Schaedler et al. (2017)
8.	Polyacrylonitrile (PAN)	Reinforced polyvinyl alcohol (PVA)	High-impact resistance materials for army combat wear, sports, and biomedical applications	Sahay et al. (2020)
9.	Thermoplastic polyurethane (TPU)	N/A	Functional parts for aerospace engineering, medical devices, sports equipment	Xu et al. (2020)
10.	Recycled polyamide 12 (PA12)	Iron oxide nanoparticles (Fe_3O_4 NPs)	Automotive parts, sports industry, leisure, and consumer goods	Saliakas (2023)
11.	Onyx (chopped, etc.)	Fiberglass, Kevlar	Progressive damage failure analysis in aerospace, automotive, manufacturing, and biomedical sectors	León-Becerra et al. (2023)
12.	High-density polyethylene (HDPE), polypropylene (PP)	Marble slurry (MS)	Fabrication of value-added products through extrusion	Chaudhary et al. (2021)

13.	Polyacrylonitrile (PAN)	Reinforced polyvinyl alcohol (PVA)	Combat army vest, sports gear, flexible piezoelectric sensors, solar photovoltaic (PV) modules	Agarwal et al. (2020)
14.	Polydimethylsiloxane (PDMS)	Nanoparticles (e.g., Graphene Oxide)	Flexible, durable materials for automotive, sports, and biomedical applications	Knothe Tate (2020)
15.	N/A	N/A	Mechanobiological properties for advanced materials in medicine, safety, transport, sports	Montgomery et al. (2021)
16.	Polylactic acid (PLA)	Carbon fiber	Flexural strength, sports equipment, aerospace, automotive applications	Mantecón (2024)
17.	Thermoplastic polyurethane (TPU)	Iron oxide nanoparticles (Fe_3O_4 NPs)	Enhanced mechanical properties, sports, medical, and automotive applications	Kumar et al. (2023)
18.	Polylactic acid (PLA)	N/A	Sports equipment, automotive parts, biomedical applications	Clares et al. (2023)
19.	Polylactic acid (PLA)	N/A	Sports mouthguards	Schewe et al. (2022)
20.	Thermoplastic polymers	Carbon fiber	Fused filament fabrication for aerospace, automotive, and sports industries	Goh (2018)
21.	Shape memory polymers (SMP)	N/A	Electrically activated sensors and actuators for sports equipment, UAVs, robotic applications	Garces and Ayranci (2020)

multiwalled carbon nanotubes (MWCNT) dispersed in polydimethylsiloxane (PDMS), was printed using the direct ink writing (DIW) technique. The sensor exhibited high sensitivity over a wide range of applied pressures, making it suitable for various sports applications where precise pressure measurement is critical. Printing high-resolution, three-dimensional shapes allows for design flexibility and structural complexity, ideal for wearable or implantable electronics in sports (Fekiri et al., 2020). In cycling, integrating sensors into gloves can significantly enhance safety and performance. A recent advancement involved developing smart gloves equipped with capacitive pressure sensors. These sensors were embedded using additive manufacturing techniques and silicon molding. The user-centered design methodology ensured the gloves accounted for the cyclist's anthropometric characteristics and sports gestures. This approach allows for precise pressure analysis on the handlebar, improving injury prevention and performance monitoring during cycling (Caporaso et al., 2022). Shape memory polymers (SMPs) and their composites have shown great potential as sensors and actuators in sports equipment. Utilizing extrusion additive manufacturing (EBAM), also known as fused filament fabrication (FFF), researchers created electrically activated Shape memory polymer composites (SMPC). These composites can be programmed to trigger motion via voltage application, making them suitable for creating adaptive and responsive sports equipment. This technology enables manufacturing intricate sensor geometries, such as those needed for structural health monitoring systems in sports gear (Garces & Ayranci, 2020). Auxetic metamaterials, which exhibit a negative Poisson's ratio, are promising for various demanding applications, including sports. Through additive manufacturing, these materials can be tailored for specific mechanical properties, such as relative density, auxetic behavior, and stiffness. A study characterized the mechanical performance of sine-based lattice structures, highlighting their potential in biomedical engineering, smart sensors, sports, and soft robotics. The ability to fine-tune these properties makes auxetic metamaterials suitable for applications that require customized mechanical responses and enhanced durability (Ramírez-Gutiérrez et al., 2023). Security in sports venues is critical, and integrating sensors for explosive detection can enhance safety. An advanced immunosensor system was developed to detect ultrasmall traces of explosives. This system uses a monolithic glass column with immobilized hapten to trap fluorescently labeled antibodies, providing rapid and empathetic detection capabilities. Although primarily designed for security, such sensors can also be adapted for use in sports environments to ensure the safety of athletes and spectators (Paul et al., 2020). Inspired by naturally occurring helicoidal structures, researchers have developed tough helicoidally architectured synthetic structural composites (HA-SSCs) using additive manufacturing. These composites, fabricated through near-field electrospinning, exhibit superior adhesion and enhanced toughness and impact properties. The ability to tune mechanical and physical properties through architectural features makes these composites ideal for flexible piezoelectric sensors in sports applications. Such sensors can be used in helmets and other protective gear, offering enhanced impact resistance and durability (Agarwal et al., 2020). Integrating sensors into sports equipment through additive manufacturing offers numerous benefits, including enhanced safety, performance monitoring, and customization. The ability to embed sensors precisely and flexibly allows for innovative applications that were previously unattainable. As additive manufacturing technologies continue to evolve, the sports industry will likely see even more advanced and multifunctional equipment designed to meet the specific needs of athletes.

Sensors are essential in many applications pertaining to sports, playing a crucial role in monitoring, analysis, and optimization of athletic performance, as well as ensuring safety and regulatory compliance. From tracking vital physiological parameters like lactate levels during exercise to detecting specific substances such as artificial sweeteners in sports drinks, sensors provide accurate, real-time data that can be used to enhance training regimens, improve recovery times, and maintain the integrity of sports through antidoping measures. Whether integrated into wearable devices, used in on-site testing, or employed in laboratory settings, these sensors enable athletes, coaches, and regulatory bodies to make informed decisions, ultimately contributing to better performance outcomes and safer sports environments. An electrochemical sensor using molecularly imprinted polymers (MIPs) combined with multiwalled carbon nanotubes (MWCNTs) for the selective detection of aspartame in sports drinks, achieving a detection limit as low as 22 nM (Su & Su, 2024). In another example, the detection of lactate, a key metabolite in sports physiology, was enhanced using a sensor based on Ag NPs@GO/GCE, providing a linear range up to 600 μM with a sensitivity of 16.5 $\mu A/cm^2/mM$, demonstrating reliable performance in real sweat samples (Wang, 2021). A notable advancement in antidoping efforts includes a ZnO/TiO2 nanocomposite sensor for methyltestosterone detection, which outperformed traditional methods in sensitivity and specificity, proving effective in synthetic urine samples (Liu & Quan, 2023). Similarly, the hydrothermal production of a Pt/GO nanocomposite has led to an electrochemical sensor for triamcinolone detection in urine, with a detection limit of 12 nM and a linear range up to 130 μM, highlighting its potential in doping control (Wang & Zhang, 2022). Further extending the scope of sports-related analyte detection, a CuN-T/GCE modified electrode has been employed for the sensitive detection of the food dye sunset yellow in sports drinks, achieving a detection limit of 13 nM and demonstrating high accuracy with recovery rates between 96.0% and 107.0% (Zhou et al., 2023). Additionally, a sensor based on Co3O4 and electropolymerized molecular imprinting polymer film exhibited a detection limit of 33 nM for tartrazine in sports drinks, providing a rapid and economical approach to monitoring this compound (Su & Su, 2023). A novel biosensor using a g-C3N4/PANI nanocomposite demonstrated exceptional sensitivity for fentanyl detection with a detection limit of 0.006 μM, showcasing its potential for real-time, on-site antidoping applications (Hu et al., 2024). The integration of these advanced sensor technologies not only enhances the detection capabilities for various analytes in sports contexts but also underscores the potential for improving the accuracy, speed, and accessibility of monitoring tools crucial for maintaining fairness and safety in sports.

15.6 Sports guards in additive manufacturing

Sports guards are crucial in protecting athletes from injuries in various high-impact sports. They are designed to absorb and dissipate energy, reducing the risk of fractures, concussions, and other severe injuries. With advancements in additive manufacturing, the production of sports guards has significantly evolved, offering enhanced customization, performance, and safety. In contact sports such as football, hockey, and boxing, using mouthguards and shin guards is essential. These protective devices are designed to shield athletes from oriental injuries and leg fractures, which are common due to the high-impact nature of these sports. Mouthguards, in particular, help prevent dental injuries by absorbing and redistributing the forces that can otherwise damage teeth and oral tissues. Shin guards protect the lower legs from impacts, crucial in sports like soccer, where kicks and collisions are frequent.

Additive manufacturing, or 3D printing, has revolutionized the design and production of sports guards. This technology allows for precise customization to fit the unique anatomical features of each athlete, enhancing comfort and effectiveness. Unlike traditional manufacturing methods, 3D printing can produce complex geometries that improve the protective performance of sports guards. For instance, the ability to layer different materials enables the creation of multidensity structures that can absorb and dissipate impact energy more effectively. A study compared additively manufactured mouthguards made from Arnitel® ID 2045 to those made using traditional thermoforming with Ethylene-vinyl acetate (EVA). The results showed that 3D-printed mouthguards led to a 10% reduction in peak force and a 25% reduction in impulse during high-impact tests, indicating superior protective performance (Lißner et al., 2023). 3D printing allows for the design and production of custom-made mouthguards that can be tailored to the athlete's specific needs. Finite element analysis (FEA) can be used to optimize the design by evaluating different variables such as layer stiffness, thickness, and space inclusion. This method enhances the mouthguard's ability to distribute load and absorb impact energy, resulting in significantly higher protection than conventional methods (Kumar et al., 2023). In addition to mouthguards, shin guards have also benefited from additive manufacturing. Using materials like polyethylene terephthalate glycol (PETG) and carbon fiber composites, researchers have developed lightweight yet strong shin guards that provide enhanced protection. The patterned structure of these 3D-printed guards reduces weight by 42% compared to traditional solid guards while maintaining high tensile, compressive, and flexural properties (Alarifi & Alarifi, 2023). Advanced materials in 3D printing have expanded the possibilities for sports guards. For example, using additive manufacturing, mouthguards made from multi-layered materials have improved protective performance. These guards incorporate softer inner layers for comfort and more complex outer layers for impact resistance. A case study demonstrated that 3D-printed multilayer mouthguards provided a clinically acceptable fit and were well-received by patients, highlighting the potential for further adoption in sports (Unkovskiy et al., 2021). Comparative studies between traditional and additive manufacturing methods reveal significant advantages of 3D printing. For instance, a study involving the evaluation of different mouthguard materials under high-impact conditions found that additively manufactured mouthguards outperformed conventional EVA mouthguards in terms of force reduction and energy dissipation (Schewe et al., 2022). Additionally, the ability to produce complex, tailored designs through additive manufacturing allows for innovations such as sensors and smart materials. These advancements can provide real-time data on impact forces and improve injury prevention strategies. The combination of digital design, simulation, and 3D printing represents a powerful toolset for developing next-generation sports guards that offer superior protection and comfort (Dias Gonçalves et al., 2024). In conclusion, the integration of additive manufacturing in the production of sports guards has led to significant advancements in customization, performance, and safety. By leveraging the capabilities of 3D printing and advanced materials; manufacturers can produce highly effective protective gear that meets the specific needs of athletes across various sports. The ongoing research and development in this field promise even more significant innovations, further enhancing the safety and performance of sports equipment.

Three-dimensional (3D) printing has brought significant advancements in medical applications, particularly in creating anatomically accurate models and implants (Fig. 15.7). This technology enables personalized healthcare solutions, including custom-fit implants, surgical planning models, and patient-specific prosthetics. For cranial surgeries and reconstructions, 3D printing allows for the precise replication of complex skull structures, supporting surgeons in planning intricate procedures. By transforming imaging data from CT or MRI scans into physical models, 3D printing addresses patient-specific anatomical challenges more accurately, improving surgical outcomes and reducing time spent in the operating room. Fig. 15.7 demonstrates the creation process of a 3D-printed model of the human skull derived from CT scan data, specifically for cranial surgery applications. In Fig. 15.7A, a reference CT scan shows essential anatomical features, such as the brain mass, skull bone, and facial structure, which help guide segmentation. Fig. 15.7B shows the segmented CAD model isolating only the cranial vault, removing other soft tissue structures to focus on the skull's bony anatomy. This segmentation step is vital for extracting the required anatomical details for accurate modeling. Fig. 15.7C presents a prepared layout of half of the skull for 3D printing. This configuration provides a physical model that surgeons can examine visually and tactilely, offering a practical reference for surgical planning. In Fig. 15.7D, the model's design is further detailed, showing the layered structure that emulates natural bone anatomy. The model comprises two cortical slabs, each 2 mm thick, surrounding a trabecular layer. Varying infill levels are applied to replicate different bone densities, with a 25% infill for the trabecular layer and 100% for the denser cortical layer. This approach ensures the model's structural integrity and creates a realistic representation of cranial bone, aiding in surgical practice and enhancing the surgeon's understanding of the patient's unique anatomy.

One of the most critical factors influencing the mechanical properties of sports guards is the choice of material. Different polymers and composites used in AM offer varying levels of tensile strength, flexibility, impact resistance, and wear resistance. For instance, thermoplastic polyurethane (TPU) is widely used in sports guards due to

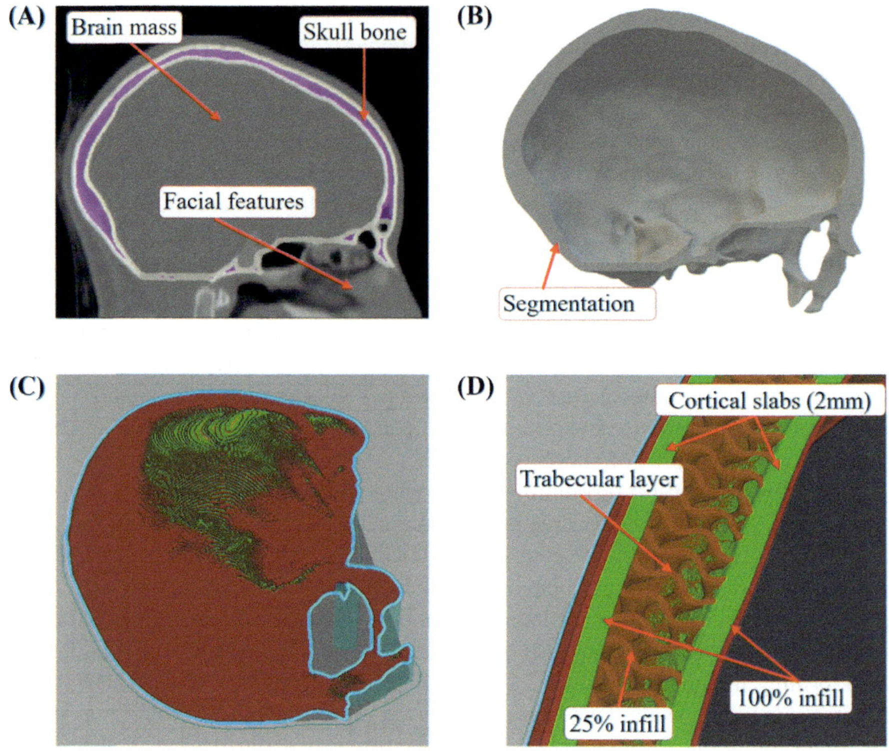

Figure 15.7 3D Model development of human skull from CT scan data for surgical planning: segmentation, CAD design, and structural infill attributes (Mantecón, 2024).
Adapted from Xu et al. (2020) under CC BY-NC-ND 4.0 license, © 2024, MDPI.

its high flexibility, excellent abrasion resistance, and good impact absorption, making it ideal for protective gear that must endure repeated impacts. On the other hand, materials like polylactic acid (PLA) provide higher rigidity and are often chosen for applications where stiffness is more important than flexibility. The inherent properties of the chosen material set the baseline for the mechanical performance of the sports guard. The orientation in which the sports guard is printed during the AM process has a significant impact on its mechanical properties. Build orientation refers to the direction in which the layers are stacked relative to the final shape of the object. Common orientations include flat, on-edge, and upright. Each orientation has distinct implications for the material's strength, stiffness, and durability. Parts printed in a flat or on-edge orientation typically exhibit higher tensile strength and greater elongation at break due to the alignment of the layers with the direction of applied stress, which allows for stronger interlayer bonding. In contrast, an upright orientation often results in weaker inter-layer bonds because the layers are perpendicular to the direction of stress, making the final product more susceptible to failure under load. Therefore, choosing the appropriate build orientation is crucial for ensuring that the sports guard performs well under the specific conditions it will face in use. Layer thickness is another crucial parameter that affects the mechanical properties of sports guards. In additive manufacturing, parts are built layer by layer, and the thickness of these layers can influence the overall strength, surface finish, and durability of the final product. Thinner layers generally lead to a smoother surface finish and finer detail, which is desirable for parts that require high precision. However, thinner layers also increase the build time and may create more opportunities for defects to occur, such as incomplete bonding between layers, which can weaken the structure. Conversely, thicker layers can reduce build time and potentially lower production costs, but they may result in a rougher surface finish and weaker inter-layer adhesion, compromising the mechanical properties of the sports guard.

The interaction between the polymer matrix and the reinforcement material is crucial for the performance of composites, as it directly affects their mechanical properties, durability, and overall functionality (Bhatt et al., 2023; Chakrabarti et al., 2022; Punetha et al., 2017, 2020; Punetha et al., 2019). The mechanism of this interaction can be better understood through scanning electron microscopy (SEM) images (Figs. 15.8 and 15.9), which reveal the microstructural details and failure patterns within the composite material. The SEM images provided depict various reinforcement patterns within a nylon matrix, which is commonly used in high-performance applications like sports guards. These patterns include voids, fiber pull-out, fiber breakage, and debonding, each representing different aspects of the interaction between the nylon matrix and the reinforcement fibers such as carbon fiber, fiberglass, and Kevlar.

Voids, as seen in several images, are regions within the composite where the polymer matrix fails to fill the space around the fibers completely. These voids can act as stress concentrators, leading to premature failure under mechanical loads. In the context of sports guards, which need to absorb and dissipate impact energy effectively, the presence of voids can significantly reduce the material's protective capabilities. This is one of the primary failure mechanisms observed in the SEM images. Fiber pull-out occurs when the bond between the polymer matrix and the reinforcement fiber is insufficient, causing the fiber to slip out of the matrix under stress. This mechanism is particularly evident in the images where carbon fibers are embedded in the nylon matrix. The effectiveness of the fiber reinforcement is compromised in such cases, leading to reduced mechanical strength and stiffness. In sports guards, where the material must withstand repeated impacts, fiber pull-out can lead to a reduction in protective performance over time. Another critical failure mechanism is fiber breakage, which occurs when the applied stress exceeds the tensile strength of the fiber. The SEM images show broken fibers within the nylon matrix, indicating that the composite has undergone significant stress. Fiber breakage can severely diminish the mechanical properties of the composite, as the load-bearing capacity of the reinforcement is lost. In the design of sports guards, minimizing fiber breakage through optimized fiber selection and processing techniques is essential for maintaining the integrity and durability of the guard under high-stress conditions. Debonding refers to the separation of the fiber from the matrix at the interface, leading to a loss of load transfer between the two components. The SEM images reveal instances of debonding, particularly in the Kevlar-reinforced composites. This

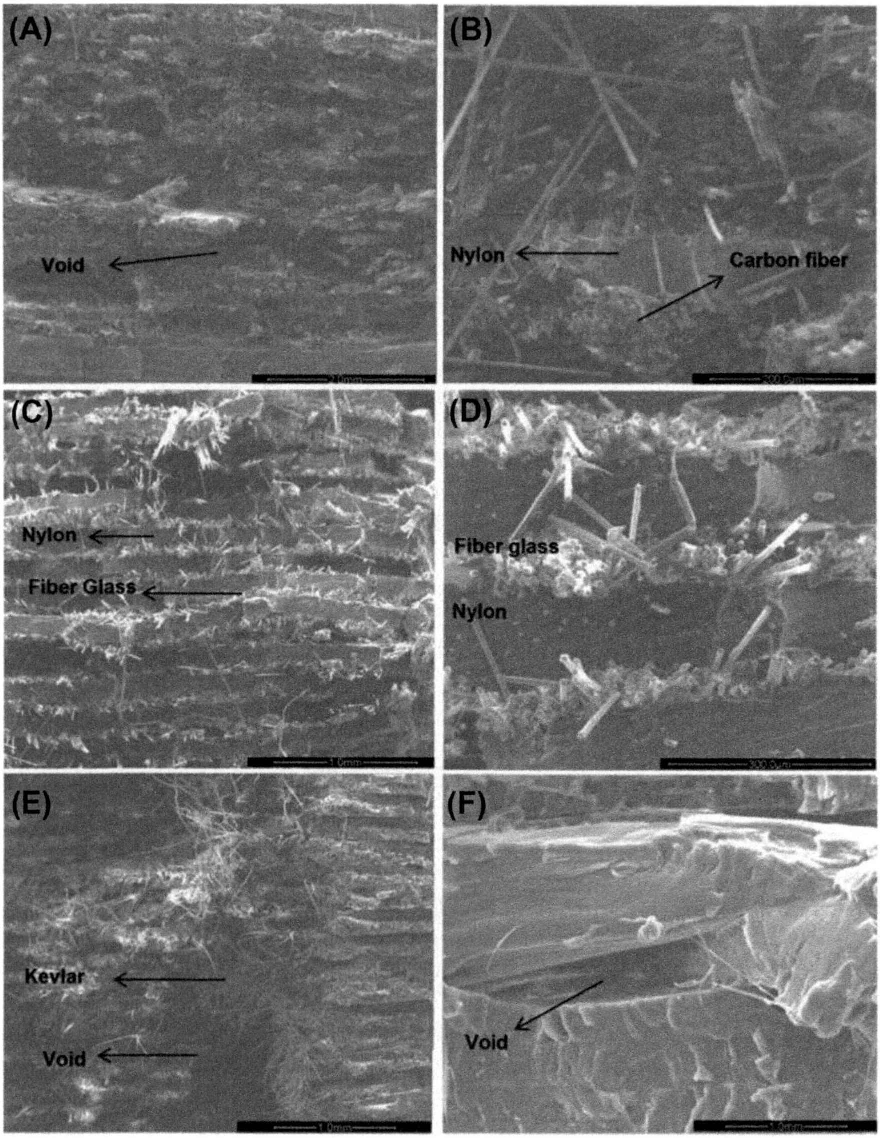

Figure 15.8 SEM images of CFRAM components posttensile testing. (A, B) Nylon-CF; (C, D) Nylon-FG; (E, F) Nylon-Kevlar.
Copyright 2019. Reproduced with permission from Elsevier (Mohammadizadeh et al., 2019).

failure mechanism can result in localized weakening of the material, which is detrimental to the overall performance of sports guards. Ensuring a strong interfacial bond between the matrix and reinforcement is critical for the longevity and effectiveness of the guard.

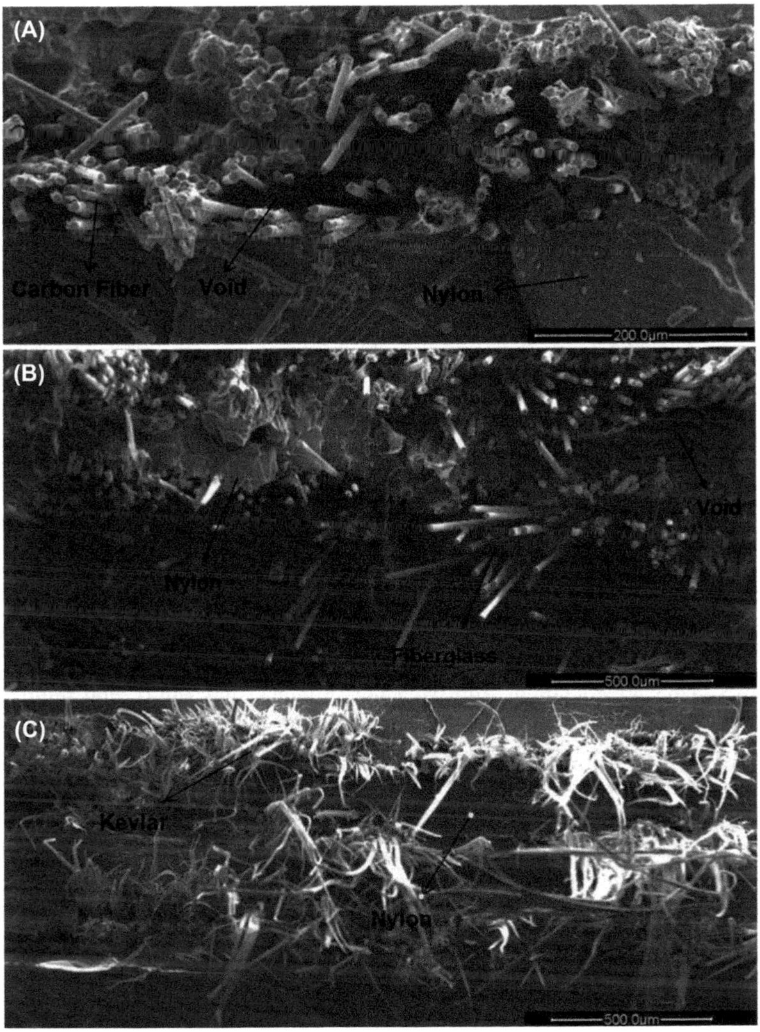

Figure 15.9 SEM images of CFRAM components postfatigue testing. (A) Nylon-CF; (B) Nylon-FG; (C) Nylon-FG.
Copyright 2019. Reproduced with permission from Elsevier (Mohammadizadeh et al., 2019).

The settings for print speed and temperature during the AM process also play a significant role in determining the mechanical properties of the sports guards. Print speed affects how quickly material is deposited, and if the speed is too high, it may lead to insufficient bonding between layers, resulting in a weaker structure. On the other hand, if the speed is too low, it can lead to excessive material build-up, increasing the risk of defects such as warping or uneven surfaces. Temperature settings are equally important; they must be optimized for the specific material being used to ensure proper melting and bonding. Inadequate temperature control can lead to poor layer adhesion, resulting in a brittle final product that lacks the necessary toughness for impact protection. Postprocessing steps, such as annealing or surface finishing, can further influence the mechanical properties of additively manufactured sports guards. Annealing, for example, can relieve internal stresses within the material, improving its overall strength and durability. Surface finishing processes can enhance the wear resistance and aesthetic qualities of the sports guard, while also sealing any minor defects that may have occurred during printing.

15.7 Sport equipment

Integrating functionally graded small-scale materials (FGSMs) in sports structures represents a groundbreaking advancement in sports engineering. FGSMs, known for their tailored material properties, significantly enhance performance and durability in sports equipment. This research emphasizes the interdisciplinary approach to designing sports equipment with FGSMs, focusing on lightweight, high-strength materials capable of withstanding dynamic loads. Advanced manufacturing techniques, such as additive manufacturing and nanotechnology, allow for precise control over material composition and microstructure. Computational modeling is employed to evaluate the mechanical behavior and performance characteristics of FGSM-based components, showcasing their potential to revolutionize sports equipment and structures. This advancement improves performance and safety and contributes to sustainability in sports engineering Developing automated sports equipment, such as ball machines, is essential for enhancing athlete training. Additive manufacturing plays a crucial role in this development, enabling the production of customized components that meet the specific needs of various sports. This study focuses on the additive manufacturing process used to create components for a ball machine prototype, which can throw balls at specific speeds and trajectories to facilitate effective workouts. Material extrusion is utilized to produce innovative, customized elements in the drive system. The design process involves using Fusion 360 software for component design and performing finite element analysis to ensure the components can withstand stress scenarios. The result is a functional prototype demonstrating the innovative potential of additive manufacturing in sports equipment design (Tunsoiu, 2022). Thermoplastic polyurethane (TPU) is a highly valued polymer in sports equipment manufacturing due to its ductility, biocompatibility, and excellent abrasion resistance. This study investigates the mechanical properties of additively manufactured TPU parts, focusing on the effects of build orientation, powder mix ratio, and postprocessing. The mix ratio of new and reused powders is critical, with new powder providing better particle quality and thermal properties. Build orientation also significantly impacts the tensile strength and deformability of TPU parts, with flat and on-edge orientations performing better than upright orientations. Additionally, postprocessing enhances the deformability of TPU parts, demonstrating the material's versatility and potential in high-performance applications across various fields (Xu et al., 2020). Auxetic structures, which exhibit a negative Poisson's ratio, are particularly suited for energy absorption applications in sports equipment. This research explores the creation of different auxetic structures using FDM 3D printing and examines their energy absorption capabilities. By experimenting with various raster angles and materials such as PLA and PLA+, the study finds that energy absorption can be significantly enhanced by transitioning from brittle to ductile materials and by grading the thickness of the structures. These findings, validated through simulation and modeling using Abaqus software, highlight the potential of auxetic structures to improve the safety and performance of protective sports gear (Hosseini & Mazaheri, 2024). Acetophenone azine, a potent allergen recently discovered in shin pads and footwear made from ethyl vinyl acetate (EVA), underscores the importance of monitoring and testing materials used in sports equipment. This compound, likely formed during manufacturing, necessitates careful patch testing to prevent allergic reactions in athletes. The discovery highlights the need for stringent quality control in the production of sports gear to ensure athlete safety (Raison-Peyron & Sasseville, 2021). The personalization of insoles using additive manufacturing, combined with 3D foot scanning and CAD systems, offers significant benefits for runners. This study evaluates the short and medium-term use of personalized insoles designed to match the plantar geometry of the feet, providing better fit and reducing discomfort. Biomechanical data show that personalized insoles lower peak pressures under the heel and improve overall fit, reducing injury risk. The ability to customize insoles through additive manufacturing demonstrates the technology's potential to enhance athlete performance and comfort (Salles & Gyi, 2013). These studies collectively highlight the transformative impact of additive manufacturing on the design and production of sports equipment. By enabling the creation of customized, high-performance components, additive manufacturing enhances athletes' safety, comfort, and overall experience across various sports disciplines.

15.8 Laser sintered sport equipment

Laser sintering is a pivotal technology in additive manufacturing (AM), revolutionizing how complex components and materials are designed and produced. This advanced manufacturing process utilizes a high-powered laser to fuse powdered material layer by layer, creating intricate and robust three-dimensional structures. The powdered material, often a polymer, metal, or ceramic, is selectively heated by the laser, causing it to sinter or bond together to form a solid mass without reaching its melting point. The integration of laser sintering into additive manufacturing has opened new avenues in the sports industry. This synergy allows for producing lightweight, durable, and precisely engineered components that were previously impossible or highly impractical to manufacture using conventional methods. Through laser sintering, additive manufacturing continues to push the boundaries of innovation, enabling the creation of parts with complex geometries, enhanced material properties, and tailored functionalities.

Exploring mechanoactive materials through advanced manufacturing techniques, including multidimensional printing and laser sintering, reveals their remarkable innovative properties under mechanical loading. These properties make them highly suitable for various applications, such as medicine, safety, transport, and sports, encompassing both internal implants and external wearables (Knothe Tate, 2020). Laser additive manufacturing, particularly laser sintering, plays a crucial role in achieving the precision and complexity required for these advanced materials. In addition to medical and safety applications, laser additive manufacturing is pivotal in customizing surfboard and stand-up paddle (SUP) board fins. By utilizing selective laser sintering (SLS), this study details a methodology for designing and testing customized fins, enhancing performance and allowing for greater experimentation within the sport (Novak, 2020). The enhanced performance of polymer lattices in sports equipment is achieved through a systematic workflow involving the design, fabrication, and evaluation of these lattices using SLS. The study illustrates how the compression properties of these lattices can be precisely tuned to maintain specific mass targets, thereby optimizing the performance of sports equipment (Clares et al., 2023). In sports footwear, laser additive manufacturing techniques, including SLS, are employed for rapid prototyping. This research assesses these techniques based on criteria such as accuracy, surface finish, material range, and build time, underscoring the advantages of AM in producing custom-tailored sports footwear (Manoharan et al., 2013). Laser sintering is also pivotal in the creation of complex lattice structures for energy-absorbent sports equipment. By examining the properties of Duraform® Flex and varying LS processing parameters, this study provides insights into optimizing the design of both solid and lightweight components to enhance their mechanical properties (Soe et al., 2014). The design of multifunctional shoe midsoles incorporates functionally gradient wave springs (FGWS) at critical foot pressure areas, utilizing materials such as PA 12 and TPU. This research evaluates the load-bearing capacity, energy absorption, stiffness, and cushioning properties, demonstrating the potential of laser additive manufacturing in creating versatile midsoles tailored for various applications (Ul Haq et al., 2022). Moreover, a design methodology for producing polyamide lattice-based lightweight fairing components using laser sintering has been developed, particularly for motorsports. This study encompasses the generation and simulation of boundary conformal non-periodic lattices, optimization of production parameters, and experimental validation, ensuring high quality and reliability in motorsports prototypes (Marschall et al., 2021).

The study on mechanoactive materials showcases the potential of advanced manufacturing methods such as multidimensional printing and laser sintering. These materials exhibit innovative properties under mechanical loading, making them ideal for medical applications, safety, transport, and sports applications. The use of the Microscopy-Aided Design And Manufacture (MADAME) approach, which incorporates principles of recursive logic, enables the design and manufacture of materials that can be used for both internal implants and external wearables, thus expanding the horizon of sports applications (Knothe Tate, 2020). In water sports, additive manufacturing has enabled the customization of surfboard and stand-up paddle (SUP) board fins. This research employs selective laser sintering (SLS) to produce fins that can be tailored to individual needs. The study details a parametric CAD system allowing surfers to modify fin geometry in real-time, enhancing performance and promoting innovation within the sport (Novak, 2020). Additive manufacturing's ability to enhance the performance of polymer lattices in sporting equipment is demonstrated through a systematic workflow involving design, fabrication, and performance evaluation.

The study uses SLS to optimize lattice designs, focusing on printed mass and compressive modulus variables. This approach ensures that the compression properties of these lattices can be finely tuned to improve sports equipment performance (Clares et al., 2023). Laser sintering (LS) produces complex lattice structures from elastomeric materials aimed at energy-absorbent sports equipment like personalized helmets and running shoes. The research examines the material properties of Duraform® Flex, highlighting how variations in LS processing parameters significantly impact the mechanical properties of the resulting parts. This knowledge is crucial for optimizing designs for both solid and lightweight components (Soe et al., 2014). The design of multifunctional shoe midsoles is another innovative application of additive manufacturing. This study integrates functionally gradient wave springs (FGWS) at critical foot pressure areas, using materials like PA 12 and TPU. The midsoles are manufactured through SLS and tested for load-bearing capacity, energy absorption, stiffness, and cushioning properties, demonstrating the versatility and performance benefits of AM in sports footwear (Ul Haq et al., 2022). Additive manufacturing is also making strides in the area of impact protection in sports technology. This study focuses on cellular structures that can be tailored for specific responses by modifying geometric design variables. The use of periodic lattice structures in protective gear is optimized through a detailed analysis of factors such as cell width and strut cross-sectional area, enhancing their impact resistance capabilities (Rossiter et al., 2020). In motorsports, laser sintering facilitates the production of personalized motorcycle structural components. By investigating the mechanical properties of thin-walled structures, this research identifies the optimal wall thickness and other parameters necessary to achieve the desired mechanical properties in lightweight components, ensuring high performance and reliability. In summary, additive manufacturing, mainly through laser sintering, offers significant advancements in sports technology by enabling the customization and optimization of equipment. This technology enhances performance and safety and opens new avenues for innovation across various sports applications.

15.9 Challenges and future directions

Additive manufacturing (AM) has brought significant advancements to the sports equipment industry, offering customized, high-performance gear that enhances athlete performance and safety. However, this technology faces several challenges that need to be addressed to fully harness its potential. Among the key challenges, the most crucial are limited technological advancements, cost considerations, regulatory and safety standards, and environmental impact. The primary challenge in using AM for sports equipment is the need for more available materials. Although progress has been made in developing high-performance polymers and nanocomposites, the range of materials that can be effectively used in AM remains limited compared to traditional manufacturing methods. The mechanical properties of AM-produced parts, such as strength, toughness, and fatigue resistance, can sometimes fall short of those produced by conventional means. Materials like PLA and PA12 offer excellent properties for specific applications. However, they may need to meet the performance requirements for more demanding uses, such as high-impact sports gear or extreme environmental conditions. Therefore, developing new materials and enhancing existing ones will be crucial for expanding the application of AM in sports. Technological challenges also exist with AM processes. The layer-by-layer construction can lead to issues such as anisotropy, where the mechanical properties of a part differ depending on the direction of the load. This can create weaker points within the structure, potentially compromising the integrity of the sports equipment. Techniques like selective laser sintering (SLS) and fused filament fabrication (FFF) need further refinement to ensure uniformity and consistency in the mechanical properties of the final products. Addressing these issues will require advancements in AM technology, including improved software for design and simulation, better control over the manufacturing process, and the development of postprocessing techniques to enhance the mechanical properties of AM parts.

Cost considerations pose a crucial challenge in the widespread adoption of AM for sports equipment. While AM can reduce material waste and enable the production of custom-fit products, the initial costs of AM machines and materials can be high. This is especially true for high-performance polymers and nanocomposites, which can be expensive to produce and process. Additionally, the cost of AM equipment, maintenance, and the expertise required to operate these machines can be prohibitive, particularly for smaller manufacturers. To overcome these barriers, there is a need to develop more affordable AM technologies and materials, as well as to increase the scalability and efficiency of AM processes. Regulatory and safety standards also pose significant challenges. The sports equipment industry is subject to stringent regulations to ensure the safety and performance of products. AM-produced sports equipment must meet these standards, which can be challenging given the technology's relatively new and evolving nature. Ensuring that AM parts are as safe and reliable as those produced by traditional methods requires rigorous testing and validation processes. Additionally, there is a need to develop specific standards and guidelines for AM-produced sports equipment to provide manufacturers with clear criteria for compliance. Collaborative efforts between industry stakeholders, regulatory bodies, and research institutions will be essential to establish these standards and ensure the safe integration of AM in sports. The environmental impact of AM is another critical area that needs attention. While AM can reduce material waste and enable the use of more sustainable materials, the energy consumption of AM processes can be high. The production and disposal of AM materials, particularly polymers, can also have environmental implications. Developing more energy-efficient AM technologies and sustainable materials that can be easily recycled or biodegraded will be crucial for minimizing the environmental footprint of AM in sports equipment. Research into alternative energy sources for AM machines and using biodegradable polymers like PLA can contribute to making AM a more sustainable manufacturing option.

Despite these challenges, the future of AM in sports equipment is promising, with several exciting directions for further development. One key area is the continued advancement of multifunctional materials and intelligent composites. These materials can offer enhanced properties such as self-healing, shape memory, and embedded sensing capabilities, significantly improving the performance and functionality of sports equipment. For instance, shape memory polymers (SMPs) can create adaptive sports gear that responds to environmental conditions or mechanical stress, providing athletes with excellent protection and comfort. Another important direction is the integration of digital technologies and data analytics with AM. Advanced design software, simulation tools, and real-time monitoring systems can enhance the precision and efficiency of AM processes, allowing for the creation of more complex and optimized designs. Additionally, integrating sensors and IoT (Internet of Things) technology into AM-produced sports equipment can provide athletes with valuable data on their performance and the condition of their gear, enabling more personalized and effective training programs. Customization and personalization will continue to be significant trends in AM applications in sports. The ability to produce custom-fit equipment tailored to individual athletes' unique needs and preferences offers significant performance, comfort, and injury prevention benefits. Advances in AM technology will further enhance the capability to create highly personalized sports gear, from custom-fit footwear and protective padding to individually tailored sportswear and accessories. Collaborative efforts between researchers, manufacturers, and athletes will be crucial for driving innovation and overcoming the challenges associated with AM in sports. Research institutions can contribute to developing new materials and technologies, while manufacturers can focus on refining AM processes and scaling up production. Athletes can provide valuable feedback on the performance and functionality of AM-produced equipment, helping guide the development of new products. By addressing these challenges and leveraging the unique capabilities of AM, we can create sports equipment that enhances performance, safety, and sustainability, ultimately benefiting athletes and the broader sports community.

References

Agarwal, K., Sahay, R., Baji, A., & Budiman, A. S. (2020). Impact-resistant and tough helicoidally aligned ribbon reinforced composites with tunable mechanical properties via integrated additive manufacturing methodologies. *ACS Appl Polym Mater, 2*(8), 3491–3504. https://doi.org/10.1021/acsapm.0c00518.

Alarifi, M. I., & Alarifi, I. M. (2023). Comprehensive structural evaluation of composite materials in 3D printed shin guards. *Journal of Materials Research and Technology, 27*, 6912–6923. https://doi.org/10.1016/j.jmrt.2023.11.131.

Ali, N., & Mostafa, E. M. (2024). Biocarbon enhanced flexible nanofiber mats for sustainable energy generation and wearable device applications. *Fibers and Polymers, 25*(3), 869–878. https://doi.org/10.1007/s12221-024-00479-7.

Al-Mohaimeed, A. M., Al Omar, S. Y., & El-Tohamy, M. F. (2023). Fast and novel multiwalled carbon nanotubes decorated with metal oxide nanoparticles for potentiometric detection of a prohibited medication in sports acebutolol hydrochloride. *Heliyon, 9*(10). https://doi.org/10.1016/j.heliyon.2023.e20997.

Bhatt, S., Punetha, V. D., Pathak, R., & Punetha, M. (2023). Two-dimensional carbon nanomaterial-based biosensors: Micromachines for advancing the medical diagnosis. *Advanced Structured Materials, 190*, 181–225. https://doi.org/10.1007/978-3-031-28942-2_9.

Caporaso, T., Bellitti, P., Grazioso, S., Serpelloni, M., Sardini, E., & Lanotti, A. (2022). Concept generation and preliminary prototyping of a tailored smart glove with capacitive pressure sensors for force grip analysis in cycling. *Comput Aided Des Appl*, 87–98. https://doi.org/10.14733/cadaps.2023.S6.87-98.

Chakrabarti, C., Khan Pathan, S., Deep Punetha, V., & Pillai, S. A. (2022). Interaction of Tetronics® micelles with stimuli and additives and a commanding aspect towards drug delivery: An overview. *J Mol Liq, 366*. https://doi.org/10.1016/j.molliq.2022.120289.

Chaudhary, R., Upadhyaya, S., & Sangal, V. K. (2021). A parametric study on the HDPE/PP and marble slurry waste utilisation using single screw extruder. *Asian Journal of Water, Environment and Pollution, 18*(4), 9–17. https://doi.org/10.3233/AJW210038.

Chen, B., Wang, J., Liu, N., & Yan, F. (2013). Synergetic effect of lubricant additive and reinforcement additive on the tribological behaviors of PEEK-based composites under seawater lubrication. *Tribology Transactions, 56*(4), 672–680. https://doi.org/10.1080/10402004.2013.779403.

Choe, C.-M., Sok, S.-H., Ri, W.-S., Yang, W.-C., Kim, U.-H., & Jong, Y.-G. (2021). Manufacture of plaster core mold for large oxygen plant components using fused deposition modeling (FDM). *International Journal of Metalcasting, 15*(4), 1275–1281. https://doi.org/10.1007/s40962-020-00549-5.

Clares, A. P., Manogharan, G., Thomas, L., & Krzeminski, D. (2023). Design-property decision-making in polymer lattices when controlling for printed mass. *Sports Engineering, 26*(1), 28. https://doi.org/10.1007/s12283-023-00418-4.

Dhapola, P. S., et al. (2024). Environment-friendly approach for synthesis of promising porous carbon: Empowering supercapacitors for a sustainable future. *Mater Adv, 5*(6), 2430–2440. https://doi.org/10.1039/d3ma00984j.

Dias Gonçalves, V. P., Fontes Vieira, C. M., & Colorado Lopera, H. A. (2024). The production and materials of mouthguards: Conventional vs additive manufacturing—A systematic review. *Heliyon, 10*(14), e34294. https://doi.org/10.1016/j.heliyon.2024.e34294.

Fajardo, J. I., Farez, M. V., & Paltán, C. A. (2023). Experimental analysis of the relationship between textile structure, tensile strength and comfort in 3D printed structured fabrics. *Polymers (Basel), 15*(1). https://doi.org/10.3390/polym15010152.

Fekiri, C., Kim, H. C., & Lee, I. H. (2020). 3D-printable carbon nanotubes-based composite for flexible piezoresistive sensors. *Materials, 13*(23), 5482. https://doi.org/10.3390/ma13235482.

Filgueira, A. K. L., Gallardo, I. D., Martins, K. Y. N., Castelo Branco, R. R., Bezerra, K. C. T., & de Morais, M. E. (2022). Manufacture of facial orthosis in ABS by the additive manufacturing process: A customized application in high performance sports. *Lecture Notes in Mechanical Engineering*, 422–432. https://doi.org/10.1007/978-3-030-79168-1_37.

Frizziero, L., et al. (2021). An innovative ford sedan with enhanced stylistic design engineering (SDE) via augmented reality and additive manufacturing. *Designs (Basel), 5*(3). https://doi.org/10.3390/designs5030046.

Gantenbein, S., et al. (2021). Spin-printing of liquid crystal polymer into recyclable and strong all-fiber materials. *Adv Funct Mater, 31*(52). https://doi.org/10.1002/adfm.202104574.

Gao, Y., Guo, W., & Baghaei, S. (2024). Ceramic based nanocomposites with alumina-carbon nanotube reinforcement for improved energy absorption in sports-related injuries: Microstructural analysis and low velocity impact. *Ceram Int, 50*(7), 11129–11137. https://doi.org/10.1016/j.ceramint.2024.01.014.

Garces, I. T., & Ayranci, C. (2018). A view into additive manufactured electro-active reinforced smart composite structures. *Manuf Lett, 16*, 1–5. https://doi.org/10.1016/j.mfglet.2018.02.008.

Garces, I. T., & Ayranci, C. (2020). Active control of 4D prints: Towards 4D printed reliable actuators and sensors. *Sensors and Actuators. A, Physical, 301*, 111717. https://doi.org/10.1016/j.sna.2019.111717.

Gisario, A., Veniali, F., Barletta, M., Tagliaferri, V., & Vesco, S. (2017). Laser transmission welding of poly(ethylene terephthalate) and biodegradable poly(ethylene terephthalate)-based blends. *Opt Lasers Eng, 90*, 110–118. https://doi.org/10.1016/j.optlaseng.2016.10.010.

Goh, G. D., et al. (2018). Characterization of mechanical properties and fracture mode of additively manufactured carbon fiber and glass fiber reinforced thermoplastics. *Materials and Design, 137*, 79–89. https://doi.org/10.1016/j.matdes.2017.10.021.

Gong, Z., et al. (2024). Flexible PVDF/SiC/FeCl3 nanofiber membrane generators with synergistically enhanced piezoelectricity. *Nano Energy, 122*. https://doi.org/10.1016/j.nanoen.2024.109290.

Górski, F., et al. (2023). Development and testing of an individualized sensorised 3D printed upper limb bicycle prosthesis for adult patients. *Applied Sciences (Switzerland), 13*(23). https://doi.org/10.3390/app132312918.

Guo, D. (2023). Repair of sports bone injury based on multifunctional nanomaterial particles. *Structural Engineering and Mechanics, 86*(4), 487–501. https://doi.org/10.12989/sem.2023.86.4.487.

Hada, T., Komagamine, Y., Kanazawa, M., & Minakuchi, S. (2024). Fabrication of sports mouthguards using a semi-digital workflow with 4D-printing technology. *J Prosthodont Res, 68*(1), 181–185. https://doi.org/10.2186/jpr.JPR_D_22_00274.

Hao, Q., Yao, Z., Choi, W. J., & Kim, H. (2024). Effect of ethylene vinyl acetate foam-graphene composite material on the mechanical properties of sports footwear. *Alexandria Engineering Journal, 96*, 142–148. https://doi.org/10.1016/j.aej.2024.04.006.

He, Q., Heidari, H., Zeze, D. A., & Hosseini, E. (2024). Highly sensitive flexible pressure sensor based on PVDF-TrFE-BaTiO3 piezoelectric nanofibers. *IEEE Sens Lett, 8*(8). https://doi.org/10.1109/LSENS.2024.3430942.

Hosseini, M., & Mazaheri, H. (2024). Experimental and simulation study of the crashworthiness behavior of thickness-graded auxetic structures of ductile and brittle materials. *Journal of Sandwich Structures & Materials*. https://doi.org/10.1177/10996362241248037.

Hu, G., Li, H., & Liu, F. (2024). A biosensor based on nanocomposite of g-C3N4 and polyaniline for detection of fentanyl as a doping agent in sports. *Alexandria Engineering Journal, 87*, 515–523. https://doi.org/10.1016/j.aej.2023.12.043.

Huanbo, C., Lijun, G., Jinhu, Z., et al. (2023). Recycling of carbon fiber and performance evaluation for its reinforced composites made of additive manufacturing. *Journal of Mechanical Engineering, 59*, 375–388. https://doi.org/10.3901/JME.2023.07.375.

Jahangir, A., Singh, P. K., Singh, R. C., Punetha, V. D., & Rawat, S. (2024). Ion conducting polyethylene oxide-based polymer electrolyte dopped with ionic liquid-trifluoromethane-sulfonic chloride. *Macromolecular Symposia, 413*(1). https://doi.org/10.1002/masy.202300139.

Joseph, A., Mahesh, V., & Mahesh, V. (2023). Effect of loading rates on the in-plane compressive properties of additively manufactured ABS and PLA-based hexagonal honeycomb structures. *Journal of Thermoplastic Composite Materials, 36*(3), 1113–1134. https://doi.org/10.1177/08927057211051416.

Kim, G., Park, K., Hwang, K.-J., & Jin, S. (2021). Highly sunlight reflective and infrared semi-transparent nanomesh textiles. *ACS Nano, 15*(10), 15962–15971. https://doi.org/10.1021/acsnano.1c04104.

Knothe Tate, M. L. (2020). Advanced design and manufacture of mechanoactive materials inspired by skin, bones, and skin-on-bones. *Frontiers in Bioengineering and Biotechnology, 8*. https://doi.org/10.3389/fbioe.2020.00845.

Kulakarni, S. P., Das, N. K., & Badhulika, S. (2024). Fabrication of high-performance triboelectric nanogenerator based on Ni3C nanosheets to self-power thermal patch for pain relief. *Nanotechnology, 35*(1). https://doi.org/10.1088/1361-6528/ad0057.

Kumar, S., et al. (2023). In-situ N/O-heteroatom enriched micro-/mesoporous activated carbon derived from natural waste honeycomb and paper wasp hive and its application in quasi-solid-state supercapacitor. *J Energy Storage, 72*. https://doi.org/10.1016/j.est.2023.108722.

Kumar, S., Teraiya, S., & Potdar, Y. (2023). Experimental investigation on porosity and flexural strength of polymer parts fabricated by fused deposition modeling. *Polymer Engineering and Science, 63*(2), 531–545. https://doi.org/10.1002/pen.26227.

León-Becerra, J., Hidalgo-Salazar, M.Á., & González-Estrada, O. A. (2023). Progressive damage analysis of carbon fiber-reinforced additive manufacturing composites. *The International Journal of Advanced Manufacturing Technology, 126*(5–6), 2617–2631. https://doi.org/10.1007/s00170-023-11256-w.

Li, J. (2022). Application of nanotubes combined with ethnic sports rehabilitation therapy in the treatment of patients with knee arthritis. *J Nanomater, 2022*. https://doi.org/10.1155/2022/8679892.

Li, J., & Lin, C. (2024). Design and fabrication of an electrochemical sensor based on nanocomposite of dendritic polymer and reduced graphene oxide to measure sports nutrition lactate. *Journal of Food Measurement and Characterization, 18*(7), 5364–5375. https://doi.org/10.1007/s11694-024-02572-0.

Li, Y., & Liu, W. (2024). Electrochemical detection of blood doping in sports: A novel biosensor based on nickel oxide/nitrogen-doped graphene oxide nanocomposite. *Alexandria Engineering Journal, 96*, 176–184. https://doi.org/10.1016/j.aej.2024.03.099.

Lißner, M., Goldberg, T., Townsend, D., Petrinic, N., & Bergmann, J. (2023). On the protectiveness of additively manufactured mouthguards. *Materials and Design, 234*, 112371. https://doi.org/10.1016/j.matdes.2023.112371.

Liu, Y., & Quan, C. (2023). A sensitive electrochemical sensor for the detection of sports stimulant methyltestosterone via ZnO/TiO2 nanocomposite. *Int J Electrochem Sci, 18*(10), 100308. https://doi.org/10.1016/j.ijoes.2023.100308.

Lu, Q., Darnal, A., Takahashi, H., Muliana, A. H., & Kim, J. (2022). User-centered property adjustment with programmable filament. *Conference on Human Factors in Computing Systems - Proceedings*. https://doi.org/10.1145/3491101.3519864.

Lunchev, A. V., Kashcheev, A., Tok, A.ling Y., & Lipik, V. (2022). Mechanical characteristics of poly (ethylene vinyl acetate) foams with graphene for the applications in sport footwear. *Polymer Testing, 113*. https://doi.org/10.1016/j.polymertesting.2022.107688.

Mahendran, G., Mageswari, M., Kakaravada, I., & Rao, P. K. V. (2024). Characterization of polyester composite developed using silane-treated rubber seed cellulose toughened acrylonitrile butadiene styrene honey comb core and sunn hemp fiber. *Polymer Bulletin, 81*(17), 15955–15973. https://doi.org/10.1007/s00289-024-05427-2.

Manoharan, V., Chou, S. M., Forrester, S., Chai, G. B., & Kong, P. W. (2013). Application of additive manufacturing techniques in sports footwear. *Virtual Phys Prototyp, 8*(4), 249–252. https://doi.org/10.1080/17452759.2013.862958.

Mantecón, R., et al. (2024). Additive manufacturing of head surrogates for evaluation of protection in sports. *Polymers (Basel), 16*(12), 1753. https://doi.org/10.3390/polym16121753.

Marschall, D., Sindinger, S.-L., Rippl, H., Bartosova, M., & Schagerl, M. (2021). Design, simulation, testing and application of laser-sintered conformal lattice structures on component level. *Rapid Prototyp J, 27*(11), 43–57. https://doi.org/10.1108/RPJ-10-2020-0232.

Mohammadizadeh, M., Imeri, A., Fidan, I., & Elkelany, M. (2019). 3D printed fiber reinforced polymer composites—Structural analysis. *Compos B Eng, 175*, 107112. https://doi.org/10.1016/j.compositesb.2019.107112.

Montgomery, S. M., Hilborn, H., Hamel, C. M., Kuang, X., Long, K. N., & Qi, H. J. (2021). The 3D printing and modeling of functionally graded Kelvin foams for controlling crushing performance. *Extreme Mech Lett, 46*, 101323. https://doi.org/10.1016/j.eml.2021.101323.

Nath, S. D., & Nilufar, S. (2020). An overview of additive manufacturing of polymers and associated composites. *Polymers (Basel), 12*(11), 2719. https://doi.org/10.3390/polym12112719.

Novak, J. (2020). A parametric method to customize surfboard and stand up paddle board fins for additive manufacturing. *Comput Aided Des Appl, 18*(2), 297–308. https://doi.org/10.14733/cadaps.2021.297-308.

Nugraha, A. D., Syahril, M., & Muflikhun, M. A. (2023). Excellent performance of hybrid model manufactured via additive manufacturing process reinforced with GFRP for sport climbing equipment. *Heliyon, 9*(3), e14706. https://doi.org/10.1016/j.heliyon.2023.e14706.

Pais, V., Navarro, M., Guise, C., Martins, R., & Fangueiro, R. (2022). Hydrophobic performance of electrospun fibers functionalized with TiO2 nanoparticles. *Textile Research Journal, 92*(15–16), 2719–2730. https://doi.org/10.1177/00405175211010669.

Park, J.-H., Goo, B., & Park, K. (2019). Topology optimization and additive manufacturing of customized sports item considering orthotropic anisotropy. *International Journal of Precision Engineering and Manufacturing, 20*(8), 1443–1450. https://doi.org/10.1007/s12541-019-00163-4.

Paul, M., Tscheuschner, G., Herrmann, S., & Weller, M. G. (2020). Fast detection of 2,4,6-trinitrotoluene (TNT) at ppt level by a laser-induced immunofluorometric biosensor. *Biosensors (Basel), 10*(8), 89. https://doi.org/10.3390/bios10080089.

Punetha, V. D., et al. (2017). Functionalization of carbon nanomaterials for advanced polymer nanocomposites: A comparison study between CNT and graphene. *Prog Polym Sci, 67*, 1–47. https://doi.org/10.1016/j.progpolymsci.2016.12.010.

Punetha, V. D., et al. (2022). Recent advancements in green synthesis of nanoparticles for improvement of bioactivities: A review. *Current Pharmaceutical Biotechnology, 23*(7), 904–919. https://doi.org/10.2174/1389201022666210812115233.

Punetha, V. D., Ha, Y.-M., Kim, Y.-O., Jung, Y. C., & Cho, J. W. (2019). Interaction of photothermal graphene networks with polymer chains and laser-driven photo-actuation behavior of shape memory polyurethane/epoxy/epoxy-functionalized graphene oxide nanocomposites. *Polymer (Guildf), 181*, https://doi.org/10.1016/j.polymer.2019.121791.

Punetha, V. D., Ha, Y.-M., Kim, Y.-O., Jung, Y. C., & Cho, J. W. (2020). Rapid remote actuation in shape memory hyperbranched polyurethane composites using cross-linked photothermal reduced graphene oxide networks. *Sensors and Actuators B, Chemical, 321*. https://doi.org/10.1016/j.snb.2020.128468.

Raison-Peyron, N., & Sasseville, D. (2021). Acetophenone Azine. *Dermatitis: contact, atopic, occupational, drug: official journal of the American Contact Dermatitis Society, North American Contact Dermatitis Group, 32*(1), 5–9. https://doi.org/10.1097/DER.0000000000000697.

Ramírez-Gutiérrez, D. L., Cuan-Urquizo, E., & Fuentes-Aguilar, R. Q. (2023). Sine-based lattice plates: Additive manufacturing and their mechanical properties when loaded out of plane. *Rapid Prototyp J, 29*(9), 1773–1787. https://doi.org/10.1108/RPJ-01-2023-0010.

Ranjan, R. K., Kumar, S., Punetha, V. D., Pandey, S. P., Singh, R. C., & Singh, P. K. (2024). Poly(vinylidene fluoride-CO-hexafluoropropylene) (PVDF-HFP) incorporated acetylene black: Electrical, optical, and structural studies. *Macromolecular Symposia, 413*(1). https://doi.org/10.1002/masy.202300120.

Raza, T., et al. (2022). Wearable and flexible multifunctional sensor based on laser-induced graphene for the sports monitoring system. *ACS Appl Mater Interfaces, 14*(48), 54170–54181. https://doi.org/10.1021/acsami.2c14847.

Xu, X., Wang, R., Zhang, D., & Feng, Y. (2022). [Retracted] Clinical application of graphene composite in internal fixation of snkle fracture in sports. *Advanced Functional Graded Materials: Processing and Applications*. https://doi.org/10.1155/2023/9865864.

Rizwee, M., & Kumar, D. (2024). Contribution of geometrical infill pattern on mechanical behaviour of 3D manufactured polylactic acid specimen: Experimental and numerical analysis. *Progress in Rubber, Plastics and Recycling Technology*. https://doi.org/10.1177/14777606241281614.

Rossiter, J. D., Johnson, A. A., & Bingham, G. A. (2020). Assessing the design and compressive performance of material extruded lattice structures. *3D Print Addit Manuf, 7*(1), 19–27. https://doi.org/10.1089/3dp.2019.0030.

Sahay, R., Agarwal, K., Subramani, A., Raghavan, N., Budiman, A. S., & Baji, A. (2020). Helicoidally arranged polyacrylonitrile fiber-reinforced strong and impact-resistant thin polyvinyl alcohol film enabled by electrospinning-based additive manufacturing. *Polymers (Basel), 12*(10), 2376. https://doi.org/10.3390/polym12102376.

Sahoo, N. G., Esteves, R. J., Punetha, V. D., Pestov, D., Arachchige, I. U., & McLeskey, J. T. (2016). Schottky diodes from 2D germanane. *Applied Physics Letters, 109*(2). https://doi.org/10.1063/1.4955463.

Saliakas, S., et al. (2023). Integrating exposure assessment and process hazard analysis: The nano-enabled 3D printing filament extrusion case. *Polymers (Basel), 15*(13), 2836. https://doi.org/10.3390/polym15132836.

Salles, A. S., & Gyi, D. E. (2013). An evaluation of personalised insoles developed using additive manufacturing. *Journal of Sports Sciences, 31*(4), 442–450. https://doi.org/10.1080/02640414.2012.736629.

Schaedler, T. A., Chan, L. J., Clough, E. C., Stilke, M. A., Hundley, J. M., & Masur, L. J. (2017). Nanocrystalline aluminum truss cores for lightweight sandwich structures. *JOM, 69*(12), 2626–2634. https://doi.org/10.1007/s11837-017-2539-8.

Schewe, P., Roehler, A., Spintzyk, S., & Huettig, F. (2022). Shock absorption behavior of elastic polymers for sports mouthguards: An in vitro comparison of thermoplastic forming and additive manufacturing. *Materials, 15*(8), 2928. https://doi.org/10.3390/ma15082928.

Shathi, M. A., Chen, M., Khoso, N. A., Rahman, M. T., & Bhattacharjee, B. (2020). Graphene coated textile based highly flexible and washable sports bra for human health monitoring. *Materials and Design, 193*. https://doi.org/10.1016/j.matdes.2020.108792.

Shi, L., Wang, L., Li, W., & Guan, J. (2024). Graphene-modified electrochemical analysis system for the detection of food-borne stimulants sibutramine in sports. *Journal of Food Measurement and Characterization, 18*(9), 7751–7761. https://doi.org/10.1007/s11694-024-02762-w.

Soe, S. P., Martindale, N., Constantinou, C., & Robinson, M. (2014). Mechanical characterisation of Duraform® Flex for FEA hyperelastic material modelling. *Polymer Testing, 34*, 103–112. https://doi.org/10.1016/j.polymertesting.2014.01.004.

Soltani, A., Noroozi, R., Bodaghi, M., Zolfagharian, A., & Hedayati, R. (2020). 3D printing on-water sports boards with bio-inspired core designs. *Polymers (Basel), 12*(1). https://doi.org/10.3390/polym12010249.

Song, S., Zhang, T., & Zhui, Z. (2023). Dynamic analysis of nanotube-based nanodevices for drug delivery in sports-induced varied conditions applying the modified theories. *Steel and Composite Structures, 49*(5), 487–502. https://doi.org/10.12989/scs.2023.49.5.487.

Srinivas, B., Basha, M. M., & Sankar, M. R. (2023). Experimental study on chemical polishing of laser powder bed fusion-based Inconel 718 features. *Lecture Notes in Mechanical Engineering,* 113–125. https://doi.org/10.1007/978-981-19-7612-4_10.

Su, J., & Su, X. (2023). Determination of tartrazine in sports drinks by a disposable electrochemical sensor modified with Co2O3. *Journal of Food Measurement and Characterization, 17*(6), 5856–5863. https://doi.org/10.1007/s11694-023-02094-1.

Su, J., & Su, X. (2024). A multi-walled carbon nanotubes-based molecular imprinting electrochemical sensor for the detection of Aspartame in sports beverage. *Journal of Food Measurement and Characterization, 18*(1), 618–624. https://doi.org/10.1007/s11694-023-02219-6.

Tang, J., Wu, Y., Ma, S., Yan, T., & Pan, Z. (2022). Flexible strain sensor based on CNT/TPU composite nanofiber yarn for smart sports bandage. *Compos B Eng, 232*. https://doi.org/10.1016/j.compositesb.2021.109605.

Tasch, D., Mad, A., Stadlbauer, R., & Schagerl, M. (2018). Thickness dependency of mechanical properties of laser-sintered polyamide lightweight structures. *Addit Manuf, 23*, 25–33. https://doi.org/10.1016/j.addma.2018.06.018.

Teng, P. S. P., et al. (2022). A methodology to design and fabricate a smart brace using low-cost additive manufacturing. *Virtual Phys Prototyp, 17*(4), 932–947. https://doi.org/10.1080/17452759.2022.2090384.

Toleos, L. R., Luna, N. J. A. B. D., Manuel, M. C. E., Chua, J. M. R., Sangalang, E. M. A., & So, P. C. (2020). Feasibility study for fused deposition modeling (FDM) 3D-printed propellers for unmanned aerial vehicles. *International Journal of Mechanical Engineering and Robotics Research, 9*(4), 548–558. https://doi.org/10.18178/ijmerr.9.4.548-558.

Toon, D., Vinet, A., Pain, M. T. G., & Caine, M. P. (2011). A methodology to investigate the relationship between lower-limb dynamics and shoe stiffness using custom-built footwear. *Proc Inst Mech Eng P J Sport Eng Technol, 225*(1), 32–37. https://doi.org/10.1177/1754337110396792.

Tunay, M. (2024). Bending behavior of 3D printed sandwich structures with different core geometries and thermal aging durations. *Thin-Walled Structures, 194*. https://doi.org/10.1016/j.tws.2023.111329.

Tunsoiu, N., et al. (2022). Additive manufacturing of components for a ball machine prototype. *Macromolecular Symposia, 404*(1). https://doi.org/10.1002/masy.202100437.

Ul Haq, M. R., Nazir, A., Lin, S.-C., & Jeng, J.-Y. (2022). Design and performance evaluation of multifunctional midsole using functionally gradient wave springs produced using multijet fusion additive manufacturing process. *Mater Today Commun, 31*, 103505. https://doi.org/10.1016/j.mtcomm.2022.103505.

Unkovskiy, A., Huettig, F., Kraemer-Fernandez, P., & Spintzyk, S. (2021). Multi-material 3D printing of a customized sports mouth guard: Proof-of-concept clinical case. *International Journal of Environmental Research and Public Health, 18*(23), 12762. https://doi.org/10.3390/ijerph182312762.

Vinet, A., & Caine, M. P. (2011). Design, manufacture, and evaluation of traction features on sprint footwear using laser sintered nylon-12 sole units. *Proc Inst Mech Eng P J Sport Eng Technol, 225*(4), 259–264. https://doi.org/10.1177/1754337111408376.

Vyavahare, S., & Kumar, S. (2021). Numerical and experimental investigation of FDM fabricated re-entrant auxetic structures of ABS and PLA materials under compressive loading. *Rapid Prototyp J, 27*(2), 223–244. https://doi.org/10.1108/RPJ-10-2019-0271.

Wang, W. (2021). Synthesis and application of Ag NPs@graphene oxide nanocomposite as an electrochemical sensor for sweat lactate monitoring in sports medicine. *Int J Electrochem Sci, 16*(12), 211232. https://doi.org/10.20964/2021.12.27.

Wang, Z., & Zhang, Y. (2022). Synthesis of Pt/GO composite as an electrochemical sensor for determination of synthetic corticosteroid triamcinolone as doping agents in sports. *Int J Electrochem Sci, 17*(7), 220769. https://doi.org/10.20964/2022.07.67.

Wu, Y., Li, Y., & Zhang, X. (2024). The future of graphene: Preparation from biomass waste and sports applications. *Molecules, 29*(8). https://doi.org/10.3390/molecules29081825.

Xhameni, A., Cheng, R., & Farrow, T. (2022). A precision method for integrating shock sensors in the lining of sports helmets by additive manufacturing. *IEEE Sens Lett, 6*(10), 1–4. https://doi.org/10.1109/LSENS.2022.3205249.

Xie, C. J., & Tian, L. (2021). Application of nano-composites in the recovery of sports ligament injury. *Indian Journal of Pharmaceutical Sciences, 83*, 235–242. https://doi.org/10.36468/pharmaceutical-sciences.spl.320.

Xu, T., Shen, W., Lin, X., & Xie, Y. M. (2020). Mechanical Properties of additively manufactured thermoplastic polyurethane (TPU) material affected by various processing parameters. *Polymers (Basel), 12*(12), 3010. https://doi.org/10.3390/polym12123010.

Yan, C., Zhang, T., Zheng, T., & Mahmoudi, T. (2024). Stability characteristic of bi-directional FG nano cylindrical imperfect composite: Improving the performance of sports bikes using carbon nanotubes. *Steel and Composite Structures, 50*(4), 459–474. https://doi.org/10.12989/scs.2024.50.4.459.

Yang, S., Larionova, T., Kobykhno, I., Klinkov, V., Shalnova, S., & Tolochko, O. (2024). Graphene-Doped thermoplastic polyurethane nanocomposite film-based triboelectric nanogenerator for self-powered sport sensor. *Nanomaterials, 14*(19). https://doi.org/10.3390/nano14191549.

Zeng, W., et al. (2023). Gradient CNT/PVDF piezoelectric composite with enhanced force-electric coupling for soccer training. *Nano Res, 16*(8), 11312–11319. https://doi.org/10.1007/s12274-023-5869-6.

Zhang, J., et al. (2018). Human motion monitoring in sports using wearable graphene-coated fiber sensors. *Sensors and Actuators. A, Physical, 274*, 132–140. https://doi.org/10.1016/j.sna.2018.03.011.

Zhao, C., et al. (2024). Polyacrylonitrile/multi-walled carbon nanotubes/polyurethane electrospun nanofiber membranes for sports equipment. *Advanced Composites and Hybrid Materials, 7*(2). https://doi.org/10.1007/s42114-024-00865-y.

Zhou, J., Lin, C., Lu, Q., & Jiang, J. (2023). Monitoring and management of sunset yellow in sports drinks by electrochemical sensor. *Journal of Food Measurement and Characterization, 17*(5), 4843–4849. https://doi.org/10.1007/s11694-023-02003-6.

Zhu, J., Zhang, Z., Liu, H., Liu, R., Ren, M., & Ma, G. (2024). Poly(vinylidene fluoride-trifluoroethylene)/graphene composite pressure sensors and their potential applications in sports training. *Alexandria Engineering Journal, 106*, 460–466. https://doi.org/10.1016/j.aej.2024.08.070.

Additive manufacturing of functional polymers and nanocomposite for electronics and optoelectronics

16

Vinay Deep Punetha, Rakshit Pathak, Shalini Bhatt, Mayank Punetha, and Parag Sanghani
2D Materials and LASER Actuation Laboratory, Centre of Excellence for Research, PP Savani University, Kosamba, Gujarat, India

16.1 Introduction

The fields of electronics and optoelectronics have their roots in the early 20th century, with foundational discoveries that have since revolutionized technology and modern life. Electronics, which deals with the control and movement of electrons in devices, began with the invention of the vacuum tube by John Ambrose Fleming in 1904. This was followed by Lee De Forest's invention of the triode in 1906, which allowed for amplification of electrical signals, a breakthrough that led to the development of radio and television (Zhou et al., 2023). The next major leap in electronics came with the invention of the transistor in 1947 by John Bardeen, Walter Brattain, and William Shockley at Bell Laboratories. The transistor replaced the bulky vacuum tubes and became the building block of modern electronic devices, leading to the rapid miniaturization and increased power of technology (These et al., 2021). Optoelectronics, a subfield of electronics that focuses on the interaction between light and electronic systems, saw significant advancements in the mid-20th century. The discovery of the photovoltaic effect by Albert Einstein in 1905 laid the groundwork for the development of solar cells. The invention of the laser in 1960 by Theodore Maiman and the subsequent development of light-emitting diodes (LEDs) in the early 1960s by Nick Holonyak marked significant milestones in optoelectronics. The integration of optoelectronic components into electronic systems led to the development of fiber optic communication in the 1970s, revolutionizing telecommunications by enabling the transmission of data over long distances with minimal loss. The advent of semiconductor lasers and photodetectors further advanced the field, allowing for the creation of high-speed, high-capacity communication networks (Kuo et al., 2022; Liu et al., 2016; Wu et al., 2024; Zhou et al., 2022).

The advent of Additive Manufacturing (AM) and the incorporation of nanomaterials have significantly reshaped the fields of electronics and optoelectronics. AM, commonly known as 3D printing, has introduced new possibilities in the design and fabrication of electronic components by allowing for precise, layer-by-layer

Additive Manufacturing of Functional Polymers and Nanocomposites. DOI: https://doi.org/10.1016/B978-0-443-26516-7.00016-X
© 2025 Elsevier Ltd. All rights are reserved, including those for text and data mining, AI training, and similar technologies.

construction of complex structures. This capability has led to the creation of customized electronic devices with intricate geometries that were previously impossible to achieve using traditional manufacturing techniques. The integration of nanomaterials, such as graphene, carbon nanotubes, and quantum dots, into these 3D-printed structures has further revolutionized the industry by enhancing the electrical, thermal, and optical properties of the devices. Nanomaterials enable the production of lightweight, flexible, and highly efficient components, paving the way for advancements in flexible electronics, high-performance sensors, and next-generation photonic devices. Together, AM and nanomaterials have not only reduced production costs and material waste but have also accelerated innovation in electronics and optoelectronics, enabling the rapid prototyping and development of cutting-edge technologies that are transforming industries ranging from healthcare to telecommunications (Kuo et al., 2022; Liu et al., 2016; Podsiadły, Matuszewski et al., 2021; Vidakis et al., 2021; Wu et al., 2024; Zhou et al., 2022).

Recent advancements in the field of AM and nanomaterials have significantly transformed electronics and optoelectronics, offering groundbreaking capabilities that were previously unattainable. For instance, the development of flexible transparent electrodes (FTEs) using polyacrylonitrile (PAN) and copper has evidently revolutionized flexible touchscreens and optoelectronic devices, demonstrating low sheet resistance and high light transmission (Zhou et al., 2023). Interestingly, the incorporation of cesium lead tribromide ($CsPbBr_3$) perovskite nanocrystals in optoelectronics has resulted in exceptional photoluminescence quantum yields and tunable optical properties, thus enhancing the performance of light-emitting diodes (LEDs) and other photonic devices (These et al., 2021). Moreover, the utilization of Molybdenum disulfide (MoS_2) in photodetectors has shown remarkable responsivity, which consequently positions it as a critical material for next-generation wearable and portable electronics (Kuo et al., 2022). Following this, the integration of siloxane and nitrogen-doped carbon dots (N-CDs) into LEDs has allowed for the fabrication of efficient blue light-emitting diodes, therefore broadening their applicability in bioimaging and optoelectronics (Zhou et al., 2022). Furthermore, the use of konjac glucomannan (KGM) in 3D-printed room temperature phosphorescence (RTP) materials has led to the development of innovative anticounterfeiting technologies, which underscores the versatility of AM in advanced security applications (Wu et al., 2024). In addition, the application of silver nanomeshes in transparent conductive electrodes has proven essential for the durability and flexibility required in touchscreens and polymer-dispersed liquid crystal (PDLC) devices, demonstrating high transmittance and low sheet resistance, which are critical for the efficiency and functionality of these optoelectronic devices (Liu et al., 2016).

This chapter is crucial as it bridges the rapidly evolving fields of electronics and optoelectronics, both of which are foundational to modern technology. By focusing on the integration of AM and nanomaterials, the chapter addresses the growing need for innovative solutions that enhance device performance, flexibility, and functionality. The significance of this work for academia lies in its potential to inspire new research avenues, providing a comprehensive overview of the latest advancements and their practical applications. For the research community, this chapter serves as a valuable resource, offering insights into cutting-edge materials and techniques that could revolutionize the development of next-generation electronic and optoelectronic devices. This synthesis of knowledge is not only relevant for ongoing academic research but also has profound implications for industrial applications, from flexible electronics to high-performance photonic systems.

16.2 Scope of AM in electronics

Electronic devices have become integral to modern life, encompassing a wide range of applications from consumer electronics like smartphones and wearable devices to advanced systems in aerospace, automotive, healthcare, and communications. These devices are characterized by their ability to process, store, and transmit information through electrical signals, enabling everything from basic data computation to complex operations in robotics and artificial intelligence (Nassar et al., 2023; Podsiadły, Matuszewski et al., 2021). The rapid miniaturization and enhanced performance of electronic devices have been driven by continuous advancements in materials science, semiconductor technology, and manufacturing techniques. Traditionally, electronic devices have been manufactured using subtractive processes, where material is removed from a larger block to achieve the desired form. This approach, while effective, often involves multiple steps, extensive material waste, and limitations in creating complex geometries. Furthermore, as electronic devices become increasingly multifunctional and integrated with various components, the need for more sophisticated manufacturing techniques has become evident. The demand for flexible, lightweight, and high-performance electronic devices has pushed the boundaries of conventional manufacturing, leading to the exploration of new methods that can meet the evolving requirements of the industry (Liu et al., 2022a; Peng et al., 2021). Interestingly, the incorporation of cesium lead tribromide ($CsPbBr_3$) perovskite nanocrystals in optoelectronics has resulted in exceptional photoluminescence quantum yields and tunable optical properties, thus enhancing the performance of light-emitting diodes (LEDs) and other photonic devices (Fig. 16.1) (These et al., 2021). The normalized absorbance spectra for crystalline $CsPbBr_3$ powder and aerosol deposition (AD) film are presented in Fig. 16.1A. The AD films were illuminated from either the front side (top of the AD film) or the backside (through the glass substrate). The observed deviation at wavelengths below 500 nm in the backside illumination is attributed to the light passing through the glass substrate. Both the $CsPbBr_3$ powder and AD film display a distinct absorption onset below 550 nm, characteristic of $CsPbBr_3$. Notably, a slight blue shift in the absorption onset is observed for the AD films, indicating an increase in bandgap.

Moreover, the utilization of Molybdenum disulfide (MoS_2) in photodetectors has shown remarkable responsivity, which consequently positions it as a critical material for next-generation wearable and portable electronics (Kuo et al., 2022). Since MoS_2 transitions from an indirect to a direct bandgap at the monolayer level, making it ideal for optoelectronic applications in the atomically thin limit, the fabrication of fully printed photodetectors was pursued to demonstrate the effectiveness of megasonically

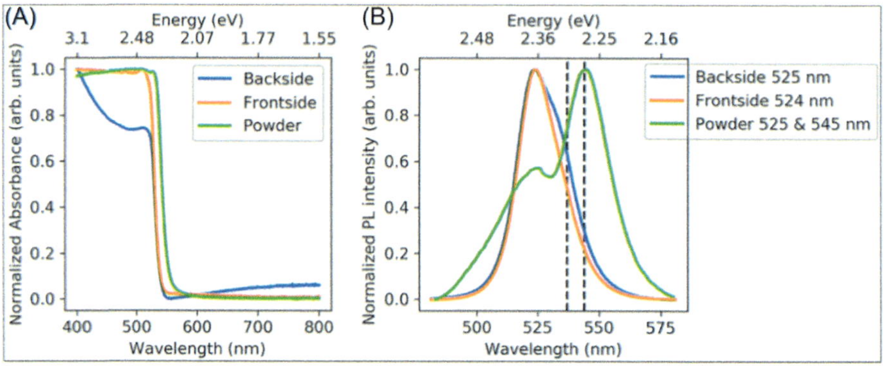

Figure 16.1 Two key characterizations of $CsPbBr_3$ materials: (A) the normalized absorbance spectra for $CsPbBr_3$ powder and AD-processed films, and (B) the normalized photoluminescence (PL) spectra of $CsPbBr_3$ powder and AD films under 402.5 nm excitation. The PL peak wavelengths are indicated in the legend, and two black dashed lines correspond to the bandgap energies, measured at 2.28 eV for the powder and 2.31 eV for the AD film, respectively. Adapted from These et al. (2021) under CC BY-NC-ND 4.0 license, © 2021, Wiley.

exfoliated MoS_2. Aerosol jet printing (AJP) was employed to deposit megasonically exfoliated MoS_2 as the channel material between AJP-printed graphene electrodes (Fig. 16.2A). Despite the significant background signal from the polyimide substrate, GIWAXS clearly reveals features corresponding to the MoS_2 (100), (101), (102), and (103) planes, indicating uniaxial texture along the c-axis and confirming the deposition of flat, stacked MoS_2 nanosheets (Fig. 16.2B). This face-on deposition of thin, laterally extensive MoS_2 nanosheets is achieved by incorporating terpineol as a high-boiling-point solvent in the AJP MoS_2 ink formulation (Fig. 16.2C). In the absence of terpineol, the MoS_2 nanosheets crumple into ball-like structures and exhibit significant overspray due to dry printing, as shown by AFM analysis (Fig. 16.2D).

AM, commonly known as 3D printing, has emerged as a transformative technology in the production of electronic devices, offering unprecedented design freedom and the ability to create complex structures with high precision. Unlike traditional subtractive manufacturing, AM builds objects layer by layer, using only the necessary amount of material, which significantly reduces waste and allows for the integration of multiple functions within a single device. This approach has opened new avenues for the fabrication of electronic components, particularly in applications where traditional methods fall short. One of the most significant impacts of AM on electronics is the ability to produce customized components with complex geometries that were previously unattainable. For example, AM enables the fabrication of flexible circuits, antennas, and sensors that can be integrated into wearable devices, biomedical implants, and other applications requiring conformal shapes and lightweight materials (Liu et al., 2024a; Okoruwa et al., 2023; Rosh-Gorsky et al., 2024). Additionally, AM has facilitated the development of multifunctional devices by allowing the combination of various materials, such as polymers, metals, and nanocomposites, within a single printing process. This capability has led to the creation of devices with enhanced

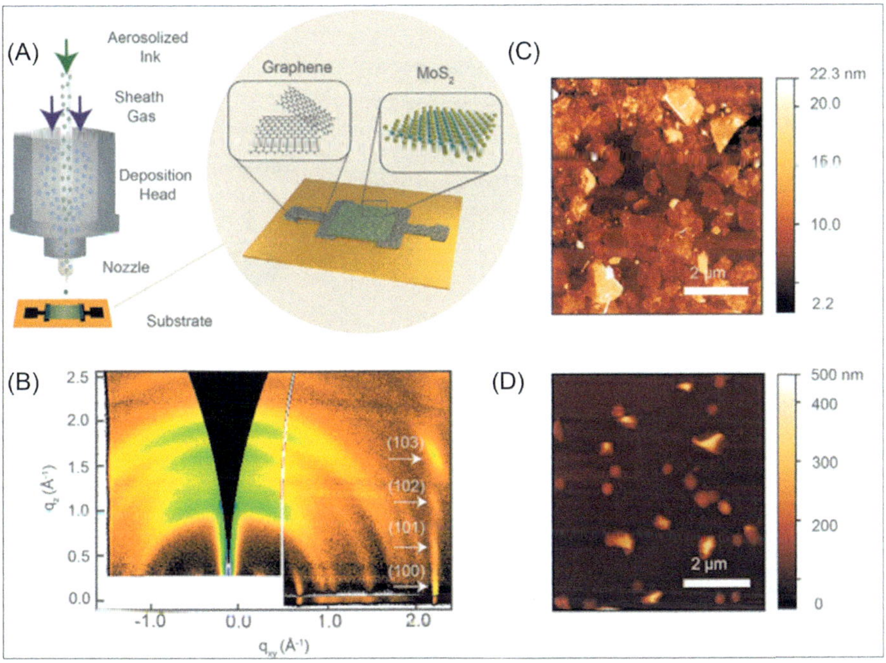

Figure 16.2 (A) Schematic illustrating aerosol jet printing (AJP) deposition and the geometry of the printed photodetector. (B) Grazing-incidence wide-angle X-ray scattering (GIWAXS) map of a printed MoS_2 film, revealing uniaxial texture along the *c*-axis, indicating flat, stacked flakes. The vertical white region represents a gap in the detector coverage, and corresponding horizontal gaps were corrected by averaging two vertically displaced images. The white area in the lower left corner is a gap in the detector (missing module), which allows transmission of the beam to a downstream small-angle X-ray scattering detector. (C) Atomic force microscopy (AFM) topography image of MoS_2 deposited by AJP with terpineol included in the ink formulation, showing flat MoS_2 flakes. (D) AFM topography image of MoS_2 deposited by AJP without terpineol, displaying crumpled MoS_2 flakes, resulting in a 20-fold wider height range. Adapted from Kuo et al. (2022) under CC BY-NC 4.0 license, © 2022, Wiley.

properties, such as improved electrical conductivity, thermal management, and mechanical strength. Moreover, AM has accelerated the prototyping and production processes in electronics, enabling rapid design iterations and reducing the time from concept to market. This speed is particularly beneficial in industries where innovation cycles are short, and the ability to quickly test and refine designs can provide a competitive advantage. AM also supports the localized production of electronic components, reducing supply chain complexities and allowing for on-demand manufacturing, which is particularly valuable in the context of personalized electronics and small-batch production (Agrawal et al., 2024; Carcione et al., 2023; Li et al., 2023; Podsiadły, Matuszewski et al., 2021). Fig. 16.3 illustrates the integration of electrochemistry and additive manufacturing, highlighting the scope of application for both conductive and nonconductive materials.

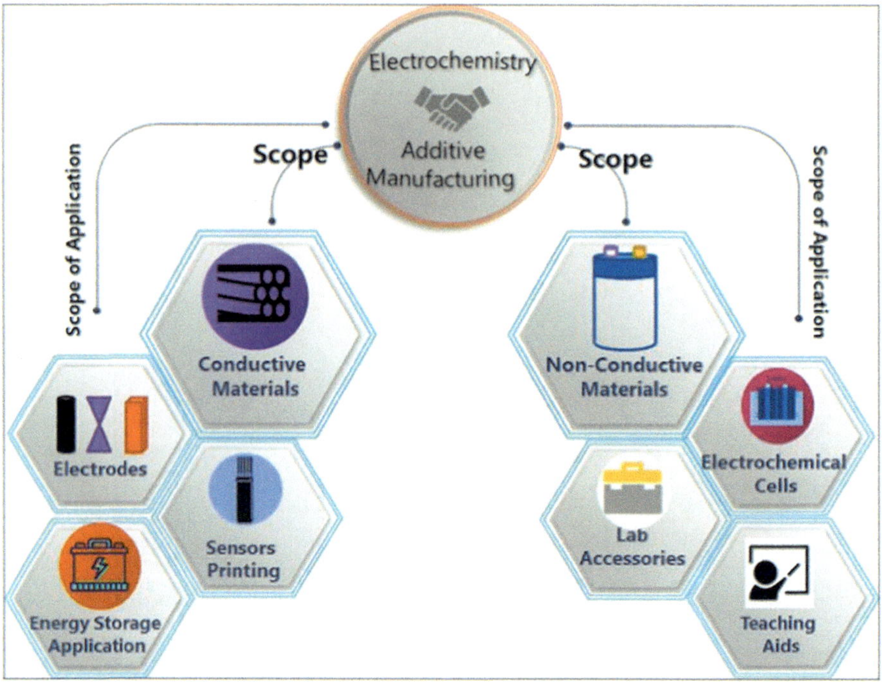

Figure 16.3 The image illustrates the integration of electrochemistry and additive manufacturing, highlighting the scope of application for both conductive and nonconductive materials. Conductive materials are applied in areas such as electrodes, sensors printing, and energy storage applications, while nonconductive materials are used for electrochemical cells, lab accessories, and teaching aids.
Adapted with permission from Rajesh & Kumawat (2024). https://doi.org/10.1007/s11581-023-05325-z, © 2023, Springer Nature.

16.2.1 Polymeric materials for additive electronics

Polymers in AM for electronics can be broadly categorized into three main types: thermoplastic polymers, thermosetting polymers, and elastomers. Each of these categories offers unique properties that make them particularly suited for specific applications within the field of electronics and optoelectronics. Thermoplastic polymers are of great significance for their versatility and recyclability, making them ideal for creating durable and complex electronic components. Thermosetting polymers, on the other hand, are prized for their high thermal stability and mechanical strength, which are essential for high-performance applications that require long-lasting and robust materials. Lastly, elastomers, with their exceptional flexibility and resilience, are indispensable in the development of wearable electronics and flexible sensors that need to withstand repeated mechanical stress. This categorization highlights the diverse roles that different types of polymers play in advancing the capabilities and functionalities of electronic devices through AM. Thermoplastic polymers are materials that soften when heated and harden upon cooling, allowing them to be reshaped and reused multiple times. These polymers are highly valued in AM due to their versatility, recyclability, and ease of processing. Polylactic acid (PLA) is one of the most used thermoplastic polymers in AM, particularly for applications in electronics and optoelectronics. PLA is derived from renewable resources such as corn starch, making it an environmentally friendly choice. Its biocompatibility and ease of processing make it ideal for producing complex structures in 3D printing. In electronics, PLA is often reinforced with materials like multiwall carbon nanotubes (MWCNTs) to enhance its mechanical properties and electrical conductivity. This combination is particularly useful for fabricating sensors, where the mechanical robustness of the polymer and the conductivity of the nanotubes are critical. The use of PLA in electronic applications highlights its potential in creating sustainable and functional devices (Vidakis et al., 2021). Acrylonitrile butadiene styrene (ABS) is another widely used thermoplastic polymer in AM. ABS is known for its toughness, impact resistance, and ability to produce durable parts, making it suitable for structural components in electronic devices. When combined with carbon nanotubes (CNTs), ABS exhibits improved tensile strength and thermal properties, which are essential for applications in structural electronics and sensor fabrication (Fig. 16.4). The compatibility of ABS with various reinforcement materials and its ability to maintain structural integrity under different conditions make it a preferred choice for developing robust electronic components (Podsiadły, Matuszewski et al., 2021).

Thermoplastic Polyurethane (TPU) is renowned for its flexibility, durability, and resistance to wear, making it an excellent material for wearable electronics and flexible sensors. TPU's ability to stretch and return to its original shape without

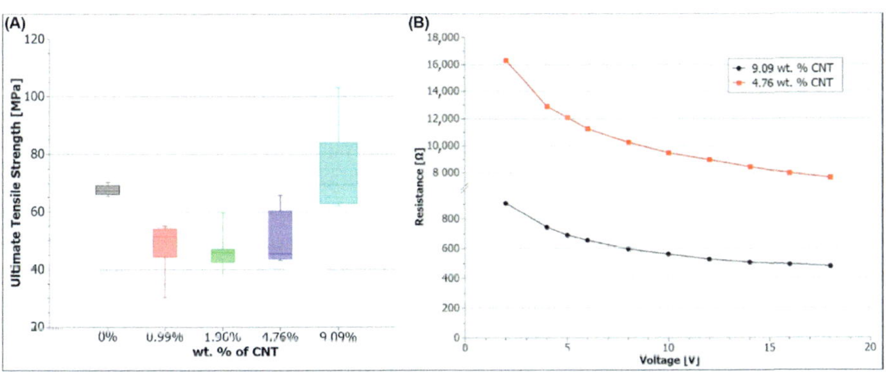

Figure 16.4 Comparison of ultimate tensile strength values for composite samples with varying CNT loadings (ranging from 0.99 to 9.09 wt.%) and ABS reference samples (0 wt.% CNTs). The tensile test results reveal that the ABS/CNT composites exhibit a decrease in tensile strength for CNT content below 5 wt.%. However, the samples with 9.09 wt.% CNTs demonstrated higher tensile strength compared to pure ABS samples.
Adapted from Podsiadły, Matuszewski, et al. (2021) under CC BY 4.0 license, © 2021, MDPI.

deformation is particularly advantageous in the development of strain sensors (Nassar et al., 2023). Its electrical conductivity is significantly enhanced, while maintaining its inherent flexibility. This property is crucial for the creation of stretchable electronics, where the material must endure repetitive mechanical stress without compromising its electrical performance. TPU's adaptability makes it a key material in the field of flexible and wearable electronics (Nassar et al., 2023). Thermosetting polymers are materials that irreversibly cure to form a solid structure when heated or chemically treated. Unlike thermoplastics, once they are set, they cannot be remelted or reshaped. These polymers are known for their high thermal stability, mechanical strength, and resistance to deformation, making them suitable for high-performance applications in electronics. Polyether Ether Ketone (PEEK) is a high-performance thermosetting polymer that is widely used in the electronics industry for applications that require exceptional thermal stability and mechanical properties. PEEK is often reinforced with carbon fibers to further enhance its strength and stiffness, making it suitable for structural components that must withstand harsh environmental conditions, such as those found in aerospace and automotive electronics. The use of PEEK in AM allows for the creation of complex, durable components that can endure high temperatures and mechanical stress, making it an invaluable material in the production of high-performance electronic devices (Kumar et al., 2024). Photopolymer epoxy resins are widely used in AM for creating high-precision components with excellent mechanical and thermal properties. These resins are typically used in stereolithography (SLA) processes, where they are cured layer by layer using UV light. The addition of nanomaterials like MWCNTs or nanoclay to epoxy resins enhances their electrical conductivity and mechanical strength, making them suitable for applications in electronics, aerospace, and medical devices. The ability of epoxy resins to form strong, durable bonds and their compatibility with various reinforcement materials make them a popular choice for producing high-performance composites in AM (Alshihabi & Kayacan, 2024).

Another polymer, polydimethylsiloxane (PDMS), which is a silicone-based thermosetting polymer known for its flexibility, biocompatibility, and resistance to extreme temperatures (Liu et al., 2022b). PDMS is often used in the fabrication of flexible and stretchable electronics, particularly in applications requiring the material to conform to irregular surfaces or undergo repeated deformation. The integration of carbon nanotubes into PDMS enhances its electrical conductivity while maintaining its flexibility, making it ideal for strain sensors and other wearable electronic components. Mechanical performance of fabricated CNT-PDMS compressive strain stressor is outlined in Fig. 16.5. Cyclic compressive testing was conducted at 50% strain for up to 100 cycles on samples with varying CNT (Carbon Nanotube) concentrations ranging from 0–5 wt%. As depicted in Fig. 16.5A, the foam-like scaffold structure demonstrated remarkable elasticity, compressing to 50% strain and fully recovering to its original shape. The stress-strain curve for the 4 wt% CNT sample, illustrated in Fig. 16.5B, revealed the presence of hysteresis loops, a characteristic seen in open-cell foams. The linear behavior observed between 0% and 25% strain corresponds to the linear elastic and plateau regions, with a Young's modulus of approximately 700 kPa. This phase is attributed to the compression and deformation of microvoid walls.

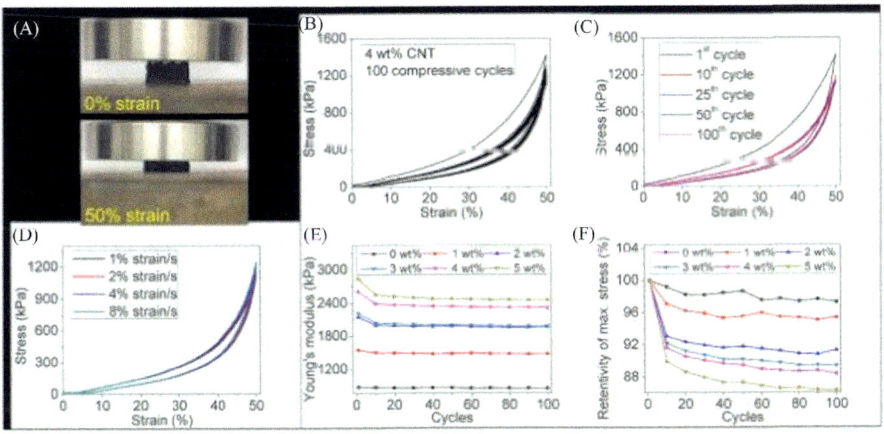

Figure 16.5 (A) Compressive testing was performed starting from the origin and at 50% strain. The stress–strain curves for the 4 wt% CNT sample are shown in (B), representing the complete 100 cycles. (C) Highlights the specific cycles for the 1st, 10th, 25th, 50th, and 100th cycles. (D) The stress–strain behavior during the 6th to 10th cycles, compressed at different strain rates of 1%, 2%, 4%, and 8% s^{-1}. (E) The graph illustrates how the Young's modulus of samples with 0–5 wt% CNT changes over the course of 100 cycles. Lastly, (F) displays the variation in the retentivity of maximum stress for 0–5 wt% CNT samples throughout 100 compressive cycles. Adapted with permission from Liu et al. (2022b). © 2021, Wiley.

Between 25% and 50% strain, a densification region with a sharply rising slope is evident, resulting from the overcompaction of microvoids. As shown in Fig. 16.5C, the hysteresis loop from the first compression cycle indicates a slight strain loss of 3.5%, likely due to plastic deformation. Subsequent cycles—10th, 25th, 50th, and 100th—show no further plastic deformation, highlighting the structural durability of the material. Fig. 16.5D further illustrates the overlapping hysteresis loops for samples tested at strain rates between 1% and 8% s^{-1}, indicating exceptional resilience during rapid deformation. The effect of CNT concentration on mechanical performance is demonstrated in Fig. 16.5E and F. Both Young's modulus and maximum stress values remained stable throughout 100 compressive cycles. The incorporation of CNT significantly enhanced the composite's mechanical strength, with Young's modulus increasing from ~800 kPa in the 0 wt% CNT sample to ~2500 kPa in the 5 wt% CNT sample. However, the maximum stress decreased from ~96% in the 0 wt% CNT sample to ~86% in the 5 wt% CNT sample, likely due to the stiffening effect introduced by the CNT structure.

PDMS's unique combination of properties makes it an essential material in the development of advanced flexible electronics that can operate reliably under a wide range of conditions (Liu et al., 2022b). Elastomers are third major type of polymers that are used in electronic devices. Elastic properties allow reversible deformation in shape. They are widely used in applications where flexibility and resilience are required, particularly in the development of wearable electronics and sensors. Polydimethylsiloxane (PDMS), as mentioned earlier, is not only a thermosetting polymer but also an elastomer due to its rubber-like flexibility. This makes it a popular choice for developing soft, flexible electronics, especially in applications involving human-machine interaction or biomedical devices. The addition of conductive fillers like carbon nanotubes enables PDMS to function as a strain sensor, capable of detecting both large- and small-scale movements with high sensitivity. The combination of PDMS's flexibility and electrical conductivity makes it an ideal material for creating stretchable electronic devices that can endure repeated mechanical stress (Liu et al., 2022b). Thermoplastic Polyurethane (TPU), while also classified as a thermoplastic, functions as an elastomer due to its exceptional elasticity and durability. TPU's ability to stretch and recover without permanent deformation makes it an excellent material for flexible and wearable electronics. Its application in strain sensors and flexible circuit elements highlights TPU's versatility and importance in the field of flexible electronics. The incorporation of silver nanoparticles or MWCNTs further enhances TPU's electrical properties, making it suitable for applications that require both flexibility and conductivity (Nassar et al., 2023) (Fig. 16.6). Natural Rubber is another elastomer used in AM for creating flexible electronic devices. Natural rubber is a renewable resource, making it an environmentally friendly choice for developing sustainable electronics. When combined with conductive fillers like acetylene black, natural rubber can be used to create piezoresistive sensors that are capable of detecting small strains, making it ideal for applications in wearable electronics and human motion monitoring. The use of natural rubber in electronic devices demonstrates the potential for creating flexible, sustainable electronics that are both functional and environmentally responsible (Georgopoulou et al., 2023).

16.2.2 Nanomaterials for additive electronics

The transformative potential of nanomaterials in various applications especially as value added composites (Jahangir et al., 2024; Kumar et al., 2023; Punetha et al., 2017, 2022; Punetha et al., 2019; Punetha et al., 2020; Ranjan et al., 2024) have shown tremendous potential in a wide range of applications (Dhapola et al., 2024). The use of nanomaterials in AM for electronics has opened new avenues for innovation, significantly enhancing the performance and capabilities of electronic devices. These nanomaterials can be broadly classified into various types, including organic, inorganic, carbon-based, and hybrid composites, each bringing unique properties that address specific challenges in electronics. Interestingly, these nanomaterials not only revolutionize traditional manufacturing techniques but also enable the creation of devices with unprecedented precision and functionality. Organic nanomaterials, such as polyaniline (PANI), have shown great promise in the development of flexible and soft electronics. For instance, PANI, when combined with polyethylene glycol diacrylate (PEGDA), forms composites that are highly conducive to applications in biomedical devices, particularly in pH monitoring sensors. This combination leverages the electrochemical activity of PANI and the flexibility of PEGDA, resulting in materials that are both functional and comfortable for wearable applications (Fig. 16.7) (Montaina et al., 2023).

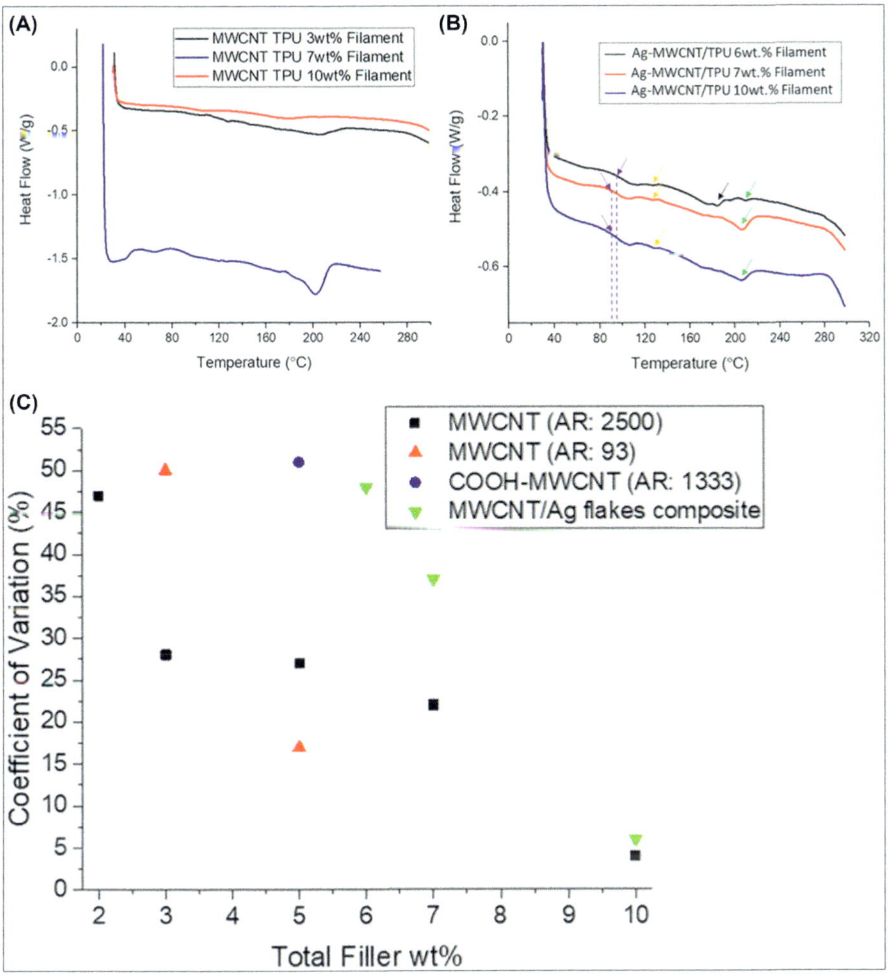

Figure 16.6 (A, B) Differential scanning calorimetry (DSC) graphs illustrating the effect of increasing filler concentration on MWCNT–TPU and Ag–MWCNT–TPU filaments. For MWCNT/TPU filaments, increasing the filler concentration had minimal impact on the temperature at which the primary endothermic events occurred. Similarly, for Ag–MWCNT/TPU filaments, higher filler concentrations slightly shifted the main endothermic events to lower temperatures. (C) Coefficient of variation (CV) for the electrical conductivity values of the filaments, averaged over five locations. A general trend of decreasing CV values is observed as filler concentration increases, indicating more uniform conductivity. The highest conductivity filaments, high aspect ratio MWCNT/TPU and 10 wt.% Ag–MWCNT/TPU, exhibited CV values of 4% and 6%, respectively, suggesting the formation of a complete filler network with conductive branches at higher filler concentrations.
Adapted from Nassar et al. (2023) under CC BY 4.0 license, © 2023, Wiley.

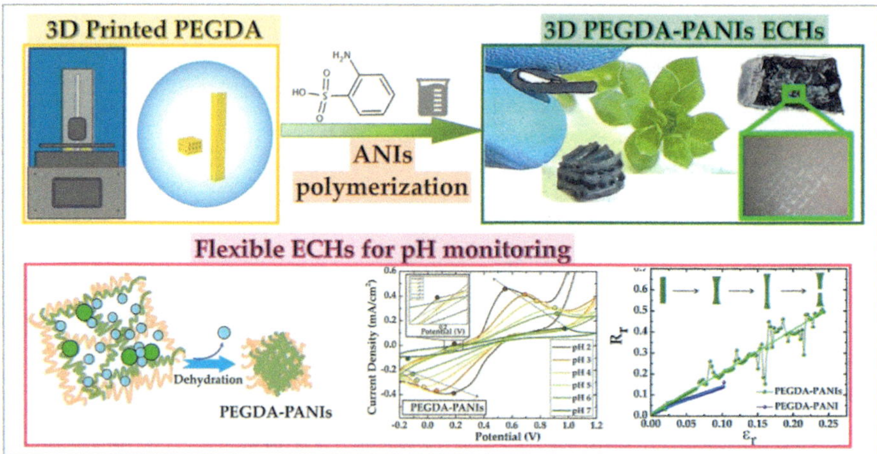

Figure 16.7 Schematic illustration of 3D-printed PEGDA structures and the subsequent polymerization of aniline derivatives (ANIs) to create flexible 3D PEGDA-PANI electrochemical hydrogels (ECHs). These flexible ECHs are designed for pH monitoring applications. The lower panels show the dehydration process in PEGDA-PANI$_s$ structures and their electrochemical performance, including current density versus potential curves at different pH levels and the relationship between relative resistance (R_t) and swelling ratio (ε_t) for PEGDA-PANI$_s$ and PEGDA-PANI.
Adapted from Carcione et al. (2023) https://doi.org/10.3390/gels9100784 under CC BY 4.0 license, © 2021, MDPI.

On the other hand, inorganic nanomaterials like tungsten and boron nitride (BN) are employed for their exceptional ability to manage radiation and heat. Tungsten's high atomic number makes it a powerful shield against radiation, while boron nitride contributes to thermal management due to its high thermal conductivity. This combination is particularly significant in the protection of microelectronic systems in space, where radiation can severely damage electronic components. Thus, the integration of these inorganic nanomaterials into electronic devices ensures both their durability and performance in extreme environments (Beck et al., 2023). In addition to organic and inorganic nanomaterials, carbon-based nanomaterials play a critical role in advancing AM for electronics. Carbon nanotubes (CNTs), for example, are frequently used to enhance the electrical and mechanical properties of polymers such as acrylonitrile butadiene styrene (ABS) and polyethylene glycol diacrylate (PEGDA. The incorporation of CNTs into these polymers leads to significant improvements in tensile strength, electrical conductivity, and thermal stability, making them suitable for a wide range of applications, from structural elements in electronics to sensors in biomedical devices (Podsiadły, Matuszewski et al., 2021) (Singh et al., 2024). Moreover, graphene, another carbon-based material, has been extensively studied for its exceptional electrical and thermal properties. The development of molecule-bridged graphene/Ag (MB-G/A) composites is a prime example of how graphene can be used to create cost-effective and highly conductive materials for printable electronics. The molecular bridges between graphene and silver enhance electron transport, resulting in inks that are not only cheaper but also more efficient for large-scale manufacturing of electronic components (Li et al., 2022). Casaluci et al. developed a method to produce graphene-based ink by exfoliating graphite in dimethylformamide using liquid-phase exfoliation (DOI; 10.1039/x0xx00000x). They employed a spray-coating technique to deposit this ink onto a transparent conductive oxide substrate, creating a counter-electrode with a large surface area of 90 cm² and a transparency of 44%, as depicted in Fig. 16.8. This graphene-based counter-electrode was then integrated into a dye-sensitized solar cell module with an active area of 43.2 cm², achieving a power conversion efficiency of 3.5%. Their research demonstrates the potential for graphene-based inks in the fabrication of fully printed, flexible, transparent, large-area, and cost-efficient photovoltaic devices on various substrates.

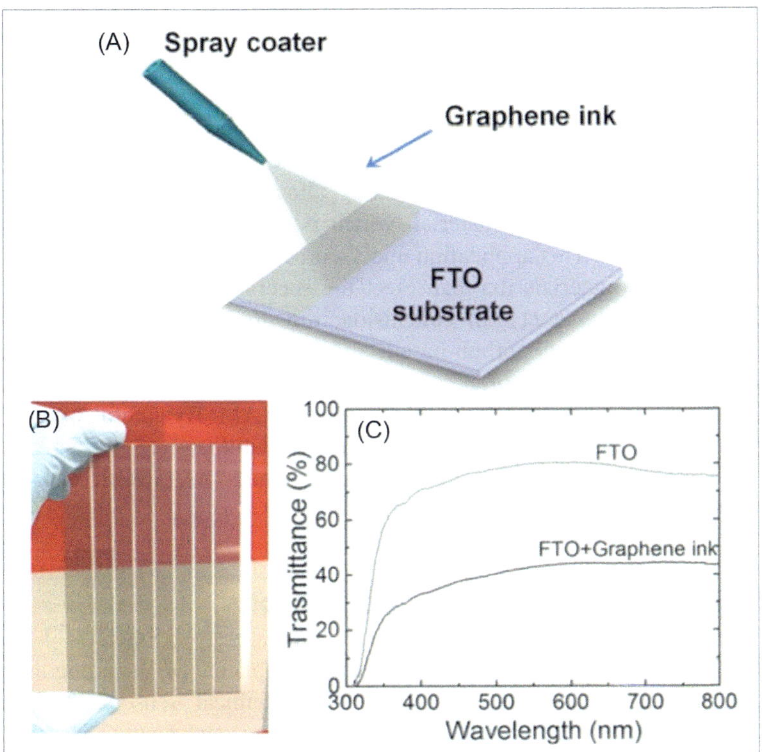

Figure 16.8 (A) Application of graphene ink via spray coating, where the ink is sprayed onto the FTO substrate using an air-assisted spray gun. (B) Optical image of the graphene-coated counter-electrode. (C) Optical transmittance comparison between the bare FTO substrate (gray curve) and the graphene-coated FTO substrate (black curve).
Adapted with permission from Yang et al. (2014) © 2016, Royal Society of Chemistry.

Hybrid nanomaterials, which combine organic, inorganic, and carbon-based components, offer a versatile approach to optimizing the properties of electronic devices. For example, nanoclay and CNTs are used together in photo-polymer epoxy resins to create composites that exhibit superior electrical and mechanical properties. The inclusion of nanoclay improves the thermal stability and mechanical strength of the resin, while CNTs enhance its electrical conductivity. This synergy between different classes of nanomaterials results in composites that are well-suited for high-performance applications in aerospace and electronics (Kasraie et al., 2021). In addition to these, functionalized submicron alumina particles are utilized in oligourethane composites to improve thermal management in printed electronics. These particles are specifically engineered to maintain a balance between mechanical integrity and thermal performance, ensuring that electronic devices can operate efficiently even under high thermal loads. This is particularly important in applications where precise thermal management is critical, such as in the cooling of electronic components (Graf et al., 2021). Silver nanowires, embedded in ethylene-co-vinyl acetate (EVA), demonstrate another application of nanomaterials in creating highly sensitive pressure sensors. The high aspect ratio of silver nanowires ensures that these sensors maintain their conductivity even under significant mechanical stress, making them ideal for robotics and health monitoring applications (Montaina et al., 2023). Finally, the use of long fibers in silicone elastomers, fabricated through Automated Fiber Embedding (AFE) techniques, showcases the potential of nanomaterials in soft robotics. These composites allow for precise control over the mechanical properties of the material, which is crucial for developing responsive and durable actuators. This application underscores the versatility of nanomaterials in creating advanced materials that can meet the specific needs of modern electronic devices (Stalin et al., 2021). In conclusion, the diverse range of nanomaterials, from organic to inorganic, carbon-based, and hybrid composites, demonstrates the extensive possibilities they offer in enhancing the performance, durability, and functionality of electronic devices produced through AM. By carefully selecting and integrating these nanomaterials into polymers, manufacturers can create highly specialized materials that meet the stringent demands of modern electronic and optoelectronic applications.

16.3 AM and electronic applications

AM has shown promising results across a range of applications, including the creation of highly conductive circuits, enhanced electrochemical sensors, and advanced materials for thermal management, radiation shielding, and energy harvesting. Key areas of impact include the fabrication of flexible and durable electronic components, the development of wearable devices with enhanced sensor capabilities, and the integration of novel materials into 3D-printed structures for improved performance in high-frequency electronics and environmental monitoring. These advancements underscore the critical role of AM in pushing the boundaries of electronic device fabrication, enabling new possibilities in fields such as biomedicine, robotics, and environmental protection. The quantitative results obtained from integrating various polymers and nanomaterials in AM for electronics reveal significant advancements in the field, with several studies offering comparative insights into material performance. Quantitative results in the study of optoelectronic devices provide essential insights into the performance and reliability of the materials and techniques used. These metrics not only offer a concrete evaluation of material properties but also signify their potential for revolutionizing the field of optoelectronics. In order to understand the quantitative results given in the table, it is important to grasp some key terms that can help in rationalizing the performance and characteristics of the materials and devices described. One such term is resistivity (Ω·m), which measures how strongly a material opposes the flow of electric current. A low resistivity value indicates excellent electrical conductivity, which is crucial for electronic applications like circuits and electrodes. For instance, the low resistivity of ultraviolet-curable resins embedded with $Cu_2(OH)PO_4$, measured at 7.5×10^{-8} Ω·m, highlights the material's efficiency in conducting electricity, making it highly suitable for advanced electronic devices (Iffelsberger et al., 2020). Another critical term is transmittance (%), which refers to the percentage of light that passes through a material. High transmittance is essential in optoelectronics, particularly for devices like flexible touch screens and transparent electrodes, where both transparency and electrical performance are needed. An example from the table is the flexible transparent electrodes made from polyacrylonitrile (PAN) and copper, which demonstrate a transmittance of 89% at 550 nm, signifying their effectiveness in maintaining optical clarity while also conducting electricity (de Oliveira et al., 2020). Photoluminescence Quantum Yield (PLQY) is also a significant term, especially for materials used in light-emitting devices like quantum dots and perovskites. PLQY measures the efficiency of photon emission from a material when it is excited by light, with higher PLQY values indicating better luminescent efficiency. For example, CsPbBr3 nanocrystals that achieve a PLQY boost from 15.4% to approximately 60% demonstrate enhanced luminescence, making them promising for high-performance LED applications (Cardoso et al., 2020). Sheet resistance (Ω/sq) is crucial for evaluating thin conductive films, which are used in flexible transparent electrodes and sensors. It represents the resistance of a square-shaped section of the material, and low sheet resistance, combined with high transmittance, is desirable. For instance, P4VP-based flexible transparent electrodes with a sheet resistance of 2.5 Ω/sq and 93% transmittance are highlighted in the table for their balance of conductivity and optical transparency, making them ideal for flexible optoelectronic devices (Koterwa et al., 2022). Photoresponsivity (A/W) is an essential parameter for photodetectors, indicating how efficiently a material converts light into electrical signals. Higher photoresponsivity means better performance in detecting light, which is vital for applications in sensors and other optoelectronic devices. Understanding these terms is fundamental in rationalizing the quantitative results presented in the table, offering insights into the material properties and potential applications of these advanced technologies (Vaghasiya et al., 2021). For instance, ultraviolet-curable resins embedded with $Cu_2(OH)PO_4$ demonstrated an impressively low resistivity of 7.5×10^{-8} Ω·m, indicative of a highly efficient conduction pathway in the manufactured circuits. Additionally, these resins achieve the highest adhesion property rating of 5B level, underscoring their excellent bonding strength and reliability for advanced electronic devices (Liu et al., 2022a). This level of performance is crucial in reducing energy losses and enhancing device efficiency, particularly in high-performance applications where reliability and durability are paramount. Similarly, the performance of polylactic acid (PLA) and thermoplastic polyurethane (TPU) reinforced with continuous carbon fibers demonstrates significant mechanical enhancements. These composites exhibit a notable tensile modulus improvement, reaching 138.5% higher than pure PEKK, along with a significant weight-saving potential of 21%, which is particularly beneficial in applications such as biomedicine, robotics, and flexible electronics (Ye et al., 2024). This can be compared to the enhancements observed in ABS polymers reinforced with carbon nanotubes (CNTs), where tensile strength improved by 12.6%, and average resistivity values decreased significantly, making these composites ideal for structural electronics and sensor applications (Podsiadły, Matuszewski et al., 2021). Both studies highlight the critical importance of enhanced mechanical properties and reduced weight in advancing flexible and robust electronic applications. Moreover, when examining composites for thermal management, various thermoplastic composites containing boron nitride (BN) and MgO particles have achieved an in-plane thermal conductivity of 1.97 W/m/K, alongside excellent electrical insulation and printability, making them ideal for lightweight 3D-printed heat sinks (Bagatella et al., 2024). Similarly, carbon-filled polymer filaments incorporating metal-filled and carbon-filled thermal conductive composites have shown superior heat dissipation performance, with a 28% improvement observed at a 90° orientation (Peng et al., 2022). The significance of these results lies in their implications for thermal management in electronics, where efficient heat dissipation is critical to maintaining performance and preventing overheating. In the realm of radiation shielding and electromagnetic interference (EMI) protection, styrene-isoprene-styrene block copolymer composites with various particles of different atomic densities have demonstrated tailored radiation attenuation for microelectronic protection in satellites (Rosh-Gorsky et al., 2024). This finding resonates with the study on nickel-plated acetate fabrics combined with carbon nanotubes and silk sericin, which exhibited EMI shielding effectiveness greater than 30 dB at the X-band frequency, alongside higher tensile strength and improved performance under strain (Duan & Lu, 2021). Both studies underscore the importance of composite materials in protecting delicate electronic components from environmental hazards, whether in space or in everyday applications. The advancements in piezoresistive sensor technology are exemplified by the development of natural rubber filled with acetylene black, which can detect small strains of less than 10% with enhanced mechano-electrical properties (Georgopoulou et al., 2023). This capability is crucial for applications in human motion monitoring and wearable electronics, where precise and reliable feedback is necessary. This can be juxtaposed with the work on flexible polyurethane (FPU) combined with multiwall carbon nanotubes (MWCNTs) and polypyrrole (PPy), which has been optimized using machine learning models to enhance the sensitivity of strain sensors (Poompiew et al., 2024). Both studies highlight the ongoing efforts to improve sensor sensitivity and durability, critical for advancing wearable technology. In the domain of dielectric composites, polyvinylidene fluoride (PVDF) combined with hollandite-like ceramics and multiwall carbon nanotubes (MWCNTs) demonstrates an optimal combination of permittivity and dielectric loss values, making them ideal for flexible electronics where maintaining device integrity and functionality is critical (Tsyganov et al., 2022). In this study Fe-doped potassium titanate with hollandite-like structure (KFTO(H)), MWCNT and PVDF were used to fabricate KFTO(H)/MWCNT/PVDF composites according to the methodology presented in Fig. 16.9. Initially, mixtures of KFTO(H) and MWCNT powders were prepared with different MWCNT concentrations (0, 0.5, 1.0, and 1.5 wt. %). These powders were homogenized and ground using a vibrating micro-mill (Fritsch Pulverisette 0) for 1 hour. The powder mixtures were then integrated into the PVDF polymer matrix at concentrations of 5.0, 7.5, 15, and 30 vol.%, calculated based on the theoretical densities of KFTO(H) (3.7 g/cm³) and PVDF (1.11 g/cm³). The procedure involved dissolving PVDF granules in DMF with a mass ratio of PVDF:DMF= 5:95. The KFTO(H)/MWCNT powder mixtures were added to this

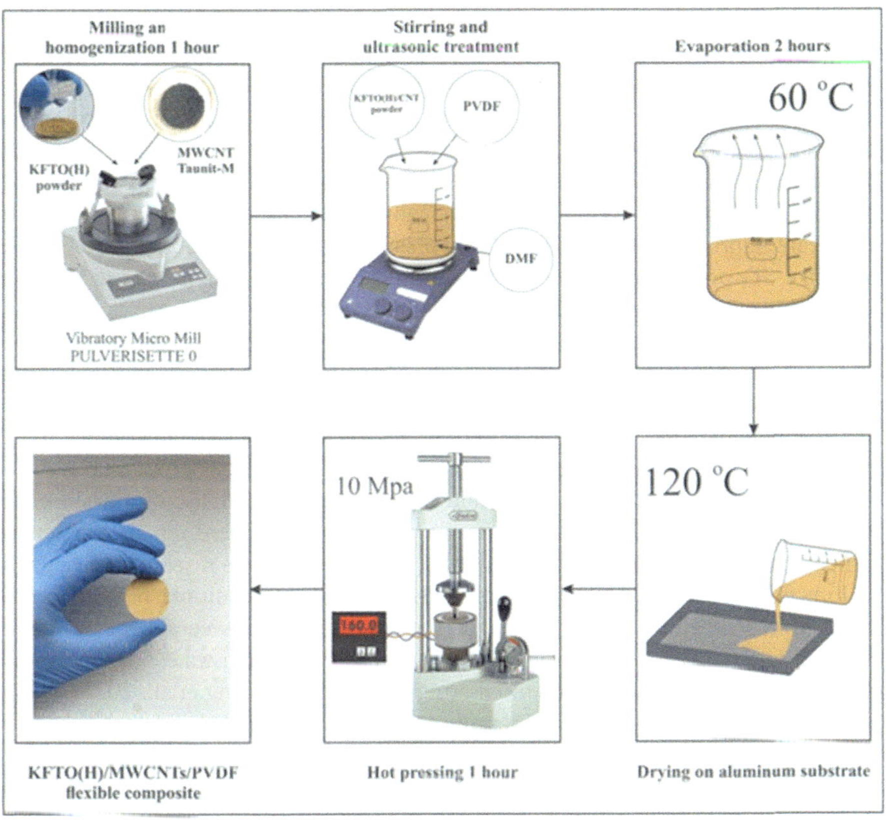

Figure 16.9 Schematic illustration of the preparation methodology of the dielectric composites consisting of Fe-doped potassium titanate with hollandite-like structure (KFTO(H)), multiwall carbon nanotubes (MWCNT) and polyvinylidene fluoride (PVDF).
Adapted from Tsyganov et al. (2022) under CC BY 4.0 license, © 2022, MDPI.

solution and stirred at 60°C for 2 hours using a magnetic stirrer, with periodic ultrasonic treatment to ensure thorough homogenization. The resulting dispersions were cast onto aluminum plates and dried at 120°C. Subsequently, the composites were hot-pressed at 160°C under 10 MPa pressure for 1 hour, resulting in flexible samples with an approximate thickness of 0.4 mm.

This is comparable to the methacrylate-terminated epoxy resins (MEP) integrated with polymerizable methacryloxypropyl cage polyhedral silsesquioxane (POSS), which achieved a remarkably low dielectric constant of 2.68 at 1 MHz, suitable for high-frequency electronic applications (Tsyganov et al., 2022). Both materials offer advancements in dielectric performance, critical for communication devices where signal integrity is paramount. Exploring energy-related applications, polydimethylsiloxane (PDMS) composites reinforced with barium titanate (BaTiO3) show enhanced voltage output and excellent electromechanical energy conversion properties, making them particularly effective for tactile imitation and mechanical energy harvesting (Wang et al., 2022).

Interestingly, 3D printing has evidently revolutionized the fabrication of electrodes and sensors due to its ability to enable novel material design, rapid prototyping, and reduced waste. For instance, fused filament modeling (FFM) has emerged as a common and accessible technique, especially in the creation of conductive composite filaments for energy storage and conversion applications (Iffelsberger et al., 2020). However, to enhance electrochemical performance, post-printing treatments like electrodeposition have been employed, which, despite their complexity, offer significant improvements. Moreover, the versatility of 3D printing extends to simpler tools like 3D pens, which have been proposed for sensor manufacturing, showing comparable performance to traditional methods but with lower costs and greater accessibility (de Oliveira et al., 2020). Furthermore, 3D printing enables the development of biosensors with high precision, demonstrated by the creation of glucose sensors and other analytical tools for real biological samples (Cardoso et al., 2020). Evidently, this versatility is further underscored by the integration of 3D-printed electrochemical devices into construction materials, providing a novel approach to energy storage within building structures (de Matos Morawski et al., 2023; Vaghasiya et al., 2021). Despite the advancements, there remain challenges, particularly in miniaturization and multiplex biosensing, where research has begun to explore fully 3D-printed microvolume cells for detecting multiple biomarkers (de Matos Morawski et al., 2023). Moreover, the development of wearable devices, such as paper-based microfluidic electrochemical sensors, demonstrates the expanding scope of 3D printing in real-time monitoring applications (Liang et al., 2021). On the contrary, while multimaterial 3D printing allows for enhanced electrochemical performance through heterolayered structures, the integration of various carbon allotropes into electrode design has shown promise in improving energy conversion and sensing applications (Palacios-Corella et al., 2023).

Moreover, 3D printing has enabled the development of novel, cost-effective, and portable devices, such as nanoparticle-based Ag/AgCl quasi-reference electrodes, which demonstrate superior stability and tactile comfort, making them highly suitable for applications in electrophysiology and glucose detection (Sun et al., 2020). In this study, nanowire-and-nanoparticle (NWP) Ag/AgCl quasi-reference electrodes (qREs) were employed as dry electrodes for monitoring cardiac electrical activity, highlighting their potential for long-term health monitoring. Unlike traditional wet electrodes, which require a conductive gel layer, these dry ECG sensors enhance patient comfort by eliminating the need for gel application (Fig. 16.10).

Interestingly, the customization potential of 3D printing is further illustrated by the creation of fibrous carbon Toray paper electrodes, which, when modified with gold nanoparticles, show impressive electrochemical performance in detecting hypochlorite, an important chemical in various industries (Mohan et al., 2022). However, the complexity of integrating multiple electrodes within a single device remains a challenge, as demonstrated in studies focusing on fused deposition modeling (FDM) to produce low-cost, disposable electrochemical devices for analyzing substances like paracetamol and cocaine (Ramos et al., 2023). In addition, the exploration of stretchable ionogels through in situ 3D printing represents a significant advancement in the field of wearable electrochemical devices, offering high stretchability, ionic conductivity, and thermomechanical stability, which are crucial for strain sensors in human motion detection (Wang et al., 2021a). Nevertheless, despite these advancements, challenges in achieving optimal conductivity and electrochemical performance persist, as seen in the development of polystyrene-based electrodes, which, although reusable, require further refinement to enhance signal-to-background responses (Rymansaib et al., 2016; Zhou et al., 2024a). Conversely, the integration of MXene with cellulose nanofibers and multiwalled carbon nanotubes in 3D printing offers promising results, particularly in enhancing the rheological properties of MXene ink and improving the electrochemical performance of printed electrodes (Zhou et al., 2024a). Furthermore, the application of metal 3D printing in fabricating bespoke electrochemical stainless-steel electrodes illustrates the potential of this technology in creating highly tailored devices for specific functions such as oxygen evolution and pH sensing (Ambrosi et al., 2016; Gevaerd et al., 2021). Moreover, innovative approaches like the use of 3D-printed microfluidic platforms for cortisol determination in saliva samples highlight the growing role of 3D printing in developing portable point-of-care devices with significant potential for real-world applications (Gevaerd et al., 2021). Finally, the development of membraneless electrochemical flow cells for hydrogen production underscores the simplicity and cost-effectiveness of 3D printing in creating modular and efficient electrolyzers, despite challenges such as product crossover and optimization of bubble evolution (O'Neil et al., 2016). Similarly, the creation of electrochemical paper based devices for detecting substances like midazolam maleate in beverages demonstrates the operational simplicity and robustness that 3D printing can offer in forensic applications, thereby opening new possibilities for on-site investigations (Rocha et al., 2021).

In the following Table 16.1, we summarize the key materials and results from recent studies on electronic devices. The table includes information on the polymers and nanomaterials used, the quantitative performance metrics reported, the types of devices investigated, and their applications.

In a similar vein, the molecule-bridged graphene/Ag (MB-G/A) composites have demonstrated a maximum conductivity of 2.0×10^5 S/m, presenting a cost-effective

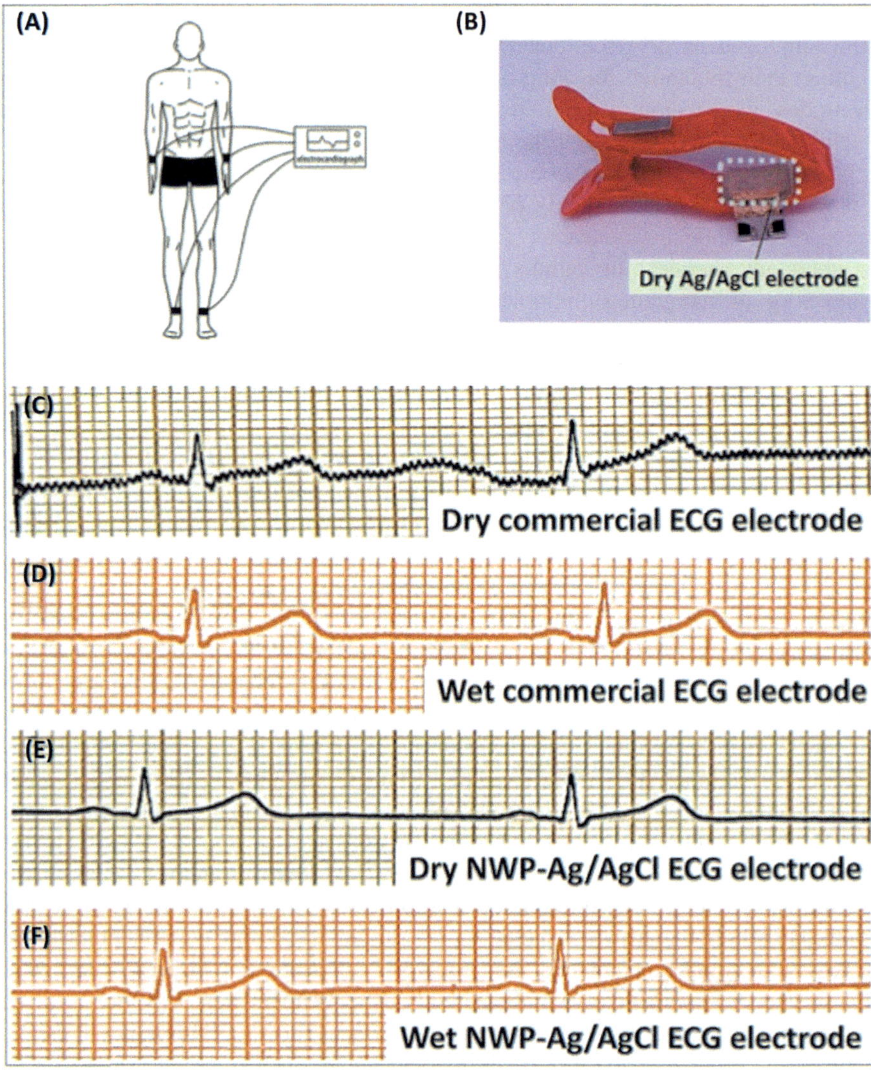

Figure 16.10 Four NWP-Ag/AgCl qREs were attached to the left and right limbs (A) and connected to commercial cardiac clips (B) to monitor heart activity. (C) The electrocardiograph (ECG) of a healthy volunteer using conventional ECG electrodes without conductive gel, where noise fluctuations are evident due to poor skin-electrode contact. The noise significantly diminished when conductive gel was applied to the same electrodes (D). For the NWP-Ag/AgCl qRE, however, no significant difference in ECG signal was observed with (F) or without (E) conductive gel, indicating their effectiveness as dry sensors. A comparison between the ECGs in (D) and (E) shows similar signal quality, suggesting that NWP-Ag/AgCl qREs can serve as reliable dry ECG sensors for long-term monitoring. Moreover, the NWP-Ag/AgCl qREs demonstrated stronger signal conductivity compared to the commercial ECG electrodes. Adopted from Sun et al. (2020) under CC BY-NC-ND 4.0 license, © 2020 Elsevier.

Table 16.1 Overview of nanomaterial-based electronic devices: Polymers, properties, performance and applications.

S. No.	Polymer (if any)	Nanomaterial (if any)	Quantitative results	Type of electronic device	Application	References
1.	Ultraviolet-curable resin	Cu2(OH)PO4	Resistivity of copper circuit: 7.5 × 10–8 Ω·m; adhesion property: highest 5B level	Circuit on composite substrate	Advanced electronic devices	Liu et al. (2022a)
2.	Liquid crystal elastomers (LCEs), resins	Conductive silver inks	Demonstrated high-speed and high-resolution printing; composites with tunable mechanical properties and multifunctionality	Circuit-embedding architectures	Soft robotics, electronics, active metamaterials	Peng et al. (2021)
3.	Polylactic acid (PLA), thermoplastic polyurethane (TPU)	Continuous carbon fibers	Tailorable gradient composites; tensile modulus: highest at 138.5% compared to pure PEKK; significant weight-saving potential of 21%	Continuous fiber-reinforced composites	Biomedicine, robotics, flexible electronics	Ye et al. (2024)
4.	Flexible polyurethane (FPU)	Multiwall carbon nanotubes (MWCNTs), polypyrrole (PPy)	Relative resistance change prediction model optimized with ML; CNT > PPy > elongation in feature importance	Strain sensors	Wearable electronics, biomedical devices	Poompiew et al. (2024)
5.	None	None	Electrode light transmission: 20 mm × 20 mm, good conductivity and mechanical stability, stable performance under different working conditions	Wearable strain sensors, heated films	Wearable multifunctional electronic devices	Liu et al. (2024a)

(*Continued*)

Table 16.1 (Continued)

S. No.	Polymer (if any)	Nanomaterial (if any)	Quantitative results	Type of electronic device	Application	References
6.	Binder materials	Fe-6.5 wt% Si particles	Saturation magnetization increase with 64 v/v% Fe-6.5 wt% Si loading; magnetic behavior characterized by VSM	Soft magnetic composite parts	Robotics, energy electronics	Okoruwa et al. (2023)
7.	Styrene-isoprene-styrene block copolymer	Various particles with different atomic densities	Radiation attenuation tailored for microelectronic protection in satellites; graded shielding capabilities	Radiation shielding materials	Microelectronic protection in satellites	Rosh-Gorsky et al. (2024)
8.	Natural rubber	Acetylene black	Piezoresistive sensor composites detect small strains $<10\%$; enhanced mechanoelectrical properties; homogeneous filler distribution achieved through AM	Piezoresistive sensors	Monitoring human motion, wearable electronics	Georgopoulou et al. (2023)
9.	Various thermoplastic composites	Boron nitride (BN), MgO particles	In-plane thermal conductivity: 1.97 W/m/K; electrical insulation, excellent printability, lightweight 3D-printed heat sinks	Thermal management solutions	Electronic devices cooling	Bagatella et al. (2024)
10.	Carbon-filled polymer filaments	Metal-filled, carbon-filled thermal conductive composites	Carbon-filled filament heat sink had best heat dissipation performance (28% higher at 90°); base temperature was the lowest in studied cases	Heat sinks	Electronics cooling	Peng et al. (2022)

11.	ABS polymer	Carbon nanotubes (CNTs)	Tensile strength improvement by 12.6% with high CNT content; average resistivity: 2.5 Ωm for 4.76 wt.% CNT, 0.15 Ωm for 9.09 wt.% CNT	Structural elements	Structural electronics, sensors	Podsiadły, Matuszewski et al. (2021)
12.	PEGDA (polyethylene glycol diacrylate)	Sulfonated polyaniline (PANIs)	High tensile strains > 40%; specific capacitances of 211 F/g; electrochemical activity suitable for pH monitoring in biofluids	pH-monitoring sensors	Biomedical, soft tissue applications	Carcione et al. (2023)
13.	Polyether ketone ketone (PEKK)	Carbon fibers	Shape fixity and recovery: CF/PEKK-A250 exhibited highest storage modulus 4438.23 MPa; 158% higher than pure PEKK; thermal degradation temperature ~568°C	Shape memory polymers (SMPs)	Self-deployable hinges, high-temperature sensors/ actuators	Kumar et al. (2024)
14.	Methacrylate-terminated epoxy resin (MEP)	Polymerizable methacryloxy-propyl cage polyhedral silsesquioxane (POSS)	Dielectric constant as low as 2.68 at 1 MHz; Td5 exceeded 320°C, indicating excellent thermal stability; suitable for 3D printing resins	Low dielectric constant composites	Communication technologies	Tsyganov et al. (2022)
15.	PEGDA (polyethylene glycol diacrylate)	Sulfonated polyaniline (PANIs)	High tensile strains > 40%; specific capacitances of 211 F/g; electrochemical activity suitable for pH monitoring in biofluids	pH-monitoring sensors	Biomedical, soft tissue applications	Carcione et al. (2023)

(*Continued*)

Table 16.1 (Continued)

S. No.	Polymer (if any)	Nanomaterial (if any)	Quantitative results	Type of electronic device	Application	References
16.	ABS polymer	Carbon nanotubes (CNTs)	Tensile strength improvement by 12.6% with high CNT content; average resistivity: 2.5 Ωm for 4.76 wt.% CNT, 0.15 Ωm for 9.09 wt.% CNT	Structural elements	Structural electronics, sensors	Podsiadły, Matuszewski et al. (2021)
17.	None	None	Demonstrated high-speed and high-resolution printing; composites with tunable mechanical properties and multifunctionality	Circuit-embedding architectures	Soft robotics, electronics, active metamaterials	Peng et al. (2021)
18.	Polydimethylsiloxane, Polydimethylsilane	Copper nanoplates	Enhanced electrical conductivity; thermal resistance; negative temperature coefficient at high temperatures	High-temperature sensing electronics	High-temperature environments	Podsiadły, Matuszewski et al. (2021)
19.	Nanocellulose, PEDOT	None	Conductive paper-based electrode with 202 S/cm; handles electric current up to 0.7 A; high specific energy of 4.05 Wh/kg, specific power of 4615 W/kg	Printed supercapacitors, flexible electronics	Supercapacitors, wearable electronics, smart packaging	Lay et al. (2023)
20.	Polylactic acid (PLA)	Multiwall carbon nanotubes (MWCNT)	Improved mechanical properties; electrothermal Joule heating performance; measurable antibacterial performance	Multifunctional nanocomposite filaments	Sensors fabrication, health monitoring devices	Vidakis et al. (2021)

21.	None	Carbon-based, ceramics, intermetallic, refractory metals	Various materials with enhanced mechanical strength, thermal stability, and resistance to harsh environments	High melting point materials	Aerospace, energy, electronics, manufacturing	Abdullah et al. (2024)
22.	Photosensitive resin	Microscopic glass, metal, polystyrene	Mechanically tunable composite geometries; structural properties enhanced via SonoPrint	3D-printed electronics, composite fabrication	Tissue engineering, biohybrid robots, aerospace	Agrawal et al. (2024)
23.	None	Graphite SLS composite	Effective absorption bandwidth of 1.04 GHz; polarization indifference; wide-angle of incidence tolerance	Radar absorbing metastructure	Stealth and electromagnetic shielding	Abdullahi and Ali (2021)
24.	Polyamide (PA12CF20)	Carbon-reinforced nylon composite	Sliding wear performance optimized; wear crack characterization via SEM; enhanced wear performance using GA-ANFIS	Functional parts, gear manufacturing	Automotive, medical, electronics	Chhabra et al. (2023)
25.	None	Tungsten, boron nitride	Capable of attenuating multiple species of radiation; favorable thermal management characteristics	Radiation shielding materials	Protection of microelectronic systems in space	Beck et al. (2023)
26.	Acrylonitrile butadiene styrene (ABS)	Multiwall carbon nanotubes (MWCNTs)	Improved tensile and thermal properties; characterized by FTIR, XRD, TGA, SEM	Structural elements, 3D printing	Aerospace, biomedical, automotive	Singh et al. (2024)

(Continued)

Table 16.1 (Continued)

S. No.	Polymer (if any)	Nanomaterial (if any)	Quantitative results	Type of electronic device	Application	References
27.	None	None	Printed copper electrodes with low resistivity of 4 × 10–4 Ω cm; compatible with soldering	3D conductive vias, embedded electronics	Modern electronics development	Rafael and Chan (2023)
28.	Oligourethane (Genomer 4247)	Functionalized submicron alumina particles	Thermal conductance of 1 W/(m·K); viscosity of around 16 mPa·s; ceramic content of 50 vol%	Jettable polymer-ceramic composites	Thermal management in printed electronics	Graf et al. (2021)
29.	Polydimethylsiloxane (PDMS)	Barium titanate (BaTiO3)	Enhanced voltage output; metamaterial-structured composites with excellent electromechanical energy conversion properties	Tactile imitation, mechanical energy harvesting	Biological signal identification, implantable self-powered electronics	Wang et al. (2022)
30.	Thermoplastic polyurethane (TPU)	Silver, multiwall carbon nanotubes (MWCNT)	Maximum electrical conductivity of 0.92 S/cm; negligible change in impedance over 1 kHz^{-1} MHz; strain sensors with distinct bending angle detection	Strain sensors, circuit elements	Embedded electronic systems, complex 3D electronics	Nassar et al. (2023)
31.	Nickel-plated acetate fabrics, silk sericin	Carbon nanotubes	EMI shielding effectiveness > 30 dB at X-band; conductive textiles with higher tensile strength and negative relative resistance changes	Wearable sensors, EMI shielding	Human-machine interaction, electromagnetic shielding	Duan and Lu (2021)

32.	Polylactic acid (PLA), polyethylene terephthalate glycol-modified (PETG), Acrylonitrile-butadiene-styrene (ABS)	None	Mechanical properties analyzed with FDM technique; failure analysis correlating mechanical behavior with microstructural characteristics	Polymer and composite components	Automotive, biomedical, aerospace, electronics	Rafael and Chan (2023)
33.	Nylon 12	Bismuth oxide (Bi_2O_3)	Improved thermal conductivity with addition of Bi_2O_3; significant tensile strength improvement	Thermal interface devices	Thermal management in electronic devices	Das et al. (2023)
34.	Epoxy	Nanoclay, carbon nanotubes (CNTs)	Enhanced electrical and mechanical properties; optimum CNT concentration for mechanical enhancement	Conductive epoxy-nanoclay nanocomposites	Aerospace structures, electronics	Kasraie et al. (2021)
35.	Nylon	Carbon fiber	Optimized nylon wall thickness and filament separation; effective voltage drop measurement to monitor resin flow	Resin frontal flow sensors	High-quality composites by monitoring resin frontal flow	Sánchez del Río et al. (2021)
36.	None	N-doped silicon	Substrate-free piezoresistive sensors; five times more sensitive than metallic ones	Strain sensors	Monitoring deformation in composite materials, rudder foil in real-time	Herry et al. (2024)

(*Continued*)

Table 16.1 (Continued)

S. No.	Polymer (if any)	Nanomaterial (if any)	Quantitative results	Type of electronic device	Application	References
37.	Thermoplastic (unspecified)	Copper, conductive thermoplastic	Electrical contact with and without copper electroplating; demonstrated a 555-timer oscillator circuit	Structural electronics	Integrating electronics into 3D printed parts	Lazarus and Tsang (2020)
38.	Polyimide (PI)	Silver nanoparticles (Ag NPs)	Electronic conductivity of 1.14×10^7 S/m after annealing at 250°C; improved interfacial bonding	Flexible and stretchable electronics	High-temperature processibility and versatility in various substrates	Yu et al. (2021)
39.	Elastomer (unspecified)	Cellulose nanocrystal (CNC)	High toughness improvement via covalently functionalized CNC; significant effects on mechanical properties at percolation threshold < 1 wt%	3D-printed elastomers	Automotive, marine, aerospace, electronics, medical, and consumer	Palaganas et al. (2023)
40.	Polydimethylsiloxane (PDMS)	Carbon nanotube (CNT)	Strain sensor with tunable performance; gauge factor (GF) of 17.4; durable over 10,000 cycles	Compressive strain sensors	Detecting large- and small-scale human motions	Liu et al. (2022b)
41.	Thermoplastic polyurethane (TPU)	Silver nanoparticles (AgNPs)	Thermal conductivity improves with AgNP content; electrical percolation threshold decreases at temperatures higher than 50°C	Flexible electronics	Medical, energy, packaging, electronic devices	Chen et al. (2021)

42.	Polyethylene glycol diacrylate (PEGDA)	Polyaniline (PANI)	Produced PEGDA-PANI composites suitable for soft electronics applications; demonstrated use as an electrode in an ECG device	Soft electrodes, biosignal monitoring	Wearable electronics, health monitoring	Montaina et al. (2023)
43.	Polyvinylidene fluoride (PVDF)	Hollandite-like ceramics, multiwall carbon nanotubes (MWCNTs)	Optimal combination of permittivity and dielectric loss (77.6 and 0.104 at 1 kHz); characterized using XRD, FTIR, SEM	Dielectric composites	Flexible electronics	Tsyganov et al. (2022)
44.	None	Ceramics, graphene-related materials (GRM)	Overview of ceramics/graphene hybrids and their applications in energy, sensors, EMI shielding, and more	Various (e.g., supercapacitors, sensors)	Energy production, storage, EMI shielding, biomaterials	Podsiadły, Skalski et al. (2021)
45.	Photo-polymer epoxy resins	Multiwall carbon nanotubes (MWCNT), titanium nitride (TiN), cubic boron nitride (c-BN)	Thermal conductivity increased by 24.21%; mechanical strength increased by 20.45%	Stereolithography nanocomposites	Electronics, aerospace, automotive, medical devices	Alshihabi and Kayacan (2024)
46.	None	Molecule-bridged graphene/Ag (MB-G/A)	Maximum conductivity of MB-G/A inks reaches 2.0×10^5 S/m; cost-effective paper-based electronics	Printable electronics	Interconnects, inductors, antennas	Li et al. (2022)
47.	Thermoplastic polyurethane (TPU)	Silver	Conductive and stretchable filament achieves 558 S/cm; maintains low resistance up to 60% strain	Stretchable electronic devices	Flexible electronics, AM	Sun et al. (2023)

(*Continued*)

Table 16.1 (Continued)

S. No.	Polymer (if any)	Nanomaterial (if any)	Quantitative results	Type of electronic device	Application	References
48.	Yarn (as a 1D material)	None	Knitted fabrics with nonlinear bulk response; constitutive model for mechanical properties	Soft robotics, wearable electronics	Soft robotics, wearable electronics, engineered tissue	Singal et al. (2024)
49.	Ethylene-co-vinyl acetate	Silver nanowires	Pressure sensor operates under low input voltage (~0.1 mV); low detection limit (12.9 Pa); high sensitivity (67.3 kPa^{-1})	Flexible pressure sensors	Robotics, health monitoring	Liu and Watt (2021)
50.	Polyamide	Microcarbon fiber (CF)	Electromagnetic shielding effectiveness (EMSE) up to 70 dB; specific shielding up to 60 $dB.cm^3/g$	Electromagnetic shielding	Electronic enclosures, EMI protection	Martins et al. (2023)
51.	None	None	Developed a three-layered multifunctional laminated composite plate of magnesium-based AZ31B-Zn-Al 1100	Biocompatible materials	Biomedical applications	Dixit et al. (2023)
52.	Silicone elastomers	Long fibers	Automated Fiber Embedding (AFE) enables fabrication of soft composites; demonstrated in soft robotic actuators	Soft robotics, actuators	Soft robotics	Stalin et al. (2021)
53.	None	Metal-polymer composite	Heat exchanger performed within 10% of a conventional CPU radiator; compact, low-cost, lightweight	Heat exchangers	Electronic cooling applications	Kailkhura et al. (2022)

solution for paper-based electronics (Li et al., 2022). Both studies highlight the innovative approaches being taken to improve energy efficiency and functionality in electronic devices, particularly in the context of self-powered and low-cost electronics. Finally, the work on photo-polymer epoxy resins reinforced with multiwall carbon nanotubes (MWCNT), titanium nitride (TiN), and cubic boron nitride (c-BN) stands out for its achievement of a 24.21% increase in thermal conductivity and a 20.45% improvement in mechanical strength, making them suitable for stereolithography in electronics, aerospace, and automotive applications (Alshihabi & Kayacan, 2024). This can be seen as part of a broader trend where advanced nanomaterials are being leveraged to enhance both the mechanical and thermal properties of polymers, thereby revolutionizing the manufacturing processes and applications of composite materials in high-tech industries.

16.4 Scope of AM in optoelectronics

Optoelectronic devices, which integrate optical and electronic functionalities, have become integral to modern technology, powering a wide array of applications from communication systems and displays to sensors and renewable energy sources. These devices, including light-emitting diodes (LEDs), photodetectors, solar cells, and laser diodes, operate by converting electrical signals into optical signals or vice versa. The seamless integration of these devices into various applications has been driven by their ability to manipulate light at both the macroscopic and microscopic levels. For instance, in telecommunications, optoelectronic devices enable the transmission of data at the speed of light, significantly enhancing bandwidth and efficiency. In consumer electronics, they are pivotal in the development of high-definition displays and advanced imaging systems. Moreover, in renewable energy, optoelectronic devices such as solar cells play a crucial role in harnessing sunlight for sustainable power generation. The continuous evolution of optoelectronic technology has thus been central to the advancement of numerous fields, making these devices indispensable in both industrial and consumer applications.

AM (AM), commonly known as 3D printing, has revolutionized the production of optoelectronic devices, enabling the creation of complex structures that were previously unattainable with traditional manufacturing techniques. AM allows for the precise deposition of materials layer by layer, which is particularly beneficial for the fabrication of intricate optoelectronic components with customized geometries. This capability has opened new avenues in the design and integration of optoelectronic devices into various systems, offering unprecedented flexibility in form factor and functionality. For example, AM techniques have enabled the development of flexible and wearable optoelectronic devices that can conform to nontraditional surfaces, expanding the potential applications of these devices in healthcare, wearable electronics, and beyond. Additionally, AM has facilitated the integration of multiple materials with distinct properties into a single device, enabling the creation of multifunctional optoelectronic systems that can perform complex tasks within a compact footprint.

The benefits of AM in the realm of optoelectronics are manifold. Firstly, AM significantly reduces material waste, as it builds structures layer by layer, using only the necessary amount of material. This contrasts with subtractive manufacturing techniques, which often involve the removal of excess material, leading to inefficiencies. Secondly, AM provides unparalleled design freedom, allowing for the creation of geometries that are either difficult or impossible to achieve with traditional manufacturing methods. This includes the fabrication of lightweight structures with optimized material distribution, which is crucial for applications where weight and material efficiency are paramount, such as in aerospace and portable electronics. Thirdly, AM supports rapid prototyping and iterative design processes, enabling faster innovation cycles. Designers can quickly produce, test, and refine prototypes, accelerating the development of new optoelectronic devices. Finally, the localized production capability of AM can lead to reduced lead times and supply chain simplification, which is particularly beneficial for the customization of optoelectronic devices tailored to specific applications or user requirements. In summary, AM not only enhances the design and production of optoelectronic devices but also paves the way for new applications and innovations in this rapidly evolving field.

16.4.1 Polymeric materials for additive optoelectronics

The versatility of polymers, combined with their inherent properties, makes them indispensable in the AM of optoelectronic devices. These materials not only provide the necessary flexibility and durability required for advanced applications but also facilitate the integration of nanomaterials, thereby enhancing the performance of these devices. Indeed, polymers have opened new avenues in the fabrication of complex structures that were previously unattainable with traditional manufacturing methods. One of the most significant polymers used in optoelectronic devices is polyacrylonitrile (PAN). PAN is particularly valued for its excellent mechanical properties and chemical resistance, making it an ideal choice for applications that require durability and flexibility. For instance, in the fabrication of Flexible Transparent Electrodes (FTEs), PAN is employed to create robust, flexible devices that can withstand mechanical stress without compromising their conductivity. Moreover, PAN's compatibility with copper nanomaterials allows for high conductivity, which is crucial for devices like flexible touch screens and other optoelectronics. The use of PAN in such applications underscores its importance in the development of advanced optoelectronic devices (Zhou et al., 2023). Fig. 16.11 illustrates the manufacturing process of high-performance flexible transparent electrodes (FTEs) based on PAN/Cu core–shell conductive mesh, which is divided into four main steps. First, a microscale 3D printing process using extrusion-based deposition (EFD) was employed to print PAN meshes containing silver ions on a flexible PET substrate. It has been demonstrated that doping silver nitrate into nonconductive polymers can generate conductive structures through a reduction reaction. PAN was selected as the carrier polymer due to its low cost and high process applicability. In the second step, hydrazine hydrate vapor was used to reduce the

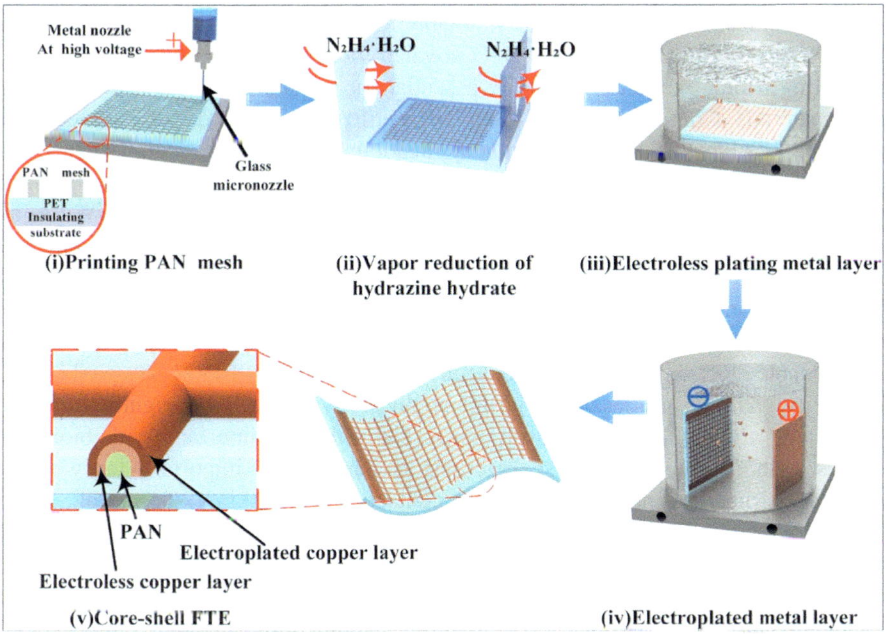

Figure 16.11 A schematic representation of the fabrication procedure for the PAN/Cu core–shell structured flexible transparent electrode.
Adapted from Zhou et al. (2023) under CC BY-NC 4.0 license © 2023, Wiley.

silver ions in the PAN/$AgNO_3$ network to elemental silver, creating a seed layer for subsequent copper deposition. The third step involved electroless copper plating on the printed PAN mesh. This additive manufacturing method, known for its simplicity and cost-effectiveness, selectively deposited copper on the PAN mesh surface, maximizing the transparency of the FTEs. Finally, an electroplating process was applied to uniformly thicken the copper layer, enhancing the electrode's environmental stability and bending resistance. This composite-plating technique endowed the FTEs with exceptional optoelectronic and physicochemical properties.

Interestingly, siloxane polymers have also found extensive use in the development of optoelectronic devices, particularly in light-emitting diodes (LEDs). Siloxane is renowned for its superior thermal stability and flexibility, qualities that are essential in ensuring the longevity and performance of LEDs. Because LEDs are often exposed to varying temperatures and mechanical stresses, the stability provided by siloxane polymers is critical. In fact, siloxane's ability to maintain its properties under these conditions makes it an excellent material for LEDs embedded with nitrogen-doped carbon dots, which are used in bioimaging and other optoelectronic applications. Thus, the use of siloxane polymers not only enhances the performance of LEDs but also extends their applicability in various fields (Zhou et al., 2022). Moreover, konjac glucomannan (KGM) has been employed in the creation of Room Temperature Phosphorescence (RTP) materials, a fascinating area of research within optoelectronics. KGM is particularly suitable for this application due to its ability to form strong hydrogen bonds and B–O covalent bonds. These bonds are crucial for protecting triplet excitons, which are essential for the phosphorescent properties of RTP materials. Consequently, KGM has been utilized in AM processes to create 3D-printed RTP objects, which are used in applications such as 3D anticounterfeiting and stereoscopic displays. The choice of KGM for these applications highlights its unique properties and its potential to revolutionize the field of optoelectronics (Wu et al., 2024). In addition to the aforementioned polymers, PEDOT (poly(3,4-ethylenedioxythiophene) polystyrene sulfonate) has also garnered significant attention in the AM of optoelectronic devices. PEDOT is a conducting polymer that is often used in the fabrication of cell-interfaced impedimetric sensors, which are employed in wearable and portable electronics. The unique electrical properties of PEDOT, coupled with its flexibility, make it an ideal material for applications where lightweight and flexible electronics are required. Moreover, PEDOT's ability to interface with biological tissues without causing adverse reactions is particularly important for wearable sensors that need to maintain prolonged contact with the skin (Wang et al., 2020). Triisopropylsilylethynyl pentacene is another polymer that has been utilized in the AM of optoelectronic devices, specifically in the fabrication of organic thin-film transistors. These transistors are crucial for flexible electronics and photonics applications, where traditional rigid materials would be unsuitable. Triisopropylsilylethynyl pentacene's ability to form high-quality, uniform thin films through laser printing has made it a material of choice for these applications. This solvent-free AM method allows for the creation of flexible substrates that can be used in a wide range of optoelectronic devices, thereby demonstrating the polymer's versatility and importance in this field (Diemer et al., 2017).

Polylactic acid (PLA), a biodegradable polymer, has also been employed in the AM of optoelectronic devices. PLA is particularly interesting because of its environmental benefits, as it is derived from renewable resources and is fully biodegradable. In the context of optoelectronics, PLA has been used to create alternating current electroluminescent (ACEL) devices. These devices, which exhibit tunable blue/green luminescence, are fabricated by incorporating metal-doped zinc sulfide (ZnS) phosphors into a PLA host matrix. The use of PLA in this context not only enhances the sustainability of the manufacturing process but also provides the necessary flexibility and biodegradability for applications in custom light-emitting devices and displays (Brubaker et al., 2019). The customized ACEL devices depicted in Fig. 16.12 were fabricated using 3D printing technology, completing the manufacturing process in under five minutes. This was achieved by depositing a single layer of PLA/phosphor filament onto an ITO slide, which was then sandwiched between conductive PLA top electrodes. The luminescent shape and behavior of the ACEL devices can be tailored by modifying the design of the conductive PLA electrode. Upon application of voltage to the printed device, luminescent activity is observed exclusively in the areas where the top electrode is in direct contact with the PLA/phosphor layer. Notably, all customized 3D-printed samples, irrespective of their shape or geometric configuration, exhibited a color transition from green to blue as the excitation frequency increased.

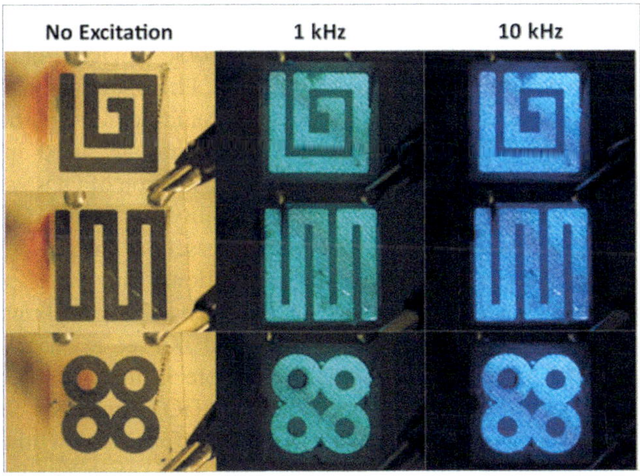

Figure 16.12 Demonstration of custom 3D-printed ACEL devices featuring various geometric patterns is presented under ambient conditions without excitation, as well as with a 245 V source at varying excitation frequencies. The devices exhibit notable visual changes in color, highlighting the impact of geometric design and excitation frequency on luminescent behavior. Under no excitation, the devices display their inherent characteristics, while the application of voltage reveals distinct color transitions, emphasizing the functionality and versatility of the 3D-printed ACEL designs.
Adapted with permission from Brubaker et al. (2019), © 2019, Royal Society of Chemistry.

Furthermore, polydimethylsiloxane (PDMS) has been utilized in the fabrication of organic photodiode arrays, particularly in applications that require the direct patterning of optoelectronic devices on curved substrates. PDMS is well-suited for these applications because of its flexibility, biocompatibility, and transparency. These properties are crucial when fabricating devices that need to conform to nontraditional surfaces or interact with biological tissues. The use of PDMS in AM enables the creation of organic photodiodes that can be integrated into a variety of optoelectronic systems, including those used in medical imaging and flexible electronics (Kim et al., 2017). Polyethyleneimine (PEI), another important polymer, is used as a precursor in the synthesis of metal oxide inks for direct ink writing in AM. PEI's ability to coordinate with metal ions makes it an excellent material for creating 3D-printed metal oxides with submicrometer resolution. These 3D microarchitectures, which exhibit strong light-matter coupling in the mid-infrared region, are used in a variety of optoelectronic applications. The versatility of PEI, combined with its ability to form stable complexes with metal ions, makes it an ideal material for the AM of functional oxides in optoelectronic devices (Chen et al., 2023). In addition to these examples, P4VP (poly(4-vinylpyridine)), when used as a core material in a conductive grid, has been shown to produce Flexible Transparent Electrodes (FTEs) with excellent mechanical stability and high transmittance. The combination of P4VP with a Cu/Ag(shell) conductive grid results in electrodes that are not only flexible but also highly durable, making them suitable for use in flexible optoelectronic applications. The use of P4VP in this context underscores its potential in the development of next-generation optoelectronic devices that require both flexibility and high performance (Zhang et al., 2023). Photopolymerizable resins have also played a crucial role in the AM of optoelectronic devices, particularly in the direct patterning of quantum dots (QDs). These resins are used in digital light processing (DLP) projectors to create isotropic encapsulation of QDs, which prevents edge ingress and degradation over time. The ability to pattern QDs with high precision and maintain their optical properties is essential for the development of miniature displays and other optoelectronic applications. The use of photopolymerizable resins in AM thus demonstrates their importance in the fabrication of high-performance, durable optoelectronic devices (Ozdemir et al., 2023). Moreover, polyacrylic acid (PAA) has been studied for its role in the AM of photocatalytic films. PAA, when combined with titanium dioxide (TiO2), enhances the rheological properties of Ti-organic inks, making them suitable for direct writing on flexible substrates. The ability to print TiO2 films with high photocatalytic activity is particularly important for applications in water purification and energy harvesting. The incorporation of PAA into these films improves their mechanical stability and adhesion, making them more effective for long-term use in environmental applications (Torres Arango et al., 2016).

16.4.2 Nanomaterials for additive optoelectronic

Nanomaterials have emerged as pivotal components in the advancement of optoelectronic devices, providing unmatched electrical, optical, and mechanical properties that have enabled the development of more efficient, compact, and versatile systems. These materials, with their unique characteristics, have been integrated into various optoelectronic applications through AM techniques, leading to significant enhancements in device performance. In this discussion, we explore several nanomaterials used in optoelectronic devices, their types, and the reasons they are favored in these cutting-edge technologies. Molybdenum disulfide (MoS_2), a transition metal dichalcogenide (TMD), is one of the most extensively researched nanomaterials for optoelectronic applications. MoS_2 exhibits a direct bandgap in its monolayer form, which is crucial for efficient light emission and absorption. This property makes MoS_2 particularly suitable for use in electroluminescent devices and photodetectors, where high sensitivity and efficiency are required. The ability of MoS_2 to maintain its optoelectronic properties even after undergoing complex solution processing further enhances its applicability in flexible, patterned, and miniaturized light sources (Rangnekar et al., 2023). Photoluminescence (PL) and electroluminescence (EL) properties of the one such prepared MoS_2 films have been shown in Fig. 16.13. The incorporation of MoS_2 into optoelectronic devices, therefore, is driven by its excellent carrier mobility, stability, and compatibility with flexible substrates, which are essential for next-generation wearable and portable electronics.

Similarly, cesium lead tribromide ($CsPbBr_3$) perovskite nanocrystals represent another class of nanomaterials that have gained significant attention in the field of optoelectronics. These nanocrystals are known for their exceptional photoluminescence quantum yield and tunable optical properties, which are crucial for applications in light-emitting

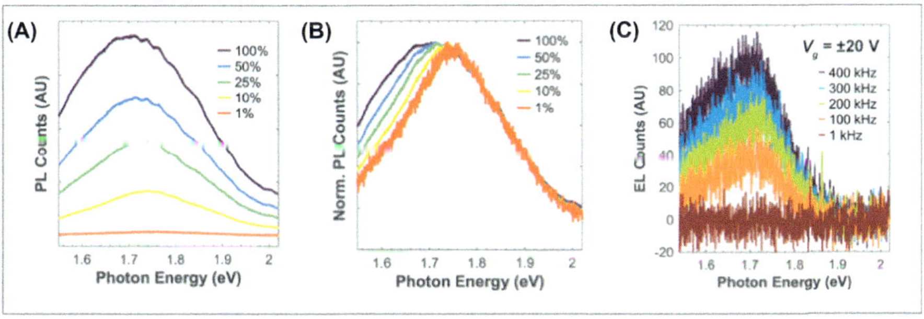

Figure 16.13. Photoluminescence (PL) and electroluminescence (EL) characteristics of the prepared MoS_2 films. (A) and (B) display the raw and normalized PL spectra, respectively, for a ~70-nm-thick AP-MoS_2 film measured under different laser power filters. The results show that as the power increases, the PL counts rise, and the emission peak exhibits a red shift from 1.75 eV to 1.70 eV. Panel (C) presents the EL spectra of AP-MoS_2 at various frequencies of the gate voltage signal, maintaining a fixed gate voltage amplitude of Vg= ± 20 V. It is noteworthy that the EL counts for the AP-MoS_2 films are significantly lower compared to those of MS-MoS_2 films of similar thickness.
Adapted with permission from Rangnekar et al. (2023), © 2023, American Chemical Society.

diodes (LEDs) and other photonic devices. The use of $CsPbBr_3$ in perovskite films, fabricated through aerosol deposition, highlights its role in enhancing the semiconducting properties and responsivity of devices (These et al., 2021). The ability to achieve dense, high-quality films at low temperatures makes $CsPbBr_3$ a preferred material for 3D surface applications, especially where efficient light emission and stability are paramount. Another significant nanomaterial used in optoelectronics is zinc sulfide (ZnS), which is often doped with other elements to tailor its optical properties. ZnS is employed in the development of alternating current electroluminescent (ACEL) devices due to its ability to emit light when subjected to an electric field. The incorporation of ZnS into a polylactic acid (PLA) host matrix for 3D printing allows the fabrication of custom-shaped, light-emitting devices with tunable luminescence (Brubaker et al., 2019). The versatility of ZnS in ACEL devices demonstrates its utility in producing bright, stable, and reproducible optical responses, which are critical for display technologies and other illumination applications. A schematic representation of the 3D-printed ACEL devices tested in this paper is shown in Fig. 16.14A. A single layer of PLA/phosphor material, measuring 100 μm in thickness, was directly deposited onto a thoroughly cleaned conductive ITO glass substrate and then sandwiched between two layers of 3D printed conductive carbon-based PLA, which served as a second electrode and had a thickness of 200 μm. Fig. 16.14B presents still images of a representative 3D-printed ACEL device subjected to a constant voltage source alongside various excitation frequency. Under these conditions, the 3D-printed samples exhibited a bright luminescence that transitioned from green to blue as the excitation frequency increased.

Nitrogen doped carbon dots (N-CDs) are another fascinating nanomaterial that has been utilized in optoelectronic devices, particularly in bioimaging and LED applications. These nanodots are known for their strong fluorescence, biocompatibility, and

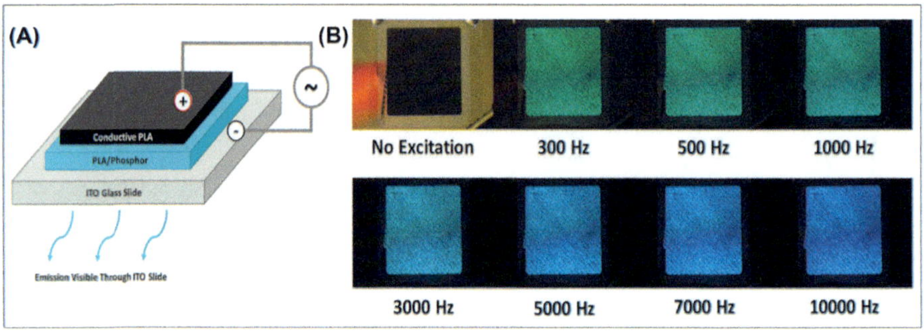

Figure 16.14 (A) Schematic diagram of the fabricated 3D-printed ACEL device, consisting of an Indium Tin Oxide (ITO) glass slide at the bottom, followed by a 3D-printed PLA/phosphor layer with a thickness of 100 μm, and topped with a 3D-printed conductive PLA electrode measuring 200 μm in thickness. The emission is observable through the bottom of the ITO glass slide. (B) Still images of the 3D-printed ACEL devices at various excitation frequencies under a constant applied voltage of 245 V.
Adapted with permission from Brubaker et al. (2019), © 2019, Royal Society of Chemistry.

ease of functionalization, making them ideal for biological sensing and imaging (Zhou et al., 2022). When integrated into siloxane matrices, N-CDs can be used to fabricate blue light-emitting diodes (LEDs), which are essential for various optoelectronic applications, including displays and sensors. The selective interaction of N-CDs with specific ions, such as Fe^{3+}, further enhances their utility in detecting and imaging applications, where sensitivity and specificity are crucial. Silver (Ag) nanomeshes and grids are also widely used in optoelectronic devices, particularly in flexible transparent electrodes (FTEs) and polymer-dispersed liquid crystal (PDLC) devices. Silver's high electrical conductivity and transparency make it an excellent material for creating electrodes that require both flexibility and durability (Liu et al., 2016; Zhou et al., 2023). The use of Ag nanomeshes in FTEs enables the development of touchscreens, displays, and other optoelectronic devices that can withstand repeated bending and mechanical stress while maintaining high optical performance. The combination of high transmittance and low resistance in these nanomeshes is critical for developing next-generation optoelectronic interfaces that are lightweight, robust, and efficient. Transition metal oxides, such as titanium dioxide (TiO_2), are also prominent in optoelectronic applications due to their excellent photocatalytic properties and ability to form highly crystalline structures at relatively low temperatures. TiO_2 is often used in the fabrication of photocatalysts and sensors, where its ability to convert light into chemical energy is harnessed for environmental and energy applications (Gerlein et al., 2020). The use of low-energy laser techniques to achieve high-yield conversion of TiO_2 precursors into crystalline films demonstrates the potential of this nanomaterial in flexible hybrid electronics and printable photovoltaics. The inherent stability and chemical inertness of TiO_2 further enhance its suitability for long-term applications in harsh environments. Fig. 16.15 illustrates the laser-induced conversion of TiO_2 to anatase and rutile phases. To demonstrate the versatile applications of this TiO_2 solution and the low-energy conversion process suitable for additive manufacturing, a

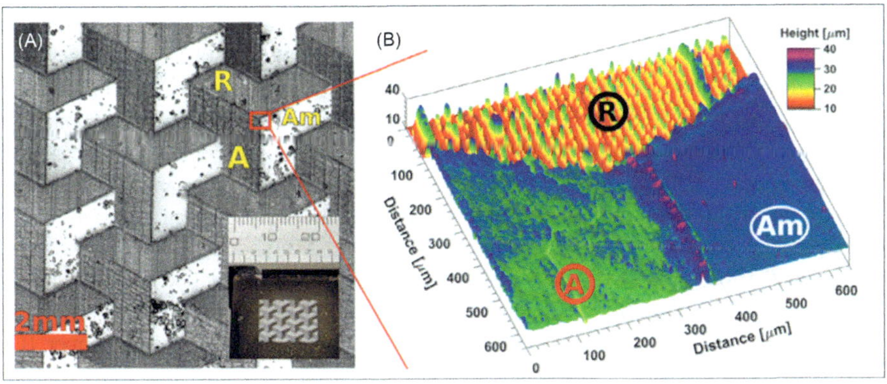

Figure 16.15 (A) Micrograph of the patterned film, showcasing the coexistence of both phases alongside an unconverted amorphous region in between. Each section is identified as either amorphous, anatase, or rutile. Inset: photograph of the patterned sample. (B) Magnified 3D reconstruction of a portion of the sample. The unconverted amorphous area (Am) is depicted in blue; the rutile area (R) appears in red-yellow, and the anatase area (A) is shown in green-blue. Adapted with permission from Gerlein et al. (2020), © 2023, Wiley.

geometric pattern was printed featuring coexisting rutile and anatase regions alongside amorphous unconverted sections (as depicted in provided figure). The initial pattern design, presented in the inset of Fig. 16.15A, displays three distinct colors: dark brown represents the amorphous film, dark gray indicates the anatase area, and light gray signifies the rutile section. Optical characterization of the sample, illustrated in Fig. 16.15B, clearly identifies the different regions, labeled as (Am) for amorphous, (A) for anatase-rich, and (R) for rutile-rich areas, which are easily distinguishable within this densely mixed film.

Another noteworthy nanomaterial is the nickel-based metal-organic framework (Ni-MOF), which is used in conjunction with other inorganic phases like BiOI and $AgVO_3$ to create hierarchical heterojunctions for photocatalytic applications. The integration of Ni-MOFs into polymer substrates with fractal geometry allows for enhanced photodegradation of pollutants, such as Rhodamine B, under visible light (Kang et al., 2023). The reduced bandgap of Ni-MOFs, achieved through innovative synthesis methods, facilitates their activation by visible light, making them highly effective in environmental remediation and chemical sensing. Additionally, transition metal dichalcogenides (TMDCs), such as tungsten diselenide (WSe_2), are increasingly used in photodetectors and other optoelectronic devices due to their unique electronic and optical properties. TMDCs offer the advantage of being fabricated with controlled morphologies and programmable geometries, which are essential for integrating these materials into flexible and conformal electronics (Hossain et al., 2024; Zhou et al., 2024b). The ability of TMDCs to maintain high photoresponsivity and other optoelectronic properties even when subjected to mechanical strain makes them ideal for applications where flexibility and durability are required, such as in wearable devices and sensors. Copper (Cu) and silver (Ag) grids, often combined with polymers such as polyacrylonitrile (PAN) and poly(4-vinylpyridine) (P4VP), are also used to create flexible transparent electrodes with ultra-fine line widths and excellent mechanical stability. These grids are particularly useful in applications where high conductivity and transparency are required, such as in flexible displays and touchscreens (Zhang et al., 2023; Zhou et al., 2023). The ability to create fine, conductive lines with low surface roughness and high environmental stability underscores the importance of these nanomaterials in developing optoelectronic devices that can operate reliably under various conditions. In summary, nanomaterials have revolutionized the field of optoelectronics, enabling the development of devices with enhanced performance, flexibility, and functionality. Molybdenum disulfide (MoS_2), cesium lead tribromide ($CsPbBr_3$), zinc sulfide (ZnS), nitrogen-doped carbon dots (N-CDs), silver (Ag) nanomeshes, titanium dioxide (TiO_2), nickel-based metal-organic frameworks (Ni-MOFs), transition metal dichalcogenides (TMDCs), and copper (Cu) grids are among the many nanomaterials that have been integrated into optoelectronic devices through AM techniques. Each of these materials brings unique properties that make them suitable for specific applications, from flexible displays and sensors to environmental remediation and energy harvesting. As research and development in nanomaterials continue to advance, we can expect to see even more innovative applications of these materials in optoelectronics, further pushing the boundaries of what is possible in this rapidly evolving field.

16.5 AM and optoelectronic applications

For instance, in the investigation of molybdenum disulfide (MoS_2), the electroluminescence spectral peaks were identified at 1.88–1.90 eV, with the emission intensity showing a direct correlation with the film thickness, which ranged from 10 to 70 nm. Electroluminescence is the phenomenon where a material emits light in response to an electric current, making it a cornerstone in the development of light-emitting devices. The increase in emission intensity with film thickness is a crucial finding, as it demonstrates that thicker films could enhance the brightness of electroluminescent devices, thereby broadening their application in flexible and miniaturized light sources (Rangnekar et al., 2023). Similarly, the development of flexible transparent electrodes (FTEs) using polyacrylonitrile (PAN) and copper (Cu) presented impressive quantitative results. The FTEs exhibited a conductivity of 1 Ω/sq, which is a measure of how easily electricity can pass through the material. This low sheet resistance is vital for ensuring high efficiency in devices such as touchscreens and displays, where maintaining conductivity while minimizing energy loss is paramount. Furthermore, the FTEs maintained 89% light transmission at 550 nm, indicating that they can allow a significant amount of light to pass through, which is crucial for display technologies. Interestingly, the resistance change rate of these electrodes was only 1.4% after undergoing 2000 bending cycles at a 3 mm radius, showcasing their remarkable mechanical flexibility and durability. These properties are essential for the next generation of flexible electronics, where materials must withstand repeated bending and flexing without degrading in performance (Zhou et al., 2023). In the case of ultraviolet-A (UVA) micro-LEDs, the quantitative results further highlight the advancements in optoelectronic devices. The micro-LEDs exhibited an in-plane compressive stress of 0.043 GPa and a dislocation density of 106 cm^{-2}. Dislocation density refers to the number of defects within a material's crystal structure, and a lower density is desirable as it leads to fewer disruptions in the material's electrical properties. This is significant because it directly impacts the efficiency and longevity of the micro-LEDs. The devices also demonstrated a minimum ideality factor of 1.43, which is a measure of how closely a diode follows the ideal diode equation, with lower values indicating better performance. Additionally, the series resistance was recorded at 4.85Ω, further underscoring the efficiency of charge carrier movement within the device. These metrics are vital for enhancing the performance of optoelectronic devices, particularly in AM, where precision and reliability are crucial (Liu et al., 2024b). Moreover, the study of triplet–triplet annihilation upconversion (TTA-UC) using $CsPbBr_3$ nanocrystals with lanthanide Ce^{3+} ions yielded high photoluminescence quantum yield (PLQY) and extended single-exciton lifetime. Photoluminescence quantum yield is a measure of the efficiency with which a material can re-emit absorbed light, and a high PLQY is indicative of a material's potential in applications such as photovoltaics and light-emitting diodes (LEDs). The extended single-exciton lifetime, which refers to the duration that an exciton (a bound state of an electron and an electron hole) exists before recombining, is significant as it enhances the efficiency of light emission processes. These properties, combined with quantum confinement and uplifted energy levels, underscore the material's suitability for advanced optoelectronic applications, including bioimaging and photocatalysis (Gong et al., 2023). Following this, the study of DNA-templated polymers in photonic crystals revealed a photonic bandgap at 450 nm and 650 nm in the visible spectrum. A photonic bandgap is a range of wavelengths where light propagation is forbidden, allowing the crystal to control light in specific ways. This characteristic is essential for applications that require precise manipulation of light, such as in structural color and AM. The ability to create such bandgaps in the visible spectrum demonstrates the potential of these materials to revolutionize optical devices by providing new ways to manage and use light (Bhagyasree et al., 2022). The use of ultrathin molybdenum disulfide (MoS_2) nanosheets in photodetectors also provided significant quantitative results. The devices exhibited a responsivity exceeding 103 A/W, a measure of the electrical output per unit of light input, indicating an exceptionally high sensitivity to light. Responsivity is a critical parameter for photodetectors, as it directly correlates with the device's ability to detect weak light signals. The nanosheets, being approximately 1.2 nm thick, also demonstrated a high aspect ratio, which is the ratio of the length to the width of the nanosheets, approximately 1 μm in this case. A high aspect ratio increases the surface area exposed to light, thereby enhancing the interaction between the light and the material, which is essential for improving the efficiency of photodetectors used in flexible optoelectronics (Kuo et al., 2022). In contrast, the study of silver (Ag) mesh in transparent conductive electrodes (TCEs) for polymer dispersed liquid crystal (PDLC) devices presented different but equally important quantitative results. The TCEs achieved a maximum transmittance of 60% in the "on" state and a minimum transmittance of 0.1% in the "off" state. Transmittance is the measure of how much light passes through a material, and these values indicate the device's effectiveness in controlling light transmission, which is crucial for display technologies like smart windows. Furthermore, the sheet resistance was found to be 5.58 Ω/sq, highlighting the

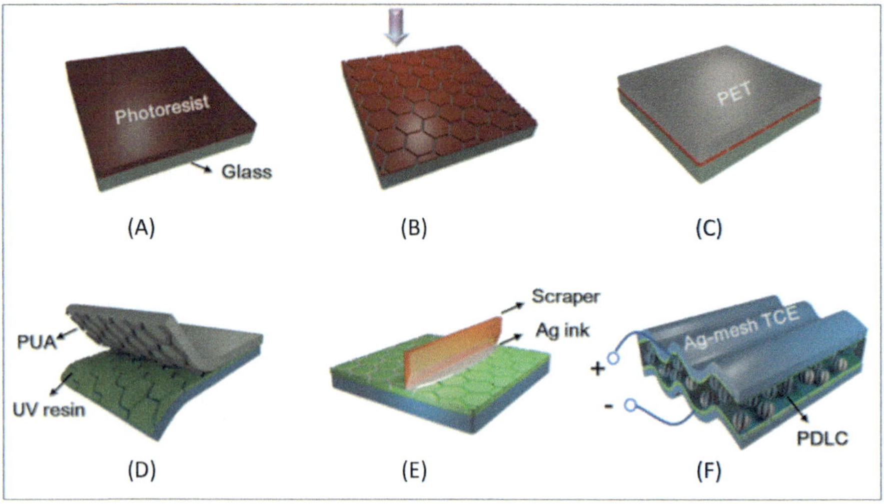

Figure 16.16 Schematic illustration of the fabrication process for a polymer-dispersed liquid crystal (PDLC) device utilizing an embedded silver mesh on a flexible substrate. The fabrication steps include: (A) spin-coating of the photoresist layer; (B) creation of patterned microgrooves on the photoresist using laser direct-writing lithography; (C, D) replication of the pattern from the photoresist to the soft mold; (E) dispersing and scratching silver ink into the microgrooves; and (F) assembly of the PDLC cell.
Adapted fromLiu et al. (2016), © Copyright 2016, Optica Publishing Group.

material's excellent conductivity. This balance of high transmittance and low resistance is vital for the efficiency and functionality of optoelectronic devices (Liu et al., 2016). Fig. 16.16 represents the fabrication process of the embedded Ag mesh TCE film.

Moreover, in the study of siloxane and nitrogen-doped carbon dots (N-CDs) used in light-emitting diodes (LEDs), the quantitative results revealed a selective interaction with Fe3+ ions and the successful fabrication of a blue LED. The N-CDs had a size of approximately 2.3 nm, a critical factor in determining the color and intensity of light emitted by the LED. The interaction with Fe^{3+} ions indicates the material's potential in bioimaging applications, where specific interactions with biological molecules are required for imaging processes. The ability to create small, efficient LEDs using these materials demonstrates their potential to revolutionize bioimaging and optoelectronics (Zhou et al., 2022). Finally, the lifetime of 2.14 seconds for room temperature phosphorescence (RTP) in 3D-printed objects made from konjac glucomannan (KGM) was another significant finding. Phosphorescence refers to the emission of light by a substance after it has absorbed light or other electromagnetic radiation, and the lifetime is the duration the light is emitted. A longer lifetime is desirable for applications in 3D anticounterfeiting and stereoscopic displays, where persistent light emission is needed. The ability to produce such materials with 3D printing technology opens new possibilities for advanced security and display applications (Wu et al., 2024).

The following Table 16.2 presents an overview of the materials and findings from recent optoelectronic device studies. It details the polymers and nanomaterials

Table 16.2 Overview of nanomaterial-based optoelectronic devices: Polymers, properties, performance and applications.

S. No.	Polymer (if any)	Nanomaterial (if any)	Quantitative results	Type of optoelectronic device	Application	References
1.	None	Molybdenum disulfide (MoS2)	Electroluminescence spectral peaks at 1.88–1.90 eV; emission intensity increases with film thickness (10–70 nm)	Electroluminescent device	Flexible, patterned, miniaturized light sources	Rangnekar et al. (2023)
2.	Polyacrylonitrile (PAN)	Copper (Cu)	Conductivity: 1 Ω/sq; 89% light transmission (at 550 nm); resistance change rate: 1.4% after 2000 bends at a 3 mm radius	Flexible Transparent Electrode (FTE)	Flexible touch screens, optoelectronics	Zhou et al. (2023)
3.	None	None	In-plane compressive stress: 0.043 GPa; dislocation density: 106 cm^{-2}; minimum ideality factor: 1.43; series resistance: 4.85Ω	Ultraviolet-A (UVA) micro-LED	AM, optoelectronics	Liu et al. (2024b)
4.	None	CsPbBr3, Lanthanide Ce^{3+} ions	High photoluminescence quantum yield; extended single-exciton lifetime; quantum confinement; uplifted energy levels	Triplet–triplet annihilation upconversion (TTA-UC)	Photovoltaics, light-emitting diodes, photocatalysis, bioimaging	Gong et al. (2023)
5.	DNA-templated polymer	None	Photonic bandgap at 450 nm and 650 nm in the visible spectrum	Photonic crystal	AM, structural color	Bhagyasree et al. (2022)

(Continued)

Table 16.2 (Continued)

S. No.	Polymer (if any)	Nanomaterial (if any)	Quantitative results	Type of optoelectronic device	Application	References
6.	Polyimide	Molybdenum disulfide (MoS2)	Responsivity exceeding 103 A/W; ultrathin nanosheets (≈1.2 nm thick); high aspect ratio (≈1 μm lateral size)	Photodetector	Flexible optoelectronics	Kuo et al. (2022)
7.	None	Silver (Ag) mesh	Maximum transmittance: 60% (on state); minimum transmittance: 0.1% (off state); low sheet resistance: 5.58 Ω/sq	Transparent Conductive Electrode (TCE)	Polymer Dispersed Liquid Crystal (PDLC) device	Liu et al. (2016)
8.	Siloxane	Nitrogen-doped carbon dots (N-CDs)	Selective to Fe^{3+}; blue light-emitting diode (LED) fabricated; size of 2.3 nm	Light-emitting diode (LED)	Bioimaging, optoelectronics	Zhou et al. (2022)
9.	Konjac glucomannan (KGM)	None	Lifetime of 2.14 seconds for room temperature phosphorescence (RTP); 3D-printed RTP objects	Room temperature phosphorescence (RTP)	3D anti-counterfeiting, stereoscopic displays	Wu et al. (2024)
10.	None	Titanium dioxide (TiO2)	High-yield conversion to highly crystalline TiO_2 at room temperature using low-energy laser; anatase or rutile phase achievable	Photocatalyst	Printable flexible hybrid electronics, photovoltaics	Gerlein et al. (2020)
11.	Polymer substrates with fractal geometry	Nickel-based metal-organic framework (Ni-MOF), BiOI, $AgVO_3$	Enhanced photodegradation of Rhodamine B (RhB) dye; reduced bandgap for visible light activation	Photocatalyst	Water depollution, chemical sensors, optical devices	Kang et al. (2023)

12.	None	None	LED-based devices for absorbance and fluorescence measurements; focus on low-cost, portable, and reliable methods	LED-based device	Absorbance and fluorescence measurements	Barreto et al. (2021)
13.	PEDOT	Silver (Ag)	Fiber arrays exhibit more than 95% transmittance (350 to 750 nm); used in 3D moisture flow sensors and respiratory sensors	Cell-interfaced impedimetric sensor	Wearable/portable sensors, optoelectronics	Wang et al. (2020)
14.	[poly(3,4-ethylenedioxythiophene) polystyrene sulfonate]	Silver (Ag)	Fiber arrays exhibit more than 95% transmittance (350 to 750 nm); used in 3D moisture flow sensors and respiratory sensors	Cell-interfaced impedimetric sensor	Wearable/portable sensors, optoelectronics	Wang et al. (2020)
15	None	None	Increased light output by a factor of 6.2; emission half-angle reduced by 50%; enhanced irradiance patterns	LED-based illuminator	Endoscopy, illumination optics	Thiele et al. (2016)
16.	Triisopropylsilylethynyl pentacene	None	First organic thin-film transistor fabricated with laser printing; solvent-free AM; flexible substrates	Organic thin-film transistor	Flexible electronics, photonics	Diemer et al. (2017)
17.	Polylactic acid (PLA)	Zinc sulfide (ZnS)	ACEL devices display tunable blue/green luminescence; emission dependent on applied voltage and frequency	Alternating current electroluminescent (ACEL)	Light-emitting devices, custom displays	Brubaker et al. (2019)

(Continued)

Table 16.2 (Continued)

S. No.	Polymer (if any)	Nanomaterial (if any)	Quantitative results	Type of optoelectronic device	Application	References
18.	Organic interposers	None	AM of organic interposers for 3D integration; state-of-the-art 3D printing enables interposers with added features like integrated vias	Organic Interposer	3D integration of optoelectronic and micromechanical components	Roshanghias et al. (2020)
19.	Sn–Pb composite	None	Film Sn6.3Pb93.7 prepared using fused-coating based AM (FCAM) method; porosity of 46.48%; decreased Sn content, increased Pb content	Film for optoelectronics	Industrial, solar, and optoelectronic applications	Dadi and Peravali (2022)
20.	None	Cesium lead tribromide ($CsPbBr_3$)	Aerosol deposition (AD) used to fabricate perovskite films; 20–35 μm thick layers; good semiconducting properties; high responsivity	Perovskite films	Optoelectronic device manufacturing on 3D surfaces	These et al. (2021)
21.	P4VP (core)	Cu/Ag(shell) conductive grid	Sheet resistance of 2.5 Ω; transmittance of 93%; ultrafine line width of 4 μm; excellent mechanical stability	Flexible transparent electrode (FTE)	Flexible optoelectronic applications	Zhang et al. (2023)
22.	Photopolymerizable resin	Quantum dots (QDs)	Direct patterning of QDs through maskless lithography; isotropic encapsulation of QDs; prevents edge ingress from lateral sides of QD films	Light emitters	Miniature displays, optoelectronics	Ozdemir et al. (2023)

23.	Nanoporous (NPo) polymeric material	Silver (Ag)	Conductive structures with sheet resistance < 2 Ω/sq; photodetector integration based on organic semiconductor	3D-printed conductive structures	Customizable electronic devices, optoelectronics	Xia et al. (2023)
24.	None	Quantum dots (QDs)	Patterning of QD layers with a feature size down to ≈1.5 μm; patterned QD films maintain ≈75% of original PLQY	QD light-emitting diode (QLED)	Displays, optoelectronics	Cho et al. (2020)
25.	None	Transition metal dichalcogenides (TMDCs)	Review of patterned growth strategies for TMDCs with programmable geometries; focus on optoelectronic and electronic applications	TMDC-based optoelectronic devices	Nano- and optoelectronics	Zhou et al. (2024b)
26.	Polydimethylsiloxane (PDMS)	None	Fabrication of hemispherical organic photodiode arrays; direct patterning on curved substrates without excessive strain	Organic photodiode	Optoelectronics on curved substrates	Kim et al. (2017)
27.	Polyethyleneimine (PEI)	Metal oxides (e.g., ZnO, CuO, In2O3, Ga_2O_3, TiO_2)	3D-printed metal oxides with submicrometer resolution; strong light-matter coupling in the mid-infrared region	3D-printed optoelectronic devices	AM, optoelectronics	Chen et al. (2023)

(Continued)

Table 16.2 (Continued)

S. No.	Polymer (if any)	Nanomaterial (if any)	Quantitative results	Type of optoelectronic device	Application	References
28.	None	None	Review of fiber laser phased arrays for dynamic beam steering; discusses strategies for nonmechanical, high-precision beam steering	Laser phased arrays	Beam steering, optoelectronics	Bowang et al. (2023)
29.	None	Cesium lead tribromide ($CsPb(Br/I)_3$)	PLQY of $CsPb(Br/I)_3$ NCs boosted from 15.4% to ≈60%; white LEDs with a luminous efficiency of 113.0 lm/W	Perovskite Nanocrystals (NCs)	High-performance LEDs, 3D printing	Wang et al. (2021b)
30.	None	None	In situ and real-time monitoring of printed electronics using an integrated high-resolution imaging system; resistance estimation accuracy ≥ 90%	Aerosol jet printed (AJP) electronics	Sensors, transistors, optoelectronics	(Ross) Salary et al. (2020)
31.	Polyacrylic acid (PAA)	Titanium dioxide (TiO2)	Study on the effect of crystalline TiO2 and polymer addition on rheological properties; photocatalytic degradation of methylene blue	Photocatalytic films	Water purification, energy harvesting	Torres Arango et al. (2016)

32.	None	WSe_2, graphene	Photoresponsivity ~0.70 A/W; strain-dependent photocurrent decreases with bending; WSe_2-graphene heterostructures used in photodetectors on flexible substrates	Photodetectors	Flexible, conformal optoelectronics	Hossain et al. (2024)
33.	Soluble polymer compounds	None	Study on the use of soluble polymer molds for composite textiles; dimensional errors optimized to produce parts with up to 0.02% error; potential for implantation of optoelectronic devices	Composite textiles	Structural health monitoring, optoelectronics	Roy and Dickens (2017)

employed, the reported quantitative performance data, the specific types of devices examined, and their associated applications.

16.6 Challenges and opportunity

One of the primary challenges in the AM of functional polymers and nanocomposites for electronics and optoelectronics lies in the development of materials with the necessary properties to meet the demanding performance requirements of these applications. The integration of nanomaterials, such as carbon nanotubes (CNTs), graphene, and metal nanoparticles, into polymers often leads to difficulties in maintaining uniform dispersion and preventing agglomeration. This lack of homogeneity can result in inconsistent electrical, thermal, and mechanical properties across the printed components, which are critical for the reliable performance of electronic devices. Another significant challenge is the limited range of printable conductive materials. While some polymers can be doped with conductive fillers to enhance their electrical properties, achieving the levels of conductivity required for high-performance electronic devices remains difficult. This limitation often necessitates postprocessing steps, such as sintering or chemical treatments, which can complicate the manufacturing process and reduce the cost-effectiveness of AM. Thermal management is also a major concern, especially in optoelectronics, where the generation of heat can lead to degradation of performance and reduced device longevity. Designing materials that can efficiently dissipate heat while maintaining electrical and optical performance is a complex task, further complicated by the need to maintain flexibility and lightweight characteristics in many applications. Despite these challenges, the opportunities presented by AM in the field of electronics and optoelectronics are vast and transformative. One of the most significant opportunities is the ability to create complex, multifunctional devices in a single manufacturing process. AM allows for the integration of various materials with different properties into a single structure, enabling the development of devices with enhanced functionality, such as flexible displays, wearable sensors, and integrated photonic circuits. The customization potential of AM is another major advantage. The ability to design and produce components tailored to specific applications and user requirements offers significant benefits in fields such as medical electronics, where personalized devices can be manufactured to meet the unique needs of individual patients. This customization extends to the development of advanced optoelectronic devices, such as custom light-emitting diodes (LEDs) and photodetectors, which can be optimized for specific wavelengths or operational conditions. While the challenges in the AM of functional polymers and nanocomposites for electronics and optoelectronics are significant, the opportunities for innovation and advancement are equally compelling. As research and development in this field continue to progress, the potential for AM to revolutionize the production of electronic and optoelectronic devices is immense.

References

Abdullahi, M. B., & Ali, M. H. (2021). Additively manufactured metastructure design for broadband radar absorption. *Beni-Suef University Journal of Basic and Applied Sciences, 10*(1), 24. https://doi.org/10.1186/s43088-021-00114-x.

Abdullah, M. D., Ahmed, I., Arafat Islam, M., Ahsan, Z., & Saha, S. (2024). Recent developments and diverse applications of high melting point materials. *Results in Engineering, 22*, 102376. https://doi.org/10.1016/j.rineng.2024.102376.

Agrawal, P., Zhuang, S., Dreher, S., Mitter, S., & Ahmed, D. (2024). SonoPrint: acoustically assisted volumetric 3D printing for composites. *Advanced Materials*. https://doi.org/10.1002/adma.202408374.

Alshihabi, M., & Kayacan, M. Y. (2024). Effect of nanosized carbon nanotubes, Titanium nitride and cubic boron nitride powders on mechanical and thermal properties of sla 3d printed resin composites. *Polym Composites, 45*, 15561–15573. https://doi.org/10.1002/pc.28850.

Ambrosi, A., Moo, J. G. S., & Pumera, M. (2016). Helical 3D-printed metal electrodes as custom-shaped 3D platform for electrochemical devices. *Adv Funct Mater, 26*(5), 698–703. https://doi.org/10.1002/adfm.201503902.

Bagatella, S., Cereti, A., Manarini, F., Cavallaro, M., Suriano, R., & Levi, M. (2024). Thermally conductive and electrically insulating polymer-based composites heat sinks fabricated by fusion deposition modeling. *Polymers (Basel), 16*(3), 432. https://doi.org/10.3390/polym16030432.

Barreto, D., et al. (2021). Performing reliable absorbance and fluorescence measurements with low budget—A tutorial for beginners. *Quimica Nova*. https://doi.org/10.21577/0100-4042.20170758.

Beck, D., et al. (2023). Additive manufacturing of multimaterial composites for radiation shielding and thermal management. *ACS Applied Materials & Interfaces, 15*(29), 35400–35410. https://doi.org/10.1021/acsami.2c22478.

Bhagyasree, G. S., Sreenilayam, S., Brabazon, D., Reena, V. N., & Nithyaja, B. (2022). Transmission characteristics of DNA templated 1D photonic crystal system for 3D printing applications: Simulation. *Results in Engineering, 16*, 100750. https://doi.org/10.1016/j.rineng.2022.100750.

Bowang, S., et al. (2023). Development of fiber laser phased array beam scanning technology. *Infrared and Laser Engineering, 52*(6), 20230250. https://doi.org/10.3788/IRLA20230250.

Brubaker, C. D., Newcome, K. N., Jennings, G. K., & Adams, D. E. (2019). 3D-Printed alternating current electroluminescent devices. *J Mater Chem C Mater, 7*(19), 5573–5578. https://doi.org/10.1039/C9TC00619B.

Carcione, R., et al. (2023). Self-standing 3D-printed PEGDA–PANIs electroconductive hydrogel composites for pH monitoring. *Gels, 9*(10), 784. https://doi.org/10.3390/gels9100784.

Cardoso, R. M., et al. (2020). 3D-Printed graphene/polylactic acid electrode for bioanalysis: Biosensing of glucose and simultaneous determination of uric acid and nitrite in biological fluids. *Sensors and Actuators. B, Chemical, 307*, 127621. https://doi.org/10.1016/j.snb.2019.127621.

Chen, H., et al. (2023). A generalized polymer precursor ink design for 3D printing of functional metal oxides. *Nanomicro Lett, 15*(1), 180. https://doi.org/10.1007/s40820-023-01147-w.

Chen, X., Li, A., & Zhang, Y. (2021). Temperature effect on thermal, mechanical, and electrical properties of Ag nanoparticle filled thermoplastic polyurethane composites *42*(11), 5952–5961. https://doi.org/10.1002/pc.26276.

Chhabra, D., et al. (2023). Analysis of fused filament fabrication parameters for sliding wear performance of carbon reinforced polyamide composite material fabricated parts using a hybrid heuristic tool. *Polymer Testing, 118*, 107910. https://doi.org/10.1016/j.polymertesting.2022.107910.

Cho, H., et al. (2020). Direct Optical patterning of quantum dot light-emitting diodes via in situ ligand exchange. *Advanced Materials, 32*(46). https://doi.org/10.1002/adma.202003805.

Dadi, V., & Peravali, S. (2022). Structural and optical analysis of film Sn6.3Pb93.7 prepared from coarse Sn63Pb37. *Optical Materials (Amst), 123*, 111908. https://doi.org/10.1016/j.optmat.2021.111908.

Das, J. P., Divakaran, N., Y, A., T P, M., & Mohanty, S. (2023). Fabrication of 3D printing extrusion-based Nylon 12/Bi2O3 composite filaments for thermal interface device application. *J Manuf Process, 85*, 885–893. https://doi.org/10.1016/j.jmapro.2022.11.061.

Dhapola, P. S., et al. (2024). Environment-friendly approach for synthesis of promising porous carbon: Empowering supercapacitors for a sustainable future. *Mater Adv, 5*(6), 2430–2440. https://doi.org/10.1039/d3ma00984j.

Diemer, P. J., et al. (2017). Laser-printed organic thin-film transistors. *Advanced Materials Technologies, 2*(11). https://doi.org/10.1002/admt.201700167.

Dixit, A. R., Srivastava, A. K., Dwivedi, S., Nag, A., & Hloch, S. (2023). An investigation on microstructural features and bonding strength of magnesium-based multifunctional laminated composite developed by friction stir additive manufacturing. *The International Journal of Advanced Manufacturing Technology, 128*(1–2), 531–546. https://doi.org/10.1007/s00170-023-11911-2.

Duan, Q., & Lu, Y. (2021). Silk Sericin as a green adhesive to fabricate a textile strain sensor with excellent electromagnetic shielding performance. *ACS Appl Mater Interfaces, 13*(24), 28832–28842. https://doi.org/10.1021/acsami.1c05671.

Georgopoulou, A., Srisawadi, S., Wiroonpochit, P., & Clemens, F. (2023). Soft wearable piezoresistive sensors based on natural rubber fabricated with a customized vat-based additive manufacturing process. *Polymers (Basel), 15*(10), 2410. https://doi.org/10.3390/polym15102410.

Gerlein, L. F., Benavides-Guerrero, J. A., & Cloutier, S. G. (2020). Laser-assisted, large-area selective crystallization and patterning of titanium dioxide polymorphs. *Advanced Engineering Materials, 22*(2). https://doi.org/10.1002/adem.201901014.

Gevaerd, A., Watanabe, E. Y., Belli, C., Marcolino-Junior, L. H., & Bergamini, M. F. (2021). A complete lab-made point of care device for non-immunological electrochemical determination of cortisol levels in salivary samples. *Sensors and Actuators. B, Chemical, 332*, 129532. https://doi.org/10.1016/j.snb.2021.129532.

Gong, N., et al. (2023). Electronic state engineering in perovskite-cerium-composite nanocrystals toward enhanced triplet annihilation upconversion. *Advanced Science, 10*(34). https://doi.org/10.1002/advs.202305069.

Graf, D., Jung, J., & Hanemann, T. (2021). Formulation of a ceramic ink for 3D inkjet printing. *Micromachines (Basel), 12*(9), 1136. https://doi.org/10.3390/mi12091136.

Herry, G., Fustec, J.-C., Le Bihan, F., & Harnois, M. (2024). Substrate-free transfer of silicon- and metallic-based strain sensors on textile and in composite material for structural health monitoring. *ACS Appl Mater Interfaces, 16*(17), 22113–22121. https://doi.org/10.1021/acsami.4c01055.

Hossain, R. F., Bandyopadhyay, A. S., Adhikari, N., Aryal, S., & Kaul, A. B. (2024). Emission spectra and strain-dependent properties of flexible inkjet printed WSe_2 –graphene heterostructure photodetectors. *J Mater Chem C Mater, 12*(20), 7176–7187. https://doi.org/10.1039/D3TC03741J.

Iffelsberger, C., Ng, S., & Pumera, M. (2020). Catalyst coating of 3D printed structures via electrochemical deposition: Case of the transition metal chalcogenide MoSx for hydrogen evolution reaction. *Appl Mater Today, 20*, 100654. https://doi.org/10.1016/j.apmt.2020.100654.

Jahangir, A., Singh, P. K., Singh, R. C., Punetha, V. D., & Rawat, S. (2024). Ion conducting polyethylene oxide-based polymer electrolyte dopped with ionic liquid-trifluoromethane-sulfonic chloride. *Macromolecular Symposia, 413*(1). https://doi.org/10.1002/masy.202300139.

Kailkhura, G., Mandel, R. K., Shooshtari, A., & Ohadi, M. (2022). Numerical and experimental study of a novel additively manufactured metal-polymer composite heat-exchanger for liquid cooling electronics. *Energies (Basel), 15*(2), 598. https://doi.org/10.3390/en15020598.

Kang, Y., et al. (2023). Controlled architectures in supported photocatalysis: Exploring the potential of cold plasma discharge, additive manufacturing, and fractal geometry to form hierarchically immobilized Ni-MOF/BiOI/AgVO $_3$. *ACS Appl Mater Interfaces, 15*(26), 31849–31866. https://doi.org/10.1021/acsami.3c05596.

Kasraie, M., Shahid, P. P., & Abadi, S. (2021). Additive manufacturing of conductive and high-strength epoxy nanoclay-carbon nanotube composites. *Addit Manuf, 46*, 102098. https://doi.org/10.1016/j.addma.2021.102098.

Kim, H., Moon, J., Lee, K., & Kanicki, J. (2017). 3D printed masks and transfer stamping process to enable the fabrication of the hemispherical organic Photodiodes. *Adv Mater Technol, 2*(9). https://doi.org/10.1002/admt.201700090.

Koterwa, A., et al. (2022). The role of electrolysis and enzymatic hydrolysis treatment in the enhancement of the electrochemical properties of 3D-printed carbon black/poly(lactic acid) structures. *Applied Surface Science, 574*, 151587. https://doi.org/10.1016/j.apsusc.2021.151587.

Kumar, S., et al. (2023). In-situ N/O-heteroatom enriched micro-/mesoporous activated carbon derived from natural waste honeycomb and paper wasp hive and its application in quasi-solid-state supercapacitor. *J Energy Storage, 72*. https://doi.org/10.1016/j.est.2023.108722.

Kumar, S., Ojha, N., Ramesh, M. R., & Doddamani, M. (2024). 4D printing of heat-stimulated shape memory polymer composite for high-temperature smart structures/actuators applications. *Polym Compos.* . https://doi.org/10.1002/pc.28844.

Kuo, L., et al. (2022). All-printed ultrahigh-responsivity MoS $_2$ nanosheet photodetectors enabled by megasonic exfoliation. *Advanced Materials, 34*(34). https://doi.org/10.1002/adma.202203772.

Lay, M., Say, M. G., & Engquist, I. (2023). Direct ink writing of nanocellulose and PEDOT:PSS for flexible electronic patterned and supercapacitor papers. *Adv Mater Technol, 8*(18). https://doi.org/10.1002/admt.202300652.

Lazarus, N., & Tsang, H. H. (2020). 3-D printing structural electronics with conductive filaments. *IEEE Trans Compon Packaging Manuf Technol, 10*(12), 1965–1972. https://doi.org/10.1109/TCPMT.2020.3038563.

Liang, B., et al. (2021). An integrated paper-based microfluidic device for real-time sweat potassium monitoring. *IEEE Sensors Journal, 21*(8), 9642–9648. https://doi.org/10.1109/JSEN.2020.3009327.

Liu, F., et al. (2022a). Selective metallization on additive manufactured polymer for fabrication of integrated device. *J Mater Sci, 57*(2), 1506–1515. https://doi.org/10.1007/s10853-021-06695-3.

Liu, C., Le, L., Zhang, M., & Ding, J. (2022b). Tunable large-scale compressive strain sensor based on carbon nanotube/polydimethylsiloxane foam composites by additive manufacturing. *Advanced Engineering Materials, 24*(6). https://doi.org/10.1002/adem.202101337.

Liu, H., et al. (2024a). Direct printing of flexible multilayer composite electrodes based on electrohydrodynamic printing. *ACS Appl Electron Mater*. https://doi.org/10.1021/acsaelm.3c01164.

Liu, Y., et al. (2024b). Low dislocation density homoepitaxy ultraviolet-A micro-LEDs scale down to 3 μm. *IEEE Electron Device Letters, 45*(4), 641–644. https://doi.org/10.1109/LED.2024.3368513.

Liu, Y., Shen, S., Hu, J., & Chen, L. (2016). Embedded Ag mesh electrodes for polymer dispersed liquid crystal devices on flexible substrate. *Optics Express, 24*(22), 25774. https://doi.org/10.1364/OE.24.025774.

Liu, W. C., & Watt, A. A. R. (2021). Solvodynamically printed silver nanowire/ethylene-*co*-vinyl acetate composite films as sensitive piezoresistive pressure sensors. *ACS Appl Nano Mater, 4*(8), 7905–7916. https://doi.org/10.1021/acsanm.1c01229.

Li, W., et al. (2022). Molecule bridged graphene/Ag for highly conductive ink. *Sci China Mater, 65*(10), 2771–2778. https://doi.org/10.1007/s40843-022-2064-8.

Li, Z., Khuje, S., Islam, A., & Ren, S. (2023). Additive manufacturing of high-temperature preceramic-derived SiOC hybrid functional ceramics. *Advanced Engineering Materials, 25*(22). https://doi.org/10.1002/adem.202300957.

Martins, L. C., Silva, C. S., Fernandes, L. C., Sampaio, Á. M., & Pontes, A. J. (2023). Evaluating the electromagnetic shielding of continuous carbon fiber parts produced by additive manufacturing. *Polymers (Basel), 15*(24), 4649. https://doi.org/10.3390/polym15244649.

de Matos Morawski, F., et al. (2023). A versatile 3D printed multi-electrode cell for determination of three COVID-19 biomarkers. *Anal Chim Acta, 1258*, 341169. https://doi.org/10.1016/j.aca.2023.341169.

Mohan, J. M., Amreen, K., Javed, A., Dubey, S. K., & Goel, S. (2022). Miniaturized 3D printed electrochemical platform with optimized Fibrous carbon electrode for non-interfering hypochlorite sensing. *Chemosphere, 302*, 134915. https://doi.org/10.1016/j.chemosphere.2022.134915.

Montaina, L., Carcione, R., Pescosolido, F., Montalto, M., Battistoni, S., & Tamburri, E. (2023). Three-dimensional-printed polyethylene glycol diacrylate-polyaniline composites by in situ aniline photopolymerization: An innovative biomaterial for electrocardiogram monitoring systems. *ACS Appl Electron Mater, 5*(1), 164–172. https://doi.org/10.1021/acsaelm.2c01181.

Nassar, H., Dahiya, A. S., & Dahiya, R. (2023). Tunable conductive composite for printed sensors and embedded circuits. *Advanced Intelligent Systems, 5*(10). https://doi.org/10.1002/aisy.202300075.

Okoruwa, L., Sameni, F., Borisov, P., & Sabet, E. (2023). 3D printing soft magnet: Binder study for vat photopolymerization of ferrosilicon magnetic composites. *Polymers (Basel), 15*(16), 3482. https://doi.org/10.3390/polym15163482.

de Oliveira, F. M., de Melo, E. I., & da Silva, R. A. B. (2020). 3D Pen: A low-cost and portable tool for manufacture of 3D-printed sensors. *Sens Actuators B Chem, 321*, 128528. https://doi.org/10.1016/j.snb.2020.128528.

Ozdemir, R., Van Avermaet, H., Erdem, O., Schiettecatte, P., Hens, Z., & Aubert, T. (2023). Quantum dot patterning and encapsulation by maskless lithography for display technologies. *ACS Appl Mater Interfaces, 15*(7), 9629–9637. https://doi.org/10.1021/acsami.2c20982.

O'Neil, G. D., Christian, C. D., Brown, D. E., & Esposito, D. V. (2016). Hydrogen production with a simple and scalable membraneless electrolyzer. *J Electrochem Soc, 163*(11), F3012–F3019. https://doi.org/10.1149/2.0021611jes.

Palacios-Corella, M., Sanna, M., Muñoz, J., Ghosh, K., Wert, S., & Pumera, M. (2023). Heterolayered carbon allotrope architectonics *via* multi-material 3D printing for advanced electrochemical devices. *Virtual Phys Prototyp, 18*(1). https://doi.org/10.1080/17452759.2023.2276260.

Palaganas, N. B., Palaganas, J. O., Doroteo, S. H. Z., & Millare, J. C. (2023). Covalently functionalized cellulose nanocrystal-reinforced photocurable thermosetting elastomer for 3D printing application. *Addit Manuf, 61*, 103295. https://doi.org/10.1016/j.addma.2022.103295.

Peng, D., Khatamifar, M., & Lin, W. (2022). Experimental and numerical studies of the orientation effect on the natural convection heat dissipation of composite polymer heat sinks. *Journal of Enhanced Heat Transfer, 29*(6), 1–26. https://doi.org/10.1615/JEnhHeatTransf.2022041768.

Peng, X., et al. (2021). Integrating digital light processing with direct ink writing for hybrid 3D printing of functional structures and devices. *Addit Manuf, 40*, 101911. https://doi.org/10.1016/j.addma.2021.101911.

Podsiadły, B., Matuszewski, P., Skalski, A., & Słoma, M. (2021). Carbon nanotube-based composite filaments for 3D printing of structural and conductive elements. *Applied Sciences, 11*(3), 1272. https://doi.org/10.3390/app11031272.

Podsiadły, B., Skalski, A., & Słoma, M. (2021). Soldering of electronics components on 3D-printed conductive substrates. *Materials, 14*(14), 3850. https://doi.org/10.3390/ma14143850.

Poompiew, N., et al. (2024). Strain sensing characteristics of 3D-printed carbon nanotubes/polypyrrole/UV-curable composites: Experimental validation and machine learning predictions. *Progress in Additive Manufacturing, 10*, 581–591. https://doi.org/10.1007/s40964-024-00642-9.

Punetha, V. D., et al. (2017). Functionalization of carbon nanomaterials for advanced polymer nanocomposites: A comparison study between CNT and graphene. *Prog Polym Sci, 67*, 1–47. https://doi.org/10.1016/j.progpolymsci.2016.12.010.

Punetha, V. D., et al. (2022). Recent advancements in green synthesis of nanoparticles for improvement of bioactivities: A review. *Curr Pharm Biotechnol, 23*(7), 904–919. https://doi.org/10.2174/1389201022666210812115233.

Punetha, V. D., Ha, Y.-M., Kim, Y.-O., Jung, Y. C., & Cho, J. W. (2019). Interaction of photothermal graphene networks with polymer chains and laser-driven photo-actuation behavior of shape memory polyurethane/epoxy/epoxy-functionalized graphene oxide nanocomposites. *Polymer (Guildf), 181*. https://doi.org/10.1016/j.polymer.2019.121791.

Punetha, V. D., Ha, Y.-M., Kim, Y.-O., Jung, Y. C., & Cho, J. W. (2020). Rapid remote actuation in shape memory hyperbranched polyurethane composites using cross-linked photothermal reduced graphene oxide networks. *Sens Actuators B Chem, 321*. https://doi.org/10.1016/j.snb.2020.128468.

Rafael, R., & Chan, P. K. L. (2023). Copper composites and laser sintering: Novel hybridization method for 3D printed electronics. *Adv Mater Technol, 8*(11). https://doi.org/10.1002/admt.202201900.

Rajesh, S., & Kumawat, A. S. (2024). Integrating additive manufacturing approaches in electrochemistry for enhanced systems — a mini review. *Ionics (Kiel), 30*(2), 677–687. https://doi.org/10.1007/S11581-023-05325-Z/METRICS.

Ramos, D. L. O., de Faria, L. V., Alves, D. A. C., Muñoz, R. A. A., dos Santos, W. T. P., & Richter, E. M. (2023). Electrochemical platform produced by 3D printing for analysis of small volumes using different electrode materials. *Talanta, 265*, 124832. https://doi.org/10.1016/j.talanta.2023.124832.

Rangnekar, S. V., et al. (2023). Electroluminescence from megasonically solution-processed MoS_2 nanosheet films. *ACS Nano, 17*(17), 17516–17526. https://doi.org/10.1021/acsnano.3c06034.

Ranjan, R. K., Kumar, S., Punetha, V. D., Pandey, S. P., Singh, R. C., & Singh, P. K. (2024). Poly(Vinylidene fluoride-CO-hexafluoropropylene) (PVDF-HFP) incorporated acetylene black: Electrical, optical, and structural studies. *Macromol Symp, 413*(1). https://doi.org/10.1002/masy.202300120.

Rocha, D. S., et al. (2021). Sandpaper-based electrochemical devices assembled on a reusable 3D-printed holder to detect date rape drug in beverages. *Talanta, 232*, 122408. https://doi.org/10.1016/j.talanta.2021.122408.

Roshanghias, A., Krivec, M., Bardong, J., & Binder, A. (2020). Additive-manufactured organic interposers. *J Electron Packag, 142*(1). https://doi.org/10.1115/1.4044279.

Rosh-Gorsky, A., et al. (2024). 3D printing of composite radiation shielding for broad spectrum protection of electronic systems. *Advanced Materials, 36*, 2403822. https://doi.org/10.1002/adma.202403822.

(Ross) Salary, R., Lombardi, J. P., Weerawarne, D. L., Tootooni, M. S., Rao, P. K., & Poliks, M. D. (2020). A sparse representation classification approach for near real-time, physics-based functional monitoring of aerosol jet-fabricated electronics. *J Manuf Sci Eng, 142*(8). https://doi.org/10.1115/1.4047045.

Roy, M., & Dickens, T. J. (2017). Additive technology of soluble mold tooling for embedded devices in composite structures: A study on manufactured tolerances. *Addit Manuf, 15*, 78–86. https://doi.org/10.1016/j.addma.2017.03.012.

Rymansaib, Z., et al. (2016). All-polystyrene 3D-printed electrochemical device with embedded carbon nanofiber-graphite-polystyrene composite conductor. *Electroanalysis, 28*(7), 1517–1523. https://doi.org/10.1002/elan.201600017.

Singal, K., Dimitriyev, M. S., Gonzalez, S. E., Cachine, A. P., Quinn, S., & Matsumoto, E. A. (2024). Programming mechanics in knitted materials, stitch by stitch. *Nat Commun, 15*(1), 2622. https://doi.org/10.1038/s41467-024-46498-z.

Singh, P., Singari, R. M., & Mishra, R. S. (2024). Enhanced mechanical properties of MWCNT reinforced ABS nanocomposites fabricated through additive manufacturing process. *Polym Adv Technol, 35*(2). https://doi.org/10.1002/pat.6308.

Stalin, T., Jain, S., Thanigaivel, N. K., Teoh, J. E. M., Raj, P. M. A., & Alvarado, P. V. Y. (2021). Automated fiber embedding for soft mechatronic components. *IEEE Robot Autom Lett, 6*(2), 4071–4078. https://doi.org/10.1109/LRA.2021.3067244.

Sun, H., Zettl, J., & Willenbacher, N. (2023). Highly conductive and stretchable filament for flexible electronics. *Addit Manuf, 78*, 103872. https://doi.org/10.1016/j.addma.2023.103872.

Sun, J., et al. (2020). A novel flexible Ag/AgCl quasi-reference electrode based on silver nanowires toward ultracomfortable electrophysiology and sensitive electrochemical glucose detection. *Journal of Materials Research and Technology, 9*(6), 13425–13433. https://doi.org/10.1016/j.jmrt.2020.09.041.

Sánchez del Río, J., Pascual-González, C., Martínez, V., Jiménez, J. L., & González, C. (2021). 3D-printed resistive carbon-fiber-reinforced sensors for monitoring the resin frontal flow during composite manufacturing. *Sens Actuators A Phys, 317*, 112422. https://doi.org/10.1016/j.sna.2020.112422.

These, A., et al. (2021). Characterization of aerosol deposited cesium lead tribromide perovskite films on interdigited ITO electrodes. *Adv Electron Mater, 7*(3). https://doi.org/10.1002/aelm.202001165.

Thiele, S., Gissibl, T., Giessen, H., & Herkommer, A. M. (2016). Ultra-compact on-chip LED collimation optics by 3D femtosecond direct laser writing. *Opt Lett, 41*(13), 3029. https://doi.org/10.1364/OL.41.003029.

Torres Arango, M. A., Valença de Andrade, A. S., Cipollone, D. T., Grant, L. O., Korakakis, D., & Sierros, K. A. (2016). Robotic deposition of TiO_2 films on flexible substrates from hybrid inks: Investigation of synthesis–processing–microstructure–photocatalytic relationships. *ACS Appl Mater Interfaces, 8*(37), 24659–24670. https://doi.org/10.1021/acsami.6b05535.

Tsyganov, A., et al. (2022). Permittivity and dielectric loss balance of PVDF/K1.6Fe1.6 Ti6.4O16/MWCNT three-phase composites. *Polymers (Basel), 14*(21), 4609. https://doi.org/10.3390/polym14214609.

Vaghasiya, J. V., Mayorga-Martinez, C. C., & Pumera, M. (2021). Smart energy bricks: Ti_3C_2 @Polymer electrochemical energy storage inside bricks by 3D printing. *Adv Funct Mater, 31*(48). https://doi.org/10.1002/adfm.202106990.

Vidakis, N., et al. (2021). Additive manufacturing of multifunctional polylactic acid (PLA)—Multiwalled carbon nanotubes (MWCNTs) nanocomposites. *Nanocomposites, 7*(1), 184–199. https://doi.org/10.1080/20550324.2021.2000231.

Wang, W., et al. (2020). Inflight fiber printing toward array and 3D optoelectronic and sensing architectures. *Sci Adv, 6*(40). https://doi.org/10.1126/sciadv.aba0931.

Wang, Z., et al. (2022). Lead-free piezoelectric composite based on a metamaterial for electromechanical energy conversion. *Adv Mater Technol, 7*(12). https://doi.org/10.1002/admt.202200650.

Wang, Z., Zhang, J., Liu, J., Hao, S., Song, H., & Zhang, J. (2021a). 3D printable, highly stretchable, superior stable ionogels based on poly(ionic liquid) with hyperbranched polymers as macro-cross-linkers for high-performance strain sensors. *ACS Appl Mater Interfaces, 13*(4), 5614–5624. https://doi.org/10.1021/acsami.0c21121.

Wang, Q., et al. (2021b). Composition optimization of multifunctional $CsPb(Br/I)_3$ perovskite nanocrystals glasses with high photoluminescence quantum yield. *Adv Opt Mater, 9*(8). https://doi.org/10.1002/adom.202002075.

Wu, P., et al. (2024). 3D printed room temperature phosphorescence materials enabled by edible natural konjac glucomannan. *Advanced Materials, 36*(31). https://doi.org/10.1002/adma.202402666.

Xia, K., et al. (2023). Electrical conductivity and photodetection in 3D-printed nanoporous structures via solution-processed functional materials. *Adv Mater Technol, 8*(23). https://doi.org/10.1002/admt.202300408.

Yang, Z., Minghan, X., Yun, L., Fengjiao, H., Feng, G., Yanjie, S., Hao, W., & Yafei, Z. (2014). Nitrogen-doped, carbon-rich, highly photoluminescent carbon dots from ammonium citrate. *Nanoscale, 6*(3), 1890–1895. https://doi.org/10.1039/C3NR05380F.

Ye, W., et al. (2024). Additive manufacture of programmable multi-matrix continuous carbon fiber reinforced gradient composites. *Addit Manuf, 89*, 104255. https://doi.org/10.1016/j.addma.2024.104255.

Yu, X., et al. (2021). Additive manufacturing of sandwich-structured conductors for applications in flexible and stretchable electronics. *Adv Eng Mater, 23*(9). https://doi.org/10.1002/adem.202100286.

Zhang, H., et al. (2023). Microscale hybrid additive manufacturing of ultra-fine, embedded Cu/Ag(shell)–P4VP(core) grid for flexible transparent electrodes. *Adv Mater Technol, 8*(8). https://doi.org/10.1002/admt.202201580.

Zhou, G., et al. (2024a). 3D printed MXene-based films and cellulose nanofiber reinforced hydrogel electrolyte to enable high-performance flexible supercapacitors. *J Mater Chem A Mater, 12*(6), 3734–3744. https://doi.org/10.1039/D3TA06925G.

Zhou, H., Zhang, C., Gao, A., Shi, E., & Guo, Y. (2024b). Patterned growth of two-dimensional atomic layer semiconductors. *Chemical Communications, 60*(8), 943–955. https://doi.org/10.1039/D3CC04866G.

Zhou, J., et al. (2023). Hybrid additive manufacturing of ultra-fine high-resolution shell–core structured conductive mesh-based flexible transparent electrodes for flexible displays. *Virtual Phys Prototyp, 18*(1). https://doi.org/10.1080/17452759.2023.2268602.

Zhou, Y., Zhang, S., Li, S., Wei, L., & Zhang, L. (2022). Multifunctional N doped carbon dots for cellular sensing, light-emitting-diode, and additive manufacturing. *Mater Lett, 325*, 132838. https://doi.org/10.1016/j.matlet.2022.132838.

Additively manufactured functional polymers and nanocomposites for electrochemical device applications

17

Vinay Deep Punetha, Rakshit Pathak, Shalini Bhatt, Mayank Punetha, and Parag Sanghani
2D Materials and LASER Actuation Laboratory, Centre of Excellence for Research, PP Savani University, Kosamba, Gujarat, India

17.1 Introduction

The rise of additively manufactured (AM) functional polymers and nanocomposites represents a transformative advancement in electrochemical device applications. This involves the precise deposition of functional polymeric material layer by layer to create complex, three-dimensional structures with tailored properties and optimized performance. Remarkable advancements have been witnessed once the extensive integration of polymers with advance 0D, 1D, 2D, and 3D nanomaterials started in last few years to fabricate sophisticated nanocomposite systems, which opened new frontiers in the design and fabrication of electrochemical devices (Memarzadeh et al., 2023; Yuan et al., 2021). The superiority of AM reflects in its capability to enable unprecedented control over matrix-reinforcement composition, structural geometry, and functional performance. It allows fabrication of complex and sophisticated architectures that were previously unattainable using conventional manufacturing techniques (Ahmad et al., 2024). This capability is particularly beneficial for electrochemical devices, where the efficiency and performance of the device are closely linked to the microstructural arrangement of the active materials. Phase controlling in energy storage devices is a beautiful example of this, such as batteries and supercapacitors, in which the spatial distribution of conductive, electroactive, and insulating phases can be meticulously controlled using AM techniques (Ngidi et al., 2024). This allows for the optimization of ion and electron transport pathways, which ultimately enhances performance parameters such as energy density, power density, and cycle life of the devices. Moreover, the integration of nanomaterials of various dimensions with polymer matrix in AM bring about remarkable improvements in mechanical strength, thermal stability, electrical conductivity, and chemical resistance; enhancing the mechanical integrity of the devices, improving the stability of the active materials, and facilitating faster electron and ion transport. For example, the incorporation of graphene nanoplatelets into polymer matrices has been shown to significantly enhance the electrical conductivity and mechanical properties of the resulting composites, making them ideal candidates for use in electrodes of energy storage devices.

Additive Manufacturing of Functional Polymers and Nanocomposites. DOI: https://doi.org/10.1016/B978-0-443-26516-7.00002-X
© 2025 Elsevier Ltd. All rights are reserved, including those for text and data mining, AI training, and similar technologies.

The advancement in devices is just not limited to the energy devices, in biosensing, 3D-printed microfluidic platforms and electrochemical sensors have demonstrated enhanced detection capabilities, improving the sensitivity and portability of devices used for detecting biomarkers and analytes such as glucose and dopamine. A 3D-printed microfluidic electrochemical sensor for dopamine detection, enhanced with a $CeO_2/BaMoO_4$ nanocomposite electrode, achieved a detection limit of 2.77 μM, demonstrating the impact of 3D printing on biosensing (Hussaini et al., 2024). Similarly, a 3D-printed fluidic sensing platform with screen-printed electrodes improved dopamine detection under flow conditions, achieving a detection limit of 0.05 μM (Contreras-Naranjo et al., 2021). In glucose monitoring, a 3D-printed electro chemiluminescent system integrated with a smartphone achieved a detection limit of 0.04 mM, highlighting the technology's potential for portable diagnostics (Kumar et al., 2023). A paper-based microfluidic biosensor further demonstrated its utility in glucose detection with a wide linear range and an LoD of 25 μM (Cao et al., 2020). Fig. 17.1 shows the fields emission scanning electron microscopy (FE-SEM) image of the fabricated screen-printed electrode. The surface morphology of bare screen-printed electrodes (SPEs), reduced graphene oxide-tetraethylene pentaamine (rGO-TEPA), and rGO-TEPA/PB modified SPEs was examined using this imaging method.

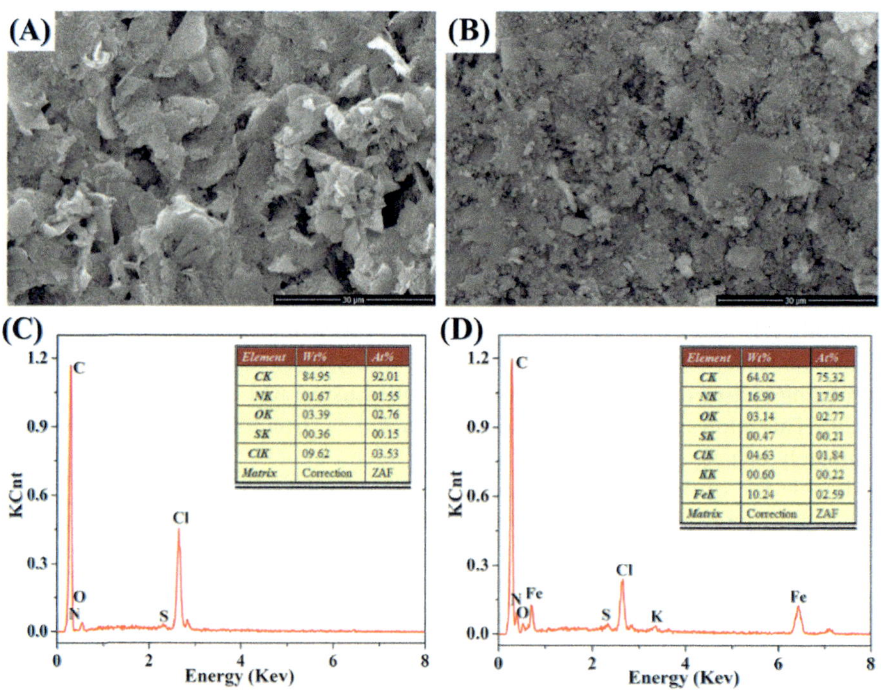

Figure 17.1 SEM images of (A) rGO-TEPA and (B) rGO-TEPA/PB modified SPEs. Energy dispersive spectrums of (C) rGO-TEPA and (D) rGO-TEPA/PB modified SPEs.
Adapted with permission from Cao et al. (2020).

By comparing the SEM images, the distribution of potassium (K) and iron (Fe)—the primary components of Prussian Blue (PB)—is clearly visible in the rGO-TEPA/PB modified paper working electrode (PWE). This observation confirms that the rGO-TEPA/PB nanocomposite was effectively distributed on the surface of the PWE.

3D printing has also made strides in cancer diagnostics, exemplified by a detachable microfluidic device with an electrochemical aptasensor that achieved a detection limit of 17 exosomes/μL, showing promise in early cancer detection (Kashefi-Kheyrabadi et al., 2020). Fig. 17.2A showing the complete analysis process of cancer detection using this technique. Fig. 17.2B showing the complete analysis system in which human plasma was taken for the cancer diagnostic. The fabricated electrochemical aptasensor integrated with a microfluidic channel showed a rapid and excellent on-site analysis of cancerous exosomes in plasma samples.

Moreover, 3D Mn_3O_4 microcubes used for nimorazole (NMZ) detection in head and neck cancer achieved a detection limit of 6 nM, further advancing cancer diagnostics (Umamaheswari et al., 2017). The primary objective of this work was to develop 3D–Mn_3O_4 microcubes as a novel electrocatalyst for the electrocatalytic reduction of NMZ. Schematic representation for the synthesis of 3D–Mn_3O_4MCs and its application towards the determination of NMZ in biological and pharmaceutical samples is shown in Fig. 17.3. The resulting amperometry sensor demonstrated significantly enhanced electrocatalytic activity, with high sensitivity and detection limits in the sub-nanomolar

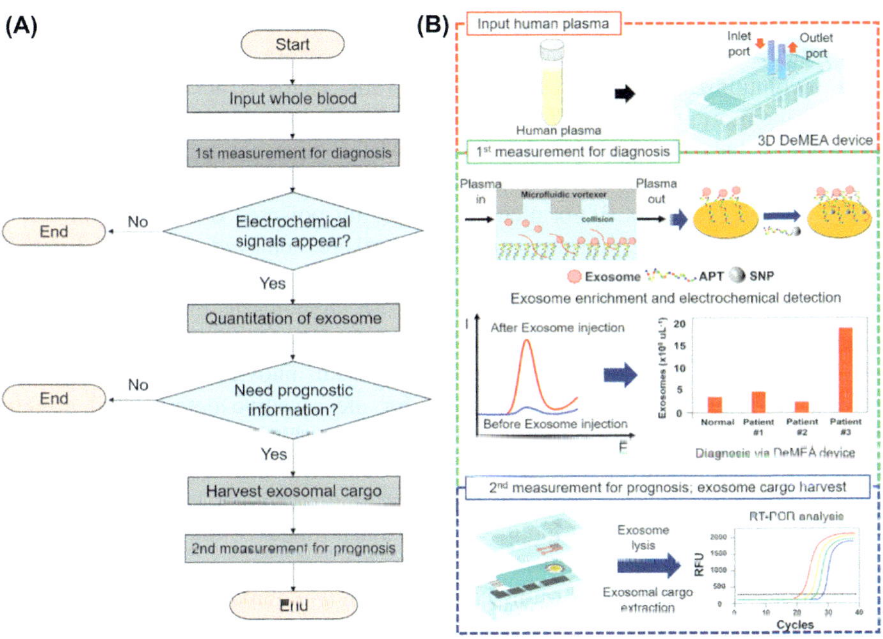

Figure 17.2 (A) The workflow chart for sequential analysis of cancerous exosomes. (B) Schematics of complete analysis system.
Adapted with permission from Kashefi-Kheyrabadi et al. (2020).

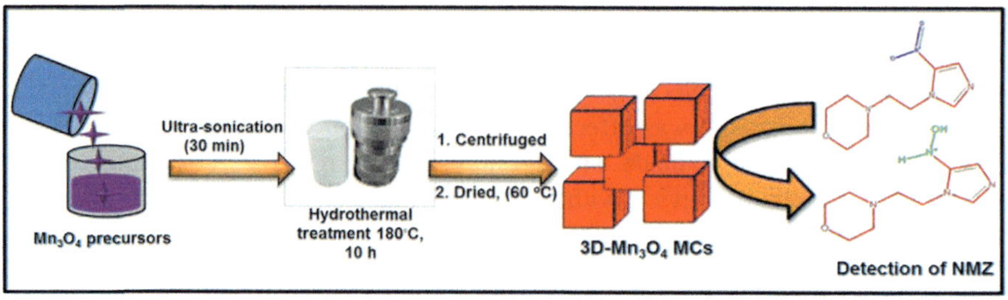

Figure 17.3 Schematic representation for the synthesis of 3D–Mn_3O_4MCs and its application towards the determination of NMZ in biological and pharmaceutical samples.
Adapted with permission from Umamaheswari et al. (2017).

range. Furthermore, this method offers additional benefits, including low overpotential, improved signal output, and rapid response time.

In energy storage, a 3D-printed flexible lithium-ion battery exhibited a volumetric energy density of 157.9 mWh/cm^3, suitable for wearable devices, while 3D-printed silicon-based electrodes showed a specific capacity of 345 mAh/ g and 96% retention after 350 cycles (Zhao et al., 2024; Beydaghi et al., 2021). In addition, 3D printing has been pivotal in creating lab-on-a-chip devices, such as a microfluidic device integrated with an indium-tin-oxide sensor for detecting tumor-derived extracellular vesicles, offering new possibilities for liquid biopsy-based cancer diagnostics (Gurudatt et al., 2023). Another 3D-printed microfluidic device enabled the simultaneous detection of multiple analytes like glucose, dopamine, and ascorbic acid, revolutionizing multiplexed biosensing (dos Santos et al., 2023). In environmental monitoring, a 3D-printed hybrid microfluidic device with disposable electrodes achieved a detection limit of 12 pM for silver nanoparticles, showcasing the potential of 3D printing in developing low-cost sensors for real-time applications (Walgama et al., 2020). 3D-printed supercapacitor-powered electro chemiluminescent immunoarray for prostate cancer biomarkers achieved detection limits of 300–500 fg/mL, highlighting the impact of 3D printing in wearable biosensing technology (Kadimisetty et al., 2016). Fig. 17.4A showing the main array and wash reservior and Fig. 17.4B showing the 3D model of the device and its working mechanism. This fabricated device demonstrated an innovative, low-cost, gravity-driven, 3D-printed, portable immunoarray for the sensitive detection of proteins. While other protein detection techniques offer comparable sensitivity, efficiency, and detection limits, but the novelty of this system lies in its affordability, enabled by 3D printing and the use of a compact, inexpensive, and rechargeable power supply.

**This chapter aims to highlight the transformative impact of additively manufactured functional polymers and nanocomposites on electrochemical device applications. The significance of the chapter lies in its comprehensive exploration of how 3D printing technology is advancing various fields. It discusses key advancements in additive manufacturing and its applications across a broad spectrum. The chapter covers the development of micro-supercapacitors, focusing on fabrication techniques that enhance their performance. It also explores 3D-printed fluidic

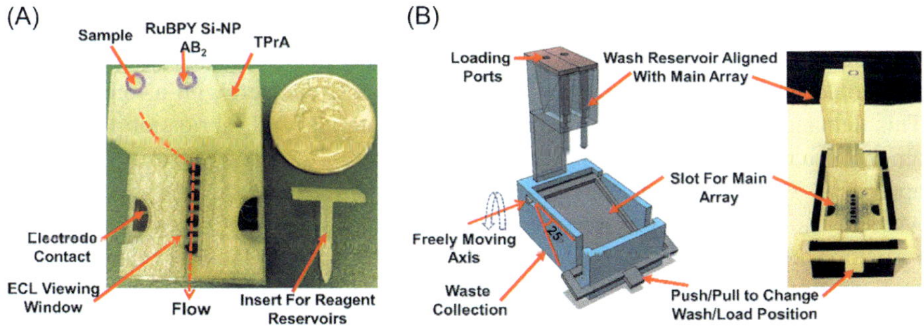

Figure 17.4 3D-printed main array and wash reservoir module. (A) The primary array, displaying three reagent reservoirs equipped with inserts, along with the flow path designed to deliver reagents to the microfluidic channel. (B) The wash reservoir module: (Left) 3D model showing a movable lever that toggles between wash and load positions, with wash reservoirs aligned to the main array; (Right) fully assembled immunoarray setup, combining the main array and wash module.
Adapted with permission from Kadimisetty et al. (2016).

devices, emphasizing their role in improving sensing and diagnostics. Further, it examines innovations in lithium-ion batteries, particularly in enhancing energy storage capacities for wearable technologies. Lab-on-a-chip systems are also discussed, with attention to their potential in streamlining medical diagnostics. Additionally, the chapter explores cancer detection devices that enable early diagnosis and concludes with a review of advancements in biomolecule detection, highlighting the precision and sensitivity of these 3D-printed devices. These developments are crucial for addressing contemporary challenges in energy sustainability, healthcare, and environmental monitoring, making this chapter a valuable resource for researchers and practitioners seeking to harness the full potential of additive manufacturing in electrochemical technologies.

17.2 Micro-supercapacitors

Supercapacitors, also known as ultracapacitors, are energy storage devices that bridge the gap between conventional capacitors and batteries by offering both high power density and moderate energy density. Their basic mechanism relies on electrostatic charge storage at the electrode-electrolyte interface through electric double layers or pseudocapacitive processes. Unlike batteries, supercapacitors store energy without undergoing chemical reactions, allowing them to charge and discharge rapidly (Goikolea & Mysyk, 2017; Castro-Gutiérrez et al., 2020). This makes them crucial for applications requiring quick bursts of energy, such as regenerative braking in electric vehicles, portable electronics, and backup power systems. Their long cycle life and fast response times highlight their growing importance in modern energy storage technologies. In recent years, the evolution of micro-supercapacitors (MSCs) has marked a significant advancement in energy storage, particularly in the context of miniaturized and portable electronics (Xu et al., 2024). Historically, the development of MSCs has been driven by the need for high-performance energy storage devices that could deliver rapid charge and discharge cycles, coupled with a long lifespan. This demand led to the exploration of various materials and fabrication techniques, among which three-dimensional (3D) printing has emerged as a revolutionary approach. The underlying physics of MSCs involves electrochemical processes where ions move between electrodes, storing and releasing energy through the formation and dissolution of electric double layers or through faradaic reactions in the case of pseudocapacitive materials. The ability to control these processes with precision at the microscale is what gives MSCs their high-power density and long cycle life.

Over the past few years, the materials science community has seen a shift towards the use of advanced materials like graphene, metal oxides, and MXenes for the fabrication of MSCs (Xu et al., 2022). However, using nanomaterials can be challenging due to their unique dimensionality and tendency to form aggregates in the matrix, which reflects in the poor performance parameters. Nanocomposites with superior properties, resonating with those of nanomaterials, can only be achieved through strong matrix-reinforcement interaction and effective dispersion. Additive manufacturing technique provides several advantages over traditional composite fabrication techniques including effective solution of material aggregation. Advance techniques such as electro-hydrodynamic-assisted fabrication can be used to create 3D structures to successfully produce advance nanocomposites with superior dispersion to fabricate stable micro-supercapacitors (MSCs). For instance, metallic molybdenum disulfide (MoS_2) in its 1 T phase has shown promise due to its capacitive performance, though its tendency to aggregate has posed challenges in maintaining its structure during electrode fabrication. The innovative use of electrohydrodynamic-assisted fabrication of crumpled MoS_2 (c-MoS_2) has overcome these challenges by producing 3D structures resistant to aggregation, which are then utilized in inkjet printing to create stable and high-performance MSCs (Shao et al., 2020). Fig. 17.5 showing the preparation and characterization of the fabricated 3D c-MoS_2. Fig. 17.5A compares the density of bulk MoS_2, ce-MoS_2 flakes, and 3D 1 T c-MoS2 powders, all in glass vials containing the same weight (4 mg). This comparison highlights the significant volume expansion of 3D 1 T c-MoS_2 following the electrohydrodynamic-assisted fabrication process. The SEM and TEM images of the as-deposited flakes (Fig. 17.5B and C) reveal uniformly dispersed structures with hierarchically porous, thin, and folded walls. Fig. 17.5D shows the bilayer thickness of 3D 1 T c-MoS_2. X-ray photoelectron spectroscopy (XPS) and Raman spectroscopy (Fig. 17.5E and F) were used to characterize the chemical states of the prepared 3D c-MoS_2. The high concentration of the 1 T phase was further confirmed by HRTEM, as shown in Fig. 17.5G. The assembled 3D 1 T c-MoS_2//rGO AMSC exhibited a wide stable working voltage of 1.75 V, excellent capacitance retention of 96% over 20,000 cycles, outstanding areal energy density of 3.85 $\mu Wh/cm^2$, and maximum power density of 12.6 mW in a $MgSO_4$ aqueous electrolyte.

Graphene and its derivatives, such as reduced holey graphene oxide (rHGO), have also been extensively studied for their role in enhancing the performance of MSCs.

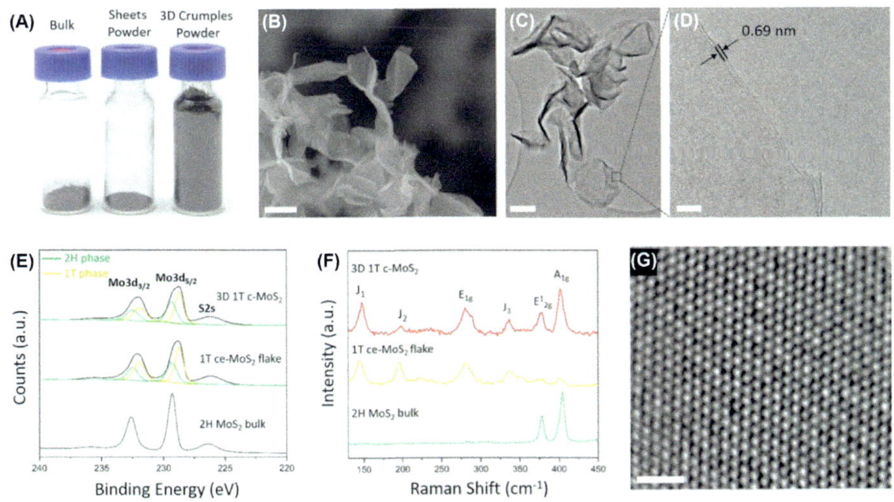

Figure 17.5 Preparation and characterization of 3D c-MoS_2 dispersion with a high concentration of the 1 T phase. (A) Photograph showing the transition from (left) bulk MoS_2 powder, (middle) ce-MoS_2 sheets, to (right) solid 3D 1 T c-MoS2 powder in vials of equal weight for comparison. The crumpling of planar 1 T ce-MoS2 sheets into 3D 1 T c-MoS2 results in notable volumetric expansion. (B) SEM image depicting the morphology of the as-printed 3D 1 T c-MoS_2 (scale bar: 200 nm). (C) Low-magnification TEM image of a single 3D 1 T c-MoS_2 (scale bar: 200 nm). (D) High-magnification TEM image of the 3D 1 T c-MoS2 edge. (E, F) High-resolution XPS spectra of Mo 3d peaks and Raman spectra comparing 3D 1 T c-MoS_2, ce-1T MoS_2 flakes, and bulk 2 H MoS_2. (G) HRTEM analysis confirming the preserved octahedral arrangement in 3D 1 T c-MoS_2.
Adapted with permission from Shao et al. (2020).

In a recent study, Tian et al. (2019) explored 3D printing techniques for rHGO-supported microelectrodes, which exhibited enhanced electrical and ionic transport due to their porous structure (Tian et al., 2019). Fig. 17.6 presents a schematic of the fabricated microelectrode. The 3D-printed microelectrodes not only facilitate electrical conductivity but also provide continuous ionic transport channels. Compared to unmodified graphene-based 3D-printed MPCs, the holey rGO-based quasi-solid-state variants show significantly improved charge storage capacity (up to 241.3 mC/cm²), superior rate capability, and excellent cycling stability, retaining 91% of their capacity after 11,000 cycles

Polypyrrole/polyaniline coaxis nanotubes have also been explored as electrode materials for MSCs, particularly for their compatibility with various substrates and their excellent processability. For instant, Zhang et al. (2018) prepared solid-state micro-supercapacitors with multilayer structure via multi-material 3D printing technique. To obtain electrode materials, they synthesized polypyrrole/polyaniline coaxis nanotubes by utilizing the self-degraded template polymerization method and in situ electrochemical polymerization. The synthesis method and characterization of electrode materials is shown in Fig. 17.7. The as-prepared aqueous composite electrode ink showed significant shear thinning behavior and excellent processability:

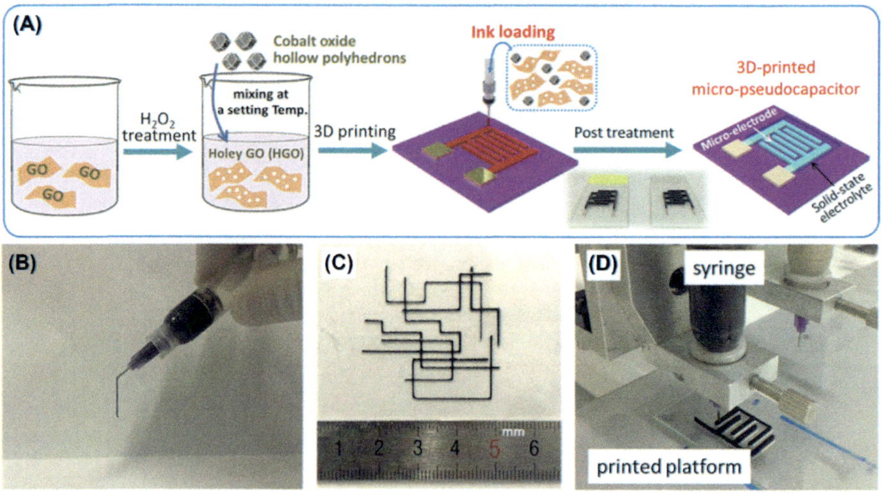

Figure 17.6 (A) Schematic illustration of the fabrication process for 3D-printed quasi-solid-state planar rHGC MPCs. (B) Optical image showing the extrusion of the HGO-Co_3O_4 composite ink from a syringe. (C) Optical image of a printed customized pattern, and (D) the printing process of microelectrodes onto an insulating substrate.
Adapted with permission from Tian et al. (2019).

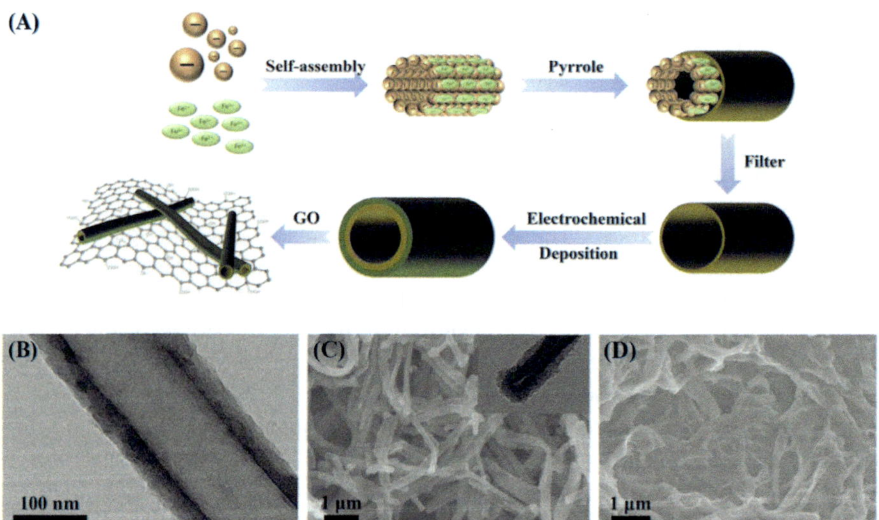

Figure 17.7 Synthesis route and characterization of electrode materials (A) Schematic illustration of the formation of the electrode ink. (B) Transmission electron microscope (TEM) image of PPy nanotubes (C) Scanning electron microscope (SEM) image of PPCANT, the inset is TEM image of PPCANT. (D) SEM image of the surface of chemically reduced electrode. Adapted with permission from Zhang et al. (2018).

a complex electrode patterning with 484 retractions was proved. In addition, the inks were well-adapted to various substrates (paper, texitle, plastic film, etc), allowing the device to be printed onto different substrates, even directly on the desired surface.

Furthermore, the incorporation of conductive materials like silver nanowires (AgNW) and phosphorus-doped graphene into 3D-printed MSCs has resulted in significant enhancements in areal capacitance and energy density, further broadening the application potential of these devices (Liu et al., 2021). The 3D-printed MSCs achieved a real capacitance of 21.6 mF/cm^2 at a scan rate of 0.01 V/s. The devices demonstrated an energy density between 0.5 and 2 $\mu Wh/cm^2$, with a peak power density of 2.5 mW/cm^2. The development of MXene-based inks for 3D printing represents another significant advancement in the field. MXenes, such as Ti_3C_2Tx, are known for their high electrical conductivity and excellent electrochemical properties, making them ideal candidates for MSCs. The ability to 3D prints these materials at room temperature has facilitated the creation of solid-state devices with exceptionally high areal capacitance, reaching up to 1035 mF/cm^2 (Orangi et al., 2020). Improving the stability and performance of MXene inks and broadening the ink composition with other 2D materials, such as graphene or MoS_2, could enhance device efficiency. In addition, efforts should focus on miniaturization and integration with self-powered systems, advancing the development of fully autonomous and flexible electronic devices that maintain high energy and power densities within a limited footprint. The benefits of using 3D printing in the fabrication of MSCs are manifold. In addition, 3D printing offers scalability and versatility, enabling the fabrication of MSCs on a variety of substrates, including flexible and wearable materials. The development of inks that can be printed onto different surfaces has expanded the application range of MSCs, making them suitable for integration into various electronic devices, from portable gadgets to large-scale energy storage systems (Chae et al., 2021; Chen et al., 2022). The process involves sequential printing of highly viscous fluids to form the metallic current collector, electrode, and electrolyte layers. The current collector, made from nickel, is produced using a combination of 3D printing and laser sintering, allowing for stable operation at voltages up to 3 V. Detailed investigations of the interaction between printed particles and laser irradiation were conducted through both experimental and simulation methods. Successive layers of graphene nanoplatelet-based electrode and ionic liquid-based electrolyte were printed on the structured current collector. The resulting 3D-printed micro-supercapacitor achieved energy densities of 25.4 to 8.5 $\mu Wh/cm^2$ and power densities ranging from 150 to 6483 $\mu W/cm^2$, nearing the performance of conventional flat-type devices (Chae et al., 2021). One of the primary challenges is the need for materials that can maintain high energy density, particularly for applications in wearable electronics. Surface modification of graphene oxide (GO) is a key approach to improve its energy density for electrochemical storage, but challenges remain in achieving both hydrophilicity and high energy density simultaneously. Traditional GO inks used in 3D printing for micro energy storage devices often fall short in these areas. The surface engineering strategy was applied to boost the performance of 3D-printed GO micro-supercapacitor chips. By treating the printed chips with a pyrrole monomer, polypyrrole was selectively anchored on the microelectrodes without obstructing the spaces between the finger electrodes. This surface modification resulted in edge-welded graphene scaffolds, leading to a significant increase in specific capacitance from 13.6 to 128.4 $mFcm^2$. This method offers a promising pathway to enhance the efficiency of GO-based micro-supercapacitors and demonstrates the practical potential of 3D printing in energy storage device fabrication. Moreover, the scalability of 3D printing techniques for industrial production remains a significant hurdle. However, ongoing research into new materials and printing methods continues to push the boundaries of what is possible in MSC technology, promising further improvements in performance and wider adoption in various fields (Orangi et al., 2020; Tang et al., 2020). The integration of advanced materials and 3D printing techniques in the fabrication of micro-supercapacitors has revolutionized energy storage, offering new opportunities for the development of high-performance, scalable, and versatile energy storage devices. These innovations not only enhance the performance of MSCs but also expand their potential applications, making them a critical component in the future of portable and wearable electronics (Wang, Tian, et al., 2020; Wang, Zhang, et al., 2020).

S. No.	Material name	Quantitative results summarized	Application	References
1.	MoS2 (1T phase)	High capacitance retention of 96% over 20,000 cycles.	Energy storage, inkjet-printed MSCs	(Shao et al., 2020)
2.	Reduced Holey Graphene Oxide (rHGO)	Specific charge storage capacity of 241.3 mC/cm^2.	Micro-supercapacitors	(Tian et al., 2019)
3.	Polypyrrole/ Polyaniline Coaxis Nanotubes	Scalable fabrication with significant shear thinning behavior.	Wearable micro-supercapacitors	(Zhang et al., 2018)
4.	MXene (Ti3C2Tx)	Areal capacitance up to ~1035 mF/cm^2.	3D-printed micro-supercapacitors	(Orangi et al., 2020)
5.	AgNW/ Phosphorus-doped Graphene	Areal capacitance of 21.6 mF/cm^2 at $0.01/Vs^{-1}$.	Micro-supercapacitors, flexible devices	(Liu et al., 2021)
6.	BCN Microspheres	Areal capacitance of 41.6 mF/cm^2, 93.3% retention after 1000 cycles.	Solid-state flexible micro-supercapacitors	(Tu et al., 2020)

7.	Fe-doped MnO2	Volumetric energy density up to 1.13×10^{-3} Wh/cm^3.	Flexible energy storage devices	(Wang et al., 2020)
8.	CuFe2O4-VACNT Composite	Area-specific energy increase of 43 times, power density increase of 15 times.	3D printed micro-supercapacitor electrodes	(Zhang et al., 2021)

17.3 Three dimentional printed fluidic devices

Fluidic devices have a rich history that traces back to the fundamental principles of fluid mechanics. Initially, microfluidic devices were developed for applications in chemical analysis, biological assays, and the manipulation of small volumes of fluids, driven by the miniaturization trends in science and technology (Convery & Gadegaard, 2019; Lisowski & Zarzycki, 2013). The underlying physics of these devices involves the precise control of fluid flow within microchannels, governed by the principles of laminar flow, diffusion, and capillary action. The advent of 3D printing technology has revolutionized the fabrication of fluidic devices, offering unparalleled flexibility and complexity in design, enabling the production of intricate geometries and integrated functionalities that were previously unattainable. One of the most widely used materials in the fabrication of 3D-printed fluidic devices is polylactic acid (PLA), which is favoured for its ease of printing and biocompatibility. Other materials like acrylonitrile butadiene styrene (ABS), poly (ethylene terephthalate) (PET), and conductive filaments doped with carbon nanotubes have also gained prominence due to their unique properties such as flexibility, chemical resistance, and electrical conductivity (Carneiro et al., 2018; Bishop et al., 2015). The incorporation of materials like graphene, reduced graphene oxide, and MXenes has further enhanced the electrochemical performance of these devices, making them highly effective for applications in biosensing, chemical detection, and energy storage. The use of 3D printing in the fabrication of fluidic devices offers several significant benefits. It allows for rapid prototyping, enabling the quick iteration of designs and the production of customized devices tailored to specific applications. Moreover, 3D printing facilitates the integration of multiple materials within a single device, allowing for the creation of complex multifunctional devices that combine fluidic channels, sensors, and electronic components. This has led to the development of innovative devices such as microfluidic chips for electrochemical detection, which have demonstrated exceptional sensitivity and selectivity in the detection of analytes like hydrogen peroxide, glucose, and dopamine (Duarte et al., 2019). For instance, Walgama et al. (2020) introduced a novel hybrid microfluidic device, hyFlow, which integrates a disposable paper electrode with a 3D-printed plastic chip for electrochemical detection of magnetic bead–silver nanoparticle (MB–AgNP) bioconjugates. The schematic flow diagram of this device is shown in Fig. 17.8. The study focused on the application of the hyFlow system for detecting MB–AgNP model conjugates. This hybrid design was developed to address challenges associated with using magnetic beads in paper-based analytical devices. The hyFlow system enhances the transport of MB conjugates to the electrochemical detection zone, resolving issues found in paper-only devices. While retaining the benefits of pressure-driven fluid transport, the system operates without the need for a pump. The device can detect both pre-formed MB–AgNP conjugates and those formed in situ, with a detection limit of 12 pM, corresponding to just 22 AgNPs per MB.

In recent years, there have been numerous advancements in the design and application of 3D-printed fluidic devices. For example, use of 3D-printed microfluidic devices with integrated electrodes for the electrochemical detection of cancer-derived exosomes, achieving ultra-low detection limits that could facilitate early-stage cancer diagnostics (Kashefi-Kheyrabadi et al., 2020). A microfluidic device with a "channel-inside-channel" configuration for the passive flow of saliva, which was used to monitor multiple salivary biomarkers, highlighting the potential of these devices in point-of-care diagnostics (Vinoth et al., 2023). Furthermore, research has shown that 3D-printed microfluidic devices can be used for complex applications such as the simultaneous detection of multiple biochemical analytes. For instance, a novel microfluidic device was developed by Dang et al. to measure the apoptotic activity of different caspase enzymes, demonstrating the device's capability to perform both sequential and parallel biochemical assays on a single platform (Dang et al., 2019). the schematic diagram of the prepared microfluidic device shown in Fig. 17.9. The fabricated multiplexed biosensing device using 3D-printed molds showed an enabling precise control over liquid sample handling. Compared to the other similar designs, this device offers a

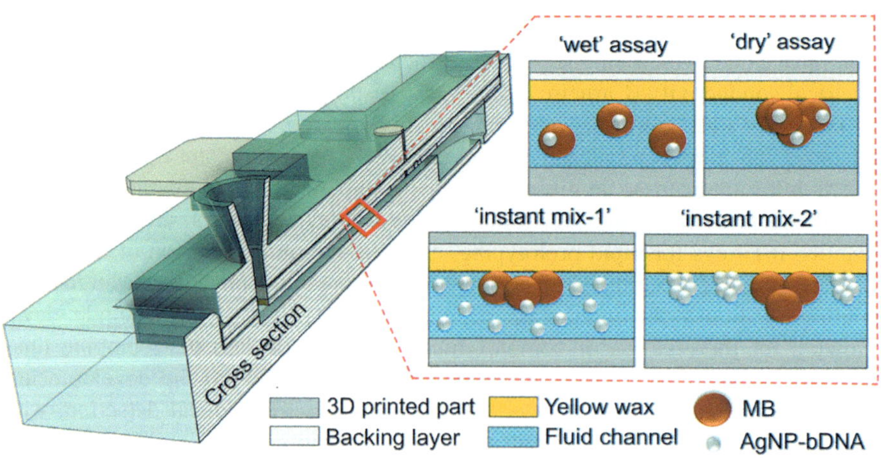

Figure 17.8 Schematic diagram of the microfluidic hyFlow device along with its capability of carrying out four types of model assays.
Adapted with permission from Walgama et al. (2020).

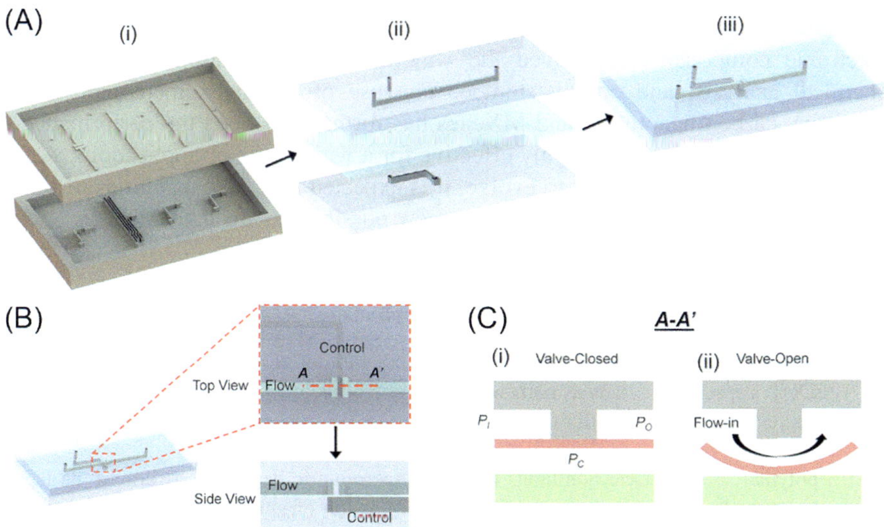

Figure 17.9 Schematic illustration of newly fabricated valve-based microfluidic devices using 3D-printed molds. (A) (i) Individual molds 3D-printed for the top and bottom layers of the valve; (ii) Soft lithography process on the 3D-printed molds to cast PDMS layers, with a thin PDMS membrane created using a spin coater at a specific speed; (iii) Schematic of the fully assembled, normally closed microfluidic actuation unit. (B) Top and side views of the valve. (C) Cross-sectional view illustrating the valve's operation mechanism.
Adapted with permission from Dang et al. (2019).

significantly simpler approach to designing, fabricating, and performing multiplexed assays. As a result, it provides a versatile platform for more straightforward and reliable multiplex detection and sensing of bioanalytes.

This level of sophistication in microfluidic device design is a testament to the versatility and potential of 3D printing in advancing analytical chemistry and biomedical engineering. In addition to their applications in diagnostics and chemical analysis, 3D-printed fluidic devices have also been utilized in energy storage. A notable example is the fabrication of micro-supercapacitors using 3D printing techniques, which have shown enhanced energy density and stability due to the use of advanced materials like MoS_2 and reduced graphene oxide (Shao et al., 2020; Tian et al., 2019) These devices are not only lightweight and flexible but also capable of operating in extreme environments, making them suitable for a wide range of applications from portable electronics to wearable devices.

Moreover, the development of hybrid microfluidic devices that combine paper-based and 3D-printed components has opened new avenues for low-cost, disposable sensors. These devices, such as the hybrid microfluidic device for the electrochemical detection of silver nanoparticles, demonstrate the feasibility of integrating complex fluidic operations with simple and cost-effective manufacturing techniques (Walgama et al., 2020). This approach is particularly beneficial for applications in resource-limited settings where access to sophisticated fabrication facilities is limited.

The ability to fabricate complex, multifunctional devices with integrated sensors and electronic components has paved the way for new applications in diagnostics, chemical analysis, and energy storage. The use of advanced materials such as graphene, carbon nanotubes, and MXenes has further enhanced the performance of these devices, making them highly effective for a wide range of applications. As 3D printing technology continues to evolve, it is expected that the capabilities of fluidic devices will continue to expand, offering new possibilities for innovation in science and technology.

Materials used	Quantitative results	Application	References
Multi-material resins (PEDOT, copper nanoparticles)	Microfluidic channels as narrow as 43 μm obtained	Fabrication of complex microfluidic devices with electrochemical detection	(Fernandes Quero et al., 2023)
ABS polymer, cotton threads, SPCEs	Detection limits for gallic acid 1.5 × 10–6 mol L, caffeic acid 8.0 × 10–7 mol L	Amperometric determination of gallic and caffeic acid in wine samples	(Carneiro et al., 2018)
3D-printed flow cell, commercial threaded reference electrodes	Improved sensitivity using pillar working electrode	In-line microfluidic amperometric detection for ATP and norepinephrine	(Hayter et al., 2020)
PLA, MEA (microelectrode array), microneedles	30 μm sized MEAs, impedance of 140 kΩ, 40 N mechanical fracture strength	Fabrication of devices for in vitro neural and cardiac electrophysiology	(Kundu et al., 2018)
PET, ABS, PBNP-modified electrodes	Detection limit of H2O2: 100 nM, linear response up to 20 μM	Electrochemical sensing of hydrogen peroxide	(Bishop et al., 2015)
Various 3D-printed polymers	Enhanced sensor response time for lactate, glucose, glutamate	High-resolution online analysis of biomarkers in microdialysate	(Samper et al., 2019)
SU-8, conducting polymers (polypyrrole)	Filtration of microparticles 60–100 μm	Electrically controlled microparticle sieve	(Tyagi et al., 2023)
PLA, oil, Span® 80	Average droplet size: 321 ± 9 μm, LOD for *E. coli*: 63.66 CFU/droplet	Label-free counting of E. coli cells in droplets	(Duarte et al., 2019)
Indium-tin-oxide	Detection range: 103	Detection of tumor-	

(ITO), specific antibodies	to 109 tdEVs/mL, linear response observed	derived extracellular vesicles (tdEVs)	(Gurudatt et al., 2023)
Pencil graphites (PGEs), 3D printed μSyringe device	Linear DA detection from 0.1 nM to 500 nM	Real-time measurement of dopamine in cerebrospinal fluid	(Alachkar et al., 2022)
PDMS membrane, 3D-printed valving units	Multiplexed biosensing, sequential and parallel assays	Multiplexed biosensing of apoptotic activity in caspase enzymes	(Dang et al., 2019)
Poly(lactic acid) filament, carbon nanotubes	Droplet length reduction from 3.4 ± 0.1 to 2.3 ± 0.1 mm	Measuring droplet size in microfluidic platforms	(Duarte et al., 2017)
Silver nanoparticles (AgNPs), multiwall carbon nanotubes (MWCNTs)	Detection limit of BPA: 0.35 μM (electrochemical), 0.01 pM (LDI-MS)	Dual detection of bisphenol A using electrochemical and LDI-MS	(Yukird et al., 2020)
Pt or Ag nanoparticles, PDMS-based microfluidic device	Sub mm size printing precision	High-throughput electrochemical screening and localized SERS	(Broccoli et al., 2023)
PolyJet 3D printing technology, platinum black deposition	Enhanced sensitivity in nitric oxide detection	Integration of electrode materials into microfluidic devices	(Castiaux et al., 2020)
Polyethylene glycol diacrylate (PEGDA) waveguides	Integrated biocompatible waveguide for chemical sensing	Use in wearable and implantable lab-on-a-chip devices	(Torres-Mapa et al., 2019)
Streptavidin-modified paramagnetic beads, CdS quantum dots (QDs)	Detection of influenza hemagglutinin via voltammetry	Detection of specific substances for infectious disease control	(Krejcova et al., 2014)
Polylactic acid (PLA), fused deposition modeling (FDM)	Resolution of ~100 μm, modularized microfluidic system	Rapid fabrication of microfluidic chips for cell-based drug screening	(Su et al., 2020)
Graphite carbon-based electrodes, 3D pen polymeric resin	Nitrite and sAA > 94 μmol L−1 and > 610 U/mL in periodontitis saliva compared to ≤ 16 μmol/L and 545 U/mL in healthy saliva	Lab-on-paper device for detecting biomarkers in saliva	(Sousa et al., 2023)
Epoxy-embedded	Linear response for	Rapid prototyping of	

gold electrodes, PDMS microfluidic components	ATP detection in flow conditions, dynamic measurement of concentration changes	electrochemical biosensors	(Belmonte & White, 2022)
Polylactic acid (PLA) filament, conductive carbon-loaded PLA filament	Detection limits for Pb(II) and Cd(II): 0.38 μg L and 0.57 μg L; Repeatability < 4.5%, Reproducibility < 9%	Sequential injection/anodic stripping voltammetry for heavy metals	(Baltima et al., 2021)
Various materials for 2D and 3D microfluidic systems	High sensitivity and reproducibility in cellular response simulation; Quantitation of multiple species with ME coupled to laser-induced fluorescence and electrochemistry	High-throughput drug screening and preclinical toxicological studies	(Caruso et al., 2020)
Graphite/resin composite, stereolithography (SLA)	Detection limits for ferrocene methanol: 4.5 μM; Linear range: 15 – 1000 μM	Reusable 3D printed separation device for electrochemical detection	(Costa et al., 2022)
Flexible screen-printed electrode (SPE), porous Mn2O3 derived from metal-organic framework (Mn-MOF)	Detection limits: 0.5 μg L for Cd (II) and 0.2 μg L for Pb (II), covering two linear ranges of 0.5 to 8 and 10 to 100 μg/L	Electrochemical detection of heavy metals in portable microfluidic device	(Hong et al., 2016)
Electrochemical aptasensor, gold nanostructures, 3D printed magnetic housing	Detection limit: 17 exosomes/μL; Wide dynamic range: 1 × 102 to 1 × 109 exosomes/μL	Early detection of cancerous exosomes in a detachable microfluidic device	(Kashefi-Kheyrabadi et al., 2020)
Glass and polymer	Used in imaging	Production of complex	

materials, ultrafast lasers for 3D internal laser writing	cells, detecting viruses and neurotransmitters, and drug transport	3D microfluidic structures	(O'Neill et al., 2014)
Paper electrode, 3D printed plastic chip, magnetic beads-silver nanoparticle (MB-AgNP) bioconjugate	Detection limit of AgNPs: 12 pM (22 AgNPs per MB)	Hybrid microfluidic device for electrochemical detection	(Walgama et al., 2020)
Covalent organic framework-based nanozymes (COF@Pt), smartphone-based 3D-printed accessory	Detection limit: 1 cell mL−1; Wide linear range: 2 to 105 cells/mL−1	Portable and sensitive device for monitoring circulating tumor cells	(Liu et al., 2022)
PDMS microfluidic channel, 3D printing pen (3D-PP), ssDNA probe	High sensitivity for SARS-CoV-2 RNA detection, fast response	Lab-on-a-chip device for SARS-CoV-2 detection using 3D printing	(Crevillen et al., 2022)
Textile threads, screen-printed electrodes (SPEs)	Sensitivity: 2.24 μA L mmol−1 for AA, 5.09 μA L mmol−1 for DA; Detection limits: 2.60 μmol/L for AA, 1.54 μmol/L for DA	Simultaneous detection of ascorbic acid and dopamine	(dos Santos et al., 2023)
Acrylonitrile butadiene styrene (ABS) filament, carbon black/polylactic acid (CB/PLA) filament, cotton threads	Detection limit for nitrite: 2.39 μmol/L; Good repeatability and reproducibility	Amperometric sensing of nitrite using microflow injection analysis	(Carvalho et al., 2021)
Thermoplastic electrodes (TPEs), polycaprolactone (PCL), polystyrene (PS)	Enhanced electrochemical detection of O_2 and H_2O_2; Introduced 3D-printed TPE template for flow-based sensor	Flow-based sensor for H2O2 detection	(Clark et al., 2024)
Screen-printed	Strong correlation	Electrochemical	

electrodes, clean-room free 3D printing, flexible porous filtration membrane	with parallel blood glucose measurements (Pearson coefficient of 0.9161)	monitoring of multiple salivary biomarkers	(Vinoth et al., 2023)

17.4 Lithium-ion batteries

Advanced 3D printing techniques have revolutionized the fabrication of energy storage devices by enabling the creation of complex, high-performance structures that were previously unattainable with traditional methods. Among these techniques, Fused Deposition Modeling (FDM) stands out for its widespread use and cost-effectiveness. FDM involves the extrusion of thermoplastic materials layer by layer to construct 3D objects. It is particularly advantageous due to its relatively low cost, making it suitable for rapid prototyping and small-scale production. Moreover, FDM can utilize various thermoplastics, including composites with conductive materials like graphene or carbon nanotubes, which enhances the functional properties of the printed structures. However, FDM has certain limitations, including lower resolution and precision compared to other advanced techniques, which can restrict the complexity of the printed structures. In addition, the reliance on thermoplastic materials may limit the types of functional materials that can be employed. An example of the use of FDM in energy storage is the fabrication of Si-based electrodes for lithium-ion batteries, where a specific capacity of up to 345 mAh/g was achieved, demonstrating the potential of this technique for flexible 3D energy storage devices (Beydaghi et al., 2021).

Another notable technique is Direct Ink Writing (DIW), which has emerged as one of the simplest yet versatile 3D printing technologies, particularly suitable for fabricating energy storage devices. DIW involves the extrusion of ink materials through a nozzle, allowing for precise control over the deposition process. This technique is highly valued for its ability to print a wide range of materials, including nanomaterial-based inks, and for achieving customized structures with high printability and desired functionality. DIW has been used to create high-performance electrodes with specific capacities close to theoretical values, such as 168 mAh/g for $LiFePO_4$ cathodes (Tian et al., 2021). The versatility and precision of DIW make it an excellent choice for developing next-generation customized and high-performance energy storage devices. In contrast, Digital Light Processing (DLP) offers a different approach by utilizing a digital light projector to cure photosensitive resins layer by layer. This technique is particularly advantageous for creating fine features and complex geometries with high resolution, making it ideal for fabricating intricate 3D architectures, such as multicomponent metal oxides used in lithium-ion battery cathodes. An example of DLP's application is the fabrication of lithium cobalt oxide (LCO) structures with micro-porous architectures. Yee et al. developed a novel method for fabricating 3D multicomponent metal oxides, unlocking new possibilities for energy storage devices like lithium-ion battery (LIB) electrodes with customizable architectures (Yee et al., 2021). Traditional methods rely on particle-based or organic–inorganic binders, limiting both resolution and chemical composition. To overcome these challenges, they used aqueous metal salt solutions as metal precursors, providing a platform for 3D printing complex metal oxides. Resin formulation and the synthesis process is schematically shown in Fig. 17.10. The 3D hydrogels were calcined to produce micro-porous, self-similar LCO structures with a resolution of approximately 100 μm. The resulting free-standing, binder- and additive-free LCO structures were integrated as cathodes into LIBs, demonstrated 76% capacity retention over 100 cycles at C/10. This innovative approach paves the way for fabricating multicomponent metal oxides with complex 3D architectures, offering a versatile solution that could extend to other materials by modifying the metal salt solutions. It mirrors the innovation in material design seen in your work on chitin and chitosan-based nanocomposites, where manipulating molecular structures opens doors to advanced functionalities in drug delivery systems.

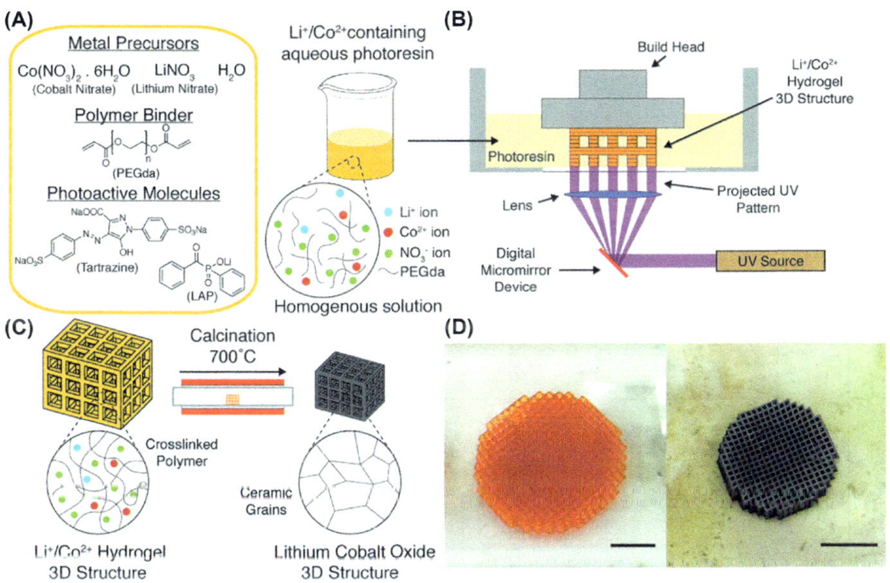

Figure 17.10 DLP printing of aqueous metal salt-containing photoresins (A) Resin formulation and schematic representation of solvated metal ions combined with PEGda in the aqueous photoresin. (B) Schematic illustration of DLP printing process for a Li^+/Co^{2+} hydrogel 3D structure. (C) The 3D Li^+/Co^{2+} hydrogel is calcined to produce a self-similar LCO structure. (D) Optical images showing the Li^+/Co^{2+} hydrogel (left) and the calcined LCO cubic lattices (right). Scale bars indicate 5 mm.
Adapted with permission from Yee et al. (2021).

S. No.	Material used	Quantitative results	3D printing content	Application	References
1.	Graphene/polylactic acid (PLA)	200-fold increase in specific capacity after chemical pre-treatment (500 mAh/ga at 40 mA/ga	Fused deposition modeling using tailored graphene/PLA filament	Li-ion batteries (3D printed freestanding anodes)	(Foster et al., 2019)
2.	Sacrificial sugar particles and polydimethylsiloxane resin	Decent electrochemical performance comparable to conventional electrodes; stable under mechanical deformations	3D-printing technology using sacrificial sugar particles	Stretchable energy storage devices for wearable electronics	(Praveen et al., 2023)
3.	Full ceramic LTO anode and LCO cathode	LTO anode: 168 mAh/ga, LCO cathode: 129 mAh/ga (96% and 94% of theoretical capacities)	Fused filament fabrication (FFF) 3D printing	Li-ion batteries with tailored geometries	(Valera-Jiménez et al., 2021)
4.	Reduced chemically modified graphene (rCMG) and Copper (Cu)	Self-standing binder-free electrode; low resistance interface and enhanced electrical properties	3D printing of multi-material structures using thermoresponsive inks	Supercapacitors and Li-ion batteries	(Rocha et al., 2017)
5.	Non-Newtonian fluid-based inks (Anode, cathode, separator layer, packaging layer)	Charge and discharge capacities: 147 and 99 mAh/g at 0.1 C rate	Multi-material integrated 3D printing method	Cylindrical Li-ion battery with core-shell structure	(Zhou et al., 2022)
6.	MXene-based anodes	Superior electrochemical performance, improved cycling stability and rate capability	Scalable printing process for 3D MXene-based anodes	High-performance Lithium-ion batteries	(Kumarov et al., 2023)
7.	Various materials including UV-curable PDMS and ETPTA	Shape-versatile ADP-LIBs with reliable extrusion, competitive electrochemical performance	Direct ink writing (DIW) 3D printing with rheological optimization	All 3D DIW printed scalable shape versatile Li-ion batteries	(Bae et al., 2023)
8.	MoS2@NiS/rGO	Reversible capacity: 289.5 mAh/ga at 0.1 A/ga, Cycling stability: 131.3 mAh/ga after 800 cycles at 1 A/ga	3D printing (3DP) technology	Sodium-ion batteries with high capacity and stability	(Han et al., 2024)

9.	Iron disulfide (FeS_2) and Lithium metal	Optimal power, uniformity, and stability; similar performance to planar configurations	Direct-ink-write (DIW) printing of FeS_2 inks	Custom-form conversion lithium batteries	(Cardenas et al., 2022)
10.	LTO/CNT-based inks	Capacity-rate linked to physical properties; limiting factor correlated with material arrangement	Comparison of various additive manufacturing techniques	Micro-energy storage devices with enhanced rate performance	(Pinilla et al., 2023)
11.	Organic solvent-based graphite anode slurry	Areal specific capacity of the printed electrode is higher than traditional coated electrodes	Pneumatic jet-based 3D printing	Li-ion batteries with improved anode performance	(Praveen et al., 2020)
12.	Poly(vinyldene fluoride) with Al2O3 filler	Comparable high rate electrochemical performance to commercial separators, better wetting and thermal stability	Dry phase inversion method and direct write extrusion	Flexible Li-ion batteries with controlled porosity	(Blake et al., 2017)
13.	Topologically-structured electrodes (TSEs)	Retains 98% of original capacity after 50 stretching cycles; superior to conventional MSEs	3D printing with topology optimization	Stretchable and flexible batteries for wearable tech	(Hu et al., 2024)
14.	Graphite anode slurry	Linear electrode pattern enhances electrochemical performance	Pneumatic printing device	Li-ion batteries with enhanced electrochemical performance	(Zhang et al., 2021)
15.	NiCo bimetallic oxyhydroxides (NixCoyOOH)	Specific capacitance: 3720 F/ga at 1 A/g, energy density: 22.8 Wh/kga at 7466 W/kga	3D printing for supercapacitors in cold environments	Supercapacitors capable of low-temperature operation	(Fan et al., 2023)
16.	LFP/LTO electrodes	92% and 88% capacity retention after 100 stretch-release cycles	3D-printed LFP/LTO electrodes	Flexible and stretchable batteries for wearable electronics	(Bao et al., 2020)
17.	PEO/LiTFSI filament	Conductivity of 2.18 Ã—10^-3 S/cm at 90 Â°C, facilitated printability	Fused deposition modeling (FDM) 3D-printer	Solid-state battery integration within 3D objects	(Maurel et al., 2020)
18.	Advanced energy storage materials with various strategies	Review of advanced battery materials and strategies	Various 3D printing methods	High performance energy storage devices including Li-ion, sodium-ion, and others	(Kurmanbayeva et al., 2021)

19.	Graphene oxide with polyvinyl alcohol/sodium metaborate	High volumetric energy density	Direct ink writing-based 3D printing	Flexible fiber lithium-ion batteries for wearable devices	(Zhao et al., 2024)
20.	Si-based electrodes with polylactic acid, carbon black-doped polypyrrole, and graphene flakes	Specific capacity up to at 20 mA g, 96% capacity retention after 350 cycles	Fused deposition modeling (FDM) method	Flexible polymeric-based 3D energy storage devices	(Beydaghi et al., 2021)
21.	Carbon nanotubes, LFP, LTO in polymer inks	Specific capacity of 110 mAh/g at 50 mA g.	3D printing with high viscous polymer inks	All-fiber LIBs for wearable electronic applications	(Wang et al., 2017)
22.	Natural graphite and $LiNi_{0.6}Co_{0.2}Mn_{0.2}O_2$ with vapor-grown carbon fibers	Prototype TYT LIB with improved electrochemical performance and flexibility	Direct ink writing-based 3D printing technology	Twisted yarn-type (TYT) LIBs for smart fabrics	(Praveen et al., 2021)
23.	Nanomaterial inks with $LiFePO_4$ and $Li_4Ti_5O_{12}$	Specific capacities of 168 mAh/g (0.5 C) and 171 mAh/g (1 C)	Direct ink writing with tailored rheological properties	Customized and high-performance energy storage devices	(Tian et al., 2021)
24.	Lithium cobalt oxide (LCO)	76% electrochemical capacity retention over 100 cycles at C/10	Digital light processing (DLP) 3D printing	Multicomponent metal oxides with complex 3D architectures	(Yee et al., 2021)
25.	MnO_2 with high mass loading	Superior properties at MnO_2 high mass loading (10 mg/cm^2 and 15 mg/cm^2)	Ultrasonic processing, freeze-drying, and 3D printing	High-performance AZIBs with grid structure 3D cathodes	(Nie et al., 2023)
26.	Vapour grown carbon fibers (VGCFs)	Prototype pouch cell with self-standing electrodes showing flexibility	3D printing to create self-standing electrodes	Flexible Li-ion batteries with integrated electrodes	(Praveen et al., 2022)
27.	Reduced graphene oxide-based printing ink	Improved conductivity and electrochemical properties due to 3D conductive network	3D printing assembly strategy	1D flexible fiber lithium-ion batteries for flexible electronics	(Zhao et al., 2023)
28.	SiNi nanoparticle anodes with CNT coatings	High capacity (500–1200 mAh/g anode) long cycle life (up to 500 cycles)	Drop-on-demand (DoD) dispensing technology	High-activity Si-based anodes for Li-ion batteries	(Ben-Barak et al., 2022)

Despite its advantages, DLP can be limited by the types of materials that can be cured using light, which may restrict its applicability in certain energy storage applications. The most widely used materials in the context of 3D printing for energy storage devices are graphene and its derivatives, carbon nanotubes (CNTs), and silicon-based materials. Graphene, including graphene oxide and reduced graphene oxide, is extensively used for its exceptional conductivity and mechanical properties, making it ideal for flexible lithium-ion batteries and 3D-printed anodes. Notable examples include the development of flexible fiber lithium-ion batteries with a volumetric energy density of 157.9 mWh/cm^3 (Zhao et al., 2024). Carbon nanotubes are another frequently utilized material, valued for their high electrical conductivity and strength. They are integrated with materials like LFP and LTO in polymer inks for fabricating flexible electrodes, as seen in the all-fiber LIBs that demonstrate high specific capacity and flexibility (Wang et al., 2017). Lastly, silicon-based materials, such as Si nanoparticles, are used for their high theoretical capacity in anodes. These materials are central to the production of advanced 3D-printed flexible energy storage devices, with notable performance improvements in applications like flexible polymeric-based 3D energy storage devices (Beydaghi et al., 2021). While both materials are used to improve the performance of energy storage devices, their specific properties lead to different applications. GO's functional groups allow for better chemical interactions and easier processing, making it a preferred choice for applications requiring good dispersibility and chemical reactivity, such as flexible and printable electrodes. On the other hand, CNTs excel in scenarios where exceptional conductivity and mechanical reinforcement are needed, such as in the fabrication of robust, flexible electrodes with high electrochemical performance. Despite their differences, both GO and CNTs are integral to the development of next-generation energy storage solutions, each offering unique benefits that cater to specific needs in the field. Advanced 3D printing techniques have significantly enhanced the development of high-performance energy storage devices, enabling the fabrication of complex and efficient structures that were previously unattainable. One such advancement is the creation of flexible fiber lithium-ion batteries using direct ink writing-based 3D printing, which achieved an impressive volumetric energy density of 157.9 mWh/cm^3 and demonstrated remarkable flexibility with the ability to withstand strains of up to 30% (Zhao et al., 2024). Similarly, the use of fused deposition modeling (FDM) to produce 3D-printed silicon-based electrodes resulted in a specific capacity of up to 345 mAh/g, with a capacity retention of 96% after 350 cycles, highlighting the potential of this technique for flexible 3D energy storage devices (Beydaghi et al., 2021). Direct ink writing (DIW) has also proven effective, particularly in fabricating high-loading nanomaterial inks, where $LiFePO_4$ cathodes and $Li_4Ti_5O_{12}$ anodes delivered specific capacities close to their theoretical values, demonstrating excellent long-term cycling stability (Tian et al., 2021).

In the realm of wearable electronics, flexible all fiber lithium-ion batteries were developed using 3D printing, achieving a specific capacity of approximately 110 mAh/g, which demonstrates their potential for integration into textile fabrics (Wang et al., 2017). In this work, Wang et al. (2017) applied a scalable, low-cost, and high-efficiency 3D printing technology to fabricate a flexible all-fiber lithium-ion battery

(LIB). To achieve this, highly viscous polymer inks containing carbon nanotubes were utilized to print lithium iron phosphate (LFP) fiber cathodes and lithium titanium oxide (LTO) fiber anodes. The schematic illustration the design concept and fabrication process of 3D printed all-fiber flexible lithium-ion batteries (LIBs) is shown in Fig. 17.11 Both fiber electrodes exhibited excellent flexibility and high electrochemical performance in half-cell configurations. The all-fiber LIB was successfully assembled by twisting the printed LFP and LTO fibers together with a gel polymer used as a quasi-solid electrolyte. This all-fiber device demonstrated a high specific capacity of approximately 110 mAh/g at a current density of 50 mA g while maintaining good flexibility in the fiber electrodes. This advancement suggests potential integration into textile fabrics for future wearable electronic applications.

In addition, MnO_2 cathodes for aqueous zinc-ion batteries have been developed using a combination of ultrasonic processing, freeze-drying, and 3D printing, resulting in superior electrochemical properties at high mass loadings (Nie et al., 2023). The integration of vapor-grown carbon fibers (VGCFs) into self-standing electrodes through 3D printing has also shown promising results, balancing flexibility with electrochemical performance in prototype pouch cells (Praveen et al., 2022). The rGO fiber electrodes were produced using a 3D printing assembly strategy, leveraging the high specific surface area of graphene oxide to enhance electronic conductivity and ion mobility (Zhao et al., 2023). Moreover, Drop-on-Demand (DoD) printing technology has been effectively used to create silicon-based anodes with high capacities and long cycle life, showcasing the technique's potential for specialized energy storage applications (Ben-Barak et al., 2022). These advancements highlight the diverse potential of 3D printing technologies in shaping the future of energy storage devices.

Despite the significant advancements and potential of 3D printing techniques in energy storage devices, several challenges need to be addressed to fully realize

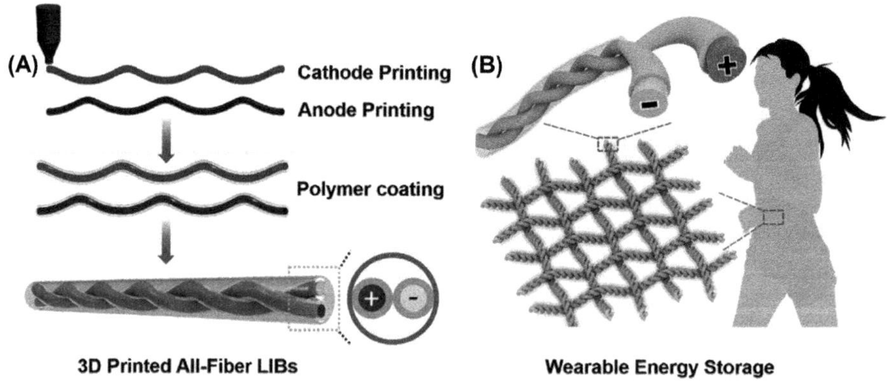

Figure 17.11 Schematic illustrating the design concept and fabrication process of 3D printed all-fiber flexible lithium-ion batteries (LIBs). (A) Overview of the 3D printing fabrication process. (B) Potential application of fiber-shaped batteries for wearable technology. Adapted with permission from Wang et al. (2017).

their benefits. One of the primary challenges is related to material limitations, particularly in terms of compatibility and printability. Not all energy storage materials are easily adaptable to 3D printing processes, as the rheological properties of inks must be precisely controlled to ensure smooth extrusion and accurate layer deposition. This often limits the range of materials that can be used, as certain high-viscosity inks necessary for effective printing may reduce the resolution and detail of the printed structures. In addition, achieving a fine balance between the mechanical properties and electrochemical performance of the printed materials can be difficult, especially when dealing with complex composites or multi-material structures. Another significant challenge is the resolution and precision of the printing techniques themselves. While methods like DLP offer high-resolution capabilities, other techniques like Fused Deposition Modeling (FDM) and Direct Ink Writing (DIW) may struggle to achieve the same level of detail, particularly when working with more viscous or heterogeneous materials. This can result in lower-quality prints with suboptimal performance characteristics, limiting the effectiveness of the final energy storage device. Moreover, scalability remains a major hurdle. While 3D printing is highly effective for prototyping and small-scale production, scaling these techniques for large-scale manufacturing while maintaining quality and performance consistency is challenging. The cost of materials, printer maintenance, and the slower printing speeds compared to traditional manufacturing processes can also hinder the broader adoption of these technologies in commercial energy storage applications. Compatibility with current technologies, ensuring long-term stability and performance under various operating conditions, and addressing issues related to the durability and lifespan of printed components are all critical considerations that must be addressed to make these techniques viable for widespread use. Despite these challenges, ongoing research and development are likely to overcome these obstacles, paving the way for more efficient, flexible, and customizable energy storage solutions in the future.

17.5 Lab on a chip

Lab-on-a-chip (LOC) technology represents a significant advancement in the miniaturization of laboratory processes, enabling complex biochemical and medical assays to be performed on a micro scale device. These devices integrate multiple laboratory functions onto a single chip, often only a few square centimeters in size, allowing for the efficient handling, analysis, and processing of small volumes of fluids. LOC systems have applications across a broad range of fields, including diagnostics, drug development, environmental monitoring, and chemical synthesis. Their ability to perform tasks that would otherwise require bulky, expensive, and time-consuming laboratory setups has revolutionized many scientific and clinical processes. The advent of 3D printing has significantly impacted the development and evolution of LOC devices. Traditionally, LOC devices were fabricated using complex, time-consuming, and expensive techniques such as photolithography and soft lithography, which limited their accessibility and widespread use. 3D printing, with its ability to rapidly prototype and fabricate intricate microfluidic structures with high precision, has democratized the development of LOC systems. It allows for the creation of customizable, complex designs that can be quickly iterated upon, enabling researchers to experiment with innovative layouts and functionalities. Furthermore, 3D printing reduces the cost of production and accelerates the transition from conceptual designs to functional prototypes, making LOC technology more accessible to a broader range of researchers and industries. Moreover, 3D printing allows for the integration of various materials and components into a single LOC device, facilitating the incorporation of sensors, electrodes, and other functional elements directly into the chip. This integration capability has led to the development of more sophisticated LOC systems that can perform complex analyses, such as simultaneous detection of multiple biomarkers or the manipulation of biological cells, with unprecedented ease and efficiency. As a result, 3D printing has not only revolutionized the fabrication process but has also expanded the possibilities and functionalities of LOC devices, driving innovation in fields that rely on micro-scale laboratory processes. Katseli et al. developed a 3D-printed electrochemical cell-on-a-chip device using a dual extruder printer (Katseli et al., 2020). Interestingly, this device costs only $0.11 in materials and is ready-to-use without any pre-activation. It demonstrated high utility for simultaneous detection of paracetamol and caffeine in pharmaceutical tablets and urine, showcasing how 3D printing can make LOC technology both cost-effective and highly functional. In contrast, the device described by Hayter et al. integrates a three-electrode setup into a microfluidic device with amperometric detection. This setup allows for in-line detection of norepinephrine and ATP, emphasizing how 3D printing can enhance the integration of complex detection systems into microfluidic devices (Hayter et al., 2020). On the other hand, the detachable microfluidic device introduced in research work showcases another significant advancement: the ability to quantify cancer-derived exosomes with ultra-sensitivity (Kashefi-Kheyrabadi et al., 2020). This case highlights how 3D printing allows for detachable components that can be separately analyzed, a feature that offers tremendous flexibility and utility in biomedical diagnostics. Moreover, the work of Durate et al. demonstrates the use of 3D printing to develop microfluidic devices with integrated electrodes for label-free counting of *Escherichia coli* cells (Duarte et al., 2019). It highlights the potential of 3D printing in microbiology, where precise control over cell detection is critical. Conversely, Gurudutt et al. explores the detection of tumor-derived extracellular vesicles (tdEVs), employing a microfluidic device integrated with an ITO sensor (Gurudatt et al., 2023). The ability to evaluate the epithelial and mesenchymal markers on tdEV surfaces highlights the versatility of 3D printing in developing complex diagnostic tools. In addition, the research Michalek et al. utilized a 3D printed device for the automated analysis of peptide-conjugated quantum dots, showcasing the automation potential that 3D printing brings to LOC systems (Michalek et al., 2016). In contrast, Walagama et al. presents a hybrid microfluidic device that integrates disposable paper electrodes with a 3D printed plastic chip, highlighting the synergy between traditional and novel materials made possible by 3D printing (Walgama et al., 2020). Another interesting case is the development of a droplet manipulation device, which employs a superhydrophobic coating on a 3D-printed microfluidic device to enable high-throughput assays (Oomen et al., 2017). This method demonstrates how 3D printing can facilitate advanced fluid dynamics in microfluidic applications. Meanwhile, Pearce et al. discusses the development of a wax 3D printer for paper-based microfluidic applications, emphasizing the role of 3D printing in cost reduction and rapid prototyping, which is especially important for scaling up research (Pearce et al., 2016). In another study, Kundu et al. presented a nontraditional fabrication of 3D microelectrode arrays capable of interfacing with 3D cellular networks (Kundu et al., 2021). This work highlights the potential of 3D printing in the development of complex biological models, contrasting with (Punjiya et al., 2018), which focuses on the creation of a paper-based analytical device with 3D fluid reservoirs for point-of-care diagnostics, underscoring the practicality and accessibility of 3D-printed LOC systems.

In similar investigation, Parekh et al. introduces a method to fabricate 3D microchannels using liquid metal as a sacrificial template (Parekh et al., 2016). This technique contrasts with the work of Najjar et al. which developed a lab-on-a-chip for multiplexed electrochemical detection of SARS-CoV-2, highlighting the application of 3D printing in rapid response to global health challenges (Najjar et al., 2022). In this work, Najjar et al. developed a 3D-printed lab-on-a-chip device designed for the rapid, accurate, and frequent detection of SARS-CoV-2 RNA and serological host antibodies, facilitating the assessment of individuals' immune status regarding COVID-19. Overview of the microfluidic chip designed for a lab-on-a-chip (LOC) sample-to-answer detection of SARS-CoV-2 RNA and antibodies in saliva is shown in Fig. 17.12. This device enables the concurrent detection of SARS-CoV-2 RNA in saliva and anti-SARS-CoV-2 immunoglobulins in saliva spiked with blood plasma, achieving results within two hours through multiplexed electrochemical outputs. The lab-on-a-chip automates the extraction, concentration, and amplification of SARS-CoV-2 RNA from unprocessed saliva. It integrates Cas12a-based enzymatic detection of the RNA using isothermal nucleic acid amplification with a sandwich-based enzyme-linked immunosorbent assay (ELISA) on electrodes functionalized with key SARS-CoV-2 antigens, including Spike S1, nucleocapsid, and receptor-binding-domain proteins. By employing inexpensive microfluidic electrochemical sensors, this innovative approach allows for multiplexed diagnostics at the point of care, potentially enabling widespread monitoring of COVID-19 infection and immunity levels in the population.

LOC technology, enhanced by 3D printing, holds immense potential across a wide range of applications. In healthcare, LOC devices can revolutionize point-of-care diagnostics by enabling rapid, on-site testing for diseases such as COVID-19, cancer, and other conditions, reducing the need for extensive lab infrastructure. The ability to detect biomarkers, including tumor-derived extracellular vesicles (tdEVs) and neurotransmitters, offers new avenues for early diagnosis and monitoring of diseases, potentially improving patient outcomes.

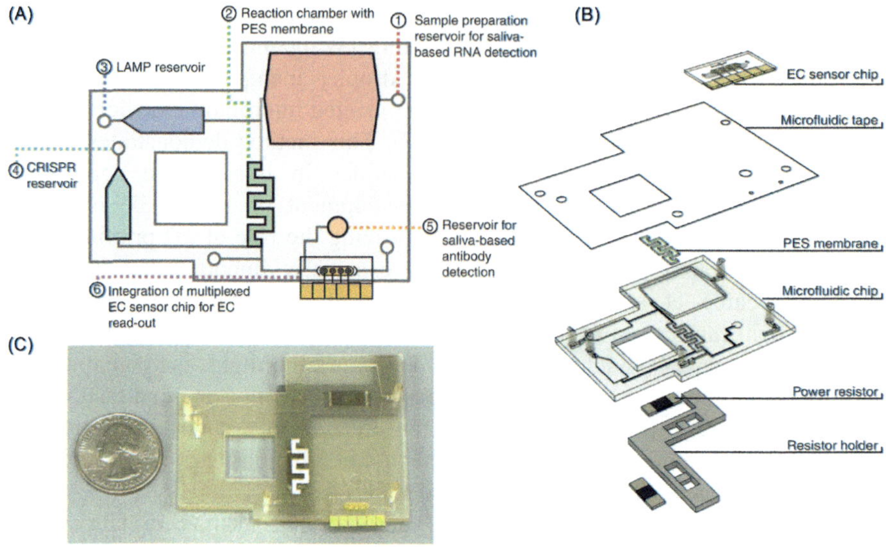

Figure 17.12 (A) Overview of the microfluidic chip designed for a lab-on-a-chip (LOC) sample-to-answer detection of SARS-CoV-2 RNA and antibodies in saliva. (1) The user adds saliva to the antibody detection reservoir and a mixture of saliva and proteinase K to the sample preparation reservoir for incubation. (2) The saliva is then pumped over the PES membrane within the reaction chamber for RNA capture, while being heated to denature potential reaction inhibitors. (3) The LAMP solution is subsequently pumped from the reservoir into the reaction chamber for incubation. (4) The CRISPR mixture is introduced into the reaction chamber, incubated, and then pumped over the electrochemical (EC) sensor chip. (5) The saliva for antibody detection is pumped over the EC sensor chip. (6) After adding polystreptavidin-HRP and TMB, the results from the EC sensor chip are read using a potentiostat. (B) Exploded view of the multiplexed system, showcasing the heater system, a sealed microfluidic chip, and the multiplexed EC sensor chip. (C) Photograph of the microfluidic system with a quarter dollar for scale.
Adapted with permission from Najjar et al. (2022).

In environmental monitoring, LOC devices can be employed to detect contaminants in water or air, offering a portable solution for real-time analysis in remote or resource-limited areas. Similarly, in food safety, these devices could be used to detect pathogens or chemical residues on-site, ensuring faster and more reliable assessments compared to traditional methods. LOC technology also has significant implications for personalized medicine, where devices can be tailored to monitor specific biomarkers in individuals, leading to more personalized and effective treatment plans. In research and drug development, LOC systems can facilitate high-throughput screening of drugs, enabling rapid evaluation of drug efficacy and toxicity with smaller sample volumes. In addition, the ability to integrate microfluidic channels with biosensors and electrodes makes LOC devices ideal for biochemical assays, cell culture, and other biological research applications, allowing researchers to conduct experiments with greater precision and efficiency. The flexibility of 3D printing in LOC fabrication also opens up possibilities in educational settings, where customized, low-cost LOC kits can be developed for teaching complex concepts in chemistry, biology, and engineering.

17.6 Devices for cancer detection

Cancer has been a persistent and formidable challenge for humanity, ranking as one of the leading causes of death worldwide. The historical battle against this disease has seen numerous advancements, yet the crux of effective cancer treatment remains rooted in early detection. Early diagnosis is crucial because it significantly enhances the chances of successful treatment, thereby reducing mortality rates. In recent years, the emergence of novel technologies has revolutionized the landscape of cancer detection. Among these, 3D printing has shown remarkable potential, offering innovative approaches to developing highly sensitive, cost-effective, and easily operable diagnostic devices. Although the application of 3D printing in cancer sensing is still in its preliminary stages, it holds great promise for the future. The advent of 3D printing technology has introduced a new paradigm in medical diagnostics, particularly in cancer detection. This technology allows for the precise fabrication of complex structures, which can be tailored to specific diagnostic needs. The ability to create intricate designs at a micro and nanoscale level has opened the door to developing novel biosensors that can detect cancer biomarkers with high sensitivity and specificity. Although this field is still evolving, the potential of 3D printing in revolutionizing cancer detection is undeniable. One of the most widely used materials in cancer detection through 3D printing is paper, particularly in the form of paper-based analytical devices (PADs). PADs offer a low-cost platform for developing point-of-need (PON) biosensors, which are crucial for cancer diagnosis, especially in resource-limited settings. These devices can be fabricated using various types of paper substrates, such as filter paper, chromatographic paper, and nitrocellulose membranes, and can be modified with different detection techniques, including colorimetry, fluorescence, chemiluminescence, and electrochemical methods. The choice of substrate, fabrication method, detection technique, assay principle, and recognition element (e.g., antibodies, aptamers, DNA, nanoparticles, enzymes) all play a critical role in determining the performance, accuracy, and sensitivity of these devices (Shalaby et al., 2023). The use of gold nanoparticles (AuNPs) in conjunction with 3D origami devices has also shown significant potential in cancer detection. For instance, Wang et al. designed a highly sensitive electrochemical (EC) immunosensor for detecting cancer antigen 125 (CA 125), a widely recognized biomarker for ovarian cancer (Wang et al., 2016). This sensor system incorporates an innovative Au nanoparticle-modified paper working electrode (Au-PWE) as the sensing platform, alongside Au nanoflowers (AuNFs) for signal amplification. As illustrated in Fig. 17.13, the porous Au-PWE was created by growing a layer of Au nanoparticles on the surfaces of cellulose fibers within the paper sample zone, significantly enhancing the conductivity and increasing the effective surface area of the bare PWE. This simple yet efficient origami EC immune-device demonstrated high sensitivity, specificity, and excellent performance in real

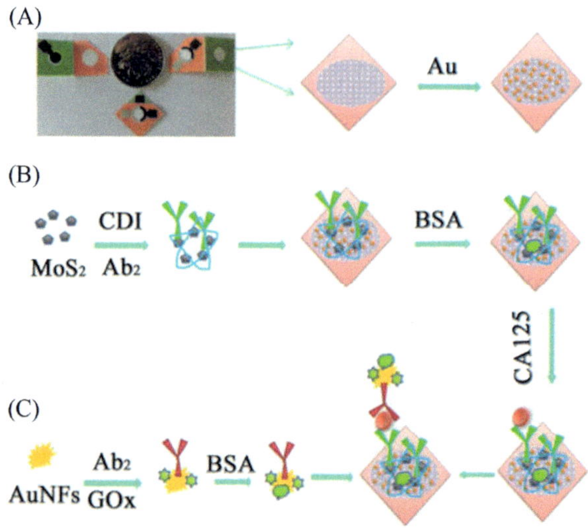

Figure 17.13 Schematic illustration of the fabrication process and assay procedure for the 3D origami electrochemical (EC) immunodevice.
Adapted with permission from Wang et al. (2016).

human serum assays, showing its potential for point-of-care testing of tumor markers. It holds promise for use in remote regions and developing countries, where access to advanced diagnostic tools may be limited.

A low-cost, sensitive, supercapacitor-powered electrochemiluminescent (ECL) protein immune array was developed using a 3D printer. This device was designed to detect three cancer biomarker proteins—prostate-specific antigen (PSA), pstate-specific membrane antigen (PSMA), and platelet factor-4 (PF-4)—in serum within 35 minutes. The immunosensor employs hand screen-printed carbon sensors with gravity flow for sample and reagent delivery. Detection limits were found to be in the range of 300–500 fg/mL for the three proteins in undiluted calf serum, demonstrating the device's potential for sensitive on-site cancer diagnostic tests in resource-limited settings (Kadimisetty et al., 2016). The synthesis of three-dimensional Mn_3O_4 microcubes (3D-Mn_3O_4 MCs) has also been explored for cancer detection, specifically for the determination of nimorazole (NMZ), a drug used in the treatment of head and neck cancer. The 3D-Mn_3O_4 MCs possess a large active area and high conductivity, which were utilized to fabricate a modified screen-printed carbon electrode (3D-Mn_3O_4MCs/SPCE). This electrode exhibited excellent electrocatalytic ability towards NMZ, with a detection limit of 6 nM. The practical feasibility of this sensor was demonstrated in human serum and NMZ tablet samples, highlighting its potential for real-time applicability in biological analysis and pharmaceutical formulations (Umamaheswari et al., 2017).

The development of bioelectrical impedance-based analysis techniques has also shown promise in cancer detection. One such technique was applied to four adherent mammalian cancer cell lines immobilized in a three-dimensional (3D) calcium alginate hydrogel matrix, mimicking in vivo tissue conditions. Paivana et al. developed a proof-of-concept bioelectrical impedance-based analysis technique to evaluate the response of various cancer cell lines to the cytostatic agent 5-fluorouracil (5-FU) (Paivana et al., 2019). Fig. 17.14 illustrates the experimental setup for the impedance measurement procedure. Two gold-coated (Au) electrodes were vertically positioned inside a custom-made, transparent, 3D-printed well, made of Polyethylene Terephthalate Glycol-Modified (PETG), which served as the cell cultivation vessel. The use of 3D printing technology in constructing the culture system provides an environment conducive to cell growth, closely mimicking physiological conditions and promoting realistic cell phenotypes. The electrodes were connected to a handheld LCR meter (Fig. 17.14B), capable of measuring impedance at three different frequencies (1 KHz, 10 KHz, and 100 KHz) to directly extract the impedance magnitude of the sample.

The study focused on four adherent mammalian cancer cell lines—SK-N-SH, HEK293, HeLa, and MCF-7—immobilized within a three-dimensional (3D) calcium alginate hydrogel matrix, simulating in vivo tissue conditions. Impedance measurements were performed to monitor cell viability in real time, with three different cell population densities (50,000, 100,000, and 200,000 cells/100 μL) used for each cell line. The aim of the study was to extract mean impedance values at various frequencies to assess how the different cancer cell lines responded to the application of 5-FU. For comparison, impedance measurements were also conducted on untreated cell lines. The results revealed that impedance values were frequency-dependent and correlated with cell population density, allowing for the identification of unique bioelectrical behavior for each cell line. This approach established specific "fingerprints" for how each cancer cell line reacts to the anticancer agent, offering valuable insights for future drug testing and cancer research. Circulating tumor cells (CTCs) represent another avenue for early cancer diagnosis. Detecting these rare cells in a patient's blood can be crucial for early diagnosis, precise cancer therapy, and monitoring therapeutic outcomes in real-time. A novel approach involved capturing and detecting CTCs using a benzoboric acid-modified gold-plated polymeric substrate with a regular 3D surface array This strategy exhibited enhanced capture efficiency compared to smooth substrates, thanks to the reversible reaction between benzoboric acid on the 3D microarray and sialic acid on

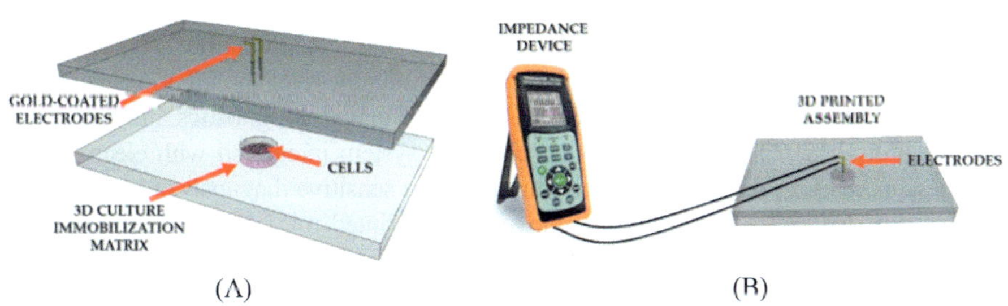

Figure 17.14 Schematic illustration of the experimental setup. (A) Illustration of the cell chamber containing the 3D cell immobilization matrix. (B) Connection of the LCR meter to the 3D-printed well.
Adapted with permission from Paivana et al. (2019).

CTCs. This method allows for the easy detachment of captured CTCs, enabling further downstream characterization of these rare cells. For this purpose, An et al. developed an efficient method for capturing and detecting rare circulating tumor cells (CTCs) from the blood of cancer patients, a critical step for early cancer diagnosis, precise therapy, and real-time monitoring of therapeutic outcomes (An et al., 2018). The approach involved a gold-plated polymeric substrate modified with benzoboric acid and featuring a regular 3D surface array. The 3D microarray design significantly improved capture efficiency, demonstrating a 3.8-fold increase compared to smooth substrates. Fig. 17.15 demonstrates the schematic design of 3-Aminophenylboronic acid (MBA) modified gold-plated polymeric substrate with regular 3D surface array used for detection and release of CTC. The benzoboric acid on the 3D microarray formed a reversible bond with the sialic acid on the surface of CTCs, allowing for the gentle detachment of captured cells without causing damage. This enables the isolated CTCs to be available for further downstream analysis. The proposed strategy offers several benefits, including enhanced capture efficiency, high sensitivity, low cost, and effective recovery of CTCs, making it a promising platform for the early detection of cancer.

In the realm of exosome-based cancer diagnostics, the quantification of cancer-derived exosomes holds great promise for minimally invasive diagnosis during the initial stages of cancer. A detachable microfluidic device implemented with an electrochemical aptasensor (DeMEA) was developed for this purpose. The aptasensor was designed to detect cancer-specific exosomes by immobilizing an aptamer against an epithelial cell adhesion molecule on the electrode surface. The integration of a microfluidic vortexer within the system increased the collision between exosomes and the sensing surface, thereby enhancing detection sensitivity. The DeMEA showed high sensitivity and specificity, with an ultra-low detection limit of 17 exosomes/μL, making it a valuable tool for early cancer detection (Kashefi-Kheyrabadi et al., 2020). Tumor-derived extracellular vesicles (tdEVs) have emerged as promising biomarkers for liquid biopsy-based cancer diagnostics. These vesicles express specific membrane proteins that reflect their cellular origin, providing insight into the epithelial-to-mesenchymal transition (EMT) in cancer—a process that is crucial in evaluating the risk of malignancy transformation. A microfluidic device integrated with an indium-tin-oxide (ITO) sensor was developed to detect tdEVs. The device utilized a multi-orifice flow-fractionation (MOFF) channel to extract pure EVs and an ITO sensor coupled with a geometrically activated surface interaction (GASI) channel for tdEV enrichment and quantification. The integrated sensors demonstrated excellent detection in the clinically relevant tdEV concentration range, offering a new avenue for liquid biopsy-based cancer prognostics and diagnostics (Gurudatt et al., 2023). A closed bipolar electrode system integrated with electrochemiluminescence (ECL) detection was introduced for the sensitive diagnosis of human breast cancer cells (MCF-7). The system employed a 3D-printed microchannel for fabricating the bipolar electrode, which was modified with the AS1411 aptamer—a specific aptamer for nucleolin. The ECL of luminol in the presence of hydrogen peroxide on the anode pole of the bipolar electrode served as the analytical signal. This aptasensor demonstrated low cost, sensitivity, and selectivity in detecting cancer cells, with detection limits as low as 10 cells (Motaghi et al., 2018). In conclusion, the integration of 3D printing technology with various sensing platforms has shown immense potential in cancer

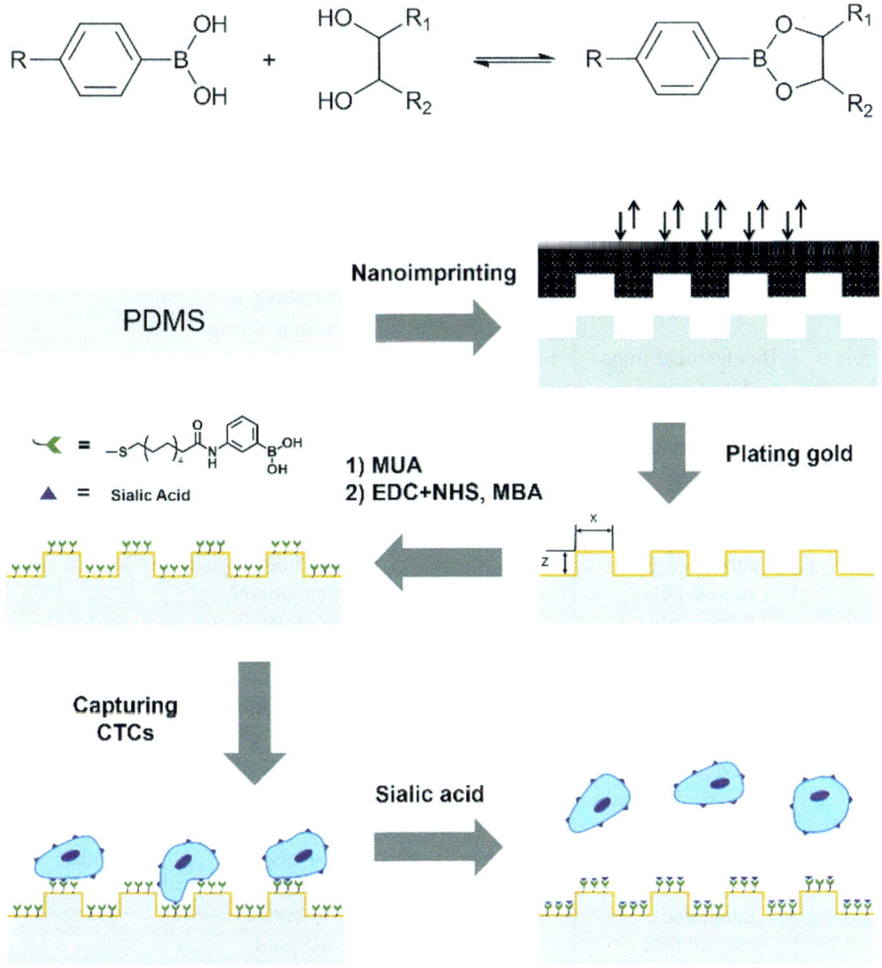

Figure 17.15 Design of an MBA-modified gold-plated polymeric substrate featuring a regular 3D surface array for the capture, detection, and release of circulating tumor cells (CTCs). Adapted with permission from An et al. (2018).

S. No.	Material used	Type of cancer/ cell	Quantitative result	DOI
1.	Paper-based analytical devices (PADs)	Cancer biomarker detection (general)	Varies based on substrate, detection method, and recognition element used.	(Shalaby et al., 2023)
2.	Au nanoparticle-modified paper working electrode (Au-PWE) and Au nanoflowers (AuNFs)	Cancer antigen 125 (CA 125) detection	Linear range: 0.001 to 50 ng/mL, Detection limit: 0.36 pg/mL.	(Wang et al., 2016)
3.	3D-printed	Prostate cancer	Detection limits:	

	supercapacitor-powered electro-chemiluminescent (ECL) protein immunoarray	biomarkers (PSA, PSMA, PF-4)	300–500 fg/mL for the 3 proteins in undiluted calf serum.	(Kadimisetty et al., 2016)
4.	3Dâ€“Mn3O4 microcubes (3Dâ€“Mn3O4MCs)	Nimorazole detection in head and neck cancer	Detection limit: 6 nM for NMZ in human serum.	(Umamaheswari et al., 2017)
5.	Bioelectrical impedance-based analysis technique in 3D calcium alginate hydrogel matrix	Four adherent mammalian cancer cell lines (SK-N-SH, HEK293, HeLa, MCF-7)	Establishes unique impedance fingerprint for each cell line treated with 5-FU.	(Paivana et al., 2019)
6.	Benzoboric acid modified gold-plated polymeric substrate with 3D surface array	Circulating tumor cells (CTCs) detection	Capture efficiency: 3.8 times higher than smooth substrate.	(An et al., 2018)
7.	Detachable microfluidic device with electrochemical aptasensor	Breast cancer-derived exosomes	Detection limit: 17 exosomes/Î¼L over a wide dynamic range.	(Kashefi-Kheyrabadi et al., 2020)
8.	Indium-tin-oxide (ITO) sensor integrated microfluidic device	Pancreatic cancer cells	Linear response between 10^3 and 10^9 tdEVs/mL.	(Gurudatt et al., 2023)
9.	Closed bipolar electrode system integrated with electrochemiluminescence (ECL) detection	Human breast cancer cells (MCF-7)	Detection limits within 10â€“100 cells and 100â€“700 cells. Limit of detection: 10 cells.	(Motaghi et al., 2018)

diagnostics. The cases discussed above illustrate the diverse approaches being explored, from paper-based analytical devices to advanced microfluidic systems. These innovations represent significant advancements in early cancer detection, offering sensitive, specific, and cost-effective solutions that could revolutionize cancer diagnostics in the future. Despite the preliminary nature of this research, the progress made so far is promising, and continued development in this field could lead to widespread clinical applications, particularly in resource-limited settings where early detection is crucial. The future of cancer diagnostics lies in the intersection of nanomaterials, 3D printing, and biosensor technology, and the ongoing research in this domain is paving the way for more accessible and effective cancer detection methods.

S. No.	Material	Type of Chip	Quantitative Results	Application	DOI
1.	Polylactic Acid, Carbon-loaded ABS	Electrochemical Cell-on-a-Chip	Cost-effective, $0.11 per device, ready-to-use without activation	Simultaneous determination of paracetamol and caffeine in tablets and urine	(Katseli et al., 2020)
2.	Threaded Reference Electrode, 3D Printed Flow Cell	Microfluidic Amperometric Detector	Increased sensitivity compared to traditional electrodes	Detection of norepinephrine and ATP nearly simultaneously by amperometry and chemiluminescence	(Hayter et al., 2020)
3.	Various for Microfluidic Channel	3D Printed Microfluidic Device	High time resolution in the detection of glutamate, glucose, and lactate	Online monitoring of biomarkers in a brain injury patient	(Samper et al., 2019)
4.	Nanocomposite, Gold Nanostructures	Electrochemical Aptasensor	Ultra-low limit of detection of 17 exosomes/μL, wide dynamic range from 1 × 102 to 1 × 109 exosomes/μL	Early detection of cancerous exosomes	(Kashefi-Kheyrabadi et al., 2020)
5.	Polylactic Acid Span® 80	Microfluidic Droplet Device	Detection limit of 63.66 CFU droplet−1 for E. coli cells	Label-free counting of E. coli cells in droplets	(Duarte et al., 2019)
6.	Indium-Tin-Oxide (ITO)	Microfluidic Device	Excellent linear response between 103 and 109 tdEVs/mL	Detection and evaluation of epithelial and mesenchymal markers on tdEV surfaces	(Gurudatt et al., 2023)
7.	γ-Fe2O3, PbS, CuS Quantum Dots	Flow-through Electrochemical Device	Results comparable with atomic absorption spectrometry	Automated analysis of HA/MT metal labels	(Michalek et al., 2016)
8.	Magnetic Bead-Silver Nanoparticle (MB-AgNP) Bioconjugate	Hybrid Microfluidic Device (hyFlow)	Detection limit of 12 pM for AgNPs	Detection of magnetic bead-silver nanoparticle bioconjugate	(Walgama et al., 2020)
9.	Ultra-Ever Dry® Coating	Droplet Manipulation Device	Compatible with human saliva, urine, rat blood, and serum; total analysis time ≈30 min	Enzyme-coupled amperometric detection of glucose	(Oomen et al., 2017)

10.	Silver Micropillars, Graphene Nanoflakes	Electrochemical Sensing Platform	Detection of dopamine at a limit of 500 attomoles	Detection of clinically relevant biomolecules at ultralow concentrations	(Ali et al., 2021)
11.	3D Printed Base Structures, Stencil Mask Evaporation	3D Microelectrode Arrays (MEAs)	Impedance decreased from 1.8 kΩ to 670 Ω for microtower electrodes, and from 55 kΩ to 39 kΩ for 30 μm electrodes	Electrical interrogation of 3D organ-on-a-chip models for pharmaceutical screening	(Kundu et al., 2021)
12.	Screen-Printed Electrodes (SPEs), Wax-Printing	Paper-Based Analytical Device	Demonstrated voltammetric, amperometric, and potentiometric measurements	Point-of-care diagnostics for dopamine, glucose, and pH	(Punjiya et al., 2018)
13.	Benzoboric Acid, Gold-Plated Polymer Substrate	3D Surface Microarray Device	3.8 times higher capture efficiency for CTCs compared to smooth substrates	Early detection and capture of circulating tumor cells (CTCs)	(An et al., 2018)
14.	Liquid Metal	Microfluidic Channels	Capability to produce robust 3D microchannels without lithography	Fabrication of microvasculature and metal features for lab-on-a-chip devices	(Parekh et al., 2016)
15.	Polylactic Acid, Cas12a Enzyme	Lab-on-a-Chip	Concurrent detection of SARS-CoV-2 RNA and antibodies within 2 hours	Monitoring of COVID-19 infection and immunity	(Najjar et al., 2022)
16.	Digital Microfluidic (DMF) Platform, Ion-Selective Membranes	DMF System	Controlled droplet manipulation for ion-selective sensing	Detection of targeted primary ions in a range from 10-6 to 1 M	(Min et al., 2019)
17.	3D Printed Reactionware Device, Arduino 3D Control	Paper-Based Microfluidic Device	Accelerated design turnover, reduced cost	Paper-based microfluidic prototyping and electrochemical detection	(Pearce et al., 2016)
18.	3D Printed Electrodes, Micro-Milling Process	AC-Induced Electric Field Biochip	Controlled 3D rotation of bovine oocytes at a rate of up to 140° s-1	3D manipulation of cells for biotechnological applications	(Benhal et al., 2014)

17.7 Biomolecule detection

The detection of biomolecules such as dopamine, glucose, and ascorbic acid (AA) has seen significant advancements due to the integration of 3D printing technologies in sensor development. This technological shift not only enhances the sensitivity and specificity of biosensors but also revolutionizes the way these essential biomolecules are monitored, offering a more efficient, cost-effective, and scalable solution for both clinical and point-of-care applications. The detection of biomolecules is of immense significance in both clinical diagnostics and research because these molecules serve as key indicators of physiological and pathological states. Accurate and timely detection of biomolecules such as neurotransmitters, sugars, and vitamins can provide critical insights into a person's health, enabling early diagnosis and monitoring of diseases, guiding treatment decisions, and even facilitating the development of personalized medicine. There are several biomolecules of great importance in clinical diagnostics and research, each playing a crucial role in maintaining physiological balance and signaling within the body. However, an in-depth discussion of all these biomolecules is beyond the scope of this chapter. Therefore, we will focus specifically on recent advancements in 3D printing technologies as they relate to the detection of three critical biomolecules: dopamine, glucose, and ascorbic acid. These molecules have been the subject of significant research and development, particularly in the context of 3D-printed biosensors, and offer a compelling look at how this technology is shaping the future of biomolecule detection. **Dopamine** is a crucial neurotransmitter involved in various neurological processes, including mood regulation, motor control, and reward-motivated behavior. Abnormal levels of dopamine are linked to several neurological disorders, such as Parkinson's disease, schizophrenia, and addiction. Therefore, the ability to accurately detect dopamine levels is vital for diagnosing and monitoring these conditions. Recent advances in 3D printing have enabled the development of highly sensitive electrochemical sensors for dopamine detection, allowing for more precise measurements and opening new avenues for early diagnosis and treatment of neurological disorders. **Glucose** detection is fundamental in managing diabetes, a chronic condition that affects millions of people worldwide. Continuous monitoring of glucose levels is essential for diabetes patients to manage their condition effectively and avoid complications such as hyperglycemia or hypoglycemia. The integration of 3D printing in glucose sensor development has led to the creation of more efficient, cost-effective, and portable devices, significantly improving the accuracy and convenience of glucose monitoring. These innovations are making it easier for patients to track their glucose levels in real-time, promoting better disease management and improving quality of life. **Ascorbic acid (AA)**, commonly known as vitamin C, plays a critical role in various physiological functions, including immune response, collagen synthesis, and antioxidant protection. Accurate detection of AA levels is important in assessing nutritional status and identifying deficiencies that could lead to health issues such as scurvy. 3D-printed devices have been developed to detect AA with high sensitivity, offering a low-cost and efficient method for monitoring vitamin C levels in various biological fluids and food products. These advancements ensure that individuals can maintain adequate levels of this essential nutrient, contributing to overall health and well-being.

Recent studies have established a tremendous scope for 3D printing in the detection of dopamine, highlighting its potential to revolutionize this area of biosensing. The unique capabilities of 3D printing, such as the ability to create highly customized and intricate sensor designs, have enabled the development of more sensitive and specific electrochemical sensors for dopamine. Hussaini et al. (2024) utilized 3D-printed electrodes modified with a $CeO_2/BaMoO_4$ nanocomposite for the electrochemical detection of dopamine hydrochloride, demonstrating the significant potential of 3D-printed biosensors in enhancing the electroactive surface area, which resulted in improved sensitivity. Specifically, the limit of detection (LoD) values were found to be 4.52 μM and 2.77 μM for CeO_2 /SPE and CeO_2 /3DPE biosensors, respectively, highlighting the superior performance of the 3D-printed electrodes (Hussaini et al., 2024). Another investigation employed a 3D-printed fluidic electrochemical sensing platform (3DP-FESP) integrated with screen-printed electrodes (SPEs) to quantitatively detect dopamine in the presence of interferents such as uric acid (UA) and ascorbic acid (AA). This setup demonstrated a significant improvement in sensitivity under flow conditions, with LoDs of 0.05 μM for dopamine detection, showcasing the enhanced performance due to the flow cell configuration (Contreras-Naranjo et al., 2021). A novel approach for the simultaneous detection of AA and dopamine involved the use of a 3D-printed microfluidic device called μTED, which utilized textile threads as microfluidic channels and an unmodified SPE as the amperometric detector. The device achieved sensitivities of 2.24 μA L/mmol for AA and 5.09 μA L/mmol for dopamine, with LoDs of 2.60 μmol/L and 1.54 μmol/L, respectively. It demonstrates the capability of 3D printing to create compact, efficient, and highly sensitive devices for the simultaneous detection of multiple biomolecules (dos Santos et al., 2023).

Moving on to glucose detection, Kumar et al., developed a 3D-printed ECL imaging system integrated with a smartphone to detect glucose and lactate. This system introduced a universal cross-platform application for real-time monitoring, with the device showing a linear range from 0.1 mM to 1 mM for glucose, and an LoD of 0.04 mM. The portability and ease of use offered by the 3D-printed system highlight the potential for such devices to revolutionize point-of-care diagnostics (Kumar et al., 2023). Moreover, a 3D paper-based microfluidic electrochemical biosensor was developed for the quantitative detection of glucose. This sensor used a prussian blue-modified reduced graphene oxide-tetraethylene pentamine (rGO-TEPA/PB) electrode for the detection of hydrogen peroxide, a product of the enzyme-catalyzed reaction in glucose detection. The device demonstrated a wide linear range of 0.1 mM–25 mM with an LoD of 25 μM, proving its utility in real-world applications such as monitoring glucose levels in human sweat and blood (Cao et al., 2020). A non-enzymatic glucose sensor was developed using a 3D hierarchical porous AuNi bimetallic aerogel, which exhibited a high sensitivity of 40.34 $\mu A/mM/cm^2$ and excellent long-term stability of 90 days. This work underscores the potential of 3D-printed materials in improving the performance and durability of glucose sensors (Li et al., 2024).

The detection of AA has also seen significant advancements with the aid of 3D printing. A low-cost 3D-printed device was developed for the detection of AA in various samples, utilizing a three-electrode sensor module prepared by screen printing technology and a microfluidic chip with a finger-actuated micropump. The device demonstrated a mixing efficiency greater than 90% and required only a microliter-scale sample volume, making it highly efficient and suitable for point-of-care testing (Liu et al., 2022). Li et al., integrated a microfluidic device with an all-in-one dual-modal immunoassay for the colorimetric and photoelectrochemical (PEC) detection of alpha-fetoprotein (AFP). The released AA from enzyme-linked immunoreactions induced significant changes in the device's signal, leading to the successful detection of AFP in serum. This dual-modality approach underscores the versatility of AA as a reporting agent in biosensing applications (Li et al., 2020). Additionally, a three-dimensional electrochemical paper-based analytical device (3D-ePAD) was developed for the simultaneous determination of AA, dopamine, and UA. The device employed a graphite paste modified with Fe_3O_4@Au-Cys/PANI on a patterned paper, achieving high sensitivity and specificity in the detection of these biomolecules. This demonstrates the capability of 3D-printed devices to handle complex sample matrices and perform simultaneous multi-analyte detection (Nontawong et al., 2018).

17.8 Prospects and challenges

The future of additively manufactured functional polymers and nanocomposites in electrochemical device applications holds significant promise, particularly in advancing technologies like micro-supercapacitors, 3D printed fluidic devices, lithium-ion batteries, LOC systems, and devices for cancer and biomolecule detection. In micro-supercapacitors, additive manufacturing allows for precise control over electrode architecture, which can significantly enhance energy density and device longevity. However, the challenge lies in optimizing the interface between nanomaterials and polymers to ensure consistent performance, along with developing scalable manufacturing processes. For 3D printed fluidic devices, the potential for integrating multifunctional materials that combine sensing, actuation, and fluid control is immense, yet achieving high precision in fluidic channel printing and ensuring material biocompatibility and chemical resistance remain critical hurdles. In the realm of lithium-ion batteries, additive manufacturing could enable the design of complex architectures that improve performance. Still, the development of printable materials that retain the necessary electrochemical properties, alongside effective thermal management strategies, is crucial to overcoming current limitations. LOC systems represent another exciting application, with the potential to revolutionize diagnostics through rapid and accurate analysis of biological samples. The integration of nanomaterials can enhance sensitivity and functionality, but ensuring reproducibility, reliability, and long-term stability of these devices remains a significant challenge, particularly when scaling up production. Similarly, additively manufactured devices for cancer detection offer the potential for highly sensitive and specific diagnostics. However, the development of cost-effective and easily manufacturable devices that provide consistent clinical performance, along with navigating regulatory pathways, are key challenges. In biomolecule detection, 3D printed devices could facilitate the development of advanced point-of-care diagnostics, but achieving high precision in printing, ensuring the stability and specificity of detection elements, and developing user-friendly interfaces are critical for widespread adoption. While the prospects for these technologies are bright, significant technical and material challenges must be addressed to fully realize their potential. The successful commercialization and integration of these advanced manufacturing techniques into mainstream applications will require close collaboration between material scientists, engineers, and industry stakeholders, emphasizing the need for continued innovation and interdisciplinary efforts to overcome existing barriers.

References

Ahmad, S., Hashmi, A. W., Singh, J., Arora, K., Tian, Y., Iqbal, F., Al-Dossari, M., & Khan, M. I. (2024). *Innovations in additive manufacturing of shape memory alloys: Alloys, microstructures, treatments, applications. Journal of Materials Research and Technology, 32*, China: Elsevier Editora Ltd, 4136–4197. https://www.sciencedirect.com/science/journal/22387854. https://doi.org/10.1016/j.jmrt.2024.08.213.

Alachkar, A., Alhassan, S., & Senel, M. (2022). Lab-In-A-Syringe: A Novel Electrochemical Biosensor for On-Site and Real-Time Monitoring of Dopamine in Freely Behaving Mice. *ACS Sensors, 7*(1), 331–337. https://doi.org/10.1021/acssensors.1c02525.

Ali, M. A., Hu, C., Yuan, B., Jahan, S., Saleh, M. S., Guo, Z., Gellman, A. J., & Panat, R. (2021). Breaking the barrier to biomolecule limit-of-detection via 3D printed multi-length-scale graphene-coated electrodes. *Nature Communications, 12*(1). https://doi.org/10.1038/s41467-021-27361-x.

An, L., Wang, G., Han, Y., Li, T., Jin, P., & Liu, S. (2018). Electrochemical biosensor for cancer cell detection based on a surface 3D micro-array. *Lab on a Chip, 18*(2), 335–342. https://doi.org/10.1039/c7lc01117b.

Bae, J., Oh, S., Lee, B., Lee, C. H., Chung, J., Kim, J., Jo, S., Seo, S., Lim, J., & Chung, S. (2023). High-performance, printable quasi-solid-state electrolytes toward all 3D direct ink writing of shape-versatile Li-ion batteries. *Energy Storage Materials, 57*, 277–288. https://doi.org/10.1016/j.ensm.2023.02.016.

Baltima, A., Panagopoulou, H., Economou, A., & Kokkinos, C. (2021). 3D-printed fluidic electrochemical microcell for sequential injection/stripping analysis of heavy metals. *Analytica Chimica Acta, 1159*, 338426. https://doi.org/10.1016/j.aca.2021.338426.

Bao, Y., Liu, Y., Kuang, Y., Fang, D., & Li, T. (2020). 3D-printed highly deformable electrodes for flexible lithium ion batteries. *Energy Storage Materials, 33*, 55–61. https://doi.org/10.1016/j.ensm.2020.07.010.

Belmonte, I., & White, R. J. (2022). 3-D printed microfluidics for rapid prototyping and testing of electrochemical, aptamer-based sensor devices under flow conditions. *Analytica Chimica Acta, 1192*, 339377. https://doi.org/10.1016/j.aca.2021.339377.

Ben-Barak, I., Schneier, D., Kamir, Y., Goor, M., Golodnitsky, D., & Peled, E. (2022). Drop-on-demand 3D-printed silicon-based anodes for lithium-ion batteries. *Journal of Solid State Electrochemistry, 26*(1), 183–193. https://doi.org/10.1007/s10008-021-05056-z.

Benhal, P., Chase, J. G., Gaynor, P., Oback, B., & Wang, W. (2014). AC electric field induced dipole-based on-chip 3D cell rotation. *Lab on a Chip, 14*(15), 2717–2727. https://doi.org/10.1039/c4lc00312h. http://pubs.rsc.org/en/journals/journal/lc.

Beydaghi, H., Abouali, S., Thorat, S. B., Del Rio Castillo, A. E., Bellani, S., Lauciello, S., Gentiluomo, S., Pellegrini, V., & Bonaccorso, F. (2021). 3D printed silicon-few layer graphene anode for advanced Li-ion batteries. *RSC Advances, 11*(56), 35051–35060. https://doi.org/10.1039/d1ra06643a. http://pubs.rsc.org/en/journals/journal/ra.

Bishop, G. W., Satterwhite, J. E., Bhakta, S., Kadimisetty, K., Gillette, K. M., Chen, E., & Rusling, J. F. (2015). 3D-Printed fluidic devices for nanoparticle preparation and flow-injection amperometry using integrated prussian blue nanoparticle-modified electrodes. *Analytical Chemistry, 87*(10), 5437–5443. https://doi.org/10.1021/acs.analchem.5b00903. http://pubs.acs.org/journal/ancham.

Blake, A. J., Kohlmeyer, R. R., Hardin, J. O., Carmona, E. A., Maruyama, B., Berrigan, J. D., Huang, H., & Durstock, M. F. (2017). 3D Printable Ceramic–Polymer Electrolytes for Flexible High-Performance Li-Ion Batteries with Enhanced Thermal Stability. *Advanced Energy Materials, 7*(14). https://doi.org/10.1002/aenm.201602920. http://onlinelibrary.wiley.com/journal/10.1002/(ISSN)1614-6840.

Broccoli, A., Vollertsen, A. R., Roels, P., van Vugt, A., van den Berg, A., & Odijk, M. (2023). Nanoparticle Printing for Microfluidic Applications: Bipolar Electrochemistry and Localized Raman Sensing Spots. *Micromachines, 14*(2), 453. https://doi.org/10.3390/mi14020453.

Cao, L., Han, G. C., Xiao, H., Chen, Z., & Fang, C. (2020). *A novel 3D paper-based microfluidic electrochemical glucose biosensor based on rGO-TEPA/PB sensitive film. Analytica Chimica Acta*, *1096*, China: Elsevier B.V, 34–43. http://www.journals.elsevier.com/analytica-chimica-acta/. https://doi.org/10.1016/j.aca.2019.10.049.

Cardenas, J. A., Bullivant, J. P., Kolesnichenko, I. V., Roach, D. J., Gallegos, M. A., Coker, E. N., Lambert, T. N., Allcorn, E., Talin, A. A., Cook, A. W., & Harrison, K. L. (2022). 3D Printing of Ridged FeS2Cathodes for Improved Rate Capability and Custom-Form Lithium Batteries. *ACS Applied Materials and Interfaces, 14*(40), 45342–45351. https://doi.org/10.1021/acsami.2c11954. http://pubs.acs.org/journal/aamick.

Carneiro, E. A., Agustini, D., Figueiredo-Filho, L. C. S., Banks, C. E., Marcolino-Junior, L. H., & Bergamini, M. F. (2018). 3D-printed Microfluidic Device Based on Cotton Threads for Amperometric Estimation of Antioxidants in Wine Samples. *Electroanalysis, 30*(1), 101–108. https://doi.org/10.1002/elan.201700579. http://onlinelibrary.wiley.com/journal/10.1002/(ISSN).

Caruso, G., Musso, N., Grasso, M., Costantino, A., Lazzarino, G., Tascedda, F., Gulisano, M., Lunte, S. M., & Caraci, F. (2020). Microfluidics as a Novel Tool for Biological and Toxicological Assays in Drug Discovery Processes: Focus on Microchip Electrophoresis. *Micromachines, 11*(6), 593. https://doi.org/10.3390/mi11060593.

Carvalho, R. M., Ferreira, V. S., & Lucca, B. G. (2021). A novel all-3D-printed thread-based microfluidic device with an embedded electrochemical detector: First application in environmental analysis of nitrite. *Analytical Methods, 13*(11), 1349–1357. https://doi.org/10.1039/d1ay00070e. http://pubs.rsc.org/en/journals/journal/ay.

Castiaux, A. D., Currens, E. R., & Martin, R. S. (2020). Direct embedding and versatile placement of electrodes in 3D printed microfluidic-devices. *Analyst, 145*(9), 3274–3282. https://doi.org/10.1039/d0an00240b. http://pubs.rsc.org/en/journals/journal/an.

Castro-Gutiérrez, J., Celzard, A., & Fierro, V. (2020). Energy Storage in Supercapacitors: Focus on Tannin-Derived Carbon Electrodes. *Frontiers in Materials, 7*. https://doi.org/10.3389/fmats.2020.00217.

Chae, C., Kim, Y. B., Lee, S. S., Kim, T. H., Jo, Y., Kim, H. S., Kim, S. K., Choi, Y., & Jeong, S. (2021). *All-3D-printed solid-state microsupercapacitors. Energy Storage Materials, 40*, South Korea: Elsevier B.V, 1–9. http://www.journals.elsevier.com/energy-storage-materials/. https://doi.org/10.1016/j.ensm.2021.04.038.

Chen, Y., Guo, M., Xu, L., Cai, Y., Tian, X., Liao, X., Wang, Z., Meng, J., Hong, X., & Mai, L. (2022). In-situ selective surface engineering of graphene micro-supercapacitor chips. *Nano Research, 15*(2), 1492–1499. https://doi.org/10.1007/s12274-021-3693-4.

Clark, K. M., Cherwin, A. E., Boes, J., Russo, M. J., & Henry, C. S. (2024). Comparing thermoplastic electrode materials: Toward enhanced sensing of O2 and H2O2 in flow devices. *Electroanalysis, 36*(11). https://doi.org/10.1002/elan.202400067. http://onlinelibrary.wiley.com/journal/10.1002/(ISSN)1521-4109.

Contreras-Naranjo, J. E., Perez-Gonzalez, V. H., Mata-Gómez, M. A., & Aguilar, O. (2021). Cost-Effective 3D-Printed-Enabled Fluidic Electrochemical Sensing Platform for Quantitative Electroanalytical Applications. *Journal of the Electrochemical Society, 168*(11), 117504. https://doi.org/10.1149/1945-7111/ac3311.

Convery, N., & Gadegaard, N. (2019). 30 years of microfluidics. *Micro and Nano Engineering, 2*, 76–91. https://doi.org/10.1016/j.mne.2019.01.003.

Costa, B. M.de C., Griveau, S., Bedioui, F., Orlye, F.d, da Silva, J. A. F., & Varenne, A. (2022). Stereolithography based 3D-printed microfluidic device with integrated electrochemical detection. *Electrochimica Acta, 407*, 139888. https://doi.org/10.1016/j.electacta.2022.139888.

Crevillen, A. G., Mayorga-Martinez, C. C., Vaghasiya, J. V., & Pumera, M. (2022). 3D-Printed SARS-CoV-2 RNA Genosensing Microfluidic System. *Advanced Materials Technologies, 7*(6). https://doi.org/10.1002/admt.202101121. http://onlinelibrary.wiley.com/journal/10.1002/(ISSN)2365-709X.

Dang, B. V., Hassanzadeh-Barforoushi, A., Syed, M. S., Yang, D., Kim, S. J., Taylor, R. A., Liu, G. J., Liu, G., & Barber, T. (2019). Microfluidic Actuation via 3D-Printed Molds toward Multiplex Biosensing of Cell Apoptosis. *ACS Sensors, 4*(8), 2181–2189. https://doi.org/10.1021/acssensors.9b01057. http://pubs.acs.org/journal/ascefj.

Duarte, L. C., Chagas, C. L. S., Ribeiro, L. E. B., & Coltro, W. K. T. (2017). 3D printing of microfluidic devices with embedded sensing electrodes for generating and measuring the size of microdroplets based on contactless conductivity detection. *Sensors and Actuators, B: Chemical, 251*, 427–432. https://doi.org/10.1016/j.snb.2017.05.011.

Duarte, L. C., Figueredo, F., Ribeiro, L. E. B., Cortón, E., & Coltro, W. K. T. (2019). *Label-free counting of Escherichia coli cells in nanoliter droplets using 3D printed microfluidic devices with integrated contactless conductivity detection. Analytica Chimica Acta, 1071*, Brazil: Elsevier B.V, 36–43. http://www.journals.elsevier.com/analytica-chimica-acta/. https://doi.org/10.1016/j.aca.2019.04.045.

Fan, J., Zheng, H., Chen, A., Gu, L., Xie, X., Fan, J., & Ding, Z. (2023). In-situ construction of amorphous–crystalline NiCo bimetallic oxyhydroxide for low-temperature supercapacitor device. *Chemical Engineering Journal, 476*, 146638. https://doi.org/10.1016/j.cej.2023.146638.

Fernandes Quero, R., Jesus, D. Pd, & Fracassi da Silva, J. A. (2023). Simple modification to allow high-efficiency and high-resolution multi-material 3D-printing fabrication of microfluidic devices. *Lab on a Chip, 23*(16), 3694–3703. https://doi.org/10.1039/d3lc00356f. http://pubs.rsc.org/en/journals/journal/lc.

Foster, C. W., Zou, G. Q., Jiang, Y., Down, M. P., Liauw, C. M., Garcia-Miranda Ferrari, A., Ji, X., Smith, G. C., Kelly, P. J., & Banks, C. E. (2019). Next-Generation Additive Manufacturing: Tailorable Graphene/Polylactic(acid) Filaments Allow the Fabrication of

3D Printable Porous Anodes for Utilisation within Lithium-Ion Batteries. *Batteries and Supercaps, 2*(5), 448–453. https://doi.org/10.1002/batt.201800148. https://onlinelibrary.wiley.com/journal/25666223.

Goikolea, E., & Mysyk, R. (2017). *Nanotechnology in Electrochemical Capacitors. Emerging Nanotechnologies in Rechargeable Energy Storage Systems*. Spain: Elsevier Inc. 131–169. http://www.sciencedirect.com/science/book/9780323429771. https://doi.org/10.1016/B978-0-323-42977-1.00004-2.

Gurudatt, N. G., Gwak, H., Hyun, K.-A., Jeong, S.-E., Lee, K., Park, S., Chung, M. J., Kim, S.-E., Hyun Jo, J., & Jung, H.-I. (2023). Electrochemical detection and analysis of tumor-derived extracellular vesicles to evaluate malignancy of pancreatic cystic neoplasm using integrated microfluidic device. *Biosensors and Bioelectronics, 226*, 115124. https://doi.org/10.1016/j.bios.2023.115124.

Han, T., He, Z., Kuang, W., Zhou, J., & Li, Y.-yun (2024). 3D printing of porous hollow nanosphere MoS2@NiS/rGO scaffolds empowering long-cycle sodium-ion batteries. *Nano Energy, 128*, 109953. https://doi.org/10.1016/j.nanoen.2024.109953.

Hayter, E. A., Castiaux, A. D., & Martin, R. S. (2020). 3D-printed microfluidic device with in-line amperometric detection that also enables multi-modal detection. *Analytical Methods, 12*(15), 2046–2051. https://doi.org/10.1039/d0ay00368a. http://pubs.rsc.org/en/journals/journal/ay.

Hong, Y., Wu, M., Chen, G., Dai, Z., Zhang, Y., Chen, G., & Dong, X. (2016). 3D Printed Microfluidic Device with Microporous Mn2O3 -Modified Screen Printed Electrode for Real-Time Determination of Heavy Metal Ions. *ACS Applied Materials & Interfaces, 8*(48), 32940–32947. https://doi.org/10.1021/acsami.6b10464.

Hu, X., Zhu, Y., Fu, Y.-F., Fan, Y., Chen, Y., Kim, D., Lu, X., Yu, B., & Chen, Y. I. (2024). 3D-printed topological-structured electrodes with exceptional mechanical properties for high-performance flexible Li-ion batteries. *Energy Storage Materials, 70*, 103560. https://doi.org/10.1016/j.ensm.2024.103560.

Hussaini, A. A., Sarilmaz, A., Ozel, F., Erdal, M. O., & Yıldırım, M. (2024). CeO2:BaMoO4 nanocomposite based 3D-printed electrodes for electrochemical detection of the dopamine. *Materials Science in Semiconductor Processing, 180*, 108587. https://doi.org/10.1016/j.mssp.2024.108587.

Kadimisetty, K., Mosa, I. M., Malla, S., Satterwhite-Warden, J. E., Kuhns, T. M., Faria, R. C., Lee, N. H., & Rusling, J. F. (2016). 3D-printed supercapacitor-powered electrochemiluminescent protein immunoarray. *Biosensors and Bioelectronics, 77*, 188–193. https://doi.org/10.1016/j.bios.2015.09.017.

Kashefi-Kheyrabadi, L., Kim, J., Chakravarty, S., Park, S., Gwak, H., Kim, S.-I., Mohammadniaei, M., Lee, M.-H., Hyun, K.-A., & Jung, H.-I. (2020). Detachable microfluidic device implemented with electrochemical aptasensor (DeMEA) for sequential analysis of cancerous exosomes. *Biosensors and Bioelectronics, 169*, 112622. https://doi.org/10.1016/j.bios.2020.112622.

Katseli, V., Economou, A., & Kokkinos, C. (2020). A novel all-3D-printed cell-on-a-chip device as a useful electroanalytical tool: Application to the simultaneous voltammetric determination of caffeine and paracetamol. *Talanta, 208*, 120388. https://doi.org/10.1016/j.talanta.2019.120388.

Krejcova, L., Nejdl, L., Rodrigo, M. A. M., Zurek, M., Matousek, M., Hynek, D., Zitka, O., Kopel, P., Adam, V., & Kizek, R. (2014). *3D printed chip for electrochemical detection of influenza virus labeled with CdS quantum dots. Biosensors and Bioelectronics, 54*, Elsevier Ltd.: Czech Republic, 421–427. http://www.elsevier.com/locate/bios. https://doi.org/10.1016/j.bios.2013.10.031.

Kumar, A., Jain, D., Bahuguna, J., Bhaiyya, M., Kumar Dubey, S., Javed, A., & Goel, S. (2023). Machine learning assisted and smartphone integrated homogeneous electrochemiluminescence biosensor platform for sample to answer detection of various human metabolites. *Biosensors and Bioelectronics, 238*, 115582. https://doi.org/10.1016/j.bios.2023.115582.

Kumarov, A., Nwaogu, E. C., Zhumadil, A., Bakenov, Z., & Nurpeissova, A. (2023). Current collector-free printed three-dimensional MXene-based anodes for lithium-ion batteries. *Electrochemistry Communications, 157*, 107621. https://doi.org/10.1016/j.elecom.2023.107621.

Kundu, A., Ausaf, T., & Rajaraman, S. (2018). 3D Printing, Ink Casting and Micromachined Lamination (3D PICLμM): A Makerspace Approach to the Fabrication of Biological Microdevices. *Micromachines, 9*(2), 85. https://doi.org/10.3390/mi9020085.

Kundu, A., McCoy, L., Azim, N., Nguyen, H., Didier, C. M., Ausaf, T., Sharma, A. D., Curley, J. L., Moore, M. J., & Rajaraman, S. (2021). Fabrication and Characterization of 3D Printed, 3D Microelectrode Arrays for Interfacing with a Peripheral Nerve-on-a-Chip. *ACS Biomaterials Science & Engineering, 7*(7), 3018–3029. https://doi.org/10.1021/acsbiomaterials.0c01184.

Kurmanbayeva, I., Mentbayeva, A., Nurpeissova, A., & Bakenov, Z. (2021). Advanced Battery Materials Research at Nazarbayev University: Review. *Eurasian Chemico-Technological Journal, 23*(3), 199. https://doi.org/10.18321/ectj1103.

Li, G., Xue, P., Fan, H., Ma, Y., Wang, H., Lu, D., Gao, J., & Wen, D. (2024). AuNi bimetallic aerogel with ultra-high stability applied in smart and portable biosensing. *Analytica Chimica Acta, 1306*, 342613. https://doi.org/10.1016/j.aca.2024.342613.

Li, X., Pan, X., Lu, J., Zhou, Y., & Gong, J. (2020). Dual-modal visual/photoelectrochemical all-in-one bioassay for rapid detection of AFP using 3D printed microreactor device. *Biosensors and Bioelectronics, 158*, 112158. https://doi.org/10.1016/j.bios.2020.112158.

Lisowski, P., & Zarzycki, P. K. (2013). Microfluidic paper-based analytical devices (μPADs) and micro total analysis systems (μTAS): Development, applications and future trends. *Chromatographia, 76*(19-20), 1201–1214. https://doi.org/10.1007/s10337-013-2413-y.

Liu, L., Lu, J. Y., Long, X. L., Zhou, R., Liu, Y. Q., Wu, Y. T., & Yan, K. W. (2021). 3D printing of high-performance micro-supercapacitors with patterned exfoliated graphene/carbon nanotube/silver nanowire electrodes. *Science China Technological Sciences, 64*(5), 1065–1073. https://doi.org/10.1007/s11431-020-1763-5. http://link.springer.com/journal/11431.

Liu, X., Wang, F., Meng, Y., Zhao, L., Shi, W., Wang, X., He, Z., Chao, J., & Li, C. (2022). Electrochemical/visual microfluidic detection with a covalent organic framework supported platinum nanozyme-based device for early diagnosis of pheochromocytoma. *Biosensors and Bioelectronics, 207*, 114208. https://doi.org/10.1016/j.bios.2022.114208.

Liu, X., Li, M., Zheng, J., Zhang, X., Zeng, J., Liao, Y., Chen, J., Yang, J., Zheng, X., & Hu, N. (2022). Electrochemical Detection of Ascorbic Acid in Finger-Actuated Microfluidic Chip. *Micromachines, 13*(9), 1479. https://doi.org/10.3390/mi13091479.

Maurel, A., Armand, M., Grugeon, S., Fleutot, B., Davoisne, C., Tortajada, H., Courty, M., Panier, S., & Dupont, L. (2020). Poly(Ethylene Oxide)–LiTFSI Solid Polymer Electrolyte Filaments for Fused Deposition Modeling Three-Dimensional Printing. *Journal of the Electrochemical Society, 167*(7), 070536. https://doi.org/10.1149/1945-7111/ab7c38.

Memarzadeh, A., Safaei, B., Tabak, A., Sahmani, S., & Kizilors, C. (2023). Advancements in additive manufacturing of polymer matrix composites: A systematic review of techniques and properties. *Materials Today Communications, 36*, 106449. https://doi.org/10.1016/j.mtcomm.2023.106449.

Michalek, P., Richtera, L., Krejcova, L., Nejdl, L., Kensova, R., Zitka, J., Kopel, P., Heger, Z., Adam, V., & Kizek, R. (2016). Bioconjugation of peptides using advanced nanomaterials to examine their interactions in 3D printed flow-through device. *Electrophoresis, 37*(3), 444–454. https://doi.org/10.1002/elps.201500345.

Min, K., Bao, C., & Kim, W. S. (2019). Additively Manufactured Digital Microfluidic Platforms for Ion-Selective Sensing. *ACS Sensors, 4*(4), 918–923. https://doi.org/10.1021/acssensors.8b01689. http://pubs.acs.org/journal/ascefj.

Motaghi, H., Ziyaee, S., Mehrgardi, M. A., Kajani, A. A., & Bordbar, A. K. (2018). *Electrochemiluminescence detection of human breast cancer cells using aptamer modified bipolar electrode mounted into 3D printed microchannel. Biosensors and Bioelectronics, 118*, Iran: Elsevier Ltd, 217–223. http://www.elsevier.com/locate/bios. https://doi.org/10.1016/j.bios.2018.07.066.

Najjar, D., Rainbow, J., Sharma Timilsina, S., Jolly, P., de Puig, H., Yafia, M., Durr, N., Sallum, H., Alter, G., Li, J. Z., Yu, X. G., Walt, D. R., Paradiso, J. A., Estrela, P., Collins, J. J., & Ingber, D. E. (2022). A lab-on-a-chip for the concurrent electrochemical detection of SARS-CoV-2 RNA and anti-SARS-CoV-2 antibodies in saliva and plasma. *Nature Biomedical Engineering, 6*(8), 968–978. https://doi.org/10.1038/s41551-022-00919-w.

Ngidi, N. P. D., Koekemoer, A. F., & Ndlela, S. S. (2024). *Recent advancement in the electrochemical performance of electrochemical capacitors based on biomass-derived porous carbon: A review. Journal of Energy Storage, 89*, South Africa: Elsevier Ltd. https://www.sciencedirect.com/science/journal/2352152X. https://doi.org/10.1016/j.est.2024.111638.

Nie, N., Wang, F., & Yao, W. (2023). Fabrication of a 3D structure MnO2 electrode with high MnO2 mass loading as the cathode for high-performance aqueous zinc-ion batteries. *Electrochimica Acta, 472*, 143423. https://doi.org/10.1016/j.electacta.2023.143423.

Nontawong, N., Amatatongchai, M., Wuepchaiyaphum, W., Chairam, S., Pimmongkol, S., Panich, S., Tamuang, S., & Jarujamrus, P. (2018). Fabrication of a three-dimensional electrochemical paper-based device (3D-ePAD) for individual and simultaneous detection of ascorbic acid, dopamine and uric acid. *International Journal of Electrochemical Science, 13*(7), 6940–6957. https://doi.org/10.20964/2018.07.65.

Oomen, P. E., Mulder, J. P. S. H., Verpoorte, E., & Oleschuk, R. D. (2017). Controlled, synchronized actuation of microdroplets by gravity in a superhydrophobic, 3D-printed device. *Analytica Chimica Acta, 988*, 50–57. https://doi.org/10.1016/j.aca.2017.08.010.

Orangi, J., Hamade, F., Davis, V. A., & Beidaghi, M. (2020). 3D Printing of Additive-Free 2D Ti3C2Tx (MXene) Ink for Fabrication of Micro-Supercapacitors with Ultra-High Energy Densities. *ACS Nano, 14*(1), 640–650. https://doi.org/10.1021/acsnano.9b07325. http://pubs.acs.org/journal/ancac3.

O'Neill, P. F., Ben Azouz, A., Vázquez, M., Liu, J., Marczak, S., Slouka, Z., Chang, H. C., Diamond, D., & Brabazon, D. (2014). Advances in three-dimensional rapid prototyping of microfluidic devices for biological applications. *Biomicrofluidics, 8*(5). https://doi.org/10.1063/1.4898632. http://scitation.aip.org/content/aip/journal/bmf.

Paivana, G., Mavrikou, S., Kaltsas, G., & Kintzios, S. (2019). Bioelectrical Analysis of Various Cancer Cell Types Immobilized in 3D Matrix and Cultured in 3D-Printed Well. *Biosensors, 9*(4), 136. https://doi.org/10.3390/bios9040136.

Parekh, D. P., Ladd, C., Panich, L., Moussa, K., & Dickey, M. D. (2016). 3D printing of liquid metals as fugitive inks for fabrication of 3D microfluidic channels. *Lab on a Chip, 16*(10), 1812–1820. https://doi.org/10.1039/C6LC00198J.

Pearce, J. M., Anzalone, N. C., & Heldt, C. L. (2016). Open-Source Wax RepRap 3-D Printer for Rapid Prototyping Paper-Based Microfluidics. *SLAS Technology, 21*(4), 510–516. https://doi.org/10.1177/2211068215624408.

Pinilla, S., Ryan, S., McKeon, L., Lian, M., Vaesen, S., Roy, A., Schmitt, W., Coleman, J. N., & Nicolosi, V. (2023). Additive Manufacturing of Li-Ion Batteries: A Comparative Study between Electrode Fabrication Processes. *Advanced Energy Materials, 13*(15). https://doi.org/10.1002/aenm.202203747.

Praveen, S., Santhoshkumar, P., Joe, Y. C., Senthil, C., & Lee, C. W. (2020). 3D-printed architecture of Li-ion batteries and its applications to smart wearable electronic devices. *Applied Materials Today, 20*, 100688. https://doi.org/10.1016/j.apmt.2020.100688.

Praveen, S., Sim, G. S., Ho, C. W., & Lee, C. W. (2021). *3D-printed twisted yarn-type Li-ion battery towards smart fabrics. Energy Storage Materials, 41*, South Korea: Elsevier B.V748–757. http://www.journals.elsevier.com/energy-storage-materials/. https://doi.org/10.1016/j.ensm.2021.07.024.

Praveen, S., Sim, G. S., Shaji, N., Nanthagopal, M., & Lee, C. W. (2022). 3D-printed self-standing electrodes for flexible Li-ion batteries. *Applied Materials Today, 26*, 100980. https://doi.org/10.1016/j.apmt.2021.100980.

Praveen, S., Kim, T., Jung, S. P., & Lee, C. W. (2023). 3D-Printed Silicone Substrates as Highly Deformable Electrodes for Stretchable Li-Ion Batteries. *Small (Weinheim an der Bergstrasse, Germany), 19*(3). https://doi.org/10.1002/smll.202205817. http://onlinelibrary.wiley.com/journal/10.1002/(ISSN)1613-6829.

Punjiya, M., Moon, C. H., Matharu, Z., Rezaei Nejad, H., & Sonkusale, S. (2018). A three-dimensional electrochemical paper-based analytical device for low-cost diagnostics. *Analyst, 143*(5), 1059–1064. https://doi.org/10.1039/c7an01837a. http://pubs.rsc.org/en/journals/journal/an.

Rocha, V. G., García-Tuñón, E., Botas, C., Markoulidis, F., Feilden, E., D'Elia, E., Ni, N., Shaffer, M., & Saiz, E. (2017). Multimaterial 3D Printing of Graphene-Based Electrodes for Electrochemical Energy Storage Using Thermoresponsive Inks. *ACS Applied Materials and Interfaces, 9*(42), 37136–37145. https://doi.org/10.1021/acsami.7b10285. http://pubs.acs.org/journal/aamick.

Samper, I. C., Gowers, S. A. N., Rogers, M. L., Murray, D. S. R. K., Jewell, S. L., Pahl, C., Strong, A. J., & Boutelle, M. G. (2019). 3D printed microfluidic device for online detection of neurochemical changes with high temporal resolution in human brain microdialysate. *Lab on a Chip, 19*(11), 2038–2048. https://doi.org/10.1039/c9lc00044e. http://pubs.rsc.org/en/journals/journal/lc.

dos Santos, V. W., Martins, G., Gogola, J. L., Kalinke, C., Agustini, D., Bergamini, M. F., & Marcolino-Junior, L. H. (2023). Simultaneous detection of dopamine and ascorbic acid by using a thread-based microfluidic device and multiple pulse amperometry. *Analytical Methods, 15*(37), 4862–4869. https://doi.org/10.1039/d3ay00693j. http://pubs.rsc.org/en/journals/journal/ay.

Shalaby, A. A., Tsao, C.-W., Ishida, A., Maeki, M., & Tokeshi, M. (2023). Microfluidic paper-based analytical devices for cancer diagnosis. *Sensors and Actuators B: Chemical, 379*, 133243. https://doi.org/10.1016/j.snb.2022.133243.

Shao, Y., Fu, J. H., Cao, Z., Song, K., Sun, R., Wan, Y., Shamim, A., Cavallo, L., Han, Y., Kaner, R. B., & Tung, V. C. (2020). 3D Crumpled Ultrathin 1T MoS2for Inkjet Printing of Mg-Ion Asymmetric Micro-supercapacitors. *ACS Nano, 14*(6), 7308–7318. https://doi.org/10.1021/acsnano.0c02585. http://pubs.acs.org/journal/ancac3.

Sousa, L. R., Silva-Neto, H. A., Castro, L. F., Oliveira, K. A., Figueredo, F., Cortón, E., & Coltro, W. K. T. (2023). "Do it yourself" protocol to fabricate dual-detection paper-based analytical device for salivary biomarker analysis. *Analytical and Bioanalytical Chemistry, 415*(18), 4391–4400. https://doi.org/10.1007/s00216-023-04581-2. https://www.springer.com/journal/216.

Su, W., Li, Y., Zhang, L., Sun, J., Liu, S., & Ding, X. (2020). Typography-Like 3D-Printed Templates for the Lithography-Free Fabrication of Microfluidic Chips. *SLAS Technology, 25*(1), 82–87. https://doi.org/10.1177/2472630319867903.

Tang, K., Ma, H., Tian, Y., Liu, Z., Jin, H., Hou, S., Zhou, K., & Tian, X. (2020). 3D printed hybrid-dimensional electrodes for flexible micro-supercapacitors with superior electrochemical behaviours. *Virtual and Physical Prototyping, 15*(sup1), 511–519. https://doi.org/10.1080/17452759.2020.1842619.

Tian, X., Tang, K., Jin, H., Wang, T., Liu, X., Yang, W., Zou, Z., Hou, S., & Zhou, K. (2019). Boosting capacitive charge storage of 3D-printed micro-pseudocapacitors via rational holey graphene engineering. *Carbon, 155*, 562–569. https://doi.org/10.1016/j.carbon.2019.08.089.

Tian, X., Wang, T., Ma, H., Tang, K., Hou, S., Jin, H., & Cao, G. (2021). A universal strategy towards 3D printable nanomaterial inks for superior cellular high-loading battery electrodes. *Journal of Materials Chemistry A, 9*(29), 16086–16092. https://doi.org/10.1039/d1ta03236d.

Torres-Mapa, M. L., Singh, M., Simon, O., Mapa, J. L., Machida, M., Günther, A., Roth, B., Heinemann, D., Terakawa, M., & Heisterkamp, A. (2019). Fabrication of a Monolithic Lab-on-a-Chip Platform with Integrated Hydrogel Waveguides for Chemical Sensing. *Sensors, 19*(19), 4333. https://doi.org/10.3390/s19194333.

Tu, D., Wu, Z., Xu, J., Zhou, Y., Yang, W., Yang, Y., Zha, X., & Shi, L. (2020). Direct Assembly of 3D-BCN Microspheres as a Microsupercapacitor Electrode for Wearable Energy Storage. *ACS Applied Materials & Interfaces, 12*(42), 47416–47424. https://doi.org/10.1021/acsami.0c11982.

Tyagi, M., Fathollahzadeh, M., Martinez, J. G., Mak, W. C., Filippini, D., & Jager, E. W. H. (2023). Radially actuating conducting polymer microactuators as gates for dynamic microparticle sieve based on printed microfluidics. *Sensors and Actuators B: Chemical, 382*, 133448. https://doi.org/10.1016/j.snb.2023.133448.

Umamaheswari, R., Akilarasan, M., Chen, S. M., Cheng, Y. H., Mani, V., Kogularasu, S., Al-Hemaid, F. M. A., Ajmal Ali, M., & Liu, X. (2017). *One-pot synthesis of three-dimensional Mn3O4 microcubes for high-level sensitive detection of head and neck cancer drug nimorazole. Journal of Colloid and Interface Science*, *505*, Taiwan: Academic Press Inc. 1193–1201. http://www.elsevier.com/inca/publications/store/6/2/2/8/6/1/index.htt. https://doi.org/10.1016/j.jcis.2017.07.006.

Valera-Jiménez, J. F., Pérez-Flores, J. C., Castro-García, M., & Canales-Vázquez, J. (2021). Development of full ceramic electrodes for lithium-ion batteries via desktop-fused filament fabrication and further sintering. *Applied Materials Today, 25*, 101243. https://doi.org/10.1016/j.apmt.2021.101243.

Vinoth, R., Sangavi, P., Nakagawa, T., Jayaraman, M., & Vinu Mohan, A. M. (2023). All-in-one microfluidic device with an integrated porous filtration membrane for on-site detection of multiple salivary biomarkers. *Sensors and Actuators B: Chemical, 379*, 133214. https://doi.org/10.1016/j.snb.2022.133214.

Walgama, C., Nguyen, M. P., Boatner, L. M., Richards, I., & Crooks, R. M. (2020). Hybrid paper and 3D-printed microfluidic device for electrochemical detection of Ag nanoparticle labels. *Lab on a Chip, 20*(9), 1648–1657. https://doi.org/10.1039/d0lc00276c. http://pubs.rsc.org/en/journals/journal/lc.

Wang, T., Tian, X., Li, L., Lu, L., Hou, S., Cao, G., & Jin, H. (2020). 3D printing-based cellular microelectrodes for high-performance asymmetric quasi-solid-state micro-pseudocapacitors. *Journal of Materials Chemistry A, 8*(4), 1749–1756. https://doi.org/10.1039/c9ta11386j.

Wang, X., Deng, W., Shen, L., Yan, M., & Yu, J. (2016). A 3D electrochemical immunodevice based on an Au paper electrode and using Au nanoflowers for amplification. *New Journal of Chemistry, 40*(3), 2835–2842. https://doi.org/10.1039/C5NJ03222A.

Wang, Y., Chen, C., Xie, H., Gao, T., Yao, Y., Pastel, G., Han, X., Li, Y., Zhao, J., Fu, K.(K.), & Hu, L. (2017). 3D-Printed All-Fiber Li-Ion Battery toward Wearable Energy Storage. *Advanced Functional Materials, 27*(43). https://doi.org/10.1002/adfm.201703140.

Wang, Y., Zhang, Y.-Z., Gao, Y.-Q., Sheng, G., & ten Elshof, J. E. (2020). Defect engineering of MnO2 nanosheets by substitutional doping for printable solid-state micro-supercapacitors. *Nano Energy, 68*, 104306. https://doi.org/10.1016/j.nanoen.2019.104306.

Xu, X., Yang, L., Zheng, W., Zhang, H., Wu, F., Tian, Z., Zhang, P., & Sun, Z. M. (2022). MXenes with applications in supercapacitors and secondary batteries: A comprehensive review. *Materials Reports: Energy, 2*(1), 100080. https://doi.org/10.1016/j.matre.2022.100080.

Xu, Y., Yu, S., Johnson, H. M., Wu, Y., Liu, X., Fang, B., & Zhang, Y. (2024). Recent progress in electrode materials for micro-supercapacitors. *iScience, 27*(2), 108786. https://doi.org/10.1016/j.isci.2024.108786.

Yee, D. W., Citrin, M. A., Taylor, Z. W., Saccone, M. A., Tovmasyan, V. L., & Greer, J. R. Hydrogel-Based Additive Manufacturing of Lithium Cobalt Oxide. http://onlinelibrary.wiley.com/journal/10.1002/(ISSN)2365-709X.

Yuan, S., Li, S., Zhu, J., & Tang, Y. (2021). Additive manufacturing of polymeric composites from material processing to structural design. *Composites Part B: Engineering, 219*, 108903. https://doi.org/10.1016/j.compositesb.2021.108903.

Yukird, J., Soum, V., Kwon, O.-S., Shin, K., Chailapakul, O., & Rodthongkum, N. (2020). 3D paper-based microfluidic device: a novel dual-detection platform of bisphenol A. *The Analyst, 145*(4), 1491–1498. https://doi.org/10.1039/c9an01738k.

Zhang, F., Wu, K., Xu, X., Wu, W., Hu, X., Yu, K., & Liang, C. (2021). 3D Printing of Graphite Electrode for Lithium-Ion Battery with High Areal Capacity. *Energy Technology, 9*(11). https://doi.org/10.1002/ente.202100628.

Zhang, H., Wang, B., & Brown, B. (2021). Aerosol-Jet-Printed CoFe2O4 Nanoparticle – Vertically Aligned Carbon Nanotube Composite for Microsupercapacitors. *The Journal of Physical Chemistry C, 125*(14), 7590–7597. https://doi.org/10.1021/acs.jpcc.1c01154.

Zhang, Y., Ji, T., Hou, S., Zhang, L., Shi, Y., Zhao, J., & Xu, X. (2018). All-printed solid-state substrate-versatile and high-performance micro-supercapacitors for in situ fabricated transferable and wearable energy storage via multi-material 3D printing. *Journal of Power Sources, 403*, 109–117. https://doi.org/10.1016/j.jpowsour.2018.09.096.

Zhao, C., Wang, R., Fang, B., Liang, H., Nie, B., Wang, R., Xu, B., Feng, S., Li, R., Li, S., Xiong, Y., Shao, Y., & Mo, R. (2023). Boosting the Lithium Storage Properties of a Flexible Li4Ti5O12/Graphene Fiber Anode via a 3D Printing Assembly Strategy. *Batteries, 9*(10), 493. https://doi.org/10.3390/batteries9100493.

Zhao, C., Wang, R., Liang, H., Fang, B., Li, R., Wu, R., & Mo, R. (2024). Autonomous self-healing strategy for flexible fiber lithium-ion battery with ultra-high mechanical properties and volumetric energy densities. *Chemical Engineering Journal, 496*, 154153. https://doi.org/10.1016/j.cej.2024.154153.

Zhou, W., Li, X., Duan, H., & Lv, P. (2022). Multi-Material Integrated Three-Dimensional Printing of Cylindrical Li-Ion Battery. *Journal of Manufacturing Science and Engineering, 144*(6). https://doi.org/10.1115/1.4052707.

18 Additively manufactured functional polymers and nanocomposites for health care and biomedical applications

Santosh Kumar[1], *and Rakesh Kumar*[2]
[1]Department of Mechanical Engineering, Chandigarh Group of Colleges, Mohali, Punjab, India, [2]Department of Regulatory Affairs & Quality Assurance, Kaushik Orthopaedic Private Limited, Sonipat, Haryana, India

18.1 Introduction

3D printing or additive manufacturing (AM) is an effective technique for fabricating a wide range of structures and complex geometries from three-dimensional (3D) model data. The procedure involves printing consecutive layers of materials on top of one another. Charles Hull invented this technique in 1986 using a procedure known as stereolithography (SLA), which was followed by advances such as powder bed fusion, fused deposition modeling (FDM), inkjet printing, and contour crafting (CC) (Ngo et al., 2018). The AM process includes creating a CAD model, converting it to STL format, transferring it to additive manufacturing equipment and manipulating the STL file, printing and building the component, removing and cleaning the manufactured part, postprocessing (if necessary), and usage (Kumar, Kumar, et al., 2021; Kumar, Ramakrishnan, et al., 2021). This approach is capable of creating devices manufactured from a variety of materials, including metals, polymers with novel geometries, ceramics, and shape-memory alloys (Cawley, 1999; Seitz et al., 2005; Vorndran et al., 2008).

One significant advantage of this technology is that it enables the production and design of complex geometries using a new macro- and microarchitecture as well as a completely linked pore network (Lam et al., 2002). The comparison of different AM processes (working, material, merits, demerits, etc.) is given in Table 18.1 (Agarwala et al., 1996; Flach et al., 1998; Fu et al., 2022; Kumar, Kumar, Chohan, et al., 2022; Kumar, Kumar, Kumar, et al., 2022; Kumar, Kushawaha, et al., 2022; Ngo et al., 2018; Parupelli & Desai, 2019; Srivastava et al., 2022; Subramanian et al., 1995; Utela et al., 2008, Wu et al., 2002; Yan & Gu, 1996; Zein et al., 2002; Ziemian & Crawn, 2001).

18.2 Additive manufacturing of polymer materials

3D polymer printing has gained popularity over the last three decades, originally for quick prototyping operations. Thermoplastics are primarily employed for their

Additive Manufacturing of Functional Polymers and Nanocomposites. DOI: https://doi.org/10.1016/B978-0-443-26516-7.00018-3
© 2025 Elsevier Ltd. All rights are reserved, including those for text and data mining, AI training, and similar technologies.

Table 18.1 Comparison of different additive manufacturing process methods.

Printing	FDM	SLA	LOM	SLS
Inventor	Scott Crump	Charles W. Hull	Helisys	Carl Deckard
Year of Invention	1990	1986	1991	1989
Principle	Matrix extrusion	Vat polymerization	Sheet lamination	Powder bed fusion
Working	Material is heated and injected through a nozzle to form a 3D part	Photosensitive thermoset polymers are cured by exposing them to ultraviolet laser light	Uses adhesives to join layers of material in sheet shapes, such as metal, plastic, or paper, to construct a component. Well-liked for prototypes	The fuses material layer by layer without ever reaching the melting temperature by using a laser beam as a heat source
Max build volume	1500 × 1100 × 1500	1500 × 750 × 550	256 x 169 x 150	81 × 330 × 460
Manufacturer and headquarter	Stratasys (USA)	Prodways (France)	-	Formlabs (USA)
Materials	Continues filaments of thermoplastic polymers (PLA, ABS), etc., continues fiber reinforcement	ABS (semiflexible materials) resin with photoactive monomers	Polymers and ceramics, paper, metal roll, polymer composites	Metal, polymers
Field of applications	Toys, RP, home applications, advanced composite components	Prototyping, medical prototyping dental applications	Foundry, nonfunctional prototypes, smart structure, paper manufacturing	Aerospace, electronics biomedical, military and complex shape object, heat exchanger, lightweight structures
Advantages	Low cost of set up, simple to operate, high speed	High quality, the resolution is fine	Less tooling and manufacturing cost, possibility to manufacture large structure.	Fine resolution, good part quality

(Continued)

Table 18.1 (Continued)

Printing	**FDM**	**SLA**	**LOM**	**SLS**
Limitation	Weak mechanical properties, only suitable for thermoplastic materials	Expensive, Applicable for limited materials, printing process is slow	Difficulty in manufacturing complex shapes	Expensive process, high porosity, low printing speed
Min. layer resolution (μm)	10–100	50–100	100	80

rigidity, high strength, and high-temperature endurance and stability. The polymers utilized in AM processes can be categorized as thermoplastics, particle-reinforced polymer composites, polymer matrix composites, fiber-reinforced polymer composites, thermoresponsive polymers, nanocomposites, and thermoplastic elastomer composites. AM makes use of a wide range of polymer materials in various forms and compositions, such as composites, nanocomposites, continuous/discontinuous fiber-reinforced thermoplastic composites, and hybrids (Fu et al., 2022; Kumar, Kumar, Chohan, et al., 2022; Kumar, Kumar, Kumar, et al., 2022; Kumar, Kushawaha, et al., 2022). The FFF technique employed a variety of polymers, including PLA, ABS, PEEK, Nylon 12/Carbon Fiber, PC, PC-ABS, HIPS, PEEK, PEI, and TPU. Aside from this, the SLS process employed powdered polycarbonate (PC), acrylic styrene (PMMA/PS), nylons (polyamide [PA]), polystyrene (PS), polyamides (PA), thermoplastic elastomers (TPE), and so on. Material jetting includes Vero, Digital ABS, Tango, and Durus, whereas multijet fusion includes elastic TPU and polyamide (PA12/Nylon 12) (Schmid & Wegener, 2016). The characteristics of the most widely used polymers are discussed in the section below.

18.2.1 ABS polymer

The commercial polymer (ABS) used in the AM process technique has tensile modulus (1627 MPa), tensile strength (22 MPa), and density (1050 kg/m^3), correspondingly (Dana et al., 2019). The mechanical characterization of thermoplastic polymer components produced using Arburg Plastic Free Forming (APF) is investigated. Furthermore, the mechanical characteristics of polymer specimens made by injection molding, a typical procedure, are investigated to assess the effect of the manufacturing process on the mechanical behavior of the components. The tensile mechanical characteristics are assessed by changing various production factors, such as the print route. Microstructural investigations were conducted to better understand the link between the production process and the final mechanical characteristics. The [0°] printing case had the weakest tensile properties in terms of mean stiffness, elongation to fracture, and UTS. Printing at 90° ([90°] and [0°/90°]) resulted in large dispersion in measured properties. Only the case [−45°/45°] provided tensile properties comparable to injection-molded specimens, both in terms of mean measures and dispersion of measures. As seen with FDM, the elongation to fracture is decreased by less than 20%. (Dana et al., 2019)

Ahn et al. (2002) discovered that the raster orientation has a significant impact on the tensile strength of an ABS product, although the printed beads' width and bed temperature had no influence.

Domingo-Espin et al. (2015) examined how, in addition to the printing orientation, additional parameters like as nozzle size, thickness, and location may have a substantial impact on the mechanical characteristics of the polycarbonate (PC) component created by additive manufacturing. Rankouhi et al. (2016) conducted comprehensive experimental research to investigate the impact of printed layer thickness and raster orientation on the mechanical characteristics of 3D-printed ABS specimens, addressing a disagreement in the literature. Their tensile test results, as well as statistical analyses of the data, clearly indicate that specimens with 0.2 mm layer thickness are stronger than specimens with 0.4 mm layer thickness, and that layer thickness and raster orientation both have a significant impact on the mechanical properties of printed specimens. The study of various authors (Casavola et al., 2015; Durgun & Ertan, 2014; Lee et al., 2005; Rankouhi et al., 2016; Rodrı´Guez et al., 2003; Tymrak et al., 2014; Wu et al., 2015; Zhang et al., 2016) uses distinct printing parameters such as printing path, layer thickness, infill degree, and feed rate.

18.2.2 Polylactic

Carothers created polylactic acid in 1932. PLA is a thermoplastic monomer made from sustainable, organic materials such as maize starch. It is classified into three subgroups: PDLLA (Poly DL-Lactic acid), PLLA (Polylactic acid), and PDLA (Poly D-lactic acid). PLA is highly suggested for medical applications because of various qualities such as low melting temperature, biocompatibility, ecofriendliness in processing and material handling, odorless ness, minimum warping, ease of printing, availability in a larger variety of colors, and blendability (Gkartzou, et al., 2017; Ilyas et al., 2021; Kumar & Kumar, 2024; Kumar & Kumar, 2025). Further, Song et al. (2017) demonstrated that polylactic acid (PLA) created via 3D printing had superior mechanical properties when compared to injection-molded polylactic acid. Biodegradable plastics are among the most promising materials for replacing traditional petroleum-based plastics, which have had a wide range of negative environmental consequences, including pollution (land, water, etc.) and climate change. Polylactic acid (PLA) is a type of biodegradable plastic that is not only commercially accessible but also safe to break down without contaminating the environment. PLA is also comparable to other traditional plastics like PP and PET in terms of mechanical, physical, biocompatibility, and processability. Because of these properties, PLA has become the most widely used biopolymer in a variety of industries, including agriculture, automobiles, and packaging (Taib et al., 2022; Torres et al., 2016). Despite the dominance of commodity plastics (e.g., PET, PP, PS, PE, etc.), natural and synthetic, bio-based and biodegradable polymer materials such as polybutylene succinate (PBS), polyhydroxyalkanoates (PHA), cellulose, and lignin have recently emerged as an important focus for the development of polymer-based material systems with functionalities that are significantly in demand by economies and the ecosystem (Daminabo et al., 2020). Fig. 18.1 illustrates the advantages and disadvantages of PLA as a material (Auras et al., 2004; Burg et al., 1999; Daminabo et al., 2020; Dorgan et al., 2001; Farah et al., 2016; Gupta et al., 2007; Jamshidian et al., 2010; Janorkar et al., 2004; Kumar & Kumar, 2022; Rasal & Hirt, 2008; Rasal et al., 2009; Sawyer, 2003; Tokiwa & Calabia, 2006).

PLA, a bioabsorbable polymer, is one of the most promising biopolymers since the monomers can be made from nontoxic renewable feedstock and is a naturally occurring organic acid. Lactic acid (2-hydroxypropionic acid, LA), a PLA component unit, occurs as two enantiomers, L- and D-lactic acid PLA also comprises stereoisomers, such as poly(L-lactide) (PLLA), poly(D-lactide) (PDLA), and poly

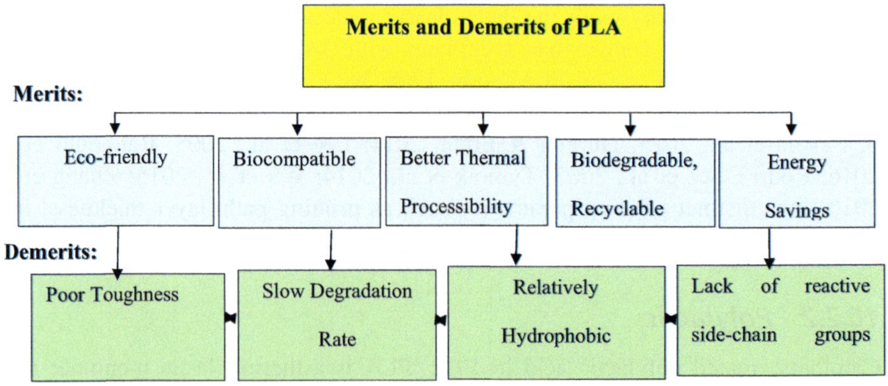

Figure 18.1 Merits and demerits of PLA material. *PLA*, Polylactic acid.

(DL-lactide) (PDLLA) (Auras et al., 2010; Farah et al., 2016; Lasprilla et al., 2011; Lopes et al., 2012). PLA exhibits physical and mechanical characteristics. In terms of physical attributes, physical attributes need knowledge of a desired part's size and weight requirements. Density may also be a significant design characteristic since high-density values indicate high transportation costs (e.g., light automobile parts minimize energy consumption). Density is commonly used to calculate "specific properties," which include dividing mechanical parameters by the appropriate density (McKeen, 2013). The physical characteristics of PLA and other biopolymers are shown in Fig. 18.2 (Van De Velde & Kiekens, 2002) and the effects of stereochemistry and crystallinity on mechanical characteristics are shown shown in Fig. 18.3 (Farah et al., 2016; Hartmann, 1998; Perego et al.,1996).

18.2.3 Polycarbonate

Polycarbonate is one of the thermoplastics often utilized in 3D printing technology. In the FDM technique, the thermoplastic polymer polycarbonate (PC) is employed (Alghamdi et al., 2021; Jasiuk et al., 2018; Ortiz-Acosta & Moore, 2019). Among the aforementioned approaches, Melted and Extruded Manufacturing (MEM) technology is an extrusion method that produces structures and models from filaments made of thermoplastic polymers, including PC (Bere et al., 2020). Further, Kumar, Kumar, et al. (2021) and Kumar, Ramakrishnan, et al. (2021) studied the influence of material composition variations on mechanical characteristics. The authors discussed the feasibility of incorporating PC (from 10 to 30 wt.%) into the ABS matrix to produce suitable materials. Scientists added 25–75 wt.% of the PC into the terpolymer. The mechanical strength and hardness of the composite improved with each increase in the material's polycarbonate (PC) content (Andrzejewski & Marciniak-Podsadna, 2020; Bano et al., 2021; Farcas et al., 2019; Kannan & Ramamoorthy, 2020; Yap et al., 2019). Several reviews have been produced, providing information on the features and characteristics of PLA. Fig. 18.4 shows the qualities of 3D items made using PC according to the data sheets of EOS GmbH, Stratasys, USA, and SYS.

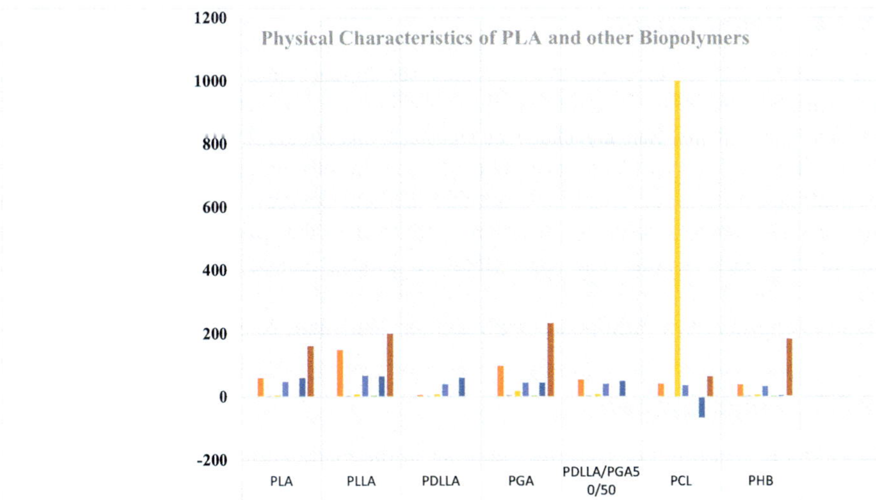

	PLA	PLLA	PDLLA	PGA	PDLLA/PGA5 0/50	PCL	PHB
Polymer Density (g/cm2)	1.25	1.3	1.27	1.71	1.4	1.146	1.262
Tensile Strength (MPa)	60	150	6.5	99.7	55.2	42	40
Tensile Modulus (Gpa)	3.5	4.14	3.45	7	4.34	0.44	4
Ultimate Strain %	6	10	10	20	10	1000	8
Specific Tensile Strength Nm/g	48	66.8	39.4	45.1	41.2	30.7	33.9
Specific Tensile Modulus KNm/g	2.8	3.85	2.36	4.51	2.14	0.38	2.97
Glass Transition Temperature	60	65	60	45	50	-65	5
Melting Temperature	162	200	0	233	0	65	182

Figure 18.2 Physical characteristics of PLA and other biopolymers. *PLA*, Polylactic acid.

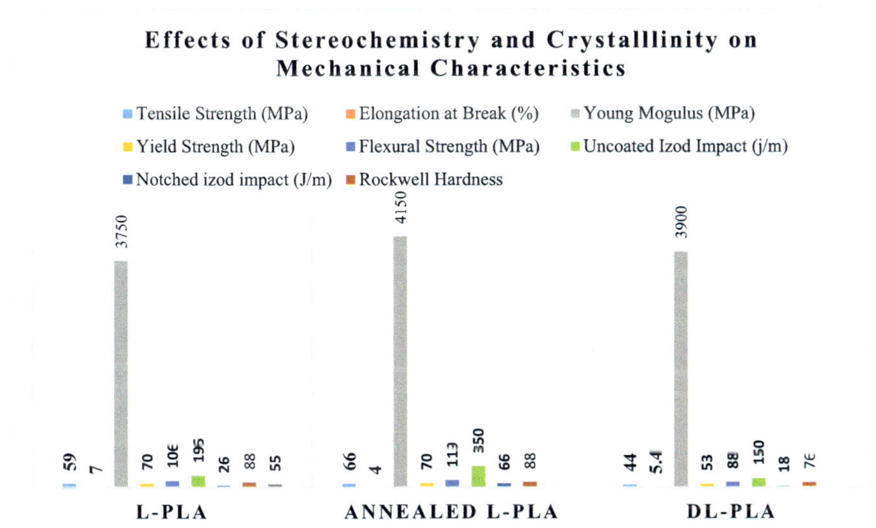

Figure 18.3 Effects of stereochemistry and crystallinity on mechanical characteristics.

18.2.4 Poly-L-lactide

It is more widely utilized however, because the fundamental component is naturally occurring. Furthermore, PLLA crystallizes easily due to its symmetric repeat unit and the flexibility of the lengthy chain (Bulanda et al., 2021; Lunt, 1998; Tsuji et al., 2006). It is a biodegradable polymer utilized in biomedical and pharmaceutical applications. It manufactures bioabsorbable medical devices, which disintegrate over time in physiological settings. Because of their historical pedigree, polylactides are one of the most simple and cost-effective biodegradable polymers for medical devices. Table 18.2 shows the thermal and mechanical properties of PLLA material, which is employed in orthopedic applications (Wuisman & Smit, 2005).

18.2.5 Polydioxanone

PDS/PDO is becoming increasingly popular in the medical industry due to its in vivo breakdown of low-toxicity monomers. PDS has a lower modulus than PLA or PGA, making it the first degradable polymer to be utilized to create a monofilament suture (Pachence & Kohn, 2000). The thermal and mechanical characteristics of PDO/PDS material are shown in Table 18.3.

18.2.6 PLLA-HA (50/50)

The PLLA-HA and PLGA-HA composites exhibited better thermal stability, higher decomposition temperature, and higher activation energy for the decomposition process than the neat polymers (Table 18.4).

18.2.7 Polydioxanone

PDS/PDO has grown in popularity in the medical industry due to its breakdown into low-toxicity monomers. In vivo, PDS has a lower modulus than PLA or PGA; hence, it was the first degradable polymer to be utilized to create a monofilament suture. The elongation at break and other parameters are maximum as seen in Table 18.5.

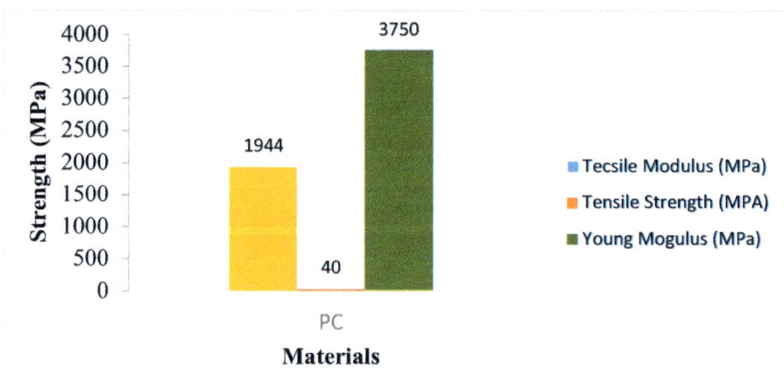

Figure 18.4 Properties of PC polymer materials. *PC*, Polycarbonate.

Table 18.2 Properties of PLLA polymer.

Materials	Melting temperature	Glass trans. temperature	Elastic modulus	Tensile strength	Elongation at beak	Recrystallization temperature	Degradation Time	References
Poly-L-lactide (PLLA)	180°C	50°C–60°C	3–4 GPa	65.0 MPa	7%	100°C	> 5.7 years	(Bergsma, 1995; Ea[illegible]wiboonthanakit et al., 2014; Farah et al., 2016; Horvath et al., 2018; Suuronen et al., 1998; Waris et al., 2004)

Table 18.3 Properties of PDO/PDS polymer.

Materials	Melting temperature	Glass trans. temperature	Elastic modulus	Tensile strength	Elongation at beak	Recrystallization temperature	Degradation Time	References
Polydioxanone (PDO/PDS)	106°C	<20 °C	798.32 ± 57.7 MPa	583.08 ± 5.3 MPa	43.7% for dia. 0.3 mm	-	Generally takes 117–240 days (Invitro)	(Albano et al., 2013; 2013; Kreszinger et al., 2018; Mäkelä et al., 2002; Martins et al., 2019; 2019)

Table 18.4 Properties of PLLA-HA (50/50) polymer.

Materials	Melting temperature	Glass trans. temperature	Elastic modulus	Tensile strength	Elongation at beak	Recrystallization temperature	Degradation Time	References
PLLA-HA (50/50)	179°C	60°C	3.5 GPa	-	-	90°C–115°C	-	(Albano et al., 2013)

Table 18.5 Properties of polydioxa none, PDO polymer.

Materials	Melting tempera-ture	Glass trans. tempera-ture	Elastic mod-ulus	Tensile stren-gth	Elongation at beak	Recrystallization temperature	Degradation Time	References
Polydioxa none (PDO)	106°C	< 20°C	798.32 ± 5 7.7 Mpa	583.08 ± 5.3 MPa	43.7% for dia. 0.3	-	Generally takes 117–240 days (In-vitro)	(Erben et al., 2022; Pachence & Kohn, 2000; Silva et al., 2015)

18.2.8 PLGA (90/10), PLGA (85 L/15 G), and PGA

PGA and PLA can be mixed to create a wide variety of PLGA polymers. Both L- and DL-lactides have been used for co-polymerization. For varied compositions, the glycolide to lactide ratio can be adjusted to regulate properties. Polyglycolide (PGA) has excellent biodegradability and biocompatibility; nevertheless, the breakdown rate of PGA is too high. Furthermore, sutures made from PGA or LA/GA copolymers can only be employed as multifilaments with a braided or twisted design to minimize the stiffness of the suture (Cohn et al., 1987) (Table 18.6).

18.3 Application of 3D-printed polymer products and composites

Additively made polymer composites are widely employed in a variety of sectors, including construction, biomedical, electronics, aerospace, and textiles. They are also utilized to create flexible components like wearable electronics and textile materials. The combination of functional polymers and nanoparticles has the potential to revolutionize the field of health care and biomedical/biomedicine. Some potential applications of this material in different sectors are as follows.

18.3.1 Biomedical applications

AM based on polymeric materials is widely employed in biomedical applications that need tailored instruments, treatments, medications, and organ replacements. Polymer-based materials utilized in biomedical applications are classed as hard or soft polymers. They are further categorized into two types: biodegradable and nonbiodegradable polymers. Biodegradable polymers are employed in situations where tissues proliferate and biomedical components within the body are no longer required once the purpose is completed. They are degraded in the human body, such as via the hydrolysis of ester bonds in polyesters. However, nonbiodegradable polymers are employed for structural implants. Some of the hard-synthetic biodegradable polymers utilized in biomedical applications include polylactic acid (PLA), polycaprolactone (PCL), polyglycolic acid (PGA), and polydioxanone (PDO). Polyaryletherketones (PAEKs) are among the nonbiodegradable rigid polymers employed. Hard polymers are used in biomedical applications such as implants, dental implants, surgical models, and prostheses. Polyethylene glycol (PEG), acrylate-based hydrogels, elastomers, and hydrogels with specific thermal characteristics, such as poloxamer pluronic F127, are examples of soft polymers utilized in biomedical applications. Tissue engineering and bioprinting, drug delivery, tissue phantoms, and soft surgical models are all examples of biomedical uses for soft polymers. Further, Ho et al. (2016) created additively built scaffolds out of polycaprolactone and CNTs for cardiac tissue engineering. Ramírez et al. (2018) also created an additive manufacturing approach for producing composite materials utilizing FFF/FDM fabrication to increase mechanical qualities. Jockusch and Özcan (2020) examined the use of additive manufacturing in dental applications, including

Table 18.6 Properties of PLGA (90/10), PLGA (85 L/15 G) and PGA polymer.

Materials	Melting temperature	Glass trans. temperature	Elastic modulus	Tensile strength	Elongation at beak	Recrystallization temperature	Degradation Time	References
PLGA (90/10)	202°C–210°C	42.0°C	3–6 GPa	45 MPa	-	-	3–4 months	(Kim et al., 2023; Lanao et al., 2013; MatWeb 2024)
PLGA (85 L/15 G)	140°C–150°C	55°C	2–4 GPa	65 MPa	310%	-	12–18 months	
PGA	215°C–225°C	40°C	6–7 GPa	95 MPa	3.67%	-	4–6 months	

research models, maxillofacial prostheses, and orthodontic equipment. The number of polymeric materials employed in biological applications has expanded dramatically during the last decade (Kohane & Langer, 2008). The main applications include resorbable sutures, medication delivery systems, and orthopedic fixation devices including pins, rods, and screws (Behravesh et al., 1999; Middleton & Tipton, 2000). The skin is the body's biggest organ, performing a variety of vital activities necessary for survival. Because they are in close touch with the external environment, they are particularly susceptible to damage and/or harm. To minimize problems, damage regeneration or repair must occur quickly (Wu et al., 2002). Polymeric biomaterials have been created during the last few decades to operate as smart skin replacements, fulfilling many of the same functions as skin. Skin replacements are a diverse set of wound-covering materials that help in wound closure and replace skin functions, either temporarily or permanently, depending on the product's properties. These compounds are used as alternatives to traditional wound care when normal therapy is not desired (Shores et al., 2007). Skin replacements are used to help heal wounds, relieve pain, and restore skin function. Skin substitutes have crucial roles in the treatment of deep dermal and full-thickness wounds of different etiologies (Halim et al., 2010). Nanocomposites of metal oxides, like zinc oxide, have been shown to have antibacterial properties and are currently being included in skin replacements. Poly(ε-caprolactone) (PCL) has received significant interest in biological applications during the last several decades. PCL provides various benefits, including cheap cost, biocompatibility, and biodegradability (Choi et al., 1999). PCL has been proposed for a variety of applications, including drug delivery systems (Allen et al., 2000; Zhong & Sun, 2001), tissue-engineered skin (plain film), and scaffolds to promote fibroblast and osteoblast development (Hutmacher et al., 2001; Ng et al., 2001), among others. Electrospinning is a good technology for producing fibers with dimensions ranging from micrometers to nanometers (Li et al., 2010). Electrospinning technology, which can readily mass-produce thin nanofibrous membranes with high conformability, might provide a solution for making skin replacements. Electrospun nanofibers mirror the inherent topographical characteristics of the extracellular matrix, potentially promoting the cell's normal activities in a biomimetic manner. Electrospun nanofibers have several properties that make them suitable as skin substitute materials, including high oxygen permeability, variable pore size, a large surface area to volume ratio, and morphological similarity to the extracellular matrix (EM) (Smith & Ma, 2004) (Zahedi et al., 2009). In addition, the author discuss technical packaging applications in the electronics, food, and beverage sectors. He suggests in the study of many authors that utilizing suitable polymers, functional nanofillers, and manufacturing procedures may be exciting approaches for developing revolutionary packaging materials with improved mechanical, thermal, and barrier capabilities (Kausar, 2019). The variety of materials that may be employed in 3D printing technology is vast, but high-performance polymer nanocomposites are a particularly significant type of material that has just lately gained popularity in additive manufacturing. High-performance polymers are a class of polymer materials that have been shown to maintain desired mechanical, thermal, and chemical qualities when exposed to hostile environments such as high temperatures, high pressure, and corrosive chemicals. When these polymers are blended with nanofillers such as carbon nanotubes, nanoclay, and graphene, their mechanical characteristics increase and they may gain features that were not there initially, such as thermal and electrical conductivity (De Leon et al., 2016).

18.3.2 Tissue engineering

Scaffolds may be created additively using functional polymer nanocomposite to offer a supporting framework for cell development and tissue formation. Researchers, for example, are creating scaffolds consisting of collagen and hydroxyapatite nanoparticles to change the way bone grows. Nanocomposite scaffolds enable numerous structural and mechanical alterations to develop novel characteristics, functionalities, and applications for tissue engineering (Wahid et al., 2018). Tissue engineering seeks to create new biofunctional tissues that can regenerate and repair damaged or sick tissue. The number of orthopedic surgical procedures conducted globally in 2017 was around 22.3 million, and iorthopat is expected to reach 28.3 million by 2022 (Esmaeili et al., 2023). Several scaffold manufacturing techniques are at the cutting edge of tissue engineering (TE), including electrospinning, freeze-drying, bioprinting, and decellularization (Wubneh et al., 2018). The different researchers used different biomaterials in bone tissue engineering that include inorganic biomaterials—metals (e.g., titanium and its alloys), inorganic biomaterials—bioceramics (e.g., HAp), β-TCP, bioactive glasses, zirconia, natural polymers: (e.g., collagen), gelatin, SF, HA, chitosan, GG, alginate, synthetic polymers (e.g., PLLA), PLGA, PCL, PU, inorganic–organic composite biomaterials (e.g., SF/β-TCP; SF/HAp; collagen/BCP), etc. (Budharaju et al., 2023; Donnaloja et al., 2020; Ginebra et al., 2018; Gunatillake, 2003; Gunja & Athanasiou, 2006; Hasan et al., 2012; Kargozar et al., 2019; Kolos & Ruys, 2015; Kundu et al., 2012; Pieralli et al., 2016; Pignatello, 2011; Purohit et al., 2019; Shaltooki et al., 2019; Silva et al., 2010; Sun & Tan, 2013; Sun et al., 2024; Tammi et al., 2008; Vieira et al., 2019; Wubneh et al., 2018). Further, Gleadall et al. (2018) describe the design of Ti6A4V scaffolds and how cell shape and porosity impact their behavior. In addition, the authors discuss the several standard processes for scaffold construction, such as freeze-drying, gas foaming, solvent casting and particle leaching, electrospinning, and thermal-induced phase separation, as well as their advantages and disadvantages. Additionally, drug delivery can be integrated into functional polymers to produce a drug delivery system. Overall, 3D printings have diverse applications due to its attractive features (Kumar & Kumar, 2024; Kumar, Kumar, et al., 2021; Kumar, Ramakrishnan, et al., 2021; Kumar, Kushawaha, et al., 2022).

18.4 Conclusion

Additive manufactured functional polymer nanocomposites represent a promising class of materials with the potential to revolutionize the field of health care and biomedicine. By combining the desired properties of the polymers with the unique function abilities of the nanoparticles, these materials can be tailored to create novel devices and implants for a wide range of biomedical applications.

References

Agarwala, M., Van Weeren, R., Bandyopadhyay, A., Whalen, P., Safari, A., & Danforth, S. (1996). Fused Deposition of Ceramics and Metals : An Overview. In *Fused Deposition of Ceramics and Metals : An Overview* , pp. 385-392. The University of Texas at Austin. https://doi.org/10.15781/t2nk36q96

Ahn, S., Montero, M., Odell, D., Roundy, S., & Wright, P. K. (2002). Anisotropic material properties of fused deposition modeling ABS. *Rapid Prototyping Journal, 8*(4), 248–257. https://doi.org/10.1108/13552540210441166.

Albano, C., González, G., Palacios, J., Karam, A., & Covis, M. (2013). PLLA-HA vs. PLGA-HA characterization and comparative analysis. *Polymer Composites, 34*(9), 1433–1442. https://doi.org/10.1002/pc.22445.

Alghamdi, S. S., John, S., Choudhury, N. R., & Dutta, N. K. (2021). Additive Manufacturing of polymer materials: Progress, promise and challenges. *Polymers, 13*(5), 753. https://doi.org/10.3390/polym13050753.

Allen, C., Han, J., Yu, Y., Maysinger, D., & Eisenberg, A. (2000). Polycaprolactone–b-poly (ethylene oxide) copolymer micelles as a delivery vehicle for dihydrotestosterone. *Journal of Controlled Release, 63*(3), 275–286. https://doi.org/10.1016/s0168-3659(99)00200-x.

Andrzejewski, J., & Marciniak-Podsadna, L. (2020). Development of thermal resistant FDM printed blends. The preparation of GPET/PC blends and evaluation of material performance. *Materials, 13*, 2057. https://doi.org/10.3390/ma13092057.

Auras, R., Harte, B., & Selke, S. (2004). An overview of polylactides as packaging materials. *Macromolecular Bioscience, 4*(9), 835–864. https://doi.org/10.1002/mabi.200400043.

Auras, R., Lim, L. T., Selke, S. E. M., & Tsuji, H. (2010). *Poly(Lactic Acid)*. Wiley eBookshttps://doi.org/10.1002/9780470649848.

Bano, S., Iqbal, T., Ramzan, N., & Farooq, U. (2021). Study of surface mechanical characteristics of ABS/PC blends using nanoindentation. *Processes, 9*(4), 637. https://doi.org/10.3390/pr9040637.

Behravesh, E., Yasko, A. W., Engel, P. S., & Mikos, A. G. (1999). Synthetic biodegradable polymers for orthopaedic applications. *Clinical Orthopaedics and Related Research, 367*, S118–S129. https://doi.org/10.1097/00003086-199910001-00012.

Bere, P., Neamtu, C., & Udroiu, R. (2020). Novel method for the manufacture of complex CFRP parts using FDM-based molds. *Polymers, 12*(10), 2220. https://doi.org/10.3390/polym12102220

Bergsma, J. (1995). Late degradation tissue response to poly(?-lactide) bone plates and screws. *Biomaterials, 16*(1), 25–31. https://doi.org/10.1016/0142-9612(95)91092-d.

Budharaju, H., Suresh, S., Sekar, M. P., De Vega, B., Sethuraman, S., Sundaramurthi, D., & Kalaskar, D. M. (2023). Ceramic materials for 3D printing of biomimetic bone scaffolds—Current state-of-the-art & future perspectives. *Materials & Design, 231*, 112064. https://doi.org/10.1016/j.matdes.2023.112064.

Bulanda, K., Oleksy, M., Oliwa, R., Budzik, G., Przeszłowski, Ł., Fal, J., & Jesionowski, T. (2021). Polymer composites based on polycarbonate (PC) applied to additive manufacturing using melted and extruded manufacturing (MEM) technology. *Polymers, 13*(15), 2455. https://doi.org/10.3390/polym13152455.

Burg, K., Holder, W., Culberson, C., Beiler, R., Greene, K., Loebsack, A., Roland, W., Mooney, D., & Halberstadt, C. (1999). Parameters affecting cellular adhesion to polylactide films. *Journal of Biomaterials Science. Polymer Edition, 10*(2), 147–161. https://doi.org/10.1163/156856299x00108.

Casavola, C., Cazzato, A., Moramarco, V., & Pappalettere, C. (2015). Orthotropic mechanical properties of fused deposition modelling parts described by classical laminate theory. *Materials & Design, 90*, 453–458. https://doi.org/10.1016/j.matdes.2015.11.009.

Cawley, J. D. (1999). Solid freeform fabrication of ceramics. *Current Opinion in Solid State and Materials Science, 4*(5), 483–489. https://doi.org/10.1016/s1359-0286(99)00055-8.

Choi, E., Kim, C., & Park, J. (1999). Synthesis and characterization of starch-g-polycaprolactone copolymer. *Macromolecules, 32*(22), 7402–7408. https://doi.org/10.1021/ma981453f.

Cohn, D., Younes, H., & Marom, G. (1987). Amorphous and crystalline morphologies in glycolic acid and lactic acid polymers. *Polymer, 28*(12), 2018–2022. https://doi.org/10.1016/0032-3861(87)90035-8.

Daminabo, S., Goel, S., Grammatikos, S., Nezhad, H., & Thakur, V. (2020). Fused deposition modeling-based additive manufacturing (3D printing): Techniques for polymer material systems. *Materials Today Chemistry, 16*, 100248. https://doi.org/10.1016/j.mtchem.2020.100248.

Dana, H. R., Barbe, F., Delbreilh, L., Azzouna, M. B., Guillet, A., & Breteau, T. (2019). Polymer additive manufacturing of ABS structure: Influence of printing direction on mechanical properties. *Journal of Manufacturing Processes, 44*, 288–298. https://doi.org/10.1016/j.jmapro.2019.06.015.

De Leon, A. C., Chen, Q., Palaganas, N. B., Palaganas, J. O., Manapat, J., & Advincula, R. C. (2016). High performance polymer nanocomposites for additive manufacturing applications. *Reactive and Functional Polymers, 103*, 141–155. https://doi.org/10.1016/j.reactfunctpolym.2016.04.010.

Domingo-Espin, M., Puigoriol-Forcada, J. M., Garcia-Granada, A. A., Llumà, J., Borros, S., & Reyes, G. (2015). Mechanical property characterization and simulation of fused deposition modeling polycarbonate parts. *Materials & Design, 83*, 670–677. https://doi.org/10.1016/j.matdes.2015.06.074.

Donnaloja, F., Jacchetti, E., Soncini, M., & Raimondi, M. T. (2020). Natural and synthetic polymers for bone scaffolds optimization. *Polymers, 12*(4), 905. https://doi.org/10.3390/polym12040905.

Dorgan, J. R., Lehermeier, H. J., Palade, L., & Cicero, J. (2001). Polylactides: Properties and prospects of an environmentally benign plastic from renewable resources. *Macromolecular Symposia, 175*(1), 55–66. https://doi.org/10.1002/1521-3900(200110)175:1.

Durgun, I., & Ertan, R. (2014). Experimental investigation of FDM process for improvement of mechanical properties and production cost. *Rapid Prototyping Journal, 20*(3), 228–235. https://doi.org/10.1108/rpj-10-2012-0091.

Eawwiboonthanakit, N., Jaafar, M., Hamid, Z. A. A., Todo, M., & Lila, B. (2014). Tensile properties of poly(L-lactic) acid(PLLA) blends. *Advanced Materials Research, 1024*, 179–183. https://doi.org/10.4028/www.scientific.net/amr.1024.179.

Erben, J., Blatonova, K., Kalous, T., Capek, L., Behalek, L., Boruvka, M., & Chvojka, J. (2022). The injection molding of biodegradable polydioxanone—A study of the dependence of the structural and mechanical properties on thermal processing conditions. *Polymers, 14*(24), 5528. https://doi.org/10.3390/polym14245528.

Esmaeili, Y., Bidram, E., Bigham, A., Atari, M., Azadani, R. N., Tavakoli, M., Salehi, S., Mirhaj, M., Basiri, A., Mirzavandi, Z., Boshtam, M., Rafienia, M., Kharazi, A. Z., Karbasi, S., Shariati, L., & Zarrabi, A. (2023). Exploring the evolution of tissue engineering strategies over the past decade: From cell-based strategies to gene-activated matrix. *Alexandria Engineering Journal, 81*, 137–169. https://doi.org/10.1016/j.aej.2023.08.080.

Farah, S., Anderson, D. G., & Langer, R. (2016). Physical and mechanical properties of PLA, and their functions in widespread applications—A comprehensive review. *Advanced Drug Delivery Reviews, 107*, 367–392. https://doi.org/10.1016/j.addr.2016.06.012.

Farcas, M. T., Stefaniak, A. B., Knepp, A. K., Bowers, L., Mandler, W. K., Kashon, M., Jackson, S. R., Stueckle, T. A., Sisler, J. D., Friend, S. A., Qi, C., Hammond, D. R., Thomas, T. A., Matheson, J., Castranova, V., & Qian, Y. (2019). Acrylonitrile butadiene styrene (ABS) and polycarbonate (PC) filaments three-dimensional (3-D) printer emissions-induced cell toxicity. *Toxicology Letters, 317*, 1–12. https://doi.org/10.1016/j.toxlet.2019.09.013.

Flach, L., Jacobs, M. A., Klosterman, D. A., & Chartoff, R. P. (1998). Simulation of Laminated Object Manufacturing (LOM) with variation of process parameters. *International Solid Freeform Fabrication Symposium*. The University of Texas at Austin. https://doi.org/10.26153/tsw/625.

Fu, X., Lin, Y., Yue, X., XunMa, N., Hur, B., & Yue, X. (2022). A review of additive manufacturing (3D printing) in aerospace: Technology, materials, applications, and challenges. *EAI/Springer Innovations in Communication and Computing,* 73–98. https://doi.org/10.1007/978-3-030-98671-1_6.

Ginebra, M., Espanol, M., Maazouz, Y., Bergez, V., & Pastorino, D. (2018). Bioceramics and bone healing. *EFORT Open Reviews, 3*(5), 173–183. https://doi.org/10.1302/2058-5241.3.170056.

Gkartzou, E., Koumoulos, E. P., & Charitidis, C. A. (2017). Production and 3D printing processing of bio-based thermoplastic filament. *Manufacturing Review, 4*(1)https://doi.org/10.1051/mfreview/2016020.

Gleadall, A., Visscher, D., Yang, J., Thomas, D., & Segal, J. (2018). Review of additive manufactured tissue engineering scaffolds: Relationship between geometry and performance. *Burns & Trauma, 6*. https://doi.org/10.1186/s41038-018-0121-4.

Gunatillake, P. (2003). Biodegradable synthetic polymers for tissue engineering. *European Cells and Materials, 5*, 1–16. https://doi.org/10.22203/ecm.v005a01.

Gunja, N. J., & Athanasiou, K. A. (2006). Biodegradable materials in arthroscopy. *Sports Medicine and Arthroscopy Review, 14*(3), 112–119. https://doi.org/10.1097/00132585-200609000-00002.

Gupta, B., Revagade, N., & Hilborn, J. (2007). Poly(lactic acid) fiber: An overview. *Progress in Polymer Science, 32*(4), 455–482. https://doi.org/10.1016/j.progpolymsci.2007.01.005.

Halim, A., Khoo, T., & Shah, J. Y. (2010). Biologic and synthetic skin substitutes: An overview. *Indian Journal of Plastic Surgery, 43*(3), 23. https://doi.org/10.4103/0970-0358.70712.

Hartmann, M. H. (1998). *High molecular weight polylactic acid polymers*. Springer eBooks367–411. https://doi.org/10.1007/978-3-662-03680-8_15.

Hasan, M. S., Ahmed, I., Parsons, A. J., Rudd, C. D., Walker, G. S., & Scotchford, C. A. (2012). Investigating the use of coupling agents to improve the interfacial properties between a resorbable phosphate glass and polylactic acid matrix. *Journal of Biomaterials Applications, 28*(3), 354–366. https://doi.org/10.1177/0885328212453634.

Ho, C. M. B., Mishra, A., Lin, P. T. P., Ng, S. H., Yeong, W. Y., Kim, Y., & Yoon, Y. (2016). 3D printed polycaprolactone carbon nanotube composite scaffolds for cardiac tissue engineering. *Macromolecular Bioscience, 17*(4). https://doi.org/10.1002/mabi.201600250.

Horvath, T., Kalman, M., Szabo, T., Roman, K., Zsoldos, G., & Kollar, M.S. (2018). The mechanical properties of polyethylene-terephthalate (PET) and polylactic-acid (PDLLA and PLLA), the influence of material structure on forming. *IOP Conference Series Materials Science and Engineering, 426*, 012018. https://doi.org/10.1088/1757-899x/426/1/012018

Hutmacher, D. W., Schantz, T., Zein, I., Ng, K. W., Teoh, S. H., & Tan, K. C. (2001). Mechanical properties and cell cultural response of polycaprolactone scaffolds designed and fabricated via fused deposition modeling. *Journal of Biomedical Materials Research, 55*(2), 203–216. https://doi.org/10.1002/1097-4636(200105)55:2.

Ilyas, R., Sapuan, S., Harussani, M., Hakimi, M., Haziq, M., Atikah, M., Asyraf, M., Ishak, Razman, Nurazzi, N., Norrrahim, M., Abral, H., & Asrofi, M. (2021). Polylactic acid (PLA) biocomposite: Processing, additive manufacturing and advanced applications. *Polymers, 13*(8), 1326. https://doi.org/10.3390/polym13081326.

Jamshidian, M., Tehrany, E. A., Imran, M., Jacquot, M., & Desobry, S. (2010). Poly-lactic acid: Production, applications, nanocomposites, and release studies. *Comprehensive Reviews in Food Science and Food Safety, 9*(5), 552–571. https://doi.org/10.1111/j.1541-4337.2010.00126.x.

Janorkar, A. V., Metters, A. T., & Hirt, D. E. (2004). Modification of poly(lactic acid) films: Enhanced wettability from surface-confined photografting and increased degradation rate due to an artifact of the photografting process. *Macromolecules, 37*(24), 9151–9159. https://doi.org/10.1021/ma049056u.

Jasiuk, I., Abueidda, D. W., Kozuch, C., Pang, S., Su, F. Y., & McKittrick, J. (2018). An overview on additive manufacturing of polymers. *JOM, 70*(3), 275–283. https://doi.org/10.1007/s11837-017-2730-y.

Jockusch, J., & Özcan, M. (2020). Additive manufacturing of dental polymers: An overview on processes, materials and applications. *Dental Materials Journal, 39*(3), 345–354. https://doi.org/10.4012/dmj.2019-123.

Kannan, S., & Ramamoorthy, M. (2020). Mechanical characterization and experimental modal analysis of 3D Printed ABS, PC and PC-ABS materials. *Materials Research Express, 7*(1), 015341. https://doi.org/10.1088/2053-1591/ab6a48.

Kargozar, S., Montazerian, M., Fiume, E., & Baino, F. (2019). Multiple and promising applications of strontium (Sr)-containing bioactive glasses in bone tissue engineering. *Frontiers in Bioengineering and Biotechnology, 7*. https://doi.org/10.3389/fbioe.2019.00161.

Kausar, A. (2019). A review of high performance polymer nanocomposites for packaging applications in electronics and food industries. *Journal of Plastic Film & Sheeting, 36*(1), 94–112. https://doi.org/10.1177/8756087919849459.

Kim, G., Gavande, V., Shaikh, V., & Lee, W. (2023). Degradation behavior of poly(lactide-Co-glycolide) monolayers investigated by langmuir technique: Accelerating effect. *Molecules (Basel, Switzerland), 28*(12), 4810. https://doi.org/10.3390/molecules28124810.

Kohane, D. S., & Langer, R. (2008). Polymeric biomaterials in tissue engineering. *Pediatric Research, 63*(5), 487–491. https://doi.org/10.1203/01.pdr.0000305937.26105.e7.

Kolos, E., & Ruys, A. (2015). Biomimetic coating on porous alumina for tissue engineering: Characterisation by cell culture and confocal microscopy. *Materials, 8*(6), 3584–3606. https://doi.org/10.3390/ma8063584.

Kreszinger, M., Toholj, B., Ačanski, A., Baloš, S., Cincović, M., Pećin, M., Lipar, M., & Smolec, O. (2018). Tensile strength retention of resorptive suture materials applied in the stomach wall—An in vitro study. *Veterinarski Arhiv, 88*(2), 235–243. https://doi.org/10.24099/vet.arhiv.170130.

Kumar, M., Ramakrishnan, R., Omarbekova, A., & R, S. K. (2021). Experimental characterization of mechanical properties and microstructure study of polycarbonate (PC) reinforced acrylonitrile-butadiene-styrene (ABS) composite with varying PC loadings. *AIMS Materials Science, 8*(1), 18–28. https://doi.org/10.3934/matersci.2021002.

Kumar, R. & Kumar S. (2025). Biomaterials and additive manufacturing. springerbriefs in applied sciences and technology. Springer Link. https://link.springer.com/book/9789819612734

Kumar, R., & Kumar, S. (2024). *Engineering aspects and medical materials of medical additive manufacturing*. Elsevier eBooks89–115. https://doi.org/10.1016/b978-0-323-95383-2.00012-3.

Kumar, R., & Kumar, S. (2022). Overview of 3D and 4D printing techniques and their emerging applications in medical sectors. *Current Materials Science, 16*(2), 143–170. https://doi.org/10.2174/2666145416666221019105748.

Kumar, R., Kumar, M., & Chohan, J. S. (2021). Material-specific properties and applications of additive manufacturing techniques: A comprehensive review. *Bulletin of Materials Science, 44*(3). https://doi.org/10.1007/s12034-021-02364-y.

Kumar, R., Kumar, M., Chohan, J. S., & Kumar, S. (2022). Effect of process parameters on surface roughness of 316L stainless steel coated 3D printed PLA parts. *Materials Today Proceedings, 68*, 734–741. https://doi.org/10.1016/j.matpr.2022.06.004.

Kumar, R., Kumar, S., Kumar, M., & Luthra, G. (2022). *Conventional and 3D printing technology for the manufacturing of metal-matrix composite*. CRC Press eBooks197–223. https://doi.org/10.1201/9781003194897-9.

Kumar, S., Kushawaha, M. K., & Kumar, R. (2022). Comprehensive study on additive manufacturing process methods, cost comparision, challenges and industrial applications. I-manager's. *Journal on Future Engineering and Technology, 17*(4), 31. https://doi.org/10.26634/jfet.17.4.18909.

Kundu, B., Rajkhowa, R., Kundu, S. C., & Wang, X. (2012). Silk fibroin biomaterials for tissue regenerations. *Advanced Drug Delivery Reviews, 65*(4), 457–470. https://doi.org/10.1016/j.addr.2012.09.043.

Lam, C., Mo, X., Teoh, S., & Hutmacher, D. (2002). Scaffold development using 3D printing with a starch-based polymer. *Materials Science and Engineering C, 20*(1–2), 49–56. https://doi.org/10.1016/s0928-4931(02)00012-7.

Lanao, R. P. F., Jonker, A. M., Wolke, J. G., Jansen, J. A., Van Hest, J. C., & Leeuwenburgh, S. C. (2013). Physicochemical properties and applications of poly(lactic-co-glycolic acid) for use in bone regeneration. *Tissue Engineering. Part B, Reviews, 19*(4), 380–390. https://doi.org/10.1089/ten.teb.2012.0443.

Lasprilla, A. J., Martinez, G. A., Lunelli, B. H., Jardini, A. L., & Filho, R. M. (2011). Poly-lactic acid synthesis for application in biomedical devices—A review. *Biotechnology Advances, 30*(1), 321–328. https://doi.org/10.1016/j.biotechadv.2011.06.019.

Lee, B., Abdullah, J., & Khan, Z. (2005). Optimization of rapid prototyping parameters for production of flexible ABS object. *Journal of Materials Processing Technology, 169*(1), 54–61. https://doi.org/10.1016/j.jmatprotec.2005.02.259.

Li, X., Xing, Y., Li, W., Jiang, Y., & Ding, Y. (2010). Antibacterial and physical properties of poly(vinyl chloride)-based film coated with ZnO nanoparticles. *Food Science and Technology International, 16*(3), 225–232. https://doi.org/10.1177/1082013209353986.

Lopes, M. S., Jardini, A., & Filho, R. M. (2012). Poly (lactic acid) production for tissue engineering applications. *Procedia Engineering, 42*, 1402–1413. https://doi.org/10.1016/j.proeng.2012.07.534.

Lunt, J. (1998). Large-scale production, properties and commercial applications of polylactic acid polymers. *Polymer Degradation and Stability, 59*(1–3), 145–152. https://doi.org/10.1016/s0141-3910(97)00148-1.

Mäkelä, P., Pohjonen, T., Törmälä, P., Waris, T., & Ashammakhi, N. (2002). Strength retention properties of self-reinforced poly l-lactide (SR-PLLA) sutures compared with polyglyconate (MaxonR) and polydioxanone (PDS) sutures. An in vitro study. *Biomaterials, 23*(12), 2587–2592. https://doi.org/10.1016/s0142-9612(01)00396-9.

Martins, J. A., Lach, A. A., Morris, H. L., Carr, A. J., & Mouthuy, P. (2019). Polydioxanone implants: A systematic review on safety and performance in patients. *Journal of Biomaterials Applications, 34*(7), 902–916. https://doi.org/10.1177/0885328219888841.

MatWeb—The Online Materials Information Resource. (n.d.). Retrieved from https://www.matweb.com/search/datasheet.aspx?MatGUID=a6ed8b3af42f4a0b9ca819009431c388&ckck=1

McKeen, L. W. (2013). *Plastics used in medical devices*. Elsevier eBooks21–53. https://doi.org/10.1016/b978-0-323-22805-3.00003-7.

Middleton, J. C., & Tipton, A. J. (2000). Synthetic biodegradable polymers as orthopedic devices. *Biomaterials, 21*(23), 2335–2346. https://doi.org/10.1016/s0142-9612(00)00101-0.

Ng, K. W., Hutmacher, D. W., Schantz, J., Ng, C. S., Too, H., Lim, T. C., Phan, T. T., & Teoh, S. H. (2001). Evaluation of ultra-thin poly(ε-caprolactone) films for tissue-engineered skin. *Tissue Engineering, 7*(4), 441–455. https://doi.org/10.1089/10763270152436490.

Ngo, T. D., Kashani, A., Imbalzano, G., Nguyen, K. T., & Hui, D. (2018). Additive manufacturing (3D printing): A review of materials, methods, applications and challenges. *Composites Part B Engineering, 143*, 172–196. https://doi.org/10.1016/j.compositesb.2018.02.012.

Ortiz-Acosta, D., & Moore, T. (2019). *Functional 3D printed polymeric materials*. IntechOpen eBooks. https://doi.org/10.5772/intechopen.80686.

Pachence, J. M., & Kohn, J. (2000). *Biodegradable polymers*. Elsevier eBooks263–277. https://doi.org/10.1016/b978-012436630-5/50026-x.

Parupelli, S. K., & Desai, S. (2019). A comprehensive review of additive manufacturing (3D printing): Processes, applications and future potential. *American Journal of Applied Sciences, 16*(8), 244–272. https://doi.org/10.3844/ajassp.2019.244.272.

Perego, G., Cella, G. D., & Bastioli, C. (1996). Effect of molecular weight and crystallinity on poly(lactic acid) mechanical properties. *Journal of Applied Polymer Science, 59*(1), 37–43. https://doi.org/10.1002/(sici)1097-4628(19960103)59:1.

Pieralli, S., Kohal, R., Jung, R., Vach, K., & Spies, B. (2016). Clinical outcomes of zirconia dental implants: A systematic review. *Journal of Dental Research, 96*(1), 38–46. https://doi.org/10.1177/0022034516664043.

Pignatello, R. (2011). Biomaterials applications for nanomedicine. *In InTech eBooks*. https://doi.org/10.5772/1957.

Purohit, S. D., Bhaskar, R., Singh, H., Yadav, I., Gupta, M. K., & Mishra, N. C. (2019). Development of a nanocomposite scaffold of gelatin–alginate–graphene oxide for bone tissue engineering. *International Journal of Biological Macromolecules, 133*, 592–602. https://doi.org/10.1016/j.ijbiomac.2019.04.113.

Ramírez, A.S., D'Amato, R., Haro, F.B., Marcos, M.I., & De Agustín Del Burgo, J.M. (2018). Composite material created by additive manufacturing techniques FFF and Robocasting for the manufacture of medical parts. In *Sixth International Conference on Technological Ecosystems for Enhancing Multiculturality, TEEM2018*, 410–415. https://doi.org/10.1145/3284179.3284249

Rankouhi, B., Javadpour, S., Delfanian, F., & Letcher, T. (2016). Failure analysis and mechanical characterization of 3D printed ABS with respect to layer thickness and orientation. *Journal of Failure Analysis and Prevention, 16*(3), 467–481. https://doi.org/10.1007/s11668-016-0113-2.

Rasal, R. M., & Hirt, D. E. (2008). Toughness decrease of PLA-PHBHHx blend films upon surface-confined photopolymerization. *Journal of Biomedical Materials Research. Part A, 88A*(4), 1079–1086. https://doi.org/10.1002/jbm.a.32009.

Rasal, R. M., Janorkar, A. V., & Hirt, D. E. (2009). Poly(lactic acid) modifications. *Progress in Polymer Science, 35*(3), 338–356. https://doi.org/10.1016/j.progpolymsci.2009.12.003.

Rodrı´Guez, J. F., Thomas, J. P., & Renaud, J. E. (2003). Design of fused-deposition ABS components for stiffness and strength. *Journal of Mechanical Design, 125*(3), 545–551. https://doi.org/10.1115/1.1582499.

Sawyer, D. J. (2003). Bioprocessing—No longer a field of dreams. *Macromolecular Symposia, 201*(1), 271–282. https://doi.org/10.1002/masy.200351130.

Schmid, M., & Wegener, K. (2016). Additive manufacturing: polymers applicable for laser sintering (LS). *Procedia Engineering, 149*, 457–464. https://doi.org/10.1016/j.proeng.2016.06.692.

Seitz, H., Rieder, W., Irsen, S., Leukers, B., & Tille, C. (2005). Three-dimensional printing of porous ceramic scaffolds for bone tissue engineering. *Journal of Biomedical Materials Research. Part B, Applied Biomaterials, 74B*(2), 782–788. https://doi.org/10.1002/jbm.b.30291.

Shaltooki, M., Dini, G., & Mehdikhani, M. (2019). Fabrication of chitosan-coated porous polycaprolactone/strontium-substituted bioactive glass nanocomposite scaffold for bone tissue engineering. *Materials Science and Engineering C, 105*, 110138. https://doi.org/10.1016/j.msec.2019.110138.

Shores, J. T., Gabriel, A., & Gupta, S. (2007). Skin substitutes and alternatives. *Advances in Skin & Wound Care, 20*(9), 493–508. https://doi.org/10.1097/01.asw.0000288217.83128.f3.

Silva, M. F., Hechenleitner, A. A. W., Irache, J. M., De Oliveira, A. J. A., & Pineda, E. A. G. (2015). Study of thermal degradation of PLGA, PLGA nanospheres and PLGA/maghemite superparamagnetic nanospheres. *Materials Research, 18*(6), 1400–1406. https://doi.org/10.1590/1516-1439.045415.

Silva, S. S., Mano, J. F., & Reis, R. L. (2010). Potential applications of natural origin polymer-based systems in soft tissue regeneration. *Critical Reviews in Biotechnology, 30*(3), 200–221. https://doi.org/10.3109/07388551.2010.505561.

Smith, L., & Ma, P. (2004). Nano-fibrous scaffolds for tissue engineering. *Colloids and Surfaces. B, Biointerfaces, 39*(3), 125–131. https://doi.org/10.1016/j.colsurfb.2003.12.004.

Song, Y., Li, Y., Song, W., Yee, K., Lee, K., & Tagarielli, V. (2017). Measurements of the mechanical response of unidirectional 3D-printed PLA. *Materials & Design, 123*, 154–164. https://doi.org/10.1016/j.matdes.2017.03.051.

Srivastava, M., Rathee, S., Patel, V., Kumar, A., & Koppad, P. G. (2022). A review of various materials for additive manufacturing: Recent trends and processing issues. *Journal of Materials Research and Technology, 21*, 2612–2641. https://doi.org/10.1016/j.jmrt.2022.10.015.

Subramanian, K., Vail, N., Barlow, J., & Marcus, H. (1995). Selective laser sintering of alumina with polymer binders. *Rapid Prototyping Journal, 1*(2), 24–35. https://doi.org/10.1108/13552549510086844.

Sun, J., & Tan, H. (2013). Alginate-based biomaterials for regenerative medicine applications. *Materials, 6*(4), 1285–1309. https://doi.org/10.3390/ma6041285.

Sun, M., Tang, L., Yang, X., Lu, J., He, H., Lin, J., He, Y., & Yu, M. (2024). Advancements of biomaterials in oral tissue engineering: Past, present, and future. *Beni-Suef University Journal of Basic and Applied Sciences, 13*(1). https://doi.org/10.1186/s43088-024-00538-1.

Suuronen, R., Pohjonen, T., Hietanen, J., & Lindqvist, C. (1998). A 5-year in vitro and in vivo study of the biodegradation of polylactide plates. *Journal of Oral and Maxillofacial Surgery, 56*(5), 604–614. https://doi.org/10.1016/s0278-2391(98)90461-x.

Taib, N. A. B., Rahman, M. R., Huda, D., Kuok, K. K., Hamdan, S., Bakri, M. K. B., Julaihi, M. R. M. B., & Khan, A. (2022). A review on poly lactic acid (PLA) as a biodegradable polymer. *Polymer Bulletin, 80*(2), 1179–1213. https://doi.org/10.1007/s00289-022-04160-y.

Tammi, R. H., Kultti, A., Kosma, V., Pirinen, R., Auvinen, P., & Tammi, M. I. (2008). Hyaluronan in human tumors: Pathobiological and prognostic messages from cell-associated and stromal hyaluronan. *Seminars in Cancer Biology, 18*(4), 288–295. https://doi.org/10.1016/j.semcancer.2008.03.005.

Tokiwa, Y., & Calabia, B. P. (2006). Biodegradability and biodegradation of poly(lactide). *Applied Microbiology and Biotechnology, 72*(2), 244–251. https://doi.org/10.1007/s00253-006-0488-1.

Torres, J., Cole, M., Owji, A., DeMastry, Z., & Gordon, A. P. (2016). An approach for mechanical property optimization of fused deposition modeling with polylactic acid via design of experiments. *Rapid Prototyping Journal, 22*(2), 387–404. https://doi.org/10.1108/rpj-07-2014-0083.

Tsuji, H., Nakano, M., Hashimoto, M., Takashima, K., Katsura, S., & Mizuno, A. (2006). Electrospinning of poly(lactic acid) stereocomplex nanofibers. *Biomacromolecules, 7*(12), 3316–3320. https://doi.org/10.1021/bm060786e.

Tymrak, B., Kreiger, M., & Pearce, J. (2014). Mechanical properties of components fabricated with open-source 3-D printers under realistic environmental conditions. *Materials & Design (1980-2015), 58*, 242–246. https://doi.org/10.1016/j.matdes.2014.02.038.

Utela, B., Storti, D., Anderson, R., & Ganter, M. (2008). A review of process development steps for new material systems in three dimensional printing (3DP). *Journal of Manufacturing Processes, 10*(2), 96–104. https://doi.org/10.1016/j.jmapro.2009.03.002.

Van De Velde, K., & Kiekens, P. (2002). Biopolymers: Overview of several properties and consequences on their applications. *Polymer Testing, 21*(4), 433–442. https://doi.org/10.1016/s0142-9418(01)00107-6.

Vieira, S., Da Silva Morais, A., Garet, E., Silva-Correia, J., Reis, R. L., González-Fernández, Á., & Oliveira, J. M. (2019). Self-mineralizing Ca-enriched methacrylated gellan gum beads for bone tissue engineering. *Acta Biomaterialia, 93*, 74–85. https://doi.org/10.1016/j.actbio.2019.01.053.

Vorndran, E., Klarner, M., Klammert, U., Grover, L. M., Patel, S., Barralet, J. E., & Gbureck, U. (2008). 3D powder printing of β-tricalcium phosphate ceramics using different strategies. *Advanced Engineering Materials, 10*(12). https://doi.org/10.1002/adem.200800179.

Wahid, F., Khan, T., Hussain, Z., & Ullah, H. (2018). *Nanocomposite scaffolds for tissue engineering: Properties, preparation and applications*. Elsevier eBooks701–735. https://doi.org/10.1016/b978-0-12-813741-3.00031-5.

Waris, E., Ashammakhi, N., Kaarela, O., Raatikainen, T., & Vasenius, J. (2004). Use of bioabsorbable osteofixation devices in the hand. *Journal of Hand Surgery (European Volume), 29*(6), 590–598. https://doi.org/10.1016/j.jhsb.2004.02.005.

Wu, G., Langrana, N. A., Sadanji, R., & Danforth, S. (2002). Solid freeform fabrication of metal components using fused deposition of metals. *Materials & Design (1980-2015), 23*(1), 97–105. https://doi.org/10.1016/s0261-3069(01)00079-6.

Wu, W., Geng, P., Li, G., Zhao, D., Zhang, H., & Zhao, J. (2015). Influence of layer thickness and raster angle on the mechanical properties of 3D-printed PEEK and a comparative mechanical study between PEEK and ABS. *Materials, 8*(9), 5834–5846. https://doi.org/10.3390/ma8095271.

Wubneh, A., Tsekoura, E. K., Ayranci, C., & Uludağ, H. (2018). Current state of fabrication technologies and materials for bone tissue engineering. *Acta Biomaterialia, 80*, 1–30. https://doi.org/10.1016/j.actbio.2018.09.031.

Wuisman, P. I. J. M., & Smit, T. H. (2005). Bioresorbable polymers: Heading for a new generation of spinal cages. *European Spine Journal, 15*(2), 133–148. https://doi.org/10.1007/s00586-005-1003-6.

Yan, X., & Gu, P. (1996). A review of rapid prototyping technologies and systems. *Computer-Aided Design, 28*(4), 307–318. https://doi.org/10.1016/0010-4485(95)00035-6.

Yap, Y. L., Toh, W., Koneru, R., Lin, K., Yeoh, K. M., Lim, C. M., Lee, J. S., Plemping, N. A., Lin, R., Ng, T. Y., Chan, K. I., Guang, H., Chan, W. Y. B., Teong, S. S., & Zheng, G. (2019). A non-destructive experimental-cum-numerical methodology for the characterization of 3D-printed materials—Polycarbonate-acrylonitrile butadiene styrene (PC-ABS). *Mechanics of Materials, 132*, 121–133. https://doi.org/10.1016/j.mechmat.2019.03.005.

Zahedi, P., Rezaeian, I., Ranaei-Siadat, S., Jafari, S., & Supaphol, P. (2009). A review on wound dressings with an emphasis on electrospun nanofibrous polymeric bandages. *Polymers for Advanced Technologies, 21*(2), 77–95. https://doi.org/10.1002/pat.1625.

Zein, I., Hutmacher, D. W., Tan, K. C., & Teoh, S. H. (2002). Fused deposition modeling of novel scaffold architectures for tissue engineering applications. *Biomaterials, 23*(4), 1169–1185. https://doi.org/10.1016/s0142-9612(01)00232-0.

Zhang, P., Liu, J., & To, A. C. (2016). Role of anisotropic properties on topology optimization of additive manufactured load bearing structures. *Scripta Materialia, 135*, 148–152. https://doi.org/10.1016/j.scriptamat.2016.10.021.

Zhong, Z., & Sun, X. S. (2001). Properties of soy protein isolate/polycaprolactone blends compatibilized by methylene diphenyl diisocyanate. *Polymer, 42*(16), 6961–6969. https://doi.org/10.1016/s0032-3861(01)00118-5.

Ziemian, C., & Crawn, P. (2001). Computer aided decision support for fused deposition modeling. *Rapid Prototyping Journal, 7*(3), 138–147. https://doi.org/10.1108/13552540110395538.

Additively manufactured functional polymers and nanocomposites for environmental applications

19

Abhinav Sharma[1], *Hendrik Faber*[1], *Abdullah N. Alodhayb*[2], *Ajit Khosla*[3], *and Thomas D. Anthopoulos*[1,4,5]

[1]King Abdullah University of Science and Technology (KAUST), KAUST Solar Center (KSC), Thuwal, Saudi Arabia, [2]Department of Physics and Astronomy, College of Science, King Saud University, Riyadh, Saudi Arabia, [3]Department of Applied Chemistry, School of Advanced Materials and Nanotechnology, Xidian University, Xi'an, P.R. China, [4]Department of Electrical and Electronic Engineering, Henry Royce Institute, The University of Manchester, United Kingdom, [5]Department of Electrical and Electronic Engineering, Photon Science Institute, The University of Manchester, United Kingdom

PLA and ABS are nonconductive thermoplastic filaments that have been employed in the production of printed electrodes and electrochemical cells. Conductive filaments are typically composed of thermoplastic insulating polymers such as ABS and PLA, as well as carbon nanomaterials such as graphene oxide, carbon black (CB), carbon nanotubes (CNTs), carbon black nanoparticles (CBNPs), and carbon nanofibers (CNFs) (Chu et al., 2021; Harris et al., 2019; Mishra et al., 2023; Shah et al., 2018). These components are utilized in manufacturing the working electrode (WE) and counter electrode (CE) for the electrochemical sensor. The rapid, cost-effective, and highly reproducible production of 3D-printed electrodes is facilitated by the easy availability of filaments in the market and the affordability of FDM 3D printers. Furthermore, certain research studies propose that conductive filaments can be developed in laboratory environments through a simple blending, mixing, and extrusion process, providing a more economical option compared to screen-printed electrodes (Honeychurch et al., 2018; Junpha et al., 2020; Rymansaib et al., 2016).

Most filaments used in 3D printing are made of plastic. The outstanding features of plastic filaments, including flexibility, reusability, rigidity, recyclability and a wide range of color options, permit widespread use in the 3D printing process. FDM 3D printers are majorly used to develop plastic items in the 3D printing process. This approach includes fusion of thermoplastic filaments and then arranged in layer by layer order to solidify into the desired forms. The ability of thermoplastics to be heated at low temperatures and molded into various shapes without undergoing chemical bonding allows the product to be recyclable and reusable. 3D-printing techniques utilize several filaments, each with distinct

Additive Manufacturing of Functional Polymers and Nanocomposites. DOI: https://doi.org/10.1016/B978-0-443-26516-7.00019-5
© 2025 Elsevier Ltd. All rights are reserved, including those for text and data mining, AI training, and similar technologies.

qualities and characteristics. The filaments usually used for FDM 3D printing are thermoplastic polyurethane (TPU), polyamide (PA), polylactic acid (PLA), polycarbonate (PC), acrylonitrile butadiene styrene (ABS), nylon, and polyethylene terephthalate glycol (PETG) (Harris et al., 2019; Liu et al., 2019; Mishra et al., 2023; Valerga et al., 2018). PLA is a commonly used material in 3D-printing due to its favorable characteristics and environmental sustainability. Derived from renewable sources like sugarcane or cornstarch, PLA is known for its biodegradability, making it an ecofriendly option for various applications including environmental, biomedical and food packaging etc. In comparison to other thermoplastics like ABS and PETG, PLA is easy to print with, featuring low warping tendencies and the ability to be printed on a cold surface. While PLA may be less sturdy and prone to deformation under extreme heat, advancements in reinforced PLA nanocomposite have enhanced its durability and heat resistance. The main advantages of using PLA in 3D printing is its ease of use, biodegradable nature, low printing temperature, which makes it compatible and prominent choice for wide range of 3D printers, including those designed for home use. Additionally, PLA does not require a heated print bed, although having one can be beneficial. PLA is known for its minimal warping during printing, allowing for more precise and accurate prints with sharper corners and features compared to other materials like ABS. Overall, the ease of printing with PLA, along with its biodegradability and availability in various colors, makes it a popular choice for both beginners and professional users in the 3D printing community.

Additive manufacturing (AM) is a cutting-edge and innovative technology that has the potential to revolutionize the manufacturing process in the future (Jayakrishna et al., 2023; Rouf et al., 2022). It has gained significant recognition in a variety of applications, which allows the development of sustainable, adaptable, and customized manufacturing processes (Rouf et al., 2022). The main feature of AM is its inherent design freedom, which allows for the creation of complex architectures and geometries that were previously impossible to achieve using traditional manufacturing techniques (Mehrpouya et al., 2019). This design flexibility opens up new opportunities for scientific and industrial improvements in a variety of fields, including aerospace, biomedical, construction, automotive, industrial engineering, and the food industry (Hawes et al., 2022; Mehrpouya et al., 2019; Mobarak et al., 2023; Rouf et al., 2022). 3D printing is a digital technology that is simple and cost-effective and capable of constructing 3D objects through a bottom-up approach. Tinkercad, Fusion 360, and Autodesk Inventor are among the software programs that are employed to designed 3D objects. Researchers have been actively investigating a variety of advanced 3D printing techniques to accomplish high-volume, precise, and customizable fabrication. This has been followed by the use of a variety of filaments, such as thermoplastic nonconductive and conductive materials (Jandyal et al., 2022). The automated layer-by-layer deposition of materials based on computer-generated models is the fundamental process of 3D printing, which is facilitated by specialized software and workstation 3D printers (Jandyal et al., 2022). This additive method facilitates the creation of prototype products that feature intricate designs and customized functionality. The production of prototypes is an optimal option due to the numerous benefits of 3D printing. These advantages encompass the ability to modify designs, the utilization of low-cost materials, minimal waste production, compact equipment, rapid manufacturing through computerized control, and the absence of a clean room facility (Chan et al., 2018; Jayakrishna et al., 2023).

19.1 Introduction

19.1.1 Additive manufacturing

Additive manufacturing (AM) has emerged as a cutting-edge technology with the potential to revolutionize traditional manufacturing processes (Jayakrishna et al., 2023; Rouf et al., 2022). This innovative approach offers sustainable, adaptable, and personalized manufacturing capabilities, attracting considerable attention for technological advancements in various industries (Rouf et al., 2022). A key distinguishing feature of AM is its intrinsic design flexibility, allow to produce complicated geometries and architectures that were not achievable using traditional manufacturing techniques (Mehrpouya et al., 2019). This design flexibility expand possibilities for scientific and industrial advancements across various sectors including aerospace, biomedical, construction, automotive, industrial engineering, food industry, etc. (Hawes et al., 2022; Mehrpouya et al., 2019; Mobarak et al., 2023; Rouf et al., 2022). Researchers have been actively exploring a range of advanced 3D printing techniques to achieve rapid, precise, and high-volume customizable fabrication. 3D printing stands out as a straightforward and economical digital technology that constructs 3D objects via a bottom-up method. The process begins with designing an object using 3D software (CAD/CAM such as Tinkercad, Fusion 360, Autodesk Inventor, etc.), followed by printing using various filaments, including thermoplastic nonconductive and conductive materials (Jandyal et al., 2022). Fundamentally, 3D printing involves the automated layer-by-layer deposition of materials based on computer-generated models, facilitated by desktop 3D printer and specialized software (Jandyal et al., 2022). This additive approach allows for the creation of solid prototype products with intricate designs and tailored functionalities. The advantages of 3D printing, including use of cost-effective material, reduced waste production, compact equipment dimensions, quick manufacturing enabled by computer control, design adaptability, and replicability, position it as a viable option for constructing device prototypes (Chan et al., 2018; Jayakrishna et al., 2023).

Additionally, AM technologies have the potential to streamline construction processes, shorten production time, and lower overall production costs, making them a natural evolution for the manufacturing industry. The integration of 3D-printing technology has significantly enhanced the capabilities of scientists, engineers, and medical professionals in the development of devices and prototypes, enabling tasks that were previously challenging. Particularly during the COVID-19 pandemic, the urgent requirement for essential items such as face shields, personal protective equipment (PPE) kits, and ventilator components surged in hospitals and public settings. The agility and versatility of 3D printing played a crucial role in rapidly addressing these critical needs by facilitating the efficient production of customized solutions to support healthcare systems and frontline workers during the crisis.

AM technology has the potential to enhance manufacturing efficiency and reduce production time in a rapid manner. The use of 3D printing has significantly raised the interest of various sectors in the automotive, industrial, and medical sectors in the ability to develop prototypes and equipment. There was an immediate need for face shields, personal protective equipment, and ventilator parts among individuals in hospitals and public locations during the COVID-19 outbreak. The versatility and flexibility of 3D printing played a vital role in quickly resolving urgent requirements. It allowed for the development of customized approaches that helped hospital systems and field workers throughout the crisis. 3D printing technologies are now being used as advanced tools to create prototype device architectures (Glasco et al., 2022; Ngo et al., 2018; Rayna & Striukova, 2016). Stereolithography was proposed by Hideo Kodama in 1981, which was the first demonstration of 3D printing. The utilization of UV light to harden a light-sensitive substance leads to the formation of a three-dimensional object. Charles Hull introduced a commercial 3D printer in 1984 that is based on the stereolithography process (Hull & Arcadia, 1984). The process involves using computer assisted design (CAD) to create a 3D model, which is then used to produce a layer-by-layer 3D object without the need for molds. Alternatively, tomographic volumetric 3D printing is a more complicated procedure than traditional layer-by-layer printing. This technique enables the printing of an entire design in a single process, eliminating the necessity for layer-by-layer slicing and printing. This streamlined procedure not only improves the efficiency of printing but also decreases both the time required for printing and the expenses associated with fabrication (Loterie et al., 2020). The worldwide market for 3D-printing technology is forecasted to experience substantial growth, with projections indicating an increase in 2021 from 12.6 billion to 34.8 billion by 2026 (Noor et al., 2022). Various 3D printing techniques have been proposed for the fabrication of biochemical sensors and energy storage devices (Sharma, AlGhamdi, et al., 2023). Notable techniques in this domain encompass fused deposition modeling (FDM), stereolithography (SLA), selective laser melting (SLM), and multijet fusion (MJF) (Gulzar et al., 2020; Hawes et al., 2022; Soares et al., 2021). Among these techniques, FDM and SLM emerge as the predominant methods for manufacturing 3D printed electrodes utilized in electrochemical sensors. This chapter aims to offer a comprehensive overview of the recent developments in 3D-printed electrochemical sensors and their sensing performance for detection of water pollutants (i.e., heavy metals), explosive/toxic substances. Furthermore, it aims to discuss the existing challenges, as well as highlight future opportunities.

19.1.2 3D printing electrode materials

Plastic is indeed the most common raw material used in 3D printing today, offering a wide range of applications from toys to household fixtures. Plastic's popularity in 3D printing is attributed to its combination of firmness, flexibility, smooth texture, and extensive color options. Additionally, plastic is a cost-effective option, making it accessible to a wide range of users. In the 3D printing process, plastic products are typically created using Fused Deposition Modeling (FDM) printers, where thermoplastic filaments are melted and layered to form the desired shapes. Thermoplastics, the more commonly used type, have the unique property of being able to undergo multiple melt and solidification cycles. This characteristic allows thermoplastics to be heated and reshaped into different forms without undergoing chemical bonding, making them recyclable and reusable. This versatility and recyclability make thermoplastics a popular choice for various 3D printing applications, as they offer the flexibility to create complex shapes and structures with the ability to adjust properties as needed.

There are numerous types of filaments utilized in 3D printing techniques, each with its unique properties and characteristics. Some common types of filaments for FDM 3D-printing include polylactic acid (PLA), polyethylene terephthalate glycol (PETG), acrylonitrile butadiene styrene (ABS), thermoplastic polyurethane (TPU), nylon, etc. (Harris et al., 2019; Liu et al., 2019; Mishra et al., 2023; Valerga et al., 2018). PLA is a commonly used material in 3D printing due to its favorable characteristics and environmental sustainability. Derived from renewable sources like sugarcane or cornstarch, PLA is known for its biodegradability, making it an ecofriendly option for various applications including environmental, biomedical and food packaging, etc. In comparison to other thermoplastics like ABS and PETG, PLA is easy to print with, featuring low warping tendencies and the ability to be printed on a cold surface. While PLA may be less sturdy and prone to deformation under extreme heat, advancements in reinforced PLA nanocomposite have enhanced its durability and heat resistance. The main advantages of using PLA in 3D printing is its ease of use, biodegradable nature, low printing temperature, which makes it compatible and prominent choice for wide range of 3D printers, including those designed for home use. Additionally, PLA does not require a heated print bed, although having one can be beneficial. PLA is known for its minimal warping during printing, allowing for more precise and accurate prints with sharper corners and features compared to other materials like ABS. Overall, the ease of printing with PLA, along with its biodegradability and availability in various colors, makes it a popular choice for both beginners and professional users in the 3D printing community.

Acrylonitrile butadiene styrene (ABS) is a thermoplastic polymer known for its toughness, durability, and versatility, making it suitable for applications that require robust parts, such as car components, household appliances, and prototypes. ABS also offers good mechanical properties, including flexibility, which allows for the creation of parts that can withstand stress and strain. Additionally, ABS has a higher melting point compared to other filaments like PLA, giving it better heat resistance and stability in high-temperature environments. Its wide availability, affordability, and variety of colors further contribute to its popularity in 3D printing. The high melting point of ABS materials makes them prone to warping when cooled during printing. To mitigate warping, it is recommended to use a heated build platform and an enclosed print chamber to maintain a consistent temperature throughout the printing process. These filaments emit toxic fumes when heated to high temperatures.

It is advisable to 3D print ABS in a well-ventilated space or with the use of a fume extraction system to minimize exposure to the emitted fumes.

19.1.3 3D printing polymer nanocomposite

Nonconductive thermoplastic filaments like PLA and ABS have been employed in the construction of printed electrochemical cells. In contrast, conductive filaments typically comprise a blend of insulating thermoplastic polymers such as PLA, ABS, and carbon-based materials like carbon black nanoparticles (CBNPs), carbon black (CB), carbon nanofibers (CNFs), carbon nanotubes (CNTs), graphite, graphene oxide among others (Chu et al., 2021; Harris et al., 2019; Mishra et al., 2023; Shah et al., 2018). These components are utilized in manufacturing the working electrode (WE) and counter electrode (CE) for the electrochemical sensor. The rapid, cost-effective, and highly reproducible production of 3D-printed electrodes is facilitated by the easy availability of filaments in the market and the affordability of FDM 3D printers. Furthermore, certain research studies propose that conductive filaments can be developed in laboratory environments through a simple blending, mixing, and extrusion process, providing a more economical option compared to screen-printed electrodes (Honeychurch et al., 2018; Junpha et al., 2020; Rymansaib et al., 2016).

19.2 3D printing techniques

19.2.1 Fused deposition modeling (FDM)

Fused deposition modeling (FDM) is a popular 3D printing method that was first introduced and patented by Scott Crump in 1989 (Sharma & Rai, 2022). FDM has become widely adopted due to its cost-effectiveness, user-friendly operation, fast fabrication speed for creating small to medium-sized objects design flexibility, ability for high-volume manufacturing, and minimal waste generation (Nga Tran et al., 2017). This technique functions by employing a layer-by-layer manufacturing approach, where thermoplastic filaments are melted to a semiliquid state and then extruded through a nozzle onto a build platform along the *X*/*Y*/*Z* axis. As the material cools down, it solidifies, allowing for the creation of successive layers to build a 3D object (Goyanes et al., 2014; Penumakala et al., 2020). Additionally, the FDM community of users actively contributes to improving hardware and software components, often sharing advancements as open-source material. The precision and reproducibility of objects produced using FDM technique rely on several key parameters. These include the printing speed, the size of the nozzle, the orientation of the nozzle/bed in relation to the *X*/*Y*/*Z* axis, the distance between the nozzle and the bed, the temperature of the nozzle/bed, and the rate of cooling. FDM is notably the most widely used printing technique for fabricating 3D printed electrodes for chemical sensors for various chemical substances (Kumar et al., 2024; Sharma, Faber, et al., 2023). The thermoplastic filament materials like PLA, ABS, polyamide (PA), and polycarbonate (PC) are commonly utilized in FDM 3D printing by commercial desktop printers (Harris et al., 2019; Liu et al., 2019; Valerga et al., 2018).

19.2.2 Selective laser melting (SLM)

Selective laser melting (SLM) is a specialized 3D printing technique specifically designed for producing metal electrodes. SLM allows for the precise fabrication of metal electrodes with high accuracy by selectively melting and fusing metal powder layer-by-layer using a high-intensity laser beam based on a 3D model (Al-Saadi et al., 2021; Yap et al., 2015). Helical-shaped stainless steel (SS) is a commonly utilized material for 3D printed electrodes in electrochemical sensors. Despite its high electrical resistivity, the surface of SS electrodes is often modified by noble metals like silver (Ag) and gold (Au) to enhance electroanalytical performance. These metal electrodes have been demonstrated for detection of various chemical substances, including water pollutants, explosives substance (Cheng et al., 2017; Lee et al., 2017; Tan et al., 2017). However, SLM technology is restricted to metal printing, necessitates careful handling of hazardous metallic particles, requires an inert environment for fabrication, and high operating and maintenance costs compared to other 3D-printing techniques.

19.2.3 Stereolithography (SLA)

Stereolithography (SLA) is a type of 3D printing technique that utilizes light-sensitive polymers to produce objects. A laser beam is utilized in SLA to cross-link the polymer at specific points, enabling for the precise construction of complicated structures (Hull & Arcadia, 1984). The interaction of the laser beam with the photosensitive liquid resin initiates polymerization at targeted points, allowing for the controlled formation of the desired 3D object (Huang et al., 2020). Following the printing of each layer, the printing platform moves in the z-direction, allowing for subsequent layers to be printed on top of one another. SLA technology has the capability to produce smooth, high-resolution structures with intricate designs at a cost-effective rate (Lakkala et al., 2023). This technology is utilized in the creation of 3D-printed ion-selective membranes (ISM), high-frequency components such as embedded resonators and filters, and various applications in biomedical engineering (Ho et al., 2022; Melchels et al., 2010). However, SLA currently has limitations such as the ability to print only one polymer at a time and the necessity for post-printing cleaning and processing steps (Quan et al., 2020).

19.2.4 3D printing pen

A 3D pen is a convenient handheld tool that facilitates the creation of prototype sensing devices. In this device, a conductive filament is loaded, heated, and extruded at a fixed temperature to produce desired shapes. Several studies have successfully utilized 3D pens to manufacture electrochemical sensors (Cardoso et al., 2020; de Oliveira et al., 2020; João, Castro, et al., 2020), offering advantages over traditional desktop 3D printers: (i) portability: 3D pens are portable, handheld, and battery-powered, allowing for the creation of 3D objects in various shapes outside of a lab setting; (ii) speed and cost: the process is rapid and more economical, with the cost of a 3D pen typically being less than 20 USD; (iii) ease of use: 3D pens are easy to operate, require minimal maintenance, and consume minimal amounts of conductive filaments; (iv) reproducibility: the use of 3D pens can lead to high manufacturing reproducibility compared to objects printed with 3D FDM printers. By simplifying the fabrication of 3DP electrodes, the adoption of 3D-printing pens has increased, offering new promises in the construction of electrochemical sensors for detection of pollutants. For example, 3D pen extruded PLA/CB filaments into a customized mold for the simultaneous detection of Pb^{2+} and Cu^{2+} in ethanol and tap water samples (João, Castro, et al., 2020; João, Squissato, et al., 2020). Additionally Cardoso et al. (2020) demonstrated the use of a 3D pen to create a reusable electrochemical sensor utilizing PLA/CB filament for detecting 2,4,6-trinitrotoluene (TNT). However, the primary limitation of the 3D pen is not suitable to process most metals and alloys due to its capability to print only selective low melting materials. Furthermore, drawbacks of this approach include low printing resolution (~5 μm), relatively high energy requirements, emission of harmful fumes, and challenges with adhesion in multi-material printing.

19.3 3D printed polymer nanocomposite for environmental monitoring

Heavy metals pose significant pollution concerns because of their toxic and nonbiodegradable nature, along with their tendency to accumulate in ecosystems (Briffa et al., 2020; Hama Aziz et al., 2023; Mitra et al., 2022; Tchounwou et al., 2012). It is essential to monitor and detect heavy metals to assess environmental health accurately. These metals can be taken in by plants and animals through a process known as bioaccumulation when they enter the environment from sources like industries and agriculture (Ali et al., 2019; Nnaji et al., 2023; Tchounwou et al., 2012). As contaminated plants, animals, or water are consumed, these metals can increase in concentration within human bodies along the food chain. It is important to note that there is currently no natural mechanism for the elimination of heavy metals from the human body (Alengebawy et al., 2021; Jaishankar et al., 2014; Singh et al., 2022). Consequently, even tiny amounts of toxic heavy metals such as lead (Pb^{2+}), cadmium (Cd^{2+}), mercury (Hg^{2+}), copper (Cu^{2+}), and others can lead to harmful effects on human health, resulting in conditions like neurological disorders and organ damage (Balali-Mood et al., 2021; Engwa et al., 2019; Jaishankar et al., 2014).

Heavy metal detection at trace amount is primarily carried out using X-ray fluorescence, inductively coupled plasma mass spectroscopy (ICP-MS) (Caroli et al., 1999; Silva et al., 2009), atomic absorption spectroscopy (Kenawy et al., 2000; Pohl, 2009), and neutron activation analysis. These techniques offer versatility across a detection of wide range of elements, high sensitivity, and low limits of detection (in femtomolar range). However, these spectroscopic methods have significant disadvantages. They require expensive chemicals/reagents, trained operators for multi-step sample processing and complex analytical procedures, making them inappropriate for real-time and on-site measurements of environmental pollutants. Moreover, these techniques can only determine total metal concentrations, and requires additional extraction and separation techniques like chromatography (Feldmann et al., 2009). These additional steps increase the risk of sample contamination during sample handling.

Electrochemical devices are known for their simple procedures, low-cost, suitability for miniaturization and reduce the risk of contamination and low sample volume measurements (Negahdary et al., 2023; Sharma et al., 2022, 2024; Sharma, AlGhamdi, et al., 2023; Sharma, Angnes, et al., 2023; Sharma et al., 2023). They also enable rapid analyses within a few minutes, real-time measurement of analytes. This capability allows for onsite monitoring of water samples. One critical aspect in heavy metal detection is selectivity. The signal of the target analyte may face interferences from different heavy metals in complex media. Surface modification with electrically conductive polymer nanocomposites and a sensitive detection method are critical for increasing the specificity and accuracy of water pollution detection. Various electrochemical techniques, including those employing square wave voltammetry (SWV), differential pulse voltammetry (DPV), square wave anodic stripping voltammetry (SWASV) and differential pulse anodic stripping voltammetry (DPASV) have been developed to address these requirements (Table 19.1). Comparison between presents the electroanalytical performance of 3D-printed chemical sensors designed for the detection of specific target analytes. Foster et al. (2020) demonstrated the electrochemical detection of Cd^{2+} and Pb^{2+} ions using a PLA/nanographite composite. Walters et al. (2020) and Baltima et al. (2021) modified PLA/carbon electrodes with bismuth (Bi) microparticles to increase the sensitivity of both Cd^{2+} and Pb^{2+} ions. In a study Foo et al. (2018), CdS nanoparticles were coated Au-PLA/graphene electrodes for Cu^{2+} ions detection. Ng et al. (2021) improved electrochemical performance by modification of PLA/graphene electrodes with Al_2O_3. Kokkinos et al. (2020) suggested a graphite electrode with a metal-organic framework as an electrode modifier for the sensitive determination of Hg^{2+} using anodic stripping voltammetry (ASV). Katseli et al. (2020) utilized in situ Au thin-film-modified PLA/carbon electrodes to improve the detection of Hg^{2+} in fish oil samples. Additionally, Giorgini Escobar et al. (2020) utilized PLA/carbon nanotubes composite for printing working electrodes. Honeychurch et al. (2018) developed a novel fully 3DP polystyrene/carbon nanofibers/graphite electrode for the trace determination of Zn^{2+} in water samples.

19.3.1 PLA/carbon nanocomposite for heavy metals detection

Heavy metal like mercury (Hg^{2+}), lead (Pb^{2+}), cadmium (Cd^{2+}), even at trace levels, are highly poisonous and pose significant threats to environment and human health (Gumpu et al., 2015; Islam et al., 2015; Jaishankar et al., 2014). In contrast, metals like manganese (Mn), zinc (Zn), and copper (Cu) are essential for various metabolic activities in human body and are only toxic at very high concentrations. Zinc is essential for enzyme function, the metabolism of nucleic acids and amino acids, DNA replication, and gene expression (McCall et al., 2000; Thompson & Thompson, 2022). Copper is necessary for iron uptake and the production of red and white blood cells (Collins et al., 2010). Brass is an alloy that contains both copper and zinc, which are widely used in water supply and plumbing systems. However, zinc that leaks from

Table 19.1 The electroanalytical performance of 3D-printed (3DP) chemical sensors designed for the detection of specific target analytes for environmental monitoring applications.

3D-printed electrode materials	3D printing techniques	Analytes	Sensing techniques	Detection range (μM)	Limit of detection (μM)	References
PLA/CB	FDM/3D-Pen	Pb^{2+}, Cu^{2+} ions	SWASV	0.048–4.2for Pb^{2+} in tap-water, 0.15–13.7 for Cu^{2+} in tap-water, 0.048–0.96 for Pb^{2+} inethanol, 0.31–6.3 for Cu^{2+} in ethanol	0.0039 for Pb^{2+}, 0.19 for Cu^{2+} in tap water, 0.0048 for Pb^{2+}, 0.19 for Cu^{2+} in ethanol	(João, Castro, et al., 2020)
PLA/CB	FDM	Pb^{2+}, Cd^{2+} ions	SWASV	0.14–1.3 for Pb^{2+}, Cd^{2+} 0.27–2.4 for Pb^{2+}, Cd^{2+}	0.012 for Pb^{2+}, 0.025 for Cd^{2+}	(Rocha, Squissato, et al., 2020)
PLA/CB	FDM/3D-pen	Pb^{2+}, Cd^{2+} ions	SWASV for Pb^{2+} & Cd^{2+}	0.1–1.57 for Pb^{2+}, 0.18–2.84 for Cd^{2+}	0.012 for Pb^{2+}, 0.10 for Cd^{2+}	(de Oliveira et al., 2020)
PLA/CB	FDM	Cu^{2+} ions	SWASV	0.15–4.7 in bioethanol	0.0015	(João, Squissato, et al., 2020)
PLA/CB	FDM	Nitrite ions	SWV	20–1500 in PBS, 25–75 in tap water and soil sample	1.96	(Pal et al., 2021)
PLA/CB	FDM	Nitrite ions	Amperometry	8–200	2.39	(Carvalho et al., 2021)
PLA/CB	FDM	Resorcinol	SWV	5–400	3.4	(Fabri et al., 2023)
PLA/CB	TES-DLP/3D-Pen	2,4,6-trinitrotoluene	SWV	5–500	1.5	(Cardoso et al., 2020)

PLA/C/Bi	FDM	Pb^{2+}, Cd^{2+} ions	SIA-ASV	0.001–0.39	0.002 for Pb^{2+}, 0.005 for Cd^{2+}	(Ealtima et al., 2021)
PLA/C/Au	FDM	Hg^{2+} ions	DPASV	0.025–0.2 in drinking water, 0.05–0.15 in (fish oil)	0.0026	(Katseli et al., 2020)
PLA/nG	FDM	Mn^{2+} ions	DPCSV	0.009–2.7 in tap and drinking water	0.0016	(Rocha, Foster, et al., 2020)
PLA/nG	FDM	Pb^{2+}, Cd^{2+} ions	SWV	0.92–22 μM for Pb^{2+}, 7–40 μM for Cd^{2+}	0.77 for Pb^{2+}, 2.84 for Cd^{2+}	(Foster et al., 2020)
PLA/G	FDM	Pb^{2+}, Sb^{3+} ions	SWASV	0.24–7.2 for Pb^{2+}, 0.41–12.31 for Sb^{3+} in gunshot residue	0.002 for Pb^{2+}, 0.015 for Sb^{3+}	(Castro et al., 2020)
PLA/G	FDM	3-chlorophenol, 4-nitrophenol, 4-chlorophenol, phenol	DPV	25–125	25	(Jyoti et al., 2021)
PLA/G	FDM	2,4,6-trinitrotoluene	SWV	1–870	0.40	(Cardoso et al., 2019)
PLA/G	FDM	Picric acid	CV	0.44–218	0.44	(Novotný et al., 2019)
PLA/G	FDM	Picric acid	CV	22–1571	22	(Manzanares Palenzuela et al., 2018)
PLA/G	FDM	Fenitrothion	DPV	15–75	15	(Jyoti et al., 2023)
PLA/G	FDM	Methyl parathion	CV	2–72	0.5	(Jyoti et al., 2023)

(*Continued*)

Table 19.1 (Continued)

3D-printed electrode materials	3D printing techniques	Analytes	Sensing techniques	Detection range (μM)	Limit of detection (μM)	References
PLA/G/Bi	FDM	Pb^{2+}, Cd^{2+}, Hg^{2+} ions	SWASV	0.1–0.5 for Pb^{2+}, 0.03–0.1 for Cd^{2+}, 0.02–0.1 for Hg^{2+}	0.011 for Pb^{2+}, 0.082 for Cd^{2+}, 0.0061 for Hg^{2+}	(Walters et al., 2020)
PLA/G/Au/ CdS NPs	FDM	Cu^{2+} ions	CA	0.01–80	0.05	(Foo et al., 2018)
PLA/CNTs	FDM	CO_2	CV		2.8×10^3	(Giorgini Escobar et al., 2020)
ABS/GO/ MWCNT	FDM	Nitrite ions	CA	10–125	4.1	(Silva-Neto et al., 2023)
PS/CNFs/Gr	FDM	Pb^{2+} ions	DPASV	2.1–258	1	(Rymansaib et al., 2016)
PS/CNFs/Gr	FDM	Zn^{2+} ions	DPASV	0.19–6.9 in DI water, 0.95–2.84 in tap water	0.131	(Honeychurch et al., 2018)
G/Ca-MOF	FDM	Hg^{2+} ions	SWASV	0.01–0.2 in drinking water, 0.001–0.15 in fish oil	0.003 in drinking water and fish oil	(Kokkinos et al., 2020)

ABS: acrylonitrile butadiene styrene, ASV: anodic stripping voltammetry, Au: gold, Bi: bismuth, CA: chronoamperomerty, Ca-MOF: calcium-metal organic framework, CB: carbon black, Cd^{2+}: cadmium, CdS NPs: cadmium sulfide nanoparticles, CNFs: carbon nanofibers, CNTs: carbon nanotubes, CO_2: carbon dioxide, Cu^{2+}: copper, CV: cyclic voltammetry, DPV: differential pulse voltammetry, DPASV: differential pulse anodic stripping voltammetry, DPCSV: differential pulse cathodic stripping voltammetry, FDM: fused deposition modeling, G: grapheme, GO: graphene oxide, G/PLA: graphene/polylactic acid, Gr: graphite, Hg^{2+}: mercury, Mn^{2+}: manganese, MWCNTs: multiwall carbon nanotubes, nG: nanographite, NH_2NH_2: hydrazine, Pb^{2+}: lead, PLA: polylactic acid, PS: polystyrene, Sb^{3+}: antimony, SIA-ASV: sequential injection-anodic stripping voltammetric, SWASV: square wave anodic stripping voltammetric, SWV: square wave voltammetry, TES-DLP: three-electrode-systems, Zn^{2+}: zinc.

plumbing fixtures can build up in tap water and harm people's health. Regulatory bodies such as the Environmental Protection Agency (EPA) set guidelines for heavy metal concentrations in groundwater to ensure human and environmental safety. Critical concentration limits are specified for various heavy metals, including 5 μg/mL for Zn, 0.6 μg/mL for Cu, 0.002 μg/mL for Hg, 0.5 μg/mL for Pb, and 0.01 μg/mL for Cd (Justino et al., 2017). Accumulation of heavy metals in organs like the liver and kidneys due to the body's incapability to metabolize them can cause health issues (Jan et al., 2015).

Lead (Pb^{2+}) is particularly toxic and widely used in applications such as batteries, gasoline, and pigments, contributing to environmental pollution and posing health risks through ingestion (Kumar et al., 2020; Wani et al., 2015). Cadmium (Cd^{2+}) is a heavy metal found naturally in fossil fuels, coal, and released during volcanic activity, leading to chronic diseases like renal failure, cancer, and osteomalacia (Alengebawy et al., 2021; Fatima et al., 2019; Rahimzadeh et al., 2017). Mercury (Hg^{2+}) is released through industrial and agricultural activities, combustion of fossil fuels, and is highly toxic to humans (Clarkson & Magos, 2006; Wu et al., 2024). Industrial waste and fuel combustion are the main contributors to heavy metal pollution in the environment, while food and water are common pathways for human exposure (Briffa et al., 2020; Tchounwou et al., 2012). Developing highly sensitive, portable, and cost-effective detection methods is essential to safeguard human health and the environment by accurately identifying trace levels of heavy metals in food and drinking water.

Rocha, Squissato, et al. (2020) utilized a 3D-printed working electrode (WE) composed of carbon black (CB) and PLA to monitor the concentration of heavy metals. The 3D-printed WE is produced by extruding CB-PLA filament using a 3D printer. In order to get a smooth surface, the 3D-printed electrode undergoes a polishing process using sandpaper. Afterward, the electrode introduced chemical and electrochemical activation using a NaOH solution to improve the electronalytical activity of the printed WE to enhance the electrical conductivity and sensing capabilities by breaking down PLA and revealing carbon black nanoparticles on the electrode surface. Furthermore, the sensing area of WE is determined by placing the rubber ring the center of electrode. The inner diameter of ring has a direct impact on the sensing area, which is critical for accurate electrochemical measurements. Fig. 19.1 shows a comprehensive presentation of a 3D printed electrochemical devices. The electrochemical device was examined using CV and EIS techniques in the presence of redox mediator. These techniques indicated the rapid electron transfer and an increased electroactive surface area on the activated electrodes. Afterward, the electrodes showed improved ability to detect Pb^{2+} andCd^{2+} in biological fluids. The detection limits for Pb^{2+} andCd^{2+} were reduced to 2.9 μg/L and 2.6 μg/L, respectively.

João, Castro et al. (2020) utilized a 3DP WE fabricated with conductive CB-PLA filaments for copper ion (Cu^{2+}) detection in fuel bioethanol. The activated 3D-printed electrodes were employed to detect copper ions in the bioethanol sample. During the process, the copper ions were reduced and deposited onto the electrode surface. Subsequently, a stripping potential was applied to the electrode, leading to the oxidation of the preconcentrated copper ions back to their ionic form. This oxidation

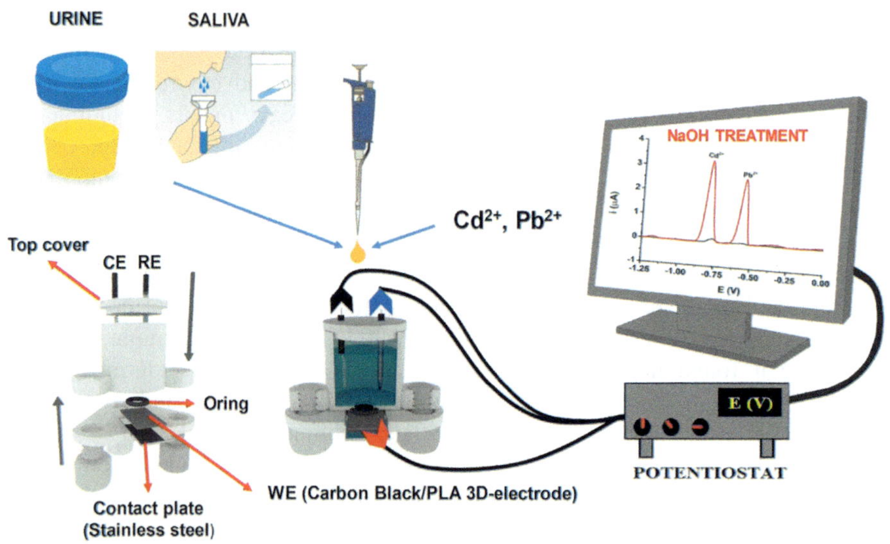

Figure 19.1 The schematic represents 3D-printed cell includes various components such as cell, top-cover, bottom-cover, and with the platinum (Pt) wire used as CE, Ag/AgCl as RE, and the 3DP WE. Additionally, contact plates and O-ring are included to make electric contact with WE. The assembled cell filled with electrolyte and established connections between potentiostat and three electrodes are shown at the middle of the diagram. The micropipette was used for injections of sample solution into the cell. The schematic represents fully 3D-printed electrochemical sensor for detection of Cd^{2+}, Pb^{2+} ions.
Reprinted with permission from Rocha, Squissato, et al. (2020), copyright 2020 Elsevier. From Rocha, D. P., Squissato, A. L., da Silva, S. M., Richter, E. M. & Munoz, R. A. A. (2020). Improved electrochemical detection of metals in biological samples using 3D-printed electrode: Chemical/electrochemical treatment exposes carbon-black conductive sites. Electrochimica Acta, 335. https://doi.org/10.1016/j.electacta.2020.135688.

process resulted in the release of a measurable current peak, which was relative to the concentration of copper ions in the sample. SWASV was utilized for the analysis of copper in the bioethanol samples using supporting electrolyte (resulting in a mixture of 30% ethanol and 70% water). The linear response for copper ions detection was found in range between 10 to 300 μg/L, and LOD was achieved 0.097 μg/L. These results demonstrate the potential of 3DP PLA/CB electrodes for sensitive and accurate analysis of copper in fuel bioethanol, highlighting their applicability in environmental monitoring and quality control in the biofuel industry. Baltima et al. (2021) introduced 3DP electrochemical microfluidic devices combined with a sequential injection analysis (SIA) system for the detection of Cd^{2+} andPb^{2+} via anodic stripping voltammetry (ASV). The fully 3D-printed microfluidic device included a 3DP microfluidic cell (made of PLA filament) integrated with three electrodes at the bottom, which were printed with PLA/CB filament via a dual extruder 3D printer in a single step. Each 3D-printed device included three conductive electrodes and a fluidic cell at a cost of approximately $0.035. Bismuth plating was employed to modify the

3DP electrodes to enhance the analytical capabilities. The study demonstrated the simultaneous determination of Cd^{2+} andPb^{2+} concentrations in honey samples ranging from 0 to 40 μg/L. The LODs was found to be 0.57 μg/L for Cd^{2+} and 0.38 μg/L for Pb^{2+} and selectivity of the 3DP electrode was assessed by examining the effect of potential interferents (Zn^{2+}, Cu^{2+}, Fe^{3+}, Sn^{2+}, and Sb^{3+}).

Katseli et al. (2020) developed a miniature 3DP device for the electrochemical detection of trace amounts of Hg^{2+} using anodic stripping voltammetry (ASV). The device was fabricated in a single-step printing process, comprises a mini cell made from a PLA nonconductive filament and three electrodes (i.e. WE, CE, RE) printed from a PLA/carbon conductive filament and electrodes were integrated at the bottom of the cell. The surface of the WE was modified with a thin layer of Au aimed to improve the sensitivity and selectivity of the electrode toward Hg^{2+}, enabling accurate and reliable detection of trace concentrations ranging from 5 to 40 μg/L. The LOD was achieved 0.52 μg/L in both fish oil and drinking water samples. Ng et al. (2021) introduced a novel method to enhance the electrochemical performance of 3D-printed nanostructured carbon (3DP-nC) electrodes for the detection of catechol, a pollutant found in natural water due to industrial waste, posing health risks to humans. The study involved coating the 3DP carbon electrodes with various thickness of Al_2O_3 insulating layer using the atomic layer deposition (ALD) process. Optimal results were achieved with an Al_2O_3 thickness of 80 ALD cycles, which enhanced charge transfer and the electrocatalysis of catechol at the Al_2O_3 surface. The surface of the 3DP-nC electrode exhibited a network of densely packed fibers. The electrooxidation activity for a catechol solution was studied using CV on pristine and Al_2O_3-coated 3DP-nC electrodes in 0.1 M $HClO_4$ electrolyte. As the thickness of the Al_2O_3 coating increased from 20 to 80 cycles, the peak current gradually increased, while the peak-to-peak separation decreased, indicating higher electron transfer and improved electrochemical activity to catechol. The CV studies of the best-performing Al_2O_3-coated 3DP-nC electrode showed a steady rising in peak current with increasing catechol concentrations, indicating the redox process involves electrocatalytic oxidation of catechol to o-benzoquinone. The calibration curve exhibited high linearity, with a LOD and limit of quantification (LOQ) approximately 7.7 μM and 44.5 μM, respectively.

19.3.2 PLA/carbon nanocomposite for phenols detection

Phenolic compounds and their derivatives pose significant environmental risks due to their high toxicity (Filik et al., 2008; Kujawski et al., 2004; Nardini, 2022). These compounds are commonly present in the effluents produced by industrial activities such as coal and petroleum processing. Additionally, they are found in various products and processes like pesticides, dyes, resins, cosmetics, plastics, antiseptic medications, and other chemical products (Bravo, 1998; Grace Pavithra et al., 2023). Elevated concentrations of phenol (>0.001 ppm) in natural water can cause harmful effects on the nervous system (Deýlová et al., 2011; Makarenko et al., 2002). The primary routes of exposure to phenolic substances are ingestion through water and food consumption or direct contact with the skin (Anku et al., 2017; Bravo, 1998).

There is a critical necessity to detect and quantify the phenolic compounds in environmental and industrial samples.

Fabri et al. (2023) developed a fully 3DP electrochemical sensors for the detection of resorcinol (RS) in water samples. Resorcinol is a phenolic compound, used in disinfectant, antiseptic in skincare products, and various industries such as organic synthesis, rubber, plastics, pharmaceuticals, and food additives. Despite its wide range of uses, resorcinol can pose environmental and health risks due to its toxic properties. It is important to monitor the presence of resorcinol to mitigate potential adverse effects on the human health and environment. The 3DP electrodes were printed with carbon black and polylactic acid filaments, and followed by electrochemically activated with a NaOH solution. The 3DP sensor exhibited a linear behavior with a concentration range from 5 to 400 μM/L, and a LOD of 3.4 μM/L. The 3DP electrochemical device offers a user-friendly and cost-effective approach for monitoring RS in the environment. Jyoti et al. (2021) demonstrated the utilization of 3DP nanocarbon electrodes (nCE) for the detection of phenols (i.e., chlorophenols and nitrophenols) through differential pulse voltammetry (DPV). These electrodes were fabricated using PLA/nC filament with optimized printing parameters and were activated using dimethylformamide (DMF) solvent to enhance their electrical conductivity. The surface of the nanocarbon electrode displayed a compact mesh-like arrangement of clustered nanocarbon with a diameter of 0.6 cm. DPV responses were recorded for varying concentrations from 25 to 125 μM of 4-chlorophenol, 3-chlorophenol, 4-nitrophenol and phenol using the 3DP-nCEs. The oxidation peak intensities were directly proportional to the concentrations of the phenolic compounds, showing an increase with higher phenol concentrations. The study highlighted the enhanced sensitivity of detecting 4-chlorophenol and phenol compared to other phenolic compounds. Additionally, DPV analysis simultaneously detected two compound mixtures, namely 3-chlorophenol with 4-nitrophenol and phenol with 4-nitrophenol. The results displayed clear anodic peaks associated with the oxidation of 3-chlorophenol at 0.545 V, phenol at 0.450 V, and 4-nitrophenol with two peaks at 0.823 V and 0.805 V. Notably, the 3DP-nCE demonstrated enhanced electrochemical reactivity toward phenolic compounds in comparison to traditional glassy carbon electrodes (GCEs).

19.3.3 PLA/carbon nanocomposite for explosive/toxic substances detection

Explosive materials often contain nitro-substituted compounds, for example, 2,4,6-trinitrophenol (TNP or Picric acid), 2,4-dinitrotoluene (DNT), and 2,4,6-trinitrotoluene (TNT). These compounds consist of both a chemical oxidant and reductant, leading to an exothermic decomposition process that result in significant environmental contamination, both in solution and vapor phases (Sun et al., 2015; Taya et al., 2018). Furthermore, these explosive materials have found applications in various industries for the production of burn cream, dyes, glass, leather, and other commercial products. The low volatility of nitroaromatic chemicals plays critical role for their detection in trace amounts, particularly in military and forensic contexts

(Stringer et al., 2010; Toal & Trogler, 2006). Electrochemical techniques are well-suited for the detection of nitroaromatic compounds by facilitating the reduction of nitro (-NO_2) groups present in the aromatic ring (Yew et al., 2016).

Cardoso et al. (2020) employed a 3D printing pen to create cost-effective and reusable electrochemical sensors for TNT detection. They designed two types of WE: three-electrode system (TES) and a single WE on a 3DP substrate. Both the single WE and TES were fabricated using 3D pen combined with two different customized techniques, that is, digital UV light processing (DLP) with UV curable acrylic resin and FDM with ABS filament. This approach resulted in the construction of four diverse electrochemical devices. The WE of both single and TES were printed using PLA/CB filaments with FDM 3D printer, followed by electrochemically activated in NaOH solution. The TES-DLP exhibited high oxidation peak potential and current density and, compared to single WE and TES-FDM system. The TES-DLP devices were selected for TNT detection, achieving a linear range from 5 to 500 μM with a LOD of 1.5 μM in a single drop analyte solution. João, Castro et al. (2020) employed a similar approach to fabricate inexpensive 3DP electrodes using a commercialized 3D pen filled with PLA/CB conductive filament. The conductive filament was heated and extruded by the 3D pen onto prefabricated templates to create 3DP electrodes. The electrodes were then subjected to electrochemical and chemical treatment for activation to enhance their sensing properties. The CV analysis conducted pre and post the chemical and electrochemical activation using the redox probe $[Fe(CN)_6]^{3-/4-}$ demonstrated a notable enhancement in electron transfer efficiency on the activated PLA/CB surface. The 3DP platform was demonstrated for the simultaneous detection of both metals copper ions (Cu^{2+}) and lead ions (Pb^{2+}) concentrations range from 1 to 870 μg/L in tap water samples. The LODs for Cu^{2+} andPb^{2+} were found to be 2 μg/L and 1 μg/L, respectively. The cost-effective and disposable 3DP platform exhibited selective monitoring of Cu^{2+} and Pb^{2+} ions against potential interferences (e.g., Hg^{2+}, Cd^{2+}, Fe^{3+}, and Zn^{2+}) commonly found in fuel ethanol.

Carvalho et al. (2021) proposed a system known as microflow injection analysis (μFIA), which utilized a 3DP microfluidic thread electroanalytical device (μTED). This system was specifically designed for the precise determination of nitrite ions present in drinking water. Nitrite ions serve as a crucial indicator in various biological and physiological processes, as well as in ensuring food safety and water quality. It is essential to monitor nitrite ion levels carefully to maintain the quality of drinking water and soil. The μTED system featured hydrophilic cotton threads as microfluidic channels, serving as passive pumps. The bottom part of the μTED, with dimensions of 95 mm in length, 60 mm in width, and 122 mm in height, was manufactured using a nonconductive ABS filament through FDM 3D printing. The electrodes dimension 45 mm in length, 3 mm in width, and 1.60 mm in height, were printed utilizing PLA/CB conductive filaments. Subsequently, the 3DP PLA/CB electrode underwent a polishing process with sandpaper in the presence of deionized water and then electrochemically activated in a NaOH solution to enhance electrical conductivity of the electrodes. The microflow channels were created using hydrophilic cotton threads. These channels were strategically placed perpendicular to the electrodes within the device. To secure the positioning of the channels and ensure proper fluid flow, the inlet and outlet reservoirs were fixed using double-sided tape. This arrangement allowed for efficient sample flow through the system, enabling accurate analysis of nitrite ions ranging from 8 to 200 μM in the water samples. The system achieved a LOD of 2.39 μM and LOQ of 7.61 μM, which were below the maximum contaminant level (MCL) considered safe for consumption by both US EPA and Brazilian standards.

19.3.4 PLA/graphene nanocomposite for heavy metals detection

Foster et al. (2020) optimized the printability and electrochemical performance of composite filaments consisting of nanographite (NG) loaded PLA for development of 3DP electrodes. They investigated different concentrations of NG (nanographite) ranging from 1% to 40% weight in PLA. Through their experimentation, they determined that a loading of 25% NG in PLA was optimal as it offered a balance between achieving sufficient conductivity and ensuring proper printability, resulting efficient 3DP electrode. The study focused on two types of 3DP electrode structures: 3D honeycomb macroporous electrodes and macroelectrodes. These electrodes were employed for the simultaneous detection of Cd^{2+} and Pb^{2+}. The 3D honeycomb macroporous electrodes exhibited higher sensitivity compared to the macroelectrodes due to their larger surface area. The LOD achieved for Cd^{2+} was approximately 300 μg/L, while for Pb^{2+} it was around 100 μg/L. The enhanced sensitivity of the 3D honeycomb structures demonstrated their potential for electrochemical sensing applications.

In this study, Walters et al. (2020) enhanced the detection sensitivity of cations Pb^{2+}, Cd^{2+}, and Hg^{2+} at trace levels by incorporating bismuth (Bi) microparticles into poly (lactic acid)/graphene (PLA/G) electrodes. The electrodes were constructed as cylindrical structures using a PLA/graphene filament. Subsequently, unmodified PLA/G electrodes were utilized to measure Hg, Pb, and Cd in acetate buffer via SWASV. Hg was successfully quantified with a LOD of 6.1 nM, while Cd and Pb did not show detectable peaks at below 400 nM concentrations. To improve the sensitivity for Cd and Pb detection, Bi microparticles were added to the PLA/G electrodes. This modification led to clear stripping peaks for both ions for Cd^{2+} and Pb^{2+} at the 20 nM level. The results show all three metals enabled quantification below the regulatory action limits set by the US Environmental Protection Agency for drinking water.

Castro et al. (2020) utilized a 3DP electrode made of PLA/G filament for the simultaneous determination of Sb^{3+} andPb^{2+} in gunshot residue samples collected by swabbing the clothes and hands of shooters using SWV. The electrochemical cell and WE were manufactured via FDM using ABS and a conductive PLA/graphene filament, respectively. Subsequently, the electrode underwent mechanical polishing and chemical treatment in dimethylformamide to expose graphene. The sensor exhibited high sensitivity with a linear range of 50 to 1500 μg/L and LOD of 1.8 μg/L and 0.5 μg/L for Sb^{3+} and $Pb^{2+,}$ respectively. Furthermore, the sensor demonstrated reusability for up to three times without compromising its electrochemical performance, achieved through simple surface renewal via mechanical polishing. Foo et al. (2018) introduced a novel 3DP photoelectrochemical (PEC) sensor using CdS nanoparticles modified 3DE/Au electrode (Fig. 19.1A). The disk shape 3DP electrode

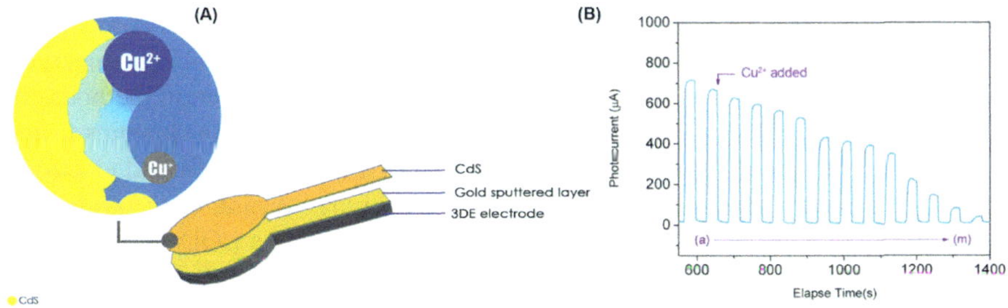

Figure 19.2 (A) Schematic representation of photoelectrochemical sensor using CdS nanoparticles modified 3DE/Au electrode. (B) Impact of Cu^{2+} on the photocurrent output of Au modified 3DE with increasing concentrations ranging from 0.01 μM to 80 μM. The inset graph displays a linear correlation between the photocurrent response ($I_0 - I$) and Cu^{2+} concentration, where I_0 and I denotes photocurrent intensity in the absence and presence of Cu^{2+}, respectively. Schematic representation of photoelectrochemical sensor using CdS nanoparticles modified 3DE/Au electrode for detection of Cu^{2+} ions.
Reprinted with permission from Foo et al. (2018), copyright 2018 Springer Nature. From Foo, C. Y., Lim, H. N., Mahdi, M. A., Wahid, M. H. & Huang, N. M. (2018). Three-Dimensional Printed Electrode and Its Novel Applications in Electronic Devices. Scientific Reports, 8(1). https://doi.org/10.1038/s41598-018-25861-3.

with a 17 mm diameter was designed using Tinkercad software. The 3D model of the electrode was printed using commercial conductive graphene filament (conductivity of 2.13 S/cm), using FDM 3D printer at temperature of 220°C. Subsequently, gold layer was sputtered onto the 3DP electrode surface, resulting in the electrode being modified with gold. The 3DP electrode design enabled the detection of trace amount of Cu^{2+} ions. Upon exposure to light, Cu^{2+} ions were reduced to Cu^{+} on the surface of Au modified 3DP electrode by capturing photo-excited electrons from the CdS nanoparticles. This process led to the formation of CuxS (where x=1, 2), which displaced the Cd^{2+} ions on the electrode surface. The change in photocurrent exhibited a consistent decrease as the Cu^{2+} ions concentrations increased from 0.01 μM to 80 μM (Fig. 19.2). The calibration plot demonstrated a good linear relationship between the change in photocurrent response and Cu^{2+} ion concentrations (inset of Fig. 19.1B), with LOD of 0.05 μM. The PEC sensor demonstrated a notable photocurrent response of approximately 724.1 μA, compared to traditional ITO/FTO electrodes.

19.3.5 PLA/graphene nanocomposite for explosive/toxic substances detection

Novotný et al. (2019) introduced a cost-effective and straightforward method to activate 3DP PLA/graphene electrodes for detecting nitroaromatic explosives through thermal annealing. The fabrication process involves creating 3DP-G/PLA disc electrodes (diameter: 8 mm, thickness: 1.6 mm) using FDM technology and a commercially available conductive filament comprising a graphene-polymer composite. Subsequently, the electrodes undergo thermal annealing in an inert atmosphere at

350°C. This thermal treatment facilitates the decomposition of the PLA component in the composite filament, leading to the formation of a well-graphitized material, a result not commonly achieved with traditional wet-chemical activation methods. The thermally annealed 3DP electrodes show enhanced electron transfer kinetics using a redox probe $[Fe(CN)_6]^{3-/4-}$ compared to untreated or chemically activated electrodes. They demonstrate improved electrochemical sensitivity toward concentrations of nitroaromatic explosives in the range of 5–50 ppm, achieving a LOD of 0.1 ppm and a LOQ of 0.3 ppm.

Cardoso et al. (2019) introduced a 3DP portable and flexible device designed for detection of trace amounts of TNT (2,4,6-Trinitrotoluene) residues on various surfaces such as gloves, metal, granite, via SWV. To fabricate the graphene/PLA electrodes, a conducting PLA/graphene (PLA/G) filament was used for the FDM 3D printing process. Prior to use, WE were polished for 30 seconds using abrasive paper to expose the nanoribbons like graphene structure within the PLA matrix, resulting an electroactive sensing surface. Fig. 19.3 illustrates the entire manufacturing process of the 3D-printed device integrated into a portable electrochemical cell designed for

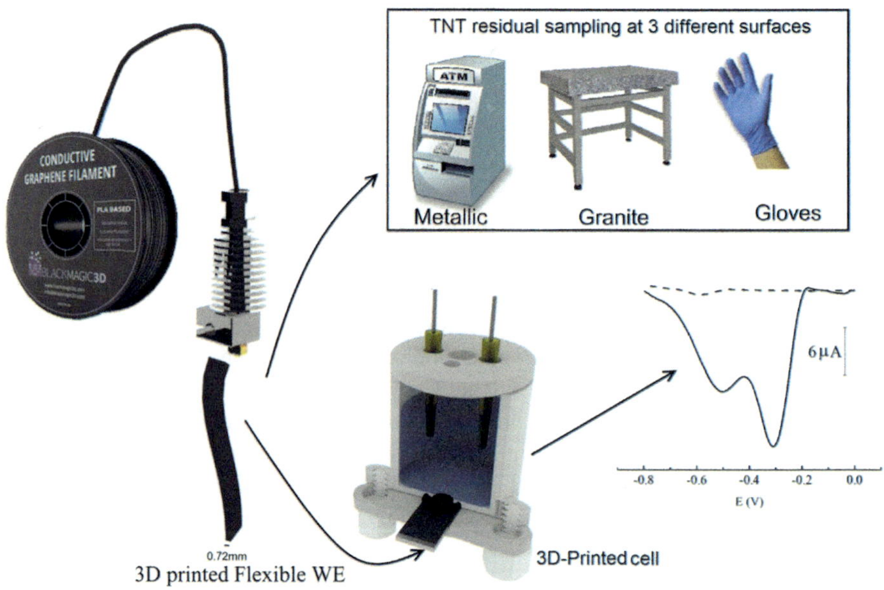

Figure 19.3 3DP electrode was printed through a 3D printing process using a PLA/grapheme filament via FDM printer. The 3DP device was assembled into a portable electrochemical cell featuring a three-electrode system to detect TNT residues on three distinct surfaces. The electrochemical sensor system for detection of TNT.
Reprinted with permission from Cardoso et al. (2019), copyright 2019 Elsevier. From Cardoso, R. M., Castro, S. V. F., Silva, M. N. T., Lima, A. P., Santana, M. H. P., Nossol, E., Silva, R. A. B., Richter, E. M., Paixão, T. R. L. C. & Muñoz, R. A. A. (2019). 3D-printed flexible device combining sampling and detection of explosives. Sensors and Actuators, B: Chemical, 292, 308–313. https://doi.org/10.1016/j.snb.2019.04.126.

SWV detection of TNT. The polished PLA/G electrode exhibited rapid response to detect TNT concentrations in the range of 1 to 870 μM within 30 seconds, with LOD of 0.40 μM. Furthermore, the sensor demonstrated the ability to detect Cu^{2+} and Pb^{2+} using SWASV from gunshot residue.

Manzanares Palenzuela et al. (2018) demonstrated the application of 3DP PLA/G electrodes for the detection of the explosive picric acid (PA) in acetate buffer. The PLA/graphene filament was utilized to create the working electrode structures in ring and disc shapes. Following the printing process, the working electrodes were treated by immersion in dimethylformamide (DMF), rinsing with ethanol, and air-drying. The detection of picric acid was achieved in the concentration range of 5 to 350 ppm by monitoring the hydroxylamine-nitroso reaction using cyclic voltammetry.

19.3.6 PLA/carbon nanotube nanocomposite for toxic substances detection

Giorgini Escobar et al. (2020) developed a fully integrated electrochemical platform that manufactured via bi-material FDM 3D printing. The platform consisted a cell made of natural polyamide and featured a WE, RE, CE fabricated from an electrically conductive PLA/CNT composite filament. The cylindrical WE was printed using the PLA/CNT composite material and activated through electrochemical anodic activation in Na_2SO_4. The electrochemical platform employed to investigate the electrochemical oxidation of aqueous hydrazine (NH_2NH_2) along with its chemical reaction with carbon dioxide (CO_2). Hydrazine, being the smallest diamine, possesses a high potential for greater absorption capacity compared to organic amines, making it an attractive sorbent for CO_2. The integration of bi-material FDM 3D printing in this study allowed for the fabrication of a customized electrochemical platform with tailored electrode materials.

19.3.7 PLA/carbon nanofiber nanocomposite for heavy metals detection

Honeychurch et al. (2018) employed 3DP polystyrene, graphite, and carbon nanofibers (PS/CNFs/Gr) electrodes for detecting trace levels of Zn^{2+} in tap-water using DPASV. To produce a composite thermoplastic filament, polystyrene pellets are mixed in chloroform under continuous stirring. Simultaneously, carbon nanofibers (CNF) and graphite flakes were sonicated in chloroform. Subsequently, the two mixture solution are mixed in an unsealed container and positioned on a heated magnetic stirrer set at 50°C within a fume hood until all the solvent evaporates. Once the solvent has completely evaporated, the resulting solid thermoplastic composite is transferred to a heated aluminum pot at 220°C. The resulting solid composite is then extruded into conductive filament at a specific temperature. The 3DP WE was initially polished using aluminum oxide solution, followed by wash with DI water. Next, repeatedly scanned using the CV was performed until a constant background current was achieved. Electrochemical studies was performed using 3DP WE with a graphite rod as a CE and pseudo-RE. The study recorded a distinct sharp anodic stripping peak, showing a linear response range from 12.7 μg/L to 450 μg/L, with a LOD of 8.6 μg/L.

Rymansaib et al. (2016) successfully developed new conductive blends of polystyrene, carbon nanofibers, graphite, and (PS/CNFs/Gr) suitable for 3D printing process. This composite material exhibited excellent surface properties and good conductivity, enabling the detection of Pb^{2+} ions at low concentrations. By incorporating graphite flake and carbon nanofibers (CNFs) into thermoplastic polystyrene polymer to produce conductive blends. The composite comprises 80% polystyrene, 10% CNFs, 10% graphite by weight, demonstrated not only good conductivity but also excellent mechanical characteristics, resulting suitable for the

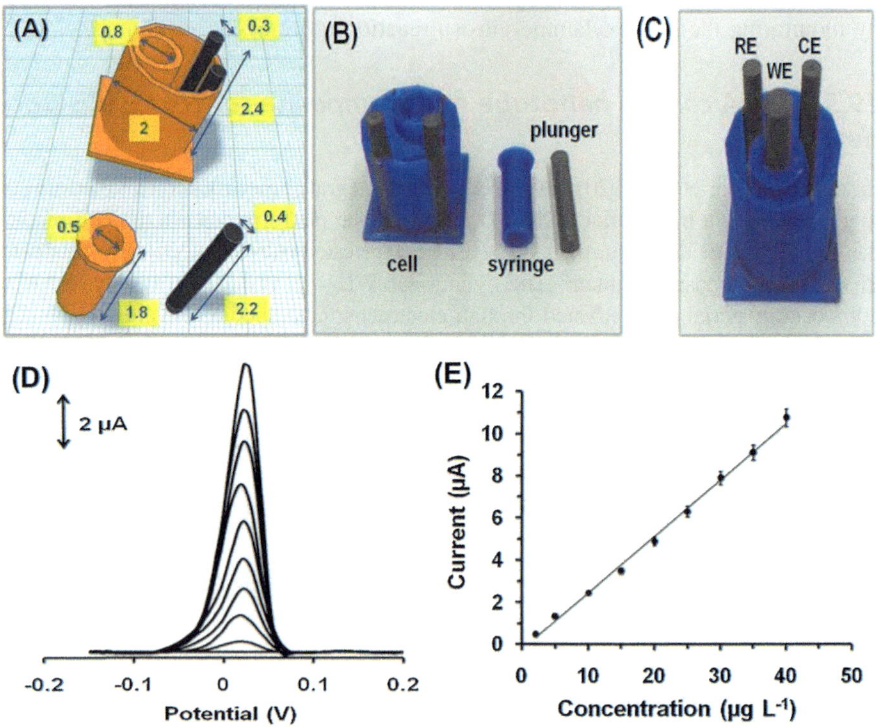

Figure 19.4 (A) Photograph shows the dimensions of the 3DP lab-in-a-syringe cell in cm. (B) Electrochemical cell, syringe, plunger of the 3DP device. (C) A complete 3DP electrochemical device imbedded with three electrodes, that is, WE, CE, RE. (D) SWV response at the Graphite/Ca-MOF electrode for increasing concentrations of Hg^{2+} from 2 to 40 μg/L with successive additions of 5 μg/L of Hg^{2+} in acetate buffer (pH 4.5). (E) The corresponding calibration plot shows the good linear current response with the Hg^{2+}concentration. A fully electrochemical sensor for detection of Hg^{2+} ion concentrations. Reprinted with permission from Kokkinos et al. (2020), copyright 2020 Elsevier. From Kokkinos, C., Economou, A., Pournara, A., Manos, M., Spanopoulos, I., Kanatzidis, M., Tziotzi, T., Petkov, V., Margariti, A., Oikonomopoulos, P. & Papaefstathiou, G. S. (2020). 3D-printed lab-in-a-syringe voltammetric cell based on a working electrode modified with a highly efficient Ca-MOF sorbent for the determination of Hg(II). Sensors and Actuators B: Chemical, 321, 128508. https://doi.org/10.1016/j.snb.2020.128508.

3D printing process. Sensors utilizing these 3DP electrodes were able to detect lead ions (Pb^{2+}) at low concentrations (1 μM) show the effectiveness of the PS/CNFs/Gr blend in electrochemical sensing applications.

19.3.8 PLA/metal-organic framework nanocomposite for heavy metals detection

Kokkinos et al. (2020) developed an innovative lab-in-a-syringe device for detecting Hg^{2+} through anodic stripping voltammetry (ASV). The device was manufactured in a single-step process, involves a small cell created from a nonconductive PLA filament and two electrodes (RE, CE) printed on the cell's sides using a conductive PLA/carbon filament (Fig. 19.4A–C). The WE was linked with a detachable 3DP syringe (made from PLA filament) and a conductive plunger (PLA/carbon) to establish electrical connectivity between the potentiostat and WE. The WE was fabricated using graphite paste modified with a metal-organic framework called Ca-MOF ([Ca(H4L)(DMA)2]•2DMA, where H4L is N,N'-bis(2,4-dicarboxyphenyl)-oxalamide and DMA is N,N-dimethylacetamide). This Ca-MOF exhibited strong sorption capability and rapid kinetics toward Hg^{2+}. The 3DP sensor demonstrated a linear correlation with the SWV current response and Hg^{2+} concentration ranges from 2 to 40 μg/L (Fig. 19.4D and E) and improved sensitivity for detection of Hg^{2+} with a LOD of 0.6 μg/L, outperforming other 3D-printed modified electrodes.

The integration of 3D printing technology and Ca-MOF as an electrode modifier offers a promising platform for highly sensitive detection of water pollutants. Assessment of the interference effects of different heavy metals (Zn^{2+}, Pb^{2+}, Cu^{2+}, and Cd^{2+}) indicated that Zn^{2+}, Pb^{2+}, and Cd^{2+} did not interfere with the Hg^{2+} analysis, while Cu^{2+} displayed a distinct stripping peak separated from the Hg peak. The sensor demonstrated the efficacy for on-site monitoring of Hg^{2+} in fish oil samples.

19.4 Conclusions, challenges, and limitations

In conclusion, the advancement of 3D printing technology has sparked significant interest in the rapid production of durable, cost-effective 3D-printed electrodes that can be customized to enhance their electrochemical capabilities. The utilization of a 3D printer enables the rapid creation of various electrode shapes and even complete analytical devices. Electrochemical sensors manufactured through 3D printing have shown reliable performance, rapid response time, and enhanced sensitivity. These advantages, coupled with the design flexibility and ease of production offered by 3D printing, pave the way for the development of point-of-care diagnostic tools that can enhance individuals' quality of life.

This chapter has outlined various 3D printed polymer nanocomposites electrodes used for detection of water pollutants, heavy metals, explosive and toxic substances for environmental monitoring. Furthermore, we have evaluated the analytical performances of these 3D-printed chemical sensors by comparing their performance in terms of limit of detection and linear range.

The majority of filaments available in the market are composed of nonconductive thermoplastic polymers like PLA and ABS, or electrically conductive carbon materials such as CB, or graphene, which are distributed within an insulating polymeric matrix. The integration of diverse carbon materials (e.g. CNTs, CNFs) into different polymers is intended to achieve sufficient electrical conductivity, a durable and mechanically stable interface that can be polished and reused, and a high signal-to-background response.

Ongoing efforts are focused on developing unique composite filaments to enhance printing performance and achieve superior electrical, mechanical, and electrochemical properties. While some studies have explored electrode modifications using noble metals like Au, Cu, or Ni through processes such as electrodeposition, there is a need for further advancements in creating polymer nanocomposite conductive filaments for printing electrodes that can deliver exceptional electrochemical performance without the need for additional activation or modification. However, before 3D-printed sensing devices can be widely accepted as diagnostic tools for environmental monitoring, several challenges and limitations need to be addressed.

References

Al-Saadi, T., Rossiter, J. A., & Panoutsos, G. (2021). Control of selective laser melting processes: Existing efforts, challenges, and future opportunities. In *2021 29th Mediterranean Conference on Control and Automation (MED)*, PUGLIA, Italy, 2021. https://doi.org/10.1109/MED51440.2021.9480258.

Alengebawy, A., Abdelkhalek, S. T., Qureshi, S. R., & Wang, M. Q. (2021). Heavy metals and pesticides toxicity in agricultural soil and plants: Ecological risks and human health implications. *Toxics 2021, 9*, 42–49. https://doi.org/10.3390/TOXICS9030042.

Ali, H., Khan, E., & Ilahi, I. (2019). Environmental chemistry and ecotoxicology of hazardous heavy metals: Environmental persistence, toxicity, and bioaccumulation. *Journal of Chemistry, 2019.* https://doi.org/10.1155/2019/6730305.

Anku, W. W., Mamo, M. A., Govender, P. P., Anku, W. W., Mamo, M. A., & Govender, P. P. (2017). Phenolic compounds in water: Sources, reactivity, toxicity and treatment methods. InTech.https://doi.org/10.5772/66927.

Balali-Mood, M., Naseri, K., Tahergorabi, Z., Khazdair, M. R., & Sadeghi, M. (2021). Toxic mechanisms of five heavy metals: Mercury, lead, chromium, cadmium, and arsenic. *Front. Pharmacol. 12*, 643972. https://doi.org/10.3389/FPHAR.2021.643972/BIBTEX.

Baltima, A., Panagopoulou, H., Economou, A., & Kokkinos, C. (2021). 3D-printed fluidic electrochemical microcell for sequential injection/stripping analysis of heavy metals. *Analytica Chimica Acta, 1159*, 338426. https://doi.org/10.1016/J.ACA.2021.338426.

Bravo, L. (1998). Polyphenols: Chemistry, dietary sources, metabolism, and nutritional significance. *Nutrition Reviews, 56*, 317–333. https://doi.org/10.1111/J.1753-4887.1998.TB01670.X.

Briffa, J., Sinagra, E., & Blundell, R. (2020). Heavy metal pollution in the environment and their toxicological effects on humans. *Heliyon, 6*, e04691. https://doi.org/10.1016/J.HELIYON.2020.E04691.

Cardoso, R. M., Castro, S. V. F., Silva, M. N. T., Lima, A. P., Santana, M. H. P., Nossol, E., Silva, R. A. B., Richter, E. M., Paixão, T. R. L. C., & Muñoz, R. A. A. (2019). 3D-printed flexible device combining sampling and detection of explosives. *Sensors and Actuators B: Chemical, 292*, 308–313. https://doi.org/10.1016/J.SNB.2019.04.126.

Cardoso, R. M., Rocha, D. P., Rocha, R. G., Stefano, J. S., Silva, R. A. B., Richter, E. M., & Muñoz, R. A. A. (2020). 3D-printing pen versus desktop 3D-printers: Fabrication of carbon black/polylactic acid electrodes for single-drop detection of 2,4,6-trinitrotoluene. *Analytica Chimica Acta, 1132*, 10–19. https://doi.org/10.1016/J.ACA.2020.07.034.

Caroli, S., Forte, G., Iamiceli, A. L., & Galoppi, B. (1999). Determination of essential and potentially toxic trace elements in honey by inductively coupled plasma-based techniques. *Talanta, 50*, 327–336. https://doi.org/10.1016/S0039-9140(99)00025-9.

Carvalho, R. M., Ferreira, V. S., & Lucca, B. G. (2021). A novel all-3D-printed thread-based microfluidic device with an embedded electrochemical detector: First application in environmental analysis of nitrite. *Analytical Methods, 13*, 1349–1357. https://doi.org/10.1039/D1AY00070E.

Castro, S. V. F., Lima, A. P., Rocha, R. G., Cardoso, R. M., Montes, R. H. O., Santana, M. H. P., Richter, E. M., & Munoz, R. A. A. (2020). Simultaneous determination of lead and antimony in gunshot residue using a 3D-printed platform working as sampler and sensor. *Analytica Chimica Acta, 1130*, 126–136. https://doi.org/10.1016/J.ACA.2020.07.033.

Chan, H. K., Griffin, J., Lim, J. J., Zeng, F., & Chiu, A. S. F. (2018). The impact of 3D Printing Technology on the supply chain: Manufacturing and legal perspectives. *International Journal of Production Economics, 205*, 156–162. https://doi.org/10.1016/J.IJPE.2018.09.009.

Cheng, T. S., Nasir, M. Z. M., Ambrosi, A., & Pumera, M. (2017). 3D-printed metal electrodes for electrochemical detection of phenols. *Applied Materials Today*, 9, 212–219. https://doi.org/10.1016/J.APMT.2017.07.005.

Chu, T., Park, S., & Fu, K. (2021). 3D printing-enabled advanced electrode architecture design. *Carbon Energy, 3*, 424–439. https://doi.org/10.1002/CEY2.114.

Clarkson, T. W., & Magos, L. (2006). The toxicology of mercury and its chemical compounds. *Critical Reviews in Toxicology, 36*, 609–662. https://doi.org/10.1080/10408440600845619.

Collins, J. F., Prohaska, J. R., & Knutson, M. D. (2010). Metabolic crossroads of iron and copper. *Nutrition Reviews, 68*, 133–147. https://doi.org/10.1111/J.1753-4887.2010.00271.X.

Deýlová, D., Yosypchuk, B., Vyskočil, V., & Barek, J. (2011). Voltammetric determination of 4-nitrophenol and 5-nitrobenzimidazole using different types of silver solid amalgam electrodes—A comparative study. *Electroanalysis, 23*, 1548–1555. https://doi.org/10.1002/ELAN.201100067.

Engwa, G.A., Ferdinand, P.U., Nwalo, F.N., Unachukwu, M.N., Engwa, G.A., Ferdinand, P.U., Nwalo, F.N., Unachukwu, M.N., 2019. Mechanism and Health Effects of Heavy Metal Toxicity in Humans. Poisoning Mod. World - New Tricks an Old Dog? https://doi.org/10.5772/INTECHOPEN.82511.

Fabri, J., Silva, L. R. G., Stefano, J. S., Pereira, J. F. S., Cocco, D. R., Muñoz, R. A. A., & Rocha, D. P. (2023). In situ electrochemical determination of resorcinol using a fully 3D printed apparatus. *Microchemical Journal, Devoted to the Application of Microtechniques in all Branches of Science, 191*, 108810. https://doi.org/10.1016/J.MICROC.2023.108810.

Fatima, G., Raza, A. M., Hadi, N., Nigam, N., & Mahdi, A. A. (2019). Cadmium in human diseases: it's more than just a mere metal. Indian Journal of Clinical Biochemistry, *34*, 371–378. https://doi.org/10.1007/S12291-019-00839-8/METRICS.

Feldmann, J., Salaun, P., & Lombi, E. (2009). Critical review perspective: elemental speciation analysis methods in environmental chemistry—Moving towards methodological integration. *Environmental Chemistry, 6*, 275–289. https://doi.org/10.1071/EN09018.

Filik, H., Hayvali, M., Kiliç, E., Apak, R., Aksu, D., Yanaz, Z., & Çengel, T. (2008). Development of an optical fibre reflectance sensor for p-aminophenol detection based on immobilised bis-8-hydroxyquinoline. *Talanta, 77*, 103–109. https://doi.org/10.1016/J.TALANTA.2008.05.045.

Foo, C. Y., Lim, H. N., Mahdi, M. A., Wahid, M. H., & Huang, N. M. (2018). Three-dimensional printed electrode and its novel applications in electronic devices. *Scientific Reports 2018, 81*(8), 1–11. https://doi.org/10.1038/s41598-018-25861-3.

Foster, C. W., Elbardisy, H. M., Down, M. P., Keefe, E. M., Smith, G. C., & Banks, C. E. (2020). Additively manufactured graphitic electrochemical sensing platforms. *Chemical Engineering Journal, 381*, 122343. https://doi.org/10.1016/J.CEJ.2019.122343.

Giorgini Escobar, J., Vaněčková, E., Nováková Lachmanová, Š., Vivaldi, F., Heyda, J., Kubišta, J., Shestivska, V., Španěl, P., Schwarzová-Pecková, K., Rathouský, J., Sebechlebská, T., & Kolivoška, V. (2020). The development of a fully integrated 3D printed electrochemical platform and its application to investigate the chemical reaction between carbon dioxide and hydrazine. *Electrochimica Acta, 360*, 136984. https://doi.org/10.1016/J.ELECTACTA.2020.136984.

Glasco, D. L., Sheelam, A., Ho, N. H. B., Mamaril, A. M., King, M., & Bell, J. G. (2022). Editors' Choice—Review—3D Printing: An innovative trend in analytical sensing. *ECS Sensors Plus, 1*, 010602. https://doi.org/10.1149/2754-2726/AC5C7A.

Goyanes, A., Buanz, A. B. M., Basit, A. W., & Gaisford, S. (2014). Fused-filament 3D printing (3DP) for fabrication of tablets. *International Journal of Pharmaceutics, 476*, 88–92. https://doi.org/10.1016/J.IJPHARM.2014.09.044.

Grace Pavithra, K., Sundar Rajan, P., Arun, J., Brindhadevi, K., Hoang Le, Q., & Pugazhendhi, A. (2023). A review on recent advancements in extraction, removal and recovery of phenols from phenolic wastewater: Challenges and future outlook. *Environmental Research, 237*, 117005. https://doi.org/10.1016/J.ENVRES.2023.117005.

Gulzar, U., Glynn, C., & O'Dwyer, C. (2020). Additive manufacturing for energy storage: Methods, designs and material selection for customizable 3D printed batteries and supercapacitors. *Current Opinion in Electrochemistry, 20*, 46–53. https://doi.org/10.1016/J.COELEC.2020.02.009.

Gumpu, M. B., Sethuraman, S., Krishnan, U. M., & Rayappan, J. B. B. (2015). A review on detection of heavy metal ions in water – An electrochemical approach. *Sensors and Actuators B: Chemical, 213*, 515–533. https://doi.org/10.1016/J.SNB.2015.02.122.

Hama Aziz, K. H., Mustafa, F. S., Omer, K. M., Hama, S., Hamarawf, R. F., & Rahman, K. O. (2023). Heavy metal pollution in the aquatic environment: Efficient and low-cost removal approaches to eliminate their toxicity: A review. *RSC Advances, 13*, 17595–17610. https://doi.org/10.1039/D3RA00723E.

Harris, M., Potgieter, J., Archer, R., & Arif, K. M. (2019). Effect of material and process specific factors on the strength of printed parts in fused filament fabrication: A review of recent developments. *Materials, 12*, 1664. https://doi.org/10.3390/MA12101664.

Hawes, G. F., Rehman, S., Rangom, Y., & Pope, M. A. (2022). Advanced manufacturing approaches for electrochemical energy storage devices. *International Materials Reviews, 68*. https://doi.org/10.1080/09506608.2022.2086388/SUPPL_FILE/YIMR_A_2086388_SM7326.DOCX.

Honeychurch, K. C., Rymansaib, Z., & Iravani, P. (2018). Anodic stripping voltammetric determination of zinc at a 3-D printed carbon nanofiber–graphite–polystyrene electrode using a carbon pseudo-reference electrode. *Sensors and Actuators B: Chemical, 267*, 476–482. https://doi.org/10.1016/J.SNB.2018.04.054.

Ho, N. H. B., Glasco, D. L., & Bell, J. G. (2022). Potentiometric analysis of benzalkonium chloride with 3D printed ion-selective membranes. *ECS Sensors Plus, 1*, 020601. https://doi.org/10.1149/2754-2726/AC8438.

Huang, J., Qin, Q., & Wang, J. (2020). A Review of Stereolithography: Processes and Systems. *Process, 8*, 1138. https://doi.org/10.3390/PR8091138.

Hull, C. W., & Arcadia, C. (1984). Apparatus for production of three-dimensional objects by stereolithography. Google Patent.

Islam, M. S., Ahmed, M. K., Raknuzzaman, M., Habibullah -Al- Mamun, M., & Islam, M. K. (2015). Heavy metal pollution in surface water and sediment: A preliminary assessment of an urban river in a developing country. *Ecological Indicators, 48*, 282–291. https://doi.org/10.1016/J.ECOLIND.2014.08.016.

Jaishankar, M., Tseten, T., Anbalagan, N., Mathew, B. B., & Beeregowda, K. N. (2014). Toxicity, mechanism and health effects of some heavy metals. *Interdisciplinary Toxicology, 7*, 60–72. https://doi.org/10.2478/INTOX-2014-0009.

Jandyal, A., Chaturvedi, I., Wazir, I., Raina, A., & Ul Haq, M. I. (2022). 3D printing—A review of processes, materials and applications in industry 4.0. *Sustainable Operations and Computers, 3*, 33–42. https://doi.org/10.1016/J.SUSOC.2021.09.004.

Jan, A. T., Azam, M., Siddiqui, K., Ali, A., Choi, I., & Haq, Q. M. R. (2015). Heavy metals and human health: mechanistic insight into toxicity and counter defense system of antioxidants. International Journal of Molecular Sciences, *16*, 29592–29630. https://doi.org/10.3390/IJMS161226183.

Jayakrishna, M., Vijay, M., & Khan, B. (2023). An overview of extensive analysis of 3D printing applications in the manufacturing sector. *Journal of Engineering*, 2023, 1–23. https://doi.org/10.1155/2023/7465737.

João, A. F., Castro, S. V. F., Cardoso, R. M., Gamela, R. R., Rocha, D. P., Richter, E. M., & Muñoz, R. A. A. (2020). 3D printing pen using conductive filaments to fabricate affordable electrochemical sensors for trace metal monitoring. *Journal of Electroanalytical Chemistry, 876*, 114701. https://doi.org/10.1016/J.JELECHEM.2020.114701.

João, A. F., Squissato, A. L., Richter, E. M., & Muñoz, R. A. A. (2020). Additive-manufactured sensors for biofuel analysis: Copper determination in bioethanol using a 3D-printed carbon black/polylactic electrode. *Analytical and Bioanalytical Chemistry, 412*, 2755–2762. https://doi.org/10.1007/S00216-020-02513-Y/TABLES/4.

Junpha, J., Wisitsoraat, A., Prathumwan, R., Chaengsawang, W., Khomungkhun, K., & Subannajui, K. (2020). Electronic tongue and cyclic voltammetric sensors based on carbon nanotube/polylactic composites fabricated by fused deposition modelling 3D printing. *Materials Science and Engineering: C, 117*, 111319. https://doi.org/10.1016/J.MSEC.2020.111319.

Justino, C. I. L., Duarte, A. C., & Rocha-Santos, T. A. P. (2017). Recent progress in biosensors for environmental monitoring: A review. *Sensors, 17*, 2918. https://doi.org/10.3390/S17122918.

Jyoti, Redondo, E., Alduhaish, O., & Pumera, M. (2023). 3D-printed electrochemical sensor for organophosphate nerve agents. *Electroanalysis, 35*, e202200047. https://doi.org/10.1002/ELAN.202200047.

Jyoti, Redondo, E., Alduhaish, O., & Pumera, M. (2021). 3D-printed nanocarbon sensors for the detection of chlorophenols and nitrophenols: Towards environmental applications of additive manufacturing. *Electrochemistry Communications, 125*, 106984. https://doi.org/10.1016/J.ELECOM.2021.106984.

Katseli, V., Thomaidis, N., Economou, A., & Kokkinos, C. (2020). Miniature 3D-printed integrated electrochemical cell for trace voltammetric Hg(II) determination. *Sensors and Actuators B: Chemical, 308*, 127715. https://doi.org/10.1016/J.SNB.2020.127715.

Kenawy, I. M. M., Hafez, M. A. H., Akl, M. A., & Lashein, R. R. (2000). Determination by AAS of some trace heavy metal ions in some natural and biological samples after their preconcentration using newly chemically modified chloromethylated polystyrene-PAN ion-exchanger. *Analytical Sciences: the International Journal of the Japan Society for Analytical Chemistry, 16*, 493–500. https://doi.org/10.2116/ANALSCI.16.493/METRICS.

Kokkinos, C., Economou, A., Pournara, A., Manos, M., Spanopoulos, I., Kanatzidis, M., Tziotzi, T., Petkov, V., Margariti, A., Oikonomopoulos, P., & Papaefstathiou, G. S. (2020). 3D-printed lab-in-a-syringe voltammetric cell based on a working electrode modified with a highly efficient Ca-MOF sorbent for the determination of Hg(II). *Sensors and Actuators B: Chemical, 321*, 128508. https://doi.org/10.1016/J.SNB.2020.128508.

Kujawski, W., Warszawski, A., Ratajczak, W., Porębski, T., Capała, W., & Ostrowska, I. (2004). Removal of phenol from wastewater by different separation techniques. *Desalination, 163*, 287–296. https://doi.org/10.1016/S0011-9164(04)90202-0.

Kumar, A., Kumar, A., Cabral-Pinto, M., Chaturvedi, A. K., Shabnam, A. A., Subrahmanyam, G., Mondal, R., Gupta, D. K., Malyan, S. K., Kumar, S. S., Khan, S. A., & Yadav, K. K. (2020). Lead toxicity: Health hazards, influence on food chain, and sustainable remediation approaches. *International Journal of Environmental Research and Public Health, 17*, 2179. https://doi.org/10.3390/IJERPH17072179.

Kumar, S., Singh, H., Singh, I., Bharti, S., Kumar, D., Siebert, G., & Koloor, S. S. R. (2024). A comprehensive review of FDM printing in sensor applications: Advancements and future perspectives. *Journal of Manufacturing Processes, 113*, 152–170. https://doi.org/10.1016/J.JMAPRO.2024.01.030.

Lakkala, P., Munnangi, S. R., Bandari, S., & Repka, M. (2023). Additive manufacturing technologies with emphasis on stereolithography 3D printing in pharmaceutical and medical applications: A review. *International Journal of Pharmaceutics: X, 5*, 100159. https://doi.org/10.1016/J.IJPX.2023.100159.

Lee, K. Y., Ambrosi, A., & Pumera, M. (2017). 3D-printed metal electrodes for heavy metals detection by anodic stripping voltammetry. *Electroanalysis, 29*, 2444–2453. https://doi.org/10.1002/ELAN.201700388.

Liu, Z., Wang, Y., Wu, B., Cui, C., Guo, Y., & Yan, C. (2019). A critical review of fused deposition modeling 3D printing technology in manufacturing polylactic acid parts. *The International Journal of Advanced Manufacturing Technology, 102*, 2877–2889. https://doi.org/10.1007/S00170-019-03332-X.

Loterie, D., Delrot, P., & Moser, C. (2020). High-resolution tomographic volumetric additive manufacturing. *Nature Communications, 11*, 1–6. https://doi.org/10.1038/s41467-020-14630-4.

Makarenko, A. A., Bezverbnaya, I. P., Kosheleva, I. A., Kuvichkina, T. N., Il'yasov, P. V., & Reshetilov, A. N. (2002). Development of biosensors for phenol determination from bacteria found in petroleum fields of West Siberia. *Applied Biochemistry and Microbiology, 38*, 23–27. https://doi.org/10.1023/A:1013240319259/METRICS.

Manzanares Palenzuela, C. L., Novotný, F., Krupička, P., Sofer, Z., & Pumera, M. (2018). 3D-printed graphene/polylactic acid electrodes promise high sensitivity in electroanalysis. *Analytical Chemistry, 90*, 5753–5757. https://doi.org/10.1021/ACS.ANALCHEM.8B00083/SUPPL_FILE/AC8B00083_SI_001.PDF.

McCall, K. A., Huang, C. C., & Fierke, C. A. (2000). Function and Mechanism of Zinc Metalloenzymes. *The Journal of Nutrition, 130*, 1437S–1446S. https://doi.org/10.1093/JN/130.5.1437S.

Mehrpouya, M., Dehghanghadikolaei, A., Fotovvati, B., Vosooghnia, A., Emamian, S. S., & Gisario, A. (2019). The potential of additive manufacturing in the smart factory Industrial 4.0: A review. *Applied Sciences, 9*, 3865. https://doi.org/10.3390/APP9183865.

Melchels, F. P. W., Feijen, J., & Grijpma, D. W. (2010). A review on stereolithography and its applications in biomedical engineering. *Biomaterials, 31*, 6121–6130. https://doi.org/10.1016/J.BIOMATERIALS.2010.04.050.

Mishra, V., Negi, S., & Kar, S. (2023). FDM-based additive manufacturing of recycled thermoplastics and associated composites. *Journal of Material Cycles and Waste Management, 25*, 758–784. https://doi.org/10.1007/S10163-022-01588-2/METRICS.

Mitra, S., Chakraborty, A. J., Tareq, A. M., Emran, T., Bin, Nainu, F., Khusro, A., Idris, A. M., Khandaker, M. U., Osman, H., Alhumaydhi, F. A., & Simal-Gandara, J. (2022). Impact of heavy metals on the environment and human health: Novel therapeutic insights to counter the toxicity. *Journal of King Saud University—Science, 34*, 101865. https://doi.org/10.1016/J.JKSUS.2022.101865.

Mobarak, M. H., Islam, M. A., Hossain, N., Al Mahmud, M. Z., Rayhan, M. T., Nishi, N. J., & Chowdhury, M. A. (2023). Recent advances of additive manufacturing in implant fabrication – A review. *Applied Surface Science Advances, 18*, 100462. https://doi.org/10.1016/J.APSADV.2023.100462.

Nardini, M. (2022). Phenolic compounds in food: characterization and health benefits. *Molecules, 27*, 783. https://doi.org/10.3390/MOLECULES27030783.

Negahdary, M., Sharma, A., Anthopoulos, T. D., & Angnes, L. (2023). Recent advances in electrochemical nanobiosensors for cardiac biomarkers. *TrAC Trends in Analytical Chemistry, 164*, 117104. https://doi.org/10.1016/J.TRAC.2023.117104.

Nga Tran, T., Bayer, Ilker, S., Alejandro Heredia-Guerrero, J., Frugone, M., Lagomarsino, M., Maggio, F., Athanassiou, A., Tran, T. N., Bayer, I. S., Heredia-Guerrero, J. A., Athanassiou Smart Materials, A., Frugone, M., Lagomarsino, M., & Maggio, F. (2017). Cocoa shell waste biofilaments for 3D printing applications. *Macromolecular Materials and Engineering,* 302. https://doi.org/10.1002/MAME.201700219.

Ngo, T. D., Kashani, A., Imbalzano, G., Nguyen, K. T. Q., & Hui, D. (2018). Additive manufacturing (3D printing): A review of materials, methods, applications and challenges. *Composites Part B: Engineering, 143*, 172–196. https://doi.org/10.1016/J.COMPOSITESB.2018.02.012.

Ng, S., Iffelsberger, C., Michalička, J., & Pumera, M. (2021). Atomic layer deposition of electrocatalytic insulator Al2O3on three-dimensional printed nanocarbons. *ACS Nano, 15*, 686–697. https://doi.org/10.1021/ACSNANO.0C06961/ASSET/IMAGES/MEDIUM/NN0C06961_M003.GIF.

Nnaji, N. D., Onyeaka, H., Miri, T., & Ugwa, C. (2023). Bioaccumulation for heavy metal removal: A review. *SN Applied Sciences, 5*, 1–12. https://doi.org/10.1007/S42452-023-05351-6/METRICS.

Noor, M. N. M., Leow, M. L., Lai, W. H., Hon, Y. K., Tiong, L. L., & Chern, P. M. (2022). Research landscape on 3D printing applications in healthcare within Southeast Asian countries: A systematic scoping review protocol. *BMJ Open, 12*, e065546. https://doi.org/10.1136/BMJOPEN-2022-065546.

Novotný, F., Urbanová, V., Plutnar, J., & Pumera, M. (2019). Preserving fine structure details and dramatically enhancing electron transfer rates in graphene 3D-printed electrodes via thermal annealing: toward nitroaromatic explosives sensing. *ACS Applied Materials & Interfaces Journal, 11*, 35371–35375. https://doi.org/10.1021/ACSAMI.9B06683/ASSET/IMAGES/LARGE/AM9B06683_0002.JPEG.

de Oliveira, F. M., de Melo, E. I., & da Silva, R. A. B. (2020). 3D Pen: A low-cost and portable tool for manufacture of 3D-printed sensors. *Sensors and Actuators B: Chemical, 321*, 128528. https://doi.org/10.1016/J.SNB.2020.128528.

Pal, A., Amreen, K., Dubey, S. K., & Goel, S. (2021). Highly Sensitive and interference-free electrochemical nitrite detection in a 3D printed miniaturized device. *IEEE Transactions on Nanobioscience, 20*, 175–182. https://doi.org/10.1109/TNB.2021.3063730.

Penumakala, P. K., Santo, J., & Thomas, A. (2020). A critical review on the fused deposition modeling of thermoplastic polymer composites. *Composites Part B: Engineering, 201*, 108336. https://doi.org/10.1016/J.COMPOSITESB.2020.108336.

Pohl, P. (2009). Determination of metal content in honey by atomic absorption and emission spectrometries. *TrAC Trends in Analytical Chemistry, 28*, 117–128. https://doi.org/10.1016/J.TRAC.2008.09.015.

Quan, H., Zhang, T., Xu, H., Luo, S., Nie, J., & Zhu, X. (2020). Photo-curing 3D printing technique and its challenges. *Bioactive Materials, 5*, 110–115. https://doi.org/10.1016/J.BIOACTMAT.2019.12.003.

Rahimzadeh, M. R., Rahimzadeh, M. R., Kazemi, S., & Moghadamnia, A. A. (2017). Cadmium toxicity and treatment: An update. *Caspian Journal of Internal Medicine, 8*, 135–145. https://doi.org/10.22088/CJIM.8.3.135.

Rayna, T., & Striukova, L. (2016). From rapid prototyping to home fabrication: How 3D printing is changing business model innovation. *Technological Forecasting and Social Change, 102*, 214–224. https://doi.org/10.1016/J.TECHFORE.2015.07.023.

Rocha, D. P., Foster, C. W., Munoz, R. A. A., Buller, G. A., Keefe, E. M., & Banks, C. E. (2020). Trace manganese detection via differential pulse cathodic stripping voltammetry using disposable electrodes: Additively manufactured nanographite electrochemical sensing platforms. *Analyst, 145*, 3424–3430. https://doi.org/10.1039/D0AN00018C.

Rocha, D. P., Squissato, A. L., da Silva, S. M., Richter, E. M., & Munoz, R. A. A. (2020). Improved electrochemical detection of metals in biological samples using 3D-printed electrode: Chemical/electrochemical treatment exposes carbon-black conductive sites. *Electrochimica Acta, 335*, 135688. https://doi.org/10.1016/J.ELECTACTA.2020.135688.

Rouf, S., Malik, A., Singh, N., Raina, A., Naveed, N., Siddiqui, M. I. H., & Haq, M. I. U. (2022). Additive manufacturing technologies: Industrial and medical applications. *Sustainable Operations and Computers, 3*, 258–274. https://doi.org/10.1016/J.SUSOC.2022.05.001.

Rymansaib, Z., Iravani, P., Emslie, E., Medvidović-Kosanović, M., Sak-Bosnar, M., Verdejo, R., & Marken, F. (2016). All-polystyrene 3D-printed electrochemical device with embedded carbon nanofiber-graphite-polystyrene composite conductor. *Electroanalysis, 28*, 1517–1523. https://doi.org/10.1002/ELAN.201600017.

Shah, S., Shiblee, M. N. I., Rahman, J. M. H., Basher, S., Mir, S. H., Kawakami, M., Furukawa, H., & Khosla, A. (2018). 3D printing of electrically conductive hybrid organic–inorganic composite materials. *Microsystem Technologies, 24*, 4341–4345. https://doi.org/10.1007/S00542-018-3781-X/TABLES/1.

Sharma, A., AlGhamdi, W. S., Faber, H., & Anthopoulos, T. D. (2022). Paper-Based microfluidics devices with integrated nanostructured materials for glucose detection. *Advanced microfluidics based point-of-care diagnostics*, 191–228. CRC Press. https://doi.org/10.1201/9781003033479-8.

Sharma, A., AlGhamdi, W. S., Faber, H., Lin, Y. H., Liu, C. H., Hsu, E. K., Lin, W. Z., Naphade, D., Mandal, S., Heeney, M., & Anthopoulos, T. D. (2023). Non-invasive, ultrasensitive detection of glucose in saliva using metal oxide transistors. *Biosensors & Bioelectronics, 237*, 115448. https://doi.org/10.1016/J.BIOS.2023.115448.

Sharma, A., Angnes, L., Sattarahmady, N., Negahdary, M., & Heli, H. (2023). Electrochemical immunosensors developed for amyloid-beta and tau proteins, leading biomarkers of Alzheimer's disease. *Biosens, 13*, 742. https://doi.org/10.3390/BIOS13070742.

Sharma, A., Faber, H., AlGhamdi, W. S., Naphade, D., Lin, Y. H., Heeney, M., & Anthopoulos, T. D. (2024). Label-free metal-oxide transistor biosensors for metabolite detection in human saliva. *Advancement of Science, 11*. https://doi.org/10.1002/ADVS.202306038.

Sharma, A., Faber, H., Khosla, A., & Anthopoulos, T. D. (2023). 3D printed electrochemical devices for bio-chemical sensing: A review. *Materials Science and Engineering: R: Reports, 156*, 100754. https://doi.org/10.1016/J.MSER.2023.100754.

Sharma, A., & Rai, A. (2022). Fused deposition modelling (FDM) based 3D & 4D Printing: A state of art review. In: *Materials Today: Proceedings, 62*, 367–372. https://doi.org/10.1016/J.MATPR.2022.03.679.

Silva-Neto, H. A., Duarte-Junior, G. F., Rocha, D. S., Bedioui, F., Varenne, A., & Coltro, W. K. T. (2023). Recycling 3D printed residues for the development of disposable paper-based electrochemical sensors. *ACS Applied Materials & Interfaces Journal, 15*. https://doi.org/10.1021/ACSAMI.3C00370/SUPPL_FILE/AM3C00370_SI_001.PDF.

Silva, E. L., Roldan, P.dos S., & Giné, M. F. (2009). Simultaneous preconcentration of copper, zinc, cadmium, and nickel in water samples by cloud point extraction using 4-(2-pyridylazo)-resorcinol and their determination by inductively coupled plasma optic emission spectrometry. *Journal of Hazardous Materials, 171*, 1133–1138. https://doi.org/10.1016/J.JHAZMAT.2009.06.127.

Singh, A., Sharma, A., Verma, R. K., Chopade, R. L., Pandit, P. P., Nagar, V., Aseri, V., Choudhary, S. K., Awasthi, G., Awasthi, K. K., Sankhla, M. S., Singh, A., Sharma, A., Verma, R. K., Chopade, R. L., Pandit, P. P., Nagar, V., Aseri, V., Choudhary, S. K., ... Sankhla, M. S. (2022). Heavy metal contamination of water and their toxic effect on living organisms. *The Toxicity of Environmental Pollutants. Intechopen*. https://doi.org/10.5772/INTECHOPEN.105075.

Soares, D. M., Ren, Z., Mujib, S. B., Mukherjee, S., Real, C. G. M., Anstine, M., Zanin, H., & Singh, G. (2021). Additive manufacturing of electrochemical energy storage systems electrodes. *Advanced Energy and Sustainability Research, 2*, 2000111. https://doi.org/10.1002/AESR.202000111.

Stringer, R. C., Gangopadhyay, S., & Grant, S. A. (2010). Detection of nitroaromatic explosives using a fluorescent-labeled imprinted polymer. *Analytical Chemistry, 82*, 4015–4019. https://doi.org/10.1021/AC902838C.

Sun, X., Wang, Y., & Lei, Y. (2015). Fluorescence based explosive detection: From mechanisms to sensory materials. *Chemical Society Reviews, 44*, 8019–8061. https://doi.org/10.1039/C5CS00496A.

Tan, C., Nasir, M. Z. M., Ambrosi, A., & Pumera, M. (2017). 3D printed electrodes for detection of nitroaromatic explosives and nerve agents. *Analytical Chemistry, 89*, 8995–9001. https://doi.org/10.1021/ACS.ANALCHEM.7B01614/ASSET/IMAGES/MEDIUM/AC-2017-016148_0006.GIF.

Taya, P., Maiti, B., Kumar, V., De, P., & Satapathi, S. (2018). Design of a novel FRET based fluorescent chemosensor and their application for highly sensitive detection of nitroaromatics. *Sensors and Actuators B: Chemical, 255*, 2628–2634. https://doi.org/10.1016/J.SNB.2017.09.073.

Tchounwou, P. B., Yedjou, C. G., Patlolla, A. K., & Sutton, D. J. (2012). Heavy metal toxicity and the environment. *EXS, 101*, 133–164. https://doi.org/10.1007/978-3-7643-8340-4_6/COVER.

Thompson, M. W., & Thompson, M. W. (2022). Regulation of zinc-dependent enzymes by metal carrier proteins. *BioMetals, 352*(35), 187–213. https://doi.org/10.1007/S10534-022-00373-W.

Toal, S. J., & Trogler, W. C. (2006). Polymer sensors for nitroaromatic explosives detection. *Journal of Materials Chemistry, 16*, 2871–2883. https://doi.org/10.1039/B517953J.

Valerga, A. P., Batista, M., Salguero, J., & Girot, F. (2018). Influence of PLA filament conditions on characteristics of FDM parts. *Mater, 11*, 1322. https://doi.org/10.3390/MA11081322.

Walters, J. G., Ahmed, S., Terrero Rodríguez, I. M., & O'Neil, G. D. (2020). Trace Analysis of heavy metals (Cd, Pb, Hg) using native and modified 3D printed graphene/poly(lactic acid) composite electrodes. *Electroanalysis, 32*, 859–866. https://doi.org/10.1002/ELAN.201900658.

Wani, A. L., Ara, A., & Usmani, J. A. (2015). Lead toxicity: A review. *Interdiscip. Toxicol. 8*, 55–64. https://doi.org/10.1515/INTOX-2015-0009.

Wu, Y. S., Osman, A. I., Hosny, M., Elgarahy, A. M., Eltaweil, A. S., Rooney, D. W., Chen, Z., Rahim, N. S., Sekar, M., Gopinath, S. C. B., Mat Rani, N. N. I., Batumalaie, K., & Yap, P. S. (2024). The toxicity of mercury and its chemical compounds: molecular mechanisms and environmental and human health implications: A comprehensive review. *ACS Omega, 9*, 5100–5126. https://doi.org/10.1021/ACSOMEGA.3C07047/ASSET/IMAGES/LARGE/AO3C07047_0008.JPEG.

Yap, C. Y., Chua, C. K., Dong, Z. L., Liu, Z. H., Zhang, D. Q., Loh, L. E., & Sing, S. L. (2015). Review of selective laser melting: Materials and applications. *Applied Physics Reviews, 2*https://doi.org/10.1063/1.4935926/123739.

Yew, Y. T., Ambrosi, A., & Pumera, M. (2016). Nitroaromatic explosives detection using electrochemically exfoliated graphene. *Scientific Reports, 61*(6), 1–11. https://doi.org/10.1038/srep33276.

Section 4

Theoretical modelling and machine learning approaches

Optimization techniques in additive manufacturing of functional polymers and nanocomposites

20

Serafeim Moustakidis, Konstantinos Stergiou, and Patrik Karlsson
AiDEAS OU, Tallinn, Estonia

20.1 Introduction

The development of nanocomposite materials indicates an important progression in the field of material science. Nanostructured materials have reshaped technological innovation in domains such as semiconductors, organic and inorganic materials, energy storage, and biotechnology (Okpala, 2013). On the other hand, the field of polymer chemistry has experienced a substantial change with the implementation of polymerization techniques. These sophisticated methods have facilitated the accurate production of polymers with clearly defined molecular weights, compositions, and structures. Functional polymers have a distinctive capability to fulfill specialized tasks that go beyond the typical structural functions of normal polymers, resulting in a more complex and sophisticated synthesis process (Gauthier et al., 2009).

Multiobjective optimization is crucial in engineering and many other fields, for addressing challenges that require simultaneous optimization for multiple objectives. The objectives often differ in units and typically contradict each other, presenting a complex challenge for optimization processes. The term used to describe this occurrence is "multiobjective optimization problems," which has been a topic of research since the 19th century, primarily originating in the field of economics (Pareto, 1896; Edgeworth, 1881). Over time, multiobjective optimization has become a fundamental aspect of operations research, progressing with the advancement of diverse mathematical programming techniques that are specifically tailored for solving these complex problems (Mardle & Miettinen, 2000). The concept of utilizing evolutionary algorithms for optimization was initially proposed in the late 1960s. However, it was not until the mid-1980s that the first practical use of a multiobjective evolutionary algorithm (MOEA) was developed. This event marks the beginning of what is presently recognized as "evolutionary multiobjective optimization" (EMO), a field that has had substantial expansion and advancement since its early days (Schaffer, 1984).

The significance of optimization in material science, as well as in other engineering fields, lies in its capacity to identify the most optimal and efficient solutions to problems that involve many, frequently conflicting objectives that need to be solved. The capacity to achieve an ideal balance is of paramount significance in the field of material science. This is because the fine-tuning of several properties, such as

Additive Manufacturing of Functional Polymers and Nanocomposites. DOI: https://doi.org/10.1016/B978-0-443-26516-7.00020-1
© 2025 Elsevier Ltd. All rights are reserved, including those for text and data mining, AI training, and similar technologies.

strength, durability, conductivity, and environmental sustainability, is essential for the advancement of novel materials and technologies. Optimization algorithms can be utilized to identify the optimal combination of components in the development of novel alloys or composite materials, hence achieving the desired properties (Jha et al., 2017; Peng et al., 2021). Optimization serves a crucial role in the domain of nanocomposites and functional polymers, (Dasgupta et al., 2017; Mazahery & Shabani, 2012). The aforementioned materials possess underlying complexity, characterized by features that display a high degree of sensitivity to their composition and structure at the nanoscale.

Throughout this chapter, readers will gain valuable insights regarding:

- The underlying concepts and many types of optimization algorithms are important to the field of material science, specifically in the nanocomposites and functional polymer regions.
- How these algorithms are applied specifically to nanocomposites and functional polymers to enhance their properties.
- Empirical instances in which the process of optimization has resulted in notable progressions in the characteristics and utilization of such materials.

As we progress, it is crucial to keep in mind the overriding objective: comprehending and utilizing the potential of optimization in expanding the limits of feasibility within the field of nanocomposites and functional polymers. A comprehensive understanding of this concept is crucial for anybody seeking to actively participate in or make informed decisions within the domain of advanced material development.

20.2 Background

The advancement of nanocomposite materials has revealed a shift from using large-scale reinforcements to utilizing nanoscale ones, which presents various challenges and requirements modifications in methodology. Carbon nanotubes (CNTs) have gained significant recognition as nanofillers in polymer matrices due to their high aspect ratio, low density, and remarkable mechanical, physical, and electrical capabilities (Iijima, 1991; Thostenson et al., 2001). The effective integration of carbon nanotubes (CNTs) into polymer matrices relies on two major factors: interfacial adhesion and dispersion (Fiedler et al., 2006). To achieve optimal material property improvements, it is important to not only include carbon nanotubes (CNTs) as fillers but also to meticulously address these crucial parameters. The distribution of carbon nanotubes (CNTs) in the matrix is a significant barrier in the development of nanotube-reinforced polymers. This issue is attributed to the nanoparticles' vast surface area which is several orders of magnitude greater than that of typical fillers. The extensive surface area plays a crucial role in distributing strain, but it also encourages the creation of aggregates (Frankland et al., 2002). Hence, to effectively harness the capabilities of carbon nanotubes (CNTs) in polymers, it is necessary to distribute them throughout the matrix, scatter any clusters, and ensure adequate adhesion to the polymer. Nanocomposites encounter a more complicated range of challenges than typical composites which are influenced by the mechanical, electrical, and optical properties of the fiber, matrix, and interface. These encompass not only the characteristics of the individual components but also the complex interactions controlled by the dynamics between particles and the matrix, as well as the characteristics of the production procedure. The optimal performance of nanocomposites depends on a comprehensive understanding and monitoring of these complex interactions. The objective of this synthesis method is to combine the unique characteristics of various constituent elements into a unified, versatile composite. Efforts have been increasingly focused on the development of conducting polymer nanocomposites, such as Polyaniline (PANI), due to their unique characteristics and ability to serve multiple applications in different domains. The synthesis of PANI/ ferrite nanocomposites involves the in situ polymerization of aniline in the presence of particular ferrite nanoparticles, such as $Zn_{0.6}$, $Cu_{0.4}$, $Cr_{0.5}$, $Fe_{1.5}$, and O_4 to produce these advanced materials (Jiang et al., 2008).

Functional polymers, a breakthrough domain of materials in modern material science, possess the remarkable ability to carry out specified functions, differentiating them from ordinary polymers. The functionalities of these structures are often a consequence of their distinct molecular and supramolecular structures, which are carefully designed to serve specific objectives in different applications (Wegner, 2000). The examination of functional polymers frequently involves the analysis of their molecular dynamics, which is determined by both the time scale and geometry. The dynamics enclose both linear and rotational mobility inside the polymeric chains, which are essential for interpreting the material's response under varying circumstances. Methods such as Nuclear Magnetic Resonance (NMR) spectroscopy are crucial in these investigations, providing valuable information on the molecular mechanisms and their dependence on different factors (Dionísio et al., 2004). Moreover, the presence of small-sized inclusions at the molecular level affects the mobility and arrangement of polymer chains near the interface between the polymer and the reinforcement. The interactions occurring at these interfaces have the potential to cause substantial variations in the properties of the material, which may include reductions in molecular mobility. Nevertheless, the precise origins and structural implications of these modifications continue to be subjects of ongoing research and have not been thoroughly grasped. The increasing fascination with these phenomena has resulted in the implementation of molecular modeling approaches to examine the arrangement and behavior at the interface between nanoparticles and polymer matrices. This approach provides a fresh understanding of the fundamental principles that regulate the behavior of functional polymers in these dynamic systems (Brown et al., 2003).

Optimization in materials science is a notable progress in the field, providing multiple advantages. Evolutionary algorithms can offer significant benefits in optimizing materials' processes, including improving material quality, reducing costs, and decreasing production time to increase productivity (Köksal et al., 2011). The implementation of these approaches in materials science has resulted in enhanced productivity and efficiency across multiple areas. These algorithms offer a methodical technique for investigating an extensive range of designs, identifying the most optimal solutions that may not be apparent through conventional empirical

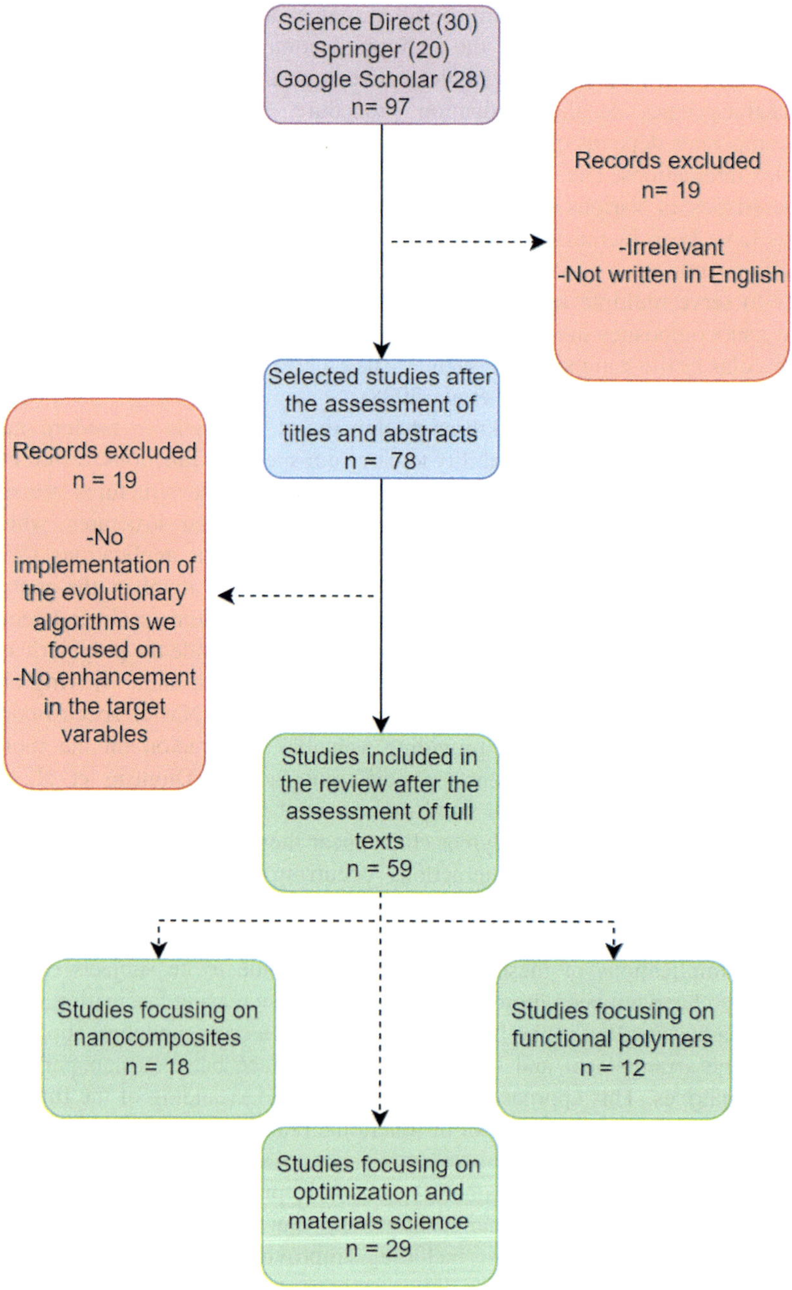

Figure. 20.1 Inclusion results based on the eligibility criteria.

methods. They can serve a role in improving the accuracy and efficiency of predictions and provide the examination of sophisticated material behaviors and interactions, resulting in improved and efficient decision-making throughout the design and production phases (Stergiou et al., 2023). Applications of evolutionary algorithms in material science can be found in the optimization of open-air perovskite solar cell manufacturing (Liu et al., 2021), carbon steels (Diao et al., 2022), zeolites and porous crystalline materials (Gandhi & Hasan, 2022), thermoplastic composites (Islam et al., 2022), and biomaterials (Suwardi et al., 2022).

Computational approaches in materials science can be found also in the nanocomposite and functional polymers field. Particularly, the dental restorative nanocomposite Filtek Z350k (Karimzadeh et al., 2019), with particle sizes ranging from 50 nm to 4 μm, was subjected to a series of nanoindentation tests to extract its hardness and elastic modulus properties. In this study experimental techniques combined with computational modeling were utilized to improve the material's mechanical behavior. The approach involved developing several finite element (FE) models to simulate the elastic-plastic deformation of the nanocomposite. The tests provided data with considerable dispersion, reflecting the local behavior of the material. The computational models aimed to link the exported data to the material's mechanical behavior and the global response of the nanocomposite, thus providing a more holistic understanding of its characteristics. Moreover, Monte Carlo multi-objective optimization algorithms (MCMO) have been utilized for the optimization of functional polymers (Mannodi-Kanakkithodi et al., 2016). Focusing on optimizing both the bandgap and dielectric constant of such polymers a dataset generated from Density Functional Theory (DFT) calculations and by the implementation of two Monte Carlo algorithms the Pareto front of the chemical space could be identified. The algorithms aimed to identify polymers that are Pareto optimal, which means that they offer the best trade-off between the objectives of bandgap and dielectric constant optimization.

20.3 Methodology

20.3.1 Reporting

The optimization chapter process, including literature search, research selection, and data extraction, was carried out independently by two authors (K.S. and S.M.). Any discrepancies were settled by a third author (P.K.).

20.3.2 Literature search

The present chapter was carried out by employing three distinct bibliographic sources (Elsevier, Springer, and Google Scholar). A supplementary manual search was conducted to identify publications referenced in the retrieved papers, as well as articles that referenced the obtained papers. The title, abstract, and conclusion of the papers were used to assess their overall eligibility. We thoroughly examined the complete texts of the initial qualifiers to determine if they met the criteria for inclusion.

20.3.3 Eligibility criteria

The inclusion criteria were as follows: (i) papers focusing on material science; (ii) studies on optimization in material science; (iii) studies on optimization in nanocomposites; (iv) studies on optimization in functional polymers; (v) ML-driven optimization approaches in material science; (vi) ML-driven optimization approaches in nanocomposites and functional polymers. Over 90 scientific papers were found in the search results, which were narrowed down according to the exclusion criteria: (i) irrelevant studies and (ii) articles not written in English.

20.3.4 Data extraction

A Microsoft Excel spreadsheet was utilized to store the extracted data through a standardized table, whereby the following information was included for each article: first author, publication year, database type, description of data, and evolutionary algorithms.

20.3.5 Analysis

The capacity of the proposed optimization algorithms was assessed using multiple assessment criteria. The findings and results were extracted, summarized, and discussed in the subsequent parts. Fig. 20.1 displays the findings of the study, which were determined according to the specified eligibility criteria. Out of the 97 papers examined, 78 of them met the first exclusion criteria, which excluded irrelevant studies, and non-English papers. Out of the total 78, only 59 met the second exclusion criteria, which involved excluding studies that did not specifically focus on evolutionary algorithms and those that did not show any improvement in the target variables. These excluded studies were not considered in the review. Out of the 59 research that satisfied both the initial and subsequent exclusion criteria, 18 were ultimately utilized for the enhancement of nanocomposites, 12 for the enhancement of functional polymers, and 29 for the enhancement of materials science in general. The study aims to conduct a comprehensive analysis of the most efficient optimization applications in the field of materials design, with a specific focus on nanocomposites and functional polymers.

20.4 Optimization algorithms: an overview

Within this section, we provide a concise overview of the prevailing evolutionary algorithms that are frequently employed. Evolutionary algorithms can be implemented using Python, C++, and R programming languages. Below we present the evolutionary algorithms most frequently utilized in the literature.

Genetic algorithm (GA) (Kumar et al., 2010): The GA, a type of evolutionary algorithm, employs techniques influenced by genetics and natural selection to tackle challenges related to optimization. By relying on bioinspired operators, they are commonly used to generate improved solutions to optimization. Genetic algorithms employ evolutionary operators such as mutation, crossover, and selection to iteratively generate a population of possible solutions to a problem across multiple generations. The goal is to select the most competent individuals from each generation to uncover the optimal solution. An individual's level of fitness is determined by their proficiency at resolving problems. Individuals with high strength are more likely to be selected for reproduction.

Bayesian Optimization (BO) (Frazier, 2018): BO is an optimization method that utilizes Bayesian inference to identify the most promising solution to a problem. The objective of this technique is to resolve the problem by identifying the most advantageous option. To achieve success, it is necessary to create a probabilistic model of the objective function and utilize this model to search for the optimal solution to the problem. Only at that point will it acquire the capacity to operate efficiently. This technique is employed to address a diverse range of optimization problems, including hyperparameter tuning and global optimization, among other frequently encountered instances. Specifically, it is utilized to create an alternative for the objective and evaluate the level of uncertainty linked to it. The sampling location is selected by utilizing an acquisition function that has been built based on the surrogate. At a certain candidate point x, the Bayesian posterior probability distribution, generated by the statistical model, which is consistently a Gaussian process, offers an explanation of the possible values for $f(x)$. The posterior distribution is updated whenever a new observation of f is made at a different location. The acquisition function aims to calculate the value acquired by evaluating the objective function at a new location, using the posterior distribution over f.

Particle swarm optimization (PSO) (Poli et al., 2007): PSO is an evolutionary algorithm that emulates the motion of particles inside a search space to discover the most effective solutions for optimization problems. Particles explore the search region and adapt their positions according to their observations and the observations of nearby particles. Through incremental progression, the particles will ultimately reach the intended result. After all particles have been relocated, the process proceeds to the next iteration. The swarm will ultimately approach the maximum of the fitness function. Particle swarm optimization can be used to address many optimization problems such as function optimization, feature selection, and clustering. Robotics, image processing, and data mining are among the various fields where it is applied.

The aforementioned algorithms are the most frequently implemented in optimization problems each with its own set of advantages and disadvantages. GA is well-suited for problems with a large and complex search space, as it can explore a wide range of solutions, and is robust against the trapping in the local optima, making it suitable for multimodal optimization problems. It can be easily implemented in parallel computing environments, enhancing its efficiency, and it can be adapted to different environments, useful in dynamic problems. On the other hand, GA can be computationally expensive due to the need for multiple generations and a large population size. A risk exists considering the premature convergence where the algorithm may converge to suboptimal solutions, while also the performance of the GA is highly dependent on the correct setting of parameters like mutation rate, crossover rate, and population size. PSO is simpler to implement and also requires fewer parameters to adjust. It is efficient for continuous problems and converges faster to a solution compared to other methods like GA. It doesn't require gradient information, which can be considered beneficial for problems where such information is not available. The disadvantages of the PSO are the risk of getting trapped into the local optima, especially in complex and multimodal spaces, and also its performance is heavily influenced by initial parameter settings. Finally, the BO is highly efficient for problems where evaluations of the objective function are computationally expensive, such as hyperparameter tuning of ML models. It can incorporate prior knowledge to improve the efficiency of the search process, and can provide also a well-balanced trade-off between the search space and exploiting information. BO is less effective in optimization problems with a very high number of dimensions, and can become computationally expensive when dealing with large datasets.

20.5 Impact of optimization in material science

The field of material science is constantly evolving in terms of optimization, specifically in the development of new materials and the improvement of existing ones. In this domain, novel methodologies may have an important impact on the final properties and practical applications of materials. "Materials by design" requires the development of methods that enable the modification of material microstructures to meet specific design criteria. Kulkarni et al. (2004) tried to optimize the microstructure of an alloy. The study focused on optimizing key properties (strength, ductility, and temperature) by focusing on the correlations of between the microstructure and these macroscopic properties. A GA was then utilized to synthesize the optimal microstructure, fulfilling the defined design requirements. For the yield strength optimization, the target was the 900 MPa. The optimization requirements were met for radius (*r*), volume fraction (*f*), and solubility of Mg (*c*) with the values of 14 nm, 13.34%, and 1.63%, respectively. The study demonstrated the variation in strength across several generations, showcasing the GA capacity to systematically investigate the range of parameters and ultimately reach an optimal solution. The optimal solution was found within the designated search space, rather than on the boundary, suggesting successful navigation and optimization by the GA. In their paper, Kou et al. (2012) provide insightful case studies on the application of optimization techniques, specifically PSO and GA in the field of functionally graded materials (FGMs). The study focuses on the optimization of thermal stresses in a 10-layer FGM composed of Ni and Al_2O_3. From the results, the best evolutionary algorithm was the PSO which outperformed GA and classical-gradient approaches, providing von Mises stress values in the tolerance range of 2.28%. Finally, Zhang et al. (2020) proposed the implementation of BO in two distinct material design cases, by utilizing the Latent Variable Expected Improvement (LV-EI) method. The implementation involves designing a high-performance light-absorbing quasi-random solar cell and conducting a combinational search for Hybrid Organic-Inorganic Perovskites (HOIPs). Both LV-EI and Monte Carlo Expected Improvement (MC-EI) methods were used, starting with 30 random initial samples and ending after 100 iterations, repeated over 20 replicates. The LV-EI method consistently outperformed

MC-EI in terms of convergence speed and solutions' quality. The optimal design variables found were a=0.0049 nm^{-1}, b=0.0035 nm^{-1}, ρ=0.875, and t_1=100.95 nm. A combined scenario where both LV-EI and MC-EI effectively identified the optimal solution but LV-EI showed superior performance with smaller variance and quicker convergence. Specifically, LV-EI found the best solution in 28 out of 30 replicates, while MC-EI found it in only 22 out of 30 replicates. The best solution identified has a binding energy of around –41.3, which was the true optimal solution determined via exhaustive simulation.

Genetic algorithms (GA), particle swarm optimization (PSO), and Bayesian optimization (BO) can be considered very important in developing material science due to their capabilities and optimization methodologies, as demonstrated in the discussed studies. Genetic Algorithms have proven to be effective in addressing complex, multivariable challenges in the field of material science. Their proficiency in efficiently exploring large and complex search spaces, as demonstrated in the optimization of alloy microstructures, emphasizes their importance in creating materials with precise properties. The evolutionary approach of GA enables a comprehensive examination of possible solutions, making it especially valuable for scenarios involving "materials by design." Particle swarm optimization, known for its emphasis on simplicity and quick convergence, has demonstrated its effectiveness in optimizing machining parameters for different materials such as brass, aluminum, and steel. The efficacy of PSO in these instances demonstrates its suitability in optimizing processes and its capacity to improve manufacturing processes to attain targeted material qualities with enhanced efficiency and accuracy. Bayesian Optimization is particularly effective in situations when experimental or computational evaluations are costly or restricted. BO's strength is in its capacity to deliver high-quality solutions with a reduced number of iterations, as demonstrated in the design of light-absorbing solar cells and the combinatorial search for Hybrid Organic-Inorganic Perovskites. BO is particularly beneficial in high-throughput screening and material informatics due to its ability to greatly reduce the time and resources needed for material discovery and optimization. Finally, GA, PSO, and BO each make distinct contributions to the area of material science, providing powerful tools for optimizing material properties and processes. Their unique capabilities in addressing complex optimization problems, optimizing processes, and conducting efficient searches in settings with limited evaluation underscore the increasing significance of computational approaches in the progress of material science and engineering.

20.6 Optimization in nanocomposites

20.6.1 Genetic algorithms (GA)

The precise control of the composite's microstructure is essential for achieving specified functional properties in nanocomposites, which is critical for "materials by design" techniques. GA is an optimization approach that has various implementations in the field of nanocomposites. In this subsection, we present an amount of GA approaches. Jafari et al. (2017) focused on enhancing the hardness of Al6061 nanocomposites reinforced with multiwall carbon nanotubes (MWCNTs). The study aimed to find the optimal conditions for achieving a maximum hardness in Al6061-MWCNT nanocomposite through a GA implementation. The GA was utilized to search for the optimal combination of parameters that would yield the optimum hardness. The parameters were sintering temperature, time, compact pressure, milling time, and via speed. The GA algorithm found that the combination of sintering temperature equal to 346°C, time equal to 0.33 hours, compact pressure of 248.82 MPA, milling time of 19.66 hours, and via speed of 310.5 rpm would provide the optimum hardness which was 87.5 micro-Vickers in the composite with 0.53 wt% CNT. The most important parameter was time and the optimal values compared with the experimental data confirmed the effectiveness of the GA-ANN model in finding optimal conditions for the nanocomposite. The study revealed the GA approach is an effective tool for optimization of mechanical alloy parameters to achieve maximum values of specific parameters, while also emphasizing the importance of the time parameter in the process.

Moghri et al. (2015) proposed a GA approach to optimize the tensile modulus and strength of the PA-6/NC nanocomposites. The authors focused on determining the optimal material and processing parameters that maximized the tensile modulus and strength by utilizing a combination of Response Surface Methodology (RSM) and GA. The RSM was utilized for modeling the mechanical properties of the nanocomposites, while the GA was implemented to optimize the RSM models and find the best combination of parameters including nano clay content, melt temperature, feeding rate, and screw speed. The maximum tensile modulus achieved from the GA was 3985 MPa, for tensile strength 93.5 MPa, while optimal processing parameters included a melt temperature of 235°C, screw speed of 577 rpm, and feeding rate of 7 rpm. To validate the significance of the models and the effect of the parameters on the tensile properties Analysis of Variance (ANOVA) has been conducted, and demonstrated the effectiveness of the PSM and GA combination approach. Dadrasi et al. (2019) utilized a central composite design and GA approach to model and optimize the mechanical performance of HA/GO/epoxy resin nanocomposites. The optimized mechanical properties were the flexural strength with an optimal value of 21.54 MPa under 5.17 wt% for HA and 0.38 wt% for GO conditions, flexural modulus with an optimal value of 25.7 GPa under 2.73 wt% for HA and 0.195 wt% for GO, compression strength with an optimal value of 23.95 MPa and 7 wt% for HA and 0.289 wt% for GO, and compression modulus with an optimal value of 690.5 MPa under 6.89 wt% for HA and a low amount of GO.

Moreover, Rabothata et al. (2021) utilized the Non-dominated Sorting Genetic Algorithm (NSGA-II) method to optimize both the design parameters and the mechanical properties of polymer-based nanocomposites. The main target of the study was to maximize the elastic modulus and the tensile strength. Specifically, interphase thickness, diameter of nanofillers, nanofiller loading, interphase modulus, interfacial shear stress, and nanofiller orientation were the considered features. From the results, it is shown that the optimization model was able to find the optimal solutions for the design variables, as the results were in agreement with the public dataset validating the effectiveness of the NSGA-II method. The parameters utilized to optimize the mechanical properties were the thickness of interphase (ti), the diameter of nanofillers (*t*), nanofiller loading (*hf*), interphase modulus (*Ei*), interfacial shear stress (*s*), and nanofiller orientation factor (*g*). The optimal values for normalized elastic modulus and normalized tensile strength were found in the ranges of 1.54 to 1.89 and 0.9, respectively.

Arefi-Oskoui et al. (2017) proposed a hybrid implementation that combined Neural Networks (NN) and GA for modeling and optimization of NLDH/PVDF ultrafiltration nanocomposite membrane. From the results it is shown that the PVDF concentration varied from 16 to 24 wt% with an optimal value of 1 wt%, the PVP 2900 concentration varied from 0 to 6 wt%, with an optimal value of 1 wt%, and the Mg-Al NLDH concentration varied from 0 to 1 wt% with an optimal value of 0.54 wt %. These optimal values were determined by utilizing the GA and the performance of the nanocomposite membrane has been experimentally validated, confirming that the NN-GA was able to model and optimize the nanocomposites. Finally, Rouhi Broujeni et al. (2021) utilized again the hybrid NN-GA implementation to optimize the input parameters for the maximization of the thorium (IV) ions (Th^{4+}) adsorption. The optimal conditions determined for the highest adsorption efficiency were pH equal to 6.5, weight of adsorbent equal to 0.2 g/L, and contact time equal to 63 minutes. Under these conditions, the NN predicted the concentration at 2 mg/L where maximum adsorption was indicated. The adsorption efficiency that was achieved under these optimized conditions was approximately 98.2% ± 0.1%, experimentally validated and matching the predictions of the NN-GA approach.

20.6.2 Particle swarm optimization (PSO)

In this subsection, we are going to present some of the PSO implementations regarding the nanocomposite field. Dehestani et al. (2021) presented a hybrid optimization approach based on NN and PSO to determine the microhardness of Ni/Al203 nanocomposite coatings. From the results, it can be observed that there are different ranges of microhardness values under different plating times. Specifically, with a plating time of 0.50 hours, the microhardness values ranged from 130 HV to 660 HV. For a plating time of 0.75 hours, the microhardness values were between 92 HV and 770 HV. When the plating time was set to 1.00 hour the microhardness values varied from 92 HV to 704 HV. The results of the study indicate that the coatings were significantly influenced by plating time among other factors, while confirming the NN-PSO approach as successful in terms of prediction and optimization of the microhardness. Nasr et al. (2020) evaluated various experimental conditions to examine the impact of machining parameters and graphene nanoplatelet (GNP) reinforcement ratio on the performance of Ti6Al4V nanocomposites. Five different conditions were investigated and from the results, the impact of varying cutting speed (V), feed rate (F), and GNP reinforcement ratio (R) on the machining characteristics of the nanocomposite was confirmed, highlighting the complexity of the process and the importance of optimizing these specific parameters for improving the performance. For all out of five conditions, the optimal value for V was equal to 75 m/minute. On the other hand, for two out of five conditions, the optimal value for the *F* parameter was found 210 mm/minute, 150 mm/minute for the other two conditions, and 90 mm/minute for one condition out of five. The optimal value for the *R* parameter in two conditions out of five was equal to 0.6 wt%, 1.2 wt% on the other two, and 0 wt% for one out of the five conditions.

Optimization approaches can be utilized also to improve the performance of ML algorithms that focus on the prediction of a particular property. For example, Liu et al. (2021) utilized PSO to improve the performance of a NN model for predicting the thermal conductivity of polymer nanocomposites. The performance of the ANN model was evaluated in terms of coefficient of determination (R^2) and root mean square error (RMSE) for both training and testing sets. From the implementation of PSO to enhance the ML model specific performance metrics extracted. For the ANN with one hidden layer, the R^2 coefficient was equal to 0.9134 and the RMSE metric was equal to 0.1939, in the training phase. In the testing phase, the R^2 was equal to 0.9017 and the RMSE was equal to 0.5338. For the ANN with five hidden layers the R^2 and RMSE were equal to 0.9669 and 0.1204, respectively, in the training phase, while in the testing phase, R^2 was equal to 0.9419 and RMSE equal to 0.2634. The PSO examined also ANNs with ten, fifteen, and twenty hidden layers, and the best performer was the model with the twenty hidden layers which achieved R^2 and RMSE equal to 0.9999 and 0.0061 for the training phase. In the testing phase, the R^2 coefficient and RMSE were equal to 0.9558 and 0.0514, respectively. In this study, a different employment of the PSO optimization algorithm has been conducted focusing on improving the implemented ML model rather than enhancing the properties of a nanocomposite.

Also, Tang et al. (2006) utilized the PSO approach to enhance the polymerization efficiency of nanocomposite thin films. The study explored several scenarios with varying distances and times, along with the corresponding polymerization efficiencies. For example, when the distance was set at 10 mm and time at 32.7 seconds, the desired polymerization efficiency was 79.9% and the varying result was 86.9%. In another scenario, for a distance equal to 16.9 mm and a time of 33.1 seconds, the PSO desired efficiency was 90.4% and the varying result showed an efficiency of 95.2%. The results showed the effectiveness of the approach in the optimization of the laser ablation process for creating matrix nanocomposite thin films. Another application of the PSO approach for the optimization of nanocomposites can be found in the following study. Tofigh et al. (2015) utilized the PSO algorithm combined with the Adaptive Neuro-Fuzzy Inference System (ANFIS) to enhance the processing parameters for AI matrix nanocomposites reinforced with Al_2O_3 nanoparticulates. The objective function was calculated by ANFIS and minimized using PSO. The final optimized process parameters obtained through this process were volume percentage equal to 1.8%, average particle size equal to 98.93 nm, casting temperature equal to 612.3°C, porosity equal to 2.3%, yield strength equal to 191.4 MPa, and tensile strength equal to 278.24 MPa.

20.6.3 Bayesian optimization (BO)

In the field of nanocomposites, BO approaches can be frequently found for the optimization of different properties. Iyer et al. (2020) presented a design framework for polymer nanocomposites by utilizing a mixed-variable BO approach. The method combined experimental data with computational simulations for microstructure characterization and property prediction. A key component of the framework is the Latent Variable Gaussian Process (LVGP), which allows for mixed-variable optimization, including both qualitative and quantitative material design variables. The optimization focused on electrical insulation properties, specifically maximizing dielectric breakdown strength while minimizing dielectric permittivity and loss. The approach confirmed the integration of sophisticated computational techniques for material design, offering insights into the complex interplay of composition, microstructure, and properties in nanocomposites. Liu et al. (2022) focused on optimizing the fused filament fabrication (FFF) process parameters to enhance surface quality in ABS/graphene nanocomposite manufacturing. Considering the key findings of this study, a nonparametric Bayesian framework was conducted by integrating Gaussian Process Regression (GPR) and BO to predict and optimize process parameters. Also, experimental results showed significant improvement in surface roughness and mechanical properties (like tensile strength) of the nanocomposites when optimized process parameters were utilized. Finally, the optimization process led to the discovery of optimal conditions that increased the ultimate tensile strength of the parts by 25–50% compared with parts made with only ABS. The research illustrated the effectiveness of data-centric approaches, based on BO, in optimizing processes for advanced materials.

Guan et al. (2020) employed BO to efficiently design nanocomposite solar absorbers. The optimization process focused on finding the ideal layer thickness for a W-SiC nanocomposite-based solar absorber. Over 20 rounds with different random seeds were conducted during the optimization process, and the goal was to maximize the spectrally averaged absorptance. All 20 rounds of optimization converged to the same optimal design by evaluating less than 130 structures out of a possible 64,000. That indicated that only 0.20% of the candidate structures were needed to identify the optimal structure, which illustrated the efficiency of BO. The optimal design consisted of a W-SiC nanocomposite layer, a middle SiC layer, and a bottom W layer with respective thicknesses of 7.69 nm, 41.03 nm, and 100 nm. This configuration yielded the highest spectrally averaged absorptance. Etefagh and Razfar (2023) proposed the BO approach to improve the printing parameters of 3D-bioprinted polycaprolactone/magnesium oxide nanocomposite scaffold and to achieve enhanced mechanical and osteoconductivity properties, printability, and print resolution. The created model examined and accelerated the optimization of printing parameters, such as air pressure, printing speed, and nozzle temperature, about their impact on printability and print resolution. Although the search areas for printing parameters were extensive, the BO model significantly decreased the number of experiments to only eleven iterations for every target width. The model-predicted values and actual values exhibited a strong agreement, with a width of 91%. Additionally, this model can be utilized to determine the optimal process parameters for printing filaments with varying widths to create 3D scaffolds with gradient properties.

Pyrialakos et al. (2021) introduced a framework that combined Bayesian inference combined with artificial NNs (ANN) and FEM for multiscale modeling, to identify and update microscale properties of carbon nanotube (CNT)/polymer interfaces in nanocomposites. The approach efficiently addressed the computational challenges and uncertainties in characterizing nanocomposites, by enabling more accurate prediction and optimization of their mechanical properties. Finally, Demirbay et al. (2020) examined the combination of BO and NN to predict the electrical conductivity of polystyrene/multiwalled carbon nanotube nanocomposite film coatings. They investigated various parameters such as surfactant and initiator concentrations, molecular weights, and particle sizes of PS latex along with MWCNT concentrations. The NN model was trained through a Bayesian algorithm with a target to predict accurately the electrical conductivity of the nanocomposites. The performance of the model was assessed using various statistical methods, illustrating its effectiveness in predicting the electrical conductivity of the nanocomposite film coatings.

20.7 Optimization in functional polymers

20.7.1 Genetic algorithms (GA)

Applications of the aforementioned optimization algorithms can also be found in the field of functional polymers. Silva and Biscaia (2003) applied GA for optimization of a batch polymerization process of free-radical polymerization reactors. The focus was on two objective functions and the initiator residue values obtained ranged from 2.9×10^{-8} to 5.5×10^{-6}, indicating variations in the concentration of initiator residue in the polymer. Considering the monomer conversion, the percentages achieved varied from 55.2% to 70.0%, demonstrating the effectiveness of the optimization process in enhancing the conversion rate of the monomer. In the study the optimal temperatures for the polymerization process were consistent, mostly around 376.4°C, confirming the stability of the parameter in the optimization process. Mohammadi et al. (2020) utilized a GA for optimizing the process parameters for the development of thin nanofibers of polysulfone polyelectrolytes through electrospinning. The GA was implemented to find the optimal combination of polymer solution concentration, voltage, feed rate, and needle-to-collector distance that yields the minimum diameter for the nanofibers. The optimization process involved evaluating a range of values for each parameter, assessing their impact on fiber diameter, and iteratively refining the parameter values by utilizing the principles of natural selection and genetics. From the results, it is shown that the GA successfully identified a set of parameters that produced nanofibers with an average diameter of approximately 97.5 nm. The results provided a significant improvement over the fiber diameters obtained through traditional methods. Moreover, Giro et al. (2002) in their paper "Designing Conducting Polymers Using Genetic Algorithms" explored the implementation of GAs to optimize the composition of conducting polymer alloys. The study focused on polyanilines and aimed to find the optimal relative concentration for disordered alloys exhibiting metallic properties. The GA identified the optimal ternary alloy composition as A77 B5 C18. It achieved convergence with 19 generations across various fitness function definitions. The GA's solutions were validated by comparing them with systematic searches in simpler cases, showing consistency and reliability.

20.7.2 Particle swarm optimization (PSO)

Another type of functional polymers are conducting copolymers, which are designed to exhibit electrical conductivity. Their electrical properties are a direct result of their specific molecular structure and functional groups. Thakral and Bakhshi (2014) presented a comprehensive study on designing conducting polymers. The authors utilized PSO to explore optimal compositions of copolymers based on thiophene, pyrrole, and furan, including their benzene-substituted analogs. The paper detailed how the PSO algorithm identified compositions with desirable electronic properties, such as low band gaps and high intrinsic conductivity. For example, in one system, the composition was optimized at P19 Q81, and in another, at M99 N1. Proton Exchange Membranes (PEMs) are usually designed to conduct protons while acting as an electronic insulator and a barrier to gases such as hydrogen and oxygen. This ability lets their integration in the operation of Proton Exchange Membrane Fuel Cells (PEMFCs), where they facilitate the electrochemical reaction by allowing protons to pass through while separating the fuel from the oxidant. Özdemir (2021) explored the use of the Cost Effective PSO (CEPSO) algorithm for optimizing Proton Exchange Membrane Fuel Cell (PEMFC) parameters. The research included several case studies where the optimization algorithm was utilized to optimize different PEMFC stacks. The algorithm was able to extract better global values in fewer iterations compared to reference studies, and this specific implementation led to a reduction in the total error of the different stacks. Nam and Ahn (2012) introduced an advanced approach for modeling Ionic Polymer Metal Composite (IPMC) actuators. The approach combined the Preisach operator and a fuzzy nonlinear autoregressive exogenous (NARX) model that was optimized through PSO. The implementation of the optimization algorithm provided improved performance in the identification of the system's dynamics and hysteresis behaviors.

20.7.3 Bayesian optimization (BO)

In the field of functional polymers, pyridinium salts possess unique properties due to the presence of a positively charged pyridinium ion, which makes them useful in a variety of applications. Dunlap et al. (2023) utilized Efficient Dynamic Bayesian Optimization (EDBO) for the optimization of bromide synthesis under continuous flow conditions. Specifically, they focused on maximizing yield and production rate, forming a well-defined Pareto front. The algorithm achieved high yield and production rates greater than 80% and 1 g/h, respectively. From the results, the optimal condition included a temperature equal to 138°C, a 21-minute residence time, and a 0.66 mole fraction of pyridine, yielding an 85.86% reaction yield and 0.90 g/h production rate. Karasuyama et al. (2020) proposed a computational framework to enhance dynamic conductivity and static stability. They combined theoretical calculations and Bayesian multiobjective optimization. They explored the diffusion of oxygen and lithium in different materials. The framework utilized the Pareto hyper-volume criterion to efficiently identify materials with high dynamic conductivity and static stability. For example, in the oxygen diffusion, the approach achieved the optimal value near the 20th iteration, while the standard sampling achieved the optimal solution still far from the 100th iteration. Also, for the lithium diffusion, the approach was still more efficient in identifying the Pareto optimal points compared to random sampling. Yamawaki et al. (2021) developed predictive models for bio-based polymers. In their study, BO employed combined with random forests (RF) and XGBoost to predict the decomposition degree of polylactic acid (PLA). The research emphasized the importance of moisture content in the compost environment as a critical factor for degradation. The ML models achieved an R^2 of 0.78 for RF and 0.73 for 1 H wide-line NMR spectra. The aforementioned optimization approaches demonstrate the efficacy of evolutionary algorithms for improving functional polymers.

20.8 Conclusion

As we wrap up this chapter on optimization in the domain of nanocomposites and functional polymers, we must consider the important implications and future possibilities of this topic. The exploration of material science, guided by the sophisticated techniques and practical applications of evolutionary algorithms, reveals a promising future full of potential and groundbreaking advancements. The field of nanocomposites, as explained in the previous sections, demonstrates the significant impact of optimization strategies. The involvement of genetic algorithms (GA), particle swarm optimization (PSO), and Bayesian optimization (BO) has proved crucial in achieving precise control over microstructures and improving mechanical properties. The process of optimization, as demonstrated in research conducted by Jafari et al. (2017) and Moghri et al. (2015), is not just a quest for material mastery but also an integrated combination of scientific principles and computational methods. The algorithms integrate precision into material science by enabling the optimization of parameters such as sintering temperature, compact pressure, and milling duration to obtain particular material qualities, such as maximum hardness or tensile modulus. The incorporation of these algorithms into the domain of nanocomposites signifies an important step toward a future where materials are not only found but also cautiously engineered. Within the domain of functional polymers, the process of optimization presents itself with similar, if not better interest and potential. The utilization of evolutionary algorithms in optimizing the batch polymerization process, as investigated by Silva and Biscaia (2003), or in the advancement of conducting polymers as observed in the research of Giro et al. (2002), highlights a crucial transition in the field of material science. This transition comprises not just modifications in production techniques but also a fundamental shift in the underlying framework of material design. Optimization goes beyond improving current materials and enters a domain where the fundamental properties of materials are tailored to fulfill particular requirements. The case studies of Pyridinium salts and their synthesis optimization using EDBO (Dunlap et al., 2023) or the dynamic conductivity enhancement in materials (Karasuyama et al., 2020) provide additional evidence of the adaptability and essential nature of these algorithms in the field of material science.

The incorporation of evolutionary algorithms in material science indicates the beginning of a novel era known as "Materials by Design." This approach surpasses conventional material development, suggesting a future where materials are designed with unparalleled precision and performance. The combination of advanced computing capacity and scientific investigation enables the development of materials that are not only more efficient and effective but also more compatible with the specific requirements of modern technology and sustainability. The investigation of these algorithms within the field of nanocomposites and functional polymers is not only an academic action but rather a significant step toward a future in which the limits of material capabilities are consistently reevaluated. The potential of these materials in several fields, ranging from energy to biotechnology, is vast and largely unexplored. The influence of these optimization strategies will increase significantly as they become more closely integrated with upcoming technologies such as artificial intelligence (AI) and machine learning (ML).

In conclusion, the goal of optimization in material science is a path that includes undiscovered and breakthrough options. It extends the boundaries of conventional material qualities and explores new possibilities in the development and use of materials. The future of material science, driven by evolutionary algorithms, is characterized by a diminishing distinction between what is possible and what is impossible, leading to groundbreaking inventions of great significance.

References

Arefi-Oskoui, S., Khataee, A., & Vatanpour, V. (2017). Modeling and optimization of NLDH/PVDF ultrafiltration nanocomposite membrane using artificial neural network-genetic algorithm hybrid. *ACS Combinatorial Science, 19*(7), 464–477. https://doi.org/10.1021/acscombsci.7b00046. http://pubs.acs.org/journal/jcchff.

Brown, D., Mélé, P., Marceau, S., & Albérola, N. D. (2003). A molecular dynamics study of a model nanoparticle embedded in a polymer matrix. *Macromolecules, 36*(4), 1395–1406. https://doi.org/10.1021/ma020951s.

Dadrasi, A., Fooladpanjeh, S., & Alavi Gharahbagh, A. (2019). Interactions between HA/GO/epoxy resin nanocomposites: Optimization, modeling and mechanical performance using central composite design and genetic algorithm. *Journal of the Brazilian Society of Mechanical Sciences and Engineering, 41*(2). https://doi.org/10.1007/s40430-019-1564-7. http://rd.springer.com/journal/40430.

Dasgupta, J., Sikder, J., & Mandal, D. (2017). Modeling and optimization of polymer enhanced ultrafiltration using hybrid neural-genetic algorithm based evolutionary approach. *Applied Soft Computing, 55*, 108–126. https://doi.org/10.1016/j.asoc.2017.02.002.

Dehestani, M., Khayati, G. R., & Sharafi, S. (2021). An improved optimization model to predict the microhardness of Ni/Al2O3 nanocomposite coatings prepared by electrodeposition: A hybrid artificial neural network-modified particle swarm optimization approach. *Measurement, 179*, 109423. https://doi.org/10.1016/j.measurement.2021.109423.

Demirbay, B., Bayram Kara, D., & Uğur, Ş. (2020). A Bayesian regularized feed-forward neural network model for conductivity prediction of PS/MWCNT nanocomposite film coatings. *Applied Soft Computing, 96*, 106632. https://doi.org/10.1016/j.asoc.2020.106632.

Diao, Y., Yan, L., & Gao, K. (2022). A strategy assisted machine learning to process multi-objective optimization for improving mechanical properties of carbon steels. *Chinese Society of Metals, China Journal of Materials Science and Technology, 109*, 86–93. https://doi.org/10.1016/j.jmst.2021.09.004. http://www.sciencedirect.com/science/journal/.

Dionísio, M., Alves, N. M., & Mano, J. F. (2004). Molecular dynamics in polymeric systems. *European Polymer Federation, Portugal E-Polymers*. https://doi.org/10.1515/epoly.2004.4.1.470. http://www.e-polymers.org/papers/mano_030704.pdf.

Dunlap, J. H., Ethier, J. G., Putnam-Neeb, A. A., Iyer, S., Lennon Luo, S.-X., Feng, H., Torres, J. A. G., Doyle, A. G., Swager, T. M., Vaia, R. A., Mirau, P., Crouse, C. A., & Baldwin, L. A. (2023). Continuous flow synthesis of pyridinium salts accelerated by multi-objective Bayesian optimization with active learning. *Chemical Science, 14*(30), 8061–8069. https://doi.org/10.1039/d3sc01303k.

Edgeworth, F. Y. (1881). Mathematical psychics. *Mind, 6*(24), 581–583.

Etefagh, A. H., & Razfar, M. R. (2023). *Bayesian optimization of 3D bioprinted polycaprolactone/magnesium oxide nanocomposite scaffold using a machine learning technique. Proceedings of the Institution of Mechanical Engineers, Part B: Journal of Engineering Manufacture*. Iran: SAGE Publications Ltd. https://journals.sagepub.com/home/PIB. https://doi.org/10.1177/09544054231202219.

Fiedler, B., Gojny, F. H., Wichmann, M. H. G., Nolte, M. C. M., & Schulte, K. (2006). Fundamental aspects of nano-reinforced composites. *Composites Science and Technology, 66*(16), 3115–3125. https://doi.org/10.1016/j.compscitech.2005.01.014.

Frankland, S. J. V., Caglar, A., Brenner, D. W., & Griebel, M. (2002). Molecular simulation of the influence of chemical cross-links on the shear strength of carbon nanotube-polymer interfaces. *Journal of Physical Chemistry B, 106*(12), 3046–3048. https://doi.org/10.1021/jp015591.

Frazier, P. (2018). Bayesian optimization. *Recent Advances in Optimization and Modeling of Contemporary Problems*.

Gandhi, A., & Hasan, M. M. F. (2022). Machine learning for the design and discovery of zeolites and porous crystalline materials. *Current Opinion in Chemical Engineering, 35*, 100739. https://doi.org/10.1016/j.coche.2021.100739.

Gauthier, M. A., Gibson, M. I., & Klok, H. A. (2009). Synthesis of functional polymers by post-polymerization modification. *Angewandte Chemie—International Edition, 48*(1), 48–58. https://doi.org/10.1002/anie.200801951Switzerland. http://www3.interscience.wiley.com/cgi-bin/fulltext/121541698/PDFSTART.

Giro, R., Cyrillo, M., & Galvão, D. S. (2002). Designing conducting polymers using genetic algorithms. *Chemical Physics Letters, 366*(1-2), 170–175. https://doi.org/10.1016/S0009-2614(02)01547-6.

Guan, Q., Alketbi, A. S., Raza, A., & Zhang, T. (2020). Accelerated development of refractory nanocomposite solar absorbers using Bayesian optimization. *MRS Advances, 5*(29-30), 1537–1545. https://www.cambridge.org/core/journals/mrs-advances. https://doi.org/10.1557/adv.2019.468.

Iijima, S. (1991). Helical microtubules of graphitic carbon. *Nature, 354*(6348), 56–58. https://doi.org/10.1038/354056a0.

Islam, F., Wanigasekara, C., Rajan, G., Swain, A., & Prusty, B.G. (2022). An approach for process optimisation of the Automated Fibre Placement (AFP) based thermoplastic composites manufacturing using Machine Learning, photonic sensing and thermo-mechanics modelling. *Manufacturing Letters, 32*, 10–14. http://www.journals.elsevier.com/manufacturing-letters/. https://doi.org/10.1016/j.mfglet.2022.01.002.

Iyer, A., Zhang, Y., Prasad, A., Gupta, P., Tao, S., Wang, Y., Prabhune, P., Schadler, L. S., Brinson, L. C., & Chen, W. (2020). Data centric nanocomposites design: Via mixed-variable Bayesian optimization. *Molecular Systems Design and Engineering, 5*(8), 1376–1390. https://doi.org/10.1039/d0me00079e. http://www.rsc.org/journals-books-databases/about-journals/molecular-systems-design-engineering/.

Jafari, M. M., Soroushian, S., & Khayati, G. R. (2017) Hardness Optimization for Al6061-MWCNT Nanocomposite Prepared by Mechanical Alloying Using Artificial Neural Networks and Genetic Algorithm. *Journal of Ultrafine Grained and Nanostructured Materials, 50*, 23–32.

Jha, R., Dulikravich, G. S., Colaço, M. J., Fan, M., Schwartz, J., & Koch, C. C. (2017). *Magnetic alloys design using multi-objective optimization. In Advanced structured materials, 33.* United States: Springer Verlag. http://www.springer.com/series/8611.

Jiang, J., Li, L., & Zhu, M. (2008). Polyaniline/magnetic ferrite nanocomposites obtained by in situ polymerization. *Reactive & Functional Polymers, 68*, 57–62.

Karasuyama, M., Kasugai, H., Tamura, T., & Shitara, K. (2020). Computational design of stable and highly ion-conductive materials using multi-objective bayesian optimization: Case studies on diffusion of oxygen and lithium. *Computational Materials Science, 184*, 109927. https://doi.org/10.1016/j.commatsci.2020.109927.

Karimzadeh, A., Koloor, S. S. R., Ayatollahi, M. R., Bushroa, A. R., & Yahya, M. Y. (2019). Assessment of nano-indentation method in mechanical characterization of heterogeneous nanocomposite materials using experimental and computational approaches. *Nature Publishing Group, Malaysia Scientific Reports, 9*(1). https://doi.org/10.1038/s41598-019-51904-4. http://www.nature.com/srep/index.html.

Kou, X. Y., Parks, G. T., & Tan, S. T. (2012). Optimal design of functionally graded materials using a procedural model and particle swarm optimization. *Computer-Aided Design, 44*(4), 300–310. https://doi.org/10.1016/j.cad.2011.10.007.

Kulkarni, A. J., Krishnamurthy, K., Deshmukh, S. P., & Mishra, R. S. (2004). Microstructural optimization of alloys using a genetic algorithm. *Materials Science and Engineering: A, 372*(1-2), 213–220. https://doi.org/10.1016/j.msea.2003.12.005. http://www.elsevier.com.

Kumar, M., Husain, M., Upreti, N., & Gupta, D. (2010). Genetic algorithm: Review and application. *SSRN Electronic Journal*, 1556–5068. https://doi.org/10.2139/ssrn.3529843.

Köksal, G., Batmaz, I., & Testik, M. C. (2011). A review of data mining applications for quality improvement in manufacturing industry. *Expert Systems with Applications, 38*(10), 13448–13467. https://doi.org/10.1016/j.eswa.2011.04.063.

Liu, B., Vu-Bac, N., & Rabczuk, T. (2021). A stochastic multiscale method for the prediction of the thermal conductivity of Polymer nanocomposites through hybrid machine learning algorithms. *Composite Structures, 273*, 114269. https://doi.org/10.1016/j.compstruct.2021.114269.

Liu, J., Ye, J., Momin, F., Zhang, X., & Li, A. (2022). Nonparametric Bayesian framework for material and process optimization with nanocomposite fused filament fabrication. *Additive Manufacturing, 54*, 102765. https://doi.org/10.1016/j.addma.2022.102765.

Liu, Z., Rolston, N., Flick, A. C., Colburn, T. W., Ren, Z., Dauskardt, R. H., & Buonassisi, T. (2021). Machine learning with knowledge constraints for process optimization of open-air perovskite solar cell manufacturing. *SSRN Electronic Journal*, 1556–5068. https://doi.org/10.2139/ssrn.3956817.

Mannodi-Kanakkithodi, A., Pilania, G., Ramprasad, R., Lookman, T., & Gubernatis, J. E. (2016). Multi-objective optimization techniques to design the Pareto front of organic dielectric polymers. *Computational Materials Science, 125*, 92–99. https://doi.org/10.1016/j.commatsci.2016.08.018.

Mardle, S., & Miettinen, K. M. (2000). Nonlinear multiobjective optimization. *The Journal of the Operational Research Society, 51*(2), 246. https://doi.org/10.2307/254267.

Mazahery, A., & Shabani, M. O. (2012). Assistance of novel artificial intelligence in optimization of aluminum matrix nanocomposite by genetic algorithm. *Metallurgical and Materials Transactions A, 43*(13), 5279–5285. https://doi.org/10.1007/s11661-012-1339-6.

Moghri, M., Shamaee, H., Shahrajabian, H., & Ghannadzadeh, A. (2015). The effect of different parameters on mechanical properties of PA-6/clay nanocomposite through genetic algorithm and response surface methods. *International Nano Letters, 5*(3), 133–140. https://doi.org/10.1007/s40089-015-0146-7.

Mohammadi, M., Mohammadi, N., & Mehdipour-Ataei, S. (2020). On the preparation of thin nanofibers of polysulfone polyelectrolyte for improving conductivity of proton-exchange membranes by electrospinning: Taguchi design, response surface methodology, and genetic algorithm. *International Journal of Hydrogen Energy, 45*(58), 34110–34124. http://www.journals.elsevier.com/international-journal-of-hydrogen-energy/. https://doi.org/10.1016/j.ijhydene.2020.09.125.

Nam, D. N. C., & Ahn, K. K. (2012). Identification of an ionic polymer metal composite actuator employing Preisach type fuzzy NARX model and Particle Swarm Optimization. *Sensors and Actuators, A: Physical, 183*, 105–114. https://doi.org/10.1016/j.sna.2012.05.012.

Nasr, M. M., Anwar, S., Al-Samhan, A. M., Ghaleb, M., & Dabwan, A. (2020). Milling of Graphene reinforced Ti6Al4V nanocomposites: An artificial intelligence based Industry 4.0 approach. *Materials, 13*(24), 5707. https://doi.org/10.3390/ma13245707.

Okpala, C.C. (2013). Nanocomposites—An overview.

Özdemir, M. T. (2021). Optimal parameter estimation of polymer electrolyte membrane fuel cells model with chaos embedded particle swarm optimization. *International Journal of Hydrogen Energy, 46*(30), 16465–16480. https://doi.org/10.1016/j.ijhydene.2020.12.203. http://www.journals.elsevier.com/international-journal-of-hydrogen-energy/.

Pareto, V. (1896). Cours D'Economie Politique (Volumes II.F). Rouge.

Peng, X., Qiu, C., Li, J., Wu, H., Liu, Z., & Jiang, S. (2021). Multiple-scale uncertainty optimization design of hybrid composite structures based on neural network and genetic algorithm. *Composite Structures, 262*, 113371. https://doi.org/10.1016/j.compstruct.2020.113371.

Poli, R., Kennedy, J., & Blackwell, T. (2007). Particle swarm optimization. *Swarm Intelligence, 1*(1), 33–57. https://doi.org/10.1007/s11721-007-0002-0.

Pyrialakos, S., Kalogeris, I., Sotiropoulos, G., & Papadopoulos, V. (2021). A neural network-aided Bayesian identification framework for multiscale modeling of nanocomposites. *Computer Methods in Applied Mechanics and Engineering, 384*, 113937. https://doi.org/10.1016/j.cma.2021.113937.

Rabothata, M., Muthu, J., & Wegner, L. (2021). Optimum design parameters and mechanical properties of polymeric nanocomposites using NSGA-II optimization method. *Journal of Composite Materials, 55*(7), 949–972. https://doi.org/10.1177/0021998320960774.

Rouhi Broujeni, B., Nilchi, A., & Azadi, F. (2021). Adsorption modeling and optimization of thorium (IV) ion from aqueous solution using chitosan/TiO2 nanocomposite: Application of artificial neural network and genetic algorithm. *Environmental Nanotechnology, Monitoring & Management, 15*, 100400. https://doi.org/10.1016/j.enmm.2020.100400.

Schaffer, J.D. (1984). Multiple objective optimization with vector evaluated genetic algorithms. In *Proceedings of the 1st International Conference on Genetic Algorithms*. ACM Digital Library.

Silva, C. M., & Biscaia, E. C. (2003). Genetic algorithm development for multi-objective optimization of batch free-radical polymerization reactors. *Computers and Chemical Engineering, 27*, 1329–1344 Elsevier Ltd., Brazil.

Stergiou, K., Ntakolia, C., Varytis, P., Koumoulos, E., Karlsson, P., & Moustakidis, S. (2023). Enhancing property prediction and process optimization in building materials through machine learning: A review. *Computational Materials Science, 220*, 112031. https://doi.org/10.1016/j.commatsci.2023.112031.

Suwardi, A., Wang, F. K., Xue, K., Han, M. Y., Teo, P., Wang, P., Wang, S., Liu, Y., Ye, E., Li, Z., & Loh, X. J. (2022). Machine learning-driven biomaterials evolution. *Advanced Materials, 34*(1). https://doi.org/10.1002/adma.202102703. http://onlinelibrary.wiley.com/journal/10.1002/(ISSN)1521-4095.

Tang, P. H., Song, R. G., Chai, G. Z., & Zhang, Q. Z. (2006). Prediction of polymerization efficiency for PDPhSM matrix nanocomposite thin film prepared by laser ablation based on artificial neural networks. *Zhongguo Jiguang/Chinese Journal of Lasers, 33*(7), 953–958.

Thakral, P., & Bakhshi, A. K. (2014). Computational atomistic blueprinting of novel conducting copolymers using particle swarm optimization. *Journal of Computer-Aided Molecular Design, 28*(2), 111–122. https://doi.org/10.1007/s10822-014-9736-0. https://www.springer.com/journal/10822.

Thostenson, E. T., Ren, Z., & Chou, T. W. (2001). Advances in the science and technology of carbon nanotubes and their composites: A review. *Composites Science and Technology, 61*(13), 1899–1912. https://doi.org/10.1016/S0266-3538(01)00094-X. http://www.journals.elsevier.com/composites-science-and-technology/.

Tofigh, A. A., Rahimipour, M. R., Shabani, M. O., & Davami, P. (2015). Application of the combined neuro-computing, fuzzy logic and swarm intelligence for optimization of compocast nanocomposites. *Journal of Composite Materials, 49*(13), 1653–1663. https://doi.org/10.1177/0021998314538871. http://jcm.sagepub.com/content/by/year.

Wegner, G. (2000). Functional polymers. *Acta Materialia, 48*(1), 253–262. https://doi.org/10.1016/S1359-6454(99)00298-0.

Yamawaki, R., Tei, A., Ito, K., & Kikuchi, J. (2021). Decomposition factor analysis based on virtual experiments throughout Bayesian optimization for compost-degradable polymers. *Applied Sciences, 11*(6), 2820. https://doi.org/10.3390/app11062820.

Zhang, Y., Apley, D. W., & Chen, W. (2020). Bayesian optimization for materials design with mixed quantitative and qualitative variables. *Nature Research, United States Scientific Reports, 10*(1). https://doi.org/10.1038/s41598-020-60652-9. http://www.nature.com/srep/index.html.

Machine learning techniques in additive manufacturing of functional polymers and nanocomposites

21

Jyotsna G. Vitore[1], Amol D. Gholap[2], Diksha Nunse[3], Ranjitsinh Pawar[4], Sanjeevani Deshkar[5], and Prabhanjan Giram[6]
[1]Department of Pharmaceutics, National Institute of Pharmaceutical Education and Research, Ahmedabad, Gujarat, India, [2]Department of Pharmaceutics, St. John Institute of Pharmacy and Research, Palghar, Maharashtra, India, [3]Department of Pharmaceutics, Sandip Institute of Pharmaceutical Science, Nashik, India, [4]Department of Pharmaceutics, Poona College of Pharmacy, Bharti Vidyapeeth, Erandwane, Pune, India, [5]Department of Pharmaceutics, Dr. D.Y. Patil Institute of Pharmaceutical Sciences & Research, Pimpri, Pune, India, [6]Department of Pharmaceutical Sciences, The State University of New York, Buffalo, NY, United States

Abbreviations

AM Additive manufacturing
ML Machine learning
3D Three-dimensional
SLA Stereolithography
FDM Fused deposition modeling
PET Polyethylene terephthalate
FFF Fused filament fabrication
SLS Selective laser sintering
BJ Binder jetting
DED Directed energy deposition
CAD Computer-aided design
PA Polyamide
FRP Fiber-reinforced polymer
EHR Electronic health records

21.1 Introduction

Additive manufacturing (AM), commonly known as 3-dimensional (3D) printing, has revolutionized the manufacturing landscape by enabling the production of complex geometries and customized parts with unprecedented flexibility and precision.

Additive Manufacturing of Functional Polymers and Nanocomposites. DOI: https://doi.org/10.1016/B978-0-443-26516-7.00021-3
© 2025 Elsevier Ltd. All rights are reserved, including those for text and data mining, AI training, and similar technologies.

Among the various materials utilized in AM, functional polymers and nanocomposites have garnered significant attention due to their unique properties and wide range of applications. These materials exhibit enhanced mechanical, thermal, and electrical characteristics, making them suitable for diverse industries such as aerospace, automotive, healthcare, and electronics. The integration of machine learning (ML) techniques into AM processes has emerged as a transformative approach to further optimize and innovate the production of functional polymers and nanocomposites. ML algorithms can analyze wide data generated during the manufacturing process, identify patterns, and make predictions that improve the quality and performance of the final products. By leveraging ML, researchers and manufacturers can achieve better control over material properties, reduce defects, and enhance the overall efficiency of the AM process.

One of the primary advantages of using ML in AM is its ability to model complex relationships between process parameters and material properties. Traditional trial-and-error methods for optimizing AM processes are time-consuming and resource-intensive. In contrast, ML models can quickly evaluate numerous variables and their interactions, providing insights that lead to more informed decision-making. For instance, supervised learning techniques can be employed to predict the mechanical strength of printed parts based on process parameters such as temperature, speed, and layer thickness. Similarly, unsupervised learning can help in clustering and identifying anomalous patterns in the data, which might indicate potential defects or areas for improvement.

Furthermore, ML techniques facilitate the development of adaptive AM systems that can adjust parameters in real time to compensate for variations in material properties or environmental conditions. This adaptive capability is particularly crucial when working with functional polymers and nanocomposites, as their properties can be highly sensitive to changes in the manufacturing environment. Reinforcement learning, a subset of ML, can be utilized to create systems that learn from experiences and continuously improve the printing process by minimizing errors and enhancing consistency. In addition to optimizing the AM process, ML can also play a pivotal role in the design and development of new functional polymers and nanocomposites. By analyzing the vast dataset of existing materials and their properties, ML algorithms can identify novel combinations and formulations that may exhibit superior performance characteristics. This data-driven approach accelerates the material discovery process and opens up new possibilities for creating advanced materials tailored to specific applications.

In summary, the application of ML techniques in the additive manufacturing of functional polymers and nanocomposites represents a significant advancement in the field of materials science and engineering. By harnessing the power of data and intelligent algorithms, researchers and manufacturers can achieve unprecedented levels of precision, efficiency, and innovation in the production of advanced materials. The synergy between AM and ML holds the potential to transform industries and drive the development of next-generation technologies. This chapter delves into the intersection of machine learning and additive manufacturing, exploring the latest advancements, methodologies, and applications in the fabrication of functional polymers and nanocomposites. Through a comprehensive review of current research and practical implementations, we aim to provide a detailed understanding of how ML is shaping the future of AM, paving the way for groundbreaking innovations in material science and engineering.

21.1.1 Fundamentals of additive manufacturing

21.1.1.1 Types of additive manufacturing techniques

There are various techniques used in additive manufacturing discussed as follows. All the techniques are illustrated in Fig. 21.1 (Gul et al., 2018; Ngo et al., 2018).

Stereolithography (SLA)

SLA is a precise 3D printing process using digital models to create customized objects, with parameters like layer thickness and exposure time significantly affecting quality and accuracy. This technology creates high-precision parts with excellent surface finishes using photo-curable polyurethane resins, analyzed for mechanical properties and composition. SLA's ability to produce parts with complex geometries, tight tolerances, and detailed features is unmatched, allowing for rapid prototyping and form and fit testing of components with small features as fine as 0.07 mm (Melchels et al., 2010). While SLA materials are primarily polymers, which may have limitations for space applications due to outgassing and charge accumulation, ongoing research focuses on assessing different materials' responses to harsh environments like vacuum and ionizing radiation, highlighting the technology's adaptability and potential for future advancements. Additionally, advancements in photo-curable polyurethane resins used in SLA technology have enabled the creation of parts with exceptional accuracy and mechanical properties suitable for both industrial and medical applications.

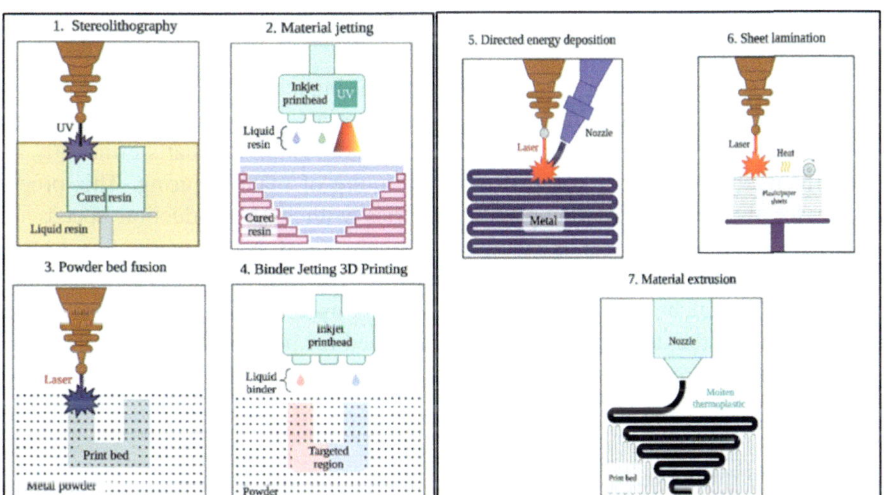

Figure 21.1 Additive manufacturing methods for 3D bioprinting involve (1) stereolithography, (2) material jetting, (3) powder bed fusion, (4) binder jetting 3D printing, (5) directed energy deposition, (6) sheet lamination, and (7) material extrusion.

Fused deposition modeling (FDM)

FDM is the same as fused filament fabrication (FFF). The term "fused deposition modeling" and the abbreviated "FDM" were trademarked by Strategy in 1991, creating the need for a second name. Many layers of melted thermoplastic material form a 3D object. In layman's terms, a typical FDM 3D printer takes a plastic filament and squeezes it through a hot end, melting it and then depositing it in layers on the print bed. These layers are fused, building up throughout the print, and eventually, they will form the finished part. This technology is cost-effective, user-friendly, and compatible with a wide range of thermoplastic polymers with diverse applications. However, its limitations include a lack of surface finishing, the strength of FDM parts, and details with resolution (Turner et al., 2014).

Selective laser sintering (SLS)

Selective laser sintering (SLS) is a layer-by-layer additive manufacturing technique that utilizes a high-powered laser to selectively fuse powdered materials, offering significant advantages over traditional manufacturing methods. Initially developed in the 1980s, SLS has evolved to accommodate various materials, including polymers, metals, ceramics, and composites, catering to diverse industrial applications such as aerospace, automotive, medical, and consumer goods. Polymers like polyamide (PA), polyethylene terephthalate (PET), and thermoplastic elastomers (TPE) dominate in prototyping and consumer goods due to cost-effectiveness and design flexibility. Medical applications include patient-specific implants and surgical guides, leveraging biocompatible materials and intricate geometries. Automotive manufacturers adopt SLS for rapid iteration of functional prototypes and lightweight parts, enhancing fuel efficiency and performance (Kruth et al., 2007).

Binder jetting (BJ)

BJ has emerged as a prominent additive manufacturing technology due to its ability to create complex geometries from a variety of materials at a lower cost compared to other AM methods. Its ability to produce complex, customized parts efficiently and cost-effectively makes it a key player in the future of manufacturing. This process allows for the creation of complex geometries with high production speeds and material versatility, making it suitable for applications across various industries, including aerospace, automotive, and medical. The built parts typically require postprocessing, such as sintering or infiltration, to enhance their mechanical properties and achieve the desired finish. Despite its need for postprocessing, it offers cost-effective production and design flexibility, making it a valuable technology in modern manufacturing (Ziaee & Crane, 2019).

Directed energy deposition (DED)

It is an advanced additive manufacturing technique that involves the focused delivery of energy, typically via a laser, electron beam, or plasma arc, to melt material as it has deposited. This process allows for the creation of intricate geometries by sequentially adding layers of material, which can be in the form of powder or wire. DED is particularly noted for its versatility in material usage, enabling the fabrication and repair of high-value components in the aerospace, automotive, and energy sectors. The precise control over the deposition process allows for the production of complex, high-performance parts, although it often requires postprocessing to achieve desired surface finishes and mechanical properties. Despite its complexity and the need for expert handling of process parameters, DED's ability to efficiently produce, repair, and enhance components underscores its significant role in modern manufacturing (Sames et al., 2016).

Sheet lamination

It is an additive manufacturing process where layers of material are bonded together to form a three-dimensional object. This technique involves stacking sheets of material, such as paper, metal, or polymer, and using an adhesive, ultrasonic welding, or other bonding methods to fuse them. Each layer is precisely cut to shape using a laser, blade, or another cutting tool before being added to the stack. The process is known for its rapid production rates and cost-effectiveness, especially for creating large, nonfunctional prototypes and visual models. However, the mechanical properties and resolution of sheet-laminated parts are generally inferior to those produced by other additive manufacturing methods like stereolithography (SLA) or selective Laser Sintering (SLS), limiting its applications to areas where these factors are less critical. Despite these limitations, sheet lamination offers significant advantages in terms of material variety and low-cost production, making it a valuable technique in specific prototyping and manufacturing scenarios (Zhang et al., 2016).

Material extrusion

This technology creates the continuous shape of the object. It involves heating a thermoplastic filament at its melting point within an extrusion nozzle. The molten feed is then deposited layer by layer onto the build platforms guided by computer-aided design (CAD) models to create three-dimensional objects. This process offers several advantages, including relative simplicity, cost-effectiveness, and the ability to use a wide range of thermoplastic materials. However, challenges such as achieving high-resolution prints and managing material properties during the printing process remain areas of ongoing research and development. Melt extrusion-based AM technologies continue to evolve, with advancements focusing on enhancing print speed, precision, and the range of printable materials, thereby expanding their applications across industries from prototyping to custom manufacturing (Turner et al., 2014).

The current literature search on machine learning is tabulated below in Table 21.1.

21.1.2 Materials used in additive manufacturing

21.1.2.1 3D printing of thermoplastics polymers

Thermoplastics are melt-processable polymers that can repeatedly solidify after cooling and soften after heating while maintaining their original characteristics

Table 21.1 Research questions posed for the systematic literature review.

Research question	Motivation
RQ1: What kind of issues in AM can be currently tackled by using technologies?	To explore state-of-the-art of AM challenges that can be tackled by using ML technologies.
RQ2: What ML algorithms have been adopted to tackle the challenges?	To present the ML algorithms adopted by previous studies to tackle challenges in AM. Answer this question will help researchers understand and select appropriate ML algorithms for dealing with similar issues in AM.
RQ3: What are the advantages and limitations of using ML algorithms to solve the issues in AM?	To study the reasons and limitations for adopting a certain ML algorithm to tackle a specific research challenge in AM. Answer this question will assist researchers to make the trade-off between ML-based methods and conventional methods, as well as exploring modified strategies.

(Chuisaca-Londa et al., 2023). The traditional methods for creating thermoplastic components are material extrusion and injection molding. However, sophisticated thermoplastic components may now be produced using the advancement of 3D printing technologies like FDM and SLS. These techniques include heating the thermoplastics to a pliable condition and then extruding them onto a platform, where solidification occurs (Burela et al., 2020). The major reasons thermoplastics are employed due to their stability and great strength at high temperatures, as well as their stiffness. Some popular thermoplastics that are utilized, include acrylic, ABS, nylon, PLA, PC, PE, and others (Gladman et al., 2019).

21.1.2.2 3D printing of polymer matrix composites

It has been used due to the inexpensive, lightweight, and easily processed, polymer materials with low melting points or in liquid condition in 3D printing. The lack of mechanical strength and utility in 3D-printed polymer items is a significant obstacle to their widespread uses, despite the materials' potential for geometric intricacy. One viable solution to these issues is to combine altered materials to accomplish desired mechanical as well as functional qualities. Thus, the creation of composite materials has already spiked the market. Numerous encouraging outcomes in the creation of novel printed composite materials reinforced by fibers, particles, or nanomaterials have shown their utility (Wang et al., 2017).

21.1.2.3 Particle-reinforced polymer composites

Recent advances in AM have made it possible to use materials other than single polymers, such as polymer composites. The need for fully functioning and load-bearing components by industries, coupled with the lack of strength and functionality of printed polymer (mostly thermoplastic) goods, led to the shift toward the usage of polymer composites for additive manufacturing. The printing polymer performs better mechanically and has more functionality when fiber reinforcement is included (Saroia et al., 2020). To create extruded filaments for printing, short fibers, and polymer pellets are combined in FDM. In this technique, polymer paste and short fibers are combined and extruded straight from the printer. Continuous carbon fibers may be printed by FDM using the mark one and mark two printers, which were recently created by Markforged. Additional additive manufacturing (AM) processes for fiber-reinforced polymer (FRP) composites include laser powder bed fusion (LPBF), which melts a powdered polymer with fibers, and selective lamination composite object manufacturing (SLCOM), which cuts and joins woven composite sheets.

21.1.2.4 Fiber-reinforced polymer composites

Its matrix can be easily made, the fiber offers a strength-to-weight ratio, and the polymer composites are less expensive. The primary structural element used in the aircraft industry in 1960 was made of fiber-reinforced polymers. FRP composites provided a versatile and affordable alternative for several production processes between 1990 and 2006. When considering standard metal materials, FRP composites have superior antiaging capabilities, a high strength-to-weight ratio, and strong antiwear qualities. For several sectors, FRP composites provide excellent performance and low weight. In fiber-reinforced polymer (FRP), thermosetting polymers include epoxy and polyurethane, whereas thermoplastics include polypropylene (PP), polyamide (PA), and polyetheretherketone (PEEK). Other characteristics of FRP composites include the use of several fiber types, including glass, carbon, aramid, and Kevlar. Fiber length parameters, such as short fibers (0.2–0.4 mm) and long fibers (10–25 mm), provide another way to categorize FRPs into subtypes. Long and short CFRPs are formed and processed using injection molding or extrusion techniques. The processing methods for the second kind of continuous fiber-reinforced plastic include pultrusion, molding, impregnating, and winding (Sonnenschein et al., 2016).

21.1.2.5 Nanocomposites

Composites are created by combining two or more monolithic polymers across a stronger and more rigid reinforcing phase (De Leon et al., 2016) The final material is categorized as a nanocomposite if any one of the two phases has one, two, or three dimensions that fall within a billionth of a meter. The kind of matrix often determines the processing approach for nanocomposite materials (Camargo et al., 2009). High shear mixing, in-situ polymerization, and solution mixing are the three primary processing techniques that can be used alone or in combination. The most common mixing techniques for the thermosetting matrix are solution processing by stirring and/or sonication, shear mixing in a three-roll mill mixer, and a combination of both. Twin-screw extruders or batch mixers are typically used for thermoplastic nanocomposite blending. The nanocomposite with superior nanofiller diffusion is produced in this kind of equipment due to extreme shear forces and complex flow (Guo et al., 2013). Hydrogels and thermoplastics that dissolve in a solvent can also be processed using a solution approach (Zhang et al., 2020). When they come together, they exhibit qualities that set them apart from every element alone. Their benefit over more conventional thermoplastics employed for 3D printing, such as PLA or ABS, is greater because of this. With these other materials and their qualities, the uses of 3D printing may be increased.

For instance, combining nylon with diced carbon fibers results in strong strength characteristics as well as chemical and thermal resistance (Clarissa et al., 2022). Compared to nonreinforced polymers, its tensile strength rose to almost six times greater. Conversely, polycarbonate (PC) is typically seen as a translucent substance. This material is tougher and more resilient than nylon (Dickson et al., 2017). This material may be utilized for plastic lenses in eyeglasses, digital discs, car components, greenhouses, and other applications since it has nearly the same internal light transmission as glass. In addition to having exceptional heat resistance, this material does not easily deteriorate under high temperatures. It is significant to note that this material may be used in operations that call for substantial transparency and high immunity to affect.

21.1.3 Current challenges in am processes

There are undoubtedly many advantages to additive manufacturing (AM), including design freedom, the capacity to print intricate structures, user-friendliness, and product customization though it has some challenges. Several problems need further study and research: the maximum part size; anisotropic mechanical properties; the formation of overhang surfaces; high costs; low manufacturing efficiency; imprecise precision; warping; cushioning; binding; voids in the upper layers; under-extrusion; layer misalignment; overextrusion; elephant foot; large-scale production; and limitations on the goods (Chen et al., 2017; Zareiyan & Khoshnevis, 2017).

21.1.3.1 Machine learning

Machine learning (ML) offers the possibility of teaching machines to handle data more skillfully. Machine learning aims to train users based on available data. The benefits of machine learning are depicted in Fig. 21.2.

21.1.3.2 Supervised learning

The best users of machine learning employ supervised learning approaches. A command dataset is used to finish the learning process of an algorithm in controlled learning. The job may be thought of as a mentor overseeing the procedure for learning, even in the case when an instructional dataset is used. The method may be used to acquire the adjustments made by the instructor and assist in making predictions about the data used in the training phase. An algorithm's conclusion could state whether it has performed to a given level or in an acceptable quantity (Caffaratti et al., 2021).

21.1.3.3 Unsupervised learning

Techniques that autonomously divide data are referred to as clustering, but unsupervised learning is another frequent term for them. These algorithms' main goal is to

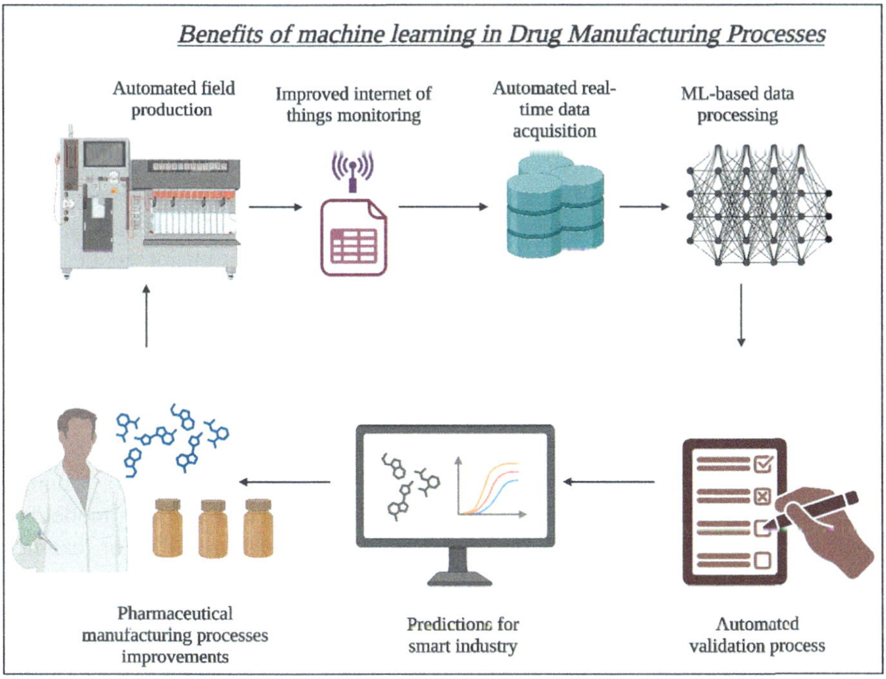

Figure 21.2 Summarizes the benefits of machine learning in drug manufacturing processes.

automatically group information so that components within one group are identical to one another as well as different from the components of other groups based on established standards. Client data, for example, might be grouped for targeted advertising efforts. The clusters may exhibit disjointness or crosses over, be arranged in hierarchical structures, be comprehensive have stochastic inclusion to classes, etc. In multiple-dimensional space, the closeness across elements is typically used to determine a comparable measure, and more metrics for similarity are employed to cluster the components. Measures like as Euclidean, Manhattan, or Murkowski are typically utilized when dealing with numerical data. For binary format, insignificant, typical, nonlinear scalars, and mixed parameters, there exist similarity measures. The measurements are often outdated metrics if the information cannot be quantified (Woergoetter & Porr, 2008).

21.1.3.4 Reinforcement learning (RL)

It is learning via engagement with the surroundings. A reinforcement-learning agent (RL agent) learns by trial and error, which means choosing its behaviors based on prior experiences (exploration) and new options (exploitation). It does this instead of overtly instructed. An action's effectiveness is encoded in a numerical reward that the RL agent gets as reinforcement. Eventually, the agent aims to learn how to choose actions that maximize the total reward. The term "reward" is employed here in an aneurism; it implies no hedonic effect, delight, or other psychic connotations (Scruggs et al., 2015).

21.1.3.5 Role of data in machine learning

Electronic health records (EHRs) are becoming used to store medical data due to the massive amount of data being collected and stored for research purposes. From massive datasets of patient information kept in electronic health records, machine learning (ML) models may automatically recognize pertinent characteristics (Chen et al., 2014). The ML model may then assist in illness identification by tracking the data and forecasting probable ailments using the pertinent attributes from the EHRs. By training from unprocessed information, machine learning and deep learning assist computer models in their ability to "adapt to new conditions and to identify and extrapolate patterns" (Tabares-Soto et al., 2020). When a model's performance metric rises on an experienced task, it is assumed to be training and whether using the model to do the task might be more beneficial (Obermeyer & Emanuel, 2016). Examples containing different characteristics would be fed into a machine learning model for a particular job, with the labels representing the intended result. For example, pathologists classify aspects from digitized photographs of human brains they have seen as well as labels indicating the presence of cancer cells in the brain (Obermeyer & Emanuel, 2016).

21.1.3.6 Key applications of machine learning in various industries

ML powers demand forecasting, segmenting consumers and recommendation systems for advertising and sales, spurring innovations, and improving user experiences as presented in Fig. 21.3.

1. Predictive analytics and intelligent decision-making: Data-driven predictive analytics is a prominent application field for machine learning in smart decision-making. To anticipate an unknown result, predictive analytics is dependent on identifying and making use of correlations amid explanations and anticipated parameters from past existences (Mahdavinejad et al., 2018).
2. Cybersecurity and threat intelligence: Cybersecurity, or the process of defending networks, systems, hardware, and data from online threats, is one of the most important aspects of Industry 4.0. Machine learning has emerged as a critical cybersecurity modern technology to find patterns, recognize malware in encrypted traffic more accurately, identify threat vectors, anticipate the locations of dangerous online neighborhoods, keep users secure while surfing, or safeguard information in the cloud by spotting unusual activity (Zimpel & Hubl, 2019).
3. Internet of Things (IoT) and smart cities: IoT is an additional necessary aspect of Industry 4.0, enabling commonplace objects to become intelligent by transmitting data and automating processes without requiring human intervention. Hence, the Internet of Things (IoT) is viewed as the great frontier that has the potential to improve nearly every aspect of our lives, including smart homes, efficient government, education, sales, retail, farming, health care, and enterprise, among many others. One of the main areas of use for IoT is the smart city, which uses technology to improve living conditions and urban amenities for citizens (Guerrero-Ibáñez et al., 2018).
4. Transportation and traffic prediction: Transportation networks are now an essential part of every nation's economic growth. However, several cities globally are witnessing an

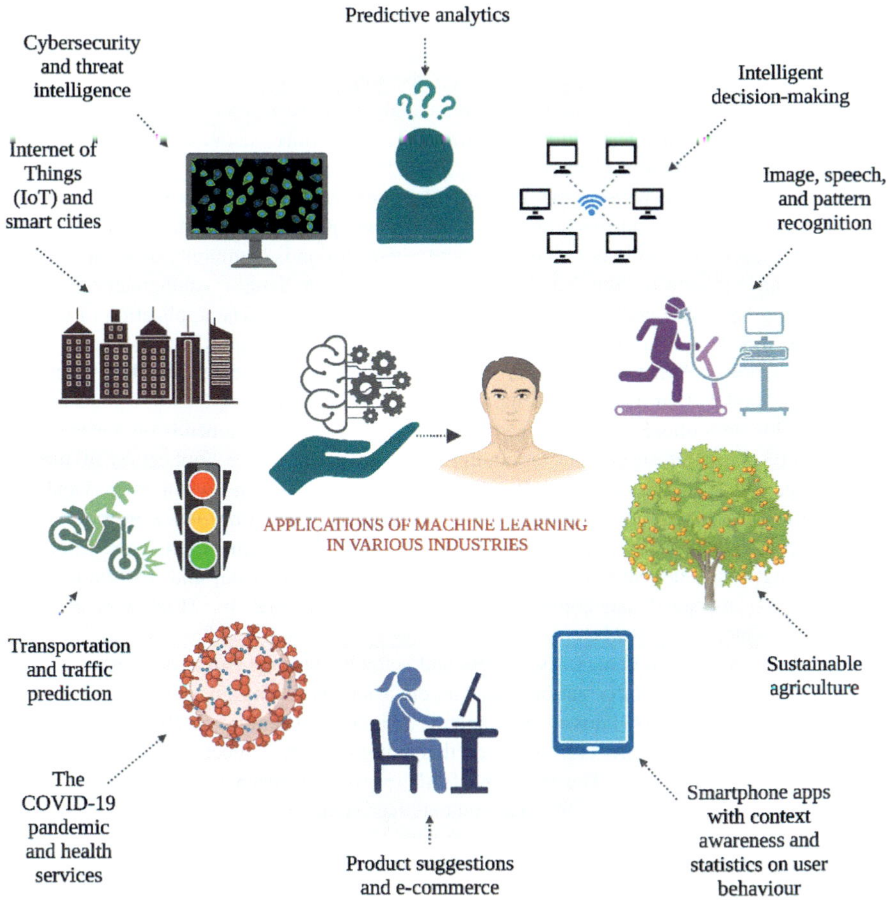

Figure 21.3 Applications of machine learning in various industries.

uncontrollably high increase in traffic volume, leading to grave problems including traffic jams, delays, elevated fuel costs, elevated carbon dioxide emissions, mishaps, crises, and a downturn in the standard of living of contemporary civilization. As a result, an intelligent transport system that forecasts traffic patterns is crucial and an essential component of a smart city. The problems can reduced with the use of precise traffic forecasts based on deep learning and machine learning models (Sroczyński & Czyżewski, 2023).

5. The COVID-19 pandemic and health services: In several medical fields, including illness anticipating, medical information extraction, finding patterns in statistics, managing patients, etc., machine learning can assist in resolving diagnostic and prognostic issues. The World Health Organisation (WHO) states that COVID-19 is a viral condition brought on by a recently identified coronavirus. In the fight against COVID-19, learning strategies have gained popularity recently (Lalmuanawma et al., 2020). The learning approaches are applied to the COVID-19 pandemic to categorize patients at greater risk, their death rate, and other abnormalities [74]. Additionally, it may be used for illness diagnosis and therapy, COVID-19 outbreak forecasting, and a deeper understanding of the virus's genesis (Mondal et al., 2021).

6. Product suggestions and e-commerce: One of the most popular and extensively utilized uses of machine learning is category suggestion, which is also among the most noticeable aspects of practically any e-commerce website on the internet today. Businesses may use the technology of machine learning to help them analyze the past purchases of their customers and provide personalized product recommendations for future purchases depending on their behavior and inclinations (Nguyen et al., 2024).
7. Image, speech, and pattern recognition: One popular and well-known application of machine learning in real-life situations is visual recognition (Sapru & Garimella, 2020), which can recognize an item as a digital picture recognition is commonly used for tasks like identifying a person or figure in a picture, classifying an X-ray as malignant or not, and suggesting tags on social networking platforms like Facebook. The application of machine learning techniques in speech recognition is also quite common. These systems, which include Google Assistant, Cortana, Siri, Alexa, and others, often employ sound and language models (Sapru & Garimella, 2020; Sarker, 2021).
8. Sustainable agriculture: The existence of all human endeavours depends on agriculture. A variety of stages of sustainable agriculture can benefit from the application of machine learning. These comprise the following stages: preproduction, when crop yield, soil properties, irrigation requirements, etc. are forecasted; production, when weather, illness and weed detection, nutrient management, livestock management, etc. are forecasted; processing, when demand is estimated and production is planned; and distribution when inventory is managed and consumer intelligence is examined, etc (Liakos et al., 2018; Sarker, Hoque, et al., 2021).
9. Smartphone apps with context awareness and statistics on user behavior: The capacity of a system to recognize information about its environment at any given time and adjust its actions accordingly is known as context awareness (Sarker, 2019). Context-aware computing uses hardware and software that automatically collect and analyze data in search of quick solutions. The strength of AI, specifically machine learning technologies and their ability to learn from data context, has significantly altered the landscape of creating mobile apps. As a result, mobile app creators may utilize machine learning to build intelligent apps that assist, amuse, and comprehend human behavior (Wu et al., 2016).

21.1.4 Applications of machine learning in additive manufacturing

21.1.4.1 Predictive modeling for material properties

Machine learning has gained remarkable potential in the prediction of mechanical properties in additive manufacturing, which helps the manufacturer optimize the design and process.

Mechanical properties prediction

Machine learning techniques have been successfully employed in predicting the mechanical properties of various materials, highlighting their efficiency and accuracy in forecasting key characteristics. Machine learning was used to develop the predictive model and to forecast the tensile strength of titanium alloy components produced through laser melting (SLM). Researchers have collected a dataset containing parameters such as laser power, scan speed, hatch spacing, etc. and various regression algorithms were used to establish the predictive model. This ML-derived predictive model demonstrated high accuracy in forecasting tensile strength, reliability, and consistency, checked by optimizing the model (Hunter et al., 2010).

This case study demonstrates, in addition, the reductive data of thermal history and mechanical properties; for example, UTS researchers have studied the final performance of developed material. It concludes that machine learning effectively enhanced the additive manufacturing process by predicting mechanical properties like tensile strength.

Thermal properties prediction

Machine learning plays a crucial role in predicting thermal properties, as evidenced by various studies. For instance, a study explores transfer learning to enhance prediction models for thermophysical properties, showing improved performance and generalization to less explored chemical spaces (Hormazabal et al., 2022). Additionally, machine learning has been utilized in geothermal exploration to generate high-resolution temperature-at-depth maps, reducing costs and time compared to conventional physics-based models (Vivas & Salehi, 2021). In the field of structural fire engineering, machine-learning techniques are applied to predict thermal responses efficiently, serving as a surrogate model for Finite Element Analysis and aiding in understanding structural behavior under fire conditions (Banerji, 2022). Furthermore, machine learning algorithms are employed to predict the thermal performance of closed-loop thermosyphon systems, with decision tree models outperforming other methods in accuracy (Birajdar & Sewatkar, 2023). These studies collectively display the diverse applications and benefits of machine learning in predicting thermal properties across different domains.

21.1.4.2 Process optimization

Machine learning is an evolving field in additive manufacturing and is being leveraged to optimize various processes at different stages. The algorithm analyses data from previous builds to predict and improve the outcomes of future prints (Suzuki et al., 2022).

Parameter optimization

ML algorithms can analyze vast amounts of data generated during AM processes. By identifying patterns and correlations, ML models can predict optimal parameters for desired outcomes, such as mechanical properties, surface finish, and structural integrity via data-driven process optimization. Predictive models can be trained to forecast the effects of different parameters (e.g., laser power, scanning speed, layer thickness) on the final product. This allows for virtual testing and reduces the need for extensive experimental trials (Suzuki et al., 2022).

Print path optimization

Print path optimization is crucial in additive manufacturing (AM) to improve quality, reduce production time, and minimize material usage. Traditional methods for print path planning often involve heuristic or rule-based approaches, which can be suboptimal and time-consuming. Machine learning (ML) offers a promising alternative by leveraging data-driven techniques to enhance print path planning (Ma et al., 2022).

21.1.4.3 Defect detection and quality control

The advancement of in situ monitoring systems allows for the collection of real-time data, which can be utilized for defect detection and closed-loop control in additive manufacturing. Spectra, photos, AE, and computed tomography (CT) are examples of real-time data that ML models can use in a variety of ways. Using experimental findings or human expertise, label the data with defects (potentially including defect kinds) or not. Using the labeled data, supervised learning models are then trained for defect detection and quality prediction in real-time, which is a common use case for machine learning classification models (Westphal & Seitz, 2021).

Image analysis for defect detection

Applying machine learning (ML) to image analysis for defect detection in additive manufacturing (AM) involves several key steps and considerations. The main components and some examples of applications. The high-quality images are captured in the AM process using a camera and sensors, which serve as primary data for training machine learning models. Data labeling involves annotating images to highlight defects, and creating a robust training dataset. Image preprocessing is performed such as Gaussian filters are applied to remove unwanted noise from the images. The image segmentation was carried out from various regions of interest, facilitating focused analysis on potential defect areas. Manual feature extractions involve various specific characteristics, such as edges, textures, or shapes where the defects can be indicative. However, deep learning approaches, particularly convolutional neural networks (CNNs), automatically learn and extract relevant features from raw image data. It is used to identify surface defects, such as cracks and scratches, by analyzing high-resolution images of printed parts. Porosity detection was performed by using deep learning models process X-ray or CT scan images to detect and quantify internal porosity within printed parts, providing insights into structural integrity. In addition, geometric defect detection was performed to compare images or 3D scans of printed parts against CAD models to detect geometric deviations, ensuring parts meet design specifications (Prabhakaran et al., 2023).

21.2 Real-time monitoring of AM processes

Machine learning-enhanced real-time monitoring systems offer significant advantages for additive manufacturing. By enabling immediate detection and correction of defects, these systems can improve the AM process's quality, reliability, and efficiency. Machine learning (ML) techniques offer powerful tools for analyzing data in real-time, enabling the detection of defects and process anomalies as they occur. Various key components are used to implement machine learning in real-time monitoring AM process including data acquisition, preprocessing, feature extraction, model development, system integration, and real-time analysis (Mohammadi & Elbestawi, 2020).

21.3 Future perspectives and challenges

It is evident that the great majority of ML applications now being used in AM research domains are heavily focused on processing-related procedures such as intraprocess surveillance and parameter optimization. However, in a few years, we may anticipate that the massive ML research efforts will pay off in the form of novel materials, a sensible production strategy, and a digital system of feedback for AM that is in process. Furthermore, we anticipate that more sophisticated ML algorithms—like XGBoost (Mylonas et al., 2019) can greatly increase operational speed and efficiency. Similar to the two faces of a coin, there may be certain drawbacks to using ML in AM design, manufacturing, or operation. The disparity of ML prediction outputs, which might lead to loss or even harm to AM practice, is a significant cause for worry. Thus, it is advised that the key to minimizing or eliminating any procedures or machine faults is to use robust, large data-driven, and transportable machine learning methods. Furthermore, it is recognized that a primary obstacle to utilizing machine learning techniques in additive manufacturing is the limited amount of data that is currently accessible (Cai et al., 2022). This is particularly true when it comes to various processes, mechanical properties and illustrations, data labeling, and extracted features, which may necessitate knowledge of both data science and materials science. Fortunately, there may be new ways to approach this problem in AM because of recently developed algorithms like single-shot learning that only need limited resources. Here, we offer a method—data reinforcement and transfer learning, that will be discussed in more detail below—to address this issue without the requirement to gather a large amount of data. The majority of AM literature using machine learning that we have covered in this article, is important to note, concentrating primarily on the technical and design components of AM. Nevertheless, there are currently very few reports on the application of ML in AM sciences and technologies, notably in the areas of microstructure analysis, novel alloy design, properties prediction, and morphology optimization.

21.4 Conclusion

Machine learning (ML) techniques have revolutionized additive manufacturing (AM) of functional polymers and nanocomposites by enhancing design, process optimization, and material performance predictions. It can work by analyzing large datasets. ML algorithms can optimize manufacturing parameters in real time, reducing material waste and costs while ensuring high-quality outputs. In particular, ML aids in developing new materials with tailored properties by modeling the relationships between composition, processing conditions, and material characteristics, which is crucial for applications in electronics, biomedical devices, and structural components. Additionally, ML improves the understanding and dispersion of nanomaterials within polymer matrices, resulting in superior mechanical strength and functionality. As ML continues to evolve, its integration in AM will drive further innovations and efficiencies, unlocking new possibilities in advanced materials and manufacturing processes.

References

Banerji, S. (2022). Using artificial intelligence to derive temperature-dependent mechanical properties of ultra-high performance concrete. In M. Z. Naser (Ed.), Leveraging artificial intelligence in engineering, management, and safety of infrastructure (pp. 403–422). CRC Press.

Birajdar, M. R., & Sewatkar, C. M. (2023). Machine learning approach to predict the thermal performance of closed-loop thermosyphon. *Heat Transfer, 52*(6), 4064–4077. https://doi.org/10.1002/HTJ.22866.

Burela, R. G., Kamineni, J. N., & Harursampath, D. (2020). Multifunctional polymer composites for 3D and 4D printing. *3D and 4D printing of polymer nanocomposite materials: processes, applications, and challenges,* 231–257. https://doi.org/10.1016/B978-0-12-816805-9.00008-9.

Caffaratti, G. D., Marchetta, M. G., Euillades, L. D., Euillades, P. A., & Forradellas, R. Q. (2021). Improving forest detection with machine learning in remote sensing data. *Remote Sensing Applications: Society and Environment, 24*, 100654. https://doi.org/10.1016/J.RSASE.2021.100654.

Cai, J., Zhang, Y., Guo, J., Zhao, X., Lv, J., & Hu, Y. (2022). ST-PN: A spatial transformed prototypical network for few-shot SAR image classification. *Remote Sensing, 14*(9), 2019. https://doi.org/10.3390/RS14092019.

Camargo, P. H. C., Satyanarayana, K. G., & Wypych, F. (2009). Nanocomposites: synthesis, structure, properties and new application opportunities. *Materials Research, 12*(1), 1–39. https://doi.org/10.1590/S1516-14392009000100002.

Chen, L., He, Y., Yang, Y., Niu, S., & Ren, H. (2017). The research status and development trend of additive manufacturing technology. *International Journal of Advanced Manufacturing Technology, 89*(9–12), 3651–3660. https://doi.org/10.1007/S00170-016-9335-4/METRICS.

Chen, M., Mao, S., & Liu, Y. (2014). Big data. *Mobile Networks and Applications, 19*(2), 171–209. https://doi.org/10.1007/S11036-013-0489-0.

Chuisaca-Londa, E., Osorio-Ordóñez, D., Dávalos-Monteiro, J., Ramirez-Cando, L., & Dávalos-Monteiro, R. (2023). Converting polymeric solutions into biomedical nanofibers through electrospinning—An overview of working parameters. *Acta Microscopica, 32*.

Clarissa, W. H. Y., Chia, C. H., Zakaria, S., & Evyan, Y. C. Y. (2022). Recent advancement in 3-D printing: nanocomposites with added functionality. *Progress in Additive Manufacturing, 7*(2), 325–350. https://doi.org/10.1007/S40964-021-00232-Z/METRICS.

Dickson, A. N., Barry, J. N., McDonnell, K. A., & Dowling, D. P. (2017). Fabrication of continuous carbon, glass, and Kevlar fiber reinforced polymer composites using additive manufacturing. *Additive Manufacturing, 16*, 146–152. https://doi.org/10.1016/J.ADDMA.2017.06.004.

Gladman, A. S., Garcia-Leiner, M., Sauer-Budge, A. F., Gladman, A. S., Garcia-Leiner, M., & Sauer-Budge, A. F. (2019). Emerging polymeric materials in additive manufacturing for use in biomedical applications. *AIMS Bioengineering, 6*(1), 1–20. https://doi.org/10.3934/BIOENG.2019.1.1.

Guerrero-Ibáñez, J., Zeadally, S., & Contreras-Castillo, J. (2018). Sensor technologies for intelligent transportation systems. *Sensors, 18*(4), 1212. https://doi.org/10.3390/S18041212.

Gul, J. Z., Sajid, M., Rehman, M. M., Siddiqui, G. U., Shah, I., Kim, K. H., et al. (2018). 3D printing for soft robotics—A review. *Science and Technology of Advanced Materials, 19*(1), 243–262. https://doi.org/10.1080/14686996.2018.1431862.

Guo, S. Z., Gosselin, F., Guerin, N., Lanouette, A. M., Heuzey, M. C., & Therriault, D. (2013). Solvent-cast three-dimensional printing of multifunctional microsystems. *Small (Weinheim an der Bergstrasse, Germany), 9*(24), 4118–4122. https://doi.org/10.1002/SMLL.201300975.

Hormazabal, R. S., Kang, J. W., Park, K., & Yang, D. R. (2022). Not from scratch: Predicting thermophysical properties through model-based transfer learning using graph convolutional networks. *Journal of Chemical Information and Modeling, 62*(22), 5411–5424. https://doi.org/10.1021/ACS.JCIM.2C00846/ASSET/IMAGES/LARGE/CI2C00846_0011.JPEG.

Hunter, P., Coveney, P. V., De Bono, B., Diaz, V., Fenner, J., Frangi, A. F., et al. (2010). A vision and strategy for the virtual physiological human in 2010 and beyond. *Philosophical Transactions of the Royal Society A: Mathematical, Physical and Engineering Sciences, 368*(1920), 2595–2614. https://doi.org/10.1098/RSTA.2010.0048.

Kruth, J. P., Levy, G., Klocke, F., & Childs, T. H. C. (2007). Consolidation phenomena in laser and powder-bed based layered manufacturing. *CIRP Annals, 56*(2), 730–759. https://doi.org/10.1016/J.CIRP.2007.10.004.

Lalmuanawma, S., Hussain, J., & Chhakchhuak, L. (2020). Applications of machine learning and artificial intelligence for Covid-19 (SARS-CoV-2) pandemic: A review. *Chaos, Solitons, and Fractals, 139*. https://doi.org/10.1016/J.CHAOS.2020.110059.

De Leon, A. C., Chen, Q., Palaganas, N. B., Palaganas, J. O., Manapat, J., & Advincula, R. C. (2016). High-performance polymer nanocomposites for additive manufacturing applications. *Reactive and Functional Polymers, 103*, 141–155. https://doi.org/10.1016/J.REACTFUNCTPOLYM.2016.04.010.

Liakos, K. G., Busato, P., Moshou, D., Pearson, S., & Bochtis, D. (2018). Machine learning in agriculture: A review. *Sensors, 18*(8), 2674. https://doi.org/10.3390/S18082674.

Ma, Z., Wan, W., Song, L., Liu, C., Liu, H., & Wu, Y. (2022). An approach of path optimization algorithm for 3D concrete printing based on graph theory. *Applied Sciences, 12*(22), 11315. https://doi.org/10.3390/APP122211315.

Mahdavinejad, M. S., Rezvan, M., Barekatain, M., Adibi, P., Barnaghi, P., & Sheth, A. P. (2018). Machine learning for internet of things data analysis: A survey. *Digital Communications and Networks, 4*(3), 161–175. https://doi.org/10.1016/J.DCAN.2017.10.002.

Melchels, F. P. W., Feijen, J., & Grijpma, D. W. (2010). A review on stereolithography and its applications in biomedical engineering. *Biomaterials, 31*(24), 6121–6130. https://doi.org/10.1016/J.BIOMATERIALS.2010.04.050.

Mohammadi, M. G., & Elbestawi, M. (2020). Real-time monitoring in L-PBF using a machine learning approach. *Procedia Manufacturing, 51*, 725–731. https://doi.org/10.1016/J.PROMFG.2020.10.102.

Mondal, M. R. H., Bharati, S., & Podder, P. (2021). Diagnosis of COVID-19 using machine learning and deep learning: A review. *Current Medical Imaging Formerly Current Medical Imaging Reviews, 17*(12), 1403–1418. https://doi.org/10.2174/1573405617666210713113439.

Mylonas, G., Triantafyllis, C., & Amaxilatis, D. (2019). An augmented reality prototype for supporting IoT-based educational activities for energy-efficient school buildings. *Electronic Notes in Theoretical Computer Science, 343*, 89–101. https://doi.org/10.1016/J.ENTCS.2019.04.012.

Ngo, T. D., Kashani, A., Imbalzano, G., Nguyen, K. T. Q., & Hui, D. (2018). Additive manufacturing (3D printing): A review of materials, methods, applications, and challenges. *Composites Part B: Engineering, 143*, 172–196. https://doi.org/10.1016/J.COMPOSITESB.2018.02.012.

Nguyen, D. N., Nguyen, V. H., Trinh, T., Ho, T., & Le, H. S. (2024). A personalized product recommendation model in e-commerce based on retrieval strategy. *Journal of Open Innovation: Technology, Market, and Complexity, 10*(2), 100303. https://doi.org/10.1016/J.JOITMC.2024.100303.

Obermeyer, Z., & Emanuel, E. J. (2016). Predicting the future—Big data, machine learning, and clinical medicine. *New England Journal of Medicine, 375*(13), 1216–1219. https://doi.org/10.1056/NEJMP1606181/SUPPL_FILE/NEJMP1606181_DISCLOSURES.PDF.

Prabhakaran, S., Annie Uthra, R., & Preetharoselyn, J. (2023). Comprehensive analysis of defect detection through image processing and machine learning for photovoltaic panels. *Lecture Notes in Electrical Engineering, 967*, 245–261. https://doi.org/10.1007/978-981-19-7169-3_23.

Sames, W. J., List, F. A., Pannala, S., Dehoff, R. R., & Babu, S. S. (2016). The metallurgy and processing science of metal additive manufacturing. *International Materials Reviews, 61*(5), 315–360. https://doi.org/10.1080/09506608.2015.1116649.

Sapru, A., & Garimella, S. (2020). Leveraging unlabeled speech for sequence discriminative training of acoustic models. *Proc. Interspeech,* 3585–3589. https://doi.org/10.21437/Interspeech.2020-2056.

Sarker, I. H. (2019). Context-aware rule learning from smartphone data: Survey, challenges, and future directions. *Journal of Big Data, 6*(1), 1–25. https://doi.org/10.1186/S40537-019-0258-4/FIGURES/2.

Sarker, I. H. (2021). Machine learning: Algorithms, real-world applications and research directions. *SN Computer Science, 2*(3), 1–21. https://doi.org/10.1007/S42979-021-00592-X/FIGURES/11.

Sarker, I. H., Hoque, M. M., Uddin, M. K., & Alsanoosy, T. (2021). Mobile data science and intelligent apps: Concepts, AI-based modeling and research directions. *Mobile Networks and Applications, 26*(1), 285. https://doi.org/10.1007/S11036-020-01650-Z.

Saroia, J., Wang, Y., Wei, Q., Lei, M., Li, X., Guo, Y., et al. (2020). A review on 3D printed matrix polymer composites: Its potential and future challenges. *International Journal of Advanced Manufacturing Technology, 106*(5–6), 1695–1721. https://doi.org/10.1007/S00170-019-04534-Z.

Scruggs, S. B., Watson, K., Su, A. I., Hermjakob, H., Yates, J. R., Lindsey, M. L., et al. (2015). Harnessing the heart of big data. *Circulation Research, 116*(7), 1115–1119. https://doi.org/10.1161/CIRCRESAHA.115.306013.

Sonnenschein, R., Gajdosova, K., & Holly, I. (2016). FRP composites and their using in the construction of bridges. *Procedia Engineering, 161*, 477–482. https://doi.org/10.1016/J.PROENG.2016.08.665.

Sroczyński, A., & Czyżewski, A. (2023). Road traffic can be predicted by machine learning equally effectively as by complex microscopic model. *Scientific Reports, 13*(1), 1–15. https://doi.org/10.1038/s41598-023-41902-y.

Suzuki, A., Shiba, Y., Ibe, H., Takata, N., & Kobashi, M. (2022). Machine-learning assisted optimization of process parameters for controlling the microstructure in a laser powder bed fused WC/Co cemented carbide. *Additive Manufacturing, 59*, 103089. https://doi.org/10.1016/J.ADDMA.2022.103089.

Tabares-Soto, R., Orozco-Arias, S., Romero-Cano, V., Bucheli, V. S., Rodríguez-Sotelo, J. L., & Jiménez-Varón, C. F. (2020). A comparative study of machine learning and deep learning algorithms to classify cancer types based on microarray gene expression data. *PeerJ Computer Science, 6*, e270.

Turner, B. N., Strong, R., & Gold, S. A. (2014). A review of melt extrusion additive manufacturing processes: I. Process design and modeling. *Rapid Prototyping Journal, 20*(3), 192–204. https://doi.org/10.1108/RPJ-01-2013-0012/FULL/XML.

Vivas, C., & Salehi, S. (2021). Real-time model for thermal conductivity prediction in geothermal wells using surface drilling data: A machine learning approach. *Stanford Geothermal Workshop*.

Wang, X., Jiang, M., Zhou, Z., Gou, J., & Hui, D. (2017). 3D printing of polymer matrix composites: A review and prospective. *Composites Part B: Engineering, 110*, 442–458. https://doi.org/10.1016/J.COMPOSITESB.2016.11.034.

Westphal, E., & Seitz, H. (2021). A machine learning method for defect detection and visualization in selective laser sintering based on convolutional neural networks. *Additive Manufacturing, 41*, 101965. https://doi.org/10.1016/J.ADDMA.2021.101965.

Woergoetter, F., & Porr, B. (2008). Reinforcement learning. *Scholarpedia, 3*(3), 1448. https://doi.org/10.4249/SCHOLARPEDIA.1448.

Wu, L., Liu, Q., Chen, E., Yuan, N. J., Guo, G., & Xie, X. (2016). Relevance meets coverage. *ACM Transactions on Intelligent Systems and Technology (TIST), 7*(3). https://doi.org/10.1145/2700496.

Zareiyan, B., & Khoshnevis, B. (2017). Interlayer adhesion, and strength of structures in Contour Crafting—Effects of aggregate size, extrusion rate, and layer thickness. *Automation in Construction, 81*, 112–121. https://doi.org/10.1016/J.AUTCON.2017.06.013.

Zhang, C., Zhang, L., Shen, W., Liu, C., Xia, Y., & Li, R. (2016). Study on constitutive modeling and processing maps for hot deformation of medium carbon Cr–Ni–Mo alloyed steel. *Materials & Design, 90*, 804–814. https://doi.org/10.1016/J.MATDES.2015.11.036.

Zhang, Y., Purssell, C., Mao, K., & Leigh, S. (2020). A physical investigation of wear and thermal characteristics of 3D printed nylon spur gears. *Tribology International, 141*. https://doi.org/10.1016/J.TRIBOINT.2019.105953.

Ziaee, M., & Crane, N. B. (2019). Binder jetting: A review of process, materials, and methods. *Additive Manufacturing, 28*, 781–801. https://doi.org/10.1016/J.ADDMA.2019.05.031.

Zimpel, T., & Hubl, M. (2019). Smart urban objects to enhance safe participation in major events for the elderly. *Proceedings of the 2019 Federated Conference on Computer Science and Information Systems, FedCSIS, 2019*, 505–514. https://doi.org/10.15439/2019F180.

Section 5

Challenges, environmental and health impact and commercialization aspects

Critical challenges, future opportunities, life cycle assessments, environmental and health impact and commercialization aspects of additively manufactured functional polymers and nanocomposites

22

Astha Bisht[1], *Mahak Kushwaha*[1], *Swati Agarwal*[2], *and Suphiya Khan*[3]
[1]Department of Bioscience and Biotechnology, Banasthali Vidyapith, Banasthali, Rajasthan, India, [2]Drumlins Water Technologies Pvt. Ltd., Jaipur, Rajasthan, India, [3]Shriram Institute for Industrial Research, Gurugram, Haryana, India

22.1 Introduction

Additive manufacturing (AM) is a revolutionary method of modern manufacturing that uses a range of digital technologies to fabricate intricate, affordable, lightweight, durable, and customized systems and components. Additive manufacturing, often referred to as 3D printing, has revolutionized the field of materials science by contributing a multipurpose and highly customizable approach to manufacturing; Fig. 22.1 mentions some game-changing approaches of additive manufacturing. Within this realm, functional polymers and nanocomposites stand out as particularly promising areas for innovation and advancement. Additively manufactured functional polymers and nanocomposites face critical challenges in terms of limited combinations of matrix and reinforcements, hardware modification requirements, and mechanical characterization difficulties (Darji et al., 2023).

Functional polymers and nanocomposites present a multitude of advantages compared to traditional polymers, such as improved mechanical performance and the ability to customize their properties. This makes them exceptionally appealing for a diverse array of advanced applications. However, the full commercialization of additively manufactured functional polymers and nanocomposites faces significant trials that must be talked about to unlock their full potential. Additive manufacturing (AM) technology enables the production of complex parts with improved lead times and reduced material wastage. Polymer composites fabricated through AM have

Additive Manufacturing of Functional Polymers and Nanocomposites. DOI: https://doi.org/10.1016/B978-0-443-26516-7.00022-5
© 2025 Elsevier Ltd. All rights are reserved, including those for text and data mining, AI training, and similar technologies.

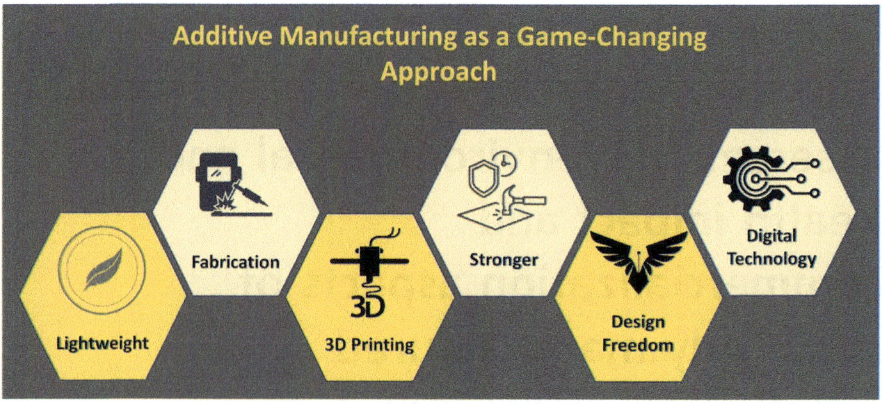

Figure 22.1 Various approaches for additive manufacturing.

compensations in the mechanical act and are suitable for useful parts (Darji et al., 2023).

One of the critical challenges in this domain revolves around the limited combinations of matrix and reinforcements available for additive manufacturing. Unlike conventional polymers, where the choices are relatively abundant, the options for additive manufacturing are more restricted, leading to constraints in material properties and performance. Additionally, modifications to hardware are often required to accommodate the unique characteristics of these materials, adding complexity and cost to the manufacturing process (Yan et al., 2023).

Mechanical characterization poses another obstacle, as accurately assessing the mechanical properties of additively manufactured functional polymers and nanocomposites can be inherently difficult. The inclusion of photocurable CNW fillers improved the thermomechanical properties of the 3D-printed nanocomposites. The functionalized chitin-based nanocomposites showed enhanced mechanical performance and high-resolution print features (Maalihan et al., 2022). Despite these challenges, the potential advantages of additively manufactured functional polymers and nanocomposites cannot be understated. However, to fully capitalize on these benefits, several key areas must be addressed.

In addition to technical hurdles, an inclusive valuation of the environmental and health ramifications of additively manufactured functional polymers and nanocomposites is imperative. Life cycle assessments (LCAs) play a pivotal role in elucidating the holistic impact of these materials, encompassing their production, utilization, and disposal phases (Alonso et al., 2020). However, further research is warranted to ascertain their release dynamics, potential human exposure pathways, and environmental fate, ensuring alignment with rigorous safety and sustainability criteria (Brouwer et al., 2019). Despite their promising prospects, the commercial deployment of additively manufactured functional polymers and nanocomposites remains constrained by multifaceted challenges. Issues such as dispersibility, scalability, cost-efficiency, and uncertainties regarding their impacts on human health and the environment impede widespread adoption (Gebler et al., 2014).

Future endeavors should focus on advancing the development of ecofriendly nanocomposites, prioritizing sustainability, and environmental stewardship (Hun et al., 2020). Opportunities for progress lie in the realm of green nanocomposite innovation, which integrates principles of sustainability into material design and production processes (Thakur et al., 2015). Embracing sustainable practices across the entire supply chain and embracing the principles of the circular economy can further augment the viability of these materials, especially in industries such as aerospace, where innovation is pivotal (Bilal et al., 2021). Additively manufactured functional polymers and nanocomposites offer immense promise, their commercialization hinges on overcoming critical challenges in mechanical characterization, materials synthesis, and environmental assessment. Nonetheless, there are worries about how polymer nanocomposites may affect the environment and human health (Qian et al., 2022). Future opportunities lie in the development of green nanocomposites, the implementation of sustainability practices, and the adoption of circular economy concepts in the aerospace industry (Tripathy & Gupta, 2024).

22.2 Methods for additive manufacturing of functional polymers and nanocomposites

Methods for additive manufacturing of functional polymers and nanocomposites encompass various techniques tailored to meet specific material requirements and application needs. These types of additive manufacturing and their uses are discussed in Table 22.1.

22.2.1 Vat polymerization

Vat polymerization is a 3D printing technology that offers high-precision fabrication and rapid printing speed. It is often used to create substances with delicate and complex structures. To overcome these limitations, researchers have integrated nanomaterials with vat polymerization-based 3D printing. This approach allows for the creation of nanocomposites with enhanced mechanical, thermal, and electrical properties. In particular, the addition of multiwalled carbon nanotubes (MWCNTs) to photopolymers has been explored to fabricate electrically conductive structures (Li et al., 2023). Vat polymerization has also been used in the biomedical industry, where considerations such as print parameters and material properties are crucial for successful prints (Iervolino et al., 2022). Overall, vat polymerization-based additive manufacturing techniques offer the potential to produce functional polymers and nanocomposites with improved properties for various applications (Kant et al., 2022).

22.2.2 Material extrusion

Material extrusion methods, such as additive manufacturing (AM) and reactive extrusion additive manufacturing (REAM), have been used to produce functional polymers and nanocomposites (Banks & Emami, 2022). These methods allow to production of multifaceted and custom geometries that are not achievable with

Table 22.1 Types of additive manufacturing and their uses.

Types of additive manufacturing	Uses	References
Fused Deposition Modeling (FDM)	Prototyping, concept modeling, low-cost production of functional parts	Gibson et al. (2014)
Stereolithography (SLA)	High-resolution prototyping, pattern making, dental applications	Deshmane et al. (2021)
Selective Laser Sintering (SLS)	Functional prototypes, end-use parts, tooling, aerospace components	Kruth et al. (2007)
Digital Light Processing (DLP)	Dental, jewelry, custom medical devices, rapid tooling	Tumbleston et al. (2015)
Directed Energy Deposition (DED)	Repairing components, adding features to existing parts, aerospace	Svetlizky et al. (2022)
Direct Metal Laser Sintering (DMLS)	Aerospace components, medical implants, automotive parts	Murr et al. (2010)
Electron Beam Melting (EBM)	Orthopedic implants, aerospace components, dental prosthetics	Thijs et al. (2010)
Selective Laser Melting (SLM)	Tooling inserts, aerospace components, dental crowns	Gu and Shen (2016)
Laminated Object Manufacturing (LOM)	Prototyping, architectural models, packaging	Dermeik and Travitzky (2020)
Ceramic 3D Printing	Dental restorations, aerospace components, electronic components	Ligon et al. (2017)

conventional manufacturing techniques (Vidakis et al., 2023). Researchers have utilized various fillers, such as carbon-based nanoparticles and iron oxide particles, to create multifunctional materials with enhanced properties (Petousis et al., 2023). The addition of fillers, such as polytetrafluoroethylene (PTFE) and silicon nitride nanoparticles, has been shown to improve the mechanical performance of the nanocomposites. Numerical modeling using computational fluid dynamics (CFD) has also been used to test the material extrusion procedure and optimize the fabrication of polymer composites. These advancements in material extrusion methods have expanded the potential applications of functional polymers and nanocomposites in areas such as sensors, structural health monitoring, and soft robotics.

22.2.3 Powder bed fusion

Powder bed fusion methods, such as selective laser sintering (SLS) and laser powder bed fusion (LPBF), are widely used for additive manufacturing of functional polymers and nanocomposites. LPBF is commonly used for metallic materials and powder-based polymers, including thermoplastics like Polyamide 12 and Polyamide 6 (Hassan et al., 2023). LPBF can also be used to fabricate complex ceramic parts by converting highly porous polymeric preforms into ceramics through preceramic polymer infiltration and pyrolysis (Tan et al., 2023). SLS, on the other hand, can be used to prepare multiscale fiber-reinforced powders, where the polymer matrix is coated on the microfiber and carbon nanotubes cover the powder surface (Pelanconi et al., 2023). These methods offer the ability to improve the mechanical properties and functionality of additively manufactured parts, such as achieving anisotropic mechanical properties for composite parts by regulating fiber orientations in the powder packing process (Schäfer et al., 2024). Additionally, powder bed fusion methods can be used to create functionally graded materials, including magneto-active composites with locally adjusted mechanical properties (Wang et al., 2023).

22.2.4 Direct light processing

Direct light processing methods, such as 3D laser nanoprinting and (DLW) direct laser writing, have been developed for additive manufacturing of functional polymers and nanocomposites. 3D laser nanoprinting allows the fabrication of 3D nanostructures with novel properties beyond the optical diffraction limit (Cazin et al., 2023). DLW technology enables the processing of functional quantum dots (QD)–polymer nanocomposites, which can be used in high-resolution displays, anti-counterfeit technology, and optical encryption (Liu et al., 2023). Additionally, DLW can be used to produce complex 3D metal nanoparticle–polymer nanocomposites, offering opportunities for applications in photonics, metamaterials, and drug delivery (Jue et al., 2023). Furthermore, direct light processing methods have been used to 3D print electrically conductive structures using photopolymers loaded with multiwalled carbon nanotubes (MWCNTs), showing potential for the development of electronic devices and microelectromechanical systems (MEMS) (Ji Hyuk et al., 2023).

22.2.5 Stereolithography

Stereolithography techniques have been applied in the additive manufacture of functional polymers and nanocomposites. These technologies take advantage of 3D printing processes to construct structures with enhanced qualities. Development of photopolymers loaded with multiwalled carbon nanotubes for 3D printing of electrically conductive structures (Iervolino et al., 2022). The fabrication of nanostructured polymer surfaces using 3D printing, combining additive manufacturing with nanolithography techniques (Martínez-García et al., 2023). Stereolithography is used to fabricate 3D structures with improved electrical behavior by incorporating metallic nanoparticles into acrylic resin (Valencia et al., 2022). Stereolithography is also used to process methacrylate-based resins containing chitin nanowhiskers, resulting in enhanced thermomechanical properties of the printed nanocomposites (Maalihan et al., 2022). These studies highlight the potential of stereolithography methods for the additive manufacturing of functional polymers and nanocomposites.

22.2.6 Direct energy deposition

Direct energy deposition (DED) methods have been used for additive manufacturing of functional polymers and nanocomposites (Woosley & Aravamudhan, 2022). These methods involve the use of 3D printing techniques to fabricate solid objects directly from computer-modeled designs. One approach is the use of Fused Deposition Modeling (FDM) 3D printing, which utilizes thermoplastic polymers to create printable composites with enhanced functionality (Kelly et al., 2021). Another approach involves the development of ink materials for 3D printing of conductive nanocomposites, such as epoxy-carbon nanotube-based nanocomposites, which exhibit improved mechanical strength and electrical conductivity (Pal & Krishnamoorthy, 2022). In addition, a new 3D printing technique called electrical field-assisted direct ink deposition (EF-DID) has been developed to create large-scale nanofilms with controllable nanostructures and tailored characteristics.

22.2.7 Sheet lamination

Sheet lamination methods have been developed for additive manufacturing of functional polymers and nanocomposites. These methods involve the lamination of sheets of materials to fabricate composite objects. The Composites-Based Additive Manufacturing (CBAM) system combines randomly oriented carbon Fiber reinforcement with thermoplastic matrix materials to fabricate composite objects (Bhatt et al., 2019). Nonetheless, there are worries about how polymer nanocomposites may affect the environment and human health (Bhatt et al., 2019). Another method combines inkjet-based printing of UV-curable polymers with the deposition of nanoscale fiber mats to fabricate multimaterial laminated nanocomposites (Spackman et al., 2015). Additionally, the Laminated Object Manufacturing (LOM) technique involves layer-by-layer lamination of paper material sheets to create parts (Ichiura et al., 2006). These sheet lamination methods offer the potential for manufacturing structures with heterogeneous materials and improved mechanical properties.

22.3 Additively manufactured functional polymers and nanocomposites

Additive manufacturing (AM) of polymers has gained significant interest in recent years due to its benefits such as easy customization, reduced material wastage, and short turnaround times (Markandan & Lai, 2023). Polymer nanocomposites, which offer superior properties compared to conventional materials, have also attracted attention in research and development. strategies for reinforcing polymer matrices during AM by adding and positioning fillers (Yaragatti & Patnaik, 2021). Additionally, the development of ink materials for 3D printing of conductive nanocomposites has been investigated, along with theoretical modeling to understand their properties (Jisun et al., 2023). Furthermore, the successful fabrication of complex 3D metal nanoparticle-polymer nanocomposites using two-photon polymerization has been reported, offering opportunities for applications in photonics, metamaterials, and drug delivery. Lastly, the creation of photopolymers with carbon nanotubes for 3D printing electrically conducting objects has demonstrated promise for the advancement of microelectromechanical systems (MEMS) and electronic devices (Iervolino et al., 2022).

22.4 Life cycle assessment of additively manufactured functional polymers and nanocomposites

Life cycle assessment (LCA) is an important tool for evaluating the environmental impact of additively manmade functional polymers and nanocomposites. LCA makes it possible to evaluate these materials' complete life cycle, from the extraction of raw materials to their disposal. Several studies have highlighted the need for LCA in this field, emphasizing the importance of considering factors such as circular economy, life cycle costs, and potential risks associated with these materials (Gil et al., 2023). LCA techniques have been used to assess the impact of polymers and polymer composites on the environment and human health, but there are still limitations to be addressed. Additionally, LCA has been applied to evaluate the environmental impacts of specific processes, such as the production of geopolymer binders for 3D printing, and using construction and demolition waste as a sustainable resource. Efforts to improve sustainability in AM include the development of bio-based polymers and recycled materials, as demonstrated by research conducted by (Domingo-Roca et al., 2022). A need for holistic assessments considering both material composition and manufacturing processes to identify opportunities for environmental improvement. (Melenka et al., 2018). A range of studies have explored the life cycle assessment (LCA) of additively manufactured functional polymers and nanocomposites (Pizza et al., 2014). LCA on graphite nanoplatelet-filled epoxy-based composites, emphasizing the substantial energy needed in their manufacturing (Schrijvers et al., 2014). The environmental impact of biodegradable agricultural films made of PBAT/LDH nanocomposites with those made of LDPE, finding that the former could be a more sustainable alternative (Pietrini et al., 2007).

LCA on composites based on poly (3-hydroxybutyrate) (PHB), indicating that they might have less of an impact on the environment than traditional petrochemical polymers, especially when sugar cane bagasse is used as a filler. The potential of high-performance polymer nanocomposites for additive manufacturing, emphasizes their improved mechanical properties and potential for use in harsh environments (De Leon et al., 2016). These studies collectively highlight the potential for additively manufactured functional polymers and nanocomposites to offer more sustainable alternatives to traditional materials. Additive manufacturing has emerged as a transformative technology, revolutionizing traditional manufacturing processes and disrupting various industries as shown in Fig. 22.1. Overall LCA studies provide valuable insights into the environmental implications of additively manufactured functional polymers and nanocomposites, guiding efforts toward more sustainable material choices and manufacturing practices.

22.5 Environmental and health impacts of additively manufactured functional polymers and nanocomposites

Additively manufactured functional polymers and nanocomposites have significant implications for both the environment and human health (Luty-Błocho et al., 2023). These materials may be used in a wide range of sectors, such as healthcare, packaging, automotive, and energy technology (Leonard & Guenin, 2023). However, there are concerns about their potential negative effects.

The commercial aspects of functional polymers and nanocomposites in additive manufacturing are tightly intertwined with considerations of their environmental and health impacts. Table 22.2 gives a general overview of different polymers and the nanocomposites made of them, highlighting the impact these materials have on the environment, human health, and application industries. Polymer-based nanocomposites can have a variation of effects on the environment; they can be decomposable in some circumstances or may contribute to plastic pollution. On the other hand, the presence of nanomaterials in the environment, such as in wastewater treatment, raises questions about their potential toxicity and the release of harmful byproducts (Bachar & Mazumder, 2022). Similarly, the use of nanoadditives in polymer nanocomposites can improve product properties but also raises concerns about their impact on humans and the environment (Bhat et al., 2022). Additive manufacturing processes typically consume energy and raw materials. While 3D printing can be more efficient than traditional subtractive manufacturing methods in certain scenarios, the overall environmental impact depends on factors such as the type of printer, materials used, and production scale (Baechler et al., 2013).

3D printing can generate waste in the form of excess material, support structures, and failed prints. Disposal of unused or failed prints can contribute to landfill waste unless suitable recycling or reclamation processes are in place. Some additive manufacturing processes, particularly those involving high-temperature printing or certain materials, can release volatile organic compounds (VOCs) and other pollutants into the atmosphere (Kreiger et al., 2017). Some polymers used in additive manufacturing, as well as the additives and solvents associated with them, can emit VOCs during printing. Prolonged exposure to VOCs may pose health risks to workers, including respiratory irritation and other health effects (Stefaniak et al., 2017). Additive manufacturing processes can generate airborne particles, including ultrafine particles and nanoparticles, which may pose respiratory health risks to workers if not properly controlled. Workers involved in additive manufacturing processes may be exposed to chemicals present in the printing materials, such as monomers, oligomers, and photoinitiators, which can have toxic effects depending on their chemical composition and concentration (Azimi et al., 2016).

Additionally, the presence of microplastics and nanoplastics in the environment, including those derived from plastic deterioration and fragmentation, raises concerns about their toxicological implications and potential harm to organisms. Overall, further research and risk assessment are needed to better understand and mitigate the

Table 22.2 Commercial aspects of functional polymers and nanocomposites.

Polymer	Nanocomposite	Effect on environment	Effect on human health	Range of sectors (application)	References
Polyethylene	Polyethylene nanocomposites	Can contribute to plastic pollution if not recycled properly	Low toxicity, but potential health risks from additives or impurities	Packaging, automotive, construction	Visakh and Lüftl (2016)
Polypropylene	Polypropylene-based nanocomposites	Non-biodegradable, can persist in the environment	Low toxicity, but may release harmful gases when burned	Packaging, textiles, automotive	Fu et al. (2022)
Polylactic Acid (PLA)	PLA-based nanocomposites	Biodegradable under certain conditions, reducing environmental impact	Generally considered safe, but potential allergic reactions	Packaging, biomedical, 3D printing	Farah and Anderson (2016)
Polyvinyl Chloride (PVC)	PVC-based nanocomposites	Environmental concerns due to chlorine content and toxic additives	Potential health risks from exposure to phthalates and other additives	Construction, automotive, healthcare	Alhassan et al. (2024)
Polyethylene Terephthalate (PET)	PET-based nanocomposites	Non-biodegradable contributes to plastic pollution	Low toxicity, but concerns about leaching of chemicals	Packaging, textiles, electronics	Li et al. (2024)
Polyurethane	Polyurethane nanocomposites	Nonbiodegradable, potential for environmental contamination	Low toxicity, but concerns about isocyanates during production	Coatings, adhesives, automotive	Hu et al. (2024)
Poly (methyl methacrylate) (PMMA)	PMMA-based nanocomposites	Nonbiodegradable, can persist in the environment	Generally considered safe, but may cause irritation or allergic reactions	Automotive, signage, medical devices	Findik et al. (2024)
Polyethylene Oxide (PEO)	PEO-based nanocomposites	Biodegradable, reducing environmental impact	Generally considered safe, but may cause gastrointestinal irritation	Pharmaceuticals, personal care, coatings	Taokaew and Chuenkaek (2024)

(*Continued*)

Table 22.2 (Continued)

Polymer	Nanocomposite	Effect on environment	Effect on human health	Range of sectors (application)	References
Polycarbonate	Polycarbonate nanocomposites	Environmental concerns due to bisphenol-A (BPA) content	Potential health risks from BPA exposure, especially to children	Electronics, automotive, medical devices	Nurazzi et al. (2024)
Polyamide (Nylon)	Nylon-based nanocomposites	Nonbiodegradable, can persist in the environment	Generally considered safe, but potential for allergic reactions	Textiles, automotive, consumer goods	Peng et al. (2024)
Polypropylene carbonate (PPC)	PPC-based nanocomposites	Biodegradable, reducing environmental impact	Generally considered safe, but limited data on long-term effects	Packaging, automotive, electronics	Su et al. (2019)
Polyvinyl Alcohol (PVA)	PVA-based nanocomposites	Biodegradable, reducing environmental impact	Generally considered safe, but potential for skin irritation	Packaging, textiles, biomedical	Park et al. (2023)
Polysulfone	Polysulfone nanocomposites	Nonbiodegradable, can persist in the environment	Generally considered safe, but potential for allergic reactions	Filtration, membranes, aerospace	Shonaike and Salami (2018)

environmental and health impacts of additively manufactured functional polymers and nanocomposites.

22.6 Commercial aspects of functional polymers and nanocomposites

The commercial aspects of functional polymers and nanocomposites encompass various factors related to their production, market demand, applications, and economic viability. Here are some key considerations:

22.6.1 Cost-effective

Functional polymers and nanocomposites offer a cost-effective solution for various commercial applications due to their unique properties and functions (Tiwari et al., 2023). Surface modifications of polymers and nanomaterials enhance their physicochemical properties, making them suitable for diverse fields like medicine, textiles, water treatment, and food packaging (Zou et al., 2008). The incorporation of nano-sized insulating particles in polymer matrices improves their thermal conductivity and fire-retardancy, creating opportunities in defense, energy storage, and structural applications (polymer nanocomposite technologies designed for biomedical applications). Nanomaterials integrated into polymers at low loading levels form nanocomposites with tailored properties, ideal for clean energy applications such as solar cells and fuel cells. By utilizing simple and abundant ingredients, cost-effective polymer composites with enhanced performance, including impact resistance and particle dispersion, can be efficiently produced.

22.6.2 Market

Polymeric nanocomposites have gained significant scientific and industrial interest due to their unique physicochemical properties and the ability to modify their structural aspects for exceptional functional advantages over bulk materials (Shivalkar et al., 2023). These materials offer improved mechanical strength, toughness, stiffness, electrical and thermal conductivity, flame retardancy, and barrier properties (Guarás & Alvarez, 2022). The use of polymeric nanocomposites has been explored in fields such as food and bioprocessing industries, where they have been utilized for biocatalysis, active packaging, waste treatment, and other functionalities. Additionally, the surface modification of polymers and nanomaterials has been investigated to enhance their physicochemical and biological properties, making them suitable for applications in the medical field, textile industry, food packaging, and water treatment (Makvandi et al., 2021).

22.6.3 Reusability

Polymeric nanocomposites have been developed and utilized in various industrial fields, including food and bioprocessing industries, due to their versatile functionalities and potential for ecofriendly and biodegradable products (Kehinde et al., 2021). Surface modifications of nanomaterials and polymers have been explored to improve their biological properties and physicochemical, leading to advancements in medical applications, water treatment, the textile industry, and food packaging (Makvandi et al., 2021). Nanocomposite membranes, which integrate inorganic and organic materials, have shown potential in different fields such as wastewater treatment, seawater desalination, and the food industry (Rasheed et al., 2020). The use of nanocomposites in food and bioprocessing industries has shown promise in improving food quality, food packaging systems, and nanoencapsulation of bioactive food compounds (Braga et al., 2021). Polymeric nanocomposites offer exceptional functional advantages over their bulk counterparts, making them suitable for various applications in healthcare, medicine, microelectronics, chemical engineering, and mechanical engineering (Shivalkar et al., 2023)

22.6.4 Flexibility

Flexible and stretchable electronic devices and systems have expanded substantial attention due to their potential applications in various fields such as robotics, healthcare, and environmental monitoring (Rishi & Beaucage, 2022). The development of flexible and stretchable electronic materials has been approached through three main strategies. One strategy involves the use of organic or polymeric materials, which can be made more flexible and elastic through chemical modification or composition formation with plasticizers or elastomers (Mahfuz et al., 2023). Another strategy is the utilization of nanometer-scale materials, which exhibit high flexibility and can be made stretchable through composition formation with elastomers (Gupta et al., 2017). The third strategy involves the use of ionogels, which are stretchable and ionically conductive (Makvandi et al., 2021). These advancements in soft functional materials have led to the development of flexible/stretchable electrodes, thermoelectric conversion materials, and polymer composites for large-area applications and tactile sensors (Tee & Ouyang, 2018).

22.7 Critical challenges

Additively manufactured functional polymers and nanocomposites present several critical challenges that need to be addressed for their widespread adoption and optimized utilization. These challenges span various aspects mentioned in Fig. 22.2 Below are some of the key critical challenges:

22.7.1 Availability

The critical challenges in the availability of a small group of polymers and metal powders are related to their scarcity, concentration in specific geographic areas, and supply uncertainties. The availability of critical metals, like earth elements, platinum group metals, and tantalum, is a strategic issue due to their crucial role in high-technology products and military systems (Lima & Jose, 2015). The concentration of

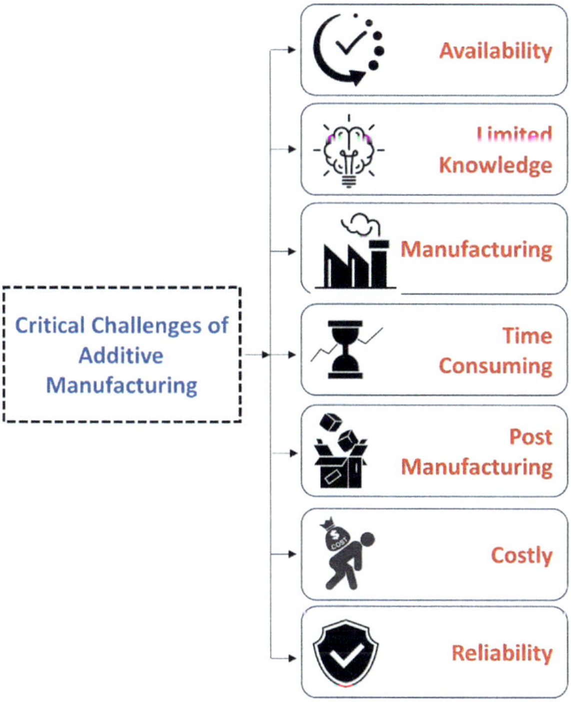

Figure 22.2 Different challenges of additive manufacturing.

metal and alloy producers further complicates the risk assessment and mitigation for industrial firms (Diallo et al., 2015). On the other hand, the availability of a small group of polymers is a challenge due to the difficulty in recognizing and separating minute structural differences in polymer structures (Slowinski et al., 2013). Most polymeric goods contain mixtures of polymers with diverse structures because conventional polymer separation techniques are not very good at controlling and differentiating polymer structures (Girtan et al., 2021). Therefore, the development of novel techniques for polymer recognition and separation is in great demand to address these challenges (Hosono & Uemura, 2021).

22.7.2 Limited knowledge

Limited knowledge of additively manufactured functional polymers and nanocomposites stems from challenges in manufacturing complex structures with high mechanical strength and shape fidelity (Latif et al., 2023). Additive manufacturing requires a workforce with specialized skills and knowledge in materials science, engineering, design, and operation of 3D printing equipment. However, there is often a shortage of adequately trained personnel in these areas, which can hinder the adoption and optimization of additive manufacturing processes (Wohlers Associates, 2021). Adequate training in safety protocols and procedures is crucial to mitigate risks associated with handling and processing additive manufacturing materials, especially those that may emit hazardous fumes or particles during printing (Masood, 2019). Additive Manufacturing of Polymer Composites (AMPCs) faces obstacles like limited material combinations and mechanical characterization challenges (Darji et al., 2023).

The evolution of smart systems using additive manufacturing involves innovative materials responsive to external stimuli, leading to 4D printing advancements (Ikram et al., 2022). Optimization of material properties through additive manufacturing often involves trial-and-error approaches due to the limited predictive capabilities stemming from insufficient knowledge of the underlying mechanisms governing property evolution during printing and postprocessing (Ning et al., 2019). Life cycle assessment (LCA) studies on nanocomposite filaments in 3D printing evaluate environmental impacts, aiding in sustainable material selection (Petrakli et al., 2022). Addressing these gaps through further research can advance the utilization of Additively Manufactured Functional Polymers and Nanocomposites for diverse applications.

22.7.3 Manufacturing

The challenges of additive functional polymers in terms of low manufacturing efficiency and optimization are being addressed through various approaches. Challenges in additive manufacturing of polymers to enhance manufacturing efficiency and overcome existing limitations. The development in additive manufacturing of polymer matrix composites (PMCs) using techniques like Fused Deposition Modeling (FDM) and Stereolithography (SLA) (Memarzadeh et al., 2023) is discussed. Emphasizing the need of creating new high-performance AM materials and refining printing techniques to broaden application areas (Zhou & Wang, 2023). Processing of polyolefins in AM, highlighting challenges and strategies for improvement (Das et al., 2023). The current research on energy efficiency in AM, proposes a comprehensive work to optimize energy consumption throughout the process (May & Psarommatis, 2023). The obstacles in AM of polymer composites and the need for mechanical characterization to enhance industrial utilization (Darji et al., 2023). These studies collectively contribute to advancing the optimization of materials and processes in additive manufacturing, aiming to enhance manufacturing efficiency and overcome existing challenges.

22.7.4 Time-consuming

Additively manufactured functional polymers and nanocomposites face critical challenges in terms of time-consuming processes. The creation and synthesis of polymer composites through additive manufacturing (AM) technologies can be hindered by various factors such as limited combinations of matrix and reinforcements, hardware modification requirements, and challenges in mechanical characterization. Additionally, the preparation of printable composite feedstock filaments and printing issues during the printing of composites can also contribute to time-consuming processes (Darji et al., 2023). Furthermore, the limited availability of different materials, inconsistent mechanical properties, and surface quality issues can restrict the functions of end-use parts in laser sintering additive manufacturing

(Kumar Penumakala et al., 2020). These difficulties point to the necessity of more study and development to get around these restrictions and improve the productivity and timeliness of the production of functional polymers and nanocomposites (Saleh Alghamdi et al., 2021).

22.7.5 Postmanufacturing

Postprocessing treatments applied to additively manufacture functional polymers and nanocomposites face a critical challenge in ensuring quality. Improving the microstructure and functional behavior of additively manufactured samples is essential as the anisotropy of AM structures is a significant problem faced by the industry. Various conditions for postprocessing treatments have been explored, leading to significant improvements in mechanical characteristics such as bending stress, stress state, impact strength, fatigue, hardness, and more. However, achieving better mechanical performance requires considering related physical characteristics such as reactivity, viscosity, processability, crystallinity, dynamic stability, heat deflection temperature, and microstructural structure (Dinița et al., 2023). Additionally, challenges in postprocessing treatments for AM include misalignment in restoration, keyhole defects, gas pores, and polycrystals in the restored parts (Sato et al., 2022). The surface quality of additively manufactured clear aligners can be improved through chemical vapor smoothing, with cyclohexanone showing the highest reduction in surface roughness. Postprocessing operations are essential to improve the surface quality and mechanical properties of AM parts, making them suitable for customers' demands. The type of postprocessing technique applied can affect the quality of multimaterial parts, with ultrasonic treatment with isopropyl alcohol having a detrimental effect on more flexible resin mixtures.

22.7.6 Costly

The critical challenge in the costly production of additively manufactured functional polymers and nanocomposites is the need to overcome hindrances in the synthesis and fabrication processes (Darji et al., 2023). These hindrances include issues related to materials synthesis, barriers, limitations, and opportunities (Markandan & Lai, 2023). Additionally, the preparation of printable composite feedstock filaments and printing issues during the printing of composites are challenges that need to be addressed (Al Rashid et al., 2021). Furthermore, the limited choice of functional precursors and versatile printing techniques in current additive manufacturing technology poses significant limitations in printing hybrid structures and devices with complex functionalities. Type of challenges encountered in the preparation process of specimens for tensile testing of thermoplastic composites reinforced with chopped carbon fibers faced difficulties in achieving consistent fiber distribution within the printed material, ensuring proper adhesion between the fibers and the matrix material, and minimizing voids or defects in the printed specimens (Majko et al., 2022). These challenges are crucial for the cost-effective production of functional polymers and nanocomposites through additive manufacturing processes.

22.7.7 Reliability

The critical challenge in the reliability of additively manufactured functional polymers and nanocomposites is the need to address technological obstacles and fill research gaps to enable widespread industrial use. Limitations in matrix and reinforcement combinations are among these problems, hardware modification requirements for existing machines, challenges in mechanical characterization, and the need for printable and reliable stretchable electronics (Raja Sekhar, 2023). The process of creating polymer composites and using additive manufacturing to fabricate them (AM) technologies also presents hindrances that need to be overcome. Additionally, the preparation of printable composite feedstock filaments and the prevention of printing defects are critical for achieving quality products in additive manufacturing. To ensure the reliability of additively manufactured functional polymers and nanocomposites, it is necessary to address these challenges and gaps in research and development (Pal et al., 2021).

22.8 Future opportunities (4D printing)

Future opportunities in 4D printing of additively manufactured functional polymers and nanocomposites include the ability to customize shape memory properties, harness desirable morphing, and create compliant structures with flexible manufacturing processes. These advancements have the potential to revolutionize fields such as smart systems, biomedical engineering, and nanomedicine (Kausar et al., 2023). The creation of intelligent materials, such as materials with form memory and stimuli-responsive polymers, along with the use of multimaterial composites, has opened up new possibilities for creating dynamic, stimuli-responsive structures. However, challenges still exist in terms of material limitations, fabrication complexities, and regulatory considerations. To fully realize the potential of 4D printing, Collaboration and multidisciplinary research in the fields of material science, biology, engineering, regulatory bodies, and business are crucial (Ikram et al., 2022). 4D printing, an extension of 3D printing, is a technology that allows materials to change shape or properties over time. It has huge potential for use in a variety of industries, including as wearable technology, medicinal, aerospace, automotive, and electronics. In the biomedical field, 4D-printed implants could help control drug delivery, while in aerospace and automotive, parts could self-assemble or morph. The intersection of 4D printing, functional polymers, and nanocomposites presents a vast landscape of opportunities across industries (Mahmood et al., 2023).

22.9 Conclusion

This chapter deliberates the most topical developments and successes in the field of additive manufacturing (AM) for polymers and their composites containing thermo-responsive, shape-memory, elastomeric, and functional materials. A summary of the present capabilities of optimization tools and additive processes shows that additive manufacturing (AM) of polymers, composites, and functional materials shows promise in transforming 3D printing starting with a prototype technique into a reliable manufacturing procedure. Innovative methods for printing plastic and composite structures have been made possible by the progress of new tools and practices in additive manufacturing (AM). These methods allow for the creation of stronger materials in more creative ways and allow for the printing of complex resins, polymer materials that have contradictory qualities, or even combining several materials to create a single product. Furthermore, occasionally by regulating the material's microstructures, printed metals can have their intrinsic strength increased.

In addition, a review is given of the several parameters that govern the mechanical characteristics, including adhesion qualities, modulus, and flexural strength. There have also been reports of AM applications in the biomedical, electronics, textile, aerospace, and construction industries. The application of various additive manufacturing techniques for polymers and their composites is reviewed in this article along with its advancements. Based on the compiled AM techniques and some of the optimization tools described, there is no question that the future of AM of polymers and their composites in 3D printing, from prototypes to the process of producing durable components, is bright. Three main benefits of 3D printing are mass customization, creative flexibility, and the ability to produce intricate structures with little waste. Methods, materials, and the condition of trending applications in several industries were investigated for 3D printing. There was also a discussion of the primary difficulties that are related to the nature of 3D printing. The recent study and development of biomaterials for prototyping intricate and customized structures with patient-specific requirements has been greatly aided by additive manufacturing (AM). However, it has to deal with obstacles like scarce resources and legal problems. The aerospace sector has made investments in additive manufacturing (AM) to create customized parts with improved strength-to-weight ratios, expedite aircraft maintenance, and enable on-demand manufacture. The world's augmented funding, research, and development would lead to a swift shift soon from conventional manufacturing techniques to 3D printing.

Acknowledgments

The study was supported by the DST major project "Low Cost-Renewable Energy Driven (LC-RED) Water Treatment Solution Centre" (Ref No. DST/TM/WTI/WIC/2K17/124), the DBT BIG-BIRAC Ref No. BIRAC/CCAMP0723/BIG-13/18, SERB-POWER major project entitled "Upgradation in MVP with Improved Adsorption, Regeneration, and Fluoride Sensing Capacities" File No. "SPF/2022/000086" and B.B. and I.D. acknowledge the support of the ISF-NSFC joint research program (grant No. 3106/19).

References

Al Rashid, A., Khan, S. A., Al-Ghamdi, S. G., & Koç, M. (2021). Additive manufacturing of polymer nanocomposites: Needs and challenges in materials, processes, and applications. *Journal of Materials Research and Technology, 14*, 910–941.

Alhassan, S., Alshammari, K., Alshammari, M., Alotaibi, T., Alshammari, A. H., Alhamazani, A., & Taha, T. A. M. (2024). Linear and nonlinear optical investigations of polyvinyl chloride modified La2O3 nanocomposite films. *Results in Physics,* 107456.

Alonso, E., Guell, C., & Fullana-i-Palmer, P. (2020). Environmental assessment of additive manufacturing. *Journal of Cleaner Production, 244*, 118631.

Azimi, P., Zhao, D., Pouzet, C., & Crain, N. E. (2016). Emissions of ultrafine particles and volatile organic compounds from commercially available desktop three-dimensional printers with multiple filaments. *Environmental Science & Technology, 50*(3), 1260–1268.

Bachar, S. C., & Mazumder, K. (2022). Environmental and health impacts of polymer nanocomposites. *Woodhead Publishing in Materials,* 547–570.

Baechler, C., DeVuono, M., & Pearce, J. M. (2013). Environmental life cycle analysis of distributed three-dimensional printing and conventional manufacturing of polymer products. *ACS Sustainable Chemistry & Engineering, 1*(12), 1511–1519.

Banks, J.D., & Emami, A. (2022, December 29). Carbon-Based Piezoresistive Polymer Nanocomposites by Extrusion Additive Manufacturing: Process, Material Design, and Current Progress. 3D Printing and Additive Manufacturing.

Bhat, M.A., Gedik, K., & Eftade, O.G. (2022) "Environmental Toxicity of Emerging Micro and Nanoplastics" CRC Press.

Bhatt, P. M., Kabir, A. M., Peralta, M., Bruck, H. A., & Gupta, S. K. (2019). A robotic cell for performing sheet lamination-based additive manufacturing. *Additive Manufacturing (Elsevier), 27*, 278–289.

Bilal, M., Zhao, Y., Rasheed, T., Li, C., Zhang, X., Iqbal, H. M. N., & Wang, W. (2021). Circular economy models for additive manufacturing of polymer composites. *Journal of Cleaner Production, 284*, 124643.

Braga, A. R. C., Lemes, A. C., & Rosso, V. V. D. (2021). Polymer Nanocomposites for Industrial Applications Chapter 7-Polymer nanocomposite's applications in food and bioprocessing industry. *Micro and Nano Technologies,* 237–250.

Brouwer, D. H., Koers-Jacquemijns, M., Bello, D., Fransman, W., & van Tongeren, M. (2019). Human exposure to nanomaterials: exploring nano-release scenarios in nanotechnology laboratories. *Journal of Hazardous Materials, 363*, 447–458.

Cazin, I., Rossegger, E., Roppolo, I., Sangermano, M., Granitzer, P., Rumpf, K., & Schlögl, S. (2023). Digital light processing 3D printing of dynamic magneto-responsive thiol-acrylate composites. *RSC Advances, 13*(26), 17536–17544.

Darji, V., Singh, S., & Mali, H. S. (2023). Mechanical characterization of additively manufactured polymer composites: A state-of-the-art review and future scope. *Polymer Composites, 44*(8), 4370–4419.

Das, A., Bryant, J. S., Williams, C. B., & Bortner, M. J. (2023). Melt-based additive manufacturing of polyolefins using material extrusion and powder bed fusion. *Polymer Reviews, 63*(4), 895–960.

Dermeik, B., & Travitzky, N. (2020). Laminated object manufacturing of ceramic-based materials. *Advanced Engineering Materials, 22*(9), 2000256.

Deshmane, S., Kendre, P., Mahajan, H., & Jain, S. (2021). Stereolithography 3D printing technology in pharmaceuticals: A review. *Drug Development and Industrial Pharmacy, 47*(9), 1362–1372.

Diallo, M.S., Baier, G., Moyer, B.A., & Hamelers, H. (2015, August 18). Critical materials recovery from solutions and wastes: Retrospective and outlook. Environmental Science & Technology, 49, 9387–9389.

Diniță, A., Neacșa, A., Portoaca, A.I., Tănase, M., Ilincă, C., & Ramadan, I. (2023, June 26). Additive manufacturing post-processing treatments, a review with emphasis on mechanical characteristics. Materials, 16(13), 4610.

Domingo-Roca, R., Asciak, L., Windmill, J. F. C., Mulvana, H., & Jackson-Camargo, J. C. (2022). Non-destructive analysis of the mechanical properties of 3D-printed materials. *Journal of Nondestructive Evaluation, 41*(1), 22.

Farah, S., & Anderson, D. G. (2016). Langer R. Physical and mechanical properties of PLA, and their functions in widespread applications—A comprehensive review. *Advanced Drug Delivery Reviews, 107*, 367–392.

Findik, Ş., Öztürk, S., Örnek, O., Köker, R., & Kösemen, A. (2024). Detailed Investigation of Plasticized PMMA Dielectric for Improved Performance of Organic Field-Effect Transistors. *Journal of Electronic Materials,* 1–8.

Fu, Q., Wang, Z., Bu, M., Su, L., Kou, H., Li, N., & Niu, D. (2022). Constitutive behaviour and modelling of hybrid basalt–polypropylene fibre-reinforced concrete considering coupling effect of fibre reinforcement and mechanical damage. *Materials and Structures, 55*(6), 155.

Gebler, M., Uiterkamp, A. J. M. S., & Visser, C. (2014). A global sustainability perspective on 3D printing technologies. *Energy Policy, 74*, 158–167.

Gibson, I., Rosen, D.W., & Stucker, B. (2014). Additive manufacturing technologies: 3D printing, rapid prototyping, and direct digital manufacturing. Springer.

Gil, L. D., de-Camargo, I., Gutierrez-Velasquez, E. I., & Colorado, H. A. (2023). Systematic overview of nanocomposites obtained by VAT photopolymerization techniques: A cost and life cycle assessment approach. *Revista UIS Ingenierías, 22*, 2.

Girtan, M., Wittenberg, A., Grilli, M. L., de Oliveira, D. P. S., Giosuè, C., & Ruello, M. L. (2021). The critical raw materials issue between scarcity, supply risk, and unique properties. *Materials, 14*(8), 1826.

Gu, D. D., & Shen, Y. F. (2016). Additive manufacturing: Technology, applications and research needs. *Frontiers of Mechanical Engineering, 11*(3), 215–243.

Guarás, M.P., & Alvarez, V.A. (2022). Consumer nanoproducts based on polymer nanocomposites matrices. Nanoproducts. Springer.

Gupta, R., Huo, D., Pancholi, M., Njuguna, J., & Pancholi, K. (2017). Insulating polymer nanocomposites for high thermal conduction and fire retarding applications. Cranfield Online Research Data (CORD).

Hassan, M.S., Delgadillo, A., Mahmud, M.S., Munoz, J., Zaman, S., Gomez, S.G., Marquez, C., Ho, J.C., & Lin, Y. (2023, January 1). Additive manufacturing of carbon fiber reinforced epoxy thermoset with improved thermomechanical properties. Journal of Composites Science, 7, 171.

Hosono, N., & Uemura, T. (2021). Metal–organic frameworks as versatile media for polymer adsorption and separation. *Accounts of Chemical Research, 54*(18), 3593–3603.

Hu, C., Jin, Y., Tian, W., Yan, K., Wang, J., &; Yuan, L. (2024). Photodegradable dynamic polyurethane networks via dithioacetals. Macromolecules, 57, 1725–1733.

Hun, M., Vaish, R., Singh, N. K., Yang, M., & Tyagi, I. (2020). Recent advancements in polymer nanocomposites for energy and environmental applications: A review. *Progress in Energy and Combustion Science, 79*, 100846.

Ichiura, H., Morikawa, M., & Fujiwara, K. (Dec 2006). Method for laminating functional material and sheet-like structure. Patent.

Iervolino, F., Aurora, B., Gaia, F., Levi, M., & Suriano, R. (2022, November 18). Additive manufacturing of electrically conductive nanocomposites filled with carbon nanotubes. Advanced Engineering Materials, 24, 2200947.

Ikram, H., Al Rashid, A., & Koç, M. (2022). Additive manufacturing of smart polymeric composites: Literature review and future perspectives. *Polymer Composites, 43*(9), 6355–6380.

Jisun I., Liu, Y., Hu, Q., Trindade, G.F., Parmenter, C., Fay, M., He, Y., Irvine, D.J., Tuck, C., Wildman, R.D., Hague, R., & Turyanska, L. (04 January 2023). Strategies for integrating metal nanoparticles with two-photon polymerization process: toward high resolution functional additive manufacturing. Advanced Functional Materials, 33, 2211920.

Jue, J., Gan, Z., Luo, Z., & Li, K. (2023). Direct laser writing of functional QD–polymer structure with high resolution. *Materials, 16*(6), 2456 -2456.

Kant, R., Patel, V.K., Bhatt, G., & Bhattacharya, S. (2022, March 2). Advances in polymer materials and composites for additive manufacturing. In *Trends in fabrication of polymers and polymer composites*. AIP Publishing.

Kausar, A., Ahmad, I., Zhao, T., Aldaghri, O., & Eisa, M.H. (2023, March 14). Polymer/graphene nanocomposites via 3D and 4D printing—design and technical potential. Processes, 11, 868.

Kehinde, B.A., Chhikara, N., Sharma, P., Garg, M.K., Panghal, A. (2021). Application of polymer nanocomposites in food and bioprocessing industries. PolymerNanocomposites for Industrial Applications.

Kelly, J.P., Elmer, J.W., Ryerson, F.J., Lee, J.R.I., & Haslam, J.J. (March 2021). Directed energy deposition additive manufacturing of functionally graded Al-W composites. Additive Manufacturing, 39, 101845.

Kreiger, M. A., & Pearce, J. M. (2013). Environmental life cycle analysis of distributed polymer recycling. *Resources, Conservation and Recycling, 73*, 145–153.

Kruth, J. P., Levy, G., Klocke, F., & Childs, T. H. (2007). Consolidation phenomena in laser and powder-bed based layered manufacturing. *CIRP Annals, 56*(2), 730–759.

Kumar Penumakala, P., Santo, J., & Thomas, A. (2020). *A critical review on the fused deposition modeling of thermoplastic polymer composites. Composites Part B: Engineering, 201*(15), 108336.

Latif, M., Jiang, Y., Song, J., & Kim, J. (2023). Additively manufactured high-concentration nanocellulose composites: Structure and mechanical properties. *Polymers, 15*(3), 669.

De Leon, A. C., Chen, Q., Palaganas, N. B., Palaganas, J. O., Manapat, J., & Advincula, R. C. (2016). High performance polymer nanocomposites for additive manufacturing applications. *Reactive and Functional Polymers, 103*, 141–155.

Leonard, E., & Guenin, E. (2023). Functional sues. CRC Press.

Li, X., Niu, D., Xu, P., Yang, W., & Ma, P. (2024). High-toughness and antistatic PET/EMAG/CNTs nanocomposites from recycled sources by reactive compatibilization. *Composites Communications, 47*, 101855.

Li, Y., Wang, W., Wu, F., & Kankala, R.K. (2023, January 5). Vat polymerization-based 3D printing of nanocomposites: A mini review. Frontiers in Materials, 9.

Ligon, S. C., Liska, R., Stampfl, J., Gurr, M., & Mülhaupt, R. (2017). Polymers for 3D printing and customized additive manufacturing. *Chemical Reviews, 117*(15), 10212–10290.

Lima, P., J., R. (March 2015), Synthesis of metal-polymer nanocomposites for fuel applications. PhD thesis, École Polytechnique de Montréal.

Liu, S.F., Hou, Z.-W., Lin, L., Li, Z., & Hong-Bo, S. (29 Jan 2023). 3D laser nanoprinting of functional materials. Advanced Functional Materials, 33, 2211280.

Luty-Błocho, M.agdalena, Hoyos, M.V., & Hessel, V. (2023). Implications of the use of functional nanomaterials for the environment and health. In *Functional nanomaterials for sensors* (pp. 289–319). CRC Press.

Maalihan, R.D., Chen, Q., Tamura, H., Agueda, J.R. H.S., Pajarito, B.B., Caldona, E.B., & Advíncula, R.C. (2022, March 29). Mechanically and thermally enhanced 3D-printed photocurable polymer nanocomposites containing functionalized chitin nanowhiskers by stereolithography. ACS Applied Polymer Materials, 2513–2526.

Mahfuz H., Lambert V., & Clements F. (2023). Functional nanocomposites in the development of flexible armor. International Journal of Molecular Sciences, 24, 5067.

Mahmood, A., Akram, T., Shenggui, C., & Chen, H. (2023). Revolutionizing manufacturing: a review of 4D printing materials, stimuli, and cutting-edge applications. *Composites Part B: Engineering,* 110952.

Majko, J., Vaško, M., Handrik, M., & Sága, M. (2022). Tensile properties of additively manufactured thermoplastic composites reinforced with chopped carbon fibre. *Materials, 15*(12), 4224.

Makvandi, P., Iftekhar, S., Pizzetti, F., et al. (2021). Functionalization of polymers and nanomaterials for water treatment, food packaging, textile, and biomedical applications: A review. Environmental Chemistry Letters, 19, 583–611.

Markandan, K., & Lai, C. Q. (2023). Fabrication, properties and applications of polymer composites additively manufactured with filler alignment control: A review. *Composites Part B: Engineering, 256*, 110661.

Martínez-García, P., Rebollar, E., Nogales, A., García-Gutiérrez, M.C., Sena-Fernández, J., & Ezquerra, T.A. (2023, June 25). 3D printing-assisted nanoimprint lithography of polymers. Advanced Engineering Materials, 25, 2300344.

Masood, S. H. (2019). Materials in additive manufacturing: current trends and future perspectives. *Virtual and Physical Prototyping, 14*(1), 1–18.

May, G., & Psarommatis, F. (2023). Maximizing energy efficiency in additive manufacturing: A review and framework for future research. *Energies, 16*(10), 4179.

Melenka, G.W., Bruni-Bossio, B.M., Ayranci, C., & Carey, J.P. (2018). Examination of voids and geometry of bio-based braided composite structures. IOP Conference Series: Materials Science and Engineering, 406(1), 012012.

Memarzadeh, A., Safaei, B., Tabak, A., Sahmani, S., & Kizilors, C. (2023). Advancements in additive manufacturing of polymer matrix composites: A systematic review of techniques and properties. *Materials Today Communications,* 106449.

Murr, L. E., Gaytan, S. M., Medina, F., Lopez, H., Martinez, E., Machado, B. I., & Wicker, R. B. (2010). Next-generation biomedical implants using additive manufacturing of complex, cellular and functional mesh arrays. *Philosophical Transactions of the Royal Society A: Mathematical, Physical and Engineering Sciences, 368*(1917), 1999–2032.

Ning, F., Cong, W., & Wei, J. (2019). Additive manufacturing of carbon fiber-reinforced polymer composites using fused deposition modeling: Effects of process parameters on tensile properties. *Journal of Composite Materials, 53*(30), 4267–4278.

Nurazzi, N.M., Norrrahim, M.N.F., Mulla, M.H., Kamarudin, S.H., Rani, M.S.A., Rushdan, A.I., & Kuzmin, A.M. (2024). Mechanical performance of seashell-reinforced polymer composites for structural applications. *Polymer Composites Derived from Animal Sources*, 243-257.

Pal, C., & Krishnamoorthy, A. (2022). Chapter 12-Direct energy deposition.

Pal, A.K., Mohanty, A.K., & Misra, M. (2021, January 1). Additive manufacturing technology of polymeric materials for customized products: recent developments and future prospective. RSC Advances, 11, 36398–36438.

Park, N., Friest, M. A., & Liu, L. (2023). Enhancing the properties of polyvinyl alcohol films by blending with corn stover-derived cellulose nanocrystals and beeswax. *Polymers, 15*(21), 4321.

Pelanconi, M., Bottacin, S., Colombo, P., & Ortona, A. (November, 2023). Powder bed fusion of polyamide powders combined with different preceramic polymers infiltration and pyrolysis to produce complex-shaped ceramics. Journal of the European Ceramic Society, 43, 5871-5881.

Peng, Z., Chen, J., Chen, Y., Chen, Y., Wang, Y., & Han, C. (2024). Progress of degradable synthetic polymers in adipose tissue engineering—A review. *Journal of Biomedical Nanotechnology, 20*(3), 415–437.

Petousis, M., Michailidis, N., Papadakis, V., Korlos, A., Mountakis, N., Argyros, A., Dimitriou, E., Charou, C., Moutsopoulou, A., & Vidakis, N. (2023, May 9). Optimizing the rheological and thermomechanical response of acrylonitrile butadiene styrene/silicon nitride nanocomposites in material extrusion additive manufacturing. Nanomaterials, 13, 1588.

Petrakli, F., Gkika, A., Vlysidis, A., Karayannis, P., & Koumoulos, E. (2022). A comprehensive methodology to support decision-making towards sustainability on nanocomposite materials in additive manufacturing sector. Research Square, 10.21203/rs.3.rs-1187608/v1.

Pietrini, M., Roes, L., Patel, M. K., & Chiellini, E. (2007). Comparative life cycle studies on poly (3-hydroxybutyrate)-based composites as potential replacement for conventional petrochemical plastics. *Biomacromolecules, 8*(7), 2210–2218.

Pizza, A., Metz, R., Hassanzadeh, M., & Bantignies, J. L. (2014). Life cycle assessment of nanocomposites made of thermally conductive graphite nanoplatelets. *The International Journal of Life Cycle Assessment, 19*, 1226–1237.

Qian, X., Ostwal, M., Asatekin, A., Geise, G. M., Smith, Z. P., Phillip, W. A., & McCutcheon, J. R. (2022). A critical review and commentary on recent progress of additive manufacturing and its impact on membrane technology. *Journal of Membrane Science, 645*, 120041.

Raja Sekhar, K.S.S. (2023). International Journal of Scientific Research.

Rasheed, I.A., Rehan, Z.A., Khalid, T., Zahid, M., & Ahmad, H. (2020). Prospects of nanocomposite membranes in commercial scale. In Nanocomposite membranes for water and gas separation (pp. 457–473). Elsevier.

Rishi, K., & Beaucage, G. (2022). Surface functionalized nanoparticles applications of nanocomposites (Pages 33). CRC Press.

Saleh Alghamdi, S., John, S., Roy Choudhury, N., & Dutta, N. K. (2021). Additive manufacturing of polymer materials: Progress, promise and challenges. *Polymers. 13*(5), 753.

Sato, N., Matsumoto, M., Ogiso, H., & Sato, H. (2022). Challenges of remanufacturing using powder bed fusion based additive manufacturing. *International Journal of Automation Technology, 16*(6), 773–782.

Schäfer, K., Lutzi, M., Khan, M. B., Schäfer, L., Dirba, I., Bruns, S., Valizadeh, I, Weeger, O., Hartmann, C., Kupnik, M., Adabifiroozjaei, E., Molina-Luna, L., Skokov, K., & Gutfleisch, O. (5January, 2024). Magneto-active composites with locally tailored stiffness produced by laser powder bed fusion. Additive Manufacturing, 79, 103905.

Schrijvers, D. L., Leroux, F., Verney, V., & Patel, M. K. (2014). Ex-ante life cycle assessment of polymer nanocomposites using organo-modified layered double hydroxides for potential application in agricultural films. *Green Chemistry, 16*(12), 4969–4984.

Shivalkar, S., Ranjan, S., Sahoo, A.K. (2023). Polymeric nanocomposites: synthesis, characterization, and recent applications. In *Nanomaterials*. Springer.

Shonaike, G. O., & Salami, A. P. (Eds.). (2018). *Handbook of engineering and specialty thermoplastics: Polyethers and polyesters*. John Wiley &; Sons.

Slowinski, G., Latimer, D., & Mehlman, S. K. (2013). Dealing with shortages of critical materials: Concerned about growing supply uncertainties, firms are taking various actions to anticipate and plan for critical materials shortages. *Research-technology Management (Industrial Research Institute Inc.), 56*(5), 18.

Spackman, C., Picha, K., Gross, G.J., Nowak, J.F., Smith, P.J., Zheng, J., Samuel, J., & Mishra, S. (2015, March 1). A novel multimaterial additive manufacturing technique for fabricating laminated polymer nanocomposite structures. Journal of Micro and Nano-Manufacturing, 3, 011008.

Stefaniak, A. B., LeBouf, R. F., Duling, M. G., Yi, J., Abukabda, A. B., & McBride, C. R. (2017). Characterization of chemical contaminants generated by a desktop fused deposition modeling 3-dimensional Printer. *Journal of Occupational and Environmental Hygiene, 14*(7), 540–550.

Su, S., Kopitzky, R., Tolga, S., & Kabasci, S. (2019). Polylactide (PLA) and its blends with poly (butylene succinate) (PBS): A brief review. *Polymers, 11*(7), 1193 Commercial aspects of functional polymers and nanocomposites.

Svetlizky, D., Zheng, B., Vyatskikh, A., Das, M., Bose, S., Bandyopadhyay, A., & Eliaz, N. (2022). Laser-based directed energy deposition (DED-LB) of advanced materials. *Materials Science and Engineering: A, 840*, 142967.

Tan, P., Liu, X., Tey, W.S., Huang, J., & Zhou, K. (2023, March 6). Numerical investigation of fiber orientations and homogeneity in powder bed fusion of fiber/polymer composites. Virtual and Physical Prototyping, 18, e2162928.

Taokaew, S., & Chuenkaek, T. (2024). Developments of core/shell chitosan-based nanofibers by electrospinning. *Techniques: A Review. Fibers, 12*(3), 26.

Tee, B. C. K., & Ouyang, J. (August 2018). Soft electronically functional polymeric composite materials for a flexible and stretchable digital future. *Advanced materials (Deerfield Beach, Fla.), 30*(47), e1802560.

Thakur, V. K., Karak, N., & Gupta, R. K. (2015). Green nanocomposites: synthesis, properties and potential applications. *Green Chemistry, 17*(11), 11 -11.

Thijs, L., Verhaeghe, F., Craeghs, T., Humbeeck, J. V., & Kruth, J. P. (2010). A study of the microstructural evolution during selective laser melting of Ti–6Al–4V. *ActaMaterialia, 58*(9), 3303–3312.

Tiwari, A., Michalska-Domanska, M., & Dhoble, S. J. (2023). *Nanocomposite-based functional materials: Synthesis, properties, and applications. Functional materials from carbon, inorganic, and organic sources (pp. 317–365)*. Woodhead Publishing.

Tripathy, D.B., & Gupta, A. (2024). Nanocomposites as sustainable smart materials: A review. Journal of Reinforced Plastics and Composites, 07316844241233162.

Tumbleston, J. R., Shirvanyants, D., Ermoshkin, N., Janusziewicz, R., Johnson, A. R., Kelly, D., & DeSimone, J. M. (2015). Continuous liquid interface production of 3D objects. *Science, 347*(6228), 1349–1352.

Valencia, L.M., Herrera, M., De La Mata, M., Hernández-Saz, J., Romero-Ocaña, I., Delgado, F., Benito, J., & Molina, S.I. (2022, December 1). Stereolithography of semiconductor silver and acrylic-based nanocomposites. Polymers, 14, 5238.

Vidakis, N., Petousis, M., Moutsopoulou, A., Papadakis, V., Spiridaki, M., Mountakis, N., Charou, C., Tsikritzis, D., & Maravelakis, E. (2023, June 22). Nanocomposites with Optimized Polytetrafluoroethylene Content as a Reinforcement Agent in PA12 and PLA for Material Extrusion Additive Manufacturing. Polymers, 15, 2786.

Visakh, P. M., & Lüftl, S. (Eds.). (2016). *Polyethylene-based biocomposites and bionanocomposites*. John Wiley &; Sons.

Wang, Z., Zhou, M., Xiao, H., & Yuan, S. (June 2023). Development and evaluation of multiscale fiber-reinforced composite powders for powder-bed fusion process. Chinese Journal of Mechanical Engineering: Additive Manufacturing Frontiers, 2, 100079

Wohlers Associates. (2021). Wohlers Report 2021: 3D printing and additive manufacturing state of the industry. Wohlers Associates, Inc.

Woosley, S., & Aravamudhan, S. (2022, July 20). Functionally modified composites for FDM 3D printing. IntechOpen.

Yan, Y., Jiang, Y., Ng, E. L. L., Zhang, Y., Owh, C., Wang, F., & Chan, B. Q. Y. (2023). Progress and opportunities in additive manufacturing of electrically conductive polymer composites. *Materials Today Advances, 17*, 100333.

Yaragatti, N., & Patnaik, A. (2021). A review on additive manufacturing of polymers composites. *Materials Today: Proceedings, 44*, 4150–4157.

Zhou, X., & Wang, J. (2023). Editorial: Additive manufacturing for polymers. *Frontiers in Materials, 10*, 1230323.

Zou, H., Wu, S., & Shen, J. (2008). Polymer/silica nanocomposites: preparation, characterization, properties, and applications. *Chemical Reviews, 108*(9), 3893–3957.

Index

Note: Page numbers followed by "*f*" and "*t*" refer to figures and tables, respectively.

B

C

I

J

M

N